T0327495

Practical Partial Discharge
Measurement on Electrical Equipment

IEEE Press
445 Hoes Lane
Piscataway, NJ 08854

Practical Partial Discharge Measurement on Electrical Equipment

Greg C. Stone
Stone Dielectrics
Toronto, Canada

Andrea Cavallini
University of Bologna
Bologna, Italy

Glenn Behrmann
Consultant
Zurich, Switzerland

Claudio Angelo Serafino
Terna S.p.A.
Torino, Italy

Published by John Wiley & Sons, Inc., Hoboken, New Jersey.
Published simultaneously in Canada.

Library of Congress Cataloging-in-Publication Data
Names: Stone, Greg C., author. | Cavallini, Andrea, 1963– author. | Behrmann, Glenn, author. | Serafino, Claudio Angelo, author.
Title: Practical partial discharge measurement on electrical equipment / Greg Stone, Andrea Cavallini, Glenn Behrmann, Claudio Serafino.
Description: Hoboken, New Jersey: Wiley, [2023] | Includes index.
Identifiers: LCCN 2023012162 (print) | LCCN 2023012163 (ebook) | ISBN 9781119833314 (hardback) | ISBN 9781119833321 (adobe pdf) | ISBN 9781119833338 (epub)
Subjects: LCSH: Electric discharges–Measurement. | Electric power systems.
Classification: LCC QC701.S76 2023 (print) | LCC QC701 (ebook) | DDC 537.5/2–dc23/eng/20230606
LC record available at https://lccn.loc.gov/2023012162
LC ebook record available at https://lccn.loc.gov/2023012163

Cover Design: Wiley
Cover image provided by Andrea Cavallini

Set in 9.5/12.5pt STIXTwoText by Straive, Pondicherry, India

This book is dedicated to the engineers and scientists who took a scientific curiosity of the 1890s and built a measurement technology that is now an essential tool for factory quality control tests on new high-voltage electrical equipment, as well as a widely used technology to help determine when maintenance is needed on such equipment.

Although thousands have worked in this field, we would like to acknowledge the seminal contributions of some of its leading engineers and scientists who are no longer with us:

Ray Bartnikas (Canada)
Steve Boggs (Canada/USA)
Frederik Kreuger (The Netherlands)
Bernd Fruth (Germany)
Jitka Fuhr (Switzerland)
George Mole (UK)
Lutz Niemeyer (Germany)

Some of us have had the privilege of working with and being guided by them over the years; they are sorely missed. This book is dedicated to them, especially their willingness to share their knowledge and inspire us.

Contents

About the Authors

Greg C. Stone was one of the developers of a widely used online partial discharge test method to evaluate the condition of the high-voltage insulation in stator windings. Except for a one-year period at Canada Wire and Cable, from 1975 to 1990, he was a Dielectrics Engineer with Ontario Hydro, a large Canadian power generation company. From 1990 to 2021 he was employed at Iris Power L.P. in Toronto Canada, a motor and generator winding condition monitoring company he helped form, and one of the largest manufacturers of partial discharge measurement equipment in the world. He has published two books on motor and generator winding maintenance, contributed the machine insulation chapters to three technical encyclopedias/handbooks, and authored >200 papers concerned with partial discharge measurement and rotating machine condition assessment. Since 1980 he has also been active in creating and updating many IEEE and IEC standards. Dr. Stone has been and continues to be active in the IEEE Dielectrics and Electrical Insulation Society for his entire career, serving as its president in 1989 and 1990. He has been chosen for the IEEE Kaufmann, Dakin and Forster Awards; the CIGRE Technical Committee Award; the IEC 1906 Award; and the US-based EPRI Principal Investigator Award. Greg Stone has BASc, MASc, and PhD degrees in electrical engineering from the University of Waterloo, is a fellow of the IEEE, a fellow of the Engineering Institute of Canada, and a licensed professional engineer in Ontario, Canada.

Andrea Cavallini is a professor at the University of Bologna. In 2000, he invented the TF map, a revolutionary method to process partial discharge signals allowing noise rejection and the individual analysis of multiple partial discharge sources simultaneously active in the same equipment. From 1999 to 2012, he cofounded and collaborated with Techimp Srl, now a part of the Altanova group. During this time, he was actively involved in testing all types of apparatus used in power systems: cables, generators, transformers, and motors. He has worked on electrical as well as ultra-high frequency detection of partial discharges; in particular, he was among the first to use ultra-high frequency detection of partial discharges on inverter-fed machines, a technique that later became the state of the art. He has published more than 250 papers, most of them in the IEEE. He is member of the IEC TC 2 Rotating Machinery MT 10 concerned with the qualification, testing and diagnostics of winding insulation systems. He is active in CIGRE where he has convened D1 working groups. He has BASc, MASc, and PhD degrees in electrical engineering from the University of Bologna in Italy and is a fellow of the IEEE.

Glenn Behrmann began work in radio-frequency design and signal processing for military applications at MIT Lincoln Laboratory, the MITRE Corporation, and Signatron. After working in EMC, he began his work in PD in 1992 at ABB Corporate Research (Baden) under Bernd Fruth and Lutz Niemeyer, first in the insulation materials lab, later on site, on transformers, HV cable accessories, and GIS, playing a key role on the first large-scale UHF PD monitoring systems for 400 kV GIS in Singapore. During a two-year stint at the Paul Scherer Institute, he worked on RF beam-diagnostics for the European Free Electron Laser (X-FEL, Hamburg). In 2008, he returned to PD diagnostics of rotating machine insulation and online monitoring systems at Alstom (GE). In 2011, he rejoined ABB, focusing on all aspects of PD detection and monitoring in GIS using both conventional and UHF techniques. This included sensor development, detailed investigations of RF signal behavior, lots of onsite PD diagnostics, and a key role in PXIPD (pulsed X-ray-induced partial discharge) for detecting voids. He has authored many papers and holds patents in the field. He is an active member of CIGRE (presently secretary of D1.66, Requirements of UHF PDM systems for GIS) and IEC TC42 (including the latest revision of IEC 60270) and is chair of the IEEE PES/SA group revising IEEE 454 on PD measurement. Glenn received a BSEE from Union College (Schenectady) in 1979. Although just retired, he remains active doing onsite PD assessments and consulting.

Claudio Angelo Serafino has more than 40 years of experience in measurements and tests on high voltage equipment. He carries out commissioning and routine tests on high-voltage equipment, including circuit breakers, disconnectors, surge arresters, gas-insulated systems, current transformers, and voltage transformers. He also has experience in commissioning and routine tests on protection systems for high-voltage plants, large power generators, and transformers. He is an expert in PD measurements on large power and high-voltage transformers, performed both in the manufacturers' test labs and onsite, with the aim to investigate faults. He gained his experience working in two companies. From 1982 until 2000, he worked at ENEL, the integrated Italian electrical utility. Since 2000 he is working in the Italian TSO Terna as an expert in medium and large power transformer tests and technical specifications.

Preface

Partial discharge (PD) testing is widely used as a quality assurance test for the electrical insulation in medium- and high-voltage equipment. Owners of high-voltage equipment such as transformers, switchgear, power cables, and rotating machines are also using PD testing as a tool to determine if there is a risk of insulation failure in equipment that has been in service. This latter application has exploded in the past decade with the availability of PD test systems that measure PD during normal operation of medium- and high-voltage apparatus.

There are now dozens of vendors of PD testing systems, and many IEC and IEEE standards have been published that present the basics of PD measurement on various types of high-voltage apparatus. Also, books have been published that go into the details of both the physics of PD and PD detection theory, in addition to thousands of technical papers. These have been mainly written for researchers on the subject.

This book has a different aim. It is written for those who work for electrical equipment manufacturers and owners of medium- and high-voltage equipment who preform PD testing as only one part of their job, and who want to understand better the information from vendors and the relevant standards. Although we discuss some basic information on why PD occurs and PD measurement theory in the first few chapters, the main focus is presenting practical information that even occasional users of PD testing need to know when using commercially available PD measurement systems. There are chapters on the most common ways to detect PD, how to reduce the influence of electrical noise and interference, as well as how PD results are analyzed in general. Then there are chapters for each type of high-voltage equipment that describe the most common PD measurement methods for that equipment, as well as what insulation problems it can detect and how to interpret the PD data. Since there is a broad range of PD system vendors for each type of high-voltage equipment, we have attempted to include the most popular methods applied to each type of equipment. Sometimes, the same PD measuring system is used for different types of electrical equipment. The final chapters are brief introductions to the rapidly evolving techniques to measure PD under DC excitation and short-risetime voltage impulses.

The authors have a diverse range of backgrounds. One of us was formerly with a PD system vendor, one is both an academic researcher and a cofounder of a different PD system vendor. The other two authors are primarily users of PD testing – one mainly working for a high-voltage equipment supplier and the other working for a transmission grid utility. Hopefully, this diversity has resulted in a practical book on PD measurement.

Acknowledgments

The authors would like to thank Ms. Resi Zarb, the founding president of Iris Power L.P. as well as a former co-editor in chief of the *IEEE Electrical Insulation Magazine*. She provided invaluable assistance in reviewing every chapter of this book.

GCS would like to thank his wife, Judy Allan, for her support during this project. He would also like to acknowledge the importance of the late Mo Kurtz and the late Steve Boggs to his comprehension of PD measurement. Many significant discussions with his former colleagues at Iris Power L.P.: Howard Sedding, Mladen Sasic, Connor Chan, Vicki Warren, and Iouri Pomelkov have contributed to his understanding of the measurement and interpretation of PD.

AC would like to thank his colleagues at the University of Bologna: Gian Carlo Montanari for directing his research efforts toward partial discharges, as well as Davide Fabiani and Gaetano Pasini for countless discussions. He is particularly indebted to Alfredo Contin and Francesco Puletti with whom he had many enlightening exchanges of ideas, Fabio Ciani (*sit tibi terra levis*) for his friendly and professional support in the lab, and all the PhD students who worked with him: Carlos Gustavo Azcarraga Ramos, Peng Wang, Fabrizio Negri, Paolo Mancinelli, Luca Lusuardi, and Alberto Rumi. Last but not least, AC thanks his fiancé Violeta Zumeta for her incredible patience.

GB wishes to thank many colleagues in the field, both past and present, for passing on their know-how (or challenging mine) during many fruitful discussions in the lab or on site, often in the evenings after work or at a conference. First and foremost, I must thank the late Bernd Fruth and Lutz Niemeyer for bringing me into this strange and fascinating field, educating me, and igniting my interest in it. Along with them, my sincere thanks to Detlev Gross, Markus Schraudolph, Stefan Neuhold, Uwe Riechert, Ralf Pietsch, Wojciech Koltunowicz, Marek Florkowski, and Maria Kosse (Hering); I'm indebted to them all for their help, wisdom, PD insights, and friendship. In my work in RF propagation, I am grateful to Jasmin Smajic and Daniel Treyer for their lucid explanations of electromagnetics, impedance, and many hours of fun discussions. To anyone I have left out: my sincere apologies.

CAS firstly thanks Fiorenzo Stevanato, Mario Cimini and Diego Bosetto; they have been my teachers and they continue to be a reference. They have the great merit for instilling passion for this profession and the pleasure of increasing knowledge about PD. Many thanks to Dario Gamba, Fabio Marinelli, and Fabio Scatiggio who were with me for long durations as we undertook onsite tests searching for PD, their causes, and their consequences in large power transformers. Finally thanks to Maria Rosaria Guarniere, Evaristo Di Bartolomeo and Roberto Spezie. They demonstrated their belief in this technology and in its benefits to transformer maintenance. They spurred me to grow and to innovate, and they are examples of excellent managers.

Acronyms

ADC	Analog-to-digital converter
AI	Artificial intelligence
AIS	Air-insulated switchgear
ASTM	American Society for Testing of Materials
BDV	Breakdown voltage
BIL	Basic impulse level (lightning impulse voltage)
CIGRE	Conseil International des Grands Réseaux Electriques; Paris-based organizer of power industry conferences and publisher of consensus technical reports
CT	Current transformer
DGA	Dissolved gas analysis (in transformer oil)
EM	Electromagnetic
EMI	Electromagnetic interference
EUT	Equipment under test
FAT	Factory acceptance test
FEM	Finite element method
FPGA	Field programable gate array (type of integrated circuit)
GIL	Gas-insulated transmission line
GIS	Gas-insulated switchgear/substation
GTEM	GigaHertz transverse electromagnetic cell
GVPI	Global vacuum-pressure impregnation (stator winding manufacturing process)
HF	High frequency (3–30 MHz)
HFCT	High-frequency current transformer
HV	High voltage
HVDC	High voltage direct current
IEC	International electrotechnical commission
IEEE	Institute of electrical and electronics engineers
IPB	Isolated phase bus
LF	Low frequency, formally 30–300 kHz, but 30 kHz to 1 MHz in this book
LV	Low voltage
MV	Medium voltage
OCP	Outer corona protection (stator windings)
OEM	Original equipment manufacturer
OIP	Oil impregnated paper
OLTC	Online tap changer (transformers)

PD	Partial discharge
PDEV	Partial discharge extinction voltage
PDIV	Partial discharge inception voltage
PE	Polyethylene
PMT	Photomultiplier tube
PRPD	Phase-resolved partial discharge (plot)
PSH	Peak sample-and-hold (electronic circuit)
PT	Potential transformer
RF	Radio frequency
RFI	Radio frequency interference
RIP	Resin impregnated paper (bushings)
RIV	Radio interference (or influence) voltage
RTS	Resonant test supply/set
SCLF	Space charge limited field
TDR	Time domain reflectometry
UHF	Ultrahigh frequency (300–3000 MHz)
VFTO	Very fast transient overvoltage (GIS)
VHF	Very high frequency (30–300 MHz)
VLF	Very low frequency (AC voltage supply)
VPI	Vacuum pressure impregnation (stator bar and coil manufacturing process)
VSC	Voltage source converter
VZC	AC voltage zero crossing
XLPE	Cross-linked polyethylene (power cables)

1

Introduction

1.1 Why Perform Partial Discharge Measurements?

This book is focused on the practical aspects of the measurement of partial discharge (PD) and corona in 50/60 Hz power system equipment such as generators, motors, power cables, air- and gas-insulated switchgear (GIS), and transformers, all usually rated 3 kV and above. Such electrical equipment uses solid electrical insulation, for example polyethylene, epoxy, and polyester, or insulation composites such as oil–paper, fiberglass-reinforced polymers, or epoxy-mica, to separate high-voltage conductors from ground or to separate one AC phase from another. If this insulation fails, the equipment experiences a phase-to-ground fault or a phase-to-phase fault, which will activate protective relays to isolate the equipment from the power system. Such a failure may manifest itself as a power outage in a residential area or hospital, a loss of electrical power production capacity, or a reduction in power system reliability. In industries such as petrochemical, cement, steel, aluminum, paper, or semiconductor fabrication, these failures can be extremely expensive because modern production processes are continuous; an electrical power failure of even a few minutes may necessitate taking the entire factory out of production for days or weeks. In addition, such insulation failures can cause collateral damage to adjacent components that can greatly increase the cost of repair. For example, a large utility generator or power transformer failure can cost millions to repair, and result in a plant shutdown that can last for months, causing tens of millions of dollars in lost production.

Partial discharges are small electrical "micro-sparks" that can occur in insulation systems operating with high electric fields. The physics of PD and how it is manifested are discussed in Chapters 2, 3, and 5. PD activity can directly lead to insulation degradation and equipment failure. PD is also sometimes a symptom of poor manufacturing and/or aging of the insulation due to high temperature, mechanical forces, contamination, etc. In this case, PD might not directly lead to failure but may indicate that insulation aging due to other mechanisms is occurring and maintenance may be needed. Thus, by measuring PD activity, equipment manufacturers can often determine that the insulation system on the equipment was properly made, and equipment owners can determine if aging is occurring that could lead to failure.

Each partial discharge is accompanied by a current pulse. As presented later in this book, these current pulses can be detected by various types of sensors and measurement instruments. In addition to measuring the PD current, PD can be detected from radio frequency (RF) radiation, light

Practical Partial Discharge Measurement on Electrical Equipment, First Edition. Greg C. Stone, Andrea Cavallini, Glenn Behrmann, and Claudio Angelo Serafino.

emissions, acoustic noise, and by chemical changes in the local environment. PD testing involves the measurement of the PD current pulses and other signals that are produced by PD.

PD testing using 50/60 Hz AC is widely employed as a factory quality assurance (QA) test for all types of high-voltage equipment. Many IEEE and IEC technical standards have been published to indicate how the PD should be measured for each type of equipment, often providing guidance on interpretation, and sometimes providing information on pass/fail criteria. The premise is that if newly manufactured equipment successfully passes the PD test, then premature insulation failure due to electrical stress is unlikely.

In recent decades, with the development of digital hardware, often with powerful disturbance suppression methods and signal processing, PD testing has increasingly been applied to high-voltage equipment that has been installed in the power system or industrial plants with a view to assess if the high-voltage insulation system is degrading and may have a high risk of failure. Thus, the purpose of PD testing, once equipment has entered service, is to help with insulation condition assessment and determining the need for maintenance. There are relatively few IEEE and IEC standards for such PD testing applications. Hence, an important function of this book is to provide information for both onsite (offline) and online PD testing/monitoring of the different types of high-voltage equipment.

In this book, for simplicity, we will use the term "high-voltage insulation system," rather than the more cumbersome "medium- and high-voltage insulation system." What voltage ratings are associated with medium voltage (MV) and high voltage (HV) depends on the type of equipment. A medium-voltage motor is usually rated between 3 and 7 kV, whereas a high-voltage motor is 11 kV or higher. In electrical power transmission systems, there is a wide variation of what is meant by medium and high voltage.

1.2 Partial Discharge and Corona

There are many definitions for partial discharge. Perhaps the most widely used definition of PD comes from IEC 60270, where it is described as "a localized electrical discharge that only partially bridges the insulation between the conductors and which can or cannot occur adjacent to a conductor." That is, PD is a localized electrical breakdown of the insulation that does not immediately progress to a complete breakdown across the insulator (e.g. between the high-voltage conductor and ground). In contrast, a "complete discharge" essentially means a phase-to-phase or phase-to-ground fault has occurred, which would typically trigger protective relaying to open-circuit breakers. As is discussed in Chapters 2 and 3, since gases (and air in particular) have a dielectric strength that is a small fraction of the dielectric strength of a solid or liquid insulation, PD tends to occur where there is a gas under high electrical stress. Thus, PD almost always occurs when there is a gas-filled void within the solid or liquid insulation, or there is gas adjacent to the solid/liquid insulation along a surface. PD can also occur in a gas adjacent to metal conductors where the electric field is rapidly decreasing the greater the distance from the metal conductor. Thus, PD can occur in all types of high-voltage apparatus, regardless of the insulation system, and may even occur at relatively low voltages if distances are small (Chapter 18).

A corona discharge is a particular type of PD. In IEC 60270, corona is described as "a form of partial discharge that occurs in a gaseous media that is around conductors that are remote from solid or liquid insulation." The most common type of corona occurs in overhead electric transmission lines, from which its distinctive crackling sound can often be heard, especially during rainy/snowy/foggy weather. Such corona is caused by localized breakdown of the air due to the high electric field adjacent to the bare aluminum conductors. The corona is very localized, since the electric field more than a few centimeters away from the high-voltage conductors is too low for electrical breakdown to occur. Thus, there is no "complete breakdown" between the transmission line conductors and ground. The term "corona" has been reserved for this type of PD since, on dark nights, the glow of the "corona" surrounding the lines can often be observed visually. To clarify, corona is often visible and caused by nonuniform electric fields in the air or gas. Corona itself does not directly damage the "electrical insulation" since, for the most part, electron and ion bombardment of gas molecules have no lasting effect, and although metals may experience some discoloration and pitting, and corona can produce by-products such as ozone, this usually does not impair the function of the HV apparatus. Also, the glass and ceramic insulators that hold up the overhead transmission lines are inorganic and extremely resistant to corona. In fact, the only real negative impact of corona is the radio and television interference they cause, as well as the energy losses due to corona on the transmission line.

A hundred years ago, the terms "ionization" and "corona" were used for what is now called PD. In the 1920s, the term corona became more popular than ionization. After the 1940s, more and more papers referred to both corona and (partial) discharges interchangeably. Once the definition of corona and partial discharge were clarified by many standard-making organizations in the 1960s, corona and PD should no longer be used as synonyms. In reviewing the literature, Europeans adapted more quickly and tended not to use the corona and PD as synonyms after the 1960s. North Americans tended to use corona and PD interchangeably well into the 1980s (and a few older persons still get mixed up). In this book, PD will refer to all types of incomplete discharges. Corona will be used to refer to a particular type of PD that is associated with highly divergent electric fields around metal conductors in air.

1.3 Categories of PD Tests

PD testing has two main purposes:

- as a factory test on new equipment; and
- as a test to determine if insulation aging is taking place in installed high-voltage equipment.

The first is an offline test (that is an external AC supply is needed to energize the equipment to the test voltage). There are subcategories of factory tests: PD tests during the development stage of new equipment; type tests on a small percentage of test objects to ensure the PD is within requirements; and routine tests (quality assurance or QA tests) done on every new piece of equipment to ensure the that each test object meets manufacturer's production standards, international or national standards, and/or customer specifications. The manufacturer's production standards may exceed the requirements of international or customer requirements.

The second category can be either offline or online testing. In online testing, the test equipment is energized from the power system.

1.3.1 Factory PD Testing

Virtually all electrical equipment that uses at least some solid or liquid insulation and that is rated above about 3 kV (phase-to-phase, rms), may be given a routine factory PD test at rated or higher voltage before the equipment is shipped. Thus, either the original equipment manufacturer (OEM) of power cables, transformers, air- and gas-insulated switchgear will voluntarily perform PD tests as part of their factory quality assurance program, or the end user (eventual owner of the equipment) may require a PD test before shipment.

As mentioned above, and as discussed in some detail in Sections 3.6 and 3.7, PD will damage organic insulation materials such as polyethylene, rubber, epoxy, and oil/paper composites. The electron and ion bombardment of organic materials leads to electrical treeing or surface electrical tracking. With sufficient time, the tree or track will cause a phase-to-ground or phase-to-phase fault, and thus equipment failure. The main purpose of a factory PD test is to ensure that HV equipment using organic insulation has no PD during normal operation, and, therefore, cannot fail prematurely due to PD. In addition, if the PD activity in a specific piece of equipment is higher than occurs in the same equipment made in the past by the OEM, even though it meets requirements, it may be an indication that the components or the manufacturing process has changed. This is a signal to the OEM to investigate the root cause of the increase in PD activity to avoid similar problems with future production. For example, if the partial discharge extinction voltage (PDEV, Sections 3.6.1, 8.7.5, and 10.2) test is lower than normal in a few reels of XLPE power cable, it may mean that the extrusion process is not using the correct pressure, flow rate, etc.; the polyethylene pellets are contaminated; the curing cycle is wrong, etc., and therefore the manufacturing process should be corrected before more cable is made.

The presence of PD-like electrical interference (Chapter 9) that can lead to false indications of high PD levels in onsite or online tests (Sections 1.3.2 and 1.3.3) tends not to be too much of a problem for factory tests. This is because the tests can often be done in an electromagnetically shielded area, use an interference-free AC test supply, and/or the source of the interference can be eliminated by doing the tests when most sources of interference are not operating (e.g. at night or on weekends).

Power cables (PE, XLPE, EPR, EPDM, as well as oil-paper insulated cables), capacitors (using polymer films impregnated with a liquid), and liquid-filled power transformers (mainly oil-paper composites) all use purely organic insulation as the main insulation material. Thus, as far back as 1926, researchers were investigating the use of PD (or as they called it "ionization" testing as a QA tool in factories) [1]. In the 1950s, what today would be recognized as factory PD tests were becoming more established, as discussed by Dakin [2]. Today, most equipment that is primarily insulated with organic insulation has associated standardized PD test procedures, often with minimum acceptable levels of PDIV or PDEV. The standards are prepared by IEEE (Institute of Electrical and Electronic Engineers), IEC (International Electrotechnical Commission), and various national standards bodies. Chapters 12–15 identify the relevant QA test procedures for each type of high-voltage equipment.

Air-insulated metalclad switchgear (AIS) and gas-insulated switchgear (GIS) use air and SF_6, respectively, as the main insulation. However, the high-voltage busbars are usually supported by organic-insulated components such as fiberglass-reinforced polyester boards (AIS) or epoxy spacers (GIS). Such switchgear may also include insulating rods to operate switches, potential transformers (PTs), and current transformers (CTs) that employ molded epoxy. PD tests on these components are essential to ensure that the switchgear does not fail in service. In addition, metallic debris may be present because of the manufacturing process that can lead to corona (and even bouncing metallic particles in GIS). Thus, PD testing has long been required for assembled AIS and GIS in most countries to ensure equipment reliability, using associated standardized tests (Chapters 13 and 14).

Rotating machines have always been in a special class for factory QA testing. As discussed in Chapter 16, the high-voltage insulation in motor and generator stator windings is a composite of mica tapes bonded together with epoxy (epoxy-mica insulation). Mica, being inorganic, is extremely resistant to PD attack, and stator windings using mica tapes have been known to withstand low and moderate levels of PD in service for many decades. As a result, even though there are IEEE and IEC standards for factory PD testing, there are no international standards for acceptable and unacceptable PD activity for new equipment. Instead, OEMs often perform PD testing on newly manufactured stator windings (especially on air-cooled motors and turbine generators), as a means of ensuring the manufacturing process has not changed, rather than as an acceptance test.

1.3.2 Onsite/Offline PD Tests

Some types of new equipment, because of their physical size, must be assembled at the utility or industrial plant where it will be used. This includes large AIS, almost all GIS, cable circuits once joints and terminations are installed, and most hydro generator stator windings. Thus, the final "factory" test or "commissioning" offline PD test must be conducted at the enduser location ("onsite") to verify the quality of assembly. This is also the case for large liquid-filled power transformers, since often the insulating liquid is added only when the transformer has been delivered to the enduser site.

However, probably the more common reason for performing PD tests at the enduser site is to determine if the electrical insulation is degrading, and maintenance may be required. This requires a baseline test (which could be the commissioning tests mentioned in the previous paragraph), followed by offline tests on the equipment over the years to detect if the PD inception voltage or the extinction voltage is decreasing; or the PD magnitude at a specific test voltage is increasing over time.

The key aspect of onsite/offline tests is that the high-voltage equipment is disconnected from the power system, and a 50/60 Hz high-voltage test supply is brought to site and used to energize the capacitance of the test object. As an alternative to 50/60 Hz voltage, sometimes the high-voltage equipment may be energized using 0.1 Hz AC or an oscillating damped wave voltage. Another alternative consists of a portable variable-frequency resonant test set, where an inductance is made resonant with the test object capacitance. For power transformers, the high-voltage winding is often energized by exciting the low-voltage winding with an external power supply operating at few hundred Hz (Section 15.8). In all cases, the HV test voltage supply must have the kVA capability to raise the voltage of at least one phase of the HV equipment to the test voltage, which often is higher than the rated line-to-ground operating voltage.

The other important requirement is that PD-like interference (also called disturbances) must be minimal to measure PD from the test object alone. With onsite/offline PD testing, the test voltage supply is expected to be interference-free, eliminating an important source of interference. However, onsite PD tests are still susceptible to RF signals coming from any other PD, arcing, or sparking elsewhere in the enduser plant/station. This may greatly increase the false indication rate or reduce the sensitivity to test object PD, compared to factory PD tests. Methods to reduce the influence of such external interference are discussed in Sections 8.4 and 9.3.

1.3.3 Online PD Testing and Continuous Monitoring

In the past few decades, online PD testing, where the high-voltage equipment is self-energized, i.e. energized from the power system, is becoming more popular. The purpose is to detect any aging that has led to an increase in PD activity, and thus a greater risk of HV equipment failure, without having to shut down the HV equipment for an offline test. Since PD is an important indicator of insulation aging or cause of failure for many types of equipment, regular online testing of the PD facilitates condition-based maintenance (CBM), a powerful method for determining when maintenance or replacement is needed.

For online PD testing, most types of PD sensors must be pre-installed during an outage (i.e. the HV equipment is disconnected from the power system) for personnel safety reasons. Online PD testing comes in two flavors: periodic testing with a portable instrument or continuous monitoring with a permanently installed instrument (Section 8.6).

The most difficult aspect of online PD testing is dealing with PD-like interference from the power system, as well as other disturbances from arcing and sparking within the plant or substation. Some of the interference can be exceptionally hard to separate, since it is actually PD or corona from other equipment in the plant or substation, plus the signal levels of such sources can exceed the level of the PD signals in the equipment of interest by several orders of magnitude. An example would be harmless PD occurring on the surface of a transformer ceramic bushing due to rain or snow, or from a sharp protrusion on an adjacent overhead line. If this PD is confused with PD from within the transformer or within the transformer bushing, an asset manager may believe the transformer windings are in trouble, and schedule costly but unnecessary maintenance. As discussed in Chapter 9, there are many hardware- and software-based methods to suppress such disturbances. Many of these are specific for the type of equipment to be tested and are discussed in detail in Chapters 12–16.

1.4 PD Test Standards

As might be suspected in a technology that has been used for more than 100 years, and where the consequences of failure due to PD may result in losses of tens of millions of dollars, there has been considerable effort over the decades to create and revise PD test standards. Perhaps the oldest standard that is directly relevant is the (USA) National Electrical Manufacturers Association (NEMA) Standard 107, "Methods of Measuring Radio Noise" in 1940 [3]. This standard was created to provide a standardized method of measuring the interference from transmission line corona on broadcast radio signals. However, it was also used as an early standardized method for researchers measuring PD in power transformers and bushings [2]. NEMA 107 is revised from time to time and still in current use.

The best-known PD standard is IEC 60270 [4], which has been adopted as a national standard by many countries. This horizontal standard specifies a general-purpose method for offline PD measurement in the low-frequency range (up to about 1 MHz), on any type of test object, and applied either in the factory or for onsite, offline testing. It was developed in the 1960s and published in 1968 (where it was originally called IEC 270) [4]. A few years later, in 1973, a very similar standard was published by the American Society for Testing Materials: ASTM D 1868 [5]. IEEE also produced a similar general-purpose PD test procedure in 1973: IEEE 454, which was subsequently withdrawn, as well as IEEE C37.301, which is IEC 60270 adopted for use in switchgear. All these standards are concerned with the measurement of PD in the 30 kHz to 1 MHz frequency range, using either "narrow band" or "wideband" frequency measurement (Section 6.5.2). The main output of the test is the magnitude of the PD pulses in terms of the apparent charge of each PD pulse. The PD sensor is most often a PD-free coupling capacitor (typically in the range of 100–1000 pF) in parallel with the test object with a detection impedance; or a high-frequency current transformer (HFCT) on the ground side of the test object. These standards also inform how to convert the detected millivolt (mV) signal to apparent charge (picoCoulombs) for capacitive test objects. IEC 60270 is discussed in detail in Chapter 6.

In 2016, the first general-purpose (applicable for all types of apparatus) guidance was published covering electrical PD measurement in the frequency range between 3 and 3000 MHz, that is well above the frequency range specified in IEC 60270. This document recognized the growing application of onsite/offline and online PD testing in high-voltage equipment. The use of higher measurement frequencies usually reduces the risk of false indications due to external electrical interference and often enables the PD sites to be located within the test object. This document, IEC Technical Specification 62478, discusses several types of PD sensors (capacitors, high-frequency current transformers, and antennas), disturbance suppression methods, and PD site location methods [6]. The standard also makes it clear that it is impossible to "calibrate" detected mV signals in terms of apparent charge (pC) in these higher frequency ranges. Chapters 6 and 7 present the differences between conventional IEC 60270 charge-based PD tests and "unconventional" PD measurements at the higher frequency ranges covered in IEC TS 62478.

In addition to the general-purpose PD standards, there are many standards for the measurement of PD for each type of equipment (e.g. power cable, transformers, switchgear, and rotating machines). These more focused standards are important since how to energize the test object, the placement and type of PD sensors, etc., can often be optimized based on the physical structure of the equipment. Also, the interference suppression methods tend to be different for each type of HV equipment and each type of equipment will have its own likely causes of PD. Each cause of PD may have a different phase resolved PD (PRPD) pattern (Section 8.7.3), and thus interpretation tends to be different for each type of equipment. Both IEC and IEEE have developed standards for each type of equipment, as indicated in Chapters 12–16.

1.5 History of PD Measurement

The history of PD testing and the equipment used goes back to the 1910s. One of the first English-language papers was by Prof. Edward Bennett where he used a coupling capacitor to detect PD currents with an oscillograph to measure PD from high-voltage transmission line equipment [7]. The oscillograph is a relatively fast responding electromechanical device like an X–Y chart recorder. The recorded public discussion of this paper shows that PD measurements were being made 15 years beforehand – i.e. 1898! Since this publication there have been many hundreds of papers published on PD measurement methods and technology. There have been over 20,000 papers on PD (and corona) measurements on HV equipment in the IEEE and IEE/IET alone, according to an IEEE Xplore search (Figure 1.1). Clearly this is a prolific field.

When a sampling of these papers is reviewed, there seems to be three eras in the development of PD measuring equipment:

- Radio interference voltage (RIV) methods
- Analog detection up to 1 MHz using oscilloscope displays
- Digital detection and computer-based processing and measurement up to the GHz frequencies

Although the word "era" is used, in fact variations of measurement methods symbolized by each era are still in use today. The following presents a summary of the developments of each era, identifies some key personalities, and some of the companies that first introduced commercial equipment.

1.5.1 RIV Test – The First Era

As mentioned above, this is the first widely applied method to measure PD, although that was not the original purpose of the test. RIV is variously defined as the radio influence voltage or the radio interference voltage. The original purpose of this test was to determine the level of corona

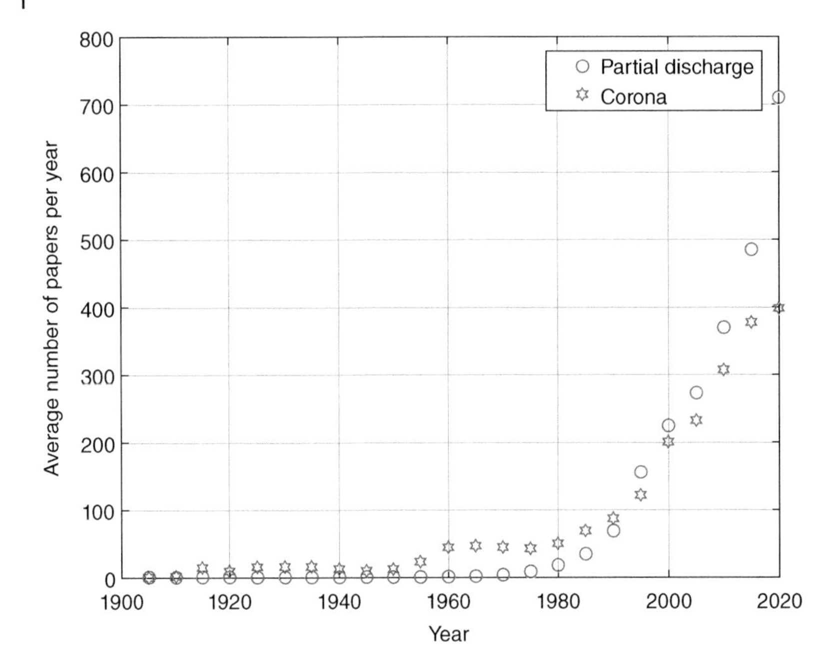

Figure 1.1 Plot of the number of papers on PD and corona vs year of publication in IEEE Xplore.

interference (in microvolts) that an overhead (outdoor) transmission line or its associated (typically glass or porcelain) insulators or transformers (typically, step-down) produces during operation. If the level is too high, complaints from the general public about poor analog radio and analog TV reception could be expected. The PD sensor was either an antenna or some type of coupling capacitor. The signals were measured by a specialized radio receiver (sometimes called a radio noise meter), usually with a center frequency about 1 MHz (i.e. within the normal AM radio broadcast band) with a narrow bandwidth of about 10 kHz. The output of the instrument was a meter that displayed the "quasi peak" – weighted level of the electrical noise produced by the PD activity. In addition, a demodulated signal from the corona could be listened to on a speaker or headphones.

Although early receivers were made by researchers, eventually commercial instruments were made for RIV measurements by Stoddart Aircraft Radio Co. in the United States and Siemens in Europe, among others. The Stoddart noise meters were manufactured beginning in 1944 by founder and IEEE Fellow Richard Stoddart. The use of Stoddart noise meters for PD measurement was a small part of the company's main business.

Since the noise meters were fundamentally to measure corona from transmission lines, researchers started using RIV methods to detect PD in oil-paper-insulated power cables and oil-filled power transformers. In 1924, Del Mar applied RIV detection to measure the PD in oil-impregnated cables to determine the maximum design electric stress for the insulation [8]. In 1965, two papers described PD measurement in power transformers for factory QA testing using the RIV method [9, 10], also referring to NEMA Standard 107 for the relevant RIV test method. The PD sensor was often the capacitance tap on the transformer bushing, normally used to measure the transformer voltage (Section 15.7.1). Meador suggested a 1000 μV limit at 1 MHz and said the main problem was electrical noise elsewhere in the factory [9]. Dr. Tom Dakin, in his chapter 6 in the Bartnikas/McMahon book on corona [11] suggested that the RIV type of PD test was still the most common type of factory PD test for transformers, up to at least 1979, when the book was written.

Although not widely recognized at the time, Mr. John Johnson of Westinghouse made a critical advance in the late 1940s with the application of PD testing to online insulation condition assessment [12]. We believe these were the first online measurements not intended to measure the radio interference from transmission lines.[1] Initially the PD pulses were detected across a resistor between the stator neutral and ground, using an early oscilloscope to measure the signal.

In the 1980s, Jim Timperley adapted the original RIV method to operating generators, together with Johnson's neutral detection [13]. Instead of measuring the PD level at a fixed frequency or using an oscilloscope, he used a specialized radio receiver that is commonly used for electromagnetic compatibility applications (i.e. measuring the RF signals emitted by electronics, power supplies, etc. to ensure they do not cause other equipment to malfunction). These commercial instruments (which are also close cousins of RF spectrum analyzers) produce a plot of RF signal magnitude (in μV) vs frequency. He initially explored the frequency ranges up to a few MHz, but later expanded the range up to 100 MHz. The PD sensor was usually a high-frequency current transformer (HFCT) mounted on the generator neutral. Timperley preferred to call this version of the RIV test the electromagnetic interference (EMI) test. There are many ways to estimate the peak PD activity, and Timperley uses the definition of quasi-peak in the IEC/CISPR 16-1 Standard. The test is still being performed by a few utilities and service companies today, although some have rebranded EMI testing as electromagnetic signature analysis (EMSA) test.

As presented in Chapter 13, EMI methods have also been applied to both offline and online PD testing of GIS.

1.5.2 Analog PD Detection Using Oscilloscopes – The Second Era

The second era of PD measurement is based on the measurement of PD using analog electronics and displaying the PD on an oscilloscope, so that the PD pulses could be seen with respect to the 50 or 60 Hz AC cycle. That is, it is a time-domain measurement, unlike first era RIV/EMI testing, which is a frequency-domain test. As is seen in Chapter 3, PD occurs in specific regions of the AC cycle depending on its cause and/or location within the insulation system. The modern (at that time) PD instrument depended on the development of better oscilloscopes. The oscilloscope can trace its history back to the development of the cathode-ray tube (CRT) by Nobel-prize winner Dr. K.F. Braun in Germany in 1897. His CRT was used by many researchers in the early 1900s to visualize the voltage and current waveforms of discharges. There were many improvements in CRTs by many researchers over the decades, but it waited until Tektronix invented the Tek 511 oscilloscope in 1946 for oscilloscopes to become externally triggered, calibrated, easy-to-use devices for PD research. The Tek 511 could record signals up to 10 MHz, which corresponds to a 30 ns pulse risetime. This led to the belief that PD pulses had risetimes of several tens of nanoseconds (instead of a few nanoseconds or less as discussed in Chapter 3). It was only logical that specialized oscilloscopes became incorporated into commercial PD instruments.

This era could be said to have started in the 1950s with the work of Dr. George Mole of the British Electrical Research Association (ERA). Mole produced a PD measuring system including a 1 nF

---

1 In a discussion with one of the authors in 1980 or so, Johnson shared an important anecdote: he said they felt they needed to perform online PD testing since Westinghouse, who had recently introduced the Thermalastic™ insulation for stator windings using mica impregnated with the synthetic polyester insulation, was suffering premature failures due to loose coils in the stator slots, leading to surface PD (what Johnson called slot discharge). Offline PD tests were not sensitive to this problem, and online testing was necessary since coil vibration in the slot only occurs when current was flowing though the coils.

Figure 1.2 Recent photograph of a still-working ERA Model 3S, manufactured by Robinson Instruments, probably in the late 1970s. The instrument was first owned and used by the former British utility the CEGB, before being acquired by Iris Power in the 1990s for its museum. The oscilloscope screen shows the typical elliptical trace of a 60 Hz sinewave with PD from a twisted pair of insulated wires superimposed on it. *Source:* Mladen Sasic, Iris Power L.P.

high-voltage PD coupler, a detection impedance using RC or RLC components, a method of synchronizing the PD to the AC cycle, analog filters, and a display based on a CRT [14]. A feature of the display was the use of an ellipse (Lissajous figure) to display the 50 or 60 Hz AC waveform. This allowed a single channel oscilloscope to display both the AC voltage and the PD in a single trace, as well as effectively doubling the sweep time-base compared to a conventional horizontal oscilloscope time-base. That is, the effective sweep speed was 1 ms/division, instead of 2 ms/division with a conventional 50 or 60 Hz sine wave, enabling the very short-duration PD pulses to be more easily seen with respect to the AC cycle. The Mole instrument, and later versions up to the ERA Model 5, were manufactured by Robinson Instruments in England. Figure 1.2 shows a recent photograph of the ERA Model 3, and the AC voltage ellipse on which the detected PD pulses are superimposed. After the commercial success of the early ERA detectors, many companies around the world made similar devices including Biddle Instruments (now part of Megger) and Hipotronics (now part of Hubbell) in the United States and Tettex Instruments (now part of the Haefely/Pfiffner Group) in Switzerland.

This generation of PD detectors worked in IEC 60270 frequency range (that is up to about 1 MHz) and in contrast to RIV methods could display the PD pulses on an oscilloscope screen with selectable "narrowband" or "wideband" frequency ranges. These analog instruments could accommodate a wide variety of coupling capacitors and test object capacitance, usually with different impedance matching units (sometimes referred to as "quadripoles" (see Section 6.3.1)) having different resistance, capacitance, and inductance (if present). The output was both an oscilloscope screen and a meter that recorded the peak (or quasi-peak) PD magnitude. Permanent recordings of the oscilloscope screen were usually made with a camera, usually a Polaroid™ instant camera, and the magnitudes estimated using a ruler.

The availability of commercial instrumentation in the 1950s that were specifically intended for PD measurement and display led to an explosion in applications to all types of high-voltage equipment. One of the pioneers of this new era was Prof. Frederik Kreuger of Delft University in The Netherlands. His PhD work led to the publication of the first English-language book about PD measurements in 1965 [15]. After a short stint at ASEA in Sweden, for most of his career Kreuger worked for the Dutch cable manufacturer Nederlandse Kabelfabriek. Kreuger, who died in 2015, investigated different PD detection methods and their sensitivity, did research on the best PD detection methods for each type of HV equipment (and especially power cables), developed what is now known as the Kreuger PD bridge to suppress disturbances, and explored how to calibrate the detected signals into apparent charge (pC) [16]. His work led directly to the development of the first international standard for application to PD measurements (IEC 270) in 1968.

Another leading researcher in this era was Dr. Ray Bartnikas. Like Kreuger, Bartnikas began his career with a cable manufacturer (Northern Electric in Canada), before continuing his research into PD measurement at the utility Hydro-Québec's Research Institute (IREQ). Bartnikas investigated optimal methods and limitations for calibrating PD in terms of apparent charge, did research into different forms of PD (including pseudo-glow discharge, Sections 3.5.3 and 4.7), and was key to the effort to develop the first American standard on PD detection, ASTM D1868, in 1973. Bartnikas also edited a book on PD measurement and interpretation, published in 1979, which is still in print [11]. As discussed in Section 1.5.3, Bartnikas, who died in 2022, was also active in the digital era with the development of PD pulse magnitude analyzers.

The research of Kreuger and Bartnikas, together with the commercial availability of relatively portable PD measuring systems, led to the widespread application of PD measurement, both in factories for QA testing of HV equipment, and also in research applications. By the end of the 1960s, virtually every manufacturer of HV equipment, plus every high voltage laboratory, had at least one of these detectors.

Another important personality of this era is Prof. Eberhard Lemke from the Technical University of Dresden, Germany. He also worked for a short time at a power cable manufacturing company, before starting his own company in 1990, Lemke Diagnostics, where he first commercialized the Lemke probe, an RF probe to locate PD sites. The company, which was eventually bought by Doble Engineering, also made IEC 60270 compliant PD instruments. Besides developing his probe, he was very active in researching the physics of PD, PD detection, and PD instrument calibration. He wrote the chapters on PD in a widely read book on high-voltage engineering [17], and chaired a CIGRE committee that prepared a technical brochure on using the 2000 version of IEC 60270, which also has a comprehensive bibliography of English- and German-language papers on the subject [18].

1.5.3 Digitizing, Ultrahigh Frequency, and Post-Processing – The Third Era

The third and current era has had three main technical focuses:

- The transition from analog electronics with an oscilloscope display to digital electronics and digital storage/display of PD data on a computer;
- With the availability of faster digital electronics (and especially analog to digital converters – ADCs), the gradual trend to measure PD at higher frequencies, into the ultrahigh frequency (UHF) range;
- Processing of captured data using digital logic devices (real time) or computer software (post-processing) to separate PD from disturbances and to identify the root cause of any detected PD.

1.5.3.1 Transition to Digital Instruments

Research into digital techniques of measuring PD can be said to have started with Bartnikas and his pulse magnitude analyzer in 1969 [19]. These early digital circuits used discrete transistors to segment the pulse magnitudes into several magnitude bins (or magnitude windows), and then count the number of pulses in each bin over a period of time. As seen in Figure 8.9, the output was a two-dimensional plot of pulse magnitude (horizontal scale) vs a (usually logarithmic) vertical scale of pulse count rate (number of pulses per second per magnitude window). Another important step was taken independently in 1976 by Dr. Andreas Kelen of ASEA in Sweden and Professors Austin and James in the United Kingdom [20, 21]. They combined home-made pulse counting electronics with the digital computers then available to count not only the number of pulses per magnitude window but also the pulses at different parts of the AC cycle. Many such research instruments that could record the number and phase position of the PD pulses were described in the 1980s. In 1988, Bernhard Fruth, Lutz Niemeyer, Marek Florkowski, and Jitka Fuhr of ABB Corporate Research in Switzerland developed a system using the IEC 60270 frequency range called the "PRPDA" – phase-resolved partial-discharge analyzer – probably the first to use the term [22]. The PRPD plot has now become an essentially quasi-standard two-dimensional "color-map" display of the three-dimensional matrix of PD pulse magnitude (vertical or *y*-axis) vs AC phase position (horizontal or *x*-axis) vs pulse count rate (the *z*-axis, represented by changes in pixel color). A concise summary of all this research was published in 1991 by Barry Ward, who chaired an IEEE working group on the subject [23].

One of the first widely used commercial IEC 60270-compliant digital PD instruments was made by Power Diagnostix, which was founded in 1992. It was developed by Dr. Detlev Gross and Dr. Bernhard Fruth, and introduced in 1993. As already mentioned above, Fruth had been an employee of ABB Corporate Research in Switzerland where he researched electrical aging and PD detection in rotating machines, HV cable, bushings, and other insulation materials. Gross had started his own electronics company in 1986 and worked with Fruth to develop what was called the ICM (Insulation Condition Monitor), again employing and further refining and popularizing the PRPD "color-map" plot of PD magnitude vs. 50/60 Hz phase cycle position vs pulse count (again displayed as color). After the Power Diagnostix (now part of Megger) ICM instrument was introduced, many companies, including Hipotronics, Lemke, Omicron, TechImp, Tettex, and many others, introduced similar IEC 60270-compliant instruments using mainly digital technology. Interestingly, Power Diagnostix also introduced an instrument that combined digital time-domain PD measurement with a spectrum analyzer [24]. Prof. Lemke, in a CIGRE brochure, outlined some of the methods used by commercial PD instruments to digitally capture and measure PD in the IEC 60270 frequency range [18]. Since 2000, very few second-era analog PD instruments were being used, due to the convenience and flexibility of digital PD instruments, as well as their ability to share data files with computers for display and data manipulation.

1.5.3.2 VHF and UHF PD Detection

The second focus of this era depended on the development of better oscilloscopes. Tektronix introduced the Tek 465 scope in 1972. Except for the CRT, it was among the first oscilloscopes to use solid-state electronics with a 100 MHz bandwidth. Of special importance for PD measurement was the introduction of a Tek 466 single-shot storage oscilloscope in 1972. The Tek 466 had a 100 MHz bandwidth, so it could clearly display a PD single pulse with a risetime as short as about 4 ns.[2] The development of the analog Tek 7104 oscilloscope in 1978 allowed the PD current pulses to be accurately recorded for the first time, since it had a bandwidth of 1 GHz (corresponding to a 0.3 ns risetime) and its microchannel image intensifier plate made clear photographic recordings of single PD current pulses possible for the first time (Figure 1.3). With each increase in oscilloscope bandwidth up to the 1 GHz range, the risetime of the PD current pulses was found to be shorter

2 One of us, who had Dr. Bartnikas as a MASc co-supervisor in 1977, recalls Bartnikas being astonished, when he was told that the measured risetime of the PD current from an electrical tree was less than 4 ns.

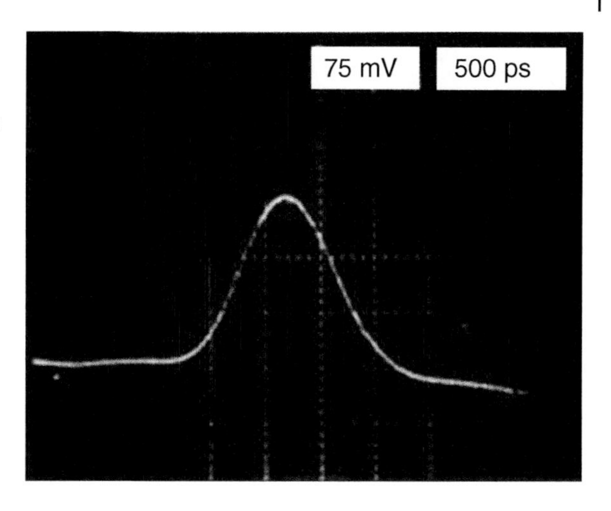

Figure 1.3 Oscilloscope photograph of a single PD current pulse through a 50-Ω resister in series with the test object (an electrical tree growing in epoxy, initiated from a razor blade) measured with a Tektronix 7104 1 GHz analog oscilloscope [25]. The vertical scale is 75 mV/division, and the horizontal scale is 500 ps/division.

than previously believed. The realization that PD created frequencies up to 1000 MHz led many researchers to investigate PD measurement in the VHF (30–300 MHz) and UHF (300–3000 MHz) frequency ranges. As discussed in Chapters 7 and 9, measuring PD in the higher frequency ranges reduced the risk of false indications from the severe electrical interference that is typically found in online PD measurement and directly led to widespread use of online PD measurement in GIS, transformers, and rotating machine stator windings (Chapters 13, 15, and 16).

As mentioned above, the development of 1 GHz oscilloscopes in the 1970s led to a tremendous amount of work on PD pulse shape. In 1982, the theoretical foundation for PD measurement above the IEC 60270 frequency range was presented by Dr. Steven Boggs, who worked for the utility Ontario Hydro in Canada (Boggs continued his research at the University of Connecticut in the United States) [25]. Boggs recognized that what he called ultrawide band (UWB) PD detection with a sensor close to the PD site would have superior ability to suppress interference, especially during online PD measurements. He and his colleagues first applied VHF and UHF detection of PD to GIS and machine stator windings. They recorded the pulse shapes from many test objects and defects. Figure 1.3 shows a single PD pulse from an electrical tree growing in an epoxy. The pulse has a risetime of about 500 ps and a pulse width of 1.5 ns (full width, half maximum). This is probably one of the first images of the true PD current pulse shape. Boggs used various types of capacitors and voltage dividers to achieve several hundred MHz bandwidth in GIS [25]. In 1991, Dr. Brian Hampton and his colleagues at the University of Strathclyde in Scotland published the design of a practical PD sensor for GIS, combined with a continuous UHF PD monitoring system [26]. The sensor was a circular plate installed on the inside surface of GIS maintenance hatch covers (i.e. inside the GIS enclosure); these acted as antennas to pick up the electric field of the PD pulse as it passed through the coaxial waveguide formed by the GIS (Section 13.8.2). Although the PD was detected in the UHF range, they used a demodulator to down-convert the UHF signal so that conventional low-frequency electronics could be used for pulse counting and determining the phase position of the PD.

In the late 1970s, Stone (who later went on to co-found Iris Power, now part of Qualitrol Corp, in 1990) and his colleagues at Ontario Hydro Research started measuring PD in the 30–300 MHz range in operating generators [27]. The advantage of the VHF frequency range is that the high-voltage PD couplers could be much smaller (80 pF) and thus fit within the generator frame; additionally, the time-of-arrival principle could be used to separate stator PD from power system

Figure 1.4 Photograph of the first commercial digital PD instrument that measured PD in the 30–300 MHz range. The PDA-H, introduced in 1986, measured the PD from a pair of permanently-installed 80 pF couplers in each phase of an operating hydro generator. The instrument was controlled by an early PC, which also served as the display device. *Source:* Mladen Sasic, Iris Power L.P.

disturbances using a pair of sensors per phase (Sections 8.4.2 and 16.9). In addition, disturbance suppression based on digital rendering of the pulse shape was possible [28].

From a commercial point of view, the extension of PD instruments to frequencies higher than 1 MHz (in what is now termed an IEC TS 62478-compliant instrument) started in 1986 when FES International (later known as Adwel and now part of Iris Power) introduced an instrument called the PDA-H to measure PD in operating hydro generators in the VHF range (Figure 1.4). In 1991, Iris Power, a spin-off from Ontario Hydro, introduced the PDA-IV, an all-digital instrument working in the VHF range that separated power system disturbances from stator winding PD on a pulse-by-pulse basis and displayed PRPD plots on a built-in LCD display. A year later Iris Power introduced the TGA-S, which worked in the UHF range with a special electromagnetic coupler (called the SSC, for *s*tator-*s*lot *c*oupler, Sections 7.4.5.4 and 16.6.2) that was installed in hydrogen-cooled turbine generators. Another UHF all-digital continuous PD monitor for GIS was introduced by a company based in Scotland called DMS. DMS, now a part of Qualitrol Corp., was founded by John Pearson, Brian Hampton, and Owen Farish of Strathclyde University in 1994, and the PD monitor was based on technology they developed at the university [26]. This was also the world's first commercial continuous online PD monitor. Today there are dozens of companies making VHF and UHF PD instruments, most of which are used for online PD monitoring.

1.5.3.3 Post-Processing of Signals

Another aspect of this era, which was facilitated by digital instruments that are easily interfaced to computers, is the development of tools to aid in the analysis of PD data. In particular, these tools use signal magnitude, phase position, count rate, and applied voltage at the time of the pulse to calculate various indicators of PD activity (quasi-peak magnitude, PD power, PD current, quadratic rate, etc. as outlined in IEC 60270), which are determined after the data has been stored in memory. Perhaps even more importantly, this post-processing can help to separate interference pulses from test object PD pulses and identify the nature of the causes of PD in a test object. This was important since there was a desire, as PD technology spread from the research/high-voltage test labs to HV equipment owners, that PD test users would be able to interpret PRPD patterns without having to be experienced PD researchers.

The first notable contribution in post-processing was made by Dr. Tatsuki Okamoto and Dr. Toshikatsu Tanaka of CRIEPI in Japan in 1986, when they started to apply statistical analysis of the PD patterns with respect to phase angle [29]. A few years later, Prof. Edward Gulski of Delft University in The Netherlands also used statistical methods based on the normal distribution to analyze PRPD patterns [30]. This work was eventually commercialized in a Haefely PD detection system. Prof. Alfredo Contin (University of Trieste) and Prof. Gian Carlo Montanari (University of Bologna) applied statistical analysis to PRPD pattern analysis using the Weibull probability distribution [31]. In all these early examples, the idea was to classify various PRPD patterns to determine the root cause of the PD. Although such techniques are not widely used today, they were the forerunners of other methods that have gained in popularity among PD test users.

In 2004, Prof. Andrea Cavallini and his colleagues at the University of Bologna and the University of Trieste in Italy were probably the first to use nonstatistical post-processing methods to suppress disturbances, as well as to identify different types of PD sources (e.g. differentiate void PD from surface PD). They developed what is known as the time-frequency (T-F) map method [32]. As described in Section 8.9.2, each pulse after A/D conversion was processed into the frequency domain at the same time as an indicator of pulse length was captured. A "map" was created with two axes (time and frequency) with the transformed pulse shape and frequency of each detected pulse. Cavallini discovered that disturbances and different types of PD sources tended to cluster in different regions of the T-F map. The clusters are identified by a skilled observer, or using specialized pattern-recognition algorithms. In many cases, there was a unique PRPD pattern for each cluster, and with experience, the patterns could be associated with different defects or disturbance sources. The technology was first applied to power cables, then spread to other types of HV equipment. This post-processing technology led to the creation by Montanari and his colleagues of a commercial company called TechImp (now part of Altanova/Doble).

Another commercial post-processing method was developed by Dr. Ronald Plath, Caspar Steineke, and Harald Emanuel at MTronix (now part of Omicron). The key feature of this post-processing method is to simultaneously capture the signals from all three phases [33, 34]. The response to an event (a PD pulse or an interference pulse) on all three phases is measured and correlated on a three-dimensional plot of the pulse magnitude in each phase (Section 8.9.3). The "3PARD" plot consists of thousands of pulses. Different types of PD and interference apparently will create clusters in different regions of the diagram. As with the T-F method, clusters are identified, and they usually have a unique PRPD pattern that identifies the nature of the interference or PD sources.

In addition to these post-processing methods, many other signal processing methods have been applied, often using various forms of artificial intelligence or fractal analysis.

1.6 The Future

We have probably entered a fourth era of PD technology. Rather than specific technical advances, the main driver has been the widespread application of continuous online PD monitoring of high-voltage equipment that started as simple research tools 40 years ago. This technology allows HV equipment owners to determine the insulation condition at any time; when PD activity appears or passes certain thresholds (amplitude and/or pulse-count), maintenance engineers are alerted that there has been a change in the insulation condition, and HV equipment maintenance may be prudent. Continuous online PD monitoring commercially started on rotating machines and GIS in the 1990s. Now tens of thousands of machines and thousands of GIS bays are being continuously monitored. A key challenge of continuous monitoring is to extract the useful information from the vast quantities of data collected. As interference separation techniques (needed to avoid false-positive indications) and data reduction methods improve, we expect continuous monitoring to expand not only to other types of HV equipment but also in the number of systems installed. The users of this equipment are not PD researchers, or even experienced high-voltage laboratory staff, but are maintenance engineers or asset managers in generating stations, substations, and industrial plants, who use the input from PD technology as only one aspect of their jobs.

One of the problems holding back further advancement in this technology is the widespread unwillingness of both OEMs and utilities (users) to share detailed information about the data gathered by online monitoring systems vs actual PD defects found in the equipment. This strong tendency toward keeping such information confidential is due, on the one hand, to obvious aspects of competition between the OEMs, but on the other hand, to the general fragmentation and compartmentalization of the power generating and distribution industry; the utilities are reluctant to release any information that may impact their SLAs (service-level agreements) and thus their business models. Without access to this information, it is very difficult to assess the effectiveness of online PDM systems and use that information to improve the technology.

1.7 Roadmap for the Book

This book is primarily intended for technicians, engineers, and scientists whose involvement with PD may be just one part of a wider range of their responsibilities, and who need a better understanding of what they are measuring and how to make and interpret the measurements accurately and effectively.

Chapters 2–4 present a summary of the physics of PD and other associated phenomena. This information may be sufficient for users of PD technology. Researchers, however, should refer to the many references found in Chapters 2–4 for a deeper understanding.

Chapters 5–7 provide some fundamental information of the main ways to detect PD including charge-based "conventional" electrical PD detection (Chapter 6), "unconventional" electromagnetic methods (Chapter 7), as well as a summary of optical, acoustic, and chemical methods. Although the term "unconventional" may imply that electromagnetic methods are less commonly used – in fact the EM methods are far more widely used today than "conventional" methods.

Chapter 8 gives users of PD test equipment some understanding how commercial PD instruments work inside. However, each manufacturer of PD instrumentation will have their own design philosophy and intellectual property that is not shared with equipment users. Thus, readers may only want to review Sections 8.6–8.9, which are important for the interpretation of PD measurements on all types of electrical equipment.

Electrical interference (also known by some as disturbances) have long caused problems during PD measurement. Interference can lead to false-positive indications of insulation problems. Thus, Chapter 9 identifies the main sources of interference and the various methods that have been developed to suppress the influence of interference.

Chapter 10 gives an overview of the basic principles used to interpret PD measurements in all types of electrical equipment. This chapter should be read before reading the interpretation section for each particular type of equipment in Chapters 12–16.

Chapters 11–16 are the heart of the book. Each chapter focuses on a particular type of electrical equipment such as power cables, transformers, etc. Over the decades, PD measurement has tended to be optimized for each type of equipment; specifically the sensors, measurement frequency range, interference suppression methods, and applicable standards are often unique to each type of electrical equipment. Also, each type of equipment tends to have a unique set of insulation issues that give rise to PD. Thus each of these chapters gives an overview of the insulation system for each type of equipment, presents the ways PD can arise due to either manufacturing or aging in service, describes the sensors used and the normal frequency ranges for both offline and online tests, outlines the standards that may be applicable, and gives an overview of interpretation, with reference to the information in Chapter 10. Each chapter presents many case studies.

The two final chapters are on rapidly evolving topics: measurement of PD in DC equipment and measurement of PD during short risetime voltage impulses, as opposed to 50/60 Hz PD measurement discussed in Chapters 6–16.

High-voltage DC systems are increasingly being applied in the transmission systems of electrical grids, especially for overhead and underground/submarine transmission lines. Also, high voltage DC is used in specialized medical and research equipment where PD has been known to occur. Since there is no alternating voltage, and the role of trapped charge (Chapter 3) is more complex, the behavior of PD under direct voltage (DC) is even more stochastic than under AC. The main tool for interpreting PD from test objects using 50/60 Hz excitation, the PRPD plot (covered in Chapter 8), is not relevant under DC conditions, since there is no AC voltage with its inherent positive- and negative-going zero crossings "modulating" the electric field. Chapter 17 introduces practical PD measurement in DC systems, but the field is also rapidly evolving, so references to further reading are presented. In addition, due to the particular behavior of moving particles (a well-known source of PD in GIS) under DC conditions, some specific aspects of PD measurement under DC are also briefly discussed in Chapter 13.

The measurement of PD during voltage impulses is becoming an increasingly important subject with the widespread adoption of power converters and semiconductor-based (e.g. IGBT) switching technologies. The short risetimes of the voltage impulses produced by such power-electronic equipment subjects solid insulation to higher electrical stress and can lead to PD in the converter modules themselves, as well as any connected equipment (power cables, transformers and machine windings, etc.). PD pulse current detection is difficult during voltage impulses, since the voltage impulses are a type of interference that can dominate the PD current pulses. Advancements are made almost daily, so this topic is only briefly discussed in Chapter 18, with many references for further reading.

References*

1 Davies, C.L. and Hoover, P.L. (1926). Ionization studies in paper insulated cables-part I. *Proceedings of the American Institute of Electrical Engineers* 45: 141–164.

2 Dakin, T.W. and Lim, J. (1957). Corona measurement and interpretation. *Transactions of the American Institute of Electrical Engineers* 76: 1059–1064.

3 NEMA 107 (1940). *Methods of Measuring Radio Noise*. (US) National Electrical Manufacturers Association.

* In this book, some common abbreviations will be used to identify publishers. IEEE is the Institute of Electrical and Electronic Engineers, AIEE is the American Institute of Electrical Engineers (a predecessor organization of the IEEE), the IET is the British Institute Engineering and Technology, IEE is the British Institute of Electrical Engineers (a predecessor of the IET), IEC – International Electrotechnical Commission – the worldwide standards organization, CIGRE – Conseil International des Grands Réseaux Electriques – is the Paris-based world-wide organization collaborating on power systems and holding biennial conference, ASTM is the American Society of Testing and Materials.

4 IEC 60270:2000 (2015). High-voltage test techniques – partial discharge measurements.
5 ASTM D1868:2020 (2020). Standard test method for detection and measurement of partial discharge (Corona) pulses in evaluation of insulation systems.
6 IEC TS 62478:2016 (2016). High voltage test techniques – measurement of partial discharges by electromagnetic and acoustic methods.
7 Bennett, E. (1913). An oscillograph study of corona. *Transactions of the American Institute of Electrical Engineers* 22: 1787–1828.
8 Del Mar, W. and Hanson, C.F. (1924). High voltage impregnated power cables. *Transactions of the American Institute of Electrical Engineers* 24 (10): 947–957.
9 Meador, J.R., Kaufman, R., and Brustle, H. (1966). Transformer corona testing. *IEEE Transactions on Power Apparatus and Systems* 84: 893–900.
10 Narbut, P. (1965). Transformer corona measurement using condenser bushing tap and resonant measuring circuits. *IEEE Transactions on Power Apparatus and Systems* 84: 652–651.
11 Bartnikas, R. and McMahon, E.J. (1979). *Engineering Dielectrics Volume 1: Corona Measurement and Interpretation*, vol. 669. ASTM Publication.
12 Johnson, J. and Warren, M. (1951). Detection of slot discharges in high voltage stator windings during operation. *Transactions of the American Institute of Electrical Engineers* 70: 1998–2000.
13 Timperley, J.E. (1983). Incipient fault identification through neutral RF monitoring of large rotating machines. *IEEE Transactions on Power Apparatus and Systems* 102: 693–698.
14 Mole, G. (1952). Design and Performance of a Portable AC Discharge Detector. ERA Report V/T 115.
15 Kreuger, F.H. (1965). *Discharge Detection in High Voltage Equipment*. American Elsevier.
16 Morshuis, P. (1995). The scientific career of Frederik Hendrik Kreuger. *IEEE Transactions on Dielectrics and Electrical Insulation* 2: 711–716.
17 Hauschild, W. and Lemke, E. (2019). *High Voltage Test and Measuring Techniques*, 2e. Springer.
18 Lemke, E. (2008). Guide for partial discharge measurements in compliance to IEC 60270. Cigre Technical Brochure, 366.
19 Bartnikas, R. and Levi, J.E. (1969). A simple pulse-height analyzer for partial discharge rate measurements. *IEEE Transactions on Instrumentation and Measurement* 18: 341–345.
20 Austin, J. and James, R.E. (1976). On-line digital computer system for measurement of partial discharges in insulation structures. *IEEE Transactions on Electrical Insulation* 11: 129–139.
21 Kelen, A. (1976). The Functional Testing of HV Generator Stator Insulation. Cigre, Paper 15-03.
22 Fruth, B. and Fuhr, J. (1990). Partial Discharge Pattern Recognition – A Tool for Diagnosis and Monitoring of Aging. Cigre Paper 15/33-12.
23 Ward, B.H. (1991). Digital techniques for partial discharge measurements. *Committee Report, Proc IEEE PES Summer Power Meeting*, Paper 91 SM 355-8 PWRD.
24 Fruth, B.A. and Gross, D.W. Combination of frequency spectrum analysis and partial discharge pattern recording. *1994 International Symposium on Electrical Insulation*, Pittsburgh, PA USA (5–8 June 1994). pp. 296–300.
25 Boggs, S.A. and Stone, G.C. (1982). Fundamental limitations in the measurement of corona and partial discharge. *IEEE Transactions on Electrical Insulation* EI-17: 143–150.
26 Pearson, J.S., Hampton, B., and Sellers, A.G. (1991). A continuous UHF monitor for gas insulated substations. *IEEE Transactions on Electrical Insulation* 26: 469–478.
27 Kurtz, M., Stone, G.C., Freeman, D. et al. (1980). Diagnostic testing of generator insulation without service interruption. Cigre Paper 11-09.
28 Campbell, S.R., Stone, G.C., and Sedding, H.G. (1992). Application of pulse width analysis to partial discharge detection. *IEEE International Symposium on Electrical Insulation*, Baltimore. https://doi.org/10.1109/ELINSL.1992.246979.

29 Okamoto, T. and Tanaka, T. (1986). Novel partial discharge measurement: computer-aided measurement systems. *IEEE Transactions on Electrical Insulation* 20: 1015–1019.

30 Gulski, E. and Kreuger, F.H. (1990). Computer-aided analysis of discharge patterns. *Journal of Physics D: Applied Physics* 23: 1569–1575.

31 Cacciari, M., Contin, A., and Montanari, G.C. (1995). Use of a mixed-Weibull distribution for the identification of PD phenomena. *IEEE Transactions on Dielectrics and Electrical Insulation* 2: 614–628.

32 Cavallini, A., Conti, M., Contin, A. et al. (2004). A new algorithm for the identification of defects generating partial discharges in rotating machines. *IEEE International Symposium on Electrical Insulation* 204–207. https://doi.org/10.1109/ELINSL.2004.1380520.

33 Plath, K.D., Plath, R., Emanuel, H., and Kalkner, W. (2002). Synchrone dreiphasige Teilentladungsmessung an Leistungstransformatoren vor Ort und im Labor. ETG-Fachtagung Diagnostik elektrischer Betriebsmittel, Beitrag O-11, Berlin.

34 Plath, R. (2005). Multi-channel PD measurements. 14th ISH, Beijing, China, Paper J-04.

2

Electric Fields and Electrical Breakdown

2.1 Electric Fields in High-Voltage Equipment

Chapters 2 and 3 taken together introduce the physics that underlie partial discharges. Chapter 2 discusses electric field theory and the concept of electric breakdown of gaseous, liquid and solid insulating materials. Chapter 3 presents the breakdown processes that lead to partial discharges. The references cited in both chapters give a more in-depth description of the PD physics.

2.1.1 Impact of Electric Field on Partial Discharges

Partial discharges (PD) cannot occur unless there are high electric fields on the surface of, or within, the electrical insulation system that can lead to a localized electrical breakdown of the insulation. Thus, to understand the physics of electrical breakdown and PD, the electric field in the insulation system must first be measured or calculated. An electric field can be created by

- a voltage between conductors, with the field determined by the geometry of the conductors and the dielectric materials separating them, along with their corresponding dielectric constant(s),
- the presence of charge carriers such as electrons and ions.

Electrical field theory is covered in all introductory texts on electromagnetism, for example Krause [1] or Hayt [2], and is, therefore, not covered in detail here. Instead, after a summary of the important relationships, the electric fields of various practical geometries are presented.

2.1.2 Basic Quantities and Equations

As the rate of change of the electrical quantities in electric power systems is slow when compared to the nanosecond-duration PD events, the quasi-stationary assumption can be invoked. This means that the equations governing the electric field within an insulation system are those applicable to electrostatic fields, summarized in Table 2.1.

In Table 2.1 and in the remainder of this book, the following symbology is adopted:

- $\boldsymbol{E}$ = electric field vector
- $\boldsymbol{D}$ = displacement vector
- $\boldsymbol{J}$ = current density vector
- V = electric potential
- ε = dielectric permittivity

Practical Partial Discharge Measurement on Electrical Equipment, First Edition. Greg C. Stone, Andrea Cavallini, Glenn Behrmann, and Claudio Angelo Serafino.

Table 2.1 Laws governing electrostatic fields.

Field equations (point form)		**Material relationships**	
$\boldsymbol{E} = -\nabla \boldsymbol{V}$	(2.1)	$\boldsymbol{D} = \varepsilon \boldsymbol{E}$	(2.4)
Gauss's equation: $\nabla \cdot \boldsymbol{D} = \varrho$	(2.2)	$\boldsymbol{J} = \sigma \boldsymbol{E}$	(2.5)
Current sources $\nabla \cdot J = -\dfrac{d\varrho}{dt}$	(2.3)		

- σ = electric conductivity
- ρ = free charge within the insulation system

Note that bold characters indicate vectors. Mathematical operators:

- $\nabla\varphi$: gradient, applied to a scalar quantity φ
- $\nabla\boldsymbol{A}$: divergence, applied to a vector quantity $\boldsymbol{A}$
- $\nabla^2\varphi = \nabla(\nabla\varphi)$: Laplacian operator, applied to a scalar quantity φ

Assuming that the current density is negligible everywhere in the insulation system (i.e. the right-hand term in Equation (2.3) can be neglected), then Equations (2.1), (2.2), and (2.4) can be combined to achieve (2.6), i.e. the Poisson equation:

$$\nabla^2 V = -\frac{\varrho}{\varepsilon} \tag{2.6}$$

The Poisson equation allows the estimation of the electric field when charge carriers such as electrons and/or ions are present. Such charges may be present from past discharges in a gas and are often referred to as space charge. In liquid and solid insulation, the charges may be relatively immobile and are referred to as trapped charge.

When the free charge within the insulation system (space or trapped charge) can be neglected, (2.6) simplifies to (2.7), i.e. the Laplace equation:

$$\nabla^2 V = 0 \tag{2.7}$$

This enables the estimation of the electric field due to geometry and the dielectric materials alone.

Except for simple geometries, the solution of either (2.6) or (2.7) is difficult to calculate analytically (see an example of a complex geometry in Figure 2.1). Several finite element method (FEM) software packages from COMSOL, Ansys, and other companies are available on the market to calculate the electric field for real geometries. In the next section, a few configurations whose solution can be calculated analytically and are of practical interest when dealing with high-voltage systems will be presented.

2.1.3 Simple Electrode Configurations

In the following, reference will be made to simple insulation systems consisting of two electrodes, i.e. two conductive surfaces whose potential difference is established by an external voltage source. In addition, charge carriers are not present. In power systems, it is common to have multi-electrode configurations (as an example, three-phase triplex power cable without concentric conductive shields on each phase, which have four electrodes: the three-phase conductors and the common shield). However, the simple configurations discussed below are very handy as they can provide an approximate model for many practical situations. As an example, Figure 2.2 shows a 1.4 MV DC

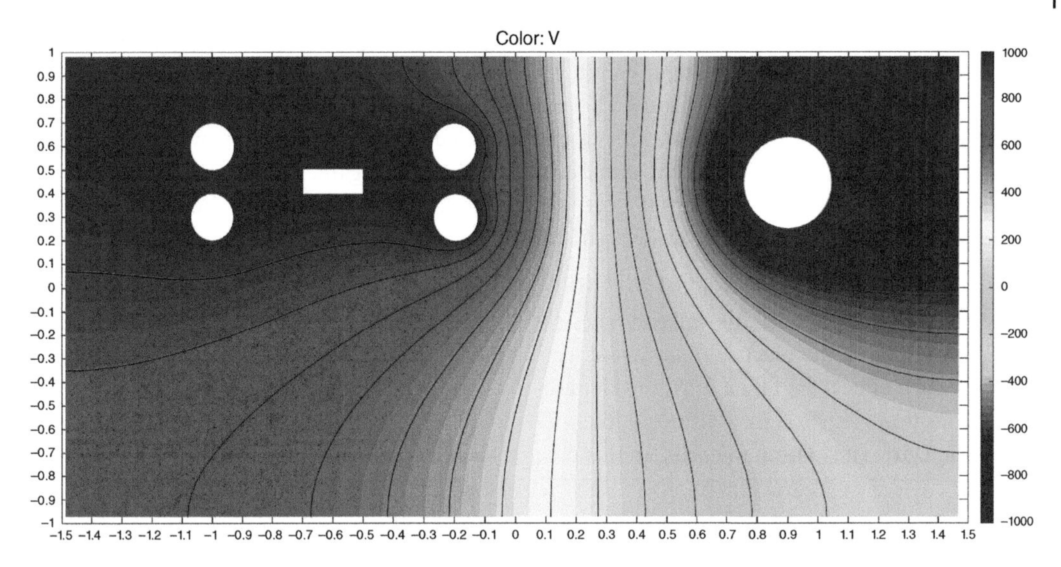

Figure 2.1 Example of equipotential lines in an electrode configuration (white shapes). The lines are equipotential lines, and the color corresponds to the positive and negative voltage magnitude at any particular location.

Figure 2.2 High-voltage test circuit: 1.4 MV HVDC generator in Terrasa, Spain.

generator at the Polytechnical University of Catalonia. The large shielding electrodes can be modeled as concentric spheres, where the inner spheres are the two donut-shaped shielding electrodes and the outer sphere is the ground. The cylindrical conductor connecting the generator with the capacitive voltage divider on the right can be modeled as coaxial cylinders, where the inner cylinder is the metal conductor shown, and the outer cylinder is, again, the ground.

It is important to observe that for all electrode configurations considered in the following, the electric fields are calculated under some assumptions:

1) The shape of the electrodes is ideal at all levels. Indeed, deviations due to, for example, machining are always found at the microscopic level; such surface irregularities can lead to local enhancement of the electric field, which can result in sub-optimal performance under high fields. Such effects are taken into consideration by designers by employing safety factors.
2) The injection of space charge is negligible.
3) The dielectric (insulating material, in this case air) is homogeneous.

2.1.3.1 Parallel Plates Capacitor

This is the simplest configuration, where two parallel metal plates are separated by a distance d. Despite its simplicity, many devices have regions of their insulation system that can be investigated assuming this type of configuration. This is true, as an example, for capacitors (away from the foil edges) and stator winding coils (away from the corners) to describe the field in the insulation between conductors.

If a potential V is applied between the electrodes, the field is constant everywhere and equal to:

$$E = V/d \tag{2.8}$$

d being the distance between the electrodes. This is an ideal situation, as the insulation is electrically stressed uniformly between the electrodes, making insulation design easier. However, there will be higher electric fields at the electrode edges.

2.1.3.2 Coaxial Cylindrical Electrodes

Coaxial cylindrical electrodes are ubiquitous in high-voltage systems. Typical examples are power cables and gas-insulated lines (note that terminations and joints cannot be represented by this model). As mentioned before, cylindrical conductors above the ground (as in an overhead transmission line) can be approximated as coaxial cylindrical electrodes if the distance between the conductor and the ground is relatively large.

Compared with the parallel plate capacitor, the field is now a function of the distance, r, from the center of the electrode system (see Figure 2.3) [3]:

$$E(r) = \frac{V}{\ln(R_2/R_1)}\frac{1}{r} \tag{2.9}$$

The maximum electric field occurs at the surface of the inner electrode, where r takes the minimum value:

$$E_{max} = E(R_1) = \frac{V}{\ln(R_2/R_1)}\frac{1}{R_1} \tag{2.10}$$

The minimum field occurs at the outer electrode. Thus, unlike the parallel plate capacitor, the field is not the same at all locations between the conductors. It is clear that the ratio of the maximum to the minimum field is R_2/R_1 and that increasing the radius of the outer cylinder decreases the maximum field through $\ln(R_2/R_1)$. Therefore, to reduce the maximum field, it is more efficient to increase the radius of the inner cylinder. It can be proved that $R_2/R_1 = e$ (e being the Natural number, 2.7183) provides the optimum field distribution, with $E_{max} = V/R_1$.

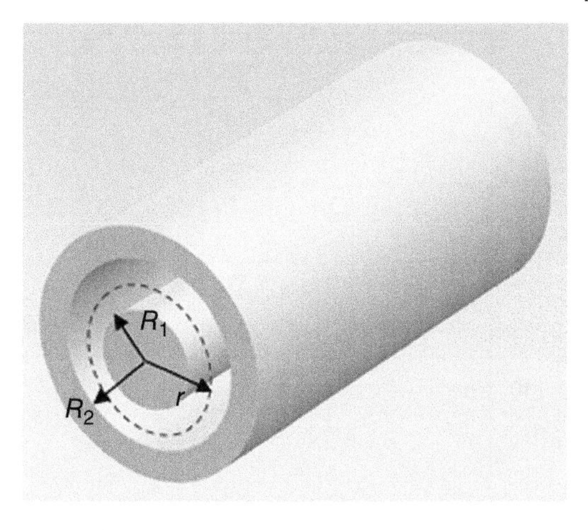

Figure 2.3 Coaxial cylindrical electrodes.

2.1.3.3 Concentric Spheres

Concentric spheres are an abstraction that can be used to represent, for example, elbows in gas-insulated switchgear or shielding electrodes at a large distance from the ground (see Figure 2.2). Following a notation like that proposed in Figure 2.3, it can be shown that the electric field is a quadratic function of the reciprocal of the distance, r, from the center of the system:

$$E(r) = \frac{V}{\left(\frac{1}{R_1} - \frac{1}{R_2}\right)} \frac{1}{r^2} \tag{2.11}$$

This indicates that this system is less efficient than the cylindrical electrode system (meaning that the maximum field is much larger than the average field V/d), thus the design of a reliable insulation system is more difficult. The maximum field is:

$$E_{max} = \frac{V}{R_1 \left(1 - R_1/R_2\right)} \tag{2.12}$$

And it can be proved that the optimal field configuration (minimum field at the inner conductor) is achieved when $R_2/R_1 = 2$.

2.1.3.4 Point/Plane Electrodes

This type of electrode configuration is often used to describe manufacturing imperfections in an insulation system (for example a protrusion on an electrode) or to carry out tests in the lab at a modest voltage level but with high electric fields. The maximum electric field (E_{max}) is at the point electrode and can be calculated in an approximate way using Mason's formula [3]:

$$E_{max} = \frac{2V}{r \ln\left(4d/r\right)} \tag{2.13}$$

where d is the distance between the point and the plane, r is the radius of the point. It is interesting to plot the behavior of this field as a function of the distance between the two electrodes and the tip radius. Figure 2.4 shows that the maximum field is not very sensitive to the distance between the electrodes, whereas the point radius has a very large influence. From a practical point of view, this indicates that to reduce the field, it is much more effective to eliminate protrusions on the surfaces rather than increasing their distance.

2.1.4 Multi-Dielectric Systems

The reader might have noted that, in the previous examples, the electric field was calculated without the permittivity of the dielectric within the electrodes. This is the case only when a single

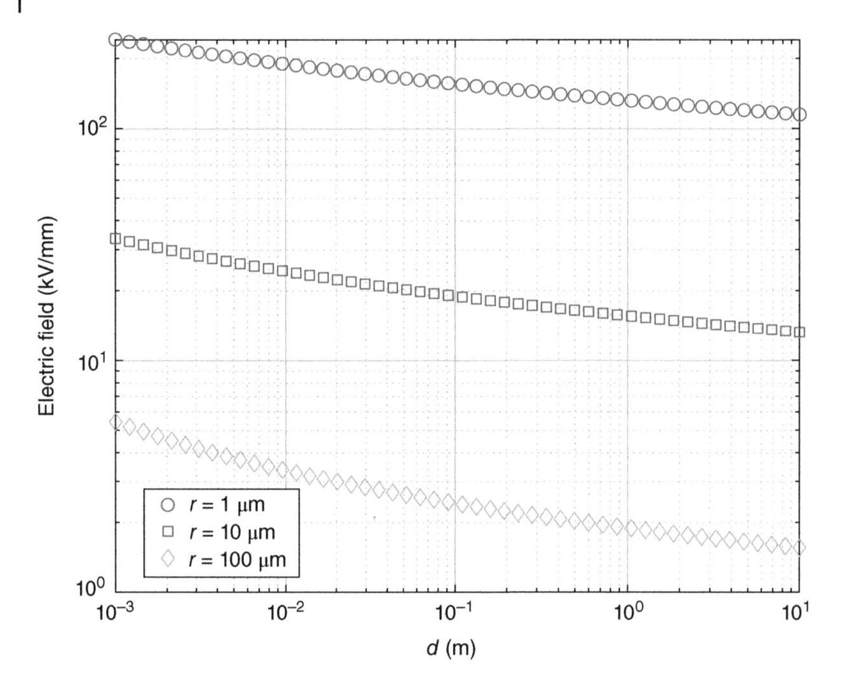

Figure 2.4 Electric field at the point electrode.

dielectric is used. When multiple dielectrics are used, this is not valid. Consider the simplest case, i.e. the parallel plate capacitor. Assuming that there are two dielectrics and that their interface follows the geometry of the system: parallel to the conducting planes.

The problem can be solved considering that in the absence of space charge or trapped charge, the component of the displacement orthogonal to the surface that separated the two dielectrics is continuous:

$$D_{1n} = D_{2n} \rightarrow \varepsilon_1 E_{1n} = \varepsilon_2 E_{2n} \tag{2.14}$$

Due to the geometry of the system, only the orthogonal component of the field will exist. Therefore, we shall drop the subscript *n* for simplicity. Since

$$d_1 E_1 + d_2 E_2 = V \tag{2.15}$$

where d_1 and d_2 are the thicknesses of the two dielectrics, one can derive

$$E_1 = \frac{\varepsilon_2 V}{d_1 \varepsilon_2 + d_2 \varepsilon_1} \quad \text{and} \quad E_2 = \frac{\varepsilon_1 V}{d_1 \varepsilon_2 + d_2 \varepsilon_1} \tag{2.16}$$

From (2.16) one observes that, if the permittivity of dielectric 1 is larger than that of dielectric 2, the field inside the dielectric 1 will be lower than that in dielectric 2. This is a general conclusion: the field in the dielectric with the lowest permittivity is generally higher than the field in the dielectric with the larger permittivity. In the next sections, some special cases of this situation are presented that are important for the reliability of high-voltage systems: cavities, interfaces, and triple points.

2.1.4.1 Cavities (Voids)

In a solid insulation system, some free space will always be present. The free space may be filled with atmospheric air, out-gassing products produced during manufacture of the solid dielectric, or other gas if the insulation system is immersed in some gaseous dielectric. In most cases, this space is in the nanometer range and does not have any implications for the reliability of the insulation. When cavities (also called voids) grow to the micrometer scale (for example in power cable insulation) or larger (e.g. stator windings, GIS insulators), they might become the site of partial discharge activity, leading to breakdown. As will be discussed at length in Chapter 3, the reasons for partial discharge inception are

1) Gases almost always have a much lower breakdown strength compared to solids.
2) Inside the cavity, the field that would exist in the dielectric in the absence of the cavity (background field) is amplified because the relative permittivity of gases is about equal to that of vacuum, whereas solids and liquids have larger relative permittivities (see some examples in Table 2.2).

Note that cavities might exist in liquids in the form of bubbles. However, bubbles tend to migrate to regions with low electrical field and/or to break into smaller bubbles due to partial discharge activity. Therefore, the partial discharge phenomena associated with bubbles are often intermittent compared to cavities within solid insulation.

Figure 2.5 sketches the three major types of cavities that can be found in insulation systems:

a) Cavity elongated in the direction of the field: This type of cavity can be due to electrical treeing as described in Section 2.4.1 (i.e. nonconductive channels eventually bridging the electrodes; these channels are the last stage of partial discharge degradation). Alternatively, it can be found at interfaces where two dielectrics come in contact and the electric field is parallel to the contact surface (e.g. in cable joints or terminations, or in the endwinding of rotating machines).
b) Flat cavity orthogonal to the field: This type of cavity is common in layered insulation as, e.g. rotating machine stator bars, conductor insulation in transformers, mass-impregnated cables.

Table 2.2 Relative permittivity of some industrial dielectrics.

State	Material	Relative permittivity (ε_r)
Gaseous dielectrics	Air	1.00058
	Sulfur hexafluoride (SF_6)	1.00204
	Methane (CH_4)	1.09690
Liquid dielectrics	Transformer oil	2.2–2.7
	Silicone oil	2.2–2.9
	Synthetic ester oil	3.1–3.2
	Natural ester oil	2.9–3.0
	Water @ 20 °C	80.2
Solid dielectrics	Polyethylene	2.2–2.4
	Polyester resin	2.8–4.5
	Epoxy resin	3.6–5
	Impregnated kraft paper	3.3
	Muscovite (mica)	6.5–8.7
	Titanium dioxide (TiO_2)	80

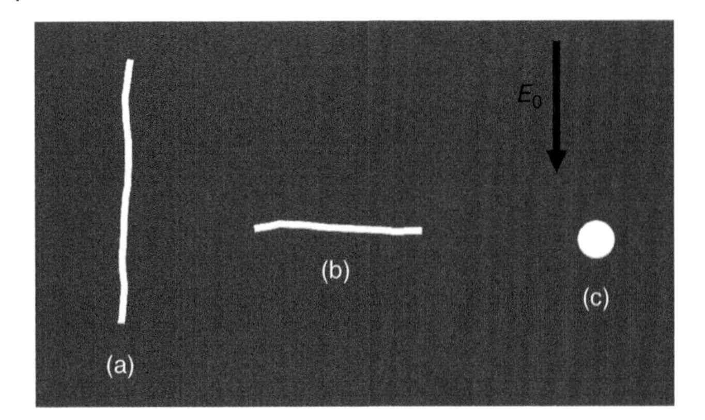

Figure 2.5 Schematic representation of the three major types of cavities: (a) cavity in the direction of the background field, (b) flat cavity orthogonal to the background field, and (c) spherical cavity. The background electric field vector is indicated with E_0.

c) Spherical cavities: These cavities are often due to the release of volatiles during the solidification of thermosetting insulation or imperfect impregnation (typically, cast resin [dry-type] transformers, epoxy spacers in GIS). They can also represent bubbles in liquids, as mentioned above.

Inside the cavity, the electric field will be the background field E_0, times an amplification factor f. Depending on the geometry of the cavity, the amplification factor is [4]:

a) Cavities elongated in the direction of the field: $f \approx 1$.
b) Flat cavities orthogonal to the field: $f \approx \varepsilon_r$
c) Spherical cavities: $f = \dfrac{3\varepsilon_r}{(1+2\varepsilon_r)}$

For (b) and (c), ε_r is the relative permittivity of the surrounding dielectric. For spherical cavities in practical dielectrics ($\varepsilon_r = 2-5$), the amplification factor takes values from 1.2 to 1.4.

2.1.4.2 Interfaces

In insulation systems, interfaces are found where materials with different dielectric constants come in contact with one another, with a tangential component of the electric field often arising as a result (as an example, high-voltage joints or cable accessories). Interfaces can be regarded as a special case of spherical cavities, in that they can be thought of as a string of cavities in the direction of the field (Figure 2.6). Interfaces are often the weakest spot in an insulation system and should be avoided

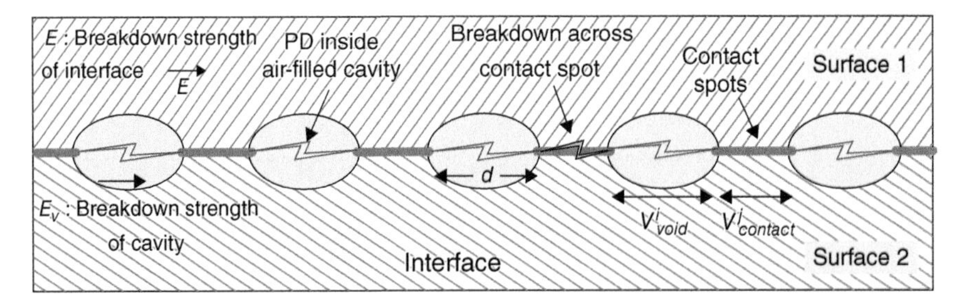

Figure 2.6 Schematic representation of an interface. *Source:* Hasheminezhad and Ildstad [5].

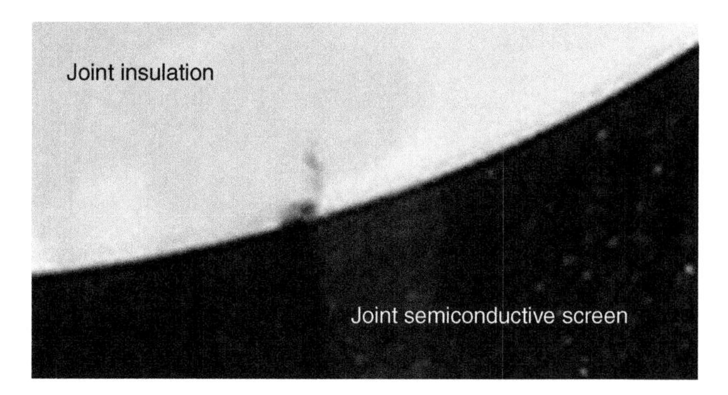

Figure 2.7 Tracking in a XLPE cable joint. *Source:* Montanari et al. [6]/IEEE.

(clearly, that is not possible in many situations). During lightning impulses, a partial discharge in a large cavity transfers the field to other cavities, possibly leading to breakdown via a domino effect. In general, however, breakdown is progressive and involves prolonged partial discharge activity, which leads to the formation of carbonized tracks on the interface, as shown in Figure 2.7.

To prevent partial discharges at interfaces, the mating surfaces are often lubricated with oil or gel to fill the cavities with a dielectric with high dielectric strength[1] and permittivity comparable to the solid.

2.1.4.3 Triple Point (Triple Junction)

Triple points (or triple junctions) are regions of an electrical insulation system where an electrode is in contact with two dielectrics having different permittivities. If the interface between the dielectrics forms an acute angle at the electrode (Figure 2.8), the electric field in the dielectric with lower permittivity can be very large (theoretically infinite at the contact point between the electrode and the dielectric). Generally, triple points can be found where electrodes are in contact with solid dielectrics (high permittivity) and gases (low permittivity). Since the dielectric strength of the gas

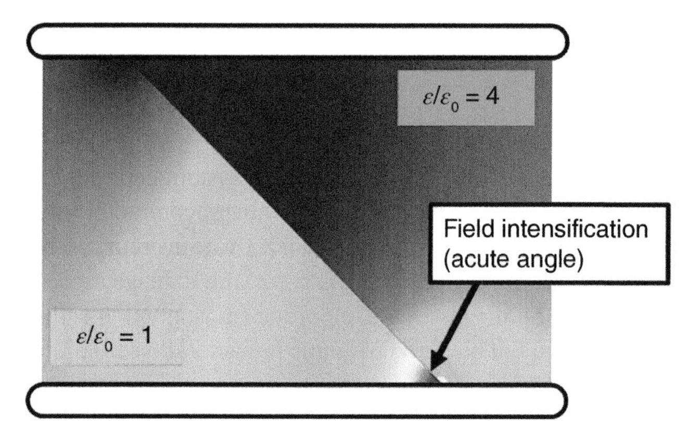

Figure 2.8 Electric field at a triple point of (for example) epoxy, air, and the electrode. The highest electric field is colored red in this FEM calculation.

1 The dielectric strength is the electric field that causes the breakdown of the gas inside the cavity. In Section 2.6, the dielectric strength will be defined in a more formal way, highlighting its dependence on several experimental parameters and, thus, its conventional definition.

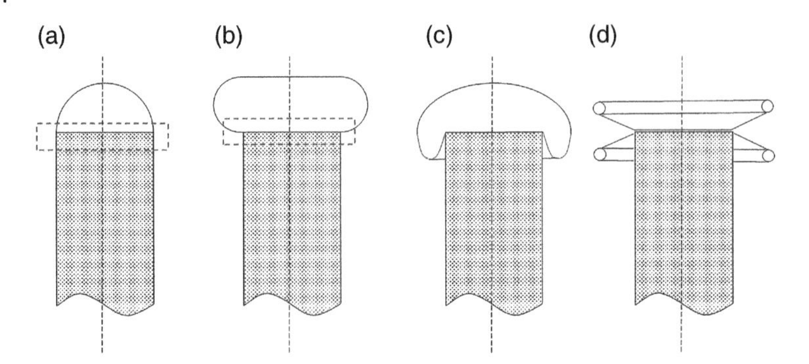

Figure 2.9 Bad, (a) and (b), versus good, (c) and (d) electrode configurations. Configurations (a) and (b) both have triple points in the regions highlighted by the dashed boxes.

is low compared with that of a solid dielectric, partial discharges, or flashover can easily occur at triple points. It is, therefore, important that all the electrodes of the insulation system are designed in a way that prevents triple points or adjusts the electric field in such a way as to reduce their effect, as shown in Figure 2.9, which compares the design of post-insulator electrodes with ((a) and (b)) and without ((c) and (d)) triple points.

2.1.5 Floating Metal Objects

Floating metal objects (also called floating masses) are not solidly connected to any electrode in the system, for example loose conductors or bolts laying on an insulating surface. Their potential is a linear combination of the potential of the system electrodes and is normally unknown. In general, floating masses are not an issue for equipment reliability, although they may create ozone that can lead to chemical aging (Section 3.4) or initiate a surface flashover under some circumstances (Section 13.4.2). Floating metal might generate partial discharges that obscure test object PD (Section 9.2.2).

2.2 Electrical Breakdown

Electrical breakdown occurs when a conductive path through the insulation system connects at least two electrodes at different potential, for example between a high-voltage conductor and ground, or between conductors in different phases. Electric breakdown will activate overcurrent or differential protection, tripping the equipment out of service. In many cases, the damage associated with breakdown can be extensive and the economic losses are important (most often, it is the energy not delivered or production losses that have the greatest economic impact, rather than the repair costs).

Extensive treatments of electrical breakdown phenomena are in [7, 8]. Except for gases, where the mechanisms are clear enough, electrical breakdown is still not completely understood for both solids and liquids. The reasons for this lack of understanding are multiple. To start with, very high fields can trigger phenomena having different nature: electronic, thermal, and mechanical phenomena can be activated, alternatively or simultaneously. Also, the composition of the dielectric is often not homogeneous, and some molecules can undergo some phenomena, whereas others do not. As an example, explanations for the streamer breakdown of liquids during lightning impulses involve hypotheses about Zener field ionization (Section 2.5), impact ionization, formation of bubbles due to evaporation, and mechanical shock waves [9]. Eventually, Zener's field ionization theory seems the most promising [10]. Zener's field ionization (i.e. the extraction of electrons from the valence band of a molecule to the vacuum level due to very intense electrical fields [11]) also explains why, depending on the field at the tip of the streamer (Section 2.3.2), different types of molecules can be ionized, thus leading to different propagation velocities.

To further complicate the topic, it is important to observe that the way the voltage is applied influences the predominant mechanisms leading to breakdown, as well as the time scale of the event. The voltage can be either ramped to some value and left constant until breakdown occurs, or continuously increased until breakdown occurs (provided the AC, DC voltage, or impulse supply can reach the breakdown voltage). Depending on the specific case, the applied voltage or the rate of rise of the voltage can have a profound impact on the time to breakdown as different breakdown mechanisms can be activated. The volt-time (V-t) curve is always used to clarify this last point. The V-t curve is typical of solids and liquids (not gases) and are discussed in Section 2.3.3. Since breakdown in gases is the simplest to explain, it will be discussed first.

As a final remark, the topic of electrical breakdown is vast and is not yet fully understood. It is a multidisciplinary field combining different branches of physics, chemistry, materials science, and electrical engineering, all of which are needed to fully explore its complexities. Therefore, the goal of this section is not to explain such an intimidating topic but to set partial discharges in a framework involving a plurality of different phenomena occurring at different electric field levels and timescales. The interested reader can probe more detailed references such as [7, 8].

2.3 Breakdown in Gases

Breakdown in gases is explained in different ways depending on the distance between the electrodes. Since the focus of this book is partial discharges, i.e. discharges occurring on a micrometer to centimeter scale, only phenomena that are relevant to these scales will be discussed here. These phenomena do not lead to the formation of hot plasmas but are the starting point of phenomena occurring at much larger scales.

2.3.1 Breakdown in Uniform Fields

Electrical breakdown in uniform fields is the simplest case. Uniform fields are constant in magnitude and direction everywhere. In practice, uniform fields can be found between parallel plate electrodes, away from the edges. It is important to note that, by increasing the voltage between bare electrodes, the conditions for the breakdown of the system will eventually be reached without significant warning signals. In contrast, if the electrodes are coated or covered by a solid dielectric, partial discharges will take place until the solid insulation is punctured. Therefore, the theory of breakdown in gases is fundamental to the understanding of partial discharges.

In the breakdown of gases, the key phenomenon is impact ionization, that is, the creation of an electron/positive ion couple upon the collision between an electron and a neutral particle (gas atom or molecule). This event occurs when the energy of an initial free electron is large enough on impact to knock an electron away from the outer orbit of the atom or molecule, i.e. an ionizing collision. (Indeed, at very high energies, multiple ionization phenomena can occur with a single collision, but we shall keep things simple.) With some simplification, one can assume that the mean energy of a free electron is equal to the product of the electric field times the average distance between two ionizing collisions (the mean free path). Since the pressure of the gas is proportional to the reciprocal of the mean free path, the ratio between the electric field and the gas pressure, E/p, is proportional to the mean energy collected by electrons between two ionization collisions. Therefore, gas discharges are normally well described using the ratio E/p, known as the reduced field.

The simplest phenomenon occurring in a gas is the Townsend avalanche. The Townsend avalanche starts from a single initial electron that is present in the gas (see later in this section). After the electron travels (statistically) the distance of the mean free path (i.e. the average distance the electron travels in the electric field between collisions with molecules), the negative electron is accelerated enough by the electric field that when it collides with a neutral atom or molecule, collision ionization occurs. That is, the avalanche consists now of two electrons and one positive ion. The two electrons are again accelerated in the direction of the field and, after traversing another mean free path, they collide with two neutral particles, creating two additional electrons and two positive ions. The process repeats itself with growing numbers of electrons in the head of the avalanche that move toward the anode. By contrast, positive ions, due to their much larger mass, remain almost still and drift very slowly toward the negative electrode (cathode). This situation is often depicted by an "ice cream cone" shape, as shown in Figure 2.10. The avalanche tail creates a positive "space charge," whereas the avalanche head, consisting of electrons, creates a negative space charge. These space charge regions create their own electric fields (Section 2.1.2).

The number of electrons in the head of the avalanche increases with 2^k where k is the number of mean free paths traveled since the initiating electron has started its travel toward the positive electrode (anode). In general, since other phenomena besides impact ionization take place (e.g. elastic collisions, impact excitation, and electron attachment, especially in electronegative gases like SF_6), the *swarm condition* is used to describe the number, n, of electrons at the head of the avalanche after a distance x from its starting point:

$$n(x) = \exp(\alpha x) \tag{2.17}$$

The parameter α is known as the effective ionization coefficient and accounts for the difference between the electrons created by impact ionization and those that get attached to neutral molecules. By differentiating (2.17), one finds that

$$\frac{dn}{dx} = \alpha n(x) \tag{2.18}$$

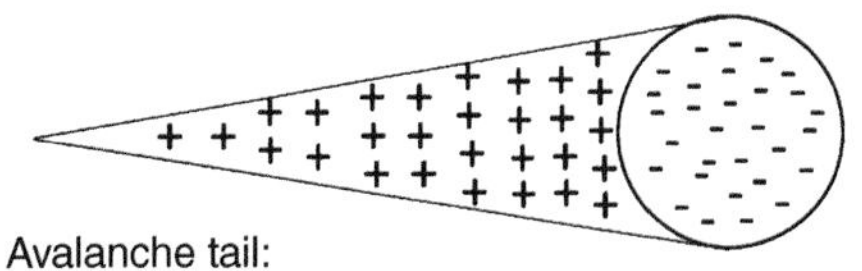

Figure 2.10 Schematic representation of a Townsend avalanche.

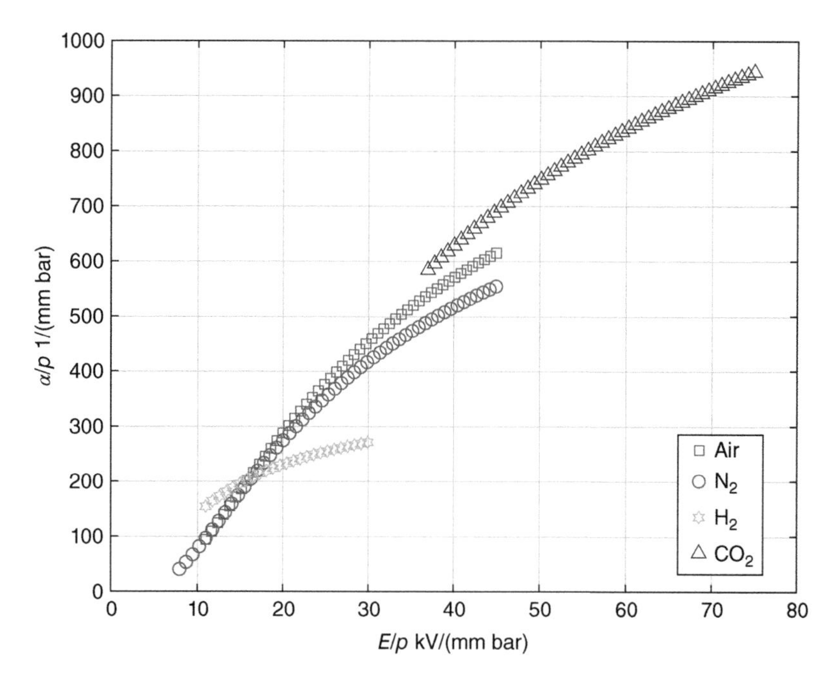

Figure 2.11 Townsend effective ionization coefficient as a function of electric field. The curves are based on Equation (2.19), with *A*, *B*, and *E*/*p* range derived. *Source:* Kuffel [7]/Newnes.

Thus, α describes, in a statistical way, the number of new electrons created through ionization by the n electrons in the head of the avalanche traveling a distance dx. It is a strong function of the electric field and of the nature of the gas. A single equation/curve can represent α/p as a function of the reduced field, E/p. As an example, Equation (2.19) is often used in high-voltage engineering (Figure 2.11):

$$\alpha / p = A\exp\left(-\frac{B}{E/p}\right) \tag{2.19}$$

From the above description, the chain of events unleashed by a Townsend avalanche is

a) the head of the avalanche grows until the anode is reached, thus rapidly increasing the current in the circuit,
b) since the avalanche grows due to collision ionization, in the process, $\exp(\alpha x) - 1$, positive ions are created,
c) the positive ions, with a lower velocity, reach the cathode, causing a slow decay of the current to zero.

How can a Townsend avalanche lead to the breakdown of the gas gap? To answer this question, it is important to understand how and where initiatory electrons are created. In principle, energetic photons from the background radiation can create free electrons in the gas. However, considering the large energy required to ionize many gases (for air, energies start from about 13 eV), this is a very rare event. On the contrary, since the energy required to extract electrons from a metal are in the order of 3–4 eV (this "work function" barrier is also reduced by the electric field), electrons come largely from the metal cathode. This is helped by microscopic protrusions on the surface of the cathode leading to local electric field intensification. After the avalanche is started, the positive ions created by the avalanche itself travel to the cathode. If the ion's energy (acquired from the electric field between the positive ions and the cathode) is about twice the energy needed to extract an electron from the cathode, the positive ion will be neutralized, and a free electron will also be released. This phenomenon is statistically accounted for by a parameter γ, which is the probability that a free electron is released at the cathode upon the impact of a positive ion. Considering that $\exp(\alpha d) - 1$,positive ions are created by an avalanche (d is the distance between the electrodes) if the starting electron has been created at the cathode, then $\gamma(\exp(\alpha d) - 1)$ free electrons are released by the impact of the positive ions created by the avalanche. Whenever

$$\gamma\left(\exp\left(\alpha d\right)-1\right)>1 \tag{2.20}$$

the phenomenon becomes self-sustaining, as each avalanche will statistically cause at least one additional avalanche. After a short time, the current in the gap will grow toward infinity, leading to the breakdown of the gap. By replacing the inequality with equality, one achieves the so-called Townsend breakdown criterion:

$$\alpha\left(E/p\right)d = \ln\left(1+1/\gamma\right) \tag{2.21}$$

Equation (2.21) can be solved assuming an analytical expression for the function $\alpha(E/p)$. Assuming (2.19), (2.21) can be rewritten in a form that puts in evidence the voltage needed to break down the gap as

$$V_{bd} = \frac{a\left(pd\right)}{\ln\left(pd\right)+b} \tag{2.22}$$

For air, $a = 4.36 \times 10^7$ V/(m atm) while $b = 12.8$. Figure 2.12 shows the so-called Paschen curve derived through Equation (2.22) (marked as Expression #1 in the figure). As can be seen, the breakdown voltage decreases as the gap length is reduced (or the pressure is decreased). The curve

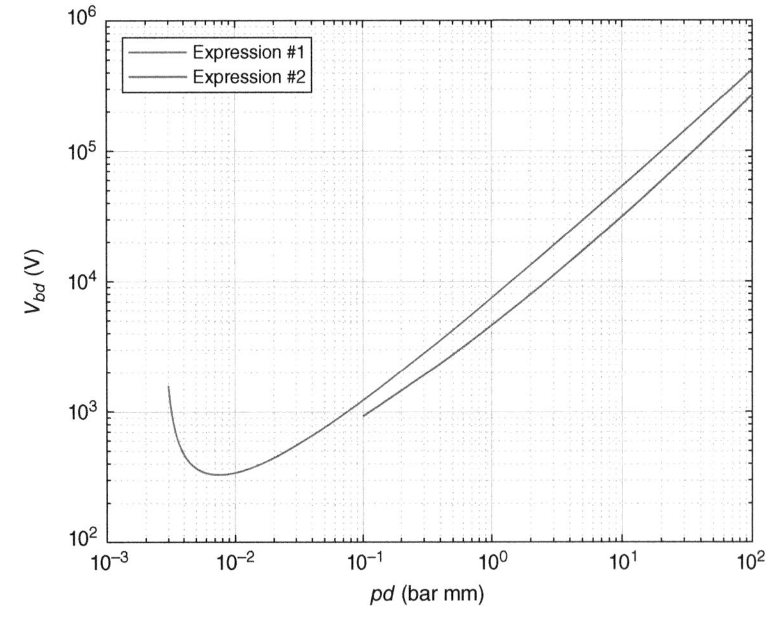

Figure 2.12 Paschen curve. Expression #1 is derived using Equation (2.22), whereas Expression #2 using Equation (2.23).

reaches a minimum breakdown voltage (327 V) for gap lengths of 7–8 μm at 100 kPa (1 bar). An alternative expression based on empirical breakdown data was developed for high-voltage systems:

$$V_{bd} = 6.72\sqrt{pd} + 24.4pd \tag{2.23}$$

Equation (2.23) assumes the pressure in bar, the gap length in cm and the voltage is in kV. The values provided by Equations. (2.22) and (2.23) are compared in Figure 2.1. As can be seen, the empirical curve tends to provide lower breakdown voltage values compared with the theoretical expression. This is related to both the quality of the air and of the electrodes (not fully controlled for (2.23)).

The Townsend discharge phenomenon is characterized by:

1) A breakdown voltage that depends on the metal and surface polish of the cathode, as well as on the nature of the gas.
2) A formative time lag as the (slow) positive ions need to travel back to the cathode to promote a new avalanche.

In 1935, Raether and Fleger observed that for pd values larger than 2.6 bar mm in air (this value was later raised, since Raether and Fleger were overvolting the gap), a different phenomenon could be observed. The phenomenon was peculiar, as the first avalanche reached the anode, it started to grow in the direction of the cathode until a plasma channel connected the two electrodes leading to breakdown. At even higher fields, avalanches started to grow in both directions from the center of the gap. In both cases, the time scale of the phenomenon was too short to allow positive ions to reach the cathode, and the breakdown voltage was independent of the cathode material and finish. Raether and Fleger thus postulated a breakdown mechanism involving a "kanal" (translated from the German as "channel" and called a "streamer" by others) that did not involve the feedback from the cathode. The mechanism assumes that when the avalanche head contains a critical number of carriers, N_{cr}, the field in the gap could be distorted (see Figure 2.13) leading to:

- Higher electric fields in the regions in front of the head of the avalanche as well as in the tail.
- Weaker fields in the center of the avalanche.

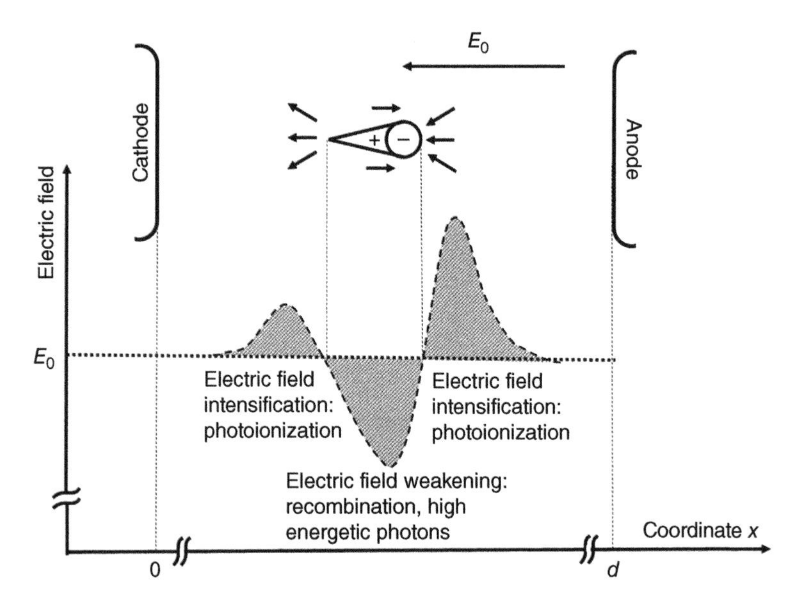

Figure 2.13 Electric field in proximity of a Townsend avalanche ($E_0 = V/d$).

The weaker field in the center of the avalanche favors the recombination of electrons and positive ions, leading to intense emission of energetic photons, including UV. Photons with energies more than the ionization potential of the gas molecules can lead to photoionization of particles in the regions where the field is stronger, giving rise to secondary avalanches. Breakdown thus occurs when the primary and secondary avalanches form a plasma channel that connects the electrodes.

Raether formulated his streamer inception criterion as [12]:

$$\exp(\alpha d) = N_{cr} \rightarrow \alpha d = k \tag{2.24}$$

where *Ncr* is a critical number of electrons in the head of the avalanche, and $k = \ln N_{cr}$. As the effective ionization coefficient is a function of the electric field, the solution of (2.24) provides the minimum field at which an avalanche grows, reaching its critical size in front of the anode. In the case of bare metal electrodes, the streamer inception criterion is also a breakdown criterion (the plasma channel between the electrodes, characterized by very low ohmic resistance, short circuits the voltage source). When at least one of the electrodes is covered by a solid dielectric, it can be considered as a partial discharge inception criterion (Section 3.6.1). Further investigation suggested that the constant k in Equation (2.24) is generally a function of the gap length, d. Raether [12] and Meek [13] both proposed more accurate methods to predict the breakdown voltage. Yet, the streamer inception criterion in Equation (2.24) remains the most popular in high-voltage engineering.

2.3.2 Breakdown in Divergent Fields

Divergent (nonuniform) electric fields can be realized using coaxial cylinders or concentric spheres, particularly when the radius of the internal electrode is much lower compared to that of the external electrode. The most important example of a divergent electric field, however, is the point/plane electrode configuration (Section 2.1.3.4). Ideally, the electric field of insulation systems should be uniform everywhere, which would ensure that the dielectric is stressed equally everywhere. Otherwise, if there are large differences between the maximum and the minimum electric field, the dielectric strength of the insulation should be large enough to prevent a localized breakdown to be initiated in the region at high stress, which then propagate into the regions at lower stress. Propagation toward regions where the field is lower might lead to breakdown or not, depending on how high the electric field is.

A classic example of localized breakdown are corona discharges, that is electron avalanches having the same nature of the discharge mechanisms discussed above: Townsend discharges, or streamers, when the distortion of the background field due to the space charge in the avalanche becomes significant. Corona discharges are typical of high-voltage overhead lines. These discharges are normally incepted at the high-voltage conductors due to the high local electric field and grow, traveling toward the ground or other parts of the system at ground potential (towers, overhead ground wires). Since the clearance between high-voltage conductors and objects at ground potential was carefully selected during the design phase, the avalanche will eventually reach regions where the field is below the minimum value for ionization and the avalanche will stop growing. As a result, breakdown will not ensue (Figure 2.14).

How can a breakdown criterion be formulated for divergent fields? It is clear from the qualitative description above that, upon reaching the regions where the Townsend ionization coefficient is lower than zero, a Townsend avalanche will stop growing. Moreover, the electrons in the head of the avalanche will attach to the gas molecules, reducing their speed. In simple words, a Townsend avalanche will "die out" without causing breakdown. Corona discharges will be different if started

Figure 2.14 Corona discharge on corona ring of 500 kV overhead power line, photographed with a conventional camera. *Source:* Nitromethane [14]/Wikimedia Commons/CC BY-SA 3.0.

from a positive or a negative electrode, as the role of space charge in the gas due to either positive or negative ions will have a profound impact on the development of these two discharge types. A thorough description can be found in [7, 15].

Since a Townsend breakdown is not feasible, breakdown in divergent fields will occur when the avalanche reaches the critical size to develop into a streamer. This can be expressed as

$$\exp\left(\int_0^{x_{lim}} \alpha\left(E\left(x\right)\right)dx\right) = N_{cr} \tag{2.25}$$

where N_{cr} has the same meaning as in (2.24). In this case, however, as the electric field is a function of the distance x as is the Townsend ionization coefficient, the number of charges in the head of the avalanche is achieved through an integral. The term x_{lim} defines the limit of the region where ionization can occur, thus $\alpha(x_{lim}) = 0$.

2.3.3 Breakdown Under Impulse Voltages – the V-t Characteristic

The phenomena described above are observed when the voltage is raised slowly, i.e. as occurs in a 50 or 60 Hz AC cycle. However, since lightning can strike transmission lines and substations, tests with impulse voltages are often carried out. Compared to the 50/60 Hz situation, it is observed that for the same electrode configuration, breakdown voltages are generally higher and more variable. This is due to two distinct phenomena. On the one hand, discharges are triggered by free electrons that might not be available when the minimum voltage for breakdown is reached. This leads to a statistical time lag. On the other hand, after the discharge is incepted, a (deterministic) formative time lag is required to build up a conductive channel leading to the complete breakdown. When the rate of rise of the voltage is short enough, i.e. the risetime of the impulse voltage is in the microsecond range, the sum of these two delays might lead to a significantly higher breakdown voltage when compared with that achieved in "static" conditions. The formative time lag is a function of the overvoltage, since the time to build up a conductive channel decreases with increasing voltages.

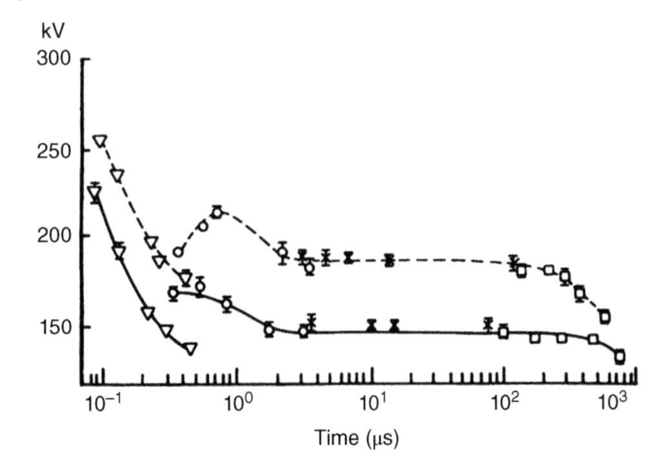

Figure 2.15 Examples of V-t characteristics for cylindrical coaxial electrodes filled with dry air (solid line) and air/1% SF_6 (dashed line) at 500 kPa. Negative polarity voltage impulses: ○ standard impulse (1 µs risetime/50 µs fall time to 50% of peak); ∇ 0.2/60 µs; * 15/1200 µs; □ 275/2500 µs. *Source:* [16].

It is possible to measure the breakdown voltage as a function of the time needed to achieve the breakdown. In general, the shorter the time, the higher the breakdown voltage. This curve is often referred to as the volt-time (V-t) characteristic. Figure 2.15 shows an example.

Since impulse voltages are unipolar, the V-t characteristic might depend on the voltage polarity. This can happen in divergent fields (nonuniform fields) if space charge formation is fast enough to influence the electric field in the gap.

2.4 Breakdown in Solids

Unlike the breakdown of gases, the breakdown of solids (and liquids) is a much more complex phenomenon that has not been completely unraveled. In general, the breakdown strength of solid insulation is 10–100 times higher than the breakdown strength of most gases at atmospheric pressure, i.e. 50–500 kV/mm.

The difference between static (essentially 50/60 Hz AC) breakdown and breakdown under lightning impulse leading to the V-t characteristic has been discussed above. For solid insulation, the V-t curve is much more complex than that found for gases. One of the striking characteristics of solids is that the V-t characteristic can be derived for impulse, ramped, and constant voltages. Depending on the type of voltage and rate of voltage rise, different phenomena will be activated:

- Electronic phenomena, which are observed when the voltage is increased rapidly or with impulses (alternatively, these phenomena can be triggered using high-power lasers). Electrons are excited to the conduction band of the dielectric. By interacting with the dielectric, they can increase its temperature and eventually lead to breakdown. When impulse voltages are used, time lags, like those observed in gas, can lead to an increase of the breakdown voltage. For this type of breakdown, the dielectric strength can be considered an intrinsic property of the material (it will depend, nevertheless, on the slew rate of the voltage). Breakdown due to electronic phenomena can occur in HV equipment during lightning impulses. When testing solid dielectric materials, specialized equipment and electrode configurations are needed to prevent anomalous breakdowns, such as flashover between the electrodes or thermal runaway. In general, under constant voltage or slowly ramped voltages, these phenomena will not occur.

- Thermal phenomena occurs mostly when the voltage is raised during engineering tests (say, hundreds of volts per second). Heat is normally produced within the dielectric by polarization and ohmic losses. When the heat due to these losses is larger than the heat that the system can exchange with the environment, the temperature of the insulation system increases up to the melting point of the insulation. This type of breakdown depends on the arrangement of the electrode system, heat exchange systems, temperature of the electrodes and of the environment. Therefore, it is a property of the insulation system, rather than a material property. For dielectric films having thickness d sandwiched between parallel-plate electrodes, the breakdown voltage tends to increase by $\sqrt{d}$. Therefore, the calculated dielectric strength will depend on $1/\sqrt{d}$, showing once more that it will not be an intrinsic property of the material.
- Aging will lead to a variety of irreversible phenomena that occur over long times in HV equipment, under moderate voltage levels. These phenomena include:
  - Partial discharges that gradually erode the insulation. PD might be active right from the start or can be activated later due to the formation of cavities within the insulation system due to electromechanical or thermal stresses that lead to delaminations, void formation, and/or the migration of materials.
  - Thermal runaway due to a progressive increase of the power losses within the insulation. This is typical of oil/paper systems when moisture content increases in the paper, due to thermal aging or water ingress caused by cracked O-rings, etc.
  - Mechanical erosion due to vibrations induced by electrodynamic forces or thermomechanical stress. Partial discharges can be incepted due to these phenomena but are a symptom of degradation, rather than a cause.

Each of these phenomena tends to occur within different time frames of voltage application (see Figure 2.16). The detailed explanations of each of these phenomena are beyond the scope of this book and are the subject of literally thousands of research papers. However, one of the most common long-term mechanisms that can lead to breakdown of HV equipment insulated with solid dielectrics is electrical treeing. PD is a fundamental aspect of electrical treeing, so more detail is discussed below.

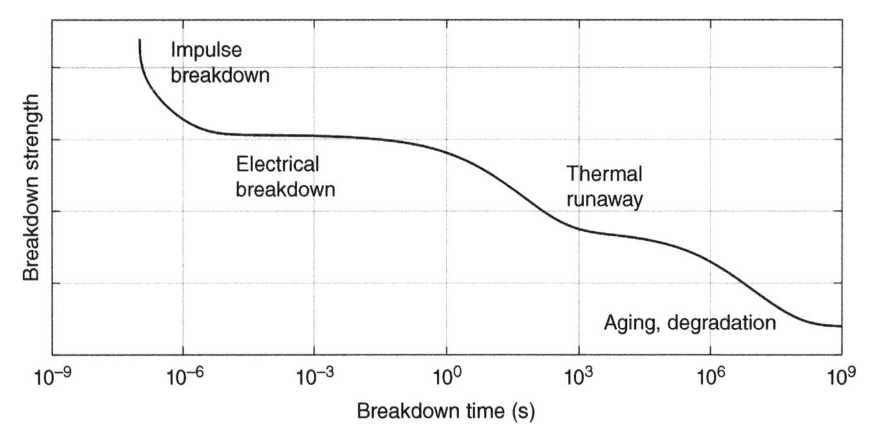

Figure 2.16 V-t characteristic of solid insulation.

2.4.1 Electrical Treeing

Electrical treeing is the formation of a network of hollow, nonconductive or partly-conductive channels within a solid dielectric. The name is derived from the appearance of the channels normally resembling the branches of a tree (or a bush) when discharges are initiated from a needle electrode inserted in a solid dielectric (Figure 2.17). Since the channels are essentially nonconductive, partial discharges occur within the tree channels, leading to the propagation of the tree until breakdown. Treeing is, therefore, the final stage of breakdown due to partial discharges.

Treeing has been studied extensively in controlled conditions in the laboratory, where typically it is artificially initiated by inserting a needle within a solid dielectric. The high field enhancement at the needle tip leads to damage of the dielectric due to injection/extraction of energetic electrons as well as the mechanical stress due to coulombic forces. Given enough time, the action of these charges will lead to the formation of a cavity large enough to support partial discharges (tree inception). At this point, partial discharges will start to erode the dielectric forming channels with a tree-like structure that will propagate toward the opposite electrode. It is interesting to note that breakdown may not immediately occur at the moment when the tree connects the two electrodes. Partial discharges need to enlarge the tree branches to a radius of about 50 μm, and only at this time do energetic arcs occur, and breakdown conditions are reached. It must be emphasized that extensive research has been

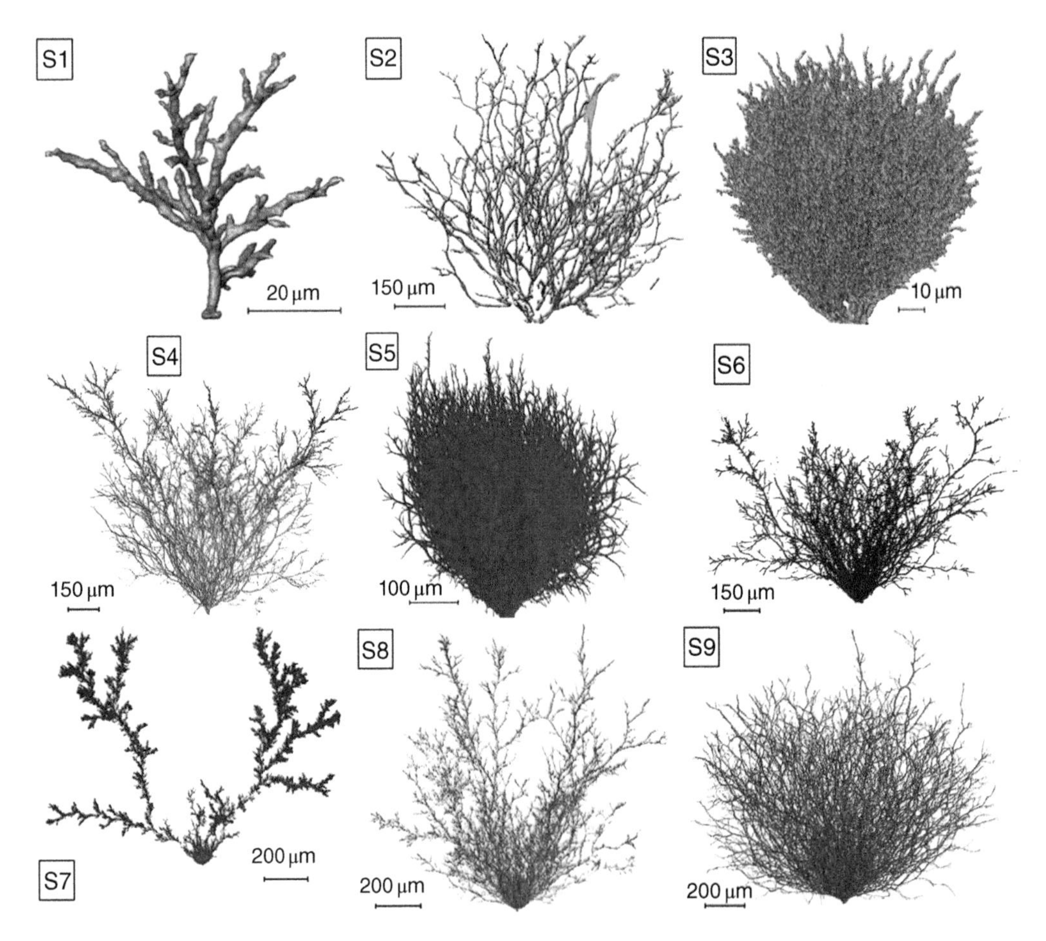

Figure 2.17 Various shapes of electrical trees grown from a needle, reconstructed using 3D X-ray computed tomography. *Source:* Schurch et al. [17]/from IEEE.

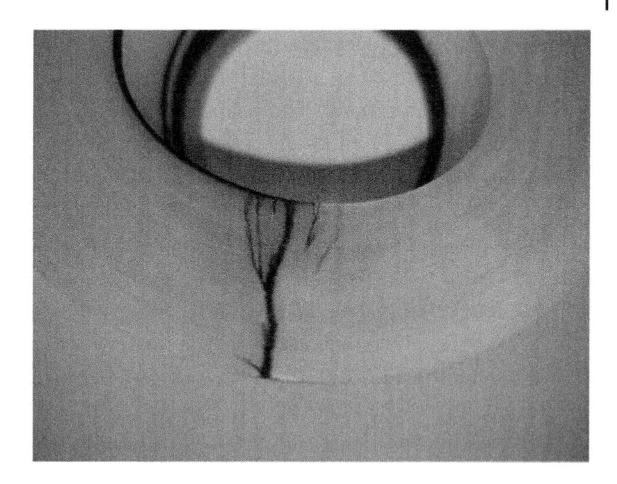

Figure 2.18 Example of an electrical treeing in a polyethylene power cable.

undertaken during the past several decades to better understand electrical treeing, its development, effects, and risk, along with trying to estimate when breakdown will occur once it starts. Much valuable work has been published, but it is still not completely understood.

The tree structure is associated with highly divergent fields. In HV equipment, highly divergent fields are avoided by design. However, a local protrusion might lead to small regions with large fields where electrical treeing can be initiated to propagate into regions where the electric field is much less divergent. This leads to less-branched structures, as seen in Figure 2.18, observed in a power cable after HV testing. Electrical trees can also grow from cavities or delamination within the insulation because of the degradation due to partial discharges at one small region of the cavity surface. An example of treeing growing from a cavity was observed by chance at the University of Bologna and is shown in Figure 2.19. The tree was initiated by intense partial discharge activity in a large (millimeter size) cavity in epoxy resin. A video was also shot of the partial discharge activity. It shows discharges both in the channel of the tree and within the cavity.

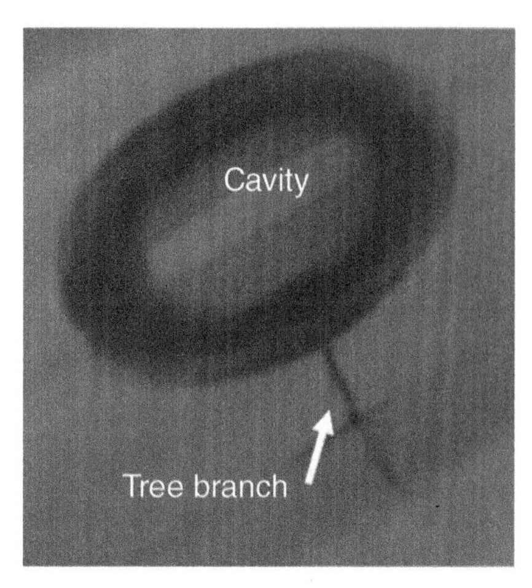

Figure 2.19 Treeing starting from a millimeter size cavity in epoxy resin.

2.5 Breakdown in Liquids

Similar to solids, liquids also have a V-t characteristic that covers a relatively long time scale. Electronic processes during voltage impulse dominate when the voltage is raised quickly enough. When the rate of rise is much lower or when the voltage is constant, "weak links" within the liquid become the site of discharge and breakdown.

a) Electronic phenomena will depend on whether the electric field is uniform or divergent along with the duration of the applied voltage:
   - Under uniform fields, electronic phenomena like those observed in solids can be observed only for very short gaps. For longer gaps, the heat released by the interaction of the highly energetic electrons with the fluid will vaporize the fluid, creating bubbles. Under these conditions, breakdown strengths up to 1000 kV/mm have been measured.

- For divergent fields and under impulse conditions, streamers will be generated starting at the point where the field is the largest. Streamers have been investigated for decades using point/plane configurations. Their shape and propagation speed depend on the polarity of the impulse voltage, and sometimes on the presence of attaching molecules (electron scavengers). Positive impulses (positive point) will have a moderately branched structure with a large propagation speed. It is speculated that positive streamers propagate as an "ionization wave," with the electric field at their tip so large that the molecules of the fluid can be directly ionized (i.e. Zener's effect, which does not involve collision with an electron). Negative streamers (from a negative polarity tip) have a more globular structure with lower propagation speed; thus, they are of less concern than positive streamers. The propagation mechanism assumed for negative streamers is that of a growing conducting sphere, with ionization occurring on the outer part of the sphere only when the sphere reaches a sufficiently large electric field in its surroundings. The charging of the sphere occurs through conduction that, given the low conductivity of the fluid, slows down the streamer. Similar to gases and solids, the breakdown can be delayed when the rate of voltage rise is short enough, resulting in higher breakdown voltages. As a final remark, the explanations reported here are probably just a part of the whole picture. The formation of low-density regions, regions where the surface tension of the fluid is highly reduced, or microbubbles can also play a part in the streamer propagation sustaining electron impact ionization. As mentioned earlier, this is still an open and fascinating topic.

b) Weak link phenomena occur on much larger time frames and are an issue in HV equipment. As for aging in solids, different phenomena can be observed caused by contaminants:
   - In power or instrument transformers using oil/paper insulation, cellulose fibers and metal particles are always present in the insulating fluid (mineral, ester, or silicone oil – see Section 15.2.2). These particles have a permittivity larger than the oil and tend, therefore, to drift toward high electric field regions due to dielectrophoretic forces. Situations where these fibers align, thus reducing the dielectric withstand strength of the oil, were recreated in the lab under moderately divergent fields, AC or DC (Figure 2.20).
   - The oil used to fill the equipment (transformers, capacitors, power modules) is generally dried and filtered before energizing, see Section 15.2.2. Its breakdown strength is usually measured at a rate of increase of 500 V/s in sphere/sphere electrode configuration, with the spheres 2.5 mm apart. Oils treated for filling power equipment have breakdown voltages normally exceeding 60 kV (about 25 kV/mm). Larger breakdown voltages, exceeding 70 kV/mm, are measured for oils tested in the laboratory, with moisture below 2–3 μl/l and filtered using a filter with pores of 1–2 μm or smaller. In time, the quality of the oil decreases due to moisture, contamination, and oxidation, leading to lower breakdown strength and favoring the former breakdown mechanisms.

For liquids, considering their regenerative (self-healing) nature, discharges are less critical than in purely organic solids, unless they involve the surface of a solid insulation where carbonized tracks, or "surface tracking" results. Partial discharges can occur in highly divergent regions, where electric fields can reach several hundreds of kV/mm. Their effect is localized, and the regenerative nature of the fluid prevents the complete breakdown of the oil gap. However, phenomena of this type are indicative of the presence of points where partial discharges can develop into streamers during over voltages such as lightning or switching impulses. Thus, it is a good idea to take countermeasures to reduce the electric field (often, in power transformers, PD in oil is incepted by mechanical tools or ropes accidently left inside the tank during maintenance operations – see

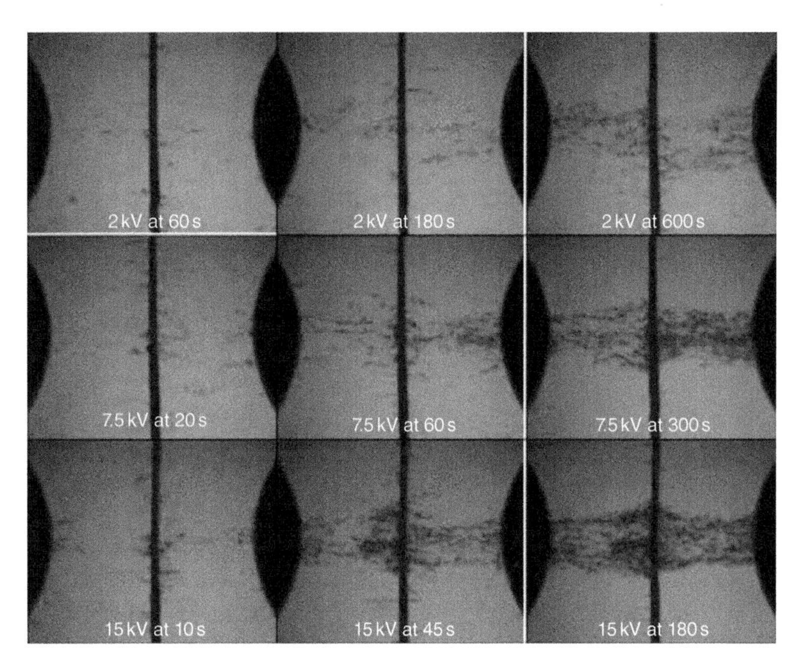

Figure 2.20 Formation of cellulose fiber bridges in mineral oil under mildly divergent fields. *Source:* Mahmud et al. [18]/IEEE.

Section 15.4.4). Partial discharges can also occur if gas bubbles occur in high-field regions. Since the permittivity of the gas is lower than that of the oil, gas bubbles will tend to migrate toward low field regions and the PD activity could eventually extinguish. However, if the reason for gassing is not removed, partial discharges can be incepted again, thus creating an erratic pattern of inception and extinction. Partial discharges are, in this case a symptom of a thermal problem since gassing is usually associated with high temperatures in the oil. While PD usually does little harm to the equipment, the reason for gassing should be removed, especially if gases associated with high temperature decomposition of the oil (acetylene and ethylene) are found when performing a chemical analysis of oil samples (Sections 5.5.2 and 15.4.6).

The V-t characteristic including all these processes is shown in Figure 2.21.

2.6 Dielectric Strength

The dielectric strength is defined conventionally as the electric field at which breakdown occurs in a parallel plate sample with thickness *d* under a uniform electric field:

$$E_{bd} = {}^{V_{bd}}\!/_{d} \tag{2.26}$$

From the above discussions, it is clear that such a definition is simplistic. For the simplest case of gases, the dielectric strength is not only a property of the gas but also depends on the gap length and gas pressure (see Figure 2.22). If streamer discharges are not activated, the characteristics of the electrodes (material, surface finish/roughness) also are important as they can influence the probability of observing secondary electrons upon the impact of positive ions on the cathode. Moreover, the larger the electrode surface area, the larger the probability of introducing weak

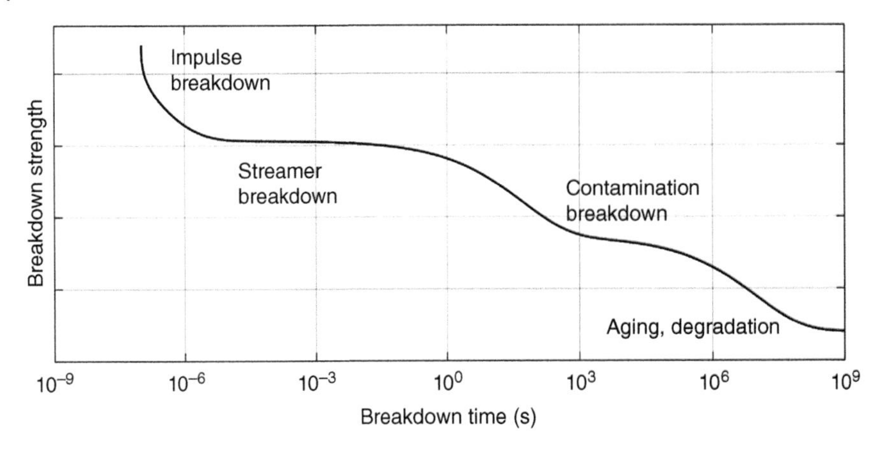

Figure 2.21 V-t characteristic of liquids. *Source:* Mahmud et al. (2014) / IEEE.

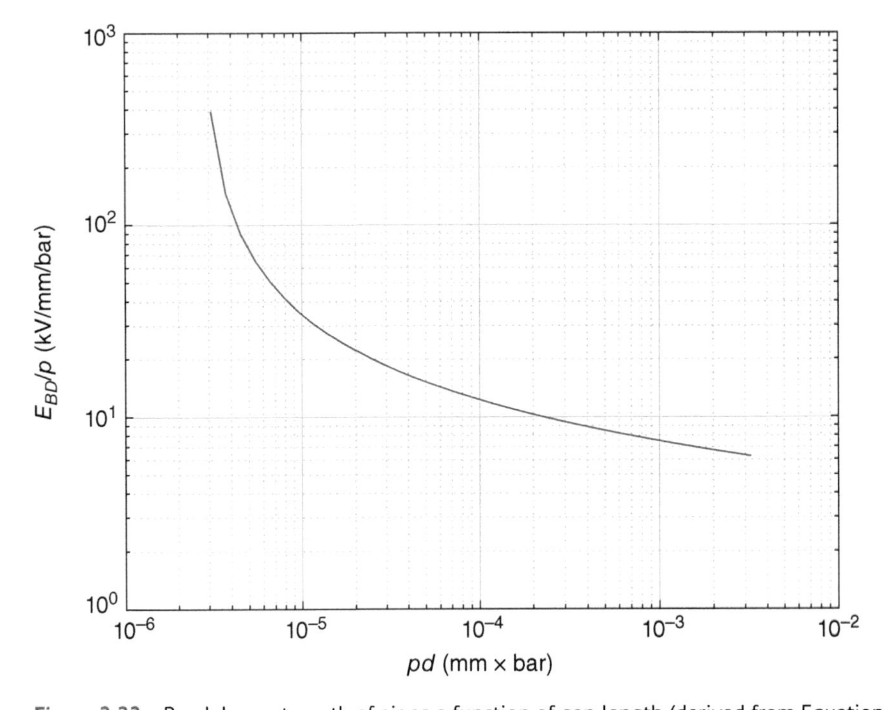

Figure 2.22 Breakdown strength of air as a function of gap length (derived from Equation (2.22) using the parameters $a = 4.36 \times 10^7$ V/(m bar) and $b = 12.8$). The curve stops at 3 mm × bar as this is about the limit for transition from Townsend to streamer breakdown.

points on the electrode surface (protrusions, microcracks, grain boundaries, oxide spots) leading to a lowering of the breakdown voltage (and thus of the calculated electric strength).

For solid insulating materials, this concept is even more complicated, since the intrinsic dielectric strength (i.e. the electric field leading to breakdown via electronic phenomena) is so high that it cannot be measured, except under very special conditions. In most cases, the breakdown of practical material samples (having thickness d) is achieved through standardized tests and is due

Table 2.3 Peak dielectric strength of some dielectrics.

Gases		Liquids		Solids	
Dielectric	Dielectric strength (kV/mm)	Dielectric	Dielectric strength (kV/mm)	Dielectric	Dielectric strength (kV/mm)
Air	2.43–3.0	Mineral oil	18	Diamond	2000
Nitrogen	2.44	Synthetic ester oil	17.5	Mica	50–250
Hydrogen	1.01	Natural ester oil	22.5	PTFE	60–173
SF_6	8.8	Silicone oil	16	XLPE	60
High vacuum	20–40 (depends on electrode)	Ethylbenzene	226	Epoxy	55–80

to thermal runaway, showing a dependence on $1/\sqrt{d}$, on electrode surface area and thickness, as well as on ambient temperature.[2] The voltage waveform applied to the insulation (DC, VLF 0.1 Hz, 50/60 Hz AC, impulse) also has a very large impact (see Figure 2.16 for solids and Figure 2.21 for liquids). Therefore, breakdown tests often provide an indirect indication of the dielectric losses and thermal conductivity and should be regarded as comparative tests rather than tests aimed at providing some intrinsic feature of the dielectric under test.

Table 2.3 provides a guide to the dielectric strength of different practical insulating materials highlighting the differences between dielectrics that are gaseous, liquid, or solid at room conditions. The dielectric strengths indicated are peak values (not rms). As can be seen, the higher the density, the higher tends to be the dielectric strength, although some notable exceptions exist as the electronic structure plays a fundamental role in defining the withstand capability of the dielectrics.

References

1 Krause, J.D. and Fleish, D.A. (1999). *Electromagnetics: with Applications*, 5e. WCB/McGraw-Hill.

2 Hayt, W.H. and Buck, J.A. (2019). *Engineering Electromagnetics*, 9e. McGraw-Hill.

3 Mason, J.H. (1955). Breakdown of solid dielectrics in divergent fields. *Proceedings of the IEE-Part C: Monographs* 102 (2): 254–263. https://doi.org/10.1049/pi-c.1955.0030.

4 Niemeyer, L. (1995). A generalized approach to partial discharge modeling. *IEEE Transactions on Dielectrics and Electrical Insulation* 2 (4): 510–528. https://doi.org/10.1109/94.407017.

5 Hasheminezhad, M. and Ildstad, E. (2012). Application of contact analysis on evaluation of breakdown strength and PD inception field strength of solid-solid interfaces. *IEEE Transactions on Dielectrics and Electrical Insulation* 19 (1): 1–7. https://doi.org/10.1109/TDEI.2012.6148496.

6 Montanari, G.C., Cavallini, A., and Puletti, F. (2006). A new approach to partial discharge testing of HV cable systems. *IEEE Electrical Insulation Magazine* 22 (1): 14–23. https://doi.org/10.1109/MEI.2006.1618967.

2 More rarely, breakdown occurs at the electrode profiles due to intense PD activity.

7 Kuffel, E., Zaengl, W.S., and Kuffel, J. (2000). *High Voltage Engineering Fundamentals*, 2e. Newnes.

8 Arora, R. and Mosch, W. (2022). *High Voltage and Electrical Insulation Engineering*. Wiley-IEEE Press.

9 Devins, J.C., Rzad, S.J., and Schwabe, R.J. (1981). Breakdown and prebreakdown phenomena in liquids. *Journal of Applied Physics* 52 (7): 4531–4545. https://doi.org/10.1063/1.329327.

10 Jadidian, J., Zahn, M., Lavesson, N. et al. (2012). Effects of impulse voltage polarity, peak amplitude, and rise time on streamers initiated from a needle electrode in transformer oil. *IEEE Transactions on Plasma Science* 40 (3): 909–918. https://doi.org/10.1109/TPS.2011.2181961.

11 Zener, C. (1934). A theory of the electrical breakdown of solid dielectrics. *Proceedings of the Royal Society of London. Series A, Containing Papers of a Mathematical and Physical Character* 145 (855): 7523–7529.

12 Raether, H. (1964). *Avalanches and Breakdown in Gases*. Butterworths.

13 Meek, J.M. and Craggs, J.D. (1953). *Electrical Breakdown of Gases*. Oxford: Clarendon Press.

14 Nitromethane. (2013). Corona discharge on corona ring of 500 kV overhead power line. https://commons.wikimedia.org/wiki/File:Corona_discharge_1.JPG (accessed 23 August 2021).

15 Kuchler, A. (2017). *High Voltage Engineering: Fundamentals – Technology – Applications*, 5e. Springer Verlag.

16 Z. Li, R. Kuffel, and Kuffel, E. (1986). "Volt-time characteristics in air, SF6/AIR mixture and w2 for coaxial cylinder and rod-sphere gaps," *IEEE Transactions on Electrical Insulation*, vol. 21, no. 2, pp. 151–155, Apr. 1986, https://doi.org/10.1109/TEI.1986.348938.

17 Schurch, R., Ardila-Rey, J., Montana, J. et al. (2019). 3D characterization of electrical tree structures. *IEEE Transactions on Dielectrics and Electrical Insulation* 26 (1): 220–228. https://doi.org/10.1109/TDEI.2018.007486.

18 Mahmud, S., Golosnoy, I.O., Chen, G. et al. (2014). Effect of kraft paper barriers on bridging in contaminated transformer oil. *2014 IEEE Conference on Electrical Insulation and Dielectric Phenomena (CEIDP)*, Des Moines, IA (19–22 October 2014), pp. 110–113. IEEE. https://doi.org/10.1109/CEIDP.2014.6995865.

3

Physics of Partial Discharge

3.1 Introduction

This chapter takes the physics described for electrical breakdown between conductors in Chapter 2 and focuses on the physics of partial discharge, which some have called partial breakdown or a localized discharge, which does not result in an electrical short between the conductors. Each partial discharge results in a current pulse that can be measured at the test object's terminals. The measurement of the PD current is called PD testing. PD testing normally determines the PD pulse magnitudes, its repetition rate (for example PD pulses per second), and the phase position of where pulses occur with respect to the AC cycle. In some cases, the polarity of the pulse current is measured, as well as the pulse shape (for example the pulse risetime or duration). Most of these PD characteristics can be shown on a single plot, called the phase-resolved PD (PRPD) plot, which is the primary method for displaying PD activity (Section 8.7.3).

This chapter also discusses the impact of PD on various insulating materials and shows how PD causes electrical aging and eventual failure of the insulation system.

3.2 Classification of Partial Discharges

The term partial discharge refers to a discharge that occurs between two surfaces, where at least one surface is not an electrode. According to [1], the term "partial discharge" refers to:

> a localized electrical discharge that only partially bridges the insulation between conductors and which can or cannot occur adjacent to a conductor.

This definition usually evokes the concept of cavities (or voids) within an insulation system, cavities that may, or may not, be bounded on one side by one of the system electrodes. Yet, cavities are just part of the story (although they play a key role when it comes to reliability). A more comprehensive classification of partial discharge phenomena is given in Figure 3.1. At least four macro-categories can be devised:

1) Internal discharges bounded or not by electrodes. In layered insulation systems, cavities are often delaminations, i.e. region where the layers of the insulation detach one from the other one. Delaminations are flat cavities orthogonal to the electric field, as discussed in Chapter 2.

Practical Partial Discharge Measurement on Electrical Equipment, First Edition. Greg C. Stone, Andrea Cavallini, Glenn Behrmann, and Claudio Angelo Serafino.

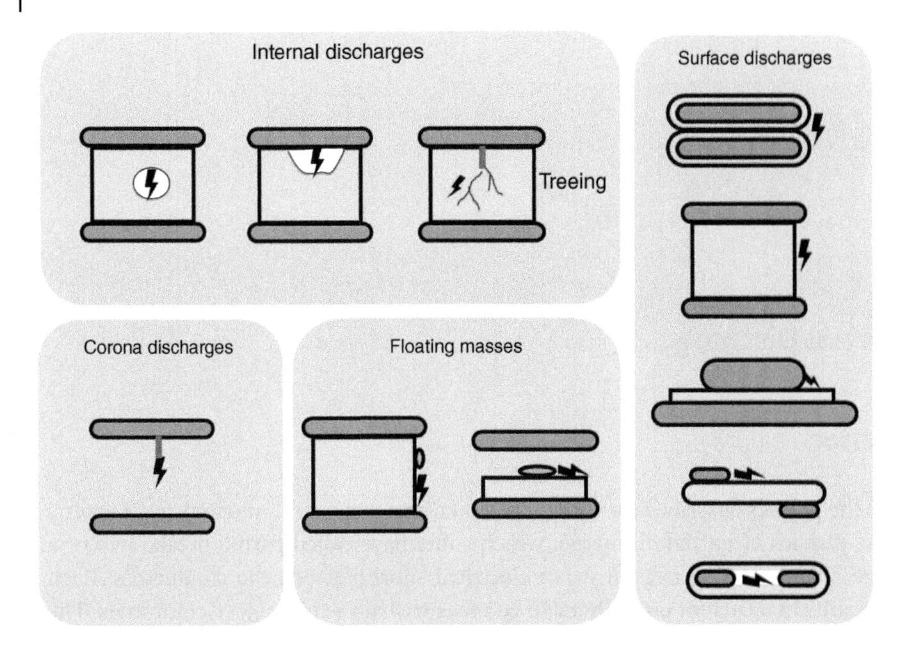

Figure 3.1 Classification of partial discharge phenomena.

Treeing can also be regarded as internal discharges. In this case, the cavities (channels of the tree) are elongated in the direction of the field. In all cases, the electric field has a non-negligible component orthogonal to the discharge surfaces.

2) Surface discharges develop on dielectric surfaces (i.e. driven by an electric field with a predominant component tangential to the dielectric surface). In contrast to some authors, we also define discharges that occur within the insulation system at the mating surface (interface) between two dielectrics (sometimes called a scarf joint) as surface discharges. Figure 2.7 shows an example of the effects of type of discharges. Triple point discharges have been classified here as surface discharges, even though, often, the component of the field normal to the surface is not negligible.
3) Corona discharges: discharge in a gaseous medium originating from a metal electrode where there is a very nonuniform electric field.
4) Floating masses originate from metallic parts that are not solidly connected to any electrode.

An alternative classification of PD sources was proposed by Niemeyer in 1995 [2] and is reported for the sake of completeness in Figure 3.2. Note that before the 1970s, English speaking countries tended to use the terms corona and PD interchangeably (Section 1.2). However, today's terminology emphasizes that corona is a specific type of PD.

3.3 PD Current Pulse Characteristics

According to the above classification, a partial discharge takes place in a gaseous region

a) between two dielectrics,
b) between an electrode and a dielectric (this includes the case of corona discharges),
c) between an electrode and a metallic floating mass (Section 2.1.5). It is assumed that the PD will not give rise to the total breakdown of the gas space between the high- and low-voltage electrodes.

In all these cases, it is possible to overvolt the gas space (i.e. exceed the minimum breakdown voltage in parts of the airgap) without breaking down the system and causing a short circuit. Thus, at low

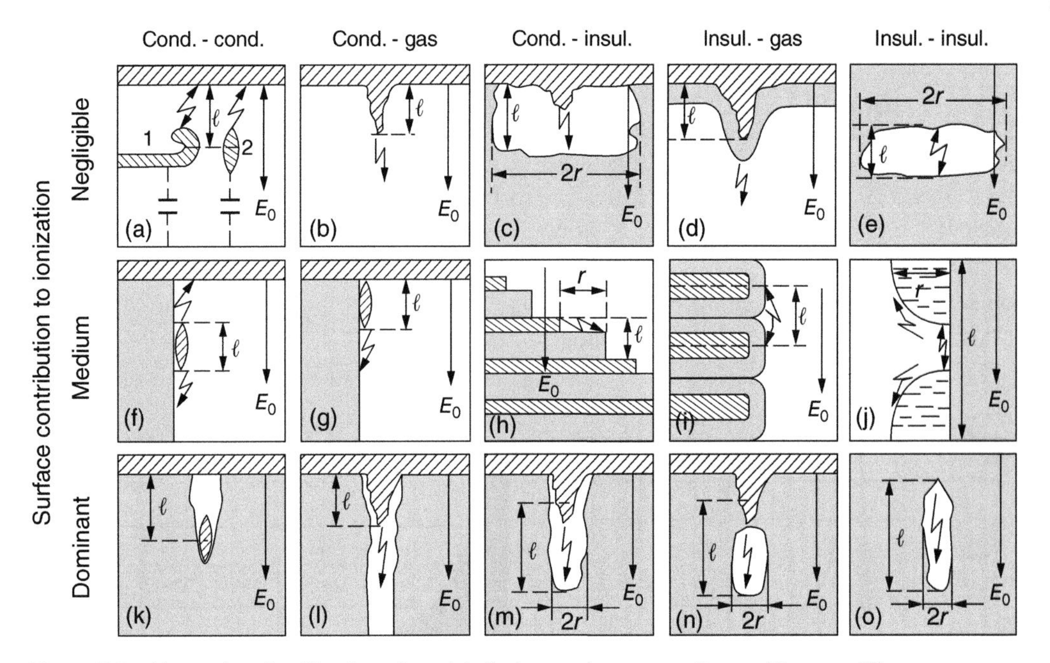

Figure 3.2 Alternative classification of partial discharge phenomena. *Source:* Niemeyer [2].

overvoltages[1] of the gas, PD will be Townsend discharges (Section 2.3.1). By increasing the applied voltage across the EUT, a gradual transition will be observed from Townsend to streamer discharges. As discussed in Section 2.3.1, the streamer will distort the field leading to secondary avalanches and the formation of a plasma channel. Compared with Townsend discharges, streamer discharges are more energetic and have a higher frequency content due to its shorter risetimes.

An extensive analysis of the properties of PD in cavities is reported in Peter Morshuis' PhD thesis [3], where a further discharge type, which he calls a pitting discharge, is observed in extremely aged cavities. Pitting discharges perhaps originate from crystals that form on the surface of the cavity due to the interactions of the weak acids created by PD activity and the polymeric matrix; these crystals tend to concentrate PD activity at some specified locations, leading to the formation of erosion pits on the opposite surface. Here, a summary of the data collected in [3] is reported. While these data cannot be generalized as they were obtained with a specific geometry and material, and with an oscilloscope having bandwidth 500 MHz [4], Table 3.1 highlights the differences between the different types of discharges. From the practical point of view, the only type of discharge with a magnitude that is likely to be detected in high- and medium-voltage equipment is the streamer discharge. Especially in cavities, the PD activity seems sometimes to extinguish, whereas it simply shifts from streamer pulse to either Townsend or pitting discharges. As an example, Figure 3.3 summarizes the differences between streamer-like, Townsend-like and pitting pulses, whereas Figure 3.4 shows pitting pulses recorded while testing a cavity in an XLPE[2] sandwich at the University of Bologna. The non-streamer pulses could only be detected after streamer pulses extinguished by using a high-sensitivity measurement cell.

---

1 Overvoltage is the ratio of the field inside the cavity to the inception field.

2 An XLPE "sandwich" consists of three layers of XLPE (or other insulating material sheets). The middle sheet has a circular hole cut into it. Thus, the sandwich has a cylindrical cavity between two layers of insulation. Such sandwiches are widely used in PD research.

Table 3.1 Main differences between Townsend and streamer discharges [3].

	Townsend	Streamer	Pitting
Waveform	"Flat Top," see Figure 3.3	Pulse like	Pulse-like, sometimes in trains of closely spaced pulses
Magnitude	<1000 μA	<150 mA	~200 μA
Rise time	<60 ns	<8 ns	<8 ns
Pulse width	<900 ns	—	—
Fall time		<30 ns	10–15 ns

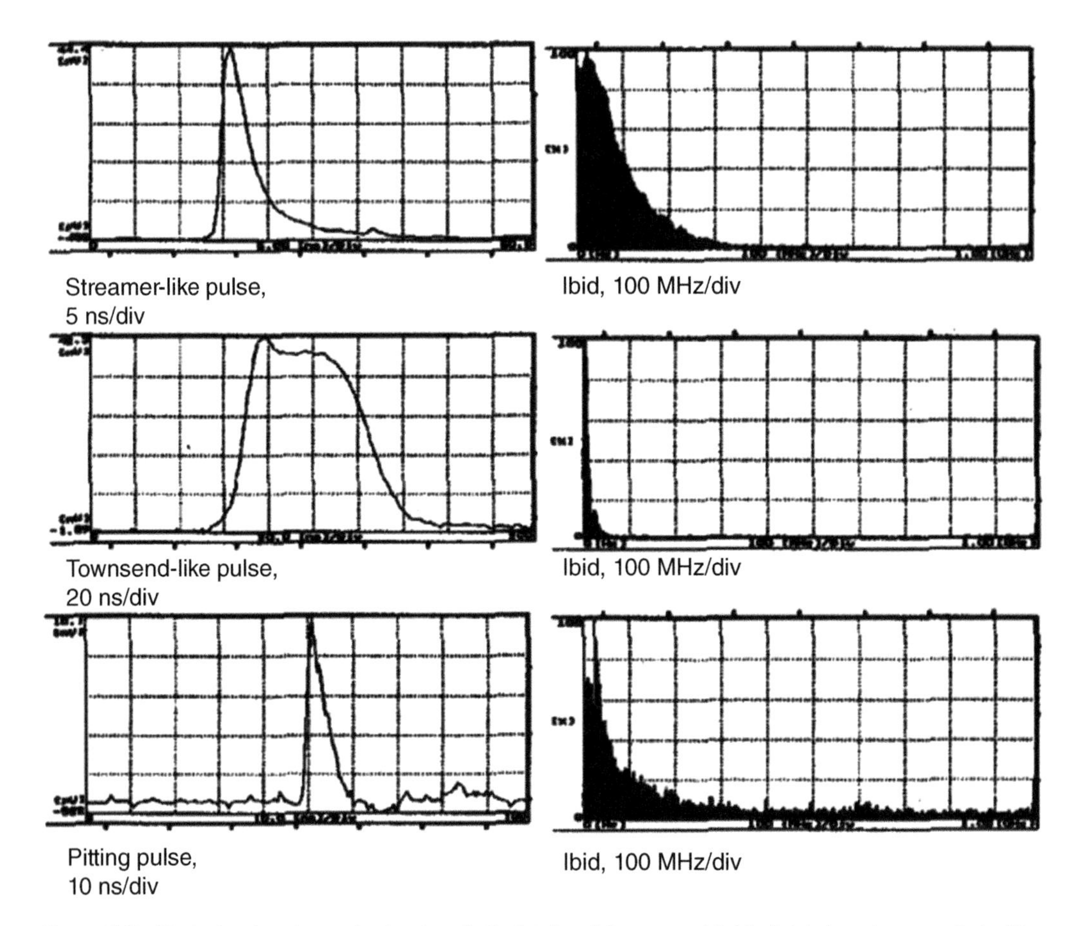

Figure 3.3 Typical pulse shapes in the time (left plots) and frequency (right plots) domains recorded with a 500 MHz digital oscilloscope. *Source:* Morshuis [4].

The pulse characteristics are also highly dependent on the type of the gas they occur in. Highly attaching gases such as SF_6 quench the discharge rapidly, leading to short duration pulses. These pulses tend to have a higher frequency content than pulses developing in non-attaching gases as air. This characteristic led to the development of UHF PD detection systems in SF_6-insulated system first (Section 13.8).

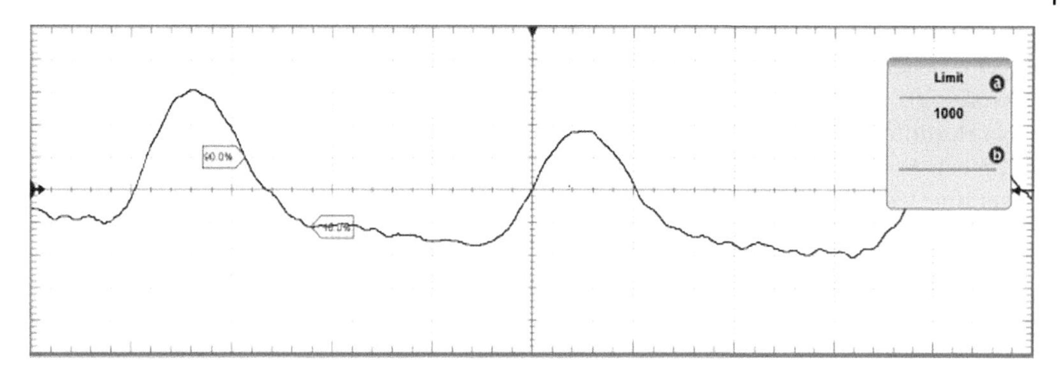

Figure 3.4 Example of pitting pulses recorded at the University of Bologna (10 ns/div, 60 μA/div).

Another factor that influences the PD pulse shape (and spectrum) is the pressure of the gas. In general, the higher the pressure, the shorter is the pulse risetime, and the higher the spectral content. As an example, Figure 3.5 shows the spectra of PD pulses recorded in air at 300, 200, and 111 mbar. As can be seen, the PD signal energy in the 300 MHz region tends to reduce as the air pressure is reduced. Since the PD is compared in the graph with the baseline

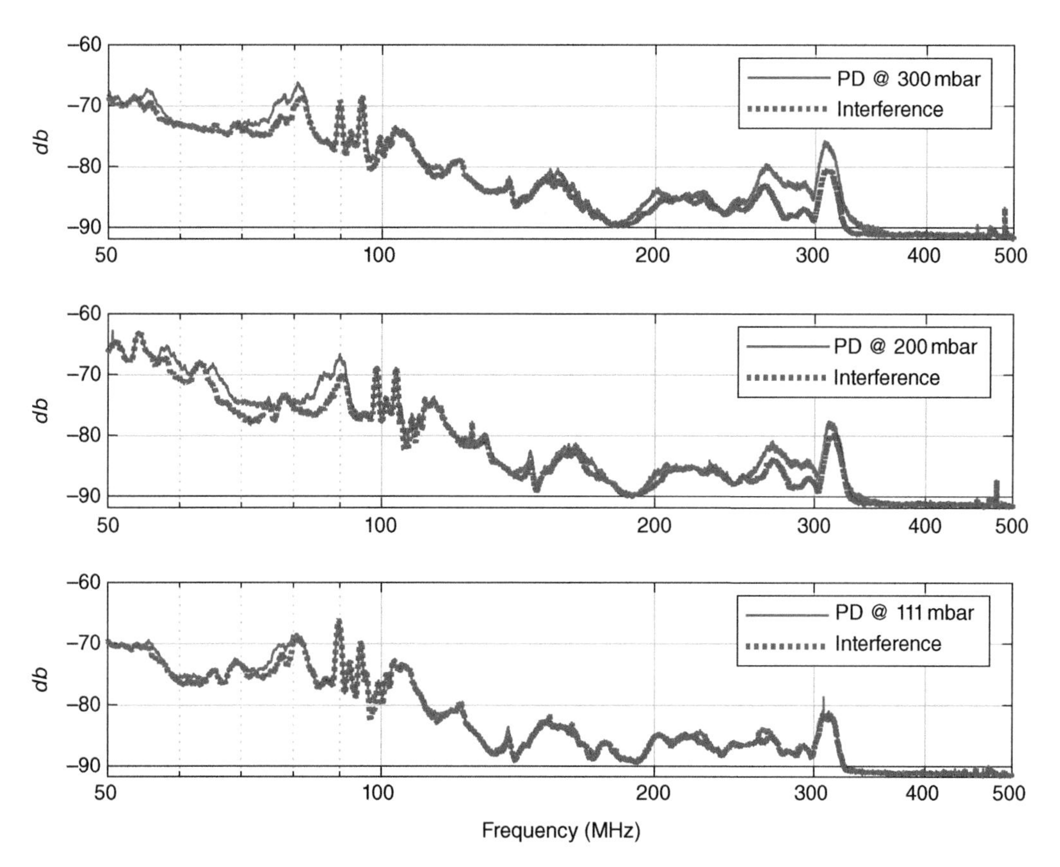

Figure 3.5 Examples of PD pulse spectra from twisted pairs at different pressure levels. The spectra were recorded using a UHF sensor and a spectrum analyzer. *Source:* Lusuardi et al. [5].

interference due to a wide bandgap inverter (Section 18.1), it appears that at the lowest pressure, separating the PD signal from the interference becomes impossible. At even lower pressures, PD loses the high-frequency content in an even more dramatical way. Figure 3.6 shows the difference between the pulses measured at standard conditions and those achieved at 12 mbar. As can be seen, the pulse measured at 12 mbar has a lower peak value and a much longer time length.

The pressure effect on PD pulse risetime and spectral content is also apparent in high-pressure hydrogen-cooled turbine generators that operate at 3–4 bar. At these pressures, the pulse risetime is typically less than 1 ns (Section 16.9.3.3). Hence, online PD measurement in such machines is often made in the UHF range.

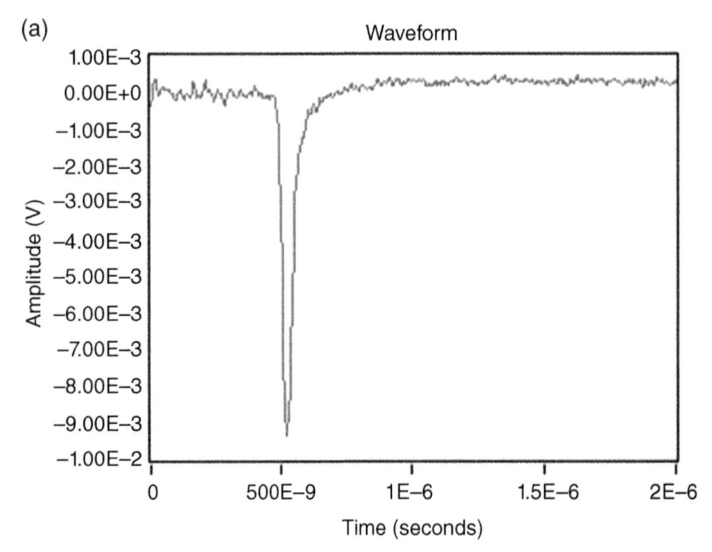

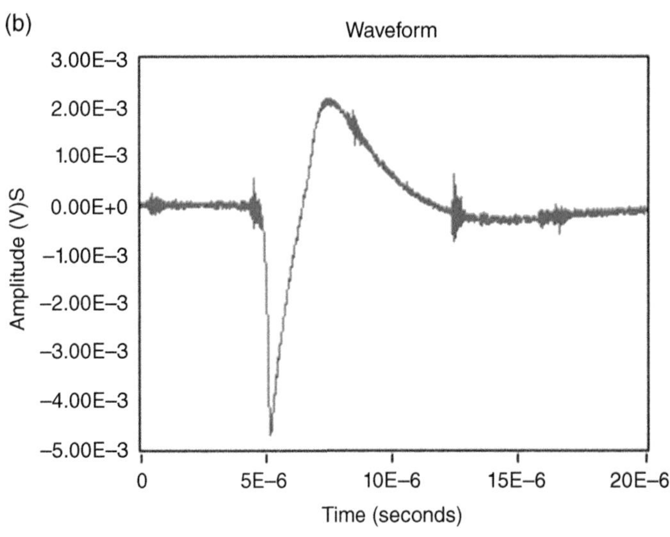

Figure 3.6 Different pulses waveform: (a) fast pulse with high-frequency content, obtained at pressures above 70 mbar in air, (b) long risetime pulses with low-frequency content obtained at 12 mbar [6]. Note the time scale in the top plot is 10 times shorter than the scale on the right. *Source:* Cavallini et al. [6].

3.4 Effects of PD

Partial discharges can interact with dielectric in different ways: through electron bombardment, chemical erosion, and thermal degradation. The first interaction is based on the so-called dissociative electron attachment (DEA). DEA occurs when an electron with sufficient energy collides with a molecule breaking the chemical bonds between some atoms in the molecule. Figure 3.7 shows the electron energy in discharges. In organic polymers, the energy required for breaking carbon–carbon or carbon–hydrogen bonds is 3–4 eV. The electrons in a discharge have energies that can be much higher than these values. Thus, electron bombardment causes some of the bonds in the polymer chain to break. Broken chemical bonds make the dielectric more brittle, and part of the material can be removed, etching the surface. As an example, Figure 3.8 shows the conditions of an epoxy slab subjected to prolonged PD activity, as well as the profile of the surface: after 1000 hours a hole of 0.3 mm is produced in the slab.

Another route for deterioration due to PD activity is corrosion by chemical by-products. Corrosion can be due to ozone that is formed in air by the PD activity (Section 5.5.1), which interacts with the dielectric surface or by free radicals created by the electron bombardment via the DEA mechanism. In [7], it was observed that nitric, formic, glycolic and glyoxylic acids, and possibly of glycolic aldehyde, glycol and glyoxal could be observed on the surface of epoxy bombarded by corona discharges. It is assumed that these chemical species could eventually lead to the formation of hydrated oxalic acid crystals. These crystals could focus PD activity at given spots eventually leading to the formation of pits, which develop gradually into tree-like structures.

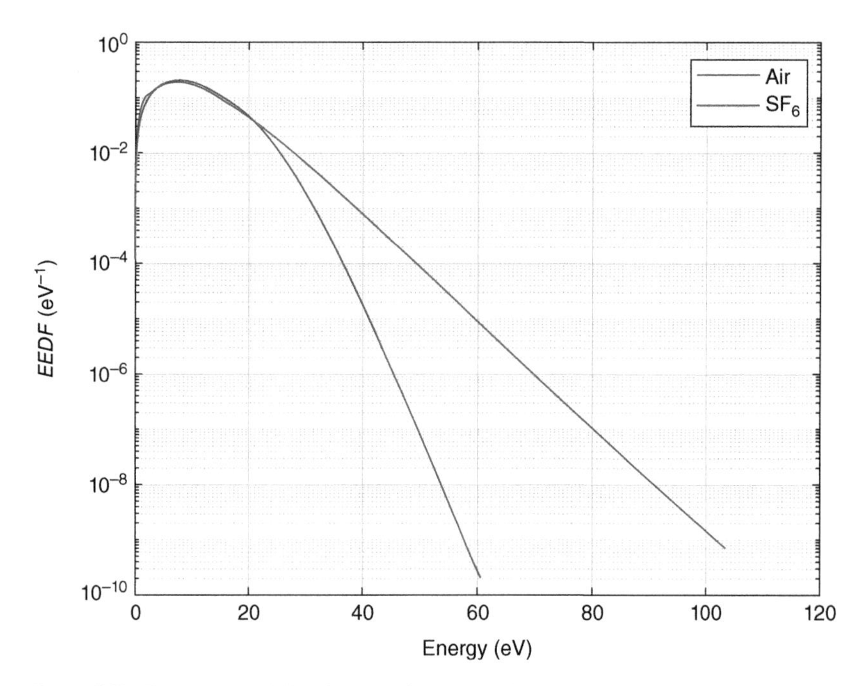

Figure 3.7 Energy probability density of electrons in air and SF_6 at 10 kV/mm. Estimate obtained using Bolsig+ software (available at https://www.bolsig.laplace.univ-tlse.fr/).

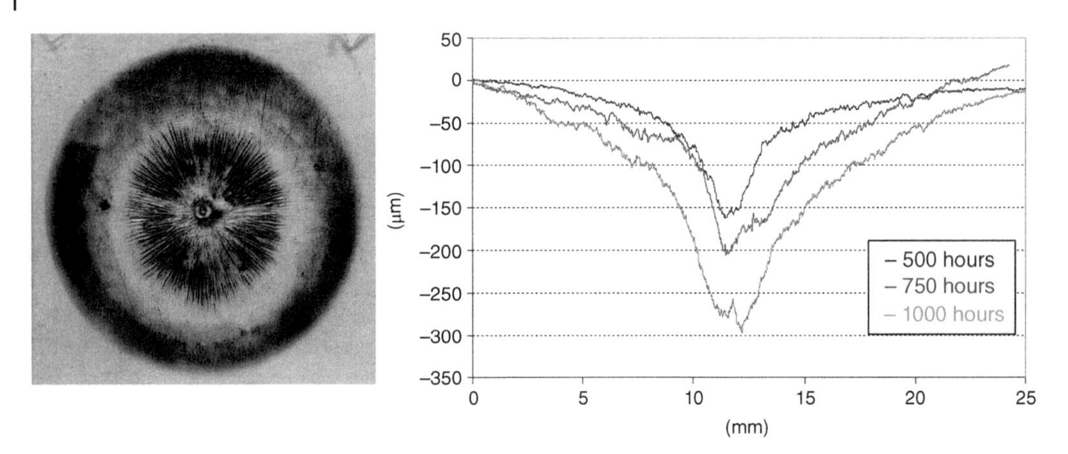

Figure 3.8 Slab of epoxy subjected to PD bombardment from a needle-plane electrode configuration: (left) picture of the conditions of the surface (right) profile of the surface.

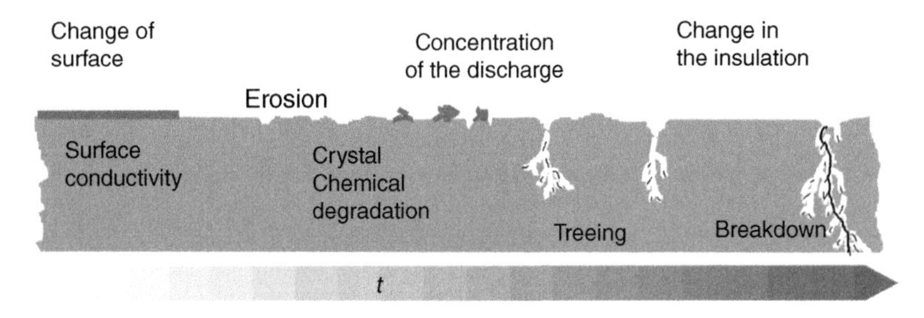

Figure 3.9 Recognized stages of PD-induced degradation as time progresses to the right. *Source:* Morshuis [3]; Temmen [8]/IOP Publishing.

Similar conclusions can be found in [3, 8]. The mechanisms leading to breakdown are summarized in Figure 3.9.

While organic materials are characterized by covalent bonds, ionic and metallic bonds are very common in inorganic materials. Ionic and metallic bonds have bond energies higher than covalent bonds (as an example, the Si-O bond is 4.7 eV, although this number might not be fully representative of complex materials having a crystalline structure as, for instance, mica). Therefore, the erosion mechanism postulated for organic materials takes much longer to occur, i.e. inorganic materials are much more resistant to deterioration than organic materials such that insulation system breakdown due to PD activity is unlikely to occur. Unfortunately, with some exceptions (glass and ceramic insulators used in overhead lines and bushings, for example) inorganic materials are not suitable as insulation materials since they lack flexibility or have too a high conductivity.

Therefore, in some types of practical HV equipment, a hybrid insulation system consisting of a polymeric matrix containing inorganic parts combines the PD resistance of inorganic materials with the flexibility of organic materials. The inorganic part of the insulation delays the development of electrical trees by forcing them to travel longer routes, see Figure 3.10. As an example, mica flakes bonded together with asphalt have been used since the 1930s to insulate stators in generators. These systems were largely replaced after the 1950s by a thermoset matrix (polyester or epoxy) containing mica (Section 16.3). More recently, winding wire enamels loaded with nanoparticles (e.g. silica, alumina or titanium oxide) were devised to slow PD-induced degradation in

Figure 3.10 Tree propagation around a 5 mm-thick mica-barrier. *Source:* Vogelsang [9]/IEEE.

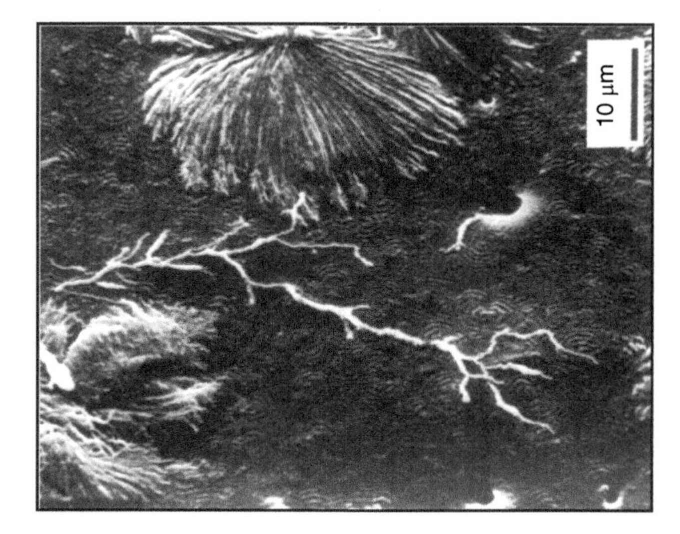

Figure 3.11 Tree channel formed at 20 °C and 15 kV in a polymer blend sample crystallized at 125 °C. *Source:* Zhao et al. [10]/The Institution of Engineering and Technology.

low-voltage machines fed by power electronics converters (Section 18.2.2). Adding inorganic barriers within a polymeric structure forces electric treeing to move around the inorganic barriers within the insulation system to fail the insulation, as seen in Figure 3.10 [9]. The result of this is that the time to failure can be considerably longer compared to that could be expected using the polymeric matrix only.

It is interesting to observe that similar phenomena were observed also in XLPE, an organic material. The propagation of electrical trees was found to move in the in the amorphous regions in XLPE samples, avoiding crystalline regions (spherulites). The likely reason is the XLPE is more densely packed in spherulites compared with the amorphous zone. Besides having a higher density, the XLPE in the spherulites also has stronger bonds between the polymeric chains due to the ordered orientation of the chains. Higher density and stronger chemical bonds tend to stop the propagation of electrical trees through spherulites, favoring the propagation in the amorphous region. See Figure 3.11.

3.5 Corona Due to Non-Uniform Electric Fields Around Conductors

As indicated previously, before the 1980s, the term "corona" was used by many English-speakers to indicate any type of PD event. Today, corona is meant to only include discharges from a metallic point (or metal wire) immersed in a gas, facing a planar electrode at a relatively large distance. This book uses the term corona with the latter meaning. Since corona phenomena are highly dependent

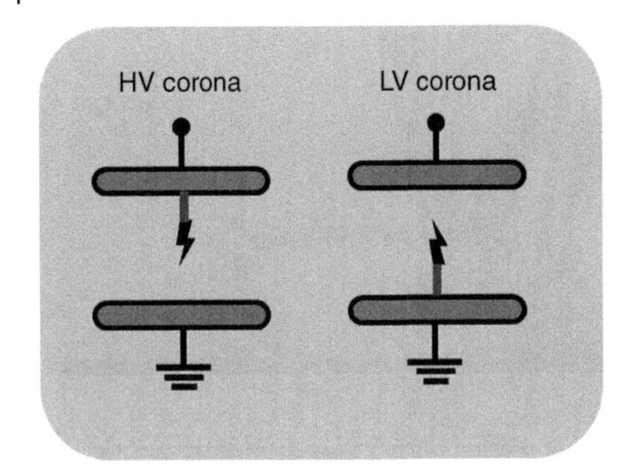

Figure 3.12 HV corona and LV corona.

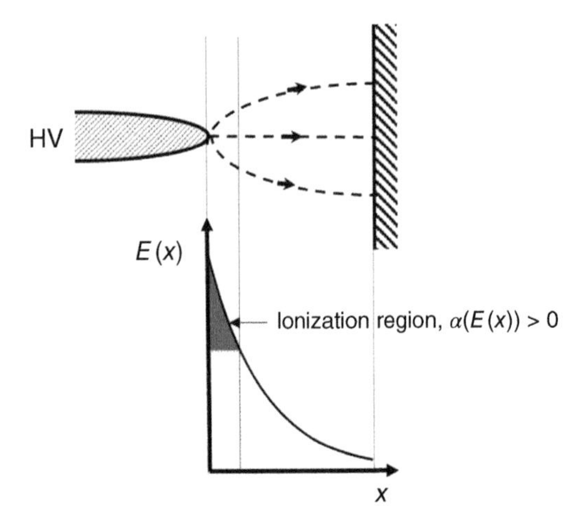

Figure 3.13 Electric field lines in a point/plane electrode configuration and electric field along the shortest line from the point to the plane.

on the polarity of the point, it is useful to distinguish between positive (or anode) from negative (cathode) corona, where positive and negative refer to the polarity of the sharp point during the discharge. Also, since the point can be connected to the high voltage or to the ground as shown in Figure 3.12, when discussing corona, it is necessary to specify whether it is HV corona or LV corona. In the following, for the sake of simplicity, we shall always refer to HV corona. LV corona will display the same characteristics of HV corona, but with the voltage polarities swapped.

3.5.1 PD and Corona Polarity

Corona PD shows striking differences between the discharges of different polarities. Observation of PD due to HV corona indicate that negative PD are incepted at lower voltages compared with positive PD. If there is a single discharge site, the magnitude of corona discharges is about constant, as is the time interval between discharges. In comparison, positive PD are incepted at higher voltages, their repetition rate is much lower, and their magnitude is much larger when compared with negative PD. The difference in magnitude can be experienced in the laboratory using AC voltages: at low overvoltages, negative PD cannot be observed nor heard, while positive PD can be both observed and heard.

These important differences between positive and negative polarity stem from the different mobilities of ions and electrons. It is important to observe that, in both cases, the prerequisite for discharges to occur is the establishment of a region in front of the sharp point tip where the field is large enough to cause ionization (see Figure 3.13). This region is normally limited to only a part of the distance between the electrodes, since the electric field reduces significantly as the distance from the sharp point increases (Section 2.1.3.4), which prevents complete breakdown of the gap. Positive and negative corona can be qualitatively explained through the following considerations:

- Positive corona (positive point): Discharges are initiated when electrons are created by photoionization of gas molecules in or in proximity to the region where the field is sufficiently large for ionization (if they are created further away, they attach to electronegative molecules such as oxygen and are not able to cause impact ionization). Since positive carriers (ions) are less easily injected than electrons, positive PD are initiated with lower probability than negative PD and are thus a much rarer event. When the PD is initiated, the electrons move to the positive point, leaving behind a space charge region of low-mobility positive ions. The low mobility positive ions tend to "focus" the discharge that normally propagates more than negative discharges, achieving larger magnitudes. The effect of the positive ions is the lowering of the field in front of the high voltage electrode, which quenches the discharge activity. Since the positive ions are attracted to the (negative) plane electrode, the positive space charge around the sharp point gradually dissipates. After some time the field at the point is again sufficient to trigger the next discharge. On the other side, when positive ions approach the plane electrode, the field might become large enough to trigger a negative PD from the (negative) plane.

- Negative corona (negative point): Due to the high electrical field, the tip emits electrons in front of it. In the region where the field is sufficiently large, the avalanche grows due to impact ionization, leaving behind positive ions. When the avalanche leaves the region where the field is sufficient for ionization, it cannot grow anymore. Furthermore, the positive ions tend to reduce the field in front of the avalanche, reducing the size of the region where ionization can occur. This explains why negative PD have small magnitudes. Since the field in the region between the positive ions and the negative point is enhanced, electrons can be emitted in this region leading to the neutralization of the positive space charge. Thus, the conditions for a new discharge are re-established quickly, explaining the large repetition rate of negative PD.

Note that if cavities occur within solid or liquid electrical insulation adjacent to the high voltage or ground electrode, a similar polarity effect occurs. An early reference to the effect of polarity on corona and PD is in [11]. Examples of PD patterns related to asymmetric electrode configurations will be discussed in the chapters dealing with measurements in the different types of equipment.

3.5.2 Corona AC Phase Position

Under 50/60 Hz AC voltages, corona phenomena are complex to understand as both positive and negative ions can stay in the gap region for times comparable with the period of the applied AC voltage. Yet, the following qualitative description of corona due to a point connected to the HV may be helpful. Increasing the voltage gradually from 0, negative PD (i.e. PD on the negative part of the AC cycle) are incepted first. These occur symmetrically around 270° of the AC cycle, with low and almost constant magnitude pulses. If the voltage is further increased, the repetition rate of the negative discharges increases notably, while the magnitudes increase only slightly. The phase range widens around 270°. When the voltage is further raised, positive PD are also incepted. Their magnitude is larger than that of negative PD and tend to occur before 90°. The latter evidence indicates that some of the positive ions in proximity of the plane electrode during the negative half cycle of the voltage have not been removed from the gap. The presence of space charge generated by positive PD eventually alters the behavior of negative PD in ways that are often difficult to explain. For corona discharges due to a point connected to the ground, the behavior of positive and negative corona are "swapped" as shown in Figure 3.14.

3.5.3 Corona Current Pulse Characteristics

Observing corona under DC voltages is easier and provides more interpretable observations than is possible for AC voltages. With DC voltage, positive PD tend to be streamer discharges at onset. When further increasing the DC voltage, the time difference between pulses decreases to a point where subsequent pulses merge into a bluish glow (see Section 4.7). The absence of pulses is more probable in electrode configurations where multiple PD sites potentially exist, such as wire/plane configurations.

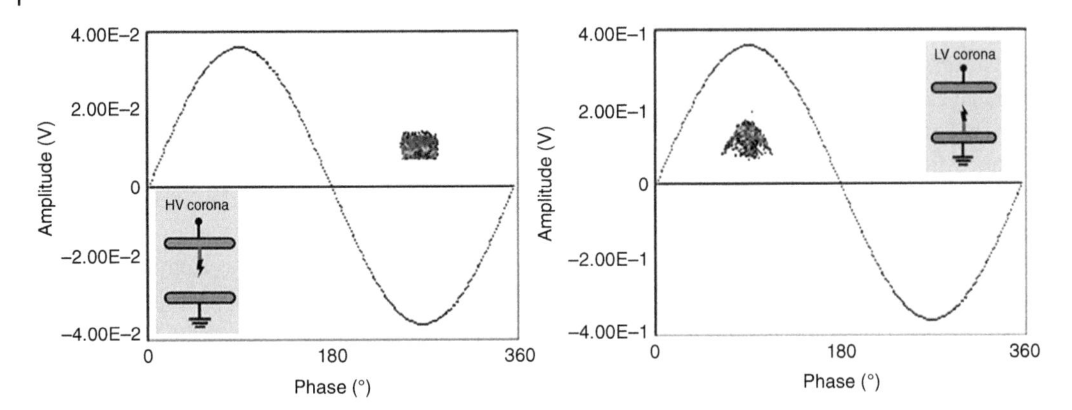

Figure 3.14 PD pattern from (left) HV and (right) LV corona. HV corona is measured using an indirect circuit. LV corona is measured using an indirect circuit.

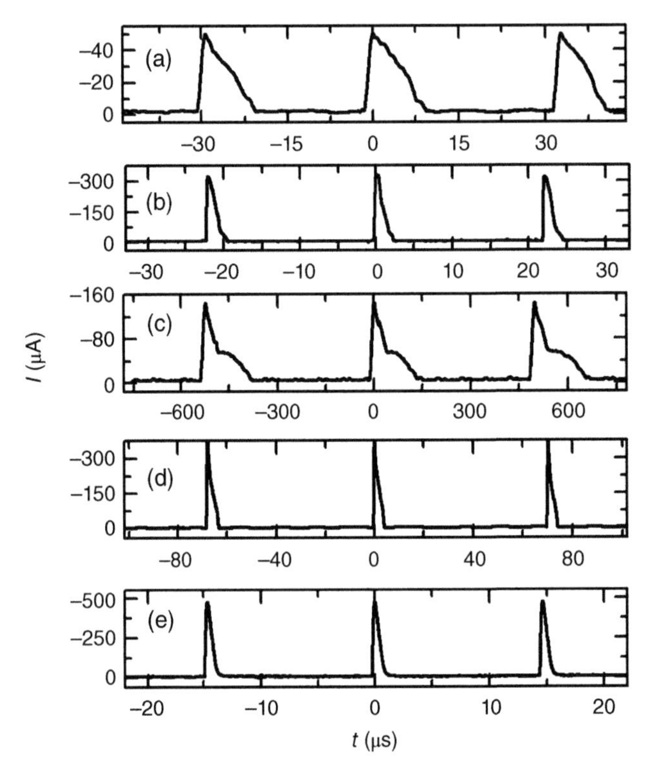

Figure 3.15 Typical waveform of Trichel pulses at p = 30 kPa in (a) pure Ar at 12 μA, (b) Ar-1% O_2 at 10 μA, (c) pure N_2 at 20 μA, (d) N_2-1% O_2 at 15 μA, and (e) air at 20 μA. *Source:* Morshuis [3]; Zhang et al. [12]/ Springer Nature/CC BY 4.0.

Negative corona PD is more complex. Initially, Townsend pulses are observed. At higher voltages, the so-called Trichel pulses can be observed. These are an intermediate stage between Townsend and glow discharges [12]. These pulses have a repetition rate that does not depend much on the distance between the electrodes. The Trichel pulses are equally spaced in time, as shown in Figure 3.15, with pulse waveforms that depend on the geometry (the rise time of the pulse depends on the point radius and gas pressure, the fall times are influenced by the presence of attaching gases). At higher voltages, a reddish glow can be observed, and the phenomenon turns, apparently, into a pulseless glow (Section 4.7), as shown in Figure 3.16.

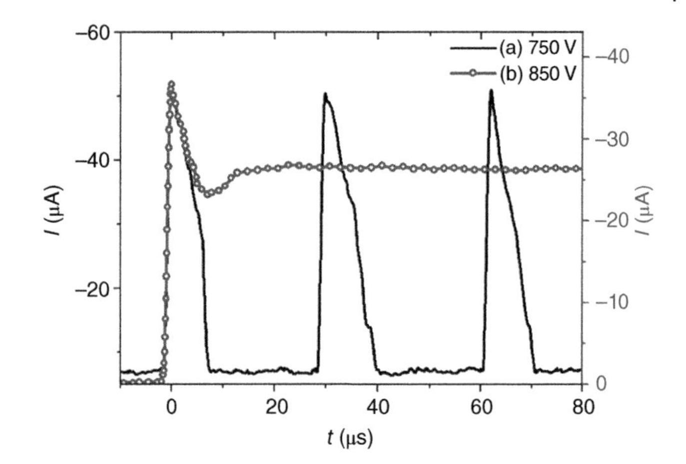

Figure 3.16 Current waveforms of (a) Trichel pulse at $V = 780\,V$ and (b) pulsed glow at $V = 850\,V$ in argon at 30 kPa. *Source:* Morshuis [3]; Zhang et al. [12]/Springer Nature/CC BY 4.0.

It is important to observe that the power dissipated by corona PD is often non-negligible. For example, widespread corona on overhead transmission lines significantly increases the power losses in transmission lines. Corona can also reduce the quality factor of resonant AC supply in an offline PD test, leading to sub-optimal performances.

3.6 Partial Discharge in Voids

PD in voids (cavities) are typically the most harmful purely electrical aging phenomenon in equipment using solid and liquid organic insulating materials. PD starting within a cavity will erode the insulation between the electrodes, following the shortest path (see Figure 3.17). Because of this, PD originating from cavities have been extensively investigated and modeled.

3.6.1 PD Inception

The electric field used to design an insulation system is a trade-off between competing requirements. On the one hand, higher design fields ensure a higher power density (which can be achieved by increasing the service voltage or by reducing the insulation thickness, thus enabling higher currents in the conductors due to better thermal conduction through the insulation). On the other hand, the electric field needs to be low enough to reduce the risk of PD occurring, thus ensuring

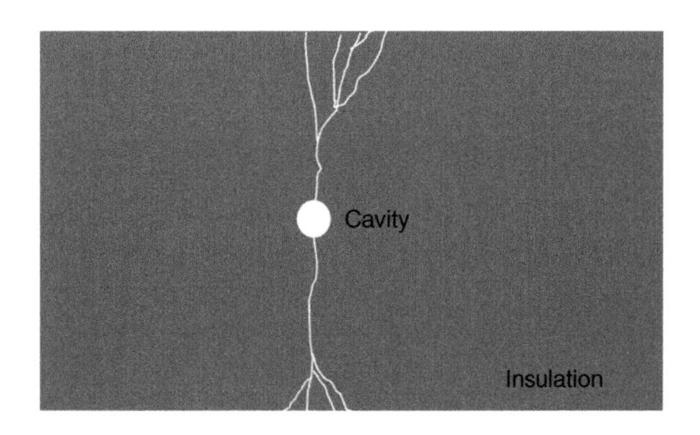

Figure 3.17 Electrical tree starting from a cavity.

HV equipment reliability. It is thus important to assess the quality of the insulation manufacturing in terms of the maximum cavity size to prevent PD in operation. Clearly, setting too stringent limits to the maximum cavity size can have very large costs in terms of manufacturing.

Assuming that the maximum cavity size compatible with the desired manufacturing process is known, then one can estimate the electrical field level that leads to partial discharge inception. To do that, the streamer inception criterion in Equation (2.24) is modified to simplify its calculation. In particular, the effective Townsend ionization coefficient can be represented in a smaller range of electric fields for most HV equipment activity as a parabolic function:

$$\frac{\alpha}{p} = C\left(\frac{E}{p} - \left(\frac{E}{p}\right)_{cr}\right)^2 \tag{3.1}$$

where $(E/p)_{cr}$ is the maximum reduced field[3] ensuring that $\alpha/p \leq 0$. Using this approximation (and assuming $k = 9$ [2] in Equation (2.24)), one gets the inception field in kV/mm:

$$E_{inc} = 2.52 \cdot p_{bar} \cdot \left(1 + \frac{27.15}{\sqrt{p_{bar} \cdot d_{\mu m}}}\right) \tag{3.2}$$

where $d_{\mu m}$ is the maximum distance that an electron can travel within the cavity (expressed in μm), and p_{bar} is the pressure of the gas inside the cavity (in bar). Considering that the field in the cavity is larger than the corresponding field in the dielectric (see Section 2.1.4.1), the design field must be further decreased by a factor that depends on the geometry of the cavity and the permittivity of the surrounding dielectric. Using these considerations, the design field is always much lower than the field that would lead to the breakdown of the insulation. As an example, in HVAC cables, 15 kV/mm is typically the upper limit of design fields, while 2.5–3.0 kV/mm is typical of rotating machines. Figure 3.18 shows the calculated electric fields needed to initiate PD in flat and spherical voids of different void diameters.

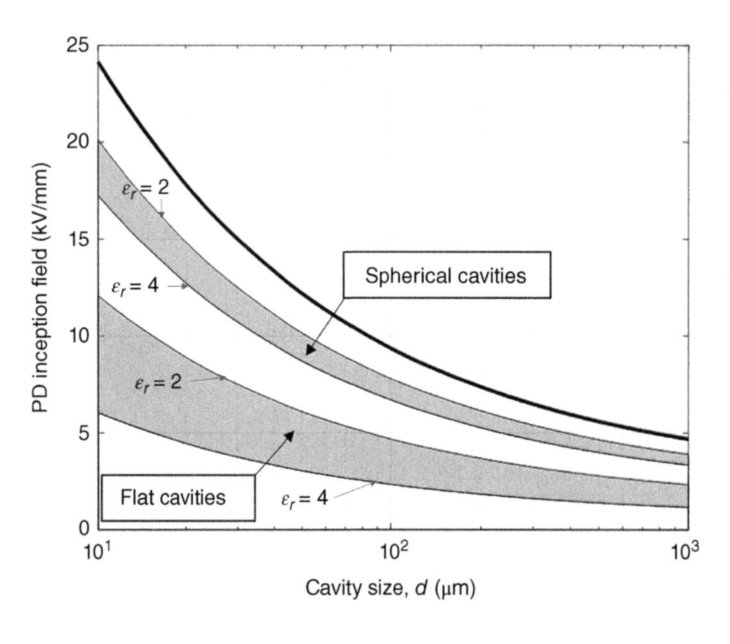

Figure 3.18 The solid curve represents the PD inception field as a function of cavity size (pressure: 1 bar). The grey areas are the corresponding fields inside the dielectric assuming either a spherical cavity or flat cavity with a solid or liquid dielectric with a permittivity in the range 2–4.

3 The reduced field is defined as the ratio E/p.

3.6.1.1 Inception Delay

If the field inside the cavity reaches the value determined through (3.2), a streamer PD can occur (note that Townsend discharges can occur whenever $E/p \geq (E/p)_{cr}$). However, nothing will happen unless a free electron becomes available to start this process. Free electrons are often assumed to be created by photoionization of the gas molecules interacting with highly energetic photons from background radiation. Niemeyer [2] reports that the average time between two starting electrons can be estimated via Equation (3.3):

$$\tau = \frac{1}{C \cdot p \cdot \underbrace{V \cdot \left(1 - \frac{1}{\sqrt[n]{\nu}}\right)}_{V_{eff}}} \tag{3.3}$$

where C is a constant that models the interaction of the background radiation, p is the pressure of the gas within the cavity, V is the volume of the cavity, ν is the ratio between the field in the cavity and E_{inc}, n a positive constant ($n \sim 2$). Clearly, if the product $p \cdot V$ decreases, so does the number of molecules in the cavity, leading to a lower photoionization probability per unit time and, therefore, longer τ values. It is interesting to notice the dependence of τ on ν. When $\nu = 1$, τ is infinite (and thus PD will not initiate). This behavior can be explained by noting that an electron avalanche develops into a streamer only when the charge is sufficiently large to distort the electric field. Depending on the field inside the cavity, the distance is equal to the largest distance between the cavity surfaces when the field equals E_{inc} (the avalanche can travel a longer distance reaching the critical size). For larger fields, this distance can be shorter as the probability of ionization increases. Therefore, when the field is equal to E_{inc}, there is only a single point inside the cavity from where the discharge can start, and thus where the initiatory electron must occur. In contrast, when the field in the void is larger, the area where a starting electron can grow into a streamer becomes larger. This is sketched in Figure 3.19.

Using Equation (3.3), Figure 3.20 shows τ as function of the cavity radius for two different values of the overvoltage. The results obtained are not reflected in actual experiments[4], but highlight that, sometimes, the difference between the time an operator has raised the voltage to a sufficient level and the time PD are detected can be very long. Thus, to measure the voltage at which PD are first incepted in HV equipment (the partial discharge inception voltage), the rate of voltage rise should be very low to allow time for radiation to occur at the critical location. Even then the measured PDIV will be higher than the actual PDIV. Note that the error will be larger for tests performed at low pressures.

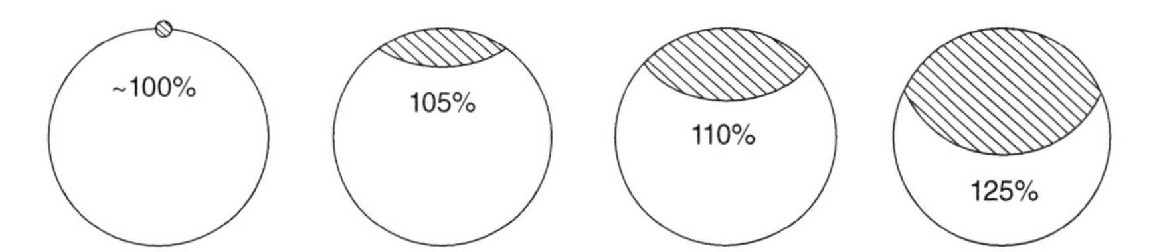

Figure 3.19 In gray, region where a starting electron must be created by the background radiation to ensure that the avalanche grows into a streamer discharge.

---

4 The authors tend to agree with prof. Christian Frank of Zurich ETH, that is the interaction of the background radiation with the polymer matrix is the most probable source of starting electrons.

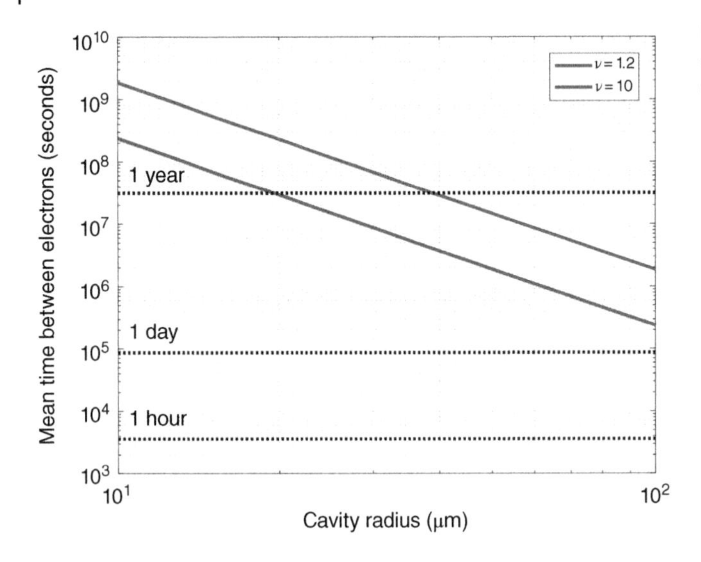

Figure 3.20 Mean time between electrons as a function of the radius (spherical cavity).

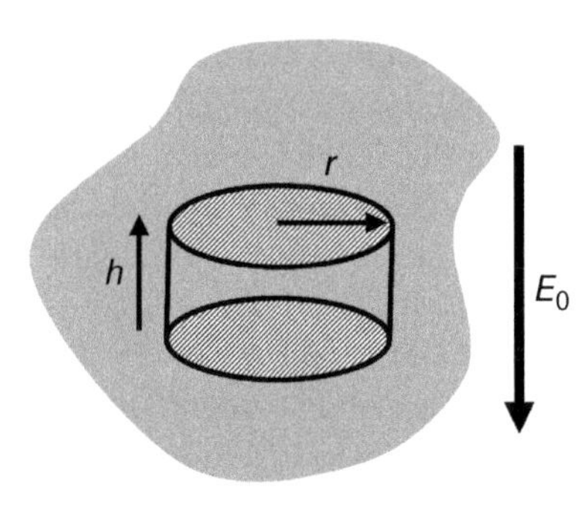

Figure 3.21 Simplified model to calculate the charge q displaced by a streamer.

If one can flood the cavity with X-rays or UV, the delay time between theoretical inception and actual inception will be much smaller. This is sometimes used to measure PD in very small cavities, for example in GIS insulators (Section 13.4.5) [13].

3.6.2 Modified Field Due to Space Charge

If the electric field is sufficient to trigger a streamer when the PD event takes place, a plasma channel (consisting of free electrons and ionized gas) will short circuit at least two points on the surface of the cavity. Assume for simplicity that the cavity is a cylinder with the parallel planes orthogonal to the field (see Figure 3.21). Furthermore, the PD will be treated as an event that effectively reduces to zero the potential difference between the parallel plates. With these assumptions, the PD can be treated as the displacement of an amount of charge that reduces the voltage across a parallel plate air gap capacitor to zero [2]. The voltage across this capacitor, V_c, is given by the product of the height of the capacitor times the field inside the gas, E_c. Thus:

$$q = C \cdot V_c = \left(\varepsilon_0 \frac{\pi \cdot r^2}{h} \right) \cdot \left(h \cdot E_c \right) = \varepsilon_0 \cdot \varepsilon_r \cdot r^2 \cdot E_c \tag{3.4}$$

For spherical cavities, the derivation of an equation like (3.4) is much more complex. Following the procedure reported in [2], the PD charge in spherical cavities having different radii is shown in Figure 3.22 as a function of the field inside the cavity. It is possible to observe that PD charge for cavities with radius lower than 100 μm have charges that do not exceed 100 pC. Discharges below 1 pC can be expected for the smallest cavities.

The space charge created by a PD tends to recombine by drifting on the cavity surface. As the field created by the space charge influences the inception and magnitude of the next discharge,

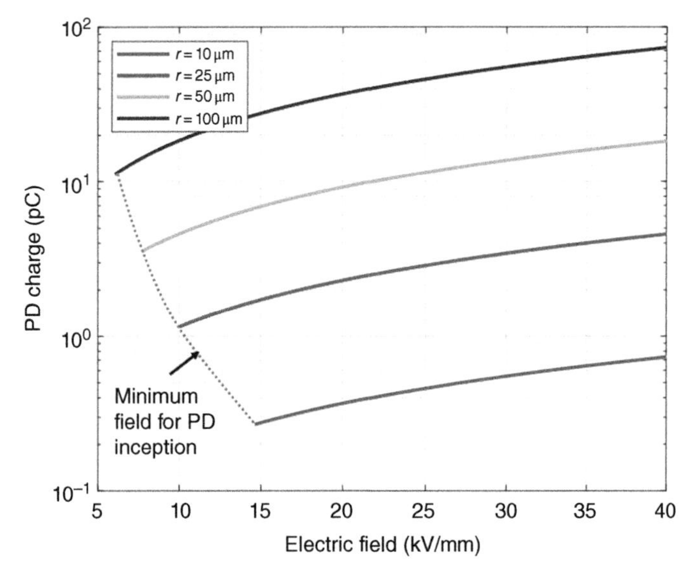

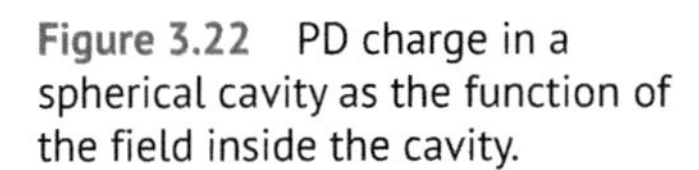
Figure 3.22 PD charge in a spherical cavity as the function of the field inside the cavity.

this charge is also a "memory" of past discharges. The theoretical condition under which the space charge does not change as a function of time will be indicated as "infinite memory." In normal conditions (finite memory), as a first approximation it can be stated that space charge follows an exponential relaxation law. As the space charge recombines, the field inside the cavity increases from almost 0 until reaching the PD inception field E_{inc}. If a free electron is available at this time, a new PD creates more plasma that reduces the field inside the cavity immediately. Otherwise, the field will continue to increase, possibly reaching its steady-state value, until a new electron becomes available. These situations are sketched in Figure 3.23 for a cavity subjected to a DC field. It is interesting to observe that, when the time constant of the charge is much larger than the average time to get a starting electron, the field drop inside the cavity is almost deterministic. Thus, according to (3.4), the magnitudes of the PD events tend to be all equal. This situation has been confirmed by experiments carried out in the lab as well as on practical objects such as power cables.

Similar considerations can be made with AC voltage, which is, however, more difficult to represent. To obtain a simple representation, it is usually assumed that (a) starting electrons are extremely abundant (so that, as soon as the internal field exceeds E_{inc} a PD is triggered immediately), and (b) the space charge due to PD does not recombine and stay constant between two different discharges (this is indicated as a system having infinite memory, i.e. the effect of a PD event is "remembered" for an indefinitely long time unless another PD event modifies the space charge). The behavior under these hypotheses is illustrated in the top of Figure 3.24: the PDs have the same magnitude and they occur at deterministic times. Moreover, negative PD can occur under positive voltages (and *vice versa*) due to the space charge. Relaxing the hypothesis of infinite memory, i.e. allowing the space charge due to PD to decay between two consecutive PD events, PDs still have the same magnitude and occur at deterministic times, but this time the phase lead between negative(positive) PD and the zero crossing of the voltage becomes less important (see Figure 3.24, center). Further reducing the memory effect, the phase lead can disappear. Eventually, if one assumes that free electrons are created at random times, bottom of Figure 3.24, the PD magnitudes and arrival phases become random.

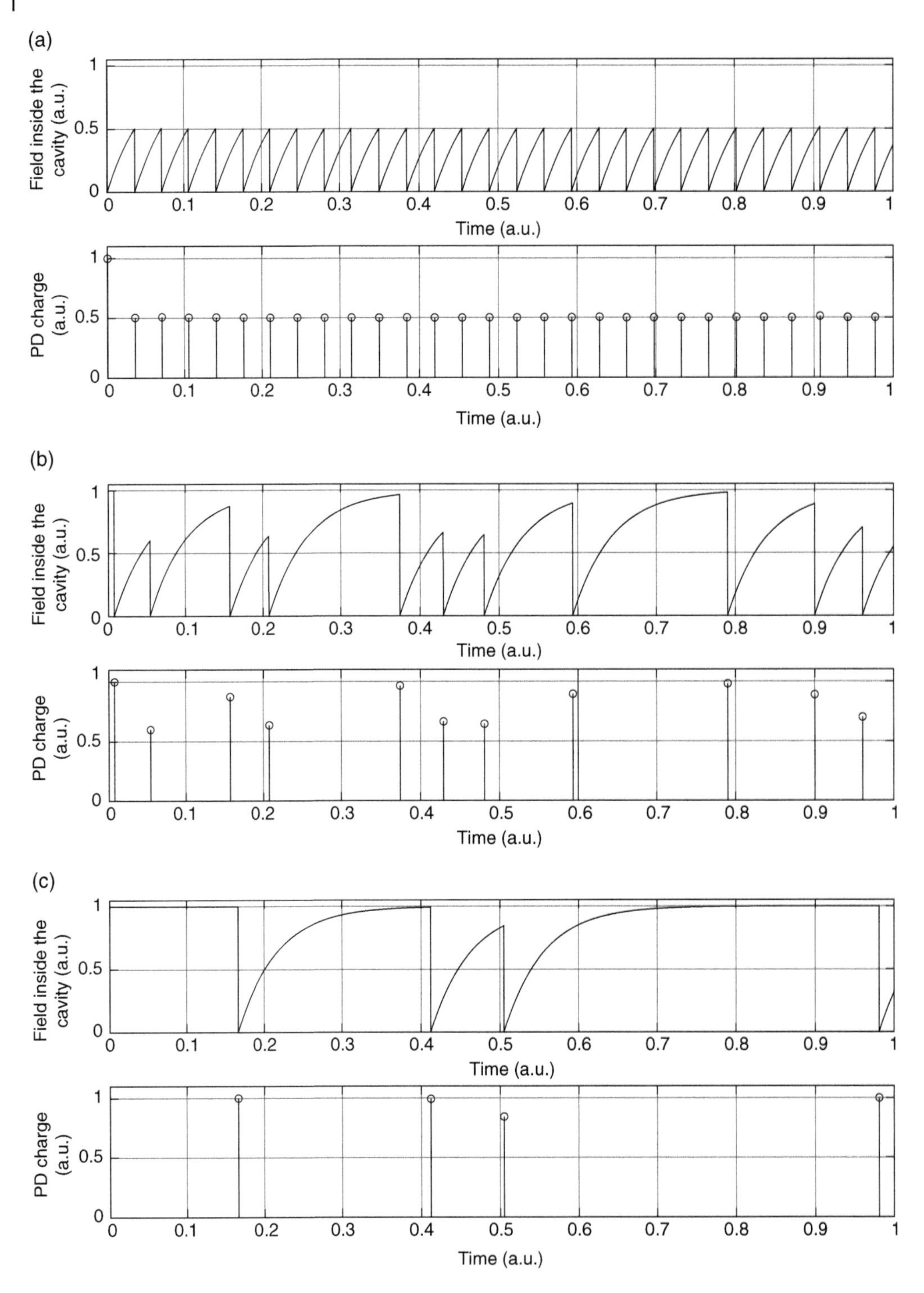

Figure 3.23 Modeling of the time behavior of the field inside a cavity and the PD pulses under DC excitation: (a) is where there are extremely abundant starting electrons- where the time constant of the charge relaxation is much longer than the average time to get a starting electron; (b) shows the condition when less abundant starting electrons are present and the time constant of the charge relaxation is comparable to the average time to get a starting electron; (c) is the case where there are scarce starting electrons, and the time constant of the charge relaxation is lower than the average time to get a starting electron.

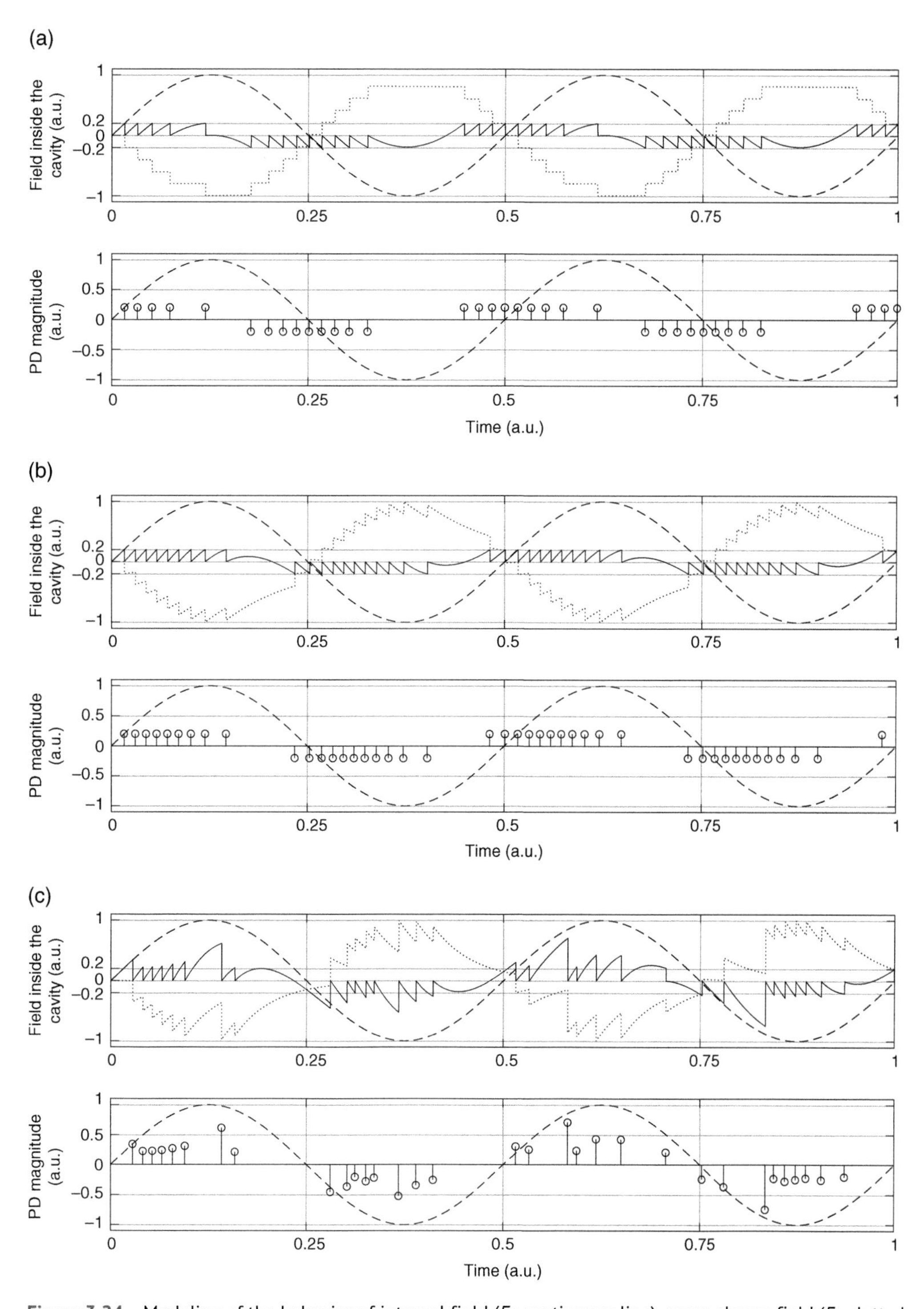

Figure 3.24 Modeling of the behavior of internal field (E_i: continuous line), space charge field (E_q: dotted line) in the presence of an external AC field (fE_o: dashed line): (a) Infinite memory, infinite availability of starting electrons. (b) Finite memory, infinite availability of starting electrons. (c) Finite memory, finite availability of starting electrons.

3.7 PD on Insulation Surfaces

3.7.1 Triple Point Junction

Triple point junctions were discussed in Section 2.1.4.3, where it was shown that at the boundary between an electrode, a high permittivity dielectric (a solid or a liquid) and a low permittivity dielectric (typically a gas) a region of high electric stress exists in the low permittivity dielectric, where an acute angle is formed between the electrode and the interface between the two dielectrics. Triple points clearly represent a threat as PD can be incepted in the low permittivity dielectric, especially if this dielectric is the gaseous state (with its relatively low breakdown strength). Depending on the applied voltage, PD might appear at a few localized spots (close to PDIV) or, above PDIV, involve the whole contour of the triple point. Further raising the voltage, if the opposite electrode is not covered by solid or liquid insulation, flashover can occur. These stages are illustrated in Figures 3.25 and 3.26 [14].

Since PDs can be incepted at several positions, these discharges have repetition rates that have a remarkable dependence on the overvoltage. For the same reason, these discharges have a large variance of PD magnitudes, with the maximum PD magnitude highly dependent on the applied voltage. Figure 3.26 shows that PD events covering different lengths (type 1: 0.3–0.5 mm, type 2: 3 mm) can be active during the same observation period at different positions of the triple point contour.

3.7.2 Electrical Tracking

The term tracking indicates the appearance of tree-like carbonized tracks on the surface of solid insulation. Tracking is normally associated with arcing phenomena and usually has a much higher energy than partial discharges. However, the arcing associated with tracking is widely measured with PD detectors, even though it is not true “PD.” As discussed in Section 4.3, arcing can develop

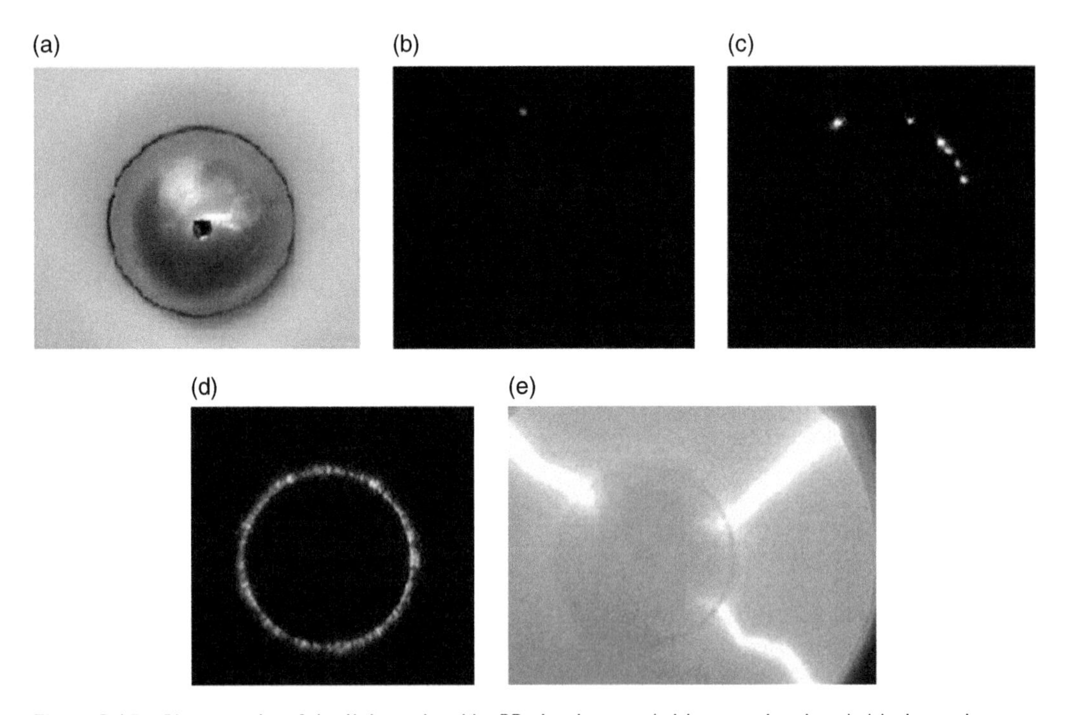

Figure 3.25 Photographs of the light emitted by PDs in nitrogen (with epoxy insulator): (a) electrode shape, (b) V < PDIV, (c) V = PDIV, (d) V > PDIV, and (e) breakdown voltage. *Source:* Tran Duy [14]/ELSEVIER.

from metal electrodes when they are separated and brought to a different potential (for example, arcs are generated in a breaker when the two electrodes are separated). A similar phenomenon occurs in large rotating machines when the conductivity of the semiconductive coating of the bars is too large, and bars can vibrate within the slots: the separation of the bar from the stator is like the separation of the electrodes of a breaker (Sections 4.4 and 16.4.3). For outdoor insulation, arcing is associated with the increase of conductivity of pollution layers due to high humidity or liquid water. The leakage current can become large enough to dry some parts (bands) of the insulator surface (Section 4.6). If the voltage potential across the "dry" band is large enough, arcing can take place across the band. This phenomenon is known as dry band arcing. This is an important aging process in stator windings (Section 16.4.7).

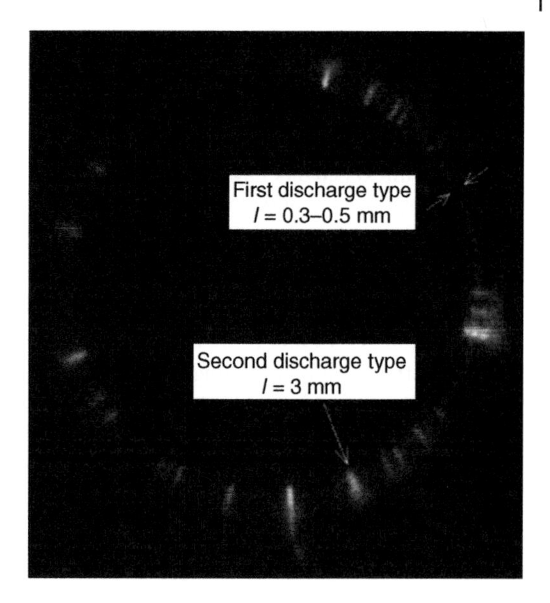

Figure 3.26 PD events above PDIV. *Source:* Tran Duy [14]/ELSEVIER.

Considering the high energy required to carbonize a polymeric insulation, true partial discharges are rarely a cause of tracking unless highly energetic PD can occur for long enough. It is however apparent that a clear-cut separation cannot be made between dry band arcing and partial discharges regarding electrical tracking formation. Considering dry band arcing, when the surface is dried, the electric field on the tracking tips can be large enough to sustain partial discharge activity. In this case, PD are more a symptom than a cause of tracking. On the other hand, tracking-like phenomena have been frequently observed in cable joints and are associated with intense PD activity [15].

3.8 Effect of Ambient Conditions and Conditioning

The behavior of partial discharges tends to evolve with time under voltage. This phenomenon can be observed routinely when testing equipment objects such as rotating machines or transformers. The cause of these trends and/or fluctuations can be twofold. On one side, changes take place in offline tests where the PD decreases over time, an effect known with the term *conditioning*. On the other hand, ambient conditions can change, altering the conditions for PD to occur or effect trapped charges by modifying the conductivity of the dielectric surfaces where the PD takes place.

3.8.1 Conditioning

The effect of prolonged PD activity in a cavity is sketched in Figure 3.9. In the first stages of PD activity, the interaction of the PD with air leads to ozone formation. Ozone is highly reactive and, coming in contact with the polymer, leads to the formation of weak acids. Acids increase the conductivity of the surface favoring the recombination of trapped charges. As mentioned in Section 3.3, this can lead to the apparent disappearance of PD.

The interaction of PD with air, leading to the formation of ozone and, eventually, to weak acids reduces the amount of oxygen within the cavity. In [16], the pressure of the gas within a cavity was observed using an indirect method. At the same time, PDIV measurements were carried out at regular times. It was found that the pressure of the gas within the cavity dropped by about 20% (more or less the same fraction of oxygen in the air), while the PDIV increased. Equation (3.2) implies that by decreasing the gas pressure, the PDIV should decrease rather than increase. A possible explanation for this unexpected behavior is that the residual gas within the cavity is nitrogen, which has ionization energy larger than that of oxygen and might require higher electric fields to sustain streamers.

Conditioning also occurs when dealing with discharges on insulating surfaces. If the surfaces are wet, energetic PD phenomena might be incepted. The PD activity can, in a matter of minutes, remove the layer of humidity, leading to a reduction of PD activity.

In fluid-filled equipment, the formation of bubbles can lead to large changes in PD activity. For instance, if the equipment voltage is raised up to a point where the field is large enough to outgas the fluid and PD are incepted, a bubble will be formed. The presence of a bubble reduces by a great deal the PDIV (in the laboratory, an 80% decrease was observed testing simple insulation models). In large transformers, bubbles are pushed by dielectrophoretic forces toward low-field regions, leading to extinction of PD. If the gas-generating defect is not removed, bubbles will be formed next, and PD will reappear after sometimes. These phenomena lead to erratic cycles of PD activity that the experts can easily trace back to bubbling within the oil.

3.8.2 Ambient/Operating Conditions

Changes in PD activity occur also as a result of changes in ambient conditions. As an example, changes in operating temperature can lead to changes in clearances between different parts of the insulation system. For example, PD is highly dependent on conductor temperature in rotating machines (the expansion of the insulation against the slots due to increased temperature can shrink the voids), transformers (the relative humidity decreases and gas solubility increases at higher temperatures), and in cables.

As mentioned above, humidity as also an important role in determining PD activity features since the PDIV normally decreases with increasing humidity. In addition, humidity influences the conductivity of the dielectrics and has a role in the removal of trapped charge and, consequently, on the PD magnitude and repetition rate.

Gas pressure can have a very important impact as it influences the PDIV largely via Equation (3.2). For power equipment such as air-cooled generators, the design of the insulation could be different depending on the altitude at which the machine will be installed. However, after installation, pressure changes due to changes in meteorological conditions will be irrelevant for PD activity. For aircraft the situation is totally different, as a commercial aircraft can go from 1000 to 150 mbar in a matter of minutes leading to a reduction of PDIV by a factor of about 50% [5].

Eventually, it should be mentioned that vibrations (for example movement of overhead lines due to wind) can lead to changes of the clearances between parts at different potential. This might result in the inception or extinction of PD activity, in some cases also PD pattern changes occurring over limited amounts of time.

3.9 Summary of Measured PD Quantities

When a PD event is measured by a PD detector, the following quantities can be measured:

- PD magnitude and polarity
- Pulse count rate
- Phase of applied voltage

- Arrival time, that is, the time elapsed from the beginning of the measurement to the time the discharge is detected

These quantities will be discussed separately in the following. One should bear in mind, however, that measuring a single PD pulse will not provide useful information about the cause and location of the PD[5]. Rather, PD should be measured for a period of time (typically a few minutes) to get a statistically stable set of quantities that represent the PD activity.

3.9.1 Magnitude

As described in Section 8.3, PD pulses are normally processed to extract information about the magnitude of the pulse (very few commercial detectors can provide the complete shape of the PD pulse). The PD magnitude may provide an indication regarding the damage caused by the PD event to the insulation. Indeed, there are some "caveats" in this respect. First, the actual PD charge cannot be measured for physical reasons explained in Sections 6.4–6.6. Next, the interaction between the electrons in the PD and the insulation depends on a number of factors including the energy (strength) and density of the bonds in the material (both changing with time under stress due to thermal aging and/or hydrolysis for instance, so the same PD can have different effects at different times), and the energy density of the electrons in the avalanche. Since the mean free path of electrons in the avalanche depends solely on the gas pressure and the gas composition, increasing the electric field at which the avalanche is generated increases the energy of the electrons leading to more damage. Thus, the information about the charge should be coupled with that of the electric field within the cavity. The electric field however can only be guessed from the voltage at which the PD takes place (neglecting the effect of trapped charge).

Alternatively, the peak of the PD pulse can be indirectly associated with the amount of the electrons reaching the surface of the cavity. Indeed, this type of measurement can be highly biased due to the detector bandwidth and attenuation phenomena (Section 6.6).

The above considerations lead to the conclusion that the PD magnitude (charge or mV) can only be used in a comparative way on similar equipment or for the same equipment over time (i.e. the trend). Besides, the information of the PD pattern with respect to the AC phase angle can be used to understand the type of the defect. As an example, in a resin-cast equipment, large PD magnitudes with a PD pattern characteristic of internal cavities will indicate that a large cavity exists, an indication that the reliability of the equipment might be compromised. However, large magnitude PD could be due to humidity on the outer surface, which will have a negligible effect. Thus, understanding whether internal or surface PD are observed is critical to assess the condition of the insulation (Section 10.4).

3.9.2 Pulse Count Rate

The PD pulse count rate (or repetition rate) can highlight different things depending on the nature of the insulation system. If a single defect gives rise to PD activity, the pulse count rate can indicate how much the defect is overvolted. As an example, Figure 3.27 shows the simplified simulation under DC excitation of PD in a single cavity at two voltage levels having a ratio 1 : 5. Simulations performed over longer times show that the ratio of the pulse count rates is about 1 : 5. Also the maximum PD magnitudes ratio is 1 : 5, confirming that PD magnitudes can also

5 Except for TDR in cables, where the measurement of a single pulse together with its reflections at the far end can be used for localization (see 12.14.1).

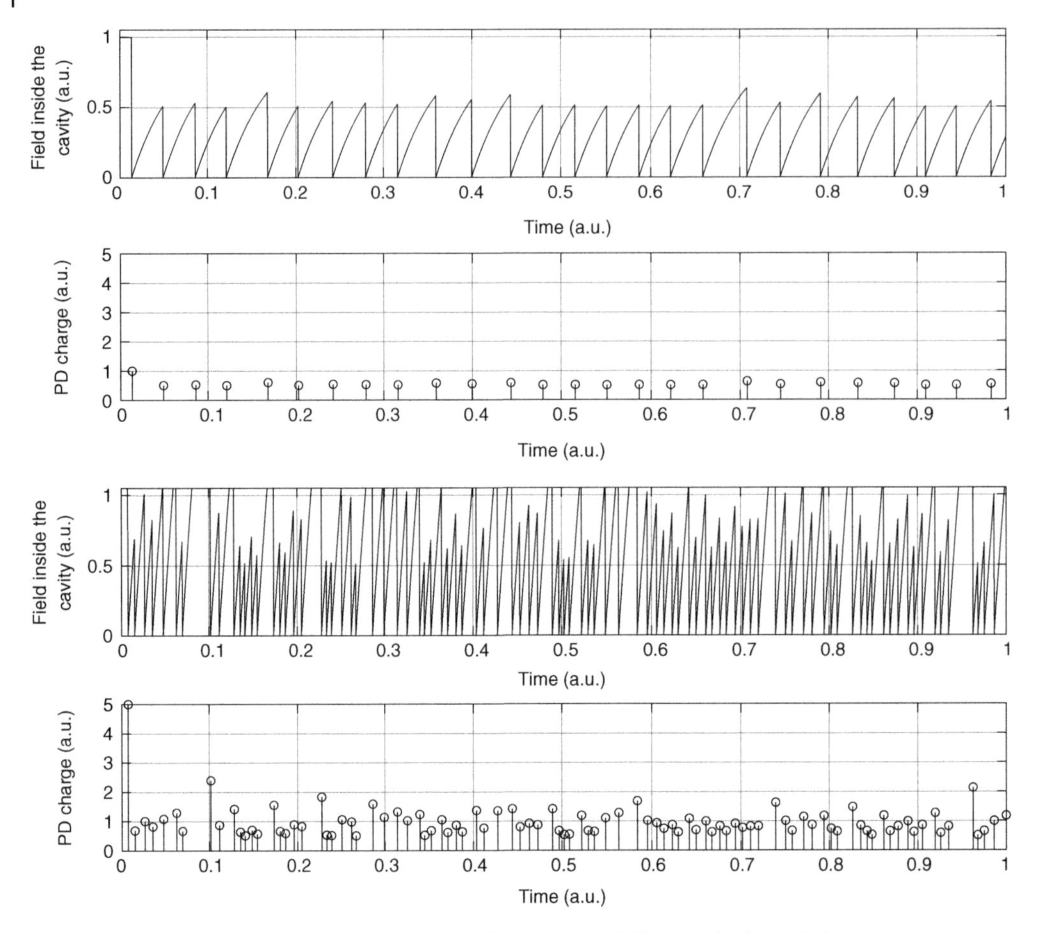

Figure 3.27 Modeling behavior of the field inside a cavity and PD magnitudes in DC.

provide an indication of the overvoltage (ratio between the applied voltage and the inception voltage).

Alternatively, for an electrical equipment with a similar cavity sizes (thus with similar PDIV for each defect), the repetition rate close to the average PDIV, where most cavities have a repetition rate of about 1 PD pulse per cycle, the global repetition rate can indicate the number of defects within the insulation system.

3.9.3 Phase Position

The analysis of where PD pulses occur with respect to the AC cycle can offer insight regarding the type of defect (Section 10.5). Consider for instance Figure 3.24. It can be observed that positive PD start before the zero crossing of the positive half-wave voltage. Negative PDs behave in the same way. The evidence that PD start before the zero crossing indicates that the defect giving rise to PD activity include dielectric surfaces able to trap charges. These charges modify the electric field at the defect allowing for inception before the zero crossing.

3.10 Understanding the PD Pattern with Respect to the AC Cycle

One of the most common mistakes in PD analysis is trying to oversimplify the analysis and jump to the conclusions based on PD magnitudes only. The impact of PD activity on insulation reliability will depend on many factors (type of materials involved, type of defect, synergies with other non-electrical stress factors). The equipment-specific chapters of this book will show that different PD sources will trigger different maintenance actions, with a wide variety of urgencies. Therefore, one of the keys to draw meaningful conclusions and schedule maintenance operations in a correct way is to perform PD source identification first.

PD source identification has usually been carried out by "PD experts" through the visual inspection of the PD pattern with respect to the AC voltage cycle (i.e. the PRPD plot – see Sections 8.7.3 and 10.5). Automatic recognition tools have been extensively proposed in literature, but these tools have not yet been widely adopted by equipment users in industry, perhaps because they are still unreliable. In the following, some concepts that can help to identify the PD source will be presented. These have general validity, i.e. they apply for all types of equipment. However, considering the large number of different situations that can arise dealing with different insulation system technologies, apparatus specific rules tend to be more effective (see Chapters 12–16). It is important to note that the criteria hold for PD from single defects. If the equipment contains many defects (which is almost always the case for stator windings), these criteria cannot be used unless a predominant defect exists.

3.10.1 Polarity Analysis

One of the first steps in PD source identification is inspecting whether there exists a polarity that is predominant with respect to the other one. Consider Figure 3.28, which deals with PD in an internal cavity. Positive PD are started from the defect surface facing the ground electrode of the system; this corresponds to conditions (a) and (d) in Figure 3.28. Negative PDs are started from the defect surface facing the high-voltage electrode of the system (conditions (b) and (c) in Figure 3.28). If other defects such as electrode-bonded cavities, surface, or corona discharges are considered, the surfaces might have substantially different electron emission capabilities, particularly when one of the defect surfaces is the electrode itself and/or the electric field is divergent.

When the defect surface facing the high-voltage electrode can provide abundant electrons (as in the example in Figure 3.29, i.e. a needle connected to the high voltage), negative PD will occur with large repetition rates, generally with low magnitude. Positive PD will be rarer event having statistically larger magnitude, as shown in Figure 3.29. The opposite will happen if the ground electrode can provide abundant electrons (see Figure 3.14).

Indeed, this analysis is simplistic. For instance, metal electrodes can be very smooth or covered with oxide layers that tend to reduce the emission of electrons. Notwithstanding, as a rule, asymmetric patterns should be regarded with great care as they might indicate that one of the defect surfaces is a metal electrode and/or a highly divergent field exist within the insulation system. Both cases correspond to the existence of a potential weak point leading to breakdown.

3.10.2 Physical Basis for PRPD Patterns

Phase resolved PD (PRPD) patterns that are collected from all the PD pulses measured for several minutes are widely used to record and display the PD activity (Sections 8.7.3 and 10.5). The following presents the physics associated with certain PRPD plots associated with different kinds of simple defects. The qualitative patterns are measured at 3 or 4 different overvoltage levels (overvoltage

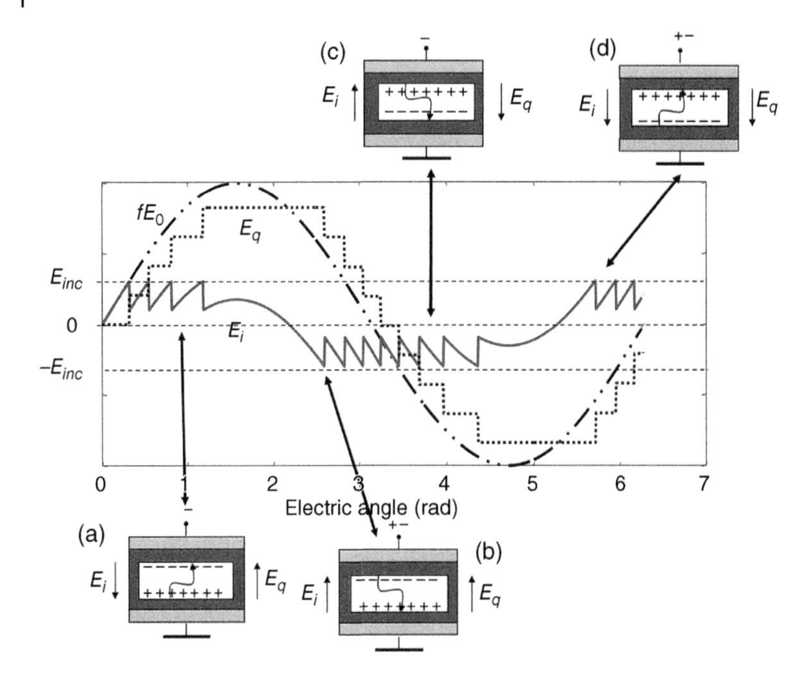

Figure 3.28 Schematic representation of PD phenomena from a single cylindrical cavity during one cycle of the fundamental voltage. The extraction of free electrons from different and differently charged surfaces is highlighted. E_0 is the Laplacian field due to the high-voltage source; E_q the field due to the space charge; E_i is the internal field, that is, the algebraic sum of E_0 and E_q.

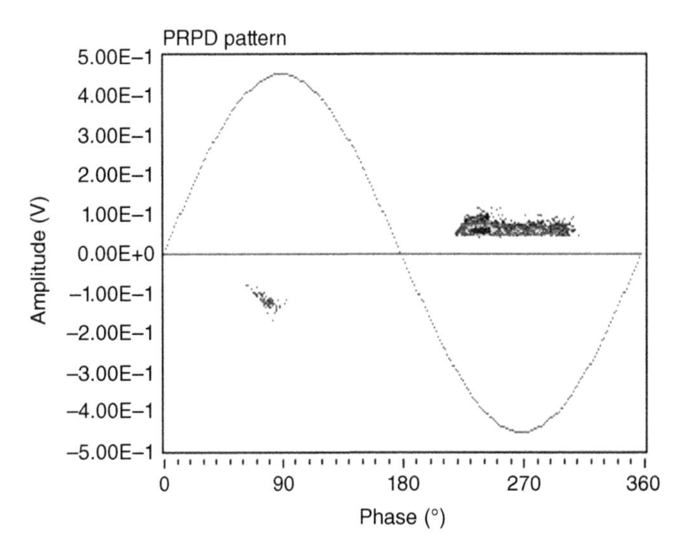

Figure 3.29 Example of corona PD from needle connected to the high voltage: negative PD are predominant.

is the ratio of the applied voltage to the PDIV). The aim of this summary is to provide the less experienced reader with insight about the features of PD from different defects. The plots show the:

1) Maximum PD magnitude
2) PD repetition rate (in pulses per cycle, PPC, represented by the color of the dots)
3) The inception phase Φ_i for both positive and negative PD (see Figure 3.30)
4) Phase angle range $\Delta\Phi$ for both positive and negative PD (see Figure 3.30)

Three overvoltage levels together with the corresponding PD patterns are used. The aim of this summary is to provide the less experienced reader with insight about the features of PD from different defects.

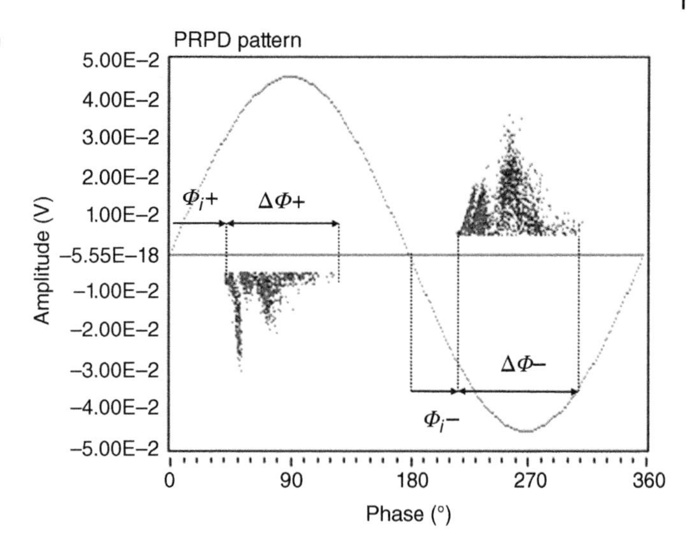

Figure 3.30 Definition of inception phase angle and phase angle range for positive and negative PD. Note, in this section positive PD are considered as those PD incepted when either the voltage or its derivative is positive. Negative PD are incepted when either the voltage or its derivative is negative. This choice was made to provide a better connection between the physics of the PD and the appearance of the PD pattern. Indirect measurement circuit.

The defects presented here are:

1) Corona PD obtained from a needle connected to the high voltage: As mentioned above, the data achieved testing high-voltage corona are also representative of those relevant to low-voltage corona, provided that the PD patterns are "swapped" (as in Figure 3.12). Dealing with statistics, those obtained from high-voltage corona can be used for low-voltage corona if statistics for positive PD are used to describe negative PD, and *vice versa*.
2) Internal PD: The defect consists of a cavity in an epoxy block containing two embedded electrodes.
3) Surface PD: This defect was achieved by cracking an insulating block containing two embedded electrodes. It is interesting to note that, for this type of defect, the PD develop at the surface crack, thus the defect surfaces are both insulating.

Starting with corona PD, it is important to observe that the pattern is characterized by a large asymmetry, see Figure 3.31. At PDIV, only negative PD are incepted. Negative PD tend to spread symmetrically around 270°, indicating that (a) inception starts as soon as the voltage exceeds the PDIV thanks to the abundance of starting electrons from the metallic tip; (b) the PD activity ceases as soon as the voltage is lower than the PDIV, indicating the absence of space charge (or memory effect). As the voltage is increased, negative PD tend to spread over wider phase ranges and positive PD are incepted. At the inception phase angle, the pattern for negative PD shows an almost vertical structure, indicative of space charge existing in front of the needle tip due to positive discharges. Indeed, depending on the geometry of the defect, at high overvoltage levels, corona can have very peculiar structures for negative PD.

For internal PD (see Figure 3.32), the pattern tends to be symmetrical. PD inception tends to occur before the voltage zero crossing due to the space charge left by previous PD. For the same reason, the largest PD magnitudes tends to occur near the inception phase angle, for both positive and negative PD. The maximum PD magnitude is not significantly influenced by the overvoltage. These features can be used to distinguish internal PD from other types of PD when tests are performed offline and the operator is at liberty to change the voltage.

Surface PD are mostly symmetrical (see Figure 3.33). Indeed, the PD pattern reported here shows a few substructures probably related to the inception of PD at different points in the crack. The most striking feature of surface PD is the remarkable dependence of the maximum PD magnitude on the overvoltage, for both polarities. This is a key feature of surface PD that can be used for recognition when tests are performed offline.

Figures 3.34–3.37 show the statistics calculated for the different defects as a function of the overvoltage. The figures highlight what was already noted qualitatively from the patterns:

1) The large dependence of PD magnitude of surface PD on the overvoltage (Figure 3.34)
2) The large amount of PD per cycle and its dependence on the overvoltage for corona PD (Figure 3.35)

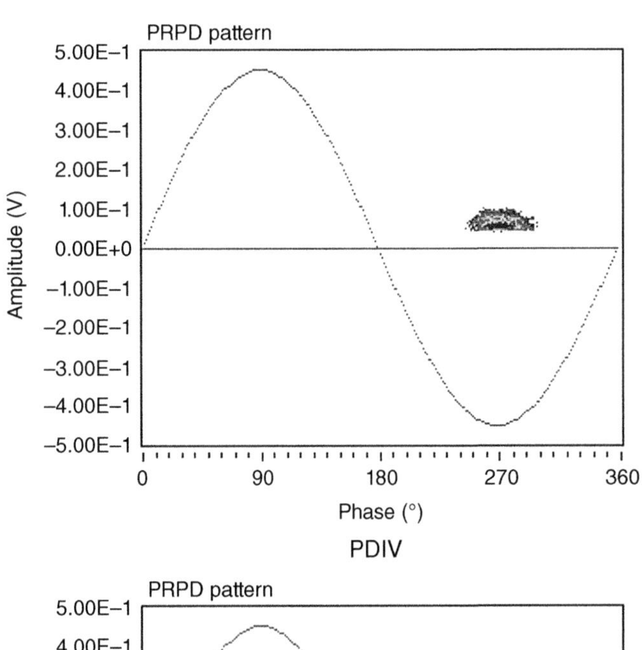

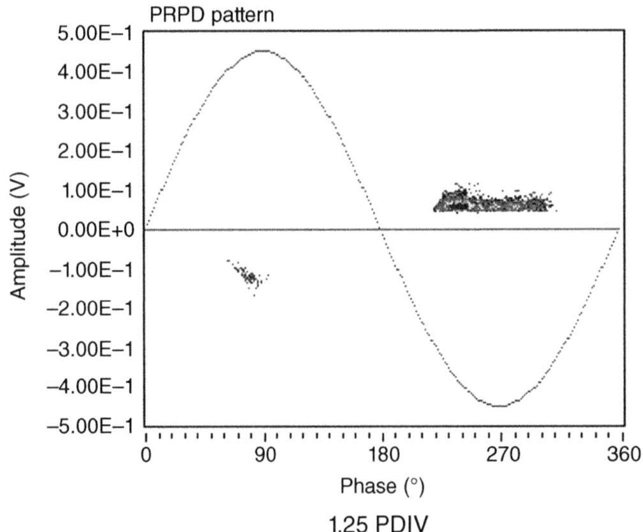

Figure 3.31 PD patterns at different overvoltages for corona PD (needle connected to the high voltage). Indirect measurement circuit.

Figure 3.31 (Continued)

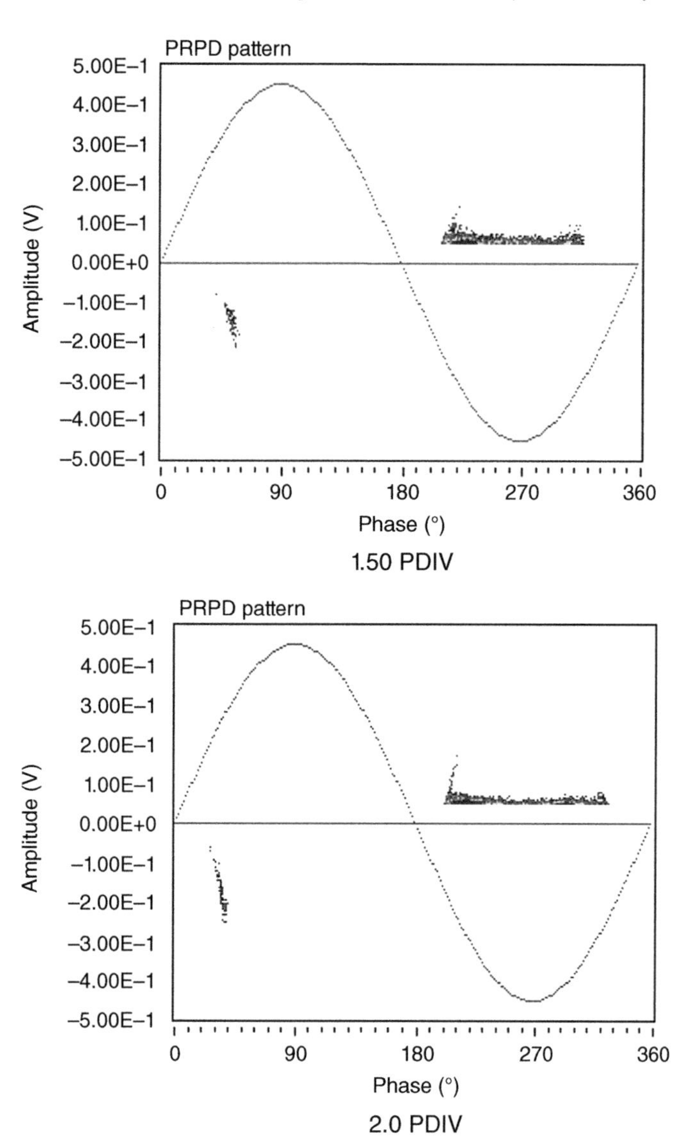

3) The negative inception phase angle for internal PD due to memory effect (Figure 3.36). The negative slope of the lines indicates that, as the voltage increases, PD tend to occur more and more to the left in the pattern.
4) The limited dependence of the phase angle range of internal PD on overvoltage (Figure 3.37). The positive slope of the lines for surface and corona PD indicate that the PD tend to spread over larger portion of the PD pattern.

These and other more complex features (not dealt here for the sake of brevity) are used by experts to distinguish the different types of PD sources in offline PD tests. Efforts were made to derive recognition software tools that could lead to identification automatically. However, these tools did

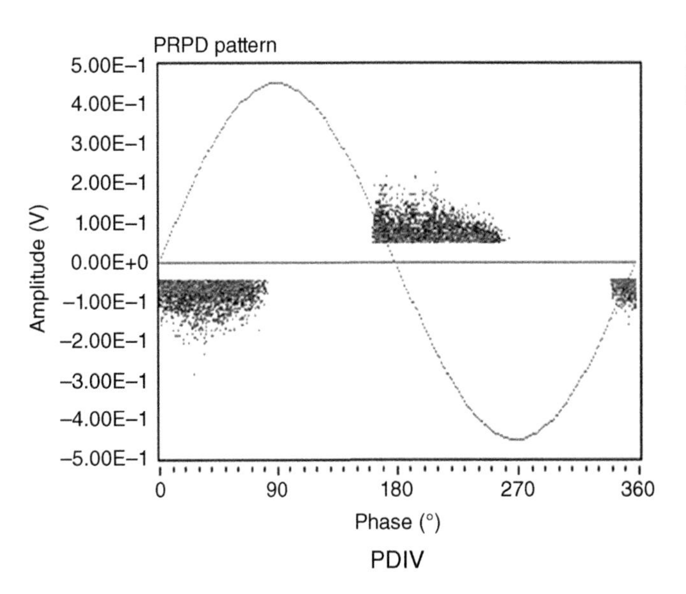

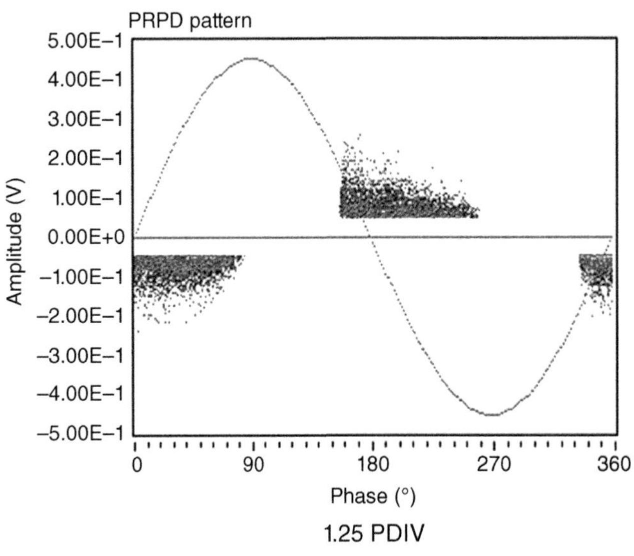

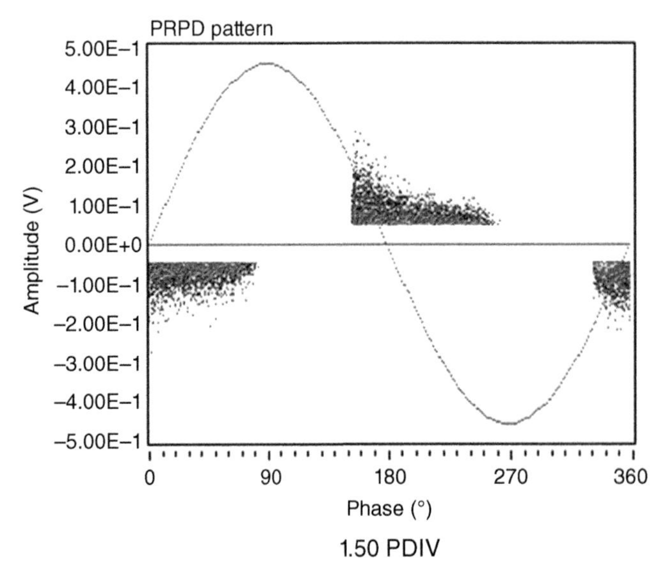

Figure 3.32 PD patterns at different overvoltages for internal PD. Indirect measurement circuit.

Figure 3.33 PD patterns at different overvoltages for surface PD (crack in epoxy casting). Indirect measurement circuit.

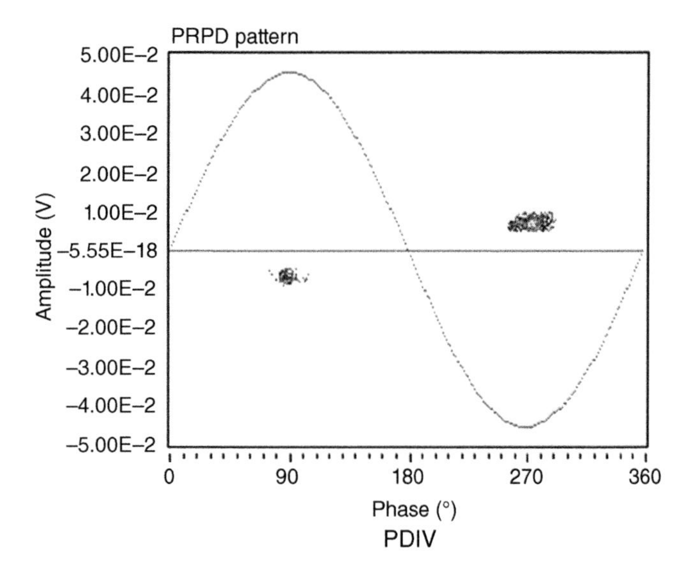

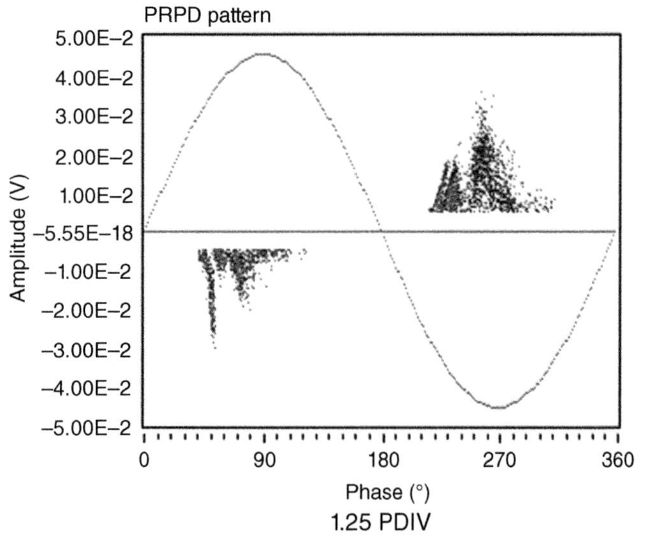

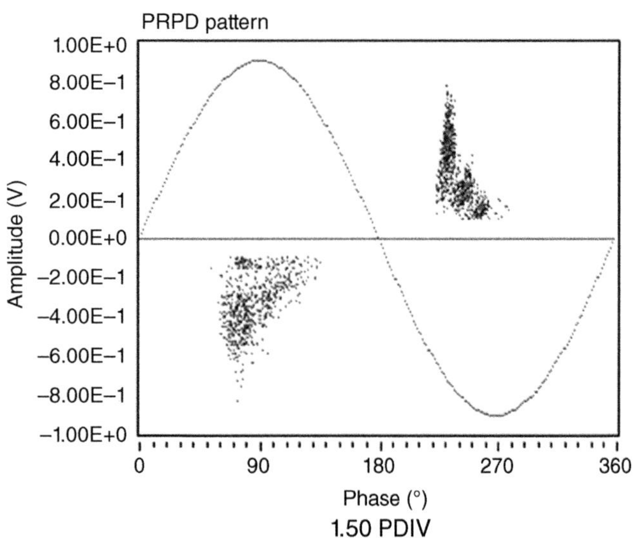

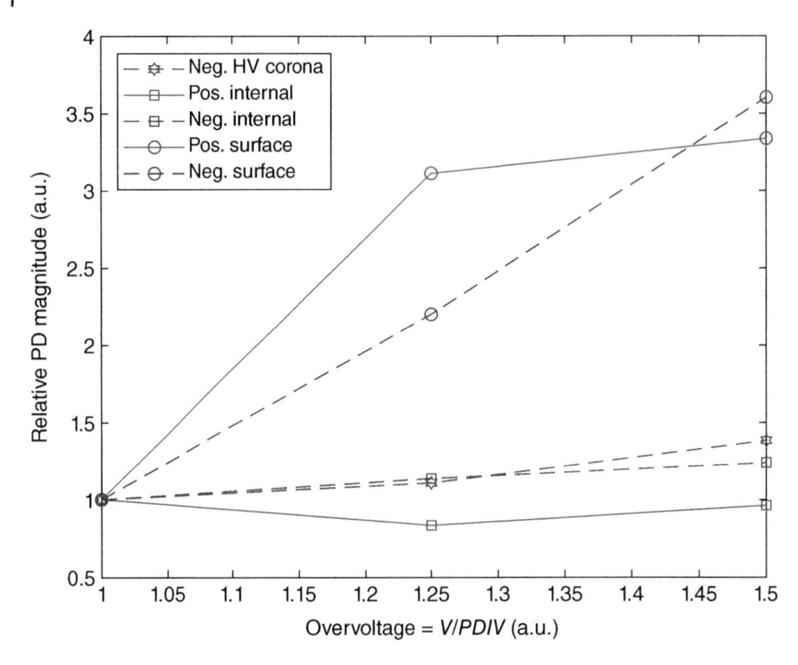

Figure 3.34 PD magnitude for corona, internal, and surface discharges. Positive and negative PDs are treated separately.

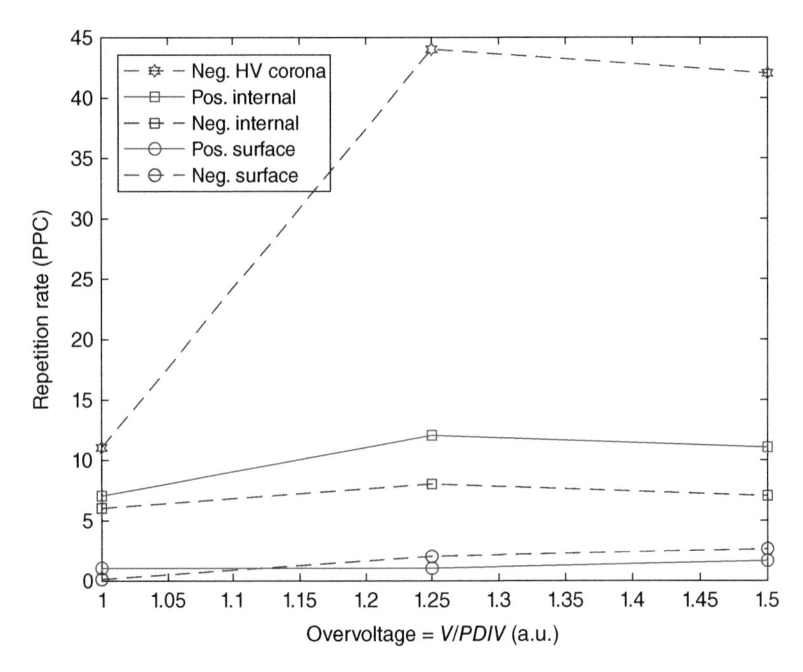

Figure 3.35 PD repetition rate for corona, internal, and surface discharges. Positive and negative PDs are treated separately.

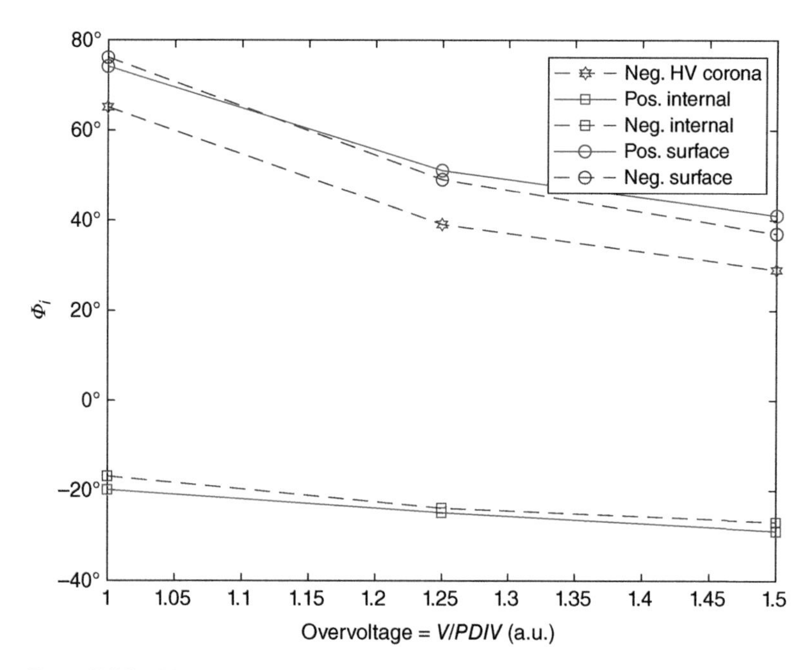

Figure 3.36 PD Inception phase angle for corona, internal, and surface discharges. Positive and negative PDs are treated separately.

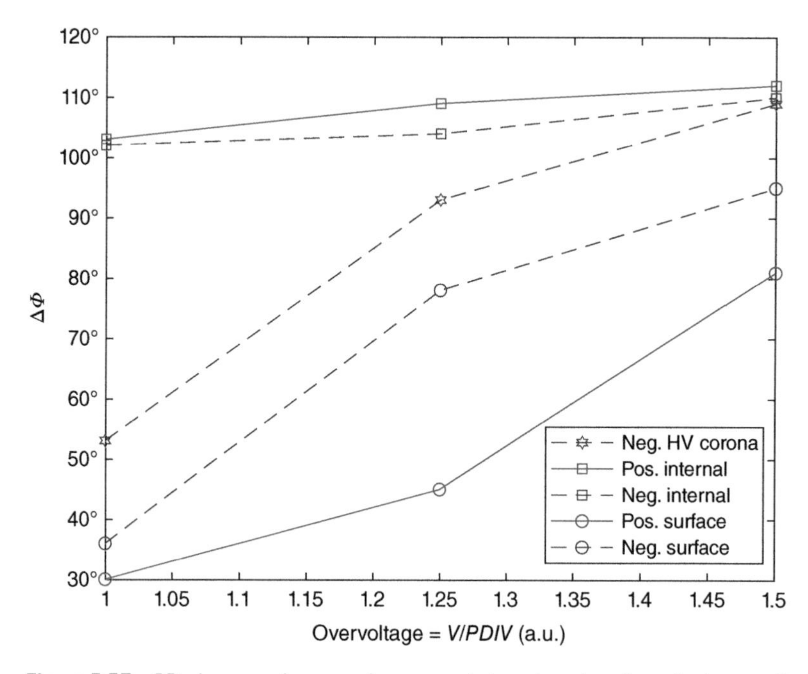

Figure 3.37 PD phase angle range for corona, internal, and surface discharges. Positive and negative PDs are treated separately.

not make their way into the market as noise and multiple PD phenomena often make the calculation of the relevant statistics very difficult.

As a final remark, it is important to warn the reader about the risks associates with generalization of PD features from a limited database not representing in a proper way the features of the possible defect types in electrical equipment. Figures 3.38–3.41 were derived using three different objects: the epoxy block with a crack; magnet wires twisted together, with one wire connected to the high voltage and the other grounded (twisted pair); and a triple point realized by placing an insulating film on a metallic plane connected to the ground and, on top, a cylindrical electrode connected to the high voltage. In three cases, one tends to use the term "surface discharges," as if these defects were behaving in the same way. However, as can be observed, the three systems behave in different ways. The triple point, owing to the bare metal facing the air gap, is notably different from the other two types of defects. Therefore, to achieve a correct diagnosis, typical PRPD patterns should be derived for each type of high-voltage equipment (cables, transformers, etc.), paying a lot of attention to the most typical defects and how to model them in the lab.

3.10.3 PD Packets

The existence in the PRPD pattern of "packets" (PD tend to concentrate at some phase positions) is often due to treeing phenomena in epoxy-insulated systems. Packets may be generated, for example, by PD discharges propagating along the tree, see Figure 3.42(left). Although packets have been found also in other phenomena (covered conductors facing each other without touching directly, see Figure 3.42(right)), the presence of packets in an epoxy-insulated system should be investigated carefully.

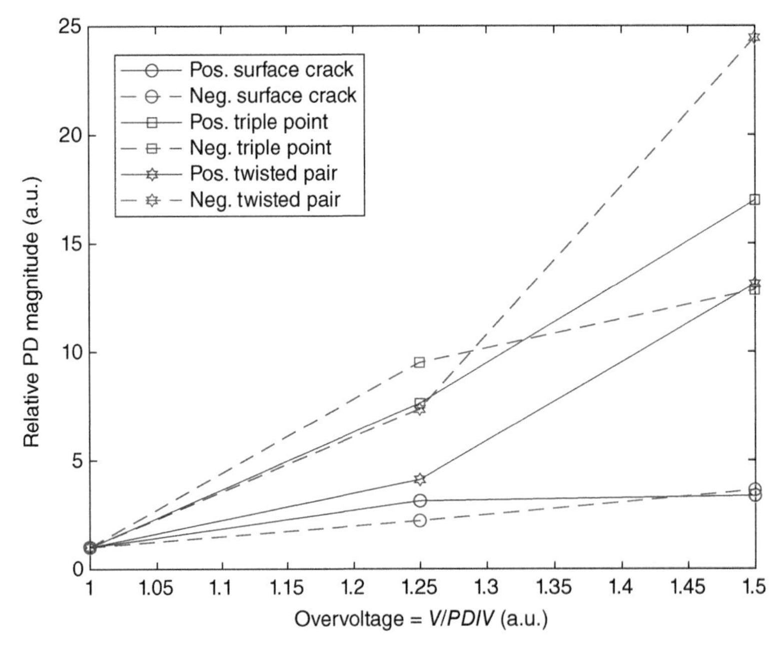

Figure 3.38 PD magnitude for three different types of surface discharges. Positive and negative PDs are treated separately.

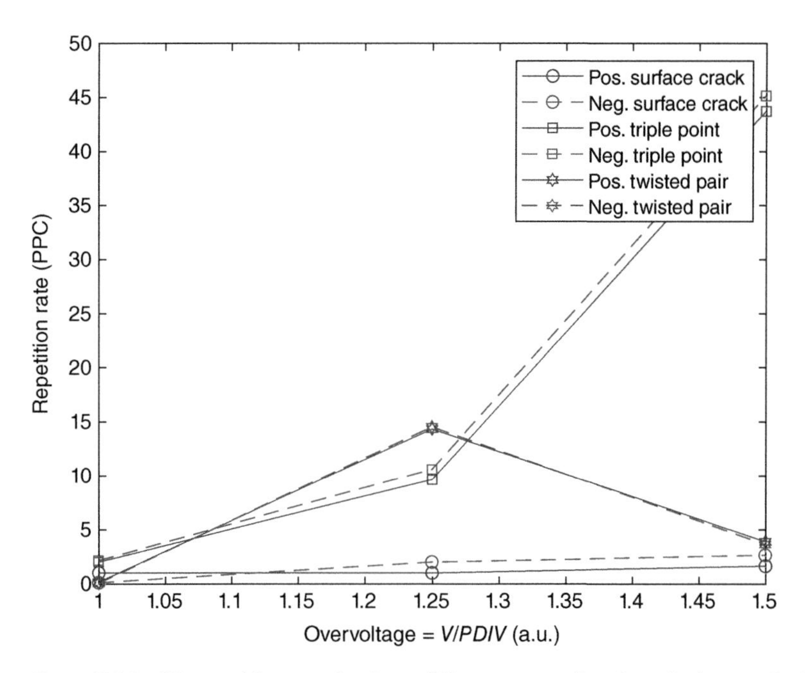

Figure 3.39 PD repetition rate for three different types of surface discharges. Positive and negative PDs are treated separately.

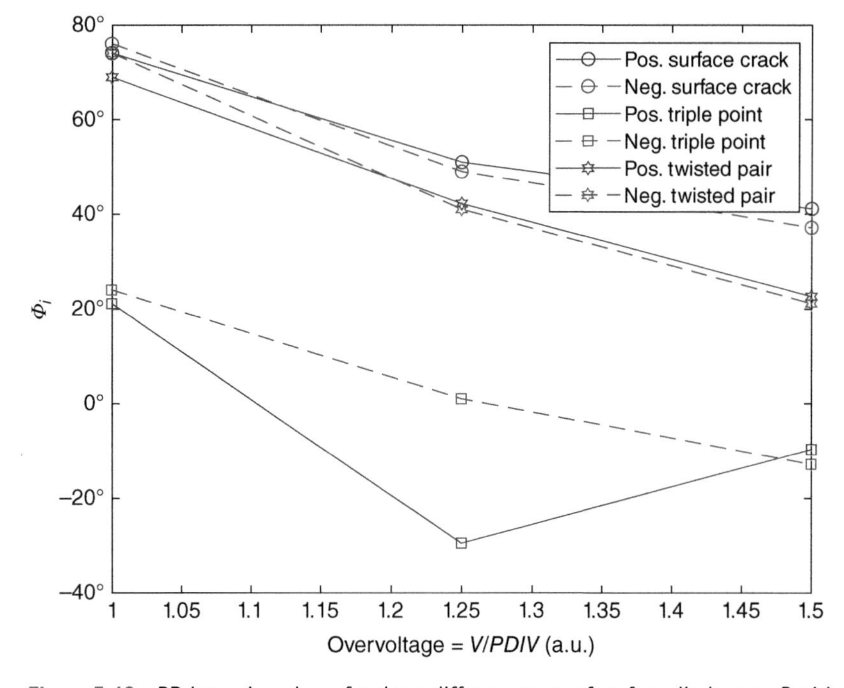

Figure 3.40 PD Inception phase for three different types of surface discharges. Positive and negative PDs are treated separately.

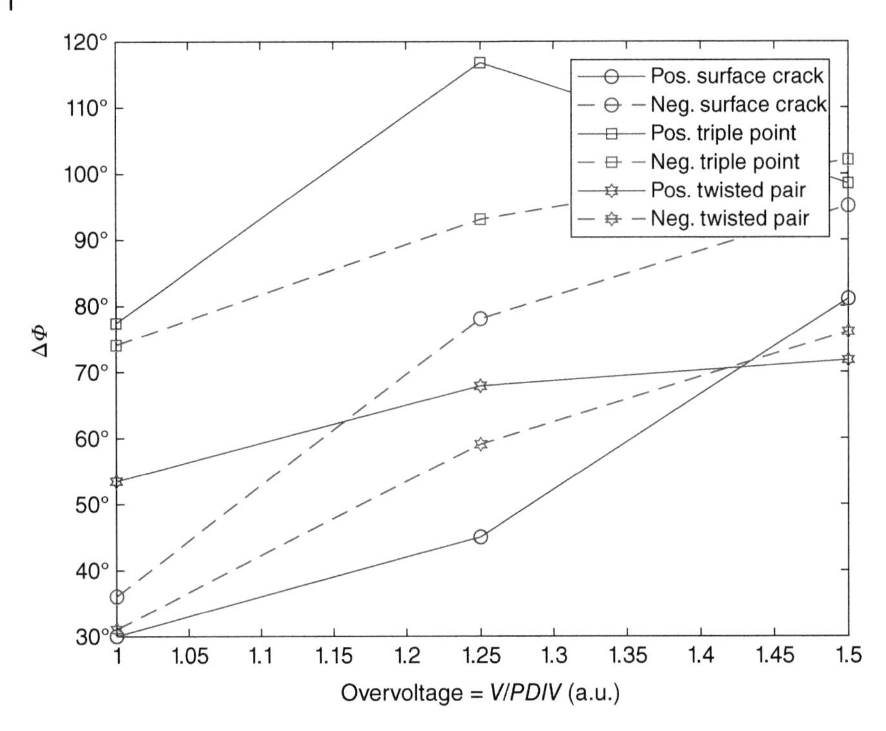

Figure 3.41 PD phase range for three different types of surface discharges. Positive and negative PDs are treated separately.

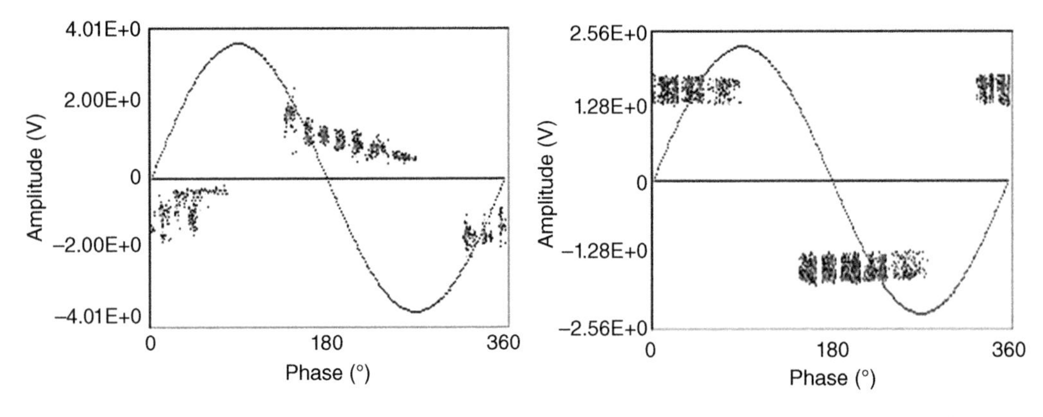

Figure 3.42 Example of packets in the PRPD pattern: (left) treeing in an epoxy slab, (right) bar-to-bar discharges in a rotating machine.

References

1 IEC Std. 60270 (2015). High-voltage test techniques – partial discharge measurements.

2 Niemeyer, L. (1995). A generalized approach to partial discharge modeling. *IEEE Transactions on Dielectrics and Electrical Insulation* 2 (4): 510–528. https://doi.org/10.1109/94.407017.

3 Morshuis, P.H.F. (1993). Partial discharge mechanisms: mechanisms leading to breakdown, analyzed by fast electrical and optical measurements. https://repository.tudelft.nl/islandora/object/uuid%3A5e39ece9-bcb1-40e0-8501-cfb963873a78 (accessed 10 September 2021).
4 Morshuis, P. (1995). Ageing of polymers studied by spectral analysis of UWB partial discharge signals *Proceedings of 1995 Conference on Electrical Insulation and Dielectric Phenomena*, pp. 226–229. https://doi.org/10.1109/CEIDP.1995.483615.
5 Lusuardi, L., Rumi, A., Cavallini, A. et al. (2021). Partial discharge phenomena in electrical machines for the more electrical aircraft. Part II: impact of reduced pressures and wide bandgap devices. *IEEE Access* 9: 27485–27495. https://doi.org/10.1109/ACCESS.2021.3058089.
6 Cavallini, A., Versari, L., and Fornasari, L. (2013). Feasibility of partial discharge detection in inverter-fed actuators used in aircrafts. *2013 Annual Report Conference on Electrical Insulation and Dielectric Phenomena*, pp. 1250–1253. https://doi.org/10.1109/CEIDP.2013.6748115.
7 Hudon, C., Bartnikas, R., and Wertheimer, M.R. (1995). Effect of physico-chemical degradation of epoxy resin on partial discharge behavior. *IEEE Transactions on Dielectrics and Electrical Insulation* 2 (6): 1083–1094. https://doi.org/10.1109/TDEI.1995.8881923.
8 Temmen, K. (2000). Evaluation of surface changes in flat cavities due to ageing by means of phase-angle resolved partial discharge measurement. *Journal of Physics D: Applied Physics* 33 (6): 603–608. https://doi.org/10.1088/0022-3727/33/6/303.
9 Vogelsang, R., Farr, T., and Frohlich, K. (2006). The effect of barriers on electrical tree propagation in composite insulation materials. *IEEE Transactions on Dielectrics and Electrical Insulation* 13 (2): 373–382. https://doi.org/10.1109/TDEI.2006.1624282.
10 Zhao, Y., Vaughan, A.S., Champion, J.V. et al. (2000). The structure of electrical trees in semi-crystalline polymers. *2000 Eighth International Conference on Dielectric Materials, Measurements and Applications (IEE Conf. Publ. No. 473)*, pp. 314–319. https://doi.org/10.1049/cp:20000525.
11 CIGRE WG 21.03 (1969). Recognition of discharges. *Electra* 11.2: 61–98.
12 Zhang, Y., Xia, Q., Jiang, Z., and Ouyang, J. (2017). Trichel pulse in various gases and the key factor for its formation. *Scientific Reports* 7 (1): 10135. https://doi.org/10.1038/s41598-017-10118-2.
13 Adili, S. and Franck, C.M. (2014). Partial discharges characterization in spherical voids using ultra-short X-ray pulses. *IEEE Transactions on Dielectrics and Electrical Insulation* 21 (2): 791–799. https://doi.org/10.1109/TDEI.2013.004180.
14 Tran Duy, C., Bonifaci, N., Denat, A. et al. (2008). Partial discharges at a triple junction metal/solid insulator/gas and simulation of inception voltage. *Journal of Electrostatics* 66 (5): 319–327. https://doi.org/10.1016/j.elstat.2008.01.011.
15 Du, B.X. and Gu, L. (2010). Effects of interfacial pressure on tracking failure between XLPE and silicon rubber. *IEEE Transactions on Dielectrics and Electrical Insulation* 17 (6): 1922–1930. https://doi.org/10.1109/TDEI.2010.5658247.
16 Wang, L., Cavallini, A., Montanari, G.C., and Testa, L. (2012). Evolution of PD patterns in polyethylene insulation cavities under AC voltage. *IEEE Transactions on Dielectrics and Electrical Insulation* 19 (2): 533–542. https://doi.org/10.1109/TDEI.2012.6180247.

4

Other Discharge Phenomena

4.1 Introduction

In Chapter 2, the electrical breakdown of gases due to the Townsend and streamer processes was discussed. In Chapter 3, the partial discharge process was described as electrical breakdown in a gas-filled void within or on the surface of solid or liquid insulation. In practical HV equipment, the solid insulation itself does not normally experience breakdown, because the short-term intrinsic electric strength of a solid (in the absence of any defects) is almost two orders of magnitude higher than the breakdown strength of gas at atmospheric pressure (Section 2.6). Sections 2.4 and 2.6 discussed that discharge pulses within the gas can degrade organic solid or liquid insulation at the void walls due to electron and ion bombardment, as well as by chemical processes. Thus, PD in solid and composite insulation is "bad" since it gradually erodes the solid insulating material by treeing (Section 2.4.1) and tracking (Section 3.7.2), eventually leading to a ground or phase-to-phase fault. To prevent premature failure of the insulation system, equipment manufacturers perform quality assurance PD tests in the factory, and end users measure PD in offline or online tests on HV equipment at site, to assure themselves that insulation system failure is not imminent.

As will be presented in detail in Chapters 6, 7, and 9, electrical PD test instruments not only detect PD but also other types of signals called "interference" or "disturbance" in the IEC 60270 and IEC 62478 standards. These interferences are pulse-like and may be confused with actual HV equipment PD. Often the interference pulses are larger and more numerous than the PD that one is trying to detect. Unless one is a skilled observer, or interference-suppression methods are employed as described in Chapter 9, then the HV equipment may be diagnosed as having high PD, when in fact the insulation system may be in good condition. This is referred to as a false positive indication.

PD and corona may occur in equipment connected to the HV equipment being PD tested. In this case, the PD is a type of interference. There are also other types of discharges that can occur, in addition to PD and corona. Some may lead to substantial interferences, others may not. The following sections discuss these different types of discharges that may lead to false positives in a PD test. A useful overview of the different types of discharges is presented by Bartnikas in references [1, 2].

Practical Partial Discharge Measurement on Electrical Equipment, First Edition. Greg C. Stone, Andrea Cavallini, Glenn Behrmann, and Claudio Angelo Serafino.

4.2 PD as Interference

One source of interference may be PD or corona in other components that are connected to the test object during the test. In the case of PD testing, PD and corona from other equipment connected to the test object may occur, especially with online PD testing. An important example of this is when corona is occurring on an overhead transmission line connected to a power transformer undergoing an online PD measurement. Such external corona and PD may be confused with test object PD, causing a false-positive indication that PD is present within the test object.

Another example is the relatively harmless PD that can occur in an isolated phase bus (IPB) – see Section 14.8 – that may overwhelm the PD being measured in either the generator or step-up transformer that is connected to the IPB. Figure 4.1 shows a support insulator in an IPB. To allow for axial expansion of the HV bus in the IPB during operation, the HV end of the support insulator has a rounded metal stud that makes a sliding contact with the HV bus. If there is a small air gap between the stud and the HV bus, it will break down when the electric field stress exceeds about 3 kV/mm peak; this will result in a floating-potential PD source being registered since the support insulator has its own capacitance to ground and is in series with the discharge. The HV bus in IPB is often painted with a black insulating varnish (to radiate heat off the bus more efficiently). Figure 4.1 shows where the discharges between the stud and the aluminum HV bus has eroded the black paint. Such PD on the IPB can be intense, making detection of generator and transformer PD very difficult due to the high level of PD interferences (both high amplitude and high pulse rate) from the IPB that connects the generator to the transformer. The IPB support insulator discharges themselves do not pose a risk to the IPB itself since the discharge is between two metal electrodes.

Figure 4.1 Photograph of a ceramic post insulator supporting the HV aluminum tube (that is painted black) in an IPB. At the sliding metal stud, there is a small (<1 mm) gap to the HV bus, creating discharges. The discharges have vaporized the black paint in the area, revealing the aluminum HV conductor. *Source:* Courtesy of A. Tabernero, Tecnatom.

4.3 Circuit Breaker Arcing

Townsend and streamer electrical breakdown processes in a gas (Chapter 2) create a spark. Once the voltage is reduced, the breakdown process stops. The discharge currents are relatively low, either due to the high impedance in the circuit, or the relatively high resistance of the spark (for example the glow discharge produced by Townsend discharge). With PD, the breakdown in the gas-filled void is usually just a spark process since the impedance of the insulation in series with the void (capacitance C_B in Figure 6.9) prevents high currents from flowing between the electrodes.

Arcing is a different type of discharge, which occurs between metal electrodes. Arcing needs a spark to initiate it. If the spark impedance is low between metal electrodes (i.e. a streamer breakdown, which creates a high density of charge carriers) and the power supply impedance is high, then a self-sustaining arc may occur even if the voltage is lowered. As discussed in Kuffel [3, section 5.2.4], arcing is a process where thermal ionization occurs within the gas. The arc will be self-sustaining when the rate of ionization due to molecular collisions and radiation exceed the recombination rate of electrons and ions. The spark will initiate an arc if the initial spark has sufficient current to raise the plasma temperature for thermal ionization to occur. The concentration of ions and electrons in an arc is so high that the arc has close to zero resistance.

Realistically, arcing will only occur as part of a breakdown process between two electrodes where there is very low series impedance in the circuit, including the power supply. A common example of arcing occurs in high-voltage circuit breakers. An arc will form once the circuit breaker contacts separate upon opening. The arc can only be interrupted when the AC voltage passes through 0 V in the AC cycle, and the ion and electron plasma are physically removed between the contacts – as occurs with air-magnetic circuit breakers and air-blast circuit breakers in substations. The arc magnitude in circuit breakers can be enormous, especially when circuit breakers open under fault conditions, where the current may be tens of thousands of amps. This arcing can create such an interference that the input electronics of continuous PD monitors can be damaged. A challenge for vendors of continuous PD monitoring systems is to protect the electronics from the huge interferences caused by circuit breaker opening under load, yet at the same time not distort the PD pulse magnitude and shape with any surge suppression circuit.

4.4 Contact Arcing and Intermittent Connections

The arcing can also be interrupted if the contacts are opened and separated by a distance large enough such that the electrical withstand strength of the insulating medium (gas, oil) is sufficient. Examples include switch or relay contact opening, or when a copper conductor carrying current fatigue cracks, and a separation occurs at the break.

In high-voltage circuit switching, an initial spark discharge can occur on both contact opening and closing, when the voltage across the gas gap is high enough that the breakdown strength of the gas gap is exceeded. With contact closing, the discharge between the contacts is short-lived since the contacts naturally come together with one another, eliminating the gas gap. With contact opening, a spark discharge across the gas gap will initiate an arc between the contacts.

In high-voltage circuits, it is easy to see that an opening of a contact will create the necessary electric breakdown across the small gas gap of the opening contacts. However, contact arcing can occur in even low-voltage circuits. In low-voltage circuits, the interruption of the current in a very short time due to increasing contact separation can generate a sufficiently high voltage across the broken contact if there is some inductance in the circuit since:

$$V = L\,\mathrm{d}i/\mathrm{d}t.$$

where di is the decrease in current to zero on opening and dt is the time it takes for the current to go to zero. Since the time to break the connection is a few microseconds, dt is small enough to create enough voltage to cause electrical breakdown of the gas between the opening contacts even if only a few tens of mA are flowing. The contact only "opens," i.e. the arc is extinguished, when the contacts are far enough apart that the voltage is not sufficient to cause electrical breakdown of the gas gap even in the presence of a plasma. Unlike PD, gas pressure does not seem to affect the arcing process, at least between 5 and 100 kPa [4].

This process can occur even with a small standard 9 V battery when a wire shorts the positive and negative terminals. Making the short does not cause the breakdown (9 V is not sufficient to exceed 3 kV/mm stress at atmospheric pressure even with a pointed wire). It is when the connection is not solid and minor movements of the wire on the battery terminal cause the connection to "make and break" – creating the spark discharge.

In industrial plants, generating stations, and utility substations, there are many places where such intermittent low-voltage connections may occur. A common example is the poor bonding of metal supports or even chain-link fencing to ground in transmission-class substations. If the metal is not solidly grounded, it will capacitively charge up in the electric fields created by nearby high-voltage conductors in the substation. If the wire bonding the metal support to ground is cracked or corrosion is present at the connection point, discharges will occur whenever the conductors separate. Separation can occur due to wind moving the wires, expansion/contraction due to diurnal temperature cycling, or general vibration that may be present from other equipment in the plant. Such discharging due to poor contacts can also occur on connection leads on the high-voltage equipment, associated secondary control and protection systems, as well as plant grounding grids. For some reason, the metal shields separating the low-voltage winding from the high-voltage conductors in molded current transformers seem to be prone to poor electrical contacts.

The arcing due to broken vibrating contacts occurs whenever the contacts are physically opened. In principle, the contact arcing can occur at any point of the 50/60 Hz AC cycle. However, due to the very small gaps that occur when a wire breaks and is vibrating, such contact discharging tends to predominate at the AC voltage zero crossings.

A special case of contact arcing occurs with slip rings in rotating machines. In spite of best efforts, the electrical connection of a carbon brush to a rotating slip ring is not perfect. The rotating ring is never perfectly round, and the shaft is often vibrating. This causes the brush to sometimes lose contact with the ring, interrupting the current and drawing an arc. This can be a powerful cause of interferences, from small motors in electric tools to the DC current supplied to the rotor winding in a 2000 MVA generator. Since the supply to the slip rings is often DC, the discharges can occur at any point of the 50/60 Hz AC cycle. A variation of this type of contact sparking occurs with overhead cranes in industrial plants and generating stations, where the supply to the traction and hoist motors comes from sliding contacts to high voltage rails.

Another source of contact arcing is within the bearings of motors and generators. Due to magnetic or static electrification, or the use of voltage inverters to supply motor windings, the shaft and rotors in rotating machines can show a potential of several hundreds of volts or more with respect to ground. This voltage is often high enough to cause electrical breakdown and contact arcing across the thin film of oil between the rotating and stationary elements of the bearings. This type of arcing may occur at any phase angle of the AC cycle.

Vibration sparking in high voltage stator windings (Section 16.4.3) is a special case of contact sparking.

4.5 Metal Oxide Layer Breakdown

A variation of poor electrical contacts can occur in bolted electrical connections. Often such connections use brass or steel bolts that are threaded into copper or aluminum busses. If the bolt is very loose, then contact sparking such as described in Section 4.2 can occur. More commonly, if connections have not been painted with antioxidation contact paste when they were made, or if the bolt threads are not tightened enough to cut through the copper oxide or aluminum oxide, electrical breakdown of the oxide will occur as the AC voltage increases from zero. Since the oxide may be only 10 nm thick, and assuming a uniform field geometry, the solid metal oxide will break down at only a few volts. Thicker oxide layers will require higher voltage to break down. Once the oxide breakdown occurs, an arc is established that may cause some local overheating for the bolted connection. Thus, this type of arcing tends to occur near the voltage zero crossings of the 50/60 Hz AC cycle and is easily identified on phase resolved PD (PRPD) plots (Sections 8.7.3 and 10.5). Depending on the metals involved along with pollutants present in the surroundings (carbon dust, sea salt, etc.), these oxide layers may become semiconducting and their sparking behavior may be strongly affected by changes in temperature or humidity.

4.6 Dry Band Arcing

Dry band arcing is a process that can occur on any exposed, polluted insulation surface where a voltage difference exists across the insulation surface between (usually) metal parts. This phenomenon was first encountered on insulators that are used to support overhead HV transmission lines [5, 6]. However, it also occurs in stator windings in the endwinding area [7], as well as in air- and gas- insulated switchgear and transformers [8, 9]. In all these types of apparatus, the dry band arcing leads to a process called electrical tracking (also known as surface tracking), as described in Section 3.7.2.

Localized dry band arcing is essentially harmless when it occurs on overhead transmission line insulators or substation post insulators that are made from glass or ceramics. These are inorganic materials and the high-temperature gas plasma created by the arcing does not cause carbonization of the glass or ceramic materials. Although the dry band arcing does not (within reason) affect the ability of glass and ceramic insulators to perform their function, this type of arcing can cause considerable interference if a PD test is being performed on HV equipment in an outdoor substation, especially if it is foggy, rainy, or snowy. If the dry band arcing is over most of the surface between high voltage and ground, then flashover may occur.

4.7 Glow (or Pulseless) Discharge

Glow discharge is included here for completeness, since it is usually not a source of interference. This type of discharge is encountered in neon tubes, fluorescent tubes, thyratrons, and plasma TVs. As discussed in Section 2.3.1, a glow discharge is a Townsend discharge process where secondary electrons originate at the cathode (or the void walls), but the density of the electrons and ions is not enough to initiate a streamer process [1, 10]. The physics of glow discharge in voids within solid and composite insulation are discussed by Bartnikas and McMahon [1]. The key features of glow discharge that have an impact on PD measurement are:

- glow discharges are more likely to occur at smaller values of pd (gas pressure x void diameter) – see Section 2.3.1
- there are essentially no short duration current pulses that accompany glow discharge

The impact of the first item is that for most HV equipment that operates at atmospheric pressure and above, glow discharge is very unlikely. If it does occur, it will be in relatively small cavities. Since most practical insulation systems will have a range of void sizes, even though some very small voids have glow discharge, there will also likely be larger voids that have the normal streamer type discharge. That is, glow discharges will normally be accompanied by streamer discharges in other voids. Glow discharges are more likely to occur in equipment operating in low pressure environments, especially below 10 kPa. Thus, electrical wiring and motor windings in aircraft flying at very high altitudes are more likely to encounter glow discharge. Glow discharges produce electron and ion bombardment of the void surfaces, as well as chemical reactions, thereby degrading the solid insulation, as discussed in Section 2.4.

Although there will be a current pulse on each half of the AC cycle when the glow discharge is initiated, this pulse will be a small fraction of the normal PD pulse magnitude. Thus, glow discharges are sometimes called pulseless discharges. Clearly, glow discharges will not be detected by conventional IEC 60270 and IEC 62478 PD detectors (Chapters 6 and 7). This becomes a limitation of conventional PD detectors. Fortunately, for most HV equipment, the insulation operates at atmospheric or above pressures, and thus glow discharges are less likely.

The only non-optical method to detect and quantify glow discharge is to measure the increase in dielectric loss that they create within an insulation system. Special bridges that could measure the dielectric loss created by the glow discharges per AC cycle were developed independently by Dakin [11] and Simons [12]. Simons' instrument was called the dielectric loss analyzer (DLA) and was manufactured by Robinson Instruments from the 1960s–1980s. Neither instrument is commercially available now, since in practice glow discharge does not seem to be an important aging factor in the conventional HV apparatus discussed in this book.

References

1. Bartnikas, R. and McMahon, E.J. (1979). *Engineering Dielectrics Volume 1: Corona Measurement and Interpretation*, vol. 669. ASTM Publication.
2. Bartnikas, R. and Novak, J. "Effect of overvoltage on the risetime and amplitude of PD pulses", *IEEE Transactions on Dielectrics and Electrical Insulation*, Aug 1995, 2 pp 557–566.
3. Kuffel, E., Zaengl, W.S., and Kuffel, J. (2000). *High Voltage Engineering: Fundamentals*. Newnes.
4. Grossjean, D. and Schweickart, D. (2021). 270 Vdc arcing at flight altitude pressures: comparison with ground level. *IEEE Electrical Insulation Conference (EIC)*. Denver, CO (7–28 June 2021). IEEE.
5. Looms, J.S.T. (1988). *Insulators for High Voltages*, IEE Power Engineering Series. Peter Peregrinus Ltd.
6. Gubanski, S.M. "Modern outdoor insulation – concerns and challenges", *IEEE Electrical Insulation Magazine*, 21 Nov 2005, pp 5–11.
7. Stone, G.C., Culbert, I., Boulter, E.A., and Dhirani, H. (2014). *Electrical Insulation for Rotating Machines*, 2e. Wiley/IEEE Press.

8 Kurtz, M. “Tracking”, *IEEE Electrical Insulation Magazine*, May 1987, 3 pp 12–14.

9 Mitchinson, P.M., Lewin, P.L., Strawbridge, B.D., and Jarman, P. “Tracking and surface discharge at the oil–pressboard interface”, *IEEE Electrical Insulation Magazine*, March 2010, 26 pp 35–41.

10 Nasser, E. (1971). *Fundamentals of Gaseous Ionization and Plasma Electronics*. Wiley.

11 Dakin, T.W. and Malinaric, P. (1960). A capacitance bridge method for measuring integrated corona charge transfer and power loss per cycle. *Transactions of the American Institute of Electrical Engineers. Part III: Power Apparatus and Systems* 79: 648–653.

12 Simons, J.S. (1966). The dielectric loss analyzer — a new tool for the measurement of total discharge energy of high-voltage insulation. *Conference on Electrical Insulation & Dielectric Phenomena - Annual Report 1966*, Pocono Manor, PA (1966), pp. 112–117. https://doi.org/10.1109/CEIDP.1966.7725204.

5

PD Measurement Overview

5.1 Introduction

Chapters 2, 3, and 4 discussed the physics of PD and related phenomena. This chapter presents an overview of how the PD can be detected and measured. When a partial discharge occurs, there will be several manifestations:

- An electrical current flows.
- This electrical current will stimulate electromagnetic (EM) radiation.
- The flow of electrons and ions in the discharge will also heat the surrounding gas or liquid, giving rise to a pressure wave that leads to audible and ultrasonic signals in air, or that travels to the walls of an enclosure.
- When the excited atoms that occur in a discharge return to their ground state, photons in the visible and ultraviolet wavelengths are emitted.
- The collisions between electrons and the gas atoms in the discharge, or the void walls made of liquid or solid insulation, may result in chemical species that can be identified.

Section 5.2 provides a brief overview of the electrical and electromagnetic detection of PD, which are then described in greater detail in Chapters 6 and 7, respectively. The remainder of this chapter provides more detail on the optical, acoustic, and chemical detection of PD. These nonelectrical methods are often used to complement the electrical/electromagnetic methods and are applied differently depending on the type of electrical equipment. Thus the nonelectrical methods are also discussed in the relevant equipment Chapters 12–16.

5.2 Charge-Based and Electromagnetic Measurement Methods

It was shown in Chapter 3 that when each PD event occurs, a current pulse is produced. A portion of the current pulse can be measured at the terminals of the HV equipment using a suitable sensor – usually a measuring impedance (quadripole) in series with a coupling capacitor that is connected in parallel to the high-voltage terminals with the equipment under test. Alternatively a high-frequency current transformer (HFCT) installed on a ground lead in series with the HV

Practical Partial Discharge Measurement on Electrical Equipment, First Edition. Greg C. Stone, Andrea Cavallini, Glenn Behrmann, and Claudio Angelo Serafino.

equipment can measure the discharge current pulse. When measured by an instrument operating at less than the 1 MHz, the current pulse is integrated to yield an indication of the apparent charge in the discharge pulse. This is the "conventional" way of measuring PD and is often called electrical or charge-based detection. The fundamentals of this measurement method are presented in Chapter 6. Charge-based PD measurements are overwhelmingly used for factory PD testing.

In addition to the current frequencies below 1 MHz, the current may contain frequencies up to the hundreds of MHz range, which can also be detected with coupling capacitors or HFCTs. There is also a radiated EM signal from each discharge pulse, which tends to predominantly occur at frequencies above 1 MHz. When the conducted or radiated signals associated with a discharge are measured at frequencies above 3 MHz, this is commonly referred to as electromagnetic detection, and sometimes referred to as "unconventional" PD detection. EM detection of PD is most commonly performed in onsite/offline PD tests and online PD tests.

Electrical and electromagnetic detection provides a wealth of quantitative information about the PD activity, including:

- the PD magnitude (in units such as pC or mV)
- the number of PD pulses per second
- the phase position of each PD pulse with respect to the 50/60 Hz AC cycle
- usually, the relative predominance of positive vs. negative polarity PD
- total energy expended by the PD in one AC cycle or one second, or other PD quantities that indicate the total or average PD activity
- phase-resolved PD (PRPD) plots (see Sections 8.7.3 and 10.5) that display in a convenient form the PD magnitude, polarity, and pulse count, as well as where the PD is occurring with respect to the 50/60 Hz AC cycle

This information is critical to the interpretation of PD in HV insulation systems. The electrical and electromagnetic methods of PD measurement are the main aim of this book, and most of the following chapters are focused on this topic.

Unfortunately, there are limits to electrical/electromagnetic PD measurements, including:

- such PD measurements generally do not precisely locate where the PD may be occurring within the HV apparatus. For many types of test objects, an electrical measurement can only locate the PD to one phase.
- As will be discussed in detail in Chapter 9, sometimes the PD signals are so overwhelmed by electrical interference that the electrical detection methods can produce a high rate of false-positive indications. If the rate is too high, users will not trust the results.
- Some endusers do not wish to pay for the cost of offline electrical/EM PD tests, or the cost of installing PD sensors for online PD testing.

In addition, even if an electrical/EM PD test is performed and high PD readings are measured, most endusers of critical HV equipment will want some type of independent confirmation that there is a problem with the insulation. This is because repairing the insulation may involve significant costs and extended shutdowns of the HV equipment.

For the above reasons, nonelectrical methods are available to measure PD. These methods may often not be possible to implement depending on the circumstances, but when they can be applied, they are very useful. The remainder of this chapter reviews the main nonelectrical/EM PD detection methods in use today.

5.3 Optical PD Detection

That discharges emit light is well known. Two common examples are lightning and static discharges. The optical emissions are primarily the result of collisions between electrons in the discharge and gas molecules that do not result in ionization, as well as electron-ion recombination. As mentioned in Section 2.3, when the molecule returns to its ground state, a photon is emitted. The energy (and thus the wavelength) of the photon depends on the initial energy of the collision, which excites the molecule. Another process that emits light occurs when a positive ion recombines with an electron. PD is easily visible with the naked eye in a low light level environment. Figure 5.1 shows the optical spectra from surface PD between two magnet wires in air that have been twisted together, with one of the wires energized and the other grounded. The spectra will change depending on the type of gas. The human eye is most sensitive to light in the wavelength range of 400–780 nm. Ultraviolet (UV) light is defined to occur in the 100–400 nm range. Figure 5.3 shows that surface PD creates both visible and UV light, and that UV wavelengths are more intense.

In the past, the human eye was most often used to detect surface PD and corona at night or in darkened rooms. In the 1960s, "night vision" cameras, also called image intensifiers, were commercialized that intensified visible light and thus made the eye more sensitive to low level PD. Corona on transmission lines and in substations at night is easily visible to the human eye that has become acclimatized to darkness, or more quickly with the aid of a night vision camera. It was also common to detect surface PD on stator windings and air-insulated switchgear when the area was dark. PD tests in the dark using the human eye are called a "blackout" or "lights-out" test [3, 4]. The great advantage with the blackout test was that the locations of surface PD could be easily determined (of course PD within the insulation is not visible unless the insulation is transparent). The problem with doing PD tests on energized high-voltage equipment in the dark is that it is more likely that an observer may inadvertently contact the high voltage in an offline test – a clear safety hazard.

Researchers sometimes use vacuum photo multiplier tubes (PMTs) to detect very low intensity optical emissions from PD. PMTs use a special surface that emits an electron when an optical photon hits it. This electron is then directed to a series of metal plates (dynodes) at progressively higher

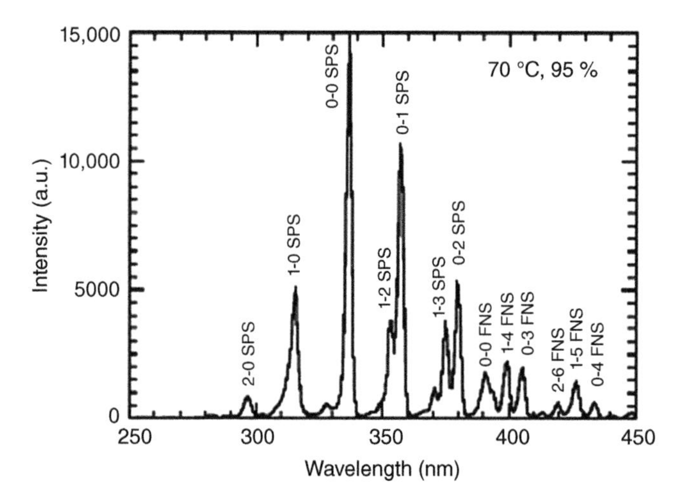

Figure 5.1 Optical spectrum measured by a spectrometer of surface PD occurring between a twisted pair of film-insulated magnet wire in atmospheric pressure air. *Source:* [1], © June (2010) IEEE.

voltages that accelerate the initial electron such that when it hits a dynode, the energy of the electron impact causes multiple electrons to be emitted (secondary emission). After several stages, a relatively high current pulse is created for each incident photon. These current pulses can be measured to give an indication of the intensity of the light from each discharge. The number of discharge events are also counted. PMTs can only be used in the dark.

In the 1990s, the UV imaging camera was invented. The first commercial device for the purpose of detecting PD was developed by Dr. Kieth Forsyth of Forsyth Electrooptics [5]. It displayed the UV light from PD while suppressing sensitivity to visible light. By filtering out the visible light, it became possible to perform a "blackout" test without turning off the lights or waiting until the sun had set, greatly improving the safety of the test. In addition, by adding a still-image camera or video camera, the PD images could be kept for later comparison and precise location assessment. This made the optical detection of surface PD much more objective.

Later developments allowed the strong visible and UV light from the sun to be filtered out while still being sensitive to the UV frequencies associated with PD and corona. This enables modern UV cameras to work outdoors in the daylight. Helicopter- and drone-mounted UV cameras have revolutionized the inspection of transmission lines and insulator strings, with inspections done in normal daylight.

A UV camera commonly used for PD and corona detection is the Daycor line made by OFIL Systems. Figure 5.2 shows a UV camera that superimposes the detected UV light from PD on a visible-light image of the test object. Note that most glass is opaque to UV light, so UV cameras cannot be used to detect PD through normal windows. Also, UV imaging cannot detect PD that is not on a surface or where direct line-of-sight to that surface cannot be obtained. Thus, UV imaging and blackout tests may miss many important sources of PD.

Another critical advance with UV imaging occurred when Dr. Claude Hudon and his colleagues at Hydro Québec discovered that a UV camera could be "calibrated" against the human eye [6]. He found that at least some commercial UV cameras had about the same sensitivity to surface PD as the human eye, when people had been in a very dark room for about 15 minutes to acclimate their

Figure 5.2 Photograph of an OFIL Scaler UV camera (lower right corner) detecting PD on a 13.8 kV stator winding. *Source:* Courtesy of M. Sasic, Iris Power-Qualitrol.

night vision. That is, the acclimatized human eye would detect almost the same PDIV as many UV cameras. This led to the development of IEEE 1799 to detect PD on the surface of stator coils and windings, a calibration procedure, and set voltage levels below which surface (both phase-to-ground and phase-to-phase) PD should not be present [3].

5.4 Acoustic Detection of PD

When PD occurs in a gas, the rapid heating in the discharge channel causes its gas volume to suddenly expand and then immediately collapse (when the surrounding gas rushes back into the partially evacuated channel). This creates a sharp transient pressure wave that radiates outward from the PD site, some of whose energy is then transmitted through the surrounding medium (gas, solid, or liquid insulation, and/or other structures) and can be detected with acoustic sensors.

Acoustic detection of PD can be traced back to the earliest days of transmission line and substation infrastructure. In one sense, the scientific investigation of acoustic PD detection started earlier than optical detection, since microphones, piezoelectric transducers, and recording devices have been available for almost a century; whereas instrument-based optical detection had to wait for the invention of practical night vision cameras in the 1960s. There are two main types of acoustic PD detection.

The first is acoustic PD detection in air. This was first used to locate corona sites in substations and on transmission lines. The purpose was to locate PD and corona sites, so they could be modified to reduce the PD and corona to decrease radio interference. The sound from surface PD or corona is transmitted through air to the human ear or microphone, enabling PD and corona detection at a safe distance from high electric fields. The ability to locate corona and surface PD in HV equipment has advanced tremendously from the time when only the human ear was available to attempt to identify PD sites. As discussed in Section 5.4.1, acoustic imaging cameras now exist that precisely locate the surface PD and corona in substations, transmission lines, bushings, stator windings, and partly disassembled air-insulated switchgear.

A different acoustic detection technology is used for HV equipment such as power transformers, power factor correction capacitor banks, and GIS. Here, a piezoelectric or fiber-optic sensor sensitive to ultrasonic pressure waves is used on the outside surface of a transformer tank, GIS enclosure, etc. Such acoustic technology dates back to Kimura and his colleagues, who in 1939 used a crystal microphone within an oil-filled power transformer to detect internal PD [7]. Such contact or immersed acoustic sensors are not only used for locating PD sites within the HV equipment, but they may also be used as the preferred PD detection method if electrical disturbances are excessive during conventional electrical testing. Unlike the airborne detection of PD and corona, where detection is remote from the sites, for this type of detection the sensors are close to the HV equipment and internal PD can be detected. This technology is discussed in Section 5.4.2, and in more detail for GIS on Section 13.7 and for transformers in Section 15.11.5.

A prolific researcher in both types of acoustic PD detection was Ron Harrold from Westinghouse R&D. He published over 30 papers until the mid-1980s on both PD detection through the air, as well as PD detection within power transformers, GIS, high-voltage capacitors, and power cable. Overviews of his and other early acoustic PD detection research are in [8, Chapter 10] and [9], although we note that the introduction of fiber-optic sensors, acoustic imaging, and powerful signal analysis methods using multiple sensors have reduced the relevance of many of his insights.

5.4.1 Acoustic Detection of PD Through the Air

The fast transient pressure waves caused by PD or corona in air is recognized as sound by human ears that are (hopefully) remote from the HV apparatus. Human hearing is sensitive to sound from about 20 Hz to 15 kHz, depending on age and many other factors. If there are few PD sources, the discharges close to the PD inception voltage will be heard in a quiet environment as discrete snaps. As the number of PD sites and PD pulse magnitudes increase, the sound progresses to something like the frying of food, i.e. a sizzling, crackling, or hissing sound.

For people working in a substation or high-voltage laboratory, the sources of PD and corona can sometimes be located simply by turning one's head. The difference in time of arrival of the sound between the two ears, and/or the difference in sound level between the ears, allows for the possibility of locating a solitary PD site. In the early 1900s, hollow insulating tubes directed from an ear toward expected PD sites were used to locate sites based on sound intensity. In the 1950s, various researchers used microphones together with parabolic reflectors to better locate PD sites in substations.

With the development of better microphones together with low noise amplifiers in the 1950s, it became apparent that the frequency range of the acoustic signals produced by PD in air extends up to almost 100 kHz [2, 8, 9]. Figure 5.3 shows the acoustic spectrum of surface PD measured with a very wide band microphone in 13.8 kV metalclad air-insulated switchgear (AIS – Chapter 14). The PD was found to have originated in the small spaces between the insulated busbar and a fiberglass-reinforced polyester board supporting the bus. Compared to the background noise when the AIS was not energized, the best PD signal to background noise response is between 30 and 50 kHz. Figure 5.4 shows another spectral response of what Harrold calls a gap discharge in air. A good signal-to-noise ratio is apparent from 10 to 80 kHz. In contrast, under similar test conditions, but in SF_6 gas, Figure 5.5 shows the signal to noise ratio was best at around 10 kHz, since SF_6 is much more proficient at absorbing acoustic signals in the ultrasonic range compared to the sonic range [9].

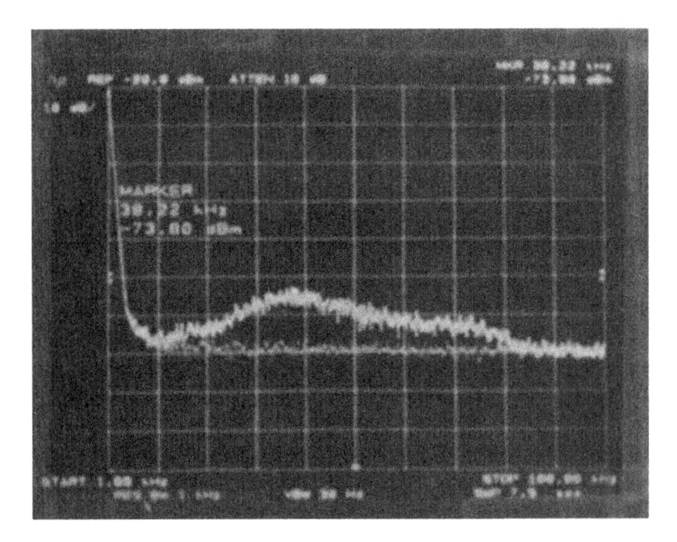

Figure 5.3 Acoustic response of surface PD vs frequency (the fainter line is the background noise when the voltage was off). *Source:* [2]/IEEE. The vertical scale is 10 dBm/div and the horizontal scale goes from 1 to 100 kHz. A B&K 4135 microphone was used whose response is flat up to 100 kHz.

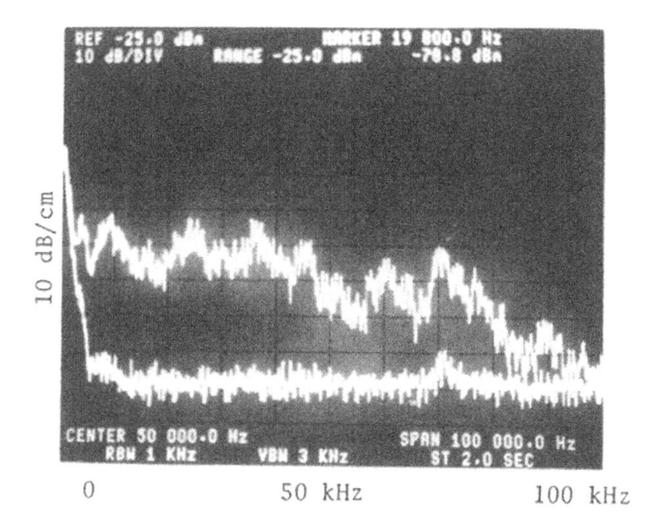

Figure 5.4 Spectrum from a "gap discharge" of about 5–10 nC in air. *Source:* [9]/IEEE.

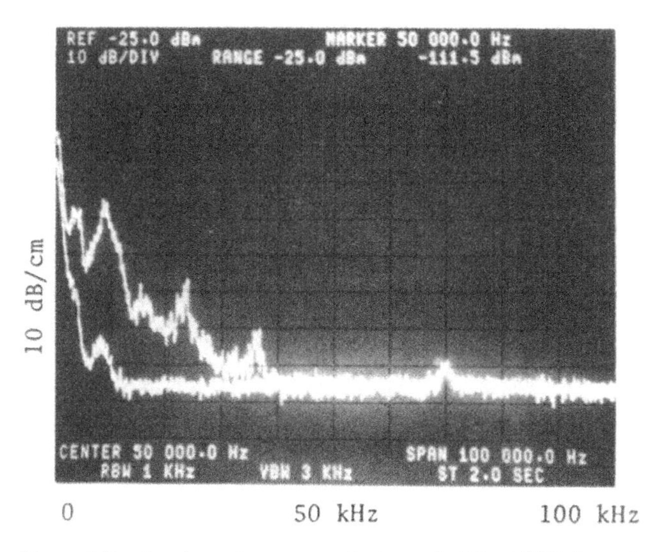

Figure 5.5 Spectrum from a gap discharge in SF_6 at 100 kPa under similar conditions to those in Figure 5.4. *Source:* [9]/IEEE.

In the 1970s, Hewlett Packard developed a commercial directional microphone and associated instrumentation that detected surface PD and corona at 40 kHz. A current commercial supplier of directional ultrasonic probes for PD detection is UE Systems, among others. The initial target application for the instrument was to detect small gas leaks from valves and pipes, but it was quickly applied to surface PD and corona location. One just pointed the directional microphone around HV equipment, e.g. in substations, the exposed insulation in rotating machines, or air-insulated switchgear, to locate the source that produced the strongest response on the sound level meter, or from the demodulated signal monitored via headphones. Studies by Harrold and others showed that 40 kHz ultrasonic probes may be sensitive to surface PD as low as 10–50 pC [9]. He also indicates that the attenuation of the ultrasonic signal in air is about 1 dB/m, due to both the inverse square law and absorption of energy due to molecular collisions, the latter depending on temperature and humidity [8].

Since hard, flat surfaces usually reflect ultrasonic signals, it was sometimes disconcerting to find high levels of 40 kHz signals coming from directions where no PD or corona could be occurring.

A revolutionary advance in acoustic PD detection occurred in 2020 when two companies (Fluke and FLIR) introduced acoustic imaging cameras. These cameras combine an array of dozens of tiny microphones with software that measures the time of arrival of a sound pulse at each microphone to produce an acoustic image of sound sources. As with recent UV cameras, the acoustic cameras superimpose this sound image on a visible light image from a normal camera. Thus, the PD sites can be easily located with respect to the HV apparatus. Some aspects of the technology used in the Fluke cameras can be found in [10].

These acoustic cameras allow the sounds to be detected in selectable frequency ranges up to as high as 100 kHz. Early experience with these cameras confirmed that PD and corona are best measured in the 30–50 kHz range and consistent with the range found with the ultrasonic probes that measure at 40 kHz [2, 11]. Unlike conventional visible light and UV cameras, as yet there is no possibility to focus or zoom in on the PD sites with acoustic cameras, so the field of view is important. As with modern UV cameras, both still images and video images of the PD are produced, the video images being the most useful.

Figure 5.6 shows the acoustic camera being used in a PD test of a motor stator winding. Figures 5.7 and 5.8 show snapshot images produced by the acoustic camera on a stator winding and post insulator in a substation, respectively. The camera also indicates the sound level (in dB) and the PD pulse count rate. In at least one brand, the sound intensity can be displayed against the AC cycle, creating a version of the PRPD plot used with conventional electrical PD tests (Section 8.7.3). It seems that the sensitivity to PD of the acoustic camera in the 35–45 kHz range is similar to that of conventional electrical (IEC 60270) detection and UV imaging [11]. An interesting observation is that the image clearly shows when reflections from hard surfaces are occurring (Figure 5.9). Such reflections caused interpretation problems with ultrasonic probes but are easily ignored with acoustic imaging.

Figure 5.6 Use of acoustic imaging camera being used on an energized 13.8 kV motor stator winding. *Source:* Courtesy of M. Sasic, Iris Power-Qualitrol.

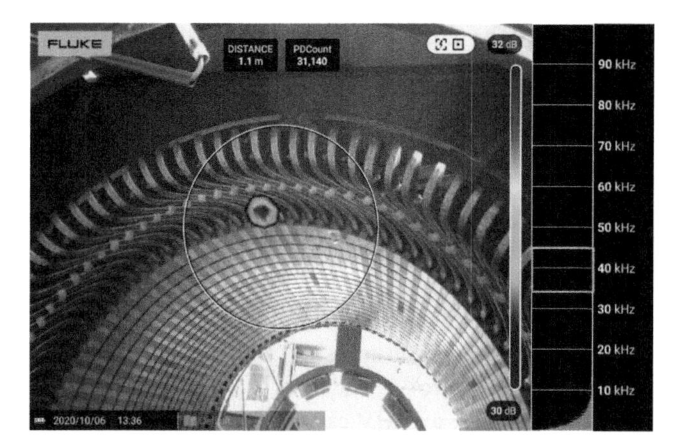

Figure 5.7 Surface PD at the stator slot exit from a motor stator winding energized at rated line-to-ground voltage. The rainbow-colored circle is where the PD is occurring. The red inner circle is where the sound level is highest. This measurement was taken in the 35–45 kHz range. *Source:* [11]/IEEE.

Figure 5.8 Corona detected from a post-insulator in a substation using an acoustic camera. *Source:* Courtesy of Fluke Corp.

Figure 5.9 Acoustic image from a point-plane corona source (described in IEEE 1799) 8 m from the camera. The rainbow-colored circle is the true corona source. The blue circle (meaning it was detected at lower sound level) on the floor is a reflection from the actual corona source. *Source:* [11]/IEEE.

5.4.2 Acoustic PD Detection Within Enclosed HV Apparatus

The acoustic techniques described in the previous section located PD sites for HV equipment operated in air, using detectors remote from the HV apparatus for reasons of safety. The acoustic signals from the PD were transmitted through the air. These methods are not effective for HV apparatus such as power transformers, power cables, gas-insulated switchgear, and HV capacitors where the electrical insulation, and any associated PD, is completely enclosed by metal. As described in [9], acoustic PD signals within an enclosure are reflected back by the enclosure (and its high acoustic impedance). Furthermore, the signal that is transmitted through the enclosure is not efficiently coupled into the air for reception by a remote microphone.

However, acoustic microphones or vibration sensors (acoustic emission (AE) or accelerometers) can be installed within the enclosure or on the enclosure's outside surface to detect internal PD. This is common for liquid-filled transformers, GIS, and HV capacitors, all of which have grounded enclosures. Acoustics were first applied to power transformers as far back as 1939. Instead of just trying to locate PD sites as in Section 5.4.1, using acoustic methods to detect PD within such apparatus may sometimes be the primary PD detection method for the equipment if UHF sensors are not installed, since LF and HF electrical methods (Chapter 6) may be severely compromised by high levels of disturbances, and optical methods are not possible since the enclosure blocks the light emission.

Piezoelectric sensors are most common for acoustic detection of PD signals in enclosed equipment; these usually employ specialized crystals that produce an electrical output signal caused by mechanical pressure transient arising from the acoustic PD signal. These devices can detect signals from a few Hz to 5 MHz, depending on the design. The sensor enclosure is metallic and grounded via the sensor output cable, therefore the sensor cannot be used in or near high electric or magnetic fields. Piezoelectric sensors are usually mounted on the grounded enclosure of the HV equipment. There are many manufacturers of piezoelectric sensors with a wide variation of design and application.

In the past 20 years, fiber-optic acoustic sensors have been developed, which use either no or only microscopic amounts of metal and thus can be installed within HV equipment such as power transformers and switchgear, and can even be attached to components operating at high voltage. These sensors use a variety of principles for operation: Fiber Bragg Grating (FBG), Fabry-Perot interferometry, and resonant vibrating reads or diaphragms. Fiber-optic sensors tend to be more expensive than piezoelectric sensors and thus are not yet widely used in practice for acoustic detection in HV equipment. However, a considerable amount of research is occurring that could change this.

The methods for detecting and locating PD within enclosures depend on the type of equipment being measured. Thus, they will be discussed in the relevant equipment chapters (Chapters 13–15). However, there are some general applications discussed below.

5.4.2.1 Power Transformers

Most high-voltage power transformers use kraft paper and cellulose pressboard as solid insulation, with the entire transformer core and winding submerged in a liquid – most commonly mineral oil (Section 15.2). PD can occur due to either manufacturing defects or aging processes (Section 15.4). The PD is commonly detected with gas-in-oil analysis (PD may create hydrogen and other gases in oil – Section 5.5.2) and/or by electrical PD detection (Section 15.4.2). However, the occurrence of PD can also be detected as acoustic signals, and PD sites can be located by an experienced engineer using acoustical methods.

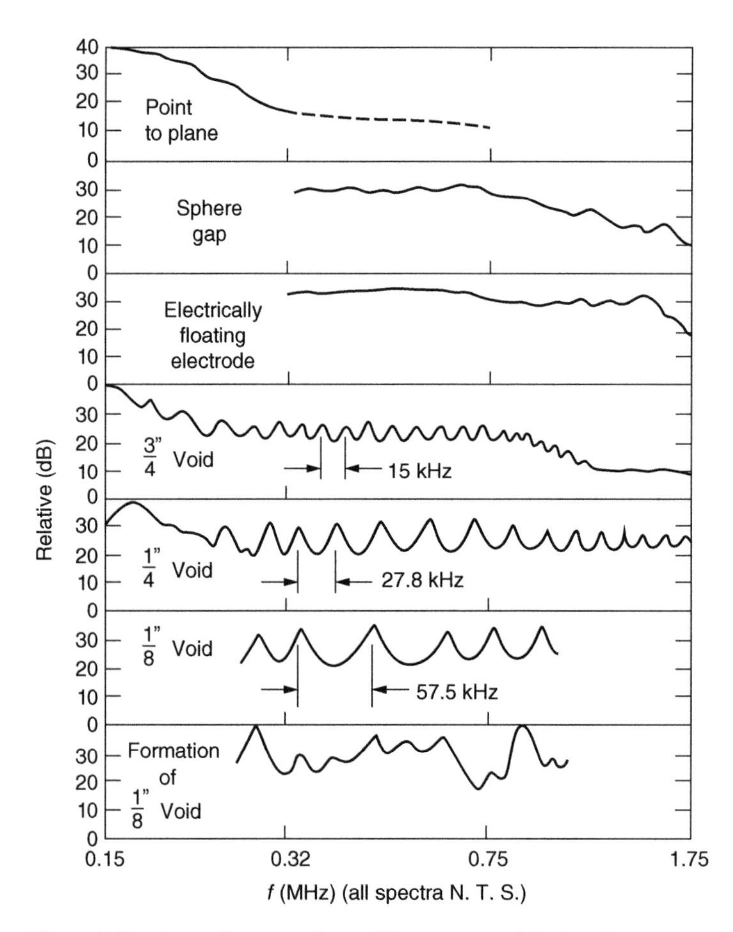

Figure 5.10 Acoustic spectra from different types of discharging sources within a tank filled with mineral oil. The piezoelectric crystal sensor was also immersed in the oil and had a flat frequency response up to 1 MHz. The sensor was placed between 10 and 100 cm from the PD source. *Source:* [12], © Dec (1975) IEEE.

The acoustic spectrum of PD and arcing within mineral oil is shown in Figure 5.10 [12]. The detected spectra can be as high as 1 MHz, much higher than that occurs with PD detected in air. This is due to:

- the PD pulses having nanosecond risetimes and durations,
- mineral oil being an excellent transmitter of acoustic signals,
- acoustic energy being efficiently coupled from the PD to the oil, and from the oil to the piezoelectric sensor mounted on the transformer enclosure.

Most practical acoustic measurements on liquid-filled power transformers used three or more movable piezoelectric sensors mounted on the outside of the transformer tank wall. Piezoelectric or fiber-optic vibration sensors mounted within the tank cannot be moved once installed, which can make localization of the PD source(s) more difficult. Sensors outside of the tank can be easily moved to help pinpoint the PD site(s). Harrold recommends the sensors detect in the 80–300 kHz range [9]. However, a European consensus CIGRE guide on acoustic PD measurement in transformers indicates that a range from 10 to 300 kHz should be used [13].

Locating PD sites can take advantage of one or more of the following approaches:

- triangulation using "time-of-flight," i.e. the time it takes for the acoustic pulse to get to each of three or more acoustic sensors at different tank wall positions (Section 8.4.2),
- relative magnitude of the detected pulses at several sensors in different tank wall positions, which assumes that pulses further from the sensor will be smaller,
- triggering the measurement using the signal from an electrical PD sensor, since compared to the velocity of acoustic transmission, the electrical signal is detected almost instantaneously.

The propagation path from the PD site to the sensor mounted on the transformer tank is usually very complicated. The most obvious path is directly through the oil to the sensor on the tank wall. However, this may not be the most likely path since the velocity of sound through mineral oil is only 25% of the velocity through steel (1400 m/s compared to 6000 m/s in steel). Thus, the acoustic pulse from a PD site may arrive at the sensor sooner by traveling to a nearby tank wall and then propagating through the steel wall to the sensor. In addition, the internal structure of the transformer is very complicated, and usually only known in detail by the manufacturer. The resulting complex signal propagation paths combined with large variations in propagation velocities in the different materials along those paths makes it difficult to locate PD sites based on differences in pulse arrival times or signal magnitude at different sensors. In practice, it seems great expertise is needed to locate a PD source within a transformer using acoustic methods, and indeed, if the PD site is close to the transformer-laminated steel core under layers of windings and insulation, it may not be detected at all. Also, if there are multiple PD sites, it is likely even an expert will not be able to locate the sources [13]. Most transformer manufacturers and a few specialized consultants can provide location services.

In principle only the sensor and a 100 MHz multichannel oscilloscope are needed to perform PD location. However, Omicron and Power Diagnostix supply piezoelectric sensors, instrumentation, and software that is optimized for PD site location in transformers (Section 15.11.5).

The field of artificial intelligence (AI) is making great progress. Since some experts can locate with a good probability a single PD site within a transformer, perhaps one day AI technology may be able to help non-experts to locate PD sites.

5.4.2.2 Gas-Insulated Switchgear and Isolated Phase Bus

GIS is now available for ratings up to 1200 kV. Most GIS and gas-insulated transmission lines (GIL) use SF_6 as the main dielectric. SF_6 is both strongly electronegative and also exhibits high electron affinity so it strongly suppresses PD activity (Sections 2.3 and 13.3.3). However, as discussed in Section 13.4, PD can occur for a variety of reasons. Thus, conventional electrical PD testing is common both as quality assurance tests in new equipment, as well as to detect problems or aging that can occur in operation. In both the case of long sections of GIL and the complex combination of many components within a GIS substation, it is very important to precisely locate the PD sites to limit the time required for GIS/GIL disassembly, repair, and reassembly to a minimum.

One method for locating a PD source in GIS, commonly called the "time-of-flight" method, determines the location based on the difference in the time-of-arrival of incoming PD pulses from two or more RF PD sensors located (hopefully) in the vicinity of the defect. This is explained further in Sections 8.4.2 and 13.10.5.

Acoustic techniques can also be used to detect and locate PD within the GIS without requiring sensors to be pre-installed, as is the case for electrical detection. The acoustic method can measure PD from sharp protrusions, floating shields or surface electrical tracking on the epoxy spacers [14–16]. As with PD in transformers, PD will produce acoustic signals to >1 MHz. Acoustic methods will usually not pick up PD within the epoxy spacers (i.e. from voids) used to support the high-voltage bus, due to absorption of the signal by the surrounding epoxy. The acoustic method is especially good at finding moving particles in GIS or GIL. Particles in GIS may lead to failure if they move to the wrong location, either by adhering to an insulator surface or triggering a "crossing" flashover (Section 13.4.1). Bouncing particles produce acoustic emissions when they impact the GIS enclosure. Generally, acoustic PD detection is done in the few 10s of kHz since SF_6 strongly attenuates signals above 40 kHz (Figure 5.5).

The simplest location technique is to use a commercial ultrasonic directional microphone, usually tuned to 40 kHz and equipped with a contact sensor instead of a microphone (Figure 13.14). The contact sensor is pressed against the outside surface of the grounded GIS enclosure to pick up the acoustic signals from inside the GIS. The closer one is to an internal PD source, the higher the signal measured.

More conventionally, one or two piezoelectric sensors are used (Section 13.7). These are temporarily placed in contact with the outer wall of the GIS enclosure. After the signals from the sensor are amplified to the mV region, the signals are displayed on a suitable multichannel oscilloscope. The closer a sensor is to the PD site, the higher is the response. Alternatively, using two sensors, one can locate the PD site using the relative time for the acoustic signal to reach sensors on either side of the PD site (Section 13.7). Location to within ± 0.5 m is possible [14].

Isolated phase bus (IPB) has a physical structure similar to GIS (Section 14.8). IPB connects a large generator to the step-up power transformer in generating stations. It is usually rated 28 kV or less, but can handle very high current (10s of kA). The high-voltage center conductor is supported by porcelain post-insulators, and the main insulation is atmospheric air, not high pressure SF_6. PD can occur within the IPB due to broken post-insulators and metal protrusions, as well as due to arcing that can occur at poor electrical contacts. Rarely will these sources lead to IPB failure. Such sources can be easily found and located using conventional electrical detection if sensors at each end of the IPB have been pre-installed. The same methods used for acoustic location in GIS can be used for the IPB (Section 14.4.4).

5.5 Chemical Detection

Partial discharge can change the chemical composition of the surrounding medium, which sometimes allows the detection of PD using chemical analysis. The changes that occur depend on the surrounding medium: air, SF_6, or oil. The most widely used chemical tests are ozone detection for PD in air; dissolved gas analysis (DGA) for PD in oil – commonly used in power transformers; and SF_6 decomposition products in the case of GIS or GIL.

5.5.1 Ozone in Air

When PD occurs in air, a gas called ozone – O_3 – is created. A discharge creates many energetic electrons (Section 2.3). An excited electron may have inelastic collisions with an oxygen molecule (O_2) in air and have sufficient energy to break the molecular bond, creating two oxygen atoms (O + O). Oxygen atoms are very chemically reactive, and if an oxygen atom is near an oxygen molecule, they will combine, creating O_3. Ozone itself is also very chemically reactive with many other molecules, often rusting metals and degrading some organic materials (Section 3.4). Thus, any ozone in the air tends to disappear in a few minutes. However, if the PD sources are continuously active, new ozone is created to replace the depleted ozone. Eventually, a steady-state level of ozone gas concentration occurs that is the balance between the creation and depletion rates. Aside from ultraviolet radiation, few other processes create ozone. Thus, the presence of ozone near HV equipment is a good indicator that PD is occurring in the air or on the surface of solid insulation in air. Continuous exposure to ozone above 0.1 ppm is deemed by many health and safety regulators to be hazardous to human beings.

Ozone can be created by PD in many types of HV equipment:

- corona from transmission lines and outdoor substation components
- PD from surface discharges in metalclad air-insulated switchgear
- PD on the surface of high-voltage stator coils/bars, both in the stator slot and in the end winding.

Generally, PD within a void enclosed within solid insulation does not produce ozone in the surrounding environment.

The amount of ozone produced by PD depends on both the number of discharge sites and the magnitude of the discharges. That is, the ozone concentration is an indicator of the total PD activity that is occurring in air. The ozone concentration in the air depends not only on the steady-state creation/depletion rate of the ozone but also on how open the air is to the surrounding environment. For example, if a lot of air is moving near the PD sites (due to high winds if outdoors, or from a fan blowing external air over an open-ventilated stator winding), the ozone concentration will be low. The PD activity may be lower in enclosed HV equipment such as totally enclosed rotating machines and metalclad air-insulated switchgear, yet show a much higher ozone concentration since little air may be circulating.

There are two main methods for measuring the ozone concentration in air:

- Chemical reaction tubes
- Electronic instruments

The ozone chemical reaction tubes are similar to the litmus test used to check acidity. If one breaks the tube open in the presence of ozone, the reagent changes color, depending on the ozone concentration. Draeger is a common supplier of such tubes.

Electronic instruments are a more convenient way to measure the ozone concentration. These devices may contain a special MOSFET sensor that changes its resistance in the presence of adsorbed ozone. Another type of sensor uses the principle that UV absorption is higher with higher ozone concentrations. The associated instruments can take spot measurements or can continuously monitor the ozone concentration.

The ozone concentration for PD detection is usually measured in parts per million (ppm). The typical ozone concentration in air is usually <0.1 ppm and is caused by thunderstorms, UV radiation, and/or certain types of industrial pollution. The human nose has a detection sensitivity of about 0.1 ppm. In open-ventilated HV equipment, readings of about 1 ppm indicate widespread surface PD activity. In enclosed HV equipment where there is little air circulation, ozone concentrations up to 25 ppm have been measured. As mentioned above, both cases may indicate the same amount of PD activity. In spite of this limitation, ozone monitoring is an excellent way to confirm high PD activity detected by electrical methods.

5.5.2 Dissolved Gas Analysis (DGA)

DGA has been used since the 1960s to determine if PD, arcing, or thermal aging of the insulation is occurring in oil-filled transformers (Section 15.7.6). When PD occurs in mineral oil, hydrogen (H_2) and various hydrocarbons are produced. These gases are normally dissolved in the oil and can be detected either by oil sampling together with chemical lab analysis, or by using specialized continuous monitors that detect these gases in the oil. Overheated organic insulation tends to produce ethylene, PD produces hydrogen, and arcing (including surface tracking) produces acetylene [17]. Thus, DGA can detect both PD and arcing, although the response time is much slower than with electrical PD detection. In fact, since PD in transformers is sometimes intermittent, and the gases in the oil can have a relatively long lifetime, it can be argued that DGA may be a more reliable method of detecting discharges than conventional electrical detection, especially when there is a high level of electrical disturbances.

There are thousands of papers on DGA by sampling or by monitoring at the transformer, and it is well beyond the scope of this book to discuss this technology in detail. A basic introduction to the topic is given by Dr. Duval of Hydro Québec in [17]. The technology for online monitoring of the gas in oil has advanced tremendously since it was introduced by Duval and others in the mid-1970s for detecting hydrogen alone. Now commercial monitors for nine or more gases dissolved in oil have been commercially introduced [18].

5.5.3 SF_6 Decomposition Products in GIS

The SF_6 used in GIS is delivered exceptionally pure and dry. However, there are usually some impurities such as H_2O and O_2 present. When PD and arcing are occurring, adjacent SF_6 will decompose, and these decomposition products then combine with H_2O and O_2 to create characteristic gases. Research has shown that the main decomposition products produced by PD and arcing include CF_4, CO_2, SOF_2, SO_2F_2, SO_2, and H_2S [19]. Although it is still not used widely, these decomposition products can be identified by sampling the gas in GIS and analyzing certain gas ratios using a gas chromatograph. As with ozone monitoring and DGA, the sensitivity of the method depends on the number of discharges sites and their magnitude. There will be a delayed response to the PD.

References

1 Kikuchi, Y., Murata, T., Fukumoto, N. et al. (2010). Investigation of partial discharge with twisted enameled wires in atmospheric humid air by optical emission spectroscopy. *IEEE Transactions on Dielectrics and Electrical Insulation* 17 (3): 839–845.

2 Van Haeren, R., Stone, G.C., Meehan, J., Kurtz, M. (1985). Preventing failure in outdoor distribution class metalclad switchgear. *IEEE Transactions on Power Apparatus and Systems* PAS-104: 2706–2712.

3 IEEE 1799 (2022). *Recommended Practice for Quality Control Testing of External Discharges on Stator Coils, Bars, and Windings*. IEEE.

4 Stone, G.C., Culbert, I., Boulter, E.A., and Dhirani, H. (2014). *Electrical Insulation for Rotating Machines*. Wiley/IEEE Press.

5 Forsyth, K. (1999). Corona detector with narrow band optical filter. US Patent 5,886,344, March 1999.

6 Tremblay, R., Hudon, C., Godin, T. et al. (2006). Improvement in requirements for stress grading systems of stator windings at Hydro-Québec. *CIGRÉ Conference*, Paris (August 2006), pp. A1-101.

7 Kimura, H., Tsumura, T., and Yokosuka, M. (1940). Corona in oil as part of commercial-frequency circuit. *Electrotechnical Journal of Japan* 4: 90–92.

8 Bartnikas, R. and McMahon, E.J. (1979). *Engineering Dielectrics Volume 1: Corona Measurement and Interpretation*, vol. 669. ASTM Publication.

9 Harrold, R.T. (1985). Acoustical technology applications in electrical insulation and dielectrics. *IEEE Transactions on Electrical Insulation* EI-20 (1): 3–19.
10 Suurmeijer, C.P., ter Keurs, J., Tsukamaki, S.M. et al. (2021). Handheld acoustic imager. US Patent D907,097, Jan 2021.
11 Stone, G.C., Sasic, M., Wendel, C., and Shaikh, A. (2021). Initial experience with acoustic imaging of PD on high voltage equipment. *IEEE Electrical Insulation Conference*, Denver, CO (7–28 June 2021).
12 Harrold, R.T. (1975). Ultrasonic spectrum signatures of under-oil corona sources. *IEEE Transactions on Electrical Insulation* EI-10 (4): 109–112.
13 Fuhr, J., S. Markalous, S. Coenen et al. (2017). Partial Discharges in Transformers. Cigre Technical Brochure 676 - WG D1.29 2017.
14 Behrmann, G.J., Neuhold, S., and Pietsch, R. (1997). Results of UHF measurements in a 220 kV substation during on-site commissioning tests. *International Symposium on High Voltage*, Montreal (August 1997).
15 Lundgaard, L.E., Skyberg, B., Schei, A., and Diessner, A. (2000). Method and Instrumentation for Acoustic Diagnosis of GIS. Cigre Paper 15-309, 2000.
16 Pietsch, R., W. Hauschild, J. Blackett et al. (2012). High Voltage On-Site PD Testing with Partial Discharge Measurement. Cigre Technical Brochure 502, June 2012.
17 Duval, M. (1989). Dissolved gas analysis: it can save your transformer. *IEEE Electrical Insulation Magazine* 5 (6): 22–27.
18 Duval, M., Atanasova-Hoehlein, I., Cyr, M. et al. (2010). Report on Gas Monitors for Oil-Filled Electrical Equipment. Cigre Technical Brochure 409, 2010.
19 Tang, J., Liu, F., Zhang, X. et al. (2012). Partial discharge recognition through an analysis of SF_6 decomposition products part 1: decomposition characteristics of SF_6 under four different partial discharges. *IEEE Transactions on Dielectrics and Electrical Insulation* 19 (1): 29–36.

6

Charge-Based PD Detection

6.1 Introduction

In this chapter, the focus is on the electrical detection of the conducted PD current at the test object terminals, and in particular the measurement of PD magnitude in terms of apparent charge, which requires the measurement to take place at frequencies less than 1 MHz (the low-frequency [LF] range). This is often referred to as an IEC 60270 detector, after the first standard on charge-based PD measuring systems. Charge-based PD measurements are often required for factory PD acceptance tests for many types of equipment and are sometimes used for onsite, offline PD tests. Other quantities measured in a PD test: pulse polarity, phase position, and pulse count rate are discussed in Sections 3.9 and 8.3. PD detection at higher frequencies than 1 MHz using various types of sensors is presented in Chapter 7.

By far the most common type of sensor used for electrical PD detection is the coupling capacitor in combination with a detection impedance, which is connected to the high-voltage terminal of the test object (or "equipment under test" – EUT). The capacitance of the coupling capacitor partly depends on the frequency range the PD is measured (Section 8.2), with values ranging from a few 10s of pF to a few nF. In all cases, the coupling capacitor is connected in parallel with the test object, and in most cases it is in series with a detection impedance (sometimes referred to as a quadripole). Alternatively, quadripoles and HFCTs are also sometimes used to measure PD currents on the ground side of EUTs. The instruments used to measure the signals from the PD sensor is discussed in Chapter 8.

This chapter also discusses the calibration of the detected mV or mA pulse signals into apparent charge for low-frequency (LF) detectors.

6.2 Basic Electrical Detection Circuits Using Coupling Capacitors

Assume that the EUT is connected only to an AC voltage source, having an inductive internal impedance. A PD takes place in fractions of microseconds. Thus, the events triggered by a PD in the circuit consist of a charge redistribution completely within the EUT, as the voltage source branch will act as an open circuit ($L\, di/dt \rightarrow \infty$). Although it will be proved later (Section 6.4), it is intuitive to conclude that the voltage drop at the EUT terminals due to the charge redistribution is so small that measuring it to infer PD events is practically impossible. In practice, electrical PD measurements are almost always carried out using a coupling capacitor (C_k) in parallel with the EUT and sensing the current flowing between the EUT and C_k using a detection impedance.

Practical Partial Discharge Measurement on Electrical Equipment, First Edition. Greg C. Stone, Andrea Cavallini, Glenn Behrmann, and Claudio Angelo Serafino.

To sense the PD current, a PD transducer is needed. To start with, it will be taken as a generic device in series with the EUT (or the coupling capacitor C_k), able to provide a suitable signal. In Section 6.3, the sensors will be described in greater detail.

In practice, the electrical detection circuit can be realized in two different ways: direct (Figure 6.1) and indirect (Figure 6.2). In these figures, Z_B is the impedance that is often connected between the AC power supply and the EUT to suppress power system noise and reduce the PD current flowing back into the power supply (Section 9.3.3). In the following, these circuits will be described having in mind what can be arranged in a HV laboratory. Note that the EUT is assumed to be primarily capacitive in nature, which is not true for transformer and rotating machine windings. Special circuits devised for cable accessories and transformers (using capacitive-graded bushings with special taps) will be described in the Chapters 12 and 15, respectively.

When coupling capacitors are used for online monitoring of the PD, the couplers need to be permanently installed. If the capacitors fail in service, then a ground fault can occur. Thus, coupling capacitors need extensive testing to ensure in-service failure will not occur over the lifetime of the high-voltage equipment. Some standards, for example IEEE 1434 and IEC 60034-27-2 for rotating machine PD testing, require many tests that must be done to assure the PD sensor itself does not lead to power system ground faults.

6.2.1 Direct Circuit

In the direct circuit, the detection impedance Z_{mi}[1] is in series with the EUT. The sensitivity achieved with the direct circuit is the largest possible as the entire external PD current will flow through the detection impedance. Discharges in the EUT are detected with the correct polarity, whereas those coming from other parts of the circuit are detected with opposite polarity (meaning that positive PD external to the EUT are detected as negative PD and vice versa). This can be readily understood by

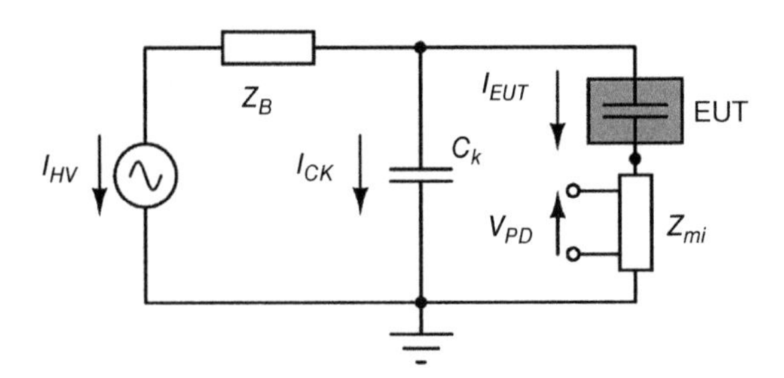

Figure 6.1 Direct measurement circuit.

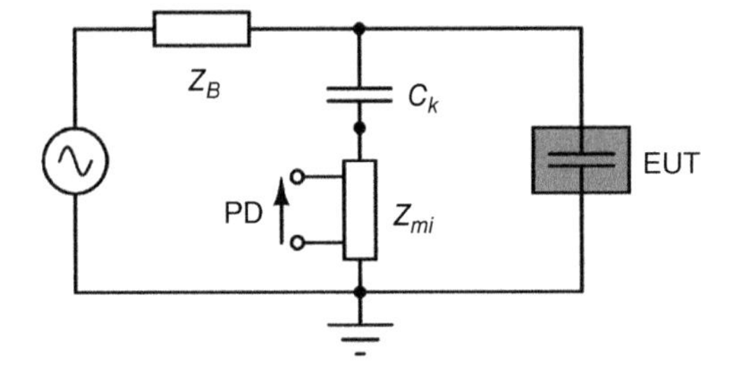

Figure 6.2 Indirect measurement circuit.

1 The detection impedance is often incorporated as part of a quadripole (a four-terminal network), that also includes a circuit to extract the 50 or 60 Hz AC waveform. The AC waveform is needed to synchronize the PD pulses with the AC voltage to create a PRPD plot (Section 8.3.2).

inspecting the flow of I_S, I_{CK}, and I_{EUT} in Figure 6.1. This simple rule can help to distinguish PD from the EUT from interference due to PD in the HV circuit (Section 9.3.5). If the stray capacitance and the capacitance of the power supply to ground is low, a capacitor C_k is used to enable the PD current external to the EUT to flow. The direct circuit can be used in the absence of C_k if the stray capacitance of the neighboring circuit is large enough (this is typically the case in HV substations).

The direct circuit has some important disadvantages. In case of breakdown of the EUT, the short circuit current flowing into the ground will induce a transient voltage within the circuitry that might destroy the PD detector. Unless fiber-optic isolation is used, test operators may also receive an electric shock if the EUT fails. Another issue is that large objects can be connected to the ground at several positions (for example, the metallic base of a transformer or a large generator), making it impossible to measure the entire PD current.

6.2.2 Indirect Circuit

With the indirect circuit, the PD is detected in series with the coupling capacitor. Therefore, the PD current polarity will be reversed for PD from the EUT (see Section 6.4.4). That is, the polarity of the detected pulses will be negative for the positive PD measured in Figure 6.1), and vice versa.

The indirect circuit has advantages and disadvantages that are complementary to those of the direct circuit. For large equipment with several ground connections, this circuit is the only one that can ensure sufficient sensitivity. Furthermore, C_k and the detection impedance Z_{mi} are usually a single piece of equipment, which also provides the AC voltage for synchronization. This arrangement simplifies the realization of the test circuit. Also, as C_k is normally more reliable than the EUT, the probability that the PD detector flashes over, or the operator is harmed due to a failure of C_k is very low.

On the other hand, a fraction of the PD current will circulate through stray capacitances between the HV conductors and ground (which are always present), reducing PD detection sensitivity. Furthermore, PD from the EUT and from the AC supply will have the same polarity, making interference rejection more complicated (Section 9.3.5).

In general, when testing objects like capacitors, GIS, AIS, transformers, and rotating machines, the indirect detection methods are more common. Direct detection is more common for online PD detection in power cables.

6.3 Measuring Impedances

6.3.1 Resistors and Quadripoles

The most obvious detection impedance is a measurement resistor, with resistance R_M. The PD current will be detected as a voltage pulse $V_{PD} = R_M \times I_{PD}$. Despite the apparent simplicity of this choice, one should pay attention to the way this measurement resistor is connected to the measuring instrument. If a resistor external to the instrument is used, the instrument should have a large input impedance (to avoid loading R_M). The instrument should also be physically close to the resistor R_M to avoid pulse reflections that might alter the magnitude and repetition rate of the PD phenomena, especially in the HF to UHF frequency ranges. Alternatively, one could use the instrument input impedance as R_M. In this case, a resistor at least 10 times R_M is connected ground to prevent an overvoltage at the C_k output if the instrument is disconnected. Most modern detectors have normally a 50 Ω input, thus the coaxial (typically RG58) cables connecting C_k with the measuring impedance will be matched to R_M, avoiding the distortion of the PD pulses by reflections. The PD instrument can thus be at some distance from C_k, reducing restrictions on the test circuit.

One can wonder what the bandwidth of this system is. The bandwidth of the resistor can be as large as 18 GHz using, e.g., *N*-type resistors (a figure vastly exceeding the needs of PD detection). The coupling capacitor is more problematic. To build a high-voltage, large capacitance C_k, some designs use a large number of large capacitors in series. However, this arrangement of the series-connected capacitors may be partly inductive, thus hampering the flow of the PD current to R_M. Gas, ceramic, or other single element capacitors have a much higher bandwidth. The bandwidth of the detection circuit will depend not only on the transfer function of the individual elements but also on the characteristics and length of the connecting cables.

Besides attenuation of the high-frequency components, the series combination of C_k and R_M can be regarded as a high-pass filter with cut-off frequency

$$f_0 = 1/\left(2\pi R_M C_k\right) \tag{6.1}$$

The larger C_k the lower the cut-off frequency, as shown in Figure 6.3. However, Equation 6.1 is derived calculating the voltage gain of the simplified circuit in Figure 6.4a. Clearly, a PD is hardly an ideal voltage generator and Thevenin or Norton representations as in Figure 6.4b, c are a more realistic choice. Therefore, the cutoff frequency calculated by Equation 6.1 should be considered as an approximation at best. Indeed, it has been observed empirically that, the larger C_k, the larger the magnitude of the lower frequency components of the detected signals. Referring again to Figure 6.4, this might indicate that, for most equipment, Z_T (or Y_N) is mainly resistive.

Due to the high-voltage supply to energize the test object, a residual 50/60 Hz component (and possibly harmonics) will be superimposed on the PD signal. Thus, PD detectors have always a high-pass filter (cutoff frequency normally a few tens of kHz) to remove these unwanted residual

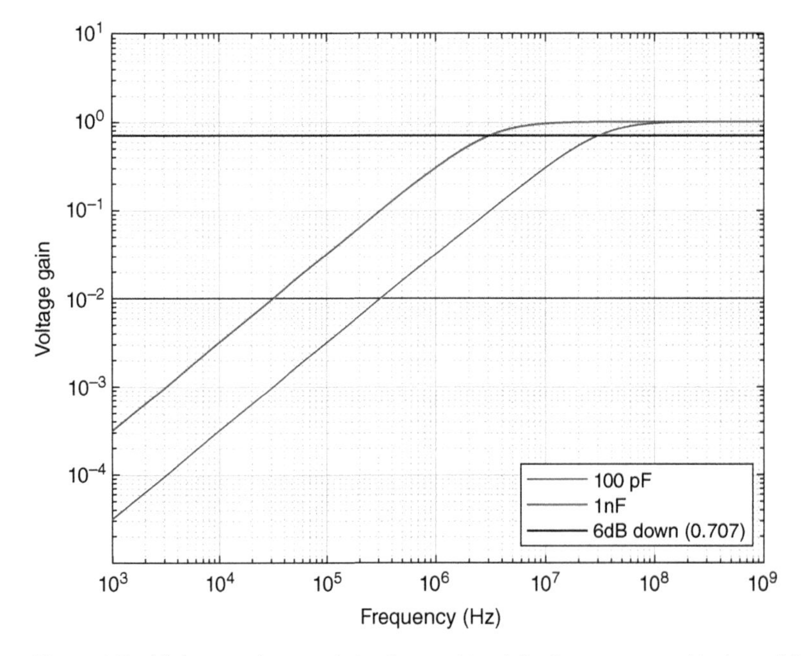

Figure 6.3 High-pass characteristic of a combined C_k–R_M sensor considering a 50 Ω resistor.

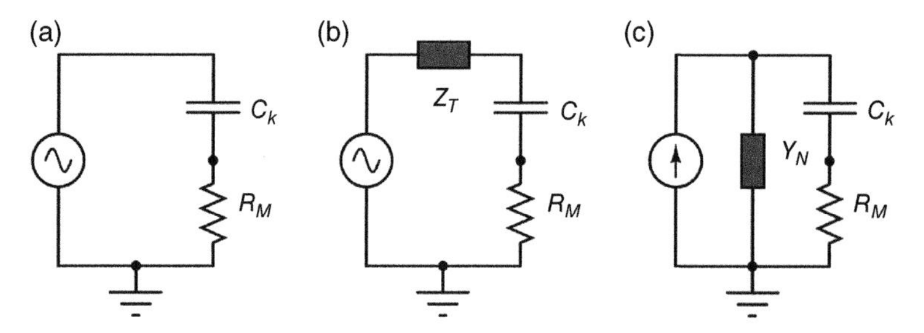

Figure 6.4 The PD source is represented as (a) an ideal voltage source, (b) a Thévenin equivalent circuit, (c) a Norton equivalent circuit.

voltage components from the source. This additional circuit is part of the "quadripole" in LF (IEC 60270 compliant) detectors.

6.3.2 AC Synchronization and Quadripoles

The residual 50/60 Hz component across Z_{mi} can be used for synchronization of the PD events with the supply voltage to create phase-resolved PD (PRPD) plots (Section 8.3.2). When Z_{mi} is a simple resistor, electronic circuitry (buffer circuits that will not load the C_k–R_M) can be used to separate the 50/60 Hz component from the PD pulses. Note that, if the test voltage is not high, C_k is too low and/or R_M is small, there may not be sufficient 50/60 Hz AC voltage across R_M to achieve a reliable AC synchronization voltage to produce PRPD plots.

Alternatively, to measure at the same time the PD pulse detection and the AC synchronizing voltage, quadripoles (a four terminal measuring impedance) are often used, especially in LF offline PD detection. With an indirect measuring circuit (Figure 6.2), quadripoles can be integrated into the coupling capacitor and will have two BNC output terminals, one for PD signal and another for AC synchronization. The typical quadripole structure is indicated in Figure 6.5. Assume that L_M is large enough to prevent any spectral component of the PD pulse to flow in the L_M–C_M branch (if not, there will be a loss of sensitivity). At 50/60 Hz, the C_k–L_M–C_M series behaves as a capacitive voltage divider since the impedance of L_M is low, thus providing a synchronization voltage between the AC and GND terminals. Quadripoles are very practical, but they need to be designed carefully. The circuit has a natural frequency $\omega_0 \cong 1/\sqrt{L_M C_M}$. If not properly designed, the natural oscillations of the system might influence the PD pulse shape, leading to longer responses and pulse superposition in the time domain.

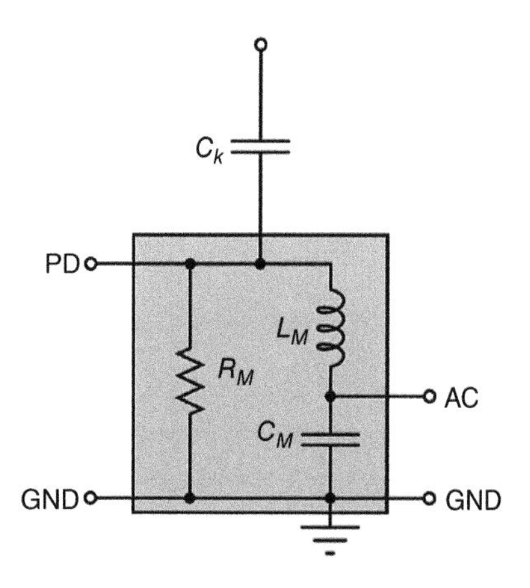

Figure 6.5 Quadripole structure.

6.3.3 High-Frequency Current Transformers

High-frequency current transformers (HFCT) can detect PD currents on the ground side of the EUT. They have either closed or split cores (see Figure 6.6). With a clamp-on core structure, they can be temporarily installed and removed easily on the ground leads of high-voltage equipment, even during service (although for safety reasons this is not recommended). Also, they may protect the PD instrument and the operator if the EUT fails during the

Figure 6.6 Closed and split core HFCTs. *Source:* Courtesy of Altanova.

test. The most important advantage of using HFCTs is that they do not affect the reliability of the power system in online testing, whereas coupling capacitors may. During online monitoring, the failure of a coupling capacitor will cause a ground fault that can cost tens or hundreds of thousands of dollars to correct. The most likely failure of an HFCT, an open circuit on the secondary winding, will have little impact, as the ferrite core at 50/60 Hz has negligible permeability leading to low voltages across the terminals of the (open) secondary winding. A shorted secondary winding, clearly, will have no impact on test object reliability.

HFCTs consist of a ferrite core and a secondary winding. The primary winding is the ground lead of the EUT itself. The secondary winding consists of a few turns. A large number of turns is generally not effective in boosting detection sensitivity since the self-inductance and the mutual capacitances between the turns tend to attenuate the high-frequency components of the PD pulse. The ferrite core is designed to match the bandwidth of the PD instrument and can be sensitive up to several tens of MHz, as illustrated in Figure 6.7. From a practical point of view, the bandwidth of the HFCT depends on the frequency at which the real and imaginary permeability of the chosen ferrite become equal. Note that the output of an HFCT does not include a significant 50/60 Hz component, so a 50/60 Hz AC synchronization signal needed for PRPD plots must come from another source (Section 8.3.2). One 50/60 Hz AC source used in combination with HFCTs is the Rogowski coil, also installed on the EUT ground lead (Figure 6.8). A Rogowski coil is an air-core current transformer with many turns to increase sensitivity to the AC current. (Rogowski coils have also been used for online detection of severe PD activity in rotating machines in petrochemical applications – see Section 16.6.2.)

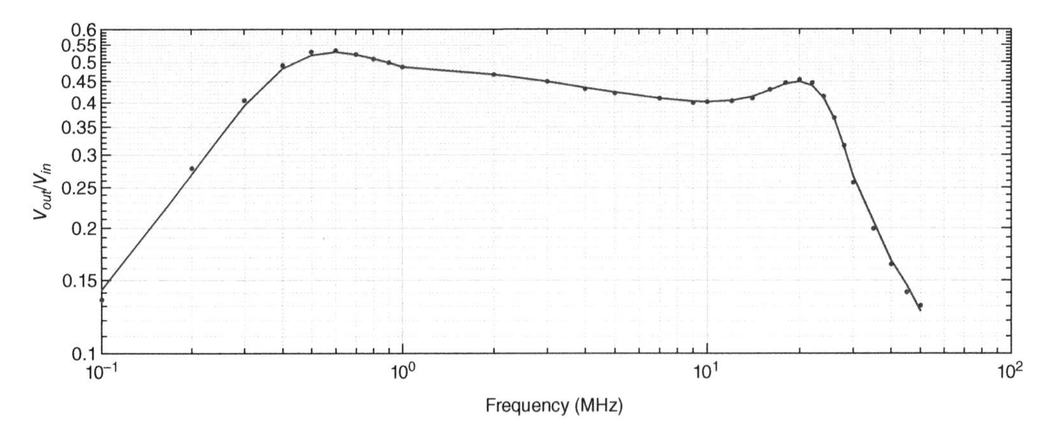

Figure 6.7 Transfer function of a commercial HFCT used for PD detection.

6.4 Electrical PD Detection Models

The models described below have the goal of assessing the effects that a partial discharge has on the circuit surrounding the defect (including the equipment under test, the coupling capacitor C_k, the voltage divider if present, the high-voltage source, and stray circuit parameters). These models relate the PD charge at the defect site (a cavity or void in this case) with what can be measured by an observer at the EUT terminals, i.e. the apparent charge if the measurement is in the low-frequency range (up to 1 MHz). The models that will be discussed are the ABC and the dipole (Pedersen's) models. The two models arrive at slightly different evaluations of the apparent charge. Indeed, the dipole model has a sound physical foundation, whereas the ABC circuit relies on approximations rarely met in practice (e.g. the dielectric in series with the cavity behaves as a capacitance separated from the remaining dielectric of the EUT). Therefore, the dipole model should be preferred to the ABC circuit. In practice, owing to its simplicity, the ABC circuit remains the most common model to evaluate the apparent charge. Therefore, both models will be explained starting from the most popular, the ABC circuit.

All models described in the following assume that all components in the test set-up are capacitive. This assumption stems from the fact that, for events happening in the nanosecond range, the impedance of the high-voltage supply (normally inductive) will behave as an open circuit. Thus, the sequence of events will be:

Figure 6.8 Terminal of a 220 kV XLPE cable with a Rogowski coil (the blue loop) and an HFCT (silver device) clamped around the ground lead. *Source:* Courtesy of Altanova.

a) The circuit is at an electrostatic equilibrium prior the PD
b) The discharge displaces the positive and negative charges on the cavity surface, altering the electrostatic equilibrium.
c) A new equilibrium is established by redistributing the charges within the system. Due to the charge of the PD, the voltage at the equipment under test terminals reduces by a very small amount.
d) The current from the power supply slowly re-establishes the voltage across the cavity.

6.4.1 ABC Model

The ABC model is by far the most accepted model. Its popularity derives on one hand from its simplicity. On the other hand, it is the oldest model, dating back at least to 1951. The equipment under test is modelled as three capacitances, as shown in Figure 6.9 where

- C_A is the capacitance of the equipment insulation, except the part that is in series with the cavity.
- C_B is the capacitance of the insulation in series with the cavity.
- C_C is the capacitance of the cavity.

The term "ABC model" is derived from the capacitor designations.

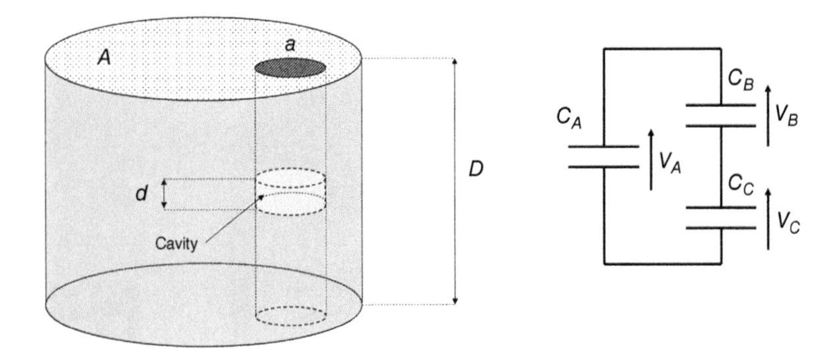

Figure 6.9 The ABC model for the equipment under test where C_C is the capacitance of the cavity.

Before continuing in the analysis, it is important to observe that this representation is somewhat general and is applicable to surface and corona PD as well. In this case, however, the boundaries between the three different capacitances are not clearly defined. Similarly, it should be noted that restricting the attention to the case of the cavity as in Figure 6.9, the capacitance C_A can also include stray capacitances that might not be readily evident.

Assuming a cylindrical cavity in Figure 6.9, the three capacitances are

$$C_A = \varepsilon_0 \varepsilon_r \frac{(A-a)}{D} \approx \varepsilon_0 \varepsilon_r \frac{A}{D} \tag{6.2}$$

$$C_B = \varepsilon_0 \varepsilon_r \frac{a}{D-d} \approx \varepsilon_0 \varepsilon_r \frac{a}{D} \tag{6.3}$$

$$C_C = \varepsilon_0 \frac{a}{d} \tag{6.4}$$

where A is the area of the electrode in the EUT, a is the area of the electrode aligned with the cavity, D is the insulation thickness, and d is the height of the cavity. Normally $A >> a$ and $D >> d$ (this might not be true dealing with corona, large voids and surface PD), then $C_A >> C_B$ and $C_C >> C_B$. Considering these relationships, and realizing that C_B and C_C form a capacitive voltage divider, the voltage across the cavity is

$$V_c \approx \varepsilon_r \frac{d}{D} V \tag{6.5}$$

$$V_B = V - V_c \approx V \tag{6.6}$$

As an example, using some values typical of stator winding rated voltage 15 kV$_{\text{rms}}$ phase to phase (or 12.2 kV peak phase to ground) and an average design electric stress of 2 kV/mm leads to an insulation thickness, D, of 6 mm. With a cylindrical cavity height (d) equal to 100 µm, a relative permittivity of the epoxy mica dielectric equal to 4, the peak voltage applied to the cavity is $4 \times 0.1/6 = 6.6\%$ of the peak phase to ground voltage, i.e. 815 V. Following a similar procedure for a 220 kV HV cable (design field 10 kV/mm, relative permittivity 2.2), with a 10 µm cavity, the voltage across the cavity is 202 V.

The analysis of the ABC circuit is straightforward. When the PD occurs in the cavity, shorting C_C, the voltage across the cavity goes to zero. Therefore, the charge in A and B must redistribute to equate V_A and V_B, with A charging B as the former was at the system voltage V, the latter at V–V_C. Eventually the voltage across the EUT will be decreased by ΔV. It is easy to prove that

$$C_B \cdot V_c = \left(C_A + C_B\right)\Delta V \approx C_{EUT} \cdot \Delta V \tag{6.7}$$

where C_{EUT} is the capacitance of the equipment under test (the sum of C_A and C_B). The meaning of Equation 6.7 is that the effects of the discharge that can be observed by measuring the change of the voltage in the system (and thus any other quantity as it will be shown) and does not depend on the charge deployed by the PD (q_{PD}) but on the so-called apparent charge (q_{app}):

$$q_{app} = C_B \cdot V_C < C_C \cdot V_C = q_{PD} \tag{6.8}$$

It is important to observe some points with regard to the ABC model.

6.4.1.1 Equivalent Circuit

Consider Equation 6.7, the change of the voltage at the EUT terminals depends on the apparent charge and on the EUT capacitance. It could be readily verified that if one or more capacitive devices are connected in parallel with the EUT, the apparent charge will not be affected (although ΔV will decrease). These findings can be summarized through the circuit of Figure 6.10. The current generator on the right is a Dirac pulse at the time of the occurrence of the PD (t_{PD}) times the apparent charge. This circuit can be used to understand the impact of a PD in a circuit comprising the EUT and other capacitive equipment such as voltage dividers and coupling capacitors.

6.4.1.2 Equivalent PD Current Generator

Any current generator connected at the EUT terminals, fast enough to simulate a PD pulse (i.e. with a frequency content exceeding the bandwidth of the detection circuit) and injecting the charge q_{app}, will provide the same effects of a PD having magnitude q_{app}. This can be inferred by inspecting Figure 6.10.

6.4.1.3 Coupling Capacitor

From the discussion of the ABC model, it is clear that the voltage drop across the EUT (and capacitive elements in parallel to it) can be very small. Suppose the apparent charge is 1 nC and the EUT capacitance 1 μF, then from Equation 6.7, the voltage drop would be 1 mV. If one wants to measure the voltage drop to detect PD, considering that the voltages are kV, it means that the precision should be 10^{-6}. Of course, this precision cannot be attained in practice (typically, HV instrument transformers for metering have a precision of the order 10^{-3}). To solve this problem, a capacitor (known as coupling capacitor and indicated as C_k) is installed in parallel to the EUT and the current flowing in the external circuit from the EUT to C_k is measured in practice. The coupling capacitor thus acts as a charge divider, and the charge in C_k is

$$q_k = \frac{1}{1 + \left(C_k / C_{EUT}\right)^{-1}} q_{app} \tag{6.9}$$

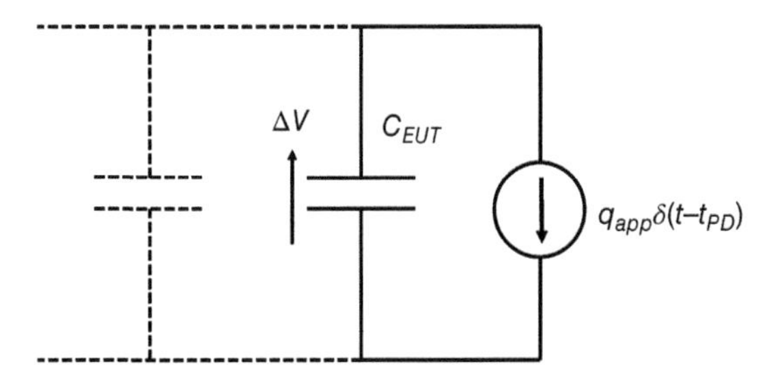

Figure 6.10 The equivalent circuit representing the ABC model. The capacitive element on the left represents any capacitive equipment in parallel with the EUT.

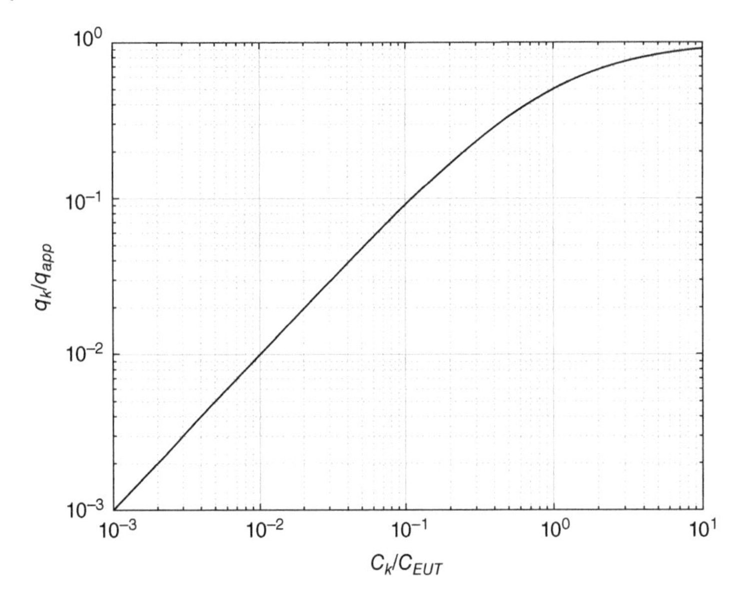

Figure 6.11 Divider ratio for the coupling capacitor.

Figure 6.11 shows the charge divider ratio of C_k as a function of the ratio C_k/C_{EUT}. It is clear that the larger the C_k the larger will be the charge flowing through the coupling capacitor, q_k. Thus, the larger the ratio C_k/C_{EUT} the higher will be the sensitivity of the detection circuit (the charge q_k must be larger than a minimum level to have a signal that exceeds the noise in the circuit).

6.4.1.4 Under Estimation of Charge

The last consequence of the ABC circuit is that the apparent charge under estimation worsens as the voltage level of the equipment increase. Rearranging Equation 6.8, one gets:

$$q_{app}/q_{PD} = C_B/C_C = \varepsilon_r \times d/D << 1 \tag{6.10}$$

Let us consider the previous examples of a rotating machine (relative permittivity 4, dielectric thickness 6 mm, and cavity thickness 100 μm) and of an HV cable (relative permittivity 2.2, dielectric thickness 18.4 mm, and cavity size 10 μm). The apparent charges are 6.6% and 0.12% the PD charge, respectively. This finding explains why limits of apparent charge reported, e.g., in equipment standards are more empirical than based upon physical considerations.

6.4.2 Dipole Model

Aage Pedersen of the Technical University of Denmark [1, 2] criticized the ABC circuit starting from the observation that the cavity is not discharged, but rather charged by the PD [2]. He highlighted also that the concept of capacitance stems from the assumption that two electrodes with equipotential surfaces are available, which is not the case of a cavity. Pedersen thus proposed a model based on electromagnetic field theory. About 20 years later, Prof. Eberhard Lemke came to the same conclusion, starting from energy balances developed in a simplified electrode configuration [3]. Compared with Pedersen's paper, Lemke's is simpler and elaborates more on the consequences of the "dipole" model, perhaps offering a better insight.

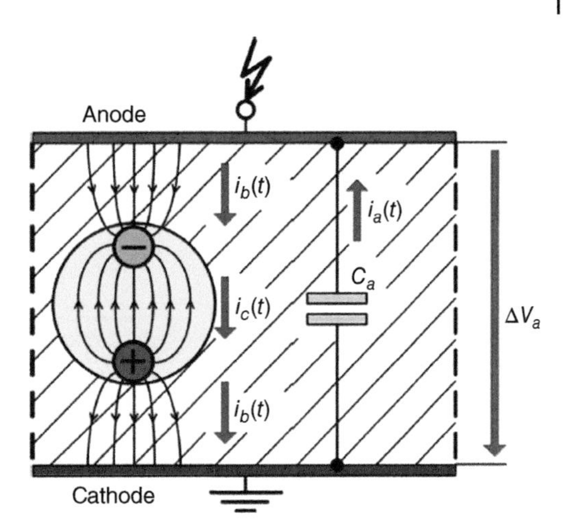

Figure 6.12 Currents within the EUT. *Source:* [3], © Nov (2012) IEEE.

According to Pedersen, the result of a PD is the establishment of a dipole within the dielectric. The dipole moment is $P = q_{PD} \times d$ where d is the height of the cavity in the direction of the field. The formation of this dipole is supported by a conduction current consisting of electrons moving toward the anode, positive ions toward the cathode. Both currents are stopped by the cavity surfaces, and, by the continuity equation, they must be balanced by displacement currents within the dielectric (see Figure 6.12). Thus, the difference of the charge on the electrodes of the system prior and after the discharge is due to polarization phenomena within the dielectric and is thus an induced charge (not an "apparent" charge as criticized in [3]). Pedersen's paper proves that the change in the charge at the EUT terminals due to the PD dipole is:

$$q_{EUT} = \Delta V \cdot C_{EUT} = -P \cdot \text{grad}\,\lambda \tag{6.11}$$

where λ is a scalar function describing the potential at all points of the insulation system assuming that is equal to 1 at the HV terminal, 0 at the ground terminal.

Lemke started from a simplified configuration and energetic balances to arrive at the same conclusions reached by Pedersen (although for a simplified defect geometry). Overall, Lemke's approach is less general but easier to follow and offers qualitative insight on the phenomena taking place during a discharge.

Let q_{EUT} be the charge on the EUT electrodes prior the PD. The variation of the charge, Δq_{EUT}, at the EUT electrodes induced by the PD (which is like the apparent charge in the ABC circuit) is:

$$\Delta q_{EUT} = \Delta V \cdot C_{EUT} = -q_{PD} \cdot \frac{d}{D} = -P/D \tag{6.12}$$

where d is the height of the cavity in the direction of the field, D is the thickness of the dielectric. Lemke argues that three cases can be envisaged:

1) $d/D \ll 1$, then $\Delta q_{EUT} = q_{PD} \cdot d/D$ is very low as little displacement field lines will involve the electrodes.
2) $d/D \approx 0.5$, then $\Delta q_{EUT} = q_{PD} \cdot d/D$ is approximately 50% of the PD charge
3) $d/D \approx 1$, then $\Delta q_{EUT} = q_{PD} \cdot d/D$ is approximately the same as the PD charge.

The three cases are illustrated in Figure 6.13. The way the displacement field lines are distributed through the sample offers a qualitative (and highly intuitive) explanation of the relationship between the PD charge and the induced charge.

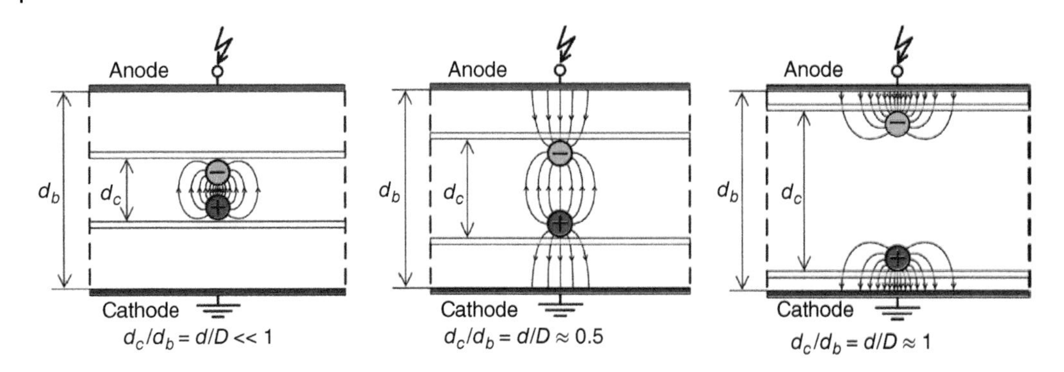

Figure 6.13 Induced charge as a function of *d/D*. *Source:* [3], © Nov (2012) IEEE. Note that the symbols used for the cavity height and insulation thickness differ from those used in this book ($d_c \equiv d$, $d_b \equiv D$).

6.4.3 Comparing the ABC Model with the Dipole Model

At this point, the reader might be confused by the description of two different models to treat the same problem. Indeed, while the authors prefer the dipole model (which reflects the physics behind a PD), the ABC model is so pervasive that is impossible to avoid it in everyday practice.

From a practical point of view, one should consider that both models arrive essentially at the same result: the ratio q_{app}/q_{PD} depends on $\varepsilon_r \times d/D$ in the ABC model (Equation 6.10), while $\Delta q_{EUT}/q_{PD}$ depends on d/D in the dipole model (Equation 6.12). The difference is in the relative permittivity, which range from 2 to 4 for most high-voltage systems. There is not much difference between the ABC and dipole models in this range of permitivities. Besides, the most important use of PD charge measurements is to determine whether a system has been manufactured in an acceptable way or not (Chapter 10). The PD charge limits are reported for each type of apparatus in technical standards (e.g. IEC, IEEE, ASTM, and local standards organizations) and were established based on testing and in-service experience acquired over the years. Thus, it is not essential to establish the real magnitude of the PD, as the limit is based on what can be measured, not on what can be inferred about the real magnitude of the event.

Any trapped charge from previous PD events within the cavity produces its own (Poisson) field (Section 2.1.2), which is superimposed on the electric field from the AC supply. The trapped charge field can be modeled as the field of a dipole. The ABC model does not include any obvious way to include the effects of trapped charge. The dipole model, on the contrary, can account for the fact that the dipole created by a PD occurring in the presence of trapped charge will be the algebraic difference between the dipole moment at the end of the discharge and the dipole moment due to the trapped charge (from previous PD).

6.4.4 Pulse Polarity

As will be discussed in Section 10.4.3, comparison of the positive and negative PD pulse polarity is sometimes useful to help determine the root cause and location of the PD. However, the PD pulse polarity depends on how the pulse is measured: in a direct circuit (Figure 6.1) or the indirect circuit (Figure 6.2). A positive PD pulse detected in Figure 6.1 would be measured as a negative pulse in Figure 6.2, and vice versa. This has given rise to two different conventions for assigning polarity to each PD pulse.

One convention is to assign pulse polarity based on the AC sinewave voltage at the instant the PD pulse is detected. That is, no matter what the measured polarity of the PD pulse is, its assigned polarity is based on the AC voltage. The advantage of this method is that pulse polarity is then independent on if the measurement circuit is direct or indirect. However, as shown in Sections 3.6 and 10.5 (and the many examples in Chapters 12–16), the PD pulse sometimes occurs before the beginning of (say) the AC positive zero crossing. This means the PD polarity would reverse as the AC voltage goes through the zero crossing. Since it is likely the same PD site is active before and after the zero crossing, this does not make technical sense. A "fix" for this approach is to assign PD pulse polarity based on the first derivative of the AC voltage. Effectively this means that PD pulse polarity is positive if the pulse occurs between −90° and +90° of the AC cycle, and vice versa, irrespective of the circuit used or the actual measured pulse polarity. As positive PD can occur beyond 90° (and negative PD beyond 270°), see Fig. 3.24, a careful combination of these two definitions is advisable.

The other convention is to assign PD pulse polarity based on the actual measured polarity of the pulse. However, the interpretation rules in Section 10.4.3 need to be reversed if a direct measurement of the PD current is made, which may be confusing to some.

Since both conventions are used in practice, PD system users need to be aware of how the software assigns polarity.

6.5 Quasi-integration in Charge-Based Measuring Systems

6.5.1 Quasi-integration Explained

As discussed in Section 3.3, if the insulation system in the test object is purely organic in nature, the assumption is that how quickly the insulation system will fail in the presence of PD depends on the charge (i.e. the number of electrons and ions) involved in each discharge. The circuits in Figures 6.1 and 6.2 measure the voltage pulse across the detection impedance. When measuring the PD at low frequencies with a charge-based measuring system, one wants to transform the PD pulses in mV to the q_{app}, so that the results can be displayed in terms of pC or nC. The detected mV can be related to the current using Ohm's law across the detection impedance. If the current is then integrated over a short period of time, the charge can be inferred. This section discusses how this is done using a "quasi-integrator." See also Section 8.8.1.

The concept of quasi-integration to obtain the PD magnitude in terms of charge can be explained referring to the Fourier transform of a PD pulse. Figure 6.14 shows two current pulses having the same integral (i.e. charge), equal to 100 pC, but with different waveforms in the time domain. For these two pulses, it is impossible to infer that their integral is the same by inspecting their oscillograms. Direct integration in the time domain of the detected PD pulses is unfeasible since, to remove the residual component from the HV AC power supply, PD detectors always have a high-pass filter as the first processing stage (Figures 6.5, 8.1–8.3). This filter has a cutoff frequency of some tens of kHz and, therefore, removes the DC component, preventing the correct calculation of the charge associated with the pulse.

In the frequency domain, the two pulses also have different spectra but at low frequencies, the spectrum components are identical. This means that focusing on low-frequency components, it is possible to compare the charge of pulses that have a different shape in the time domain.

A simple analog integrator is a low-pass filter. The principle of quasi-integration is thus to define the cutoff frequency of a low-pass filter that will process only low-frequency components having values like the DC component, as shown in Figure 6.15. Since the first stage of the detector is a

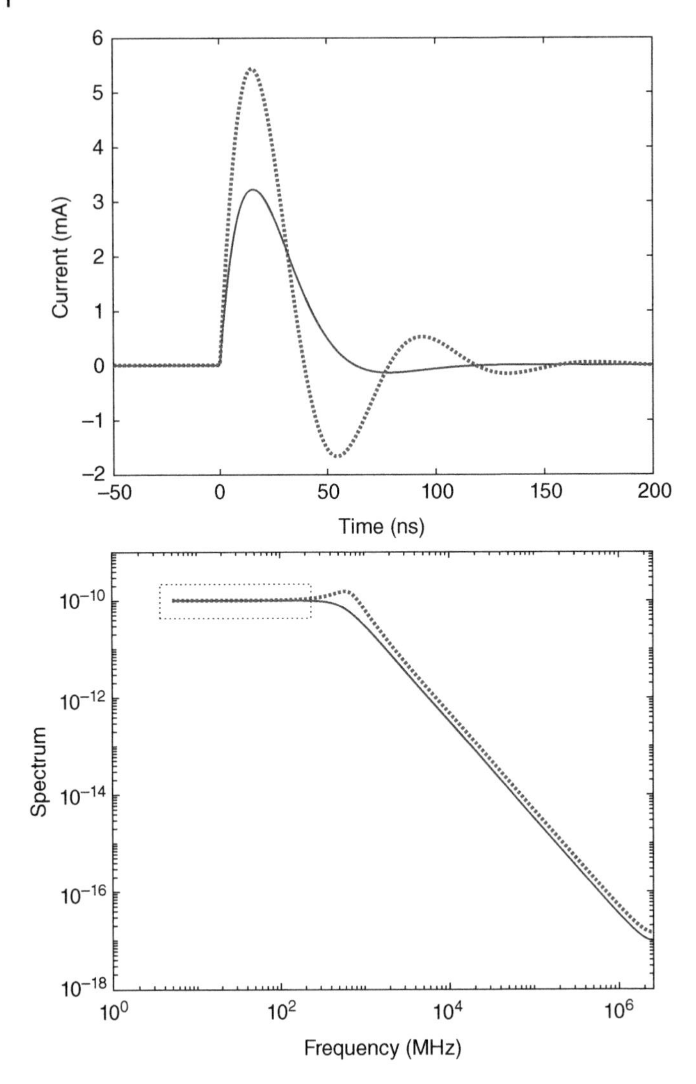

Figure 6.14 Time and frequency domain representation of two pulses with different waveshapes but equal charge (100 pC).

high-pass filter (C_k and R_M), the low-pass and the high-pass filters combine to make a band-pass filter. The output of the band-pass filter will be a pulse that will bear no information of the original PD pulse. Its output will be its impulse response scaled by the DC component of the input pulse. Thus, any quantity obtained from this pulse will provide information regarding the DC component. Usually, the peak value of the pulse is taken, since (a) it is very easy to measure with analog or digital circuitry, and (b) the signal-to-noise ratio is the best possible. Thus, the output of the quasi-integration filter is fed into peak detector, for both analog and digital detection systems (Section 8.3), and the PD measurement is "condensed" in the output of the peak detector.

6.5.2 Frequency Range of Charge-Based PD Detectors

The frequency range of PD charge-based PD measuring systems is standardized in IEC 60270 [4]. Two different types of measuring systems are conceived: wideband and narrowband

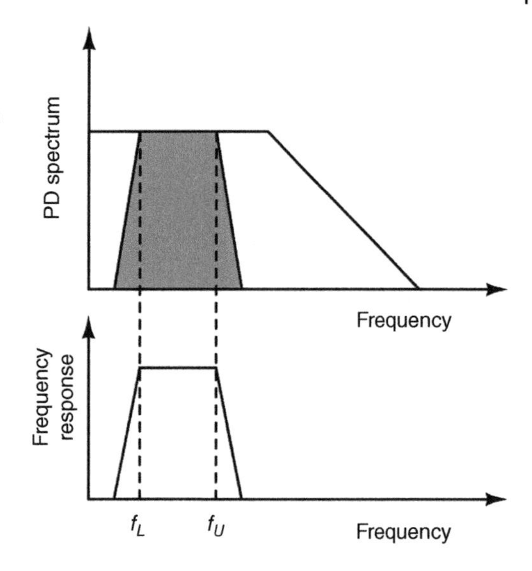

Figure 6.15 Principle of quasi-integration. The output of the filter is proportional to the DC component of the pulse as the frequency response of the filter differs from zero in a range of frequencies where the spectrum of the pulse is approximately equal to its DC component.

Table 6.1 Frequency range for quasi-integration.

Type of system	Filter characteristics
Wideband systems	• 30 kHz ≤ lower limit frequency ≤ 100 kHz • upper limit frequency ≤ 1000 kHz • 100 kHz ≤ bandwidth ≤ 900 kHz
Narrowband systems	• 9 kHz ≤ bandwidth ≤ 30 kHz • 100 kHz ≤ mid-band frequency ≤ 1000 kHz
Ultra-wideband systems	See IEC 62478

Source: Data from [4]/IEC.

detectors, the broad specifications of the filters are reported in Table 6.1. As can be seen, the upper frequency limit for quasi-integration is up to 1 MHz. It is assumed that, if the upper limit for quasi-integration is extended above this limit, the spectral components of the pulses might differ notably from the DC component. This is not the case for the example reported in Figure 6.14 where the differences between the pulses appear above 100 MHz. These pulses were, however, simulated without taking into account the bandwidth of electrical circuits and attenuation phenomena that drastically reduce the frequency content of the detected pulses in practical situation. It can be also observed that the lower cutoff frequency is above 30 kHz to prevent interference from the HV supply voltage (and higher-order voltage harmonics of 50/60 Hz, if present).

6.5.2.1 Pros and Cons of the Narrowband vs Wideband Systems

The frequency ranges for quasi-integration filters (Table 6.1) were decided having in mind both the evaluation of the apparent charge and the suppression of interference from AM radio stations (having a minimum frequency of 540 kHz) that were observed, for instance, when measuring power cables [5]. The latter point is reflected in the former editions of [4], where the upper limit for apparent charge estimation was 500 kHz. Apparently, [4] was modified to reflect the advances of interference suppression techniques by signal processing in modern PD detectors (Sections 8.4 and 8.9).

To understand the behavior of the narrowband and wideband systems, the response of two different band-pass filters (wideband and narrowband) to a train of PD pulses were simulated and are shown in Figures 6.16 and 6.17. Both filters were simulated using 4th-order Butterworth filters. The wideband filter had lower and upper cutoff frequencies of 100 kHz and 500 kHz, respectively, whereas for narrowband they are 290 kHz and 310 kHz. Figure 6.16 highlights that wideband filters preserve pulse polarity and have minor superposition effects even with PD pulses separated by only 1 μs. Using a narrowband filter, on the contrary, the polarity is lost (which makes it impossible to separate EUT PD from PD elsewhere in the circuit – Section 6.4.4) and strong superposition is observable when the pulses are spaced by 10 μs or less, i.e. if the PD repetition rate is high (Section 12.6). In this case, the PD magnitude becomes five times larger than the actual magnitude, and it is impossible to resolve the individual PD pulses, so that the PD repetition cannot be estimated. The advantage of narrowband detectors with tunable mid-band frequency is that they can help to detect PD in the case of high interference (Section 9.3.4).

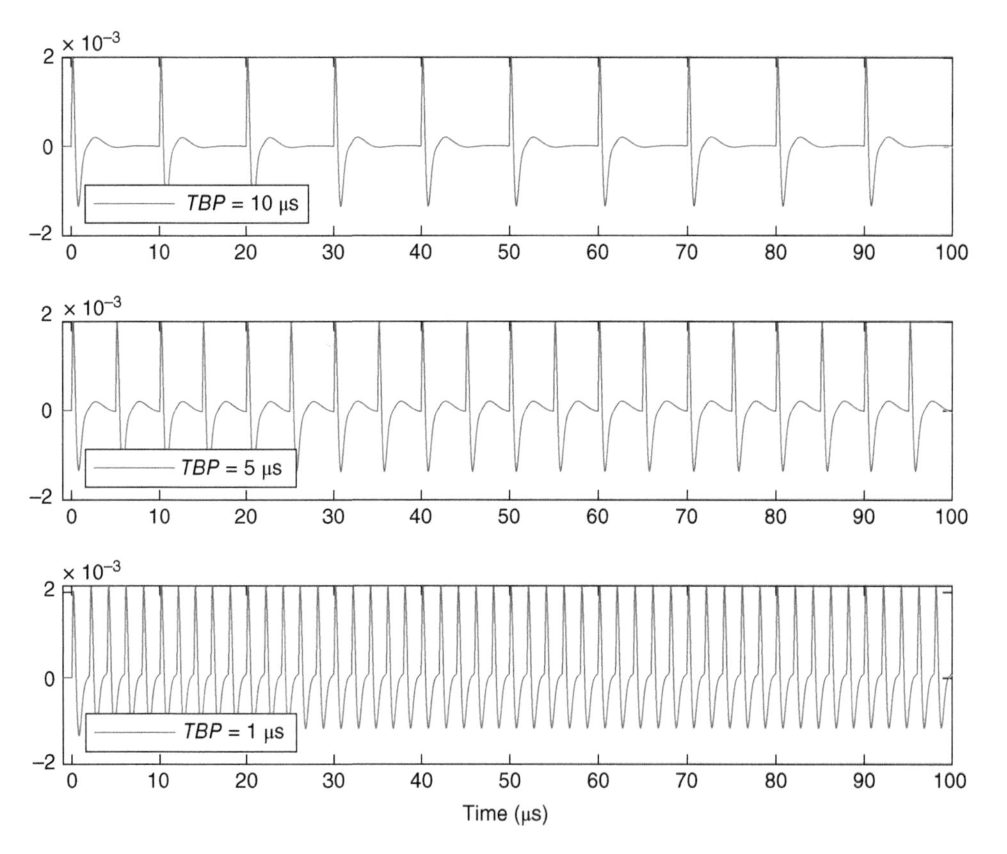

Figure 6.16 Response (in V) of a wide-band PD detector to a train of PD pulses (TBP is Time Between Pulses).

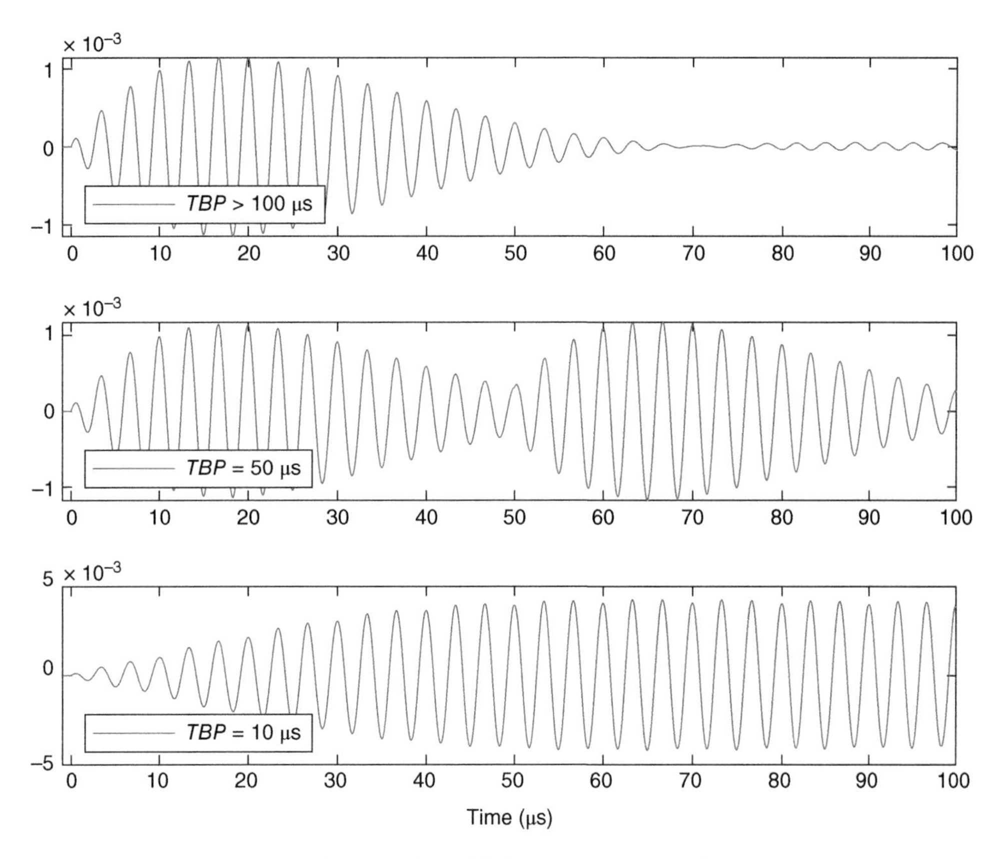

Figure 6.17 Response (in V) of a narrowband PD detector to a train of PD pulses. Note the different magnitude scales. (TBP is Time Between Pulses).

6.6 Calibration into Apparent Charge

6.6.1 Capacitive Test Objects

For charge-based PD measuring systems (IEC 60270), the quasi-integration procedure provides an estimate of the DC component of the pulse. In general, since a voltage will be measured from the detector, the DC component achieved by quasi-integration will be in mV (or in mA knowing the input resistance of the detector). Will this information be useful? In general, one should be able to compensate for the transfer function of the sensor. This might not be as straightforward as for HFCTs, where the transfer function is specified by measurements performed in the frequency domain, as in Figure 6.7, and thus not suitable to deal with the transients in the time domain associated with PD pulses. Besides, if one is using the indirect circuit (Figure 6.2), it is not clear what the impact of stray capacitances will be since they alter the EUT/C_k divider ratio in an unknown way. To solve these problems, the calibration procedure was devised. The calibration procedure starts from the equivalent circuit shown in Figure 6.10 which shows that, for a capacitive EUT, a PD can be simulated by injecting a pulse having the same apparent charge at the EUT terminals. Thus, the calibration procedure follows the steps:

1) A pulse of known charge Q_{Cal} is injected at the EUT terminals to simulate a PD pulse having apparent charge equal to Q_{Cal}. Note that, for both direct and indirect circuits, the calibrator is connected to the EUT terminals as in Figure 6.18. Except for special types of calibrators, fully isolated from the HV (usually using fiber optic cable), the HV supply is disconnected during calibration.
2) The calibrator pulse is coupled by the coupling capacitor and detection impedance, and processed by the quasi-integrator filter. The peak detector provides a voltage V_{Cal} proportional to the charge injected in the circuit.
3) The calibrator is disconnected from the circuit.
4) During the PD test, a PD pulse is processed in the same way as the calibration pulse providing the voltage V_{PD} (see Figure 6.19).
5) The calibration constant is evaluated as $k = Q_{Cal}/V_{Cal}$.
6) The apparent charge of the PD is inferred from V_{PD} and the calibration constant: $Q_{PD} = V_{PD} \times k$

The calibration charge should be generally twice the expected maximum charge of the PD from the EUT or twice the limit for acceptance/rejection in case a quality control test is performed.

It is important to know that if the frequency range of the PD detector is changed for any reason (perhaps to avoid a fixed frequency interference like a radio station), then the calibration process must be repeated at the new frequency range.

6.6.2 Distributed Test Objects

Distributed test objects do not behave as lumped capacitances but as lossy transmission lines. Typically, distributed test objects are cable circuits as well as gas-insulated substations (GIS) and transmission lines (GITL). There are important differences between the measurements performed

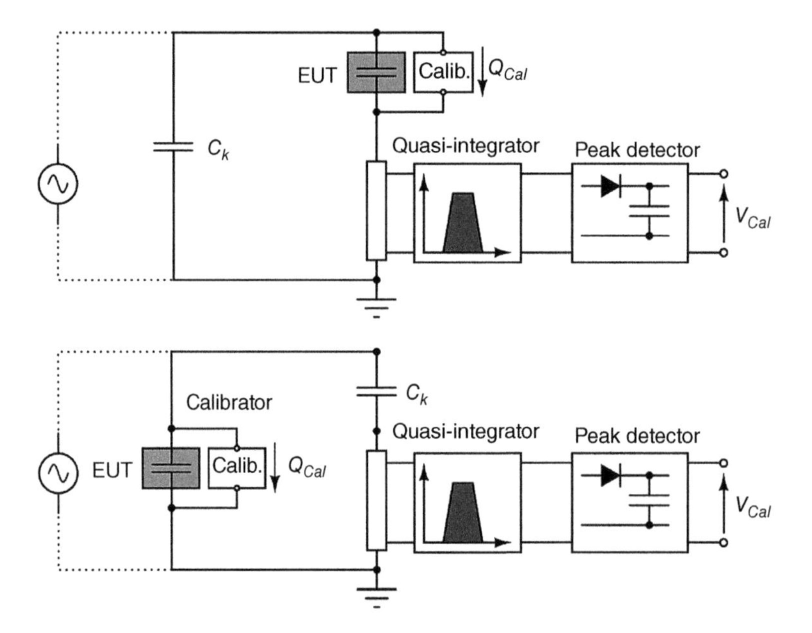

Figure 6.18 Calibration for the direct (top) and indirect (bottom) measuring systems. A pulse of known charge, Q_{Cal}, is injected at the EUT terminals (the high-voltage supply is disconnected), it is coupled by the sensors, and processed by the quasi-integrator filter. The peak detector provides a voltage V_{Cal} proportional to the charge injected in the circuit.

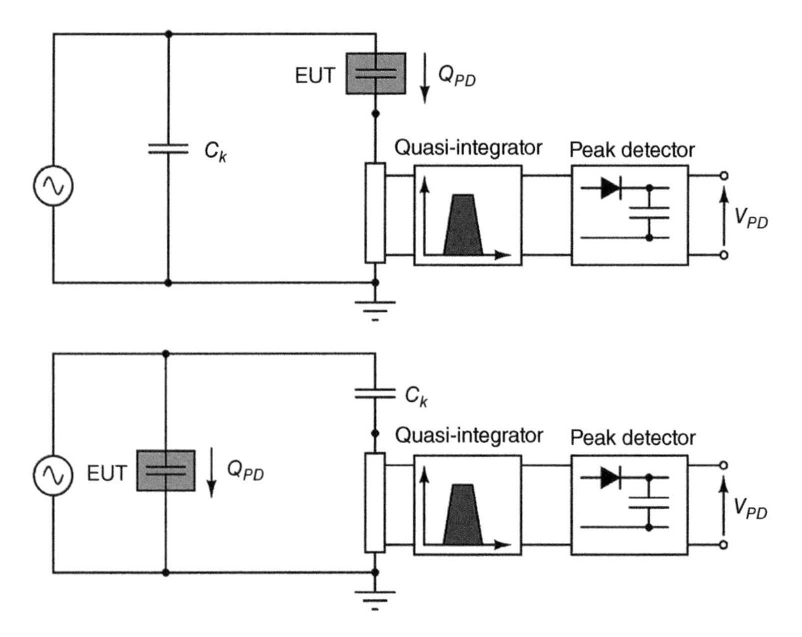

Figure 6.19 Detection stage. A PD pulse is processed in the same way as the calibration pulse in Figure 6.18 providing the voltage V_{PD}. The apparent charge of the PD is inferred from V_{PD} and the voltage provided by the calibrator V_{CAL} through linearity.

on objects behaving as transmission lines and those that can be represented using the lumped ABC or dipole models. While all phenomena will be present at the same time, it is convenient to discuss first the propagation of PD pulses in a lossless line first, then the effects of attenuation and dispersion. Additional information is in Sections 12.6 (for cables) and 13.8.2 (for GIS).

6.6.2.1 PD Pulse Splitting and Reflections

Figure 6.20 presents the behavior of a PD pulse injected in the EUT at a distance x from the near end (i.e. the end of the line where the PD coupler is installed). The PD pulse "sees" the characteristic impedance of the cable, Z_0, toward both the near and far end. As a consequence, the PD pulse will split in two identical pulses, each traveling toward one of the ends of the cable (with 50% of the PD **charge** each). Let v be the propagation speed of the PD pulses. The order of magnitude of the propagation speed can be estimated as $v \approx c/\sqrt{\varepsilon_r}$ where c is the speed of light in a vacuum, and ϵ_r is the relative permittivity of the insulation. Thus, in m/μs one gets $v \approx 300/\sqrt{\varepsilon_r}$

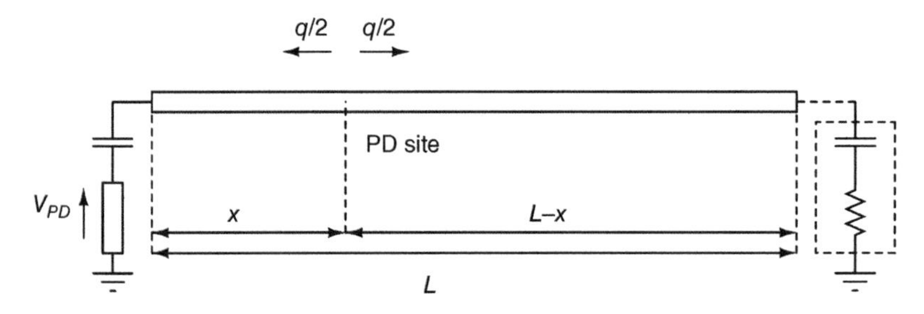

Figure 6.20 A PD pulse injected at some point of a transmission line splits into two pulses, each having 50% of the original pulse charge, traveling in opposite directions. The HV source is not represented as it can be located at either end of the transmission line.

which depending on the permittivity of the material ranges from 150 to 300 m/μs. For transmission class cables, typical propagation speeds are in the range 150–160 m/μs.

If L is the length of the transmission line, the pulse directed toward the near end will be detected after a time x/v. The other pulse will arrive at the far end after a time $(L-x)/v$. Only rarely is the far end terminated through a high-voltage blocking capacitor to a resistor matching the characteristic impedance of the cable. Normally, the far end is open-circuited. In the first case, there will be no reflection and the detector will receive only one pulse with charge $q/2$. In the second, more likely case, the PD pulse at the far end will be reflected toward the near end, where it will arrive after a total time from the PD event equal to $(2L-x)/v$. The time between the arrival of the first and second pulse will be $\Delta t = 2(L-x)/v$. When the discharge is close to the far end and x is comparable with L, Δt will be small and, depending on the detector type (wideband or narrowband) with a LF measuring instrument, there will be a significant risk that the two pulse responses will superpose at the detector quasi-integration filter output. This is illustrated through a simulation in Figure 6.21. In the simulation, the distance between the defect and far end is only 10 m. Since the propagation speed of the line is 183 m/μs assuming a polyethylene dielectric, the delay between the direct and the reflected pulse is $20/183 = 0.11$ μs. With a LF charge-based measuring system, this time is short enough to allow the output due to the direct pulse to superpose with the output of the reflected pulse, almost doubling the peak of the response. Indeed, the behavior of the estimated charge depends in ways that are difficult to predict due to the pulse response of the quasi-integrator filter, see for example Figure 6.22. In general, it will be necessary to evaluate the detection characteristic starting from the impulse response of the quasi-integrator filter.

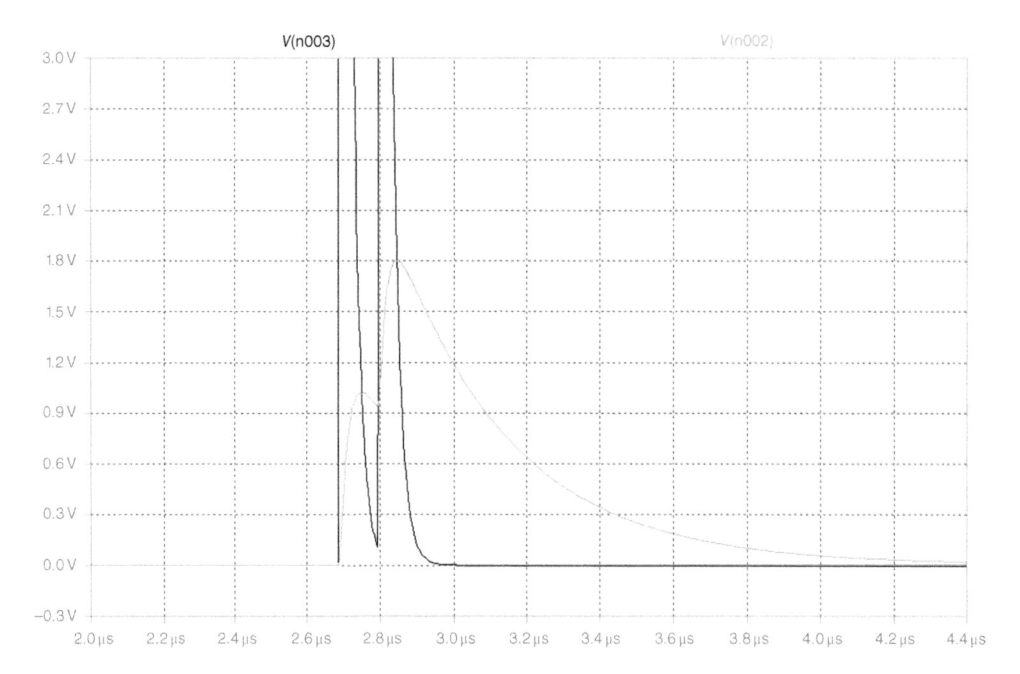

Figure 6.21 Electric pulses, V(n003)-black and detector output V(n002)-green for a PD originated at a distance of 490 m from the detector in a 500 m long transmission line. The cable is open ended at the far end. The detector is simulated by a first-order low-pass filter with cutoff frequency equal to 500 kHz. Simulations performed using LT Spice. The vertical scale is from 0.3 to 3 V (linear) and the horizontal scale is from 2 to 4.4 μs (linear).

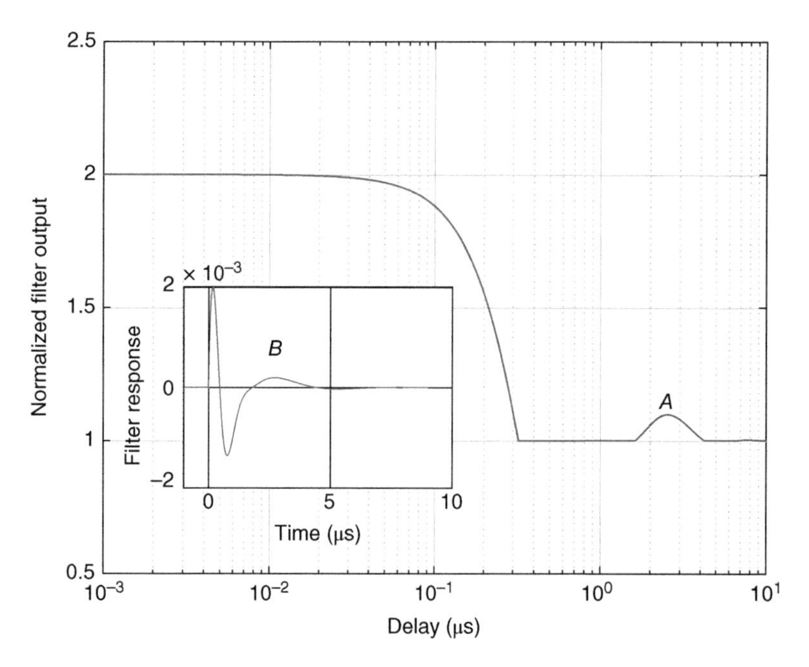

Figure 6.22 Output of the quasi-integration filter at the near end as a function of the delay of the reflected pulse. The output is normalized to the filter output in the absence of the reflected pulse. Quasi-integrator filter simulated through a Butterworth filter having order equal to 4. The filter has a lower and upper cutoff frequencies of 100 and 500 kHz. The peak indicated with A is due to the superposition of the main peak of the reflected impulse with the secondary peak (indicated with B) of the direct pulse, see inset.

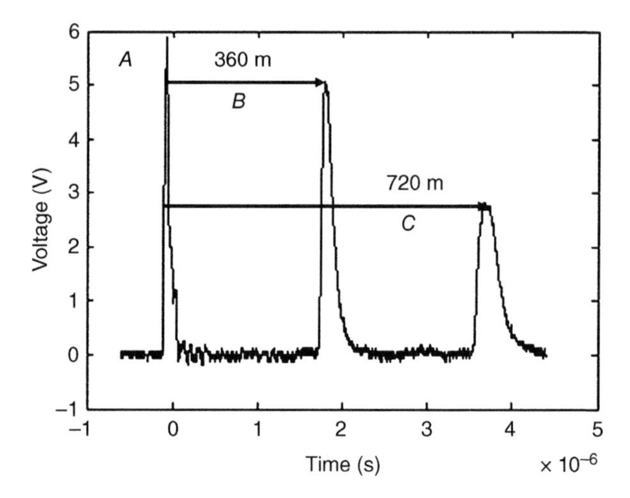

Figure 6.23 Propagation of a pulse in a cable 180 m long. The cable is open ended both at the near and far end.

6.6.2.2 Attenuation and Dispersion

Pulse attenuation and dispersion (group delay) are phenomena typical of lossy transmission lines, i.e. those that are not using a gas dielectric. Before presenting the mathematical relationships that describe attenuation and dispersion, it is worth to observe the way these phenomena influence pulses propagating in a transmission line. Figure 6.23 shows an impulse injected in a cable (A).

The cable is long 180 m. After traveling the full cable length is reflected to the near end, where it arrives traveling about 360 m (*B*). The process is then repeated, and another pulse is received at the near end (*C*) after traveling a total of 720 m. The effect of attenuation and dispersion can be understood inspecting the pulses *A*, *B*, and *C*: the high-frequency components are attenuated more than the low-frequency components. As a result, the peak value of the pulse is reduced, and the rise time is increased. The effect of dispersion is more subtle. The high- and low-frequency components travel at different speeds along the cable (group delay). Since low-frequency components are slower, they arrive later than high-frequency ones. As a result, the pulse tends to spread in the time domain.

From a mathematical point of view, attenuation and dispersion are described through the functions $\alpha(\omega)$ and $\beta(\omega)$ in Equation 6.13.

$$X(\omega,z)=X(\omega,0)\cdot\exp(-\alpha(\omega)z)\cdot\exp(j\beta(\omega)z)=X(\omega,0)\cdot\exp(-\alpha(\omega)z)\left(\cos(\beta(\omega)z)+j\cdot\sin(\beta(\omega)z)\right) \tag{6.13}$$

The real exponential, $\exp(-\alpha(\omega)z)$, describes how the different spectral components of the pulse get attenuated by traveling along the line. The imaginary part describes dispersion, i.e. how the phase of the pulse spectral components are modified along the line.

Regarding attenuation, measurements were carried out on XLPE and EPR cables [6, 7]. The results, reported in Figures 6.24 and 6.25, show that high high-frequency components are attenuated more than low-frequency components. The moderate attenuation occurring below 1 MHz is due to the skin effect. Above 1 MHz, the attenuation is due to power cable's semiconductive layers (Section 12.2.1) and becomes much stronger. In [8], these phenomena were reproduced fairly accurately by the simple model reported in Figure 6.26. Indeed, the equivalent parameters of the semiconductive layers in Figure 6.26 are frequency-dependent and the parallel connection between G_i and C_i ($i = 1, 2$) is functional to the frequency-dependent parameter estimation using impedance or vector network analyzers.

The result of attenuation due to semiconductive layers is that the magnitude and frequency content of PD pulses decrease rapidly as the pulses travel along the cable. Figure 6.27 shows the output

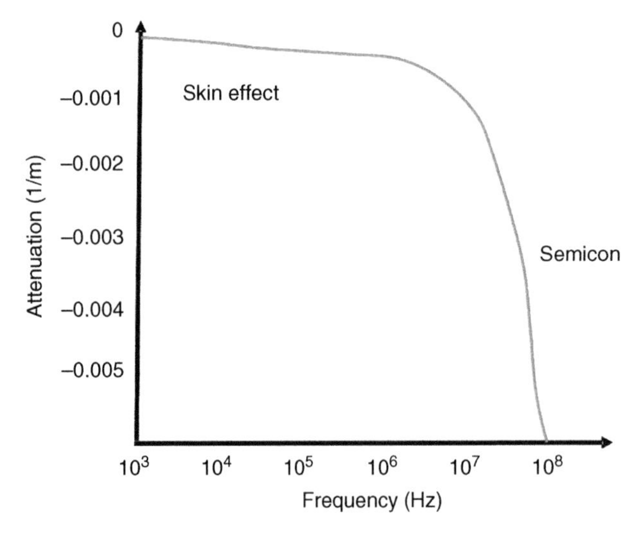

Figure 6.24 Measured attenuation in XLPE cables. *Source:* [6], © Feb (1992) IEEE.

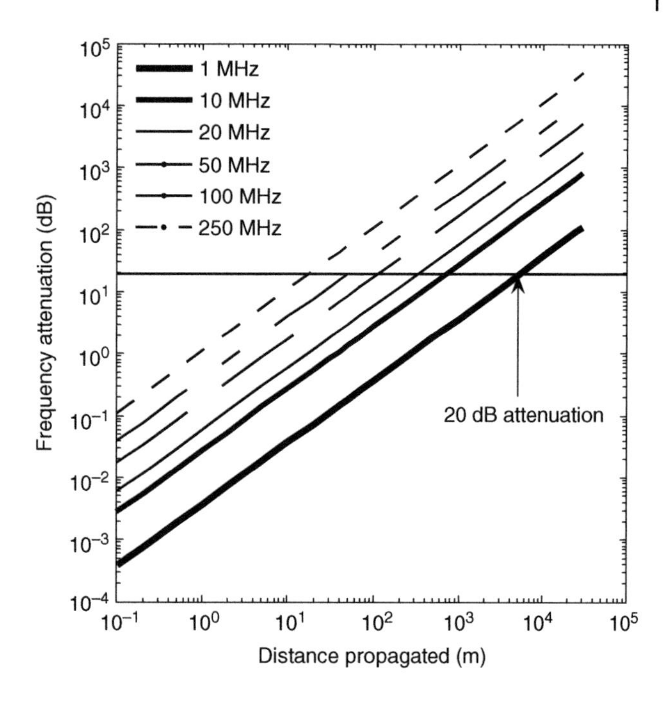

Figure 6.25 Attenuation of PD pulse spectral components as a function of distance between the PD source and the detection point. Attenuation data extracted from measurements on a 15 kV EPR distribution class cables. *Source:* [7]/M. Tozzi.

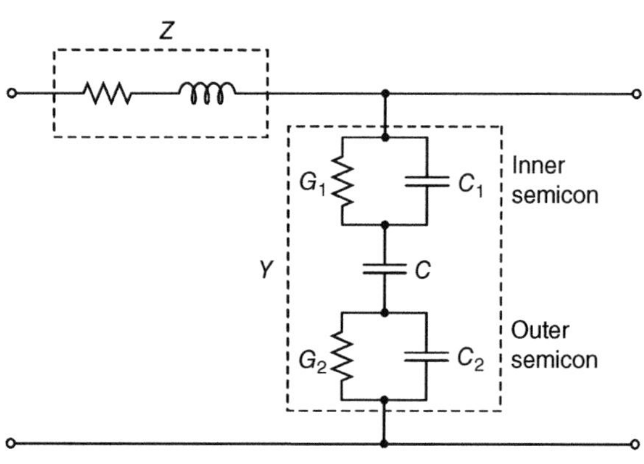

Figure 6.26 Power cable equivalent circuit. *Source:* [8], © Oct (1982) IEEE.

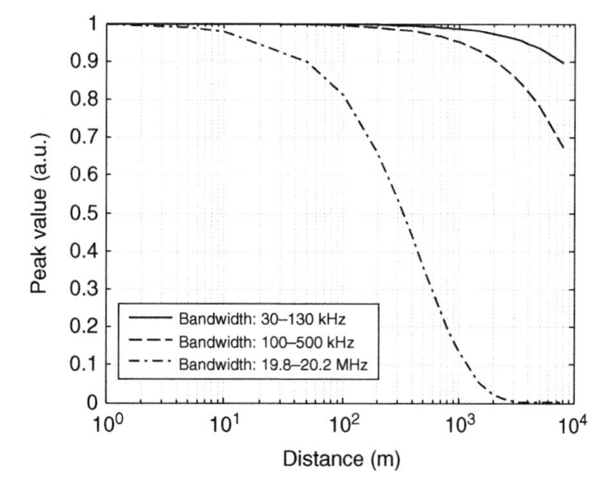

Figure 6.27 Simulation of a detector output as a function of distance between the PD source and the detection point. Cable attenuation data reported in Figure 6.25.

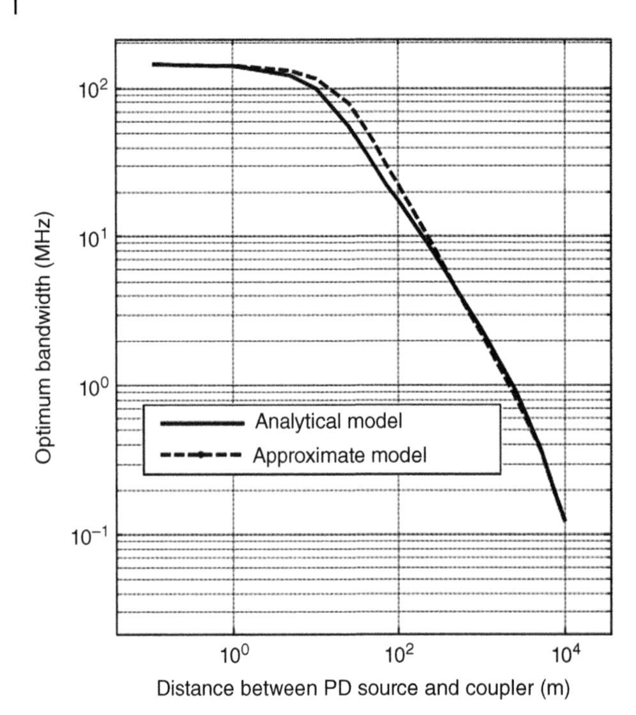

Figure 6.28 Optimum detection bandwidth. Cable attenuation data reported in Figure 6.25. *Source:* Tozzi [7].

of detectors having different filters as a function of the distance between the PD source and the detection point. The results are reported in per unit of the output of the detector when the PD source is at a negligible distance and were obtained by simulation using the attenuation values described in Figure 6.25. As can be seen, filters that operate below 1 MHz, when only skin effect losses influence the PD pulses, suffer a moderate reduction of the peak value of the pulse at the quasi-integrator output. As an example, the output reduction after 10 km for a filter with a lower and upper cutoff frequencies of 30 kHz and 130 kHz is only 10%. It is already 30% for the lower and upper cutoff frequencies of 100 kHz and 500 kHz. For a filter centered around 20 MHz, the output is negligible after 1–2 km. Figure 6.28 shows the optimum detection bandwidth as a function of the cable length [7]. As expected, this quantity is highly dependent on the distance between the PD source and the detectors, going from 100 to 150 MHz when the sensor is very close to the PD source to about 100 kHz if the cable is 10 km long.

The above considerations suggest that detectors with upper cutoff frequency of tens of MHz (i.e. in the HF range) can be highly effective when the detector can be placed in proximity of the defect. This is normally possible when the cable joints have embedded sensors or when PD detection can be performed through the link box (this latter point will be clarified in Chapter 12). However, if a long distribution cable or a submarine cable needs be measured from its terminals, lower bandwidths will ensure a better signal-to-noise ratio. Indeed, for very long cables PD detection might be impossible as the noise of the detector and the interference due to radio stations, etc. might bury the PD signal, which is already highly attenuated.

6.6.3 Inductive-Capacitive Test Objects

Inductive-capacitive test objects are devices characterized by the presence of windings and include, for instance, rotating machines, liquid-filled, and resin-cast transformers. At frequencies starting from a few tens of kHz, the equivalent circuit of these devices is very complex, as it must account for:

1) Turn self-inductance
2) Mutual inductances between turns
3) Mutual inductances between coils
4) Capacitance between turns and ground
5) Capacitance between turns
6) Mutual capacitances between coils

Accordingly, the frequency response of these devices is very complex and often displays several resonances and antiresonances at frequencies of hundreds of kHz (see Figure 6.29 for a stator winding). These natural resonances often occur at frequencies within the passband over which quasi-integration is being performed by an IEC 60270 compliant PD measurement system. To understand how this complex behavior can affect apparent charge measurements, the EUT can be modeled as a two-port network. Referring to Figure 6.30, one port (1) consists of the terminals, as an example, the high-voltage connection and the ground. The calibrator must be connected to this port. Another port (2), which is an abstraction, is where the PD takes place. Since the two ports are not coincident, the same pulse will give rise to different voltages at port 1, thus at the sensor port to which the calibrator is connected. These differences, particularly in the presence of system resonances and antiresonances

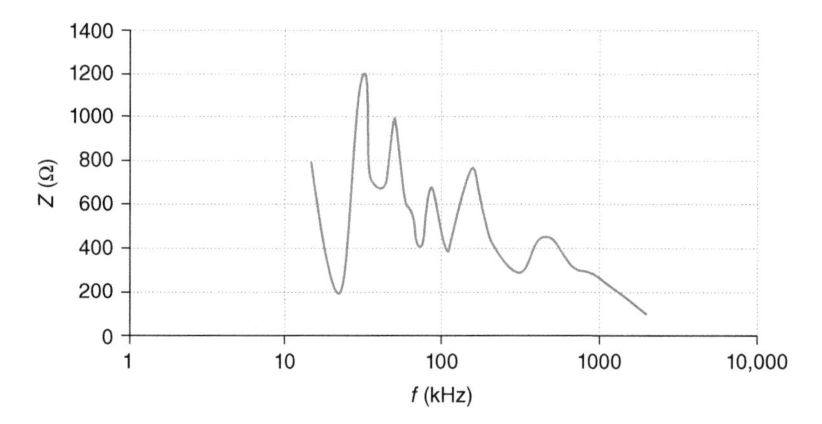

Figure 6.29 Impedance vs frequency for a 13.2 kV, 7000 HP, 1200 rpm motor stator winding in the LF range. *Source:* [9], © Sep (2021) IEEE.

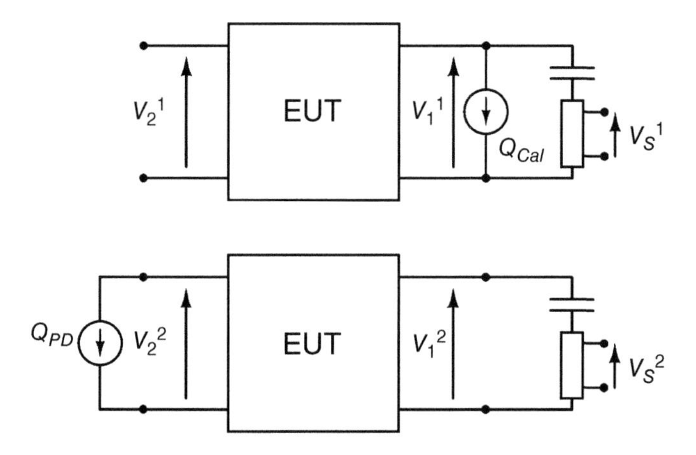

Figure 6.30 Two-port schematization of the EUT.

at frequencies in the quasi-integrator filter bandwidth, affect PD apparent charge measurements. Consequently, the use of apparent charge and calibration is not relevant for PD measurement in rotating machine windings. Indeed, the same effect occurs with transformers, where PD detection limits are still indicated in terms of apparent charge, when perhaps they should not be.

6.6.4 Practical Calibrators

Besides converting mV into pC, calibrators are also used for the sensitivity check of PD measurements carried out on inductive-capacitive test objects, as described in the next section. It is worthwhile observing that a sensitivity check is normally carried out for capacitive objects also to assess the noise floor (in pC) of the test.

The calibrator structure is shown in Figure 6.31. It consists of a voltage source V_0 in series with a charge injection capacitor C_0. To inject a known amount of charge, the series connection of C_0 and C_{EUT} should have a capacitance close to C_0, in this way $Q_{Cal} = C_0 V_0$, thus $C_0 << C_{EUT}$. Since V_0 is generally realized using electronic sources from a few to a few hundred volts, the injection capacitance is related to the amount of charge to be injected in the circuit.

Calibrators must satisfy, among other things, the following requirements on the voltage shape (see Figure 6.32) to be compliant with [4]:

- Rise time, $t_{rise} \leq 60\,\text{ns}$
- Step voltage duration, $t_d \geq 5\,\mu\text{s}$
- Fall time, $t_{fall} \geq 100\,\mu\text{s}$

Since the calibrator voltage is applied to a capacitive circuit, the current in the circuit will be made of a positive pulse having large magnitude but short duration, a negative pulse having low magnitude but long duration (both pulses have the same charge, $C_0 V_0$). The requirements on the rise and fall time are associated with the detectability of these current pulses. If the frequency content is estimated through the approximate relationship ~$0.3/T$ (where T is, alternatively, the rise or the fall time), the frequency content of the positive pulse will exceed 5 MHz, whereas that of the negative pulse will be lower than 3 kHz. Therefore, the positive pulse will be detected, whereas the negative pulse will be removed by the high-pass filter at the detector input.

An additional constraint is that the repetition period of the pulses must be controlled. Especially for narrowband detectors, injecting pulses with too high a repetition rate can lead to the superposition of the pulses at the output of the quasi-integrator filter. As an example, Figure 6.17 shows that, if the time between the pulses is 10 μs, the peak of the quasi-integrator filter will be five

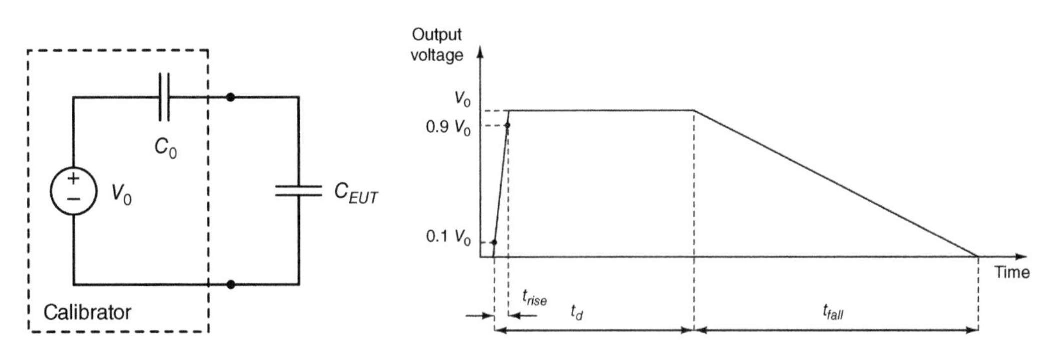

Figure 6.31 Calibrator structure.

Figure 6.32 Ideal calibrator pulse.

times larger compared to the value obtained in the absence of superposition. This indicates that, potentially, there is a risk that the magnitude of individual discharges is underestimated by a factor equal to five.

In the field, in the presence of high noise, it is convenient to synchronize the calibrator pulses with the 50/60 Hz voltage. In this way, the PD pattern associated with the calibrator will consist of unipolar pulses having constant magnitude located at equally spaced phase intervals. In the past, calibrators were synchronized to the 50/60 Hz using a photodiode that could detect the varying light flux of fluorescent lamps. To date, the change to LED lighting has made this approach useless.

References

1 ETHW Oral-History:Aage Pedersen – Engineering and Technology History Wiki. https://ethw.org/Oral-History:Aage_Pedersen#ASEA_and_Use_of_Epoxy_for_Insulation (accessed 23 December 2021).

2 Pedersen, A. (1989). On the electrical breakdown of gaseous dielectrics-an engineering approach. *IEEE Transactions on Electrical Insulation* 24 (5): 721–739. https://doi.org/10.1109/14.42156.

3 Lemke, E. (2012). A critical review of partial-discharge models. *IEEE Electrical Insulation Magazine* 28 (6): 11–16. https://doi.org/10.1109/MEI.2012.6340519.

4 IEC (2015). IEC Std. 60270. High-voltage test techniques – partial discharge measurements.

5 Mashikian, M.S., Palmieri, F., Bansal, R., and Northrop, R.B. (1992). Location of partial discharges in shielded cables in the presence of high noise. *IEEE Transactions on Electrical Insulation* 27 (1): 37–43. https://doi.org/10.1109/14.123439.

6 Steiner, J.P., Reynolds, P.H., and Weeks, W.L. (1992). Estimating the location of partial discharges in cables. *IEEE Transactions on Electrical Insulation* 27 (1): 44–59. https://doi.org/10.1109/14.123440.

7 Tozzi, M. (2010). Partial discharges in power distribution electrical systems: pulse propagation models and detection optimization. PhD thesis. University of Bologna, Bologna. https://amsdottorato.unibo.it/2308/1/Tozzi_Marco_Tesi.pdf (accessed 14 January 2022).

8 Stone, G.C. and Boggs, S.A. (1982). Propagation of partial discharge pulses in shielded power cable. *Conference on Electrical Insulation Dielectric Phenomena – Annual Report 1982*, Amherst, MA (17–21 October 1982), pp. 275–280. https://doi.org/10.1109/CEIDP.1982.7726545.

9 Stone, G., Sedding, H., and Veerkamp, W. (2021). What medium and high voltage stator winding partial discharge testing can – and cannot – tell you. *IEEE IAS Petroleum and Chemical Industry Technical Conference (PCIC)*, San Antonio, TX (13–16 September 2021), pp. 293–302. https://doi.org/10.1109/PCIC42579.2021.9728995.

7

Electromagnetic (RF) PD Detection

7.1 Why Measure Electromagnetic Signals from PD

The notion of using radiated electromagnetic signals for PD detection dates back to the 1940s, when radio interference voltage (RIV) systems were first used for corona and (later) PD detection (Section 1.5.1). These systems were operated at relatively low frequencies (up to a few MHz) [1]. With the improvement of measurement hardware, meaningful measurement at higher frequencies became more and more affordable. Around the beginning of the 1990s, gas insulated substations (GIS) started to be equipped with detection systems working at frequencies up to 1800 MHz [2, 3]. Later, rotating machines were equipped with stator slot couplers (SSCs) working up to 1 GHz [4]. The reason for this is that the risetime of PD pulses in SF_6 is extremely short, on the order of ~24 ps [3]. Using the approximate formula for the −3 dB upper cutoff frequency of the pulse:

$$f_{-3\,\mathrm{dB}} = 0.35/t_r \tag{7.1}$$

a PD pulse with a t_r of 24 ps transforms to RF frequencies in excess of 14 GHz, well past the upper boundary of the UHF band (3 GHz) [5]. Similarly, PD occurring within high pressure hydrogen in large turbine generators has a risetime of about 500 ps [4], corresponding to an upper cutoff of over 1 GHz. PD in air at 100 kPa typically has a risetime of a few ns, corresponding to about 100 MHz.

Specifically for GIS (Chapter 13), the low attenuation factor (~2 dB/m in straight sections) and closed metallic enclosure of GIS and GIL favors signal transmission over relatively large distances, thus favoring PD detection using a moderate number of sensors. Figure 7.1 shows the RF spectrum from PD in a GIS.

Later, detection of radiated PD signals also became more and more popular in rotating machines, especially for inverter-fed machines, and transformers. It is also becoming clear that PD sources in air can radiate EM fields at frequencies that are much higher than previously expected. Compared with previous expectations, recent measurements performed at the University of Bologna on a compact test cell seem to indicate that spectral components can have frequencies at 3 GHz or above (Figure 7.2). Indeed, the indications that discharges in air can reach 100 MHz is achieved using Equation (7.1), which refers to an attenuation of the spectrum equal to 3 dB under the hypothesis of an exponential decay of the pulse. However, Equation (7.1) does not imply that higher frequency components cannot exist, although with a lower energy.

The advantages offered by detection of EM-radiated signals in the upper limits of the VHF range or in the UHF range (see Table 7.1) stem from the way electrical equipment acts as an antenna [6]. The coupling between an antenna and an EM field from PD is optimal when the antenna length is

Practical Partial Discharge Measurement on Electrical Equipment, First Edition. Greg C. Stone, Andrea Cavallini, Glenn Behrmann, and Claudio Angelo Serafino.

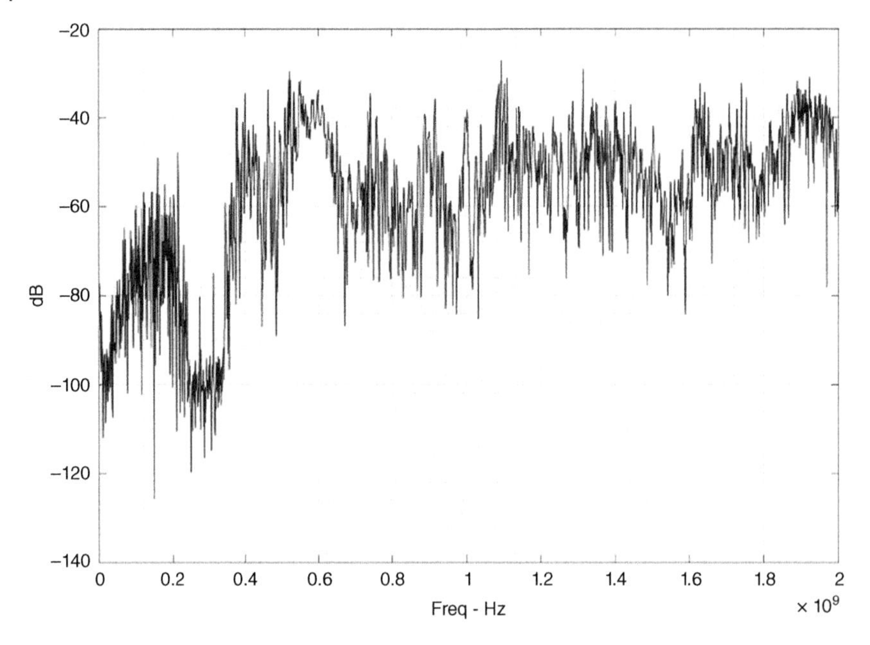

Figure 7.1 Measured spectrum from PD in GIS, measured with a UHF sensor.

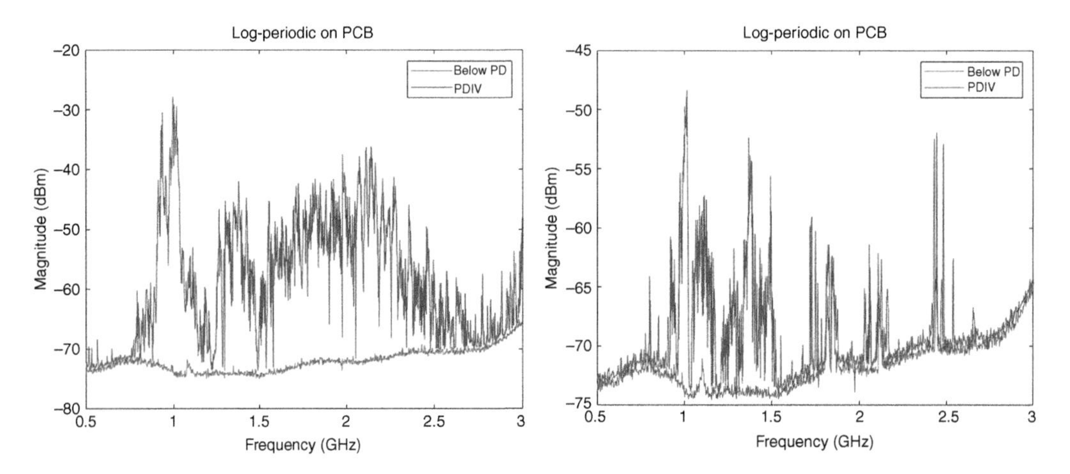

Figure 7.2 Spectrum of EM fields irradiated from PD in air from a pair of magnet wires twisted together and subjected to voltage impulses having large slew rate: (left) spectrum recorded at 15 cm from the source, (right) spectrum recorded at 150 cm from the source.

Table 7.1 Classification of radiated EM signals.

	Low frequency (LF)	Medium frequency (MF)	High frequency (HF)	Very high frequency (VHF)	Ultrahigh frequency (UHF)
Frequency range (MHz)	0.03–0.3	0.3–3	3–30	30–300	300–3000
Wavelength range (m)	10,000–1000	1000–100	100–10	10–1	1–0.1

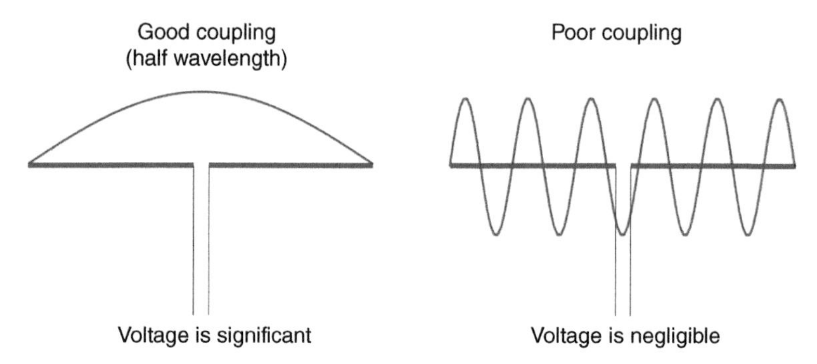

Figure 7.3 Due to the reciprocal cancellation of positive and negative half waves, compared with the antenna dimensions, the shorter the wavelength, the lower the signal at the antenna terminals.

half the wavelength of the EM field as shown in Figure 7.3. For frequencies above 200 MHz, the wavelength is shorter than 1.5 m. As most high-voltage electrical equipment is larger than 1.5 m, their coupling with interference tend to be less important above 200 MHz, favoring PD detection. Furthermore, at frequencies above 100 MHz, the skin depth is only on the order of a few micrometers, so that the equipment enclosure acts as a fairly effective Faraday cage, shielding internal PD from external interference. Even though external radio-frequency interference (RFI) can still enter the equipment via externally connected conductors, the more obtrusive levels of external interference within metallic enclosures are normally low, above 100–200 MHz. As an example, Figure 7.4 shows PD spectra measured within a small metallic tank ($26 \times 24 \times 24\,cm^3$) equipped with a medium-voltage bushing on the top. A circular aperture having a diameter equal to 7.5 cm was cut in the upper wall of the box to insert the bushing. Figure 7.4 shows that, for PD having comparable magnitudes, the shielding effect can be ~4–5 dB.

7.2 Terminology

Before venturing into technical topics, it is important to clarify the terminology that will be used in the following.

In this chapter, the term radio frequency (RF) detection will be used instead of ultrahigh frequency (UHF) detection, a term that is pervasively used in literature. As a matter of fact, PD RF detection techniques are not limited to the UHF range, that is 300–3000 MHz as defined by the International Telecommunications Union and IEC 62478, but they often extend into the very high frequency (VHF) range, that is 30–300 MHz and the HF (3–30 MHz) range (Table 7.1).

The second point worth explaining is the use of terms "noise," "interference," and "disturbance" (Chapter 9). The term "noise" will be used here to indicate the unwanted variability of the measurement that is intrinsic to the measurement system. In practice, "noise" will indicate all phenomena that are typical of electronic circuits such as thermal noise, shot noise, and $1/f$ noise. It must be clear that this type of variability of the signal cannot be eliminated (it can be reduced if more advanced and costlier equipment is used) and can be observed when the measurement system is disconnected from everything else (in practice, one would expect a signal equal to zero, but some continuous random signal is observed instead). The only way to remove noise is to turn off the instrument (not very helpful, though). Generally, noise levels are μV and are not so concerning when it comes to PD measurements.

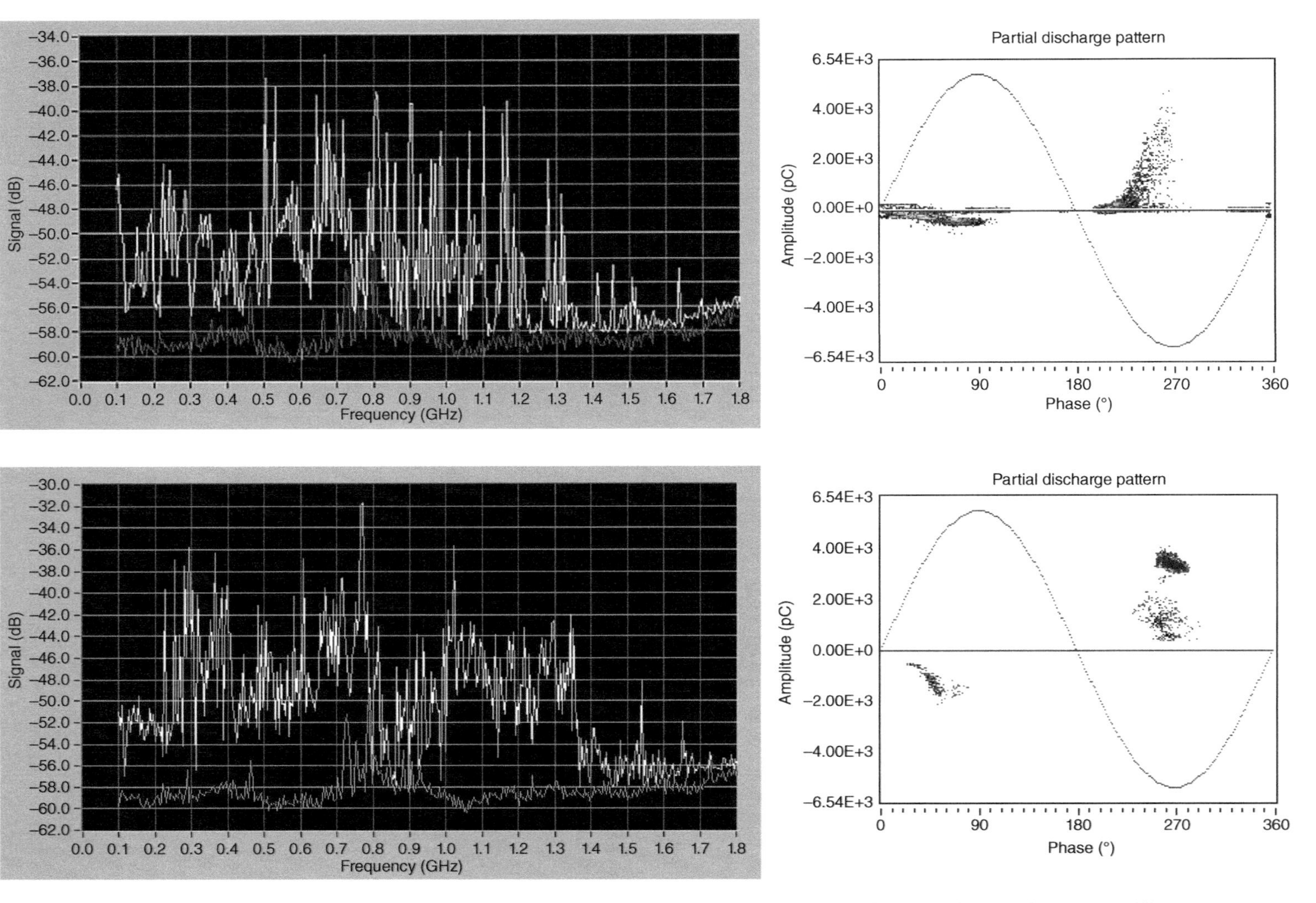

Figure 7.4 PD spectra measured within a metallic tank equipped with a medium-voltage bushing installed through a circular aperture (diameter 75 mm) in the tank upper wall. Spectrum and PRPD pattern for a PD source (top) within the tank, and outside the tank, on the surface of the bushing (bottom). White lines correspond to the PD spectra, red lines to background noise.

When the circuit external to the measurement system is connected, the signal in the measurement system tends to increase abruptly, even in the absence of applied voltage in the HV circuit (sometimes it is sufficient to connect, say, 10 cm of unterminated BNC coaxial cable). These additional, unwanted, signals are "interference": EMI, or RFI. Interference is much more relevant than noise when it comes to PD measurement as it can reach levels of a few V, hiding the PD signals. The sources of interference are man-made devices as radio stations, radars, and other electric and electronic devices of various nature (power electronic drives, DC motor collectors, welding machines, etc.). Interference due to these sources tend to fluctuate in time (think for example of a police radio station, interference will be active only when somebody is speaking on the radio), but it cannot be controlled by the operator attempting to measure PD. The only way to improve detection sensitivity is to shield the measurement circuit as much as possible.

The term "disturbance," also used frequently in technical literature as a synonym for "interference," will not be used here.

7.3 Basic Electrical Detection Circuits

A measurement system for electromagnetic PD detection can be split into three parts, as shown in Table 7.2:

- The transmission path from the PD source to the coupler (depends on the geometry and materials used to manufacture the equipment).
- The coupler, including the link with the measurement system.
- The measurement system itself.

7.3.1 Transmission Path

The EM field can be roughly subdivided into two regions: the near field and the far field. In the far-field region, the EM field waves are planar and can be derived from that of a dipole. Antenna features are calculated assuming the characteristics of planar EM waves. In the near field, the interactions between electric and magnetic fields are much more complex and difficult to analyze as they must be modeled using multi-dipole configurations. These differences are summarized in

Table 7.2 Different components of a VHF/UHF detection systems.

Transmission path	Sensors	PD measuring instruments
• PD source/sensor distance • Propagation characteristic of dielectric material(s) and geometry: – Dispersion – Attenuation – Resonance – Reflection – Diffraction	• Sensor types – Capacitive couplers – Magnetic couplers – External antennas • Connection link to the measuring instrument – RF coaxial cable – Fiber-optic cable	• Oscilloscope • Spectrum analyzer (max hold mode) • Spectrum analyzer in zero-span mode coupled with PD detector

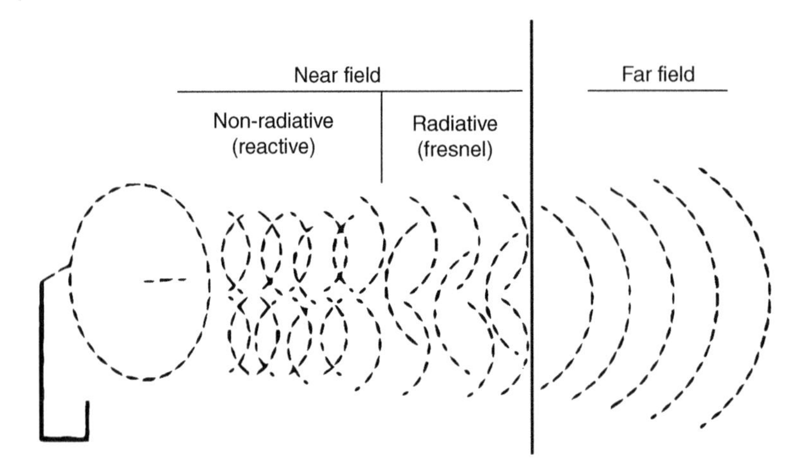

Figure 7.5 EM far- and near-field patterns. *Source:* [7].

Figure 7.5. The separation between the near- and far-field region is not precisely defined; the distance from the emitting body at which the near field turns into far field is generally calculated using the Fraunhofer distance, d_f, given by:

$$d_f = \frac{2D^2}{\lambda} = \frac{2fD^2}{c} \tag{7.2}$$

where D is the size of the emitting body, λ and f the wavelength and frequency of the EM wave, respectively. In general, if the sensor is located at a distance closer than d_f, it is said to be in the near field and thus behaves as either a purely capacitive or purely inductive sensor. If the distance is larger than d_f, the sensor can be considered to be in the far field and behaving as an antenna As a rule of thumb, 10–20 wavelengths are normally assumed in RF engineering as an estimate for d_f. As can be seen in Figure 7.6, the Fraunhofer distance is a linear function of the frequency but depends in a quadratic way on the size of the emitting body. Thus, relatively small differences in the dimensions of the emitting body can lead to a significant change in the structure of the EM waves. It is interesting to see these effects at play in Figure 7.2. The graph on the left corresponds to the spectrum measured at a very short distance from the PD source. The spectrum is approximately constant over a relatively large range of frequencies. The figure on the left is the spectrum achieved by placing the sensor 150 cm away from the PD source. Since the PD source was placed within a cylindrical tank with a diameter of 20 cm, one can estimate that the transition occurs around 1.4 GHz. Indeed, an intermediate situation can be observed as the PD source was close to the aperture: below 1.5 GHz, the spectrum is like what was previously observed (relatively constant), but above 1.5 GHz, one starts to observe the typical radiation pattern of an antenna (i.e. the spectral energy tends to concentrate at discrete frequencies). One of the problems in estimating the Fraunhofer distance is that the dimensions of the emitting body are not always clear. Is it only the region where the PD takes place, the entire EUT as it is crossed by the displacement current, or (more likely) a fraction of the entire EUT?

The calculation of the transmission path is generally unmanageable from an analytical point of view. Referring to a three-phase gas-insulated line (GIL), Figure 7.7 shows the calculated EM field distribution and its intensity on the outer (ground) conductor as a function of the angular position [8]. The GIL behaves as a very poor waveguide, and it can be observed that depending on the frequency of the signal, different transmission modes can be activated: Transverse Electromagnetic

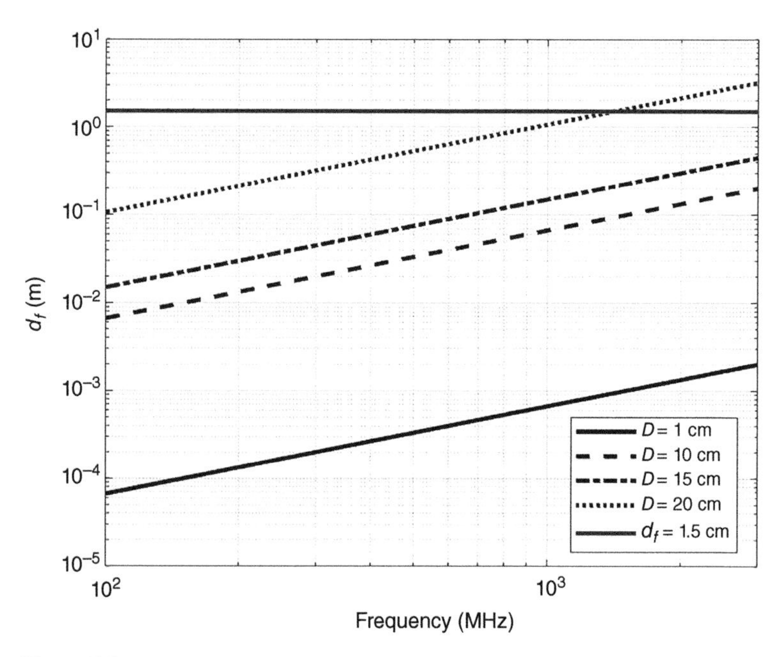

Figure 7.6 Fraunhofer distance as a function of EM wave frequency and emitting body dimensions. The blue line is at $d_f = 1.5\,\text{m}$.

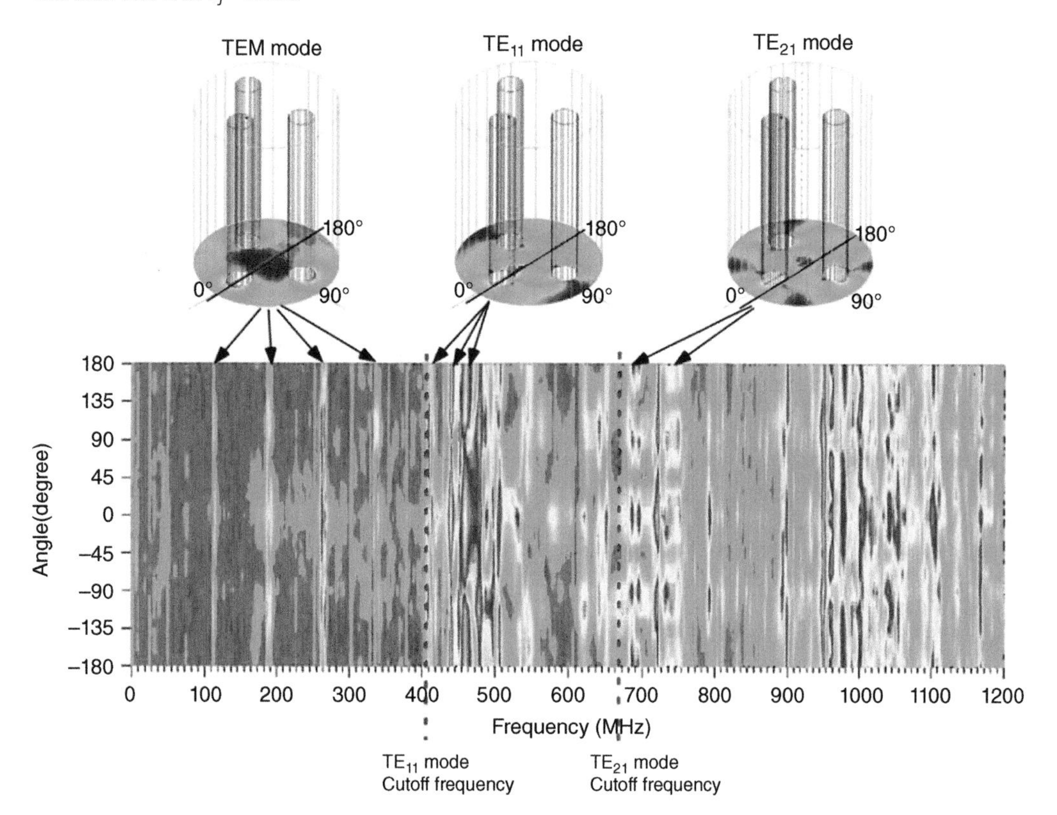

Figure 7.7 Frequency-dependent EM wave intensity in a three-phase gas-insulated line (GIL). *Source:* Okabe et al. [8].

(TEM_{00}), Transverse Electric 11 (TE_{11}), and Transverse Electric 21 (TE_{21}). Each mode (thus, each spectral component of the source) gives rise to a different intensity pattern. It is, therefore, clear that the signal coupled by a sensor will depend on the position of the PD source (on one of the three high-voltage conductors or on the external grounded enclosure) and the frequency response of the measurement system. In general, this extreme variation of the output signal prevents the system being calibrated in terms of charge (the pC/mV ratio would depend on the position of the PD source).

7.3.2 Sensors

The sensors will be presented in more detail in Section 7.4. However, it is worth observing that most sensors for PD detection will behave as a near-field sensor. The uncertainty arises since it depends on the distance of the sensor from the PD source, the frequency range where measurements are performed, and the radiation pattern (or D in Equation (7.2)) of the EUT. Many sensors (for instance, transformer sensors inside or on dielectric apertures in the tank) operate as capacitive sensors.

Besides the sensors themselves, a circuit for matching the sensor impedance with that of the amplifiers or the transmission line will often be required. These circuits are known as balancing units or, "baluns." A suitable amplifier will often be required to transmit the signal to the measurement system or, alternatively, to power the photodiode or laser used to convert the detected analog signal to an optical signal suitable to be transmitted through fiber-optic cables. Eventually, one should also consider the attenuation due to the transmission line loss between the sensors to the measurement system, which can become significant at frequencies in the UHF range (see Figure 7.8).

The final structure of the system is sketched in Figure 7.9.

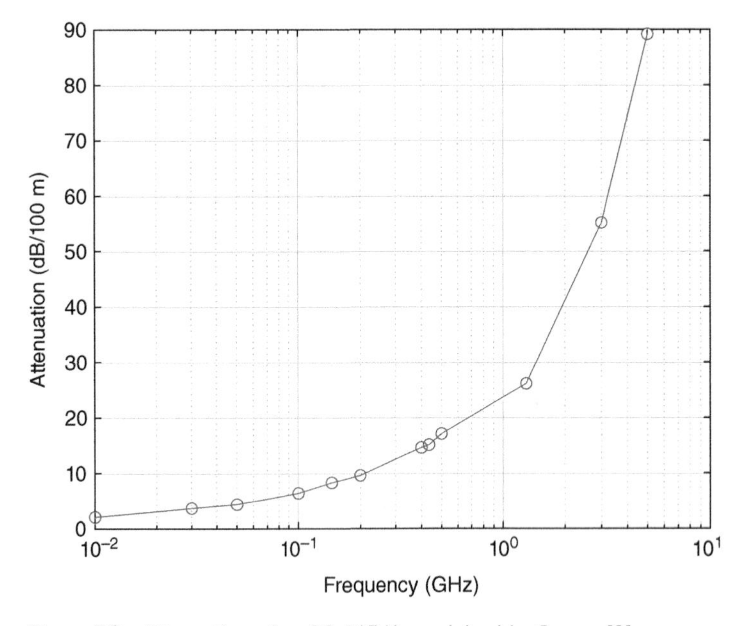

Figure 7.8 Attenuation of an RG-213 U coaxial cable. *Source:* [9].

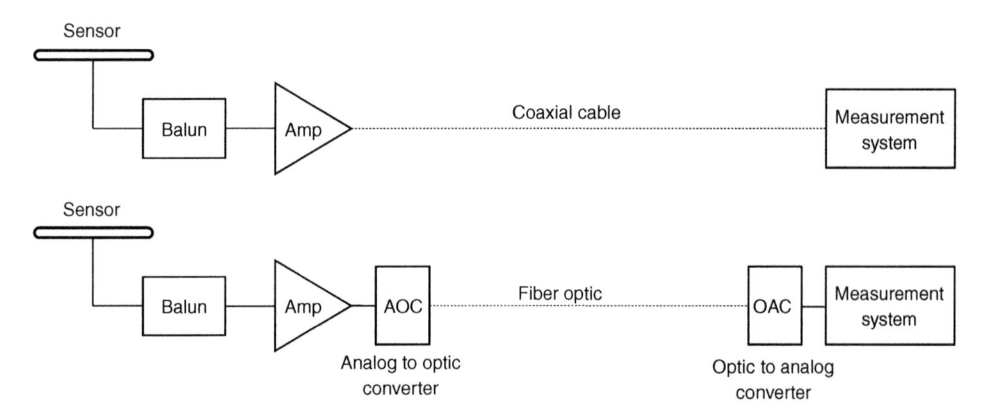

Figure 7.9 Connection link structures.

7.3.3 Time and Frequency Domain Measurement

Signals from RF PD sensors can measure in the time or in the frequency domain. In the time domain, the following options are available:

- Digital sampling oscilloscopes (DSO). Pulse acquisition in the oscilloscope is triggered by the sensor signal. The waveform on the signal provided by the sensor is thus recorded, as shown in Figure 7.10.
- PD detector connected to (typically) HF/VHF sensor. The Prysmian Prycam is an example of this type of system. The sensor embedded in the Prycam works in a frequency range that spans from the HF to the VHF range, i.e. from 0.1 to 100 MHz. Therefore, the Prycam can often pick up the

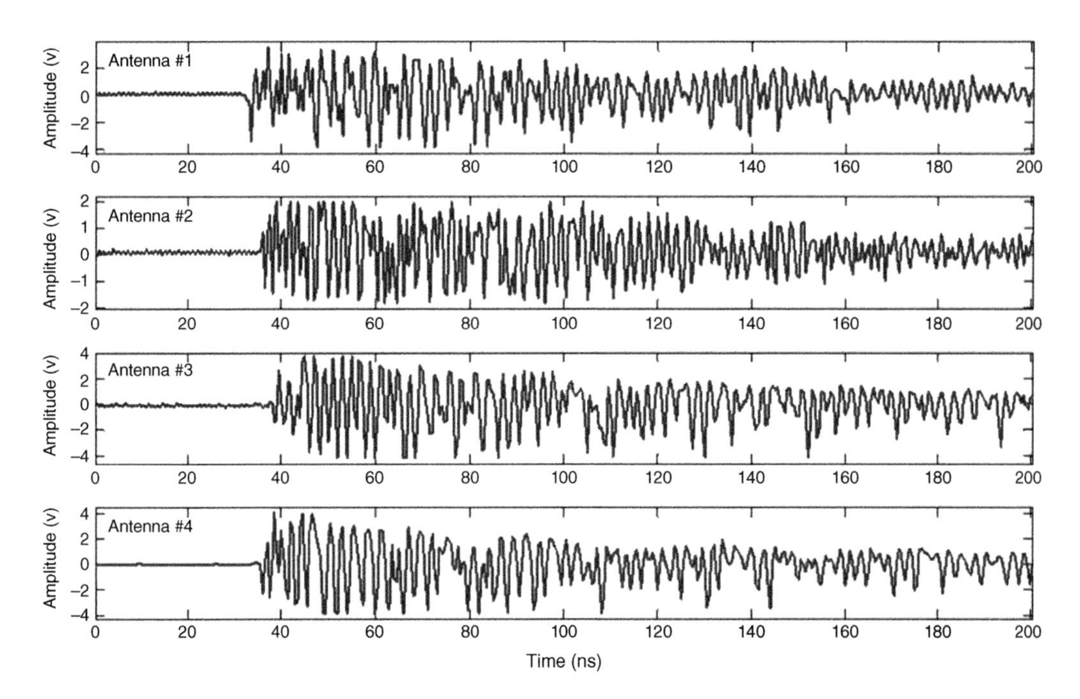

Figure 7.10 Example of readings from a four-channel oscilloscope connected to UHF sensors installed within a transformer at different distances from the PD source. *Source:* Akbari et al. [10].

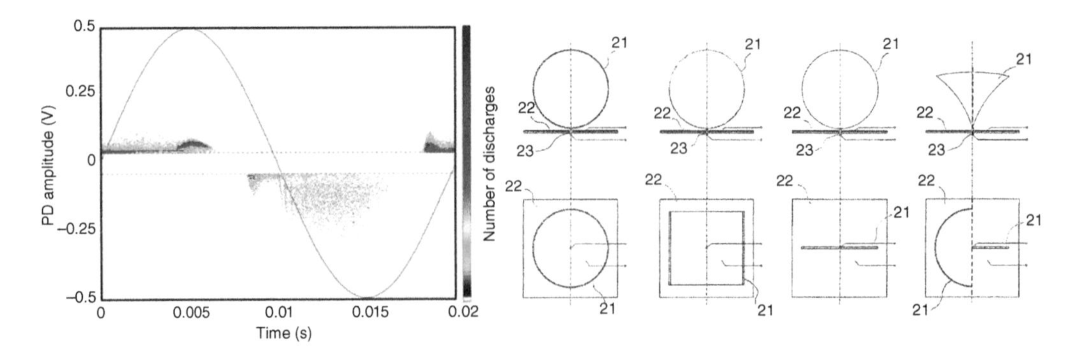

Figure 7.11 The Prysmian Prycam detector is an example of peak detector operating in the HF–VHF range. (left) PD detected in proximity of a power cable. *Source:* Romano et al. [11]/MDPI/CC BY-4.0. (right) Structure of the Prycam sensor. *Source:* Di et al. [12]/U.S. Patent/Public Domain.

PD polarity consistently. Besides, the same sensor can be used to synchronize to the 50/60 Hz voltage, thus providing the PRPD pattern (Figure 7.11).

- PD detector connected to VHF or UHF sensor through an amplitude demodulator (Section 8.3). The amplitude demodulator consists of a fast diode connected to an RC circuit that extracts the envelope of the PD pulse. Amplifiers working in the frequency band of the signals provided by the sensor are normally required to increase the voltage amplitude, thereby enabling the operation of the diode. To create PRPD plots, the detector needs to be synchronized to the 50/60 Hz (Section 8.3.2). Due to the presence of the diode, the PD are recorded only with the positive polarity (Figure 7.12). This is the approach followed by Techimp-Altanova, DMS-Qualitrol, and several other suppliers.

In the frequency domain, the typical solution has been to work with a swept-frequency RF spectrum analyzer utilizing an analog RF super-heterodyne architecture. Digital spectrum analyzers (for example, as implemented in a digital oscilloscope), in contrast, are cheaper and faster and make frequency domain measurements more accessible. However, they have a limited dynamic range, which may result in PD being suppressed in the presence of very high levels of interference.

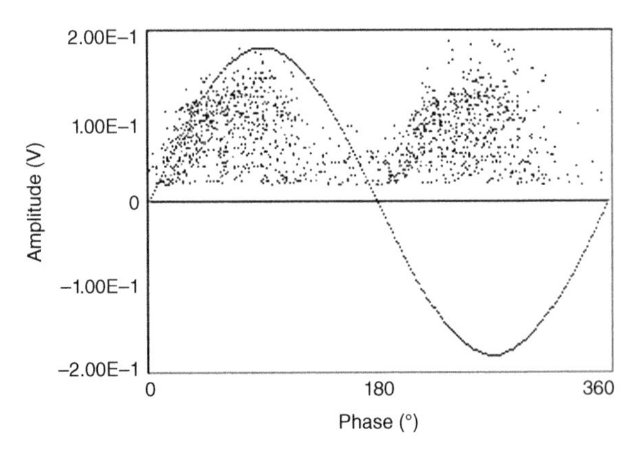

Figure 7.12 Partial discharges in a power transformer detected using an UHF antenna connected to a PD detector (Techimp PDBase II, bandwidth 40 MHz) via an envelope demodulator.

To avoid going into the details of their operation, for our purposes a spectrum analyzer uses a variable-width bandpass filter to analyze its input signal in a specifically selected range of frequencies.[1] There are typically two ways to use the spectrum analyzer for PD detection:

- The spectrum analyzer is used in the sweep frequency mode. Since PD pulses are of extremely short duration (typically ~1 ns duration), multiple sweeps are repeated with the spectrum analyzer set in the "max-hold" mode (that is at any given frequency, the spectral density displayed is the maximum accumulated signal amplitude acquired during the complete measurement interval). This type of measurement does not provide a PRPD pattern that can help to discriminate between PD and interference. Therefore, a noise baseline must be established first, and the signals exceeding that baseline are assumed to be PD activity, as in Figure 7.2 and the left images in Figure 7.4. The baseline can be established with the voltage applied to the EUT equal to zero or, if the voltage can be controlled, set to some value that does not pose the risk of incepting PD. The latter is preferable since, in the first case, the voltage source will not be connected to the EUT, and the system could be affected by interferences that differ from those observed during the test.
- The spectrum analyzer is used in the zero-span mode, that is the center frequency of the variable-bandwidth filter "window" is held constant. This is a typical feature of analog superheterodyne units, which normally provide the analog ("video") output of the center frequency "window." If this output is available, one can connect a PD detector to the output and record the PRPD pattern. In practice, it is better to use this type of measurement after having performed the first one to determine where in the RF spectrum the PD signal exhibits the best signal-to-noise ratio (SNR). Given the latter is 10 dB or more, this should enable the PD detector to record a PD pattern that is clean enough to be used for PD source recognition (Section 8.7.3). Alternatively, similar results can be achieved using an oscilloscope synchronized with the line voltage through a TTL signal. Figure 7.13 shows an example of the output achieved using an oscilloscope synchronized with the 50/60 Hz supply voltage. Indeed, since the analysis filter has normally a narrow bandwidth compared with the central frequency, the impulse response of the filter is highly oscillatory and the PD polarity is often interpreted incorrectly. Thus, it is customary to report a unipolar PRPD pattern as in Figure 7.13.

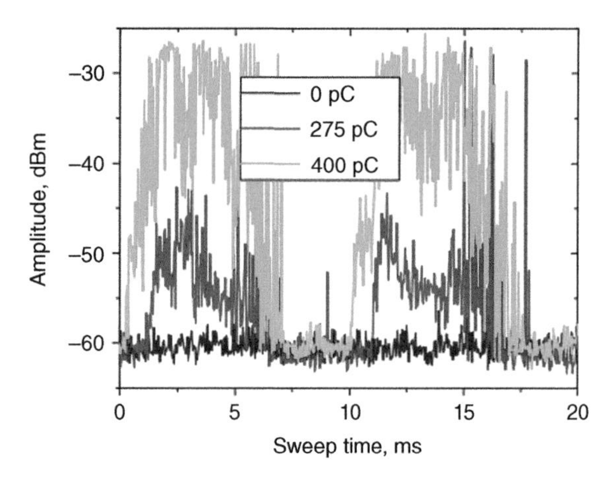

Figure 7.13 Example of PD pattern reconstructed using a (time-domain) oscilloscope connected to a spectrum analyzer in zero-span mode. *Source:* Denissov et al. [13].

1 Note that, technically speaking, this is also a measurement in the time domain, as the output of the filter is received and analyzed in the time domain. However, it has become customary to indicate it as frequency-domain measurement.

7.4 Types of RF Sensors

Sensors can be very different, as different coupling principles can be adopted (electric field, magnetic field, planar electromagnetic waves). For antennas, as mentioned above, whether the sensor will work in the near field or in the far field will depend on the distance between the PD source, metallic objects (if any), and the sensor. Yet, some sensors are designed as capacitive sensors (thus, they basically consist of metal plates insulated from ground potential), others are designed referring to specific antenna topologies with the aim of exploiting some specific antenna features (directional or omnidirectional, for example). Besides, even when the same coupling principle is adopted, the aspect can be very different due to the need to adapt the sensor to the equipment.

This section aims at providing some basic ideas. More specific information will be provided in the chapters dealing with the various equipment.

7.4.1 Ferrite Antennas

Ferrite antennas are solenoids wound on a cylindrical ferrite core. Their use for PD detection was conceived for stator winding applications in the 1960s [14]. The operating frequency of these sensors is constrained by the choice of the ferrite core. Looking at the data sheets of different ferrites, one can establish the bandwidth of the sensor by checking the frequency at which the real and imaginary permeability are equal. The choice of the bandwidth is generally weighted against the static permeability of the ferrite (a higher static permeability favors a higher detection sensitivity). In [15], detection was carried out using a spectrum analyzer with an upper frequency of 100 MHz. These sensors are highly directional sensors, thus they can be used to locate PD sources during offline tests. On the other hand, if used for monitoring, but if not properly oriented, they might not be able to couple any signal at all.

7.4.2 Magnetic Loops

Magnetic loops are placed in proximity of the ground leads as illustrated in Figure 7.14. This type of sensor was commercialized by Lemke, later described in detail and optimized by researchers at the University Carlos III in Madrid [16]. With AC PD detection applications, split core high-frequency current transformers (HFCTs) provide a better sensitivity in situations such as that described in Figure 7.14, but to detect PD within inverter-fed machines (Chapter 18), this technique can limit the problem of interference from the inverter (that prevents the use of HFCTs). These sensors might be useful also for PD detection if embedded in printed circuit boards (PCBs).

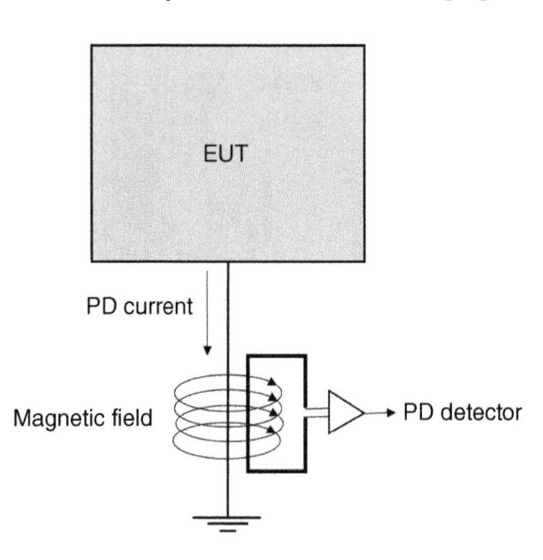

Figure 7.14 Magnetic loop operation principle.

7.4.3 Transient Earth Voltage (TEV) Sensors

The TEV sensors are used mostly for air-insulated switchgear (Chapter 14) and distribution-class (<60 kV) GIS. These sensors were pioneered by EA Technologies at the beginning of the 1990s [17, 18].

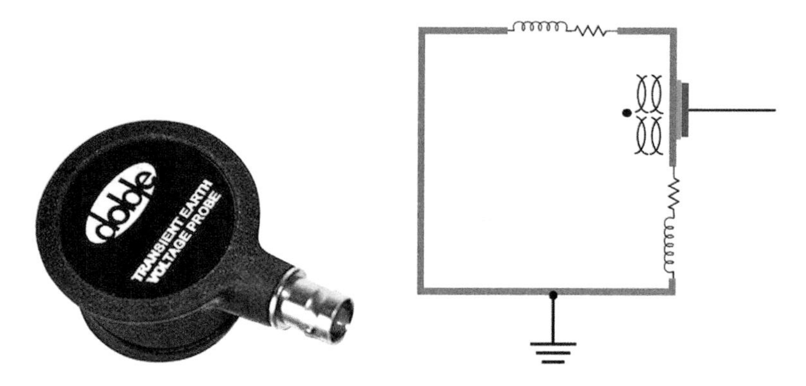

Figure 7.15 Doble TEV sensor (left) picture of the sensor. *Source:* [19]/Doble Engineering Company, (right) operation principle. *Source:* Photo courtesy of Doble.

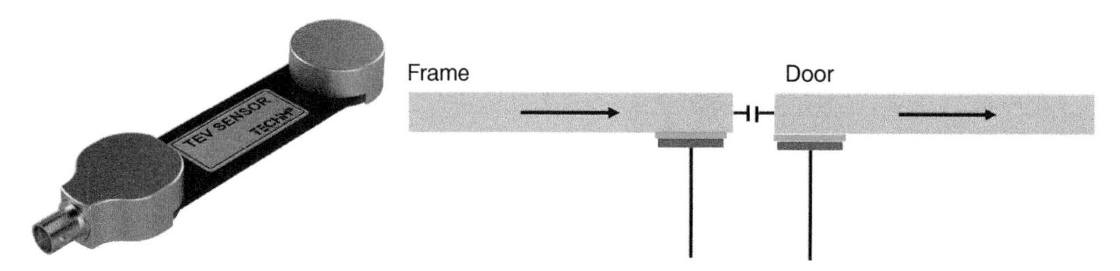

Figure 7.16 Techimp-Altanova TEV sensor (left) picture of the sensor. *Source:* [20]/Altanova Group, (right) operation principle. *Source:* Photo courtesy of Techimp-Altanova.

The principle of TEV sensors is shown in Figure 7.15, the PD source couples capacitively with the metallic enclosure of the switchgear. At the point where the capacitive coupling is maximum, the enclosure exhibits a higher potential with respect to (a remote) ground. The TEV sensor detects changes in the potential of the enclosure, including being able to detect partial discharge phenomena. Since the change in the potential of the enclosure is due to the PD current flowing to ground in the enclosure itself, the bandwidth of these sensors is normally limited to about 100 MHz.

In the same vein, but using a slightly different principle, Techimp-Altanova developed a sensor (Figure 7.16) that is made of two electrodes. These electrodes can be placed where the impedance of the enclosure increases, typically between the access doors and the enclosure frame. These sensors have also been used in GIS across the flange junctions.

7.4.4 Internal or Tank-Mounted UHF Sensors

For equipment with a metallic enclosure (GIS and transformers, for example), the UHF signals are typically detected by means of internal couplers, which are usually of similar design to capacitive couplers. The most classical example is the mushroom-like sensors developed for GIS (Chapter 13), shown in Figure 7.17, probably one of the first sensors developed for this purpose. A sensor having a similar logic was later developed for transformers, consisting of an UHF sensor that can be slid down the oil drain valve of transformer, as shown in Figure 7.18. Different sensors can be installed in the active part: horn type, helical type, or spiral type sensors [23]. In general, PD detection is performed over a large range of frequencies (from 100 to 2000 MHz or more).

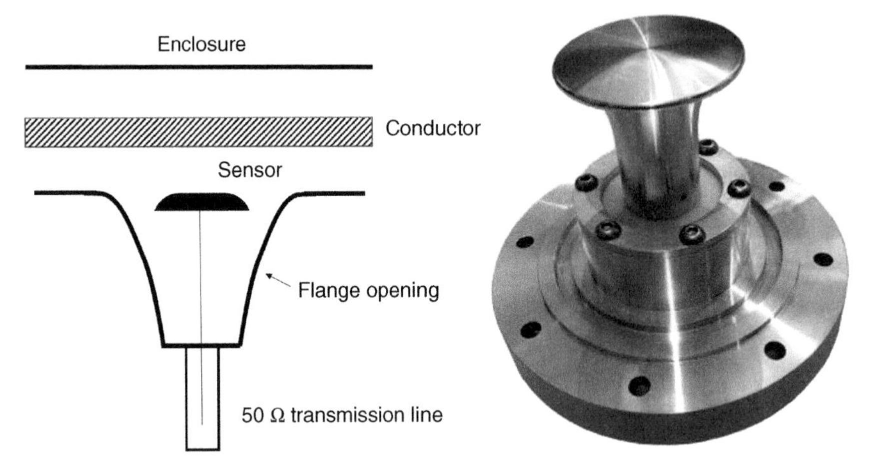

Figure 7.17 GIS electric field sensors: diagram (left) and example (right). *Source:* Uwe Riechert [21]/John Wiley & Sons (right).

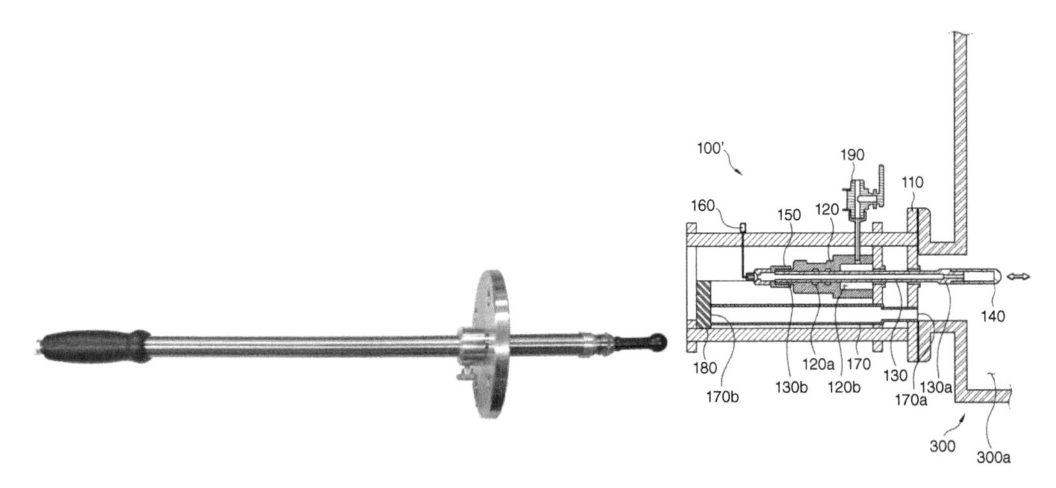

Figure 7.18 Transformer capacitive sensors: left UHF half-cone drain valve sensor. *Source:* Siegel [22]/IEEE, and right, principle of operation [23]. According to [23], different sensor types can be installed: horn type, helical type, or spiral type sensors. *Source:* Jun et al. [23].

Alternatively, it is possible to use external couplers on dielectric windows used for looking inside the equipment or hatches used to access the interior of enclosures, see examples for GIS and transformers in Figure 7.19 (and in Chapters 13 and 15).

7.4.5 Antennas

Practically all types of antennas proposed in the specialized literature have been evaluated for PD detection. The discussion below is far from exhaustive, since many other antenna (such as Vivaldi, conical, biconical, TEM, spiral, log-periodic, Hilbert, circular patch, loop, and bow tie) have been used to detect PD are not included.

7.4.5.1 Monopole

The monopole antenna is an omnidirectional antenna that can be manufactured in the simplest possible way by removing the screen from a short length of RG-213 (or other type of) coaxial cable.

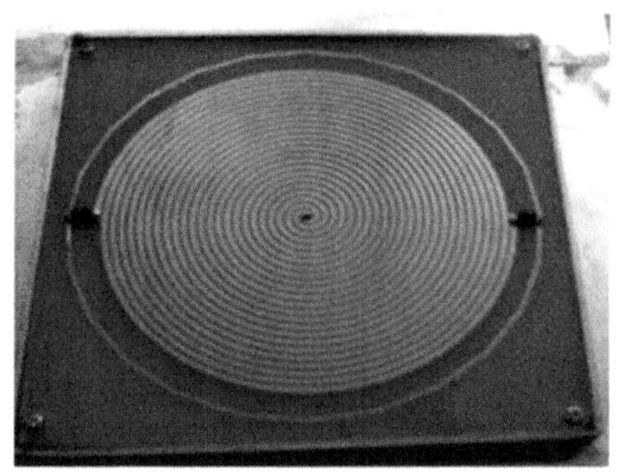

Figure 7.19 (left) GIS UHF flange sensor. *Source:* Zhang [24]/IEEE, (right) UHF PD dielectric window plate sensor (often indicated as spiral antenna). *Source:* Akbari et al. [10]/IEEE.

The exposed center conductor of the cable will correspond to half the wavelength of the EM wave at which the antenna is most sensitive. If the antenna is mounted perpendicular to a highly conductive plane (i.e. by passing the exposed center-conductor of the coaxial cable through a hole in the conductive ground-plane), it will behave as a dipole antenna but will have half the sensitivity. The monopole antenna, being extremely simple to manufacture and omnidirectional, has been used widely as a "quick and dirty" way of detecting partial discharges, but its gain is quite low.

7.4.5.2 Patch (Microstrip) Antenna

The patch antenna is one of the possible topologies of microstrip antennas, that is antennas realized on a PCB with the ground plane on one side of the PCB and the antenna's active elements printed on the other side. The non-grounded part is made with at least one patch of conductor, although arrays of patch antennas can be realized to achieve more complex sensors (see Figure 7.20). The patch antenna resonates with EM waves having wavelength (in free space)

$$\lambda_0 = \frac{L\sqrt{\varepsilon_r}}{0.49} \tag{7.3}$$

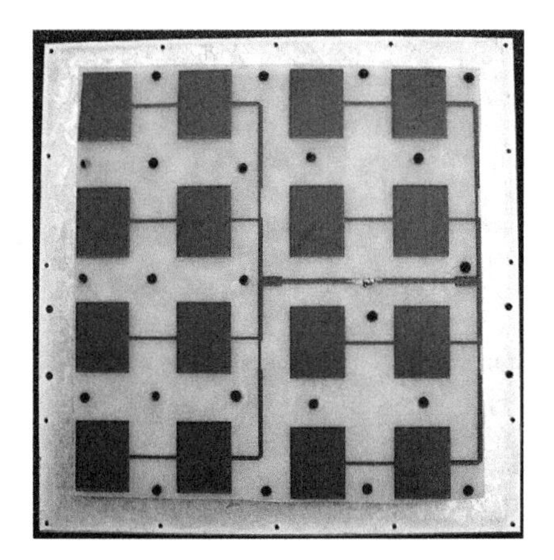

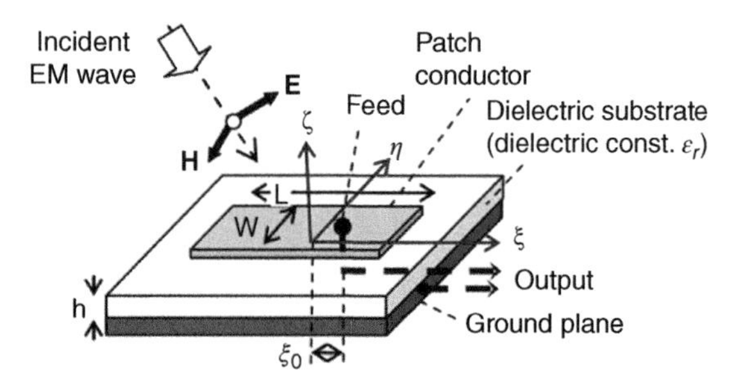

Figure 7.20 Patch antenna: (left) photograph of a 2.4 GHz patch antenna. *Source:* Serge Nueffer: https://en.wikipedia.org/wiki/Patch_antenna#/media/File:Antenne_patch_2.4_GHz.JPG. (right) Operation principle. *Source:* Shibuya et al. [25]/Wikimedia Commons/CC BY-SA 3.0.

where L shown in Figure 7.20. Typically, the characteristic impedance is on the order of a few hundreds of Ohms when the signal is collected at the PCB edge opposite to the direction of the incident wave. Therefore, arrays of patch antennas can be used to adapt the output to a 50-Ω transmission line. The antenna is highly directional and can thus be used for aiming at PD sites to locate them. Due to their structure and size, patch antennas are normally suited to detect PD signals with broadband frequency content. In [25], their use as a tool to detect PDs in the inter-turn insulation of inverter-fed machines is advocated, whereas their successful application in GIS is discussed in [26].

7.4.5.3 Horn Antenna

The use of a horn antenna for PD detection is described in detail in [27]. The antenna consists of an enclosure that acts as a waveguide consisting of pitcher-shaped copper combined with a dielectric lens as shown in Figure 7.21. The shape of the copper pitchers was investigated finding that an exponential profile was the best solution. The antenna was investigated for both near- and far-field coupling, finding that the antenna is sensitive starting from around 500 MHz, going on to form an excellent capacitive response in the near-field region over the 500–1400 MHz range of frequencies. The horn antenna has been used to detect PD in a medium-voltage power transformer by sensing the EM waves radiated by the bushing apertures on the tank of the transformer [29].

7.4.5.4 Stator Slot Couplers

The stator slot couplers (SSC, Section 16.6.2) yields an output pulse from each end whenever the electromagnetic wave produced by a PD pulse occurring within at stator slot propagates along the length of the SSC. The characteristic impedance of the 50 cm long SSC is 50 Ω, in order to match the 50 Ω impedance of the two output cables, making it non-resonant. The dual-port nature of the SSC permits determination of the direction of PD pulse travel using instrumentation that can measure which end of the SSC detected the signal first. The SSC should not be considered as either a purely capacitive, or inductive, coupler, rather, it can be understood in terms of distributed parameters. It has been primarily used in hydrogen-cooled turbine generator stator windings (Figure 7.22).

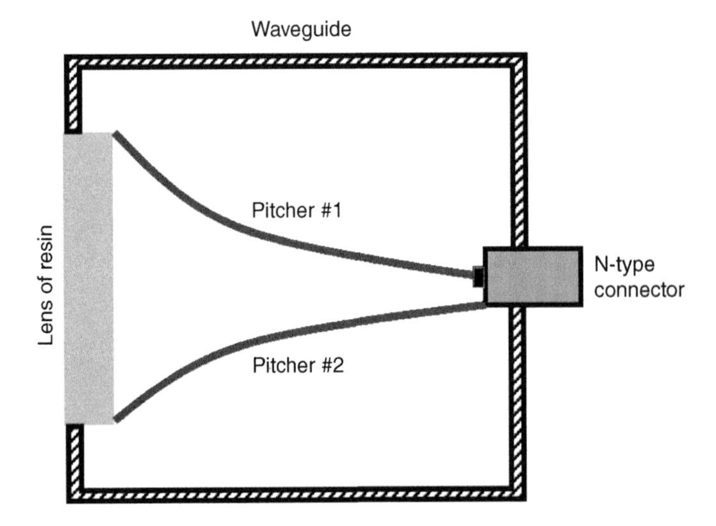

Figure 7.21 Horn antenna structure.

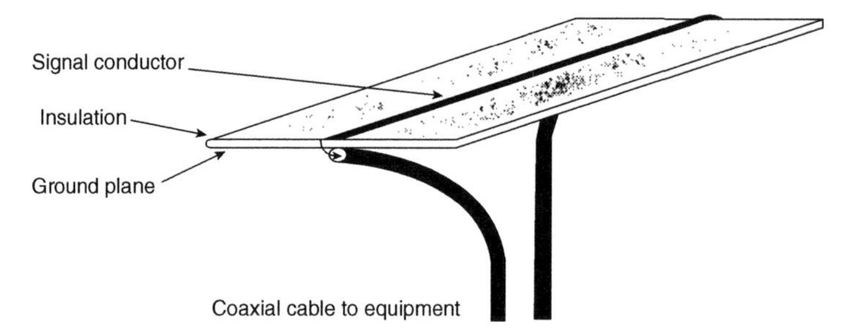

Figure 7.22 Simplified schematics of SSC. *Source:* Sedding et al. [4].

7.5 Measuring Instruments

The PD signal can be represented as a waveform with its amplitude expressed in mV if an oscilloscope is used. When using RF spectrum analyzers, signal amplitudes are normally expressed in units of RF power referenced to a standard impedance (usually 50 Ω) such as dBm (0 dBm = 1 mW of RF power into 50 Ω), dBuV (1 uV at 50 Ω), and so on. The time evolution and the features of the spectrum are controlled by selecting the bandwidth and sweep parameters whose interdependence and effect on the displayed signal are extremely complex and outside the scope of this book to explain in detail.[2] As an example, one might be inclined to assume that very narrow resolution bandwidths will lead to an accurate determination of the PD signal spectrum. However, this may not be the case because PD is by nature a (very short) pulsed signal; indeed, although counterintuitive, narrow resolution bandwidths will generally not work to reveal PD. In general, resolution bandwidths of several (e.g. 3, 5) MHz are preferable. However, the spectral lines formed by each individual PD event will be very much nonstationary; therefore, it is normal practice to use the "max-hold" mode to accumulate a so-called envelope spectrum. After multiple sweeps, the envelope of the maxima of the energies observed in each frequency band will be available.

For RIV measurements, it is customary to use a quasi-peak system to observe the evolution of the signal magnitudes. This is achieved by sending the signal to an analog peak detector. The time constant of the RC circuit is tuned in a way that highly repetitive signals will result in a larger quasi-peak output reading.

Except in tightly controlled, purposely designed laboratory test set-ups, calibration of PD measurements using RF techniques in terms of charge (pC) is impossible because PD-induced EM waves can excite the equipment's inherent internal resonances in different ways depending on the exact position of the PD site and the sensor's position (see for example GIS in Figure 7.23). Indeed, PD-induced EM waves can arrive from unknown locations with wide differences in apparent signal strength. Therefore, as discussed in IEC 62478 [6], the apparent charge in pC should never used to report PD measurements in the VHF or UHF range and PD severity assessment is often based on trending the PD measurements or empirical thresholds derived from experience on a specific device.

7.6 Performance and Sensitivity Check

As discussed in the previous section, VHF/UHF PD measurements cannot be calibrated in terms of charge (pC). What can be done in practice are performance and sensitivity checks. They are summarized and explained in Table 7.3, and outlined in IEC 62478 [6]. In most cases, the

2 It is recommended the reader refer to references such as the HP/Agilent/Keysight application note 150 and take a training course to learn how to use such instruments.

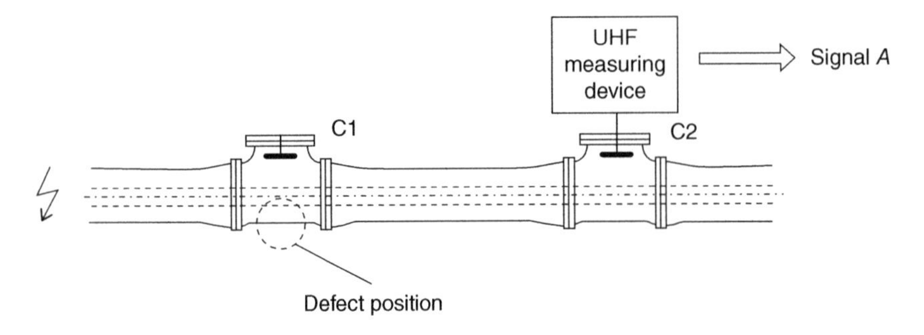

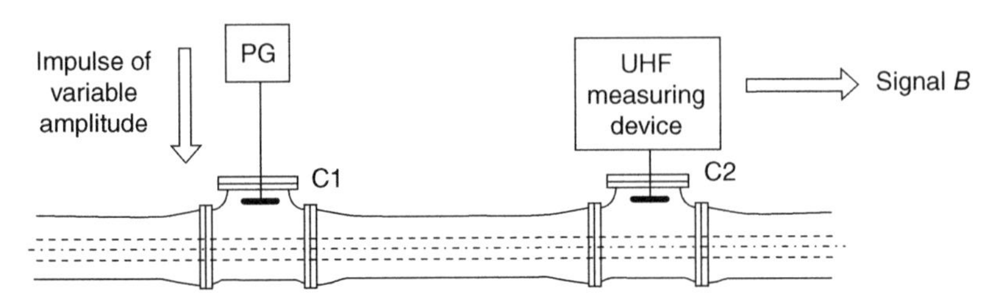

Figure 7.23 Principle of sensitivity check: (top) detection at C2 during operation, (bottom) detection at C2 during sensitivity check. *Source:* Schichler et al. [30]. See also Section 13.8.

Table 7.3 Performance and sensitivity checks of VHF/UHF detection systems.

Type of check	Equipment	Type	Note	Units
Performance check	Transformers	Single port	Injection of artificial pulses in the same sensor	(V)
	Stator windings	Dual port	Injection of artificial pulses in other sensors	(V)
Sensitivity check	Cables	Laboratory setup	PD source with known apparent charge	(pC)
	GIS		Injection of artificial pulses	(V)
	Transformers	Onsite	Sensitivity check of sensor arrangement is demonstrated as a performance check (single port or, better, dual port)	(V)

dual-port sensitivity check is the best option. In practice, a special impulse generator capable of working in the UHF range is needed. A preliminary measurement must first be performed to determine the emission spectrum and relative signal strength from an actual PD source whose charge value is known by using the method in IEC 60270 (calibrated). The sensitivity check can then be performed injecting an impulse with the exact same characteristics into the system being assessed.

As an example, Figure 7.23 shows the sensitivity check procedure used for GIS/GIL. First, a defect generating PD at a known (calibrated according to IEC 60270) charge level (e.g. a 5 pC hopping particle) is set up in the GIS enclosure containing sensor C1. The voltage is raised, and the PD signal is measured side-by-side using a UHF detector at sensor C2, for instance an RF spectrum analyzer whose spectrum at C2 is recorded. In the next step, the high voltage is turned off, and a fast risetime/high-frequency artificial pulse is injected at the sensor C1. The output voltage level of the pulse generator is adjusted to the level that best matches the spectrum due to the 5 pC particle PD. Figure 7.24 shows that, setting the voltage of the generator to 2 V, a good match was reached.

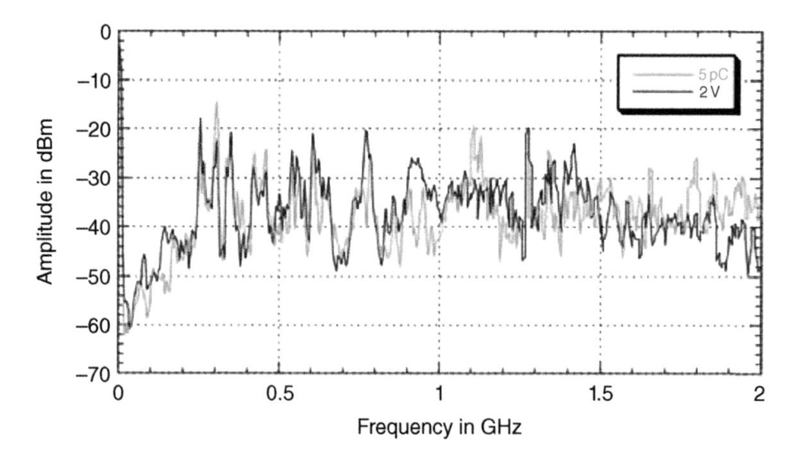

Figure 7.24 Comparison of the spectra in the UHF range for a 5 pC pulse and for a 2 V pulse injected by a pulse generator suitable for sensitivity check in the UHF range. *Source:* Schichler et al. [30].

Thus, a 2 V pulse will be used on site to verify the sensitivity of the detection system at each sensor location. See also Section 13.8.2.

7.7 PD Source Location

Localization of PD sources using VHF/UHF systems is based on measurements performed in the time domain. The "time of flight" or "time of arrival" approach (Section 8.4.2) is used to locate PD sites in power cables (Section 12.14.2), GIS (Section 13.10.5), and isolated phase bus (Section 14.8) and used as an interference separation method in rotating machines (Section 16.9.3.1). The arrival times of the signals related to a PD event at different sensors are used for this purpose (any delay due to different coaxial cable lengths should be corrected; if possible and to avoid confusion, it is strongly recommended to use all cables of the same length). Looking for instance at Figure 7.10 the PD pulses arrive at sensor #1 first, then at sensors #2 and #4 at about the same time, eventually at sensor #3. This evidence indicates that the PD source is closer to #1 than to #2 and #4, and that #3 is the sensor most distant from the PD source. On theoretical grounds, if the PD signal develops in space as a spherical wave, with a propagation velocity equal to

$$v_p = \frac{c}{\sqrt{\varepsilon_r}} \tag{7.4}$$

(ε_r being the permittivity of the medium in which the wave propagates, and c is the speed of light in vacuum), one can calculate the coordinates of the PD source starting from the following system of equations:

$$\begin{aligned} t_1 - t_2 &= \frac{D(\boldsymbol{P}_1 - \boldsymbol{P}_s)}{v_p} - \frac{D(\boldsymbol{P}_2 - \boldsymbol{P}_s)}{v_p} \\ t_1 - t_3 &= \frac{D(\boldsymbol{P}_1 - \boldsymbol{P}_s)}{v_p} - \frac{D(\boldsymbol{P}_3 - \boldsymbol{P}_s)}{v_p} \\ t_1 - t_4 &= \frac{D(\boldsymbol{P}_1 - \boldsymbol{P}_s)}{v_p} - \frac{D(\boldsymbol{P}_4 - \boldsymbol{P}_s)}{v_p} \end{aligned} \tag{7.5}$$

where t_i $i = 1, ..., 4$ are the arrival times of the signals at the PD sensor, $\boldsymbol{P}_i$ $i = 1, ..., 4$ are the vectors representing the position of the sensors, $\boldsymbol{P}_s$ represents the position of the source, and $D(\boldsymbol{A})$ represents the Euclidean length of a vector $\boldsymbol{A}$. In practice, one must solve the following system of equations:

$$\begin{aligned}
&(x_1 - x_s)^2 + (y_1 - y_s)^2 + (z_1 - z_s)^2 - (x_2 - x_s)^2 - (y_2 - y_s)^2 - (z_2 - z_s)^2 = (v_p(t_1 - t_2))^2 \\
&(x_1 - x_s)^2 + (y_1 - y_s)^2 + (z_1 - z_s)^2 - (x_3 - x_s)^2 - (y_3 - y_s)^2 - (z_3 - z_s)^2 = (v_p(t_1 - t_3))^2 \\
&(x_1 - x_s)^2 + (y_1 - y_s)^2 + (z_1 - z_s)^2 - (x_4 - x_s)^2 - (y_4 - y_s)^2 - (z_4 - z_s)^2 = (v_p(t_1 - t_4))^2
\end{aligned} \tag{7.6}$$

This method has been applied in air-insulated substations (where the assumption that the EM wave is spherical seems logical) but also to transformers. In the latter case, some inaccuracies can arise because of the complicated internal geometry and the different materials along the different signal paths that can strongly affect the electromagnetic wave propagation.

In GIS/GIL, the problem is solved following the same approach, but as the system is one-dimensional, it suffices to observe at which sensor the signal arrives first (Section 13.10.5).

References

1 Golinski, J., Malewski, R., and Train, D. (1979). Measurements of RIV on large EHV apparatus in high voltage laboratory. *IEEE Transactions on Power Apparatus and Systems* PAS-98 (3): 817–824. https://doi.org/10.1109/TPAS.1979.319294.

2 Petterson, K., Baumgartner, R., Lanz, W. et al. (1992). Development of methods for on site testing as a part of the total quality assurance testing program for GIS. *International Conference on Large High Voltage Electric Systems* 1: 23–202.

3 Reid, A.J. and Judd, M.D. (2012). Ultra-wide bandwidth measurement of partial discharge current pulses in SF6. *Journal of Physics D: Applied Physics* 45 (16): 165203. https://doi.org/10.1088/0022-3727/45/16/165203.

4 Sedding, H.G., Campbell, S.R., Stone, G.C., and Klempner, G.S. (1991). A new sensor for detecting partial discharges in operating turbine generators. *IEEE Transactions on Energy Conversion* 6 (4): 700–706. https://doi.org/10.1109/60.103644.

5 Okubo, H., Hayakawa, N., and Matsushita, A. (2002). The relationship between partial discharge current pulse waveforms and physical mechanisms. *IEEE Electrical Insulation Magazine* 18 (3): 38–45. https://doi.org/10.1109/MEI.2002.1014966.

6 IEC TS 62478 (2016). High voltage test techniques – measurement of partial discharges by electromagnetic and acoustic methods.

7 Djuknic, G.M. (2003). Method of measuring a pattern of electromagnetic radiation. US Patent 6657596.

8 Okabe, S., Kaneko, S., Yoshimura, M. et al. (2009). Propagation characteristics of electromagnetic waves in three-phase-type tank from viewpoint of partial discharge diagnosis on gas insulated switchgear. *IEEE Transactions on Dielectrics and Electrical Insulation* 16 (1): 199–205. https://doi.org/10.1109/TDEI.2009.4784568.

9 BCAR (2020). Coax cable attenuation details. https://web.archive.org/web/20200220132743/http://www.bcar.us/cablespec.htm (accessed 21 February 2022).

10 Akbari, A., Werle, P., Akbari, M., and Mirzaei, H.R. (2016). Challenges in calibration of the measurement of partial discharges at ultrahigh frequencies in power transformers. *IEEE Electrical Insulation Magazine* 32 (2): 27–34. https://doi.org/10.1109/MEI.2016.7414228.

11 Romano, P., Imburgia, A., and Ala, G. (2019). Partial discharge detection using a spherical electromagnetic sensor. *Sensors* 19 (5): https://doi.org/10.3390/s19051014.

12 Di, S.A, Candela, R., Giaconia, G.C., and Fiscelli, G. (2011). Portable partial discharge detection device. US2011156720A1.

13 Denissov, D., Kohler, W., Tenbohlen, S. et al. (2008). Wide and narrow band PD detection in plug-in cable connectors in the UHF range. *2008 International Conference on Condition Monitoring and Diagnosis*, Beijing (21–24 April 2008), pp. 1056–1059. IEEE. https://doi.org/10.1109/CMD.2008.4580464.

14 Dakin, T.W., Works, C.N., and Johnson, J.S. (1969). An electromagnetic probe for detecting and locating discharges in large rotating-machine stators. *IEEE Transactions on Power Apparatus and Systems* PAS-88 (3): 251–257. https://doi.org/10.1109/TPAS.1969.292314.

15 Ozaki, T., Abe, K., and Umemura, T. (1992). Partial discharge detection using ferrite antenna. *Conference Record of the 1992 IEEE International Symposium on Electrical Insulation*, Baltimore, MD (7–10 June 1992), pp. 371–374. IEEE. https://doi.org/10.1109/ELINSL.1992.246986.

16 Robles, G., Martinez-Tarifa, J.M., Mó, V. et al. (2009). Inductive sensor for measuring high frequency partial discharges within electrical insulation. *IEEE Transactions on Instrumentation and Measurement* 58 (11): 3907–3913. https://doi.org/10.1109/TIM.2009.2021239.

17 Mackinlay, R.R. (1990). Discharge measurements on high voltage distribution plant. *IEE Colloquium on Developments Towards Complete Monitoring and In-Service Testing of Transmission and Distribution Plant*, Chester (31 October 1990), p. 9. IEEE.

18 Jennings, E. and Collinson, A. (1993). A partial discharge monitor for the measurement of partial discharges in a high voltage plant by the transient earth voltage technique. *1993 International Conference on Partial Discharge*, Canterbury (28–30 September 1993), pp. 90–91. IEEE.

19 DOBLE TEVProbe_web.pdf. https://www.doble.com/wp-content/uploads/TEVProbe_web.pdf (accessed 2 February 2022).

20 Altanova TEV antenna – Altanova group. https://www.altanova-group.com/en/products/partial-discharge-tests/sensors-and-accessories/tev-antenna (accessed 21 February 2022).

21 Kurrer, R. and Feser, K. (1998). The application of ultra-high-frequency partial discharge measurements to gas-insulated substations. *IEEE Transactions on Power Delivery* 13 (3): 777–782. https://doi.org/10.1109/61.686974.

22 Siegel, M., Beltle, M., Tenbohlen, S., and Coenen, S. (2017). Application of UHF sensors for PD measurement at power transformers. *IEEE Transactions on Dielectrics and Electrical Insulation* 24 (1): 331–339: https://doi.org/10.1109/TDEI.2016.005913.

23 Jun, J.H., Son, H.K., and Yun, Y.J. (2013). Apparatus for detecting partial discharge. KR101264548B1.

24 Zhang, J.-N., Zhu, M-X., Liu, Q. et al. (2016). Design and development of internal UHF sensor for partial discharge detection in GIS. *2016 International Conference on Condition Monitoring and Diagnosis (CMD)*, Xi'an (25–28 September 2016), pp. 709–712. IEEE. https://doi.org/10.1109/CMD.2016.7757922.

25 Shibuya, Y., Matsumoto, S., Tanaka, M. et al. (2010). Electromagnetic waves from partial discharges and their detection using patch antenna. *IEEE Transactions on Dielectrics and Electrical Insulation* 17 (3): 862–871. https://doi.org/10.1109/TDEI.2010.5492260.

26 Hikita, M., Ohtsuka, S., Teshima, T. et al. (2007). Electromagnetic (EM) wave characteristics in GIS and measuring the EM wave leakage at the spacer aperture for partial discharge diagnosis. *IEEE Transactions on Dielectrics and Electrical Insulation* 14 (2): 453–460. https://doi.org/10.1109/TDEI.2007.344624.

27 Tozzi, M. (2010). *Partial Discharges in Power Distribution Electrical Systems: Pulse Propagation Models and Detection Optimization*. Bologna: University of Bologna. https://amsdottorato.unibo.it/2308/1/Tozzi_Marco_Tesi.pdf.

28 Altanova UHF horn antenna – Altanova group. https://www.altanova-group.com/en/products/partial-discharge-tests/sensors-and-accessories/horn-antenna (accessed 21 February 2022).

29 Tozzi, M., Saad, H., Montanari, G.C., and Cavallini, A. (2009). Analysis on partial discharge propagation and detection in MV power transformers. *2009 IEEE Conference on Electrical Insulation and Dielectric Phenomena*, Virginia Beach, VA (18–21 October 2009), pp. 360–363. IEEE. https://doi.org/10.1109/CEIDP.2009.5377860.

30 Schichler, U., Koltunowicz, W., Gautschi, D. et al. (2016). UHF partial discharge detection system for GIS: application guide for sensitivity verification: CIGRE WG D1.25. *IEEE Transactions on Dielectrics and Electrical Insulation* 23 (3): 1313–1321. https://doi.org/10.1109/TDEI.2015.005543.

8

PD Measurement System Instrumentation and Software

8.1 Introduction

This chapter attempts to give PD system users an appreciation of the technologies (and trade-offs) used by developers of instruments specifically designed to measure PD. Chapter 6 discussed the basic principles of electrical PD detection using either capacitive or HFCT sensors, together with the other major components that are part of an electrical PD detection system. In this chapter, we delve into the electronic hardware for measuring the PD in more detail and the main features of the software that displays and/or analyzes the measurement results. All commercial PD detection systems made since 2000 use digital electronics and display the measurements using software, and this is what will be presented here. For readers interested in analog PD detection systems, please refer to the references in Chapter 1, consult Kreuger's latest book [1] or the CIGRE Technical Brochure mainly authored by Lemke [2].

Virtually every PD system vendor has their own proprietary hardware implementation, so the discussion in Section 8.3 is somewhat general, but applies to all frequency ranges from LF to UHF as defined in IEC 60270 and IEC 62478, as well as Section 7.1. As a minimum, the purpose of the PD measuring system is to count each detected pulse above a threshold, determine its magnitude, and record it with the associated phase position in the AC cycle.

Software is then used to create and display the phase-resolved PD (PRPD) plots and determine other derived quantities such as peak PD magnitude, quasi-peak magnitude, PD current, and many other calculated quantities based on the basic information. Instruments working in the HF, VHF, and UHF (IEC 62478) range usually also record the polarity of each PD pulse, although the pulse polarity is usually lost with narrow band LF (IEC 60270) detectors. The pulse magnitude, count rate, and polarity information must be displayed in a way that is useful to the user of the PD measuring system.

Many PD measuring systems also come with calibrators to yield results in terms of charge (IEC 60270 instruments) or pulse generators to perform sensitivity and hardware checks (IEC 62478 instruments). In addition, provisions are sometimes made for disturbance suppression by either hardware methods or using post-processing (i.e. after the PD pulses have been digitally stored). This is also discussed in Section 9.3.

Practical Partial Discharge Measurement on Electrical Equipment, First Edition. Greg C. Stone, Andrea Cavallini, Glenn Behrmann, and Claudio Angelo Serafino.

8.2 Frequency Range Selection

The most fundamental specification of PD system instrumentation is its measurement frequency range. Most commercial PD systems work in one or more of the following ranges:

- Low frequency, LF, 30 kHz to 1 MHz
- High frequency, HF, 3–30 MHz
- Very high frequency, VHF, 30–300 MHz
- Ultrahigh frequency, UHF, 0.3–3 GHz

The electronics required depends on the frequency range in which it will work. LF electronics will not be able to respond to very short pulses in the VHF range and thus will be less sensitive. Conversely, VHF electronics may produce more electronic (thermal) noise and cannot measure the apparent charge of the PD (Section 6.5).

LF PD systems are mainly used in factory quality assurance testing of HV equipment like power cable, transformers, stator coils, switchgear, etc. and measure the PD in apparent charge (pC, nC). Most brands of LF systems measure the PD in a narrow band (typically about 10 kHz) range, as well as the wideband (typically a few hundred kHz) range below 1 MHz. Some models of LF PD systems permit choosing center frequencies within the LF range with adjustable lower and upper cutoff frequencies to avoid certain frequencies that may have high interference levels and/or to choose frequencies where the system is more sensitive to PD in the test object. The latter is usually most applicable for test objects that have some inductance – for example stator windings [3]. The filters can be implemented using analog electronics or in firmware for digitized signals.

HF, VHF, and UHF PD systems are mainly used with onsite/off-testing and for online monitoring (Section 1.3), since they are less sensitive to PD-like pulse interference from the power system, and interference suppression beyond simple filtering tends to be easier to implement. Usually, the frequency range is set by the PD sensor and the upper frequency limits of the electronics. For HF, VHF, UHF systems additional filtering may not be implemented, i.e. they are inherently "wide-band," or what some have called "ultra-wide band."

8.3 PD Detector Hardware Configurations

As described in Section 1.5.3, the early days of digital detection used simple comparators to determine the number of pulses in each magnitude window [4, 5]. Later, Austin and James [6] and Kelen [7] combined early analog to digital converters (ADCs) and the digital computers available in the 1970s to store the magnitude/number/phase position/polarity information of each pulse. For example, Austin used a peak sample and hold device with an 11-bit ADC that took 30 μs to convert the pulse magnitude to a digital memory location in a PDP 8 computer.

Today, commercially available electronics is very fast with ADCs that can sample every 0.5 ns (i.e. a 2 GHz sampling rate) and field-programmable gate arrays (FPGAs) that can process the different pulse categories at the same speed using flexible firmware to define the pulse categories. This has allowed direct digital processing for PD pulses measured even in the UHF range.

The types of digital methods used in PD systems can be broadly classed into three basic approaches:

- systems where the input pulse from the detection impedance Z_{mi} in Figures 6.1 and 6.2 (sometimes called a quadripole if there is a provision for capturing the AC voltage) is filtered and then integrated in an analog fashion into apparent charge (Section 6.6), which is then digitized using a peak, sample, and hold (PSH) ADC (Figure 8.1);

- systems where the input signal magnitude is immediately digitized using an ADC, and in parallel an FPGA processes the pulses (Figure 8.2);
- systems where the entire input signal is immediately digitized with an ADC that dumps the entire pulse waveform into memory (as a digital oscilloscope does) for later analysis using post-processing firmware or software (Figure 8.3).

For some UHF systems, the input signal is first demodulated (frequency downshifted) using analog electronics (Figure 8.4). Usually the amplitude modulation (am) demodulator design is set for the LF range. All pulse polarity and waveshape information is lost with this approach, and the peak magnitude cannot formally be calibrated into apparent charge, even if the rest of the electronics is in the LF range and compliant with IEC 60270. The rest of the PD system hardware can be any of the above systems.

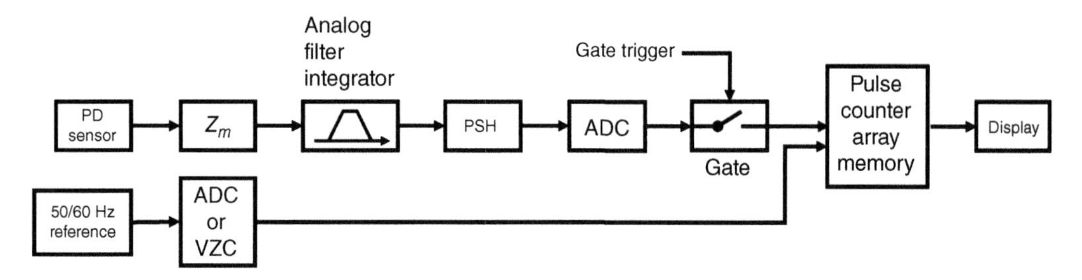

Figure 8.1 Block diagram of a mixed analog and digital detector.

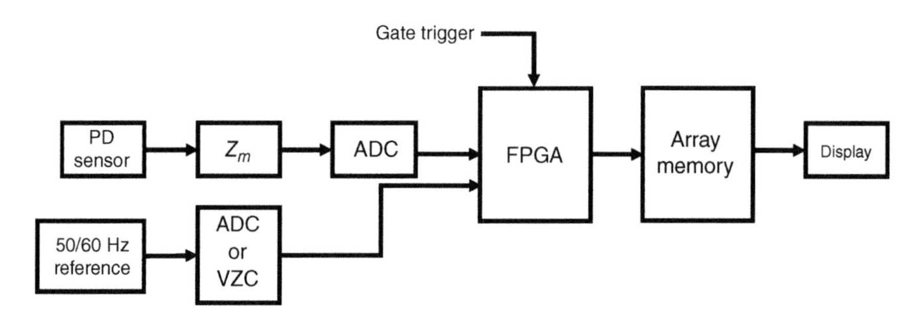

Figure 8.2 Block diagram of a typical all-digital PD detector where the origin of the pulse is determined at the time of acquisition.

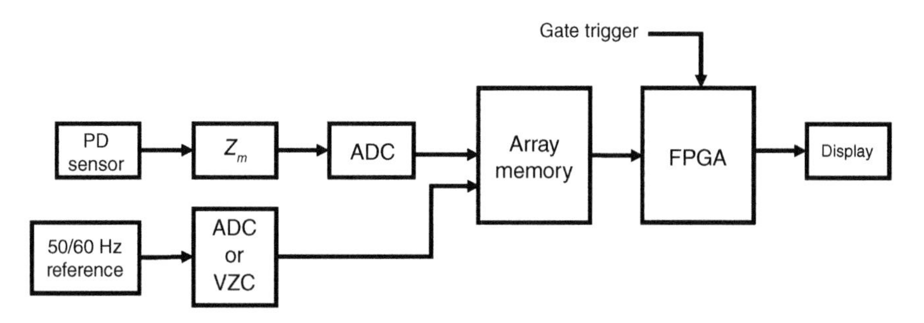

Figure 8.3 All digital detector where the waveform of each pulse is stored, and after acquisition of all pulses, the stored data is analyzed.

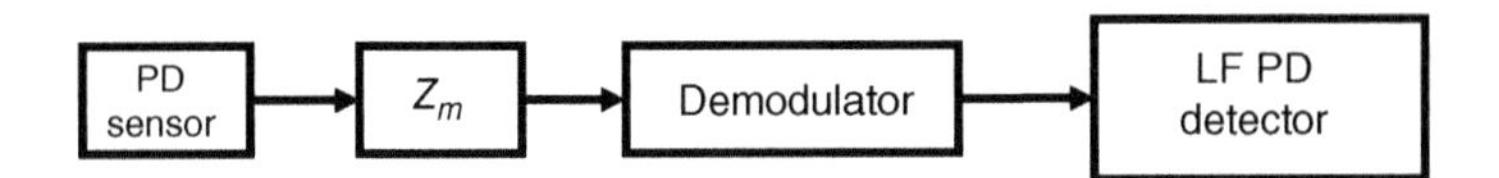

Figure 8.4 Block diagram of a UHF PD sensor that can be connected to an LF or HF detector, by using a demodulator (low-pass filter).

Each of these systems will be presented below, but first we discuss other concepts that are important in most PD measurement systems.

The cost of the electronics in modern digital PD systems is usually a small percentage of the actual selling price of the system. The major cost of PD systems is the R&D effort that went into creating the firmware embedded in FPGAs or other electronics that processes the pulses. Major vendors keep this intellectual property close to themselves, and rarely publish or patent the key details of their systems. This is to prevent "clone-makers" taking advantage of other's R&D efforts. As a result, the hardware descriptions outlined in this section are necessarily somewhat vague due to the lack of published information.

8.3.1 Minimum Threshold and Processing Time

Most digital PD systems require each recorded input pulse to be above a certain magnitude threshold. The reason is to prevent processing very small-magnitude pulses, which are likely to be internal electronic noise or minor interference detected by the PD sensor. The threshold is dependent on the measurement scale. For example, if one is trying to measure 1 pC discharges in a power cable, the threshold would normally be 0.1 pC or lower. Measuring 10 nC PD in rotating machines would require a threshold >100 pC. For most vendors, this threshold is automatically set by the instrumentation, depending on the PD magnitude measurement scale. However, some vendors allow the threshold to be adjusted by the user. In this case, users should be aware that adjusting the threshold too low may desensitize the system to important PD pulses. For measurements done on the same test object over time, the same threshold should be used to ensure comparable results.

For the systems in Figures 8.1 and 8.2, when the threshold is exceeded, the input electronics usually measures (or records) any signal appearing at the input for a fixed time duration. The duration is set by the expected time it takes for PD or interference pulses to decay below the threshold, or it can be a preset time based on the vendor's experience. For narrow band measurements in the LF range, the PD pulse may oscillate for several tens of microseconds (Figure 6.17). For VHF systems, the damped oscillation associated with each PD pulse may last only a microsecond (Figure 6.16). The time window during which the measurement is acquired is sometimes called the processing time, since in the early days of digital systems, the electronics was slow enough that it was longer than the time for the pulse to decay below the threshold.

Clearly the processing time has an impact on the number of PD pulses that can be measured by the system. If the processing time is 1 μs (a VHF or UHF detector), then in principle one million pulses can be measured per second. If the processing time is 50 μs, then the maximum pulse repetition rate will be 20,000 pulses per second, i.e. 20 kpps.

In UHF and VHF systems, the concept of processing time may be extended. In these frequency ranges, when a PD event occurs, the first pulse is the true response to the PD, and all the subsequent oscillations are due to pulse reflections and local oscillations that depend on the test object's construction. It is possible for the subsequent oscillations to be higher magnitude than the first response. This error can easily be averted by shutting down processing after the initial response. This shut down or "deadtime" would normally be the expected duration of the pulse including all oscillations.

Generally, the duration of the PD pulses in power cables, GIS, transformers, and stator windings will be different, and thus the processing time may be different. Users should be aware that PD systems intended for one type of HV equipment may not respond properly when used on a different type of HV equipment due to these factors.

8.3.2 AC Voltage Measurement and Synchronization

All conventional PD measuring systems require some way to detect the test object AC high voltage. The AC voltage is usually 50 or 60 Hz from the power system. However, 0.1 Hz (also called very low frequency or VLF) is sometimes used for high-capacitance test objects like long power cables or stator windings. For power transformer testing, AC supplies operating up to 400 Hz are usually used to energize the test object.

AC voltage detection is required to classify each pulse as to when it occurred with respect to the AC cycle phase angle. This synchronization facilitates the creation of PRPD plots (Section 8.7.3). For some detectors, it is needed to determine the AC voltage across the test object at the time of each PD pulse. As discussed later, this is needed for calculation of the power in each PD pulse (and the average power due to PD activity), as well as facilitating the measurement of the PD inception voltage (PDIV) and the PD extinction voltage (PDEV) in offline PD tests.

If there is no AC voltage synchronization signal, PD systems that produce PRPD plots will not function, and an error message will be displayed.

Some PD systems intended for offline PD testing will measure the instantaneous test voltage at the time of each PD event. These will require a conventional ADC to measure the voltage from the AC voltage source (Figures 8.1–8.3). However, many online PD measuring systems will not measure the actual voltage at each PD event, but just track when the PD pulse occurs with respect to the AC cycle (usually the positive voltage zero crossing). The latter option implies that accurate measurement of the applied voltage, which involves the installation of special voltage dividers or connections to potential transformers, is not needed. Instead, the synchronization signal can be obtained from a capacitive PD sensor (where used), and a simple voltage zero crossing (VZC) detector can be used instead of a full ADC.

For offline tests, most conventional PD systems will employ a capacitive voltage divider or resistive voltage divider that is usually a part of the variable voltage AC supply. This makes it easy to measure the AC rms voltage needed for the PDIV and PDEV, as well as the positive zero crossing to obtain a synchronizing signal for the start of each AC cycle, and thus the AC phase angle of each PD event. Alternatively, the AC voltage may be derived from the PD detection capacitor in a quadripole that separates the PD signal from the AC signal [2].

For online PD tests, the AC voltage may come from a variety of sources. Details for each type of HV equipment are presented in Chapters 12–16. However, it is usually not feasible to use a separate capacitive or resistive voltage divider since this would require their permanent installation. Asset managers and plant managers are very reluctant to add unnecessary equipment to the power system since it can fail. Accurate AC voltage measurement can be made from metering outputs of any potential (voltage) transformers that are already installed for protection or metering reasons. The only problem with this is that wires would have to be installed between the potential transformers and the PD measuring system at the HV equipment under test. In many plants and stations, this cable run can be hundreds of meters long and can cost more than the PD monitoring system itself to permanently install them!

If online capacitors are used as PD sensors, there will be a small 50/60 Hz current that will flow through the capacitors. For example, if a 1 nF PD sensor is used on an 11 kV (phase-to-phase) 50 Hz network, 2 mA of AC will flow, although it will be phase shifted 90°. This current flowing through a 1 kΩ resistance will produce 2 V rms, which is plenty of signal for AC synchronization. The AC voltage measured in this way my not be accurate due to the higher order harmonics present (recall the capacitor and resistor form a high-pass filter that accentuates higher frequencies). Using the PD coupling capacitance as an AC synchronization source will require clever changes to the PD detection impedance (Z_{mi} in Figures 6.1 and 6.2) to make sure that the 1 kΩ load does affect the PD magnitude or detection frequency; or conversely Z_{mi} does not short out the 1 kΩ resistance. Figure 6.5 shows one implementation of circuit that produces both a PD and AC synchronization signals. Such a device is sometimes called a quadripole, since it is a four-terminal device.

If an HFCT is used as the PD sensor (Section 6.3.3), then an AC synchronization signal cannot be obtained from it (HFCTs usually use a ferrite magnetic core, and these are insensitive to 50/60 Hz). In this case, it is best to obtain the synchronization signal from a potential transformer. Alternatively, if the capacitance to ground of the test object is large, for example a power cable where the HFCT is placed around the lead from the power cable sheath to ground, a power frequency CT (using a conventional steel core) or Rogowski coil (Figure 6.8) may provide enough signal for synchronization.

As a last resort, a 50/60 Hz synchronization signal can be obtained from the 120/220 V AC supply or an antenna in the plant. However, the phase relationship to the actual AC voltage is usually unknown due to the different types of transformers used between the high-voltage system and the plant's 120/220 AC voltage.

8.3.3 Combined Analog–Digital Systems

Figure 8.1 shows the block diagram of a measurement system that mixes analog and digital PD measurement and is particularly suitable for LF (i.e. IEC 60270) detection. The front end is a traditional detection impedance (Z_{mi}), analog band-pass filter that can allow narrowband or wide-band detection and also functions as an analog integrator that converts the mV pulses into apparent charge (pC or nC). The pulse magnitude is then digitized, usually with a peak sample and hold (PSH) circuit to capture the peak magnitude of each PD pulse (which may be as simple as a diode and a capacitor – Figure 6.19), and a slow ADC to digitize the PSH output magnitude. Sometimes an analog circuit (Figure 6.19) is used to measure the quasi-peak magnitude (Q_{IEC}) with the traditional discharge time constant of 0.44 s (IEC 60270),[1] to avoid the problems with the digital determination of Q_{IEC} (Section 8.8.1).

The memory for this type of system can be thought of as having an array structure that has memory cells (bins) that are addresses for each combination digitized magnitude window and digitized phase window. That is, the magnitude and phase constitute the address for a specific bin in the array. For example, there may be 1024 magnitude windows, and 256 phase windows from 0 to 360°. The contents of each magnitude/phase angle bin are the number of pulses measured during the total acquisition time (typically 60–600 s). The display software will usually normalize this to the number of PD pulses per second. If the system allows for the detection of pulse polarity (usually HF, VHF, and UHF systems, and sometimes LF systems in the wide-band mode), then the polarity may also be recorded by adding another dimension to the array (i.e. a 3D array).

This type of measurement system does not record complete pulse waveforms, so the *TF* map type of post-processing (Section 8.9.2) cannot be applied. However, the statistical and three-phase

1 Q_{IEC} is the quasi-peak magnitude of the PD in pC or nC. IEC 60270:2000 and its amendments do not formally define Q_{IEC}, but the next edition will define it as the quasi-peak PD magnitude.

synchronous measurement post-processing methods in Sections 8.9.1 and 8.9.3, respectively, can be implemented (the latter only if three parallel ADCs are present).

8.3.4 Digital System to Measure Pulse Magnitude and Selected Pulse Characteristics

The block diagram of a system that immediately digitizes the pulse from the detection impedance Z_m is shown in Figure 8.2. A digital filter (usually within an FPGA) after the ADC can select the lower and upper frequencies. The FPGA determines the magnitude of each pulse and the AC phase position is often from a VZC detector for online PD systems. The pulse magnitude and phase position define a memory bin to count the number of pulses, as described in the preceding section. Such systems became possible when ADCs with clock speed of 5 MHz (LF systems), 50 MHz (HF), and 500 MHz (VHF) became cost-competitive. The peak magnitude (in mV or pC) is determined by the FPGA. In addition to pulse polarity and magnitude, other characteristics of the PD pulse may also be established by the FPGA, such as the pulse risetime, pulse width, and/or the main pulse oscillation frequencies. Such pulse characteristics can sometimes distinguish PD from interference pulses, or different sources of PD. This system can have three separate channels to enable three-phase synchronized detection, perhaps with time stamping using the GPS system for cable PD measurement (Section 12.14.2).

Variations of this system have been used in all frequency ranges, as well as for off- and online PD detection. It is also straight-forward to add circuits and firmware to determine the location of PD or interference sources based on the time-of-flight method (Section 8.4.2).

8.3.5 Systems to Facilitate Waveform Post-Processing

Another approach is to record the entire PD waveform from Zm, much like a digital oscilloscope would, with a suitable ADC (Figure 8.3). That is, once the measurement is triggered by a pulse above the threshold, the ADC records the pulse waveform every (say) 100 ns (i.e. with a 10 M Samples/second ADC sampling rate) and dumps the recorded waveform into memory for the set duration of the pulse. 10 Mega Samples per second is suitable for an LF detector. Higher ADC clocking rates are needed for HF, VHF, and UHF detectors. Once the waveform data for each pulse is in memory, post-processing logic in an FPGA or a computer can analyze the contents at leisure. With this approach, digital algorithms can be used to filter the stored pulses with any desirable lower and upper cutoff frequencies, as well as perform digital integration of the mV pulses for conversion to charge (pC) in IEC 60270 instruments.

With the pulses stored in memory, extensive post-processing of the waveforms is possible. One of the most widely applied post-processing algorithms creates the time-frequency map whereby the duration and frequency content of each pulse is determined (Section 8.9.2). A two-dimensional map of time vs. frequency tends to show clusters of pulses in different time/frequency combinations on the map. These different clusters may point to different causes of PD and interferences.

If three different independent channels are provided (corresponding to one per phase), post-processing can be applied to determine which pulses are occurring at the same time (and their relative magnitude) – the so-called three-phase synchronous detection (Section 8.9.3). Although apparently not realized in a commercial form, wavelet or AI processing can also be performed on the recorded waveforms to suppress the effect of interferences.

Compared to the approaches in 8.2.3 and 8.2.4, the processing time to display results can take longer. To date, this approach has been mainly applied to the LF and HF ranges due to the cost of the hardware at higher sampling rates; however, commercial systems in the VHF and UHF range can be expected eventually.

8.4 Hardware-Based Interference Suppression and PD Source Identification

Section 9.1 outlines the need for suppressing interference, especially with online testing. Chapter 9 also presents various methods of interference suppression. Some of the methods are primarily based on hardware implementation. Thus, the following information is about these hardware-based approaches (filtering is not included since how it works is obvious). How these hardware methods are implemented often depends on the type of HV equipment. For example, the implementation of the time-of-flight (or time of arrival method is different between GIS, power cables, transformers, and rotating machines). Hence, further information is also in Chapters 12–16.

8.4.1 Hardware-Based Gating

The use of gates to suppress interference was first introduced in the 1980s, when digital PD instruments were first being designed. The idea is to prevent an interference pulse detected by the PD sensor from being counted as PD, based on either of the following events:

- interference is known to occur at a fixed position of the AC cycle,
- interference is detected by a "noise" sensor.

When either event occurs, a digital gate between the ADC and the memory or pulse counter array (Figures 8.1–8.3) is opened, preventing any pulse detected about the same time by the PD sensor (presumably due to the interference) from being permanently recorded.

Gating based on AC phase position is most-commonly applied when certain types of electronic power supplies such as battery chargers and DC excitation systems for generators, are operating near the test object. Such DC supplies create two or six pulses per AC cycle, caused by thyristor or transistor switching. This type of interference tends to be phase-locked to the AC, at least when the DC supply output power is constant. Triggering the gate is usually done from the AC reference voltage using the positive voltage zero crossing (VZC) circuit, in concert with a delay trigger circuit. The blocking time depends on the duration of the pulse created by the interference.

The most common signal to trigger the gate comes from an antenna or other sensor that detects when an interference pulse is occurring. The sensor is usually physically close to the source of the interference. DC power supplies, inverters, or arcing from slip rings are common applications where such gating may be useful. The antenna is fed to its own channel that usually has a threshold to ignore all signals less than a user-defined setting. The blocking time is set based on the expected duration of each interference pulse.

The use of gating can cause several problems for PD detection, including blocking legitimate PD from being counted when the gate is open. It also makes accurate PD trending over time difficult unless the trigger threshold and blocking time is consistent in all measurements. An experienced user is needed to make the trigger and blocking time adjustments at the time of the test. These aspects are discussed in more detail in Section 9.3.9.

Gating can also be employed in the reverse manner, i.e. to only allow pulses to be recorded when a gate signal is triggered. For example, as described in Section 15.12.4, a UHF sensor within a power transformer can be used as a trigger to record the signals from the transformer bushing taps on each phase. Bushing taps can then be the primary PD sensors on each phase. UHF sensors within the transformer are not susceptible to external interference, whereas bushing tap sensors are sensitive to external interference. Thus, when a UHF sensor detects a signal, it is most likely from the winding PD, thus the signals from the bushing tap sensors are recorded. When the UHF sensor does not detect a pulse, the signals from the bushing taps are ignored.

The gating feature can also be implemented during software-based post-processing, as mentioned in Section 8.9.4.

8.4.2 Time-of-Flight (or Time of Arrival) Method

This method makes use of the fact that it takes time for electrical pulses (and acoustic pulses) to travel through a test object. The time of "flight" method is widely used both to suppress external interference and to locate sources of PD within the HV equipment. This method usually requires at least two PD sensors per phase, and its implementation is greatly dependent on the nature of the HV equipment – see Chapters 12–16 for more details.

The method depends on the time it takes an electromagnetic pulse from PD (or interference) to travel between the sensors. The velocity of electrical propagation is $3x10^8$ m/s (or 0.3 m/ns) in air and gases, which is the same as the velocity of light. The velocity in other insulating materials is velocity in air divided by the square root of the material's dielectric constant, since magnetic permeability is 1 at PD detection frequencies. Figure 8.5 shows a simplified sketch of either a GIS bus or the isolated phase bus (IPB). If the distance between the two PD sensors (C_0 and C_c) is X m, and assuming the dielectric is a gas (with a velocity v of 0.3 m/ns), it will take:

$$t = X / v$$

for a pulse to travel between the sensors. For example, if the distance X is 8 m, then it takes 26.7 ns for the pulse travel time from one sensor to the other.

If each sensor is connected to a fast electronic comparator, the output of the comparator will go high when the pulse is detected above a selectable threshold. Digital logic can then be used to determine where the pulse came from. For example, if the pulse comes from the bus on the left of the left sensor (Figure 8.5), a comparator connected to C_0 will go high 26.7 ns before the gate associated with sensor C_C goes high. If a pulse originates from the right of sensor C_C, the reverse occurs – comparator C_C goes high 26.7 ns before that of sensor C_0. If the pulse originates between the pair of sensors, then both comparators go high less than 26.7 ns apart. Thus, the pulse can be decoded as coming from three locations: left of C_0, right of C_C or between C_0 and C_C. In a VHF detection system with 1 ns resolution, one can locate the pulse source between the sensors to within 0.3 m based on the difference in travel times.

Figure 8.6 shows an example of the PRPD patterns captured using two sensors on a section of isolated phase bus (IPB) between a generator and the power transformer. PRPD plots are

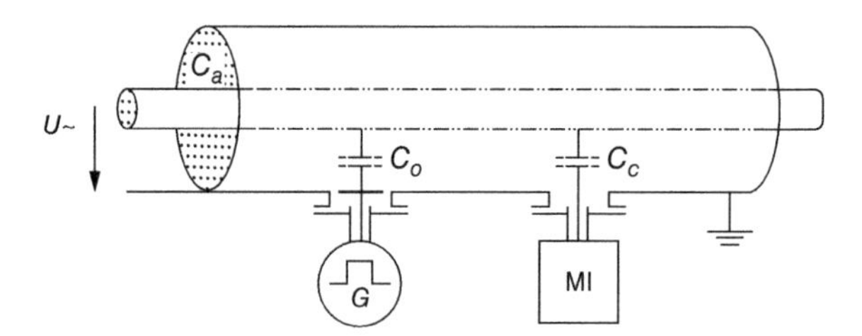

Figure 8.5 Schematic of two sensors per phase using the time-of-flight method to determine the location of PD or interference pulses.

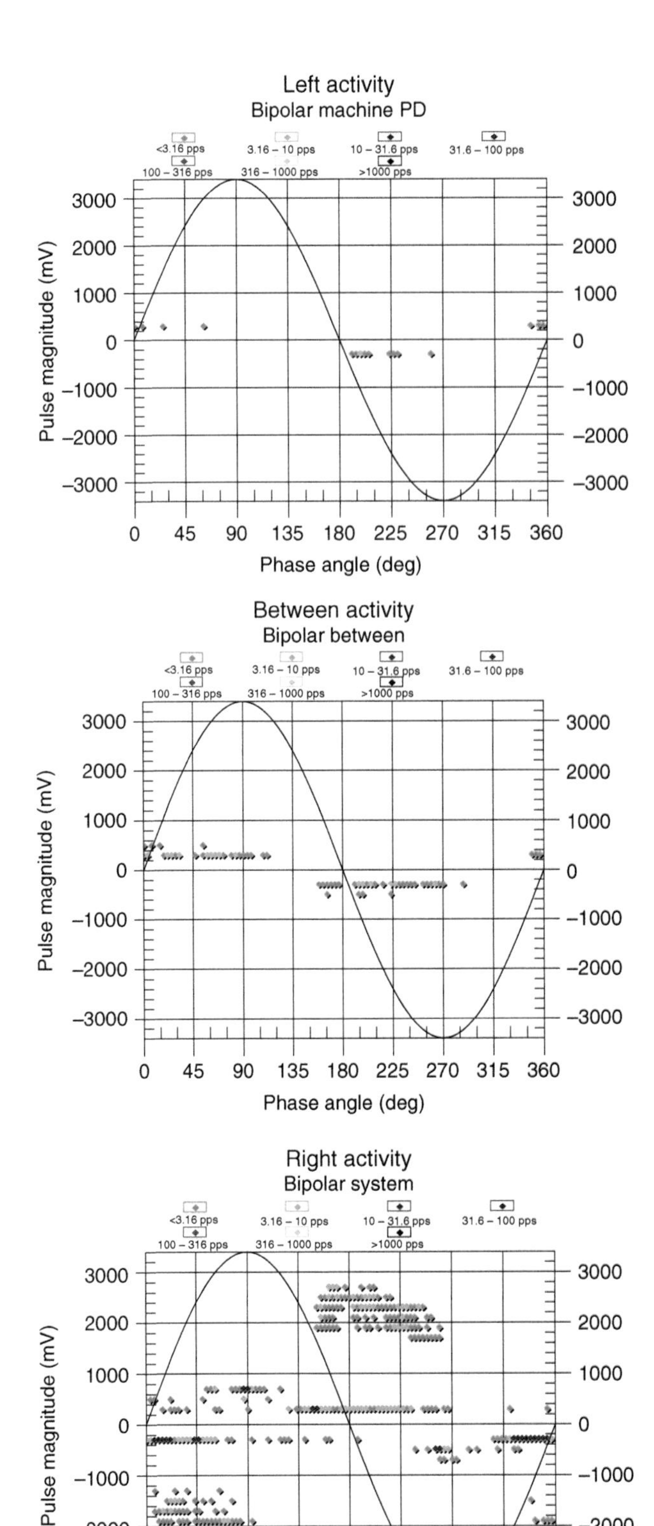

Figure 8.6 Example of VHF PRPD plots collected online from Phase A of a generator, with two sensors per phase on the generator output bus. The high activity on the right-hand plot is due to PD on the IPB beyond the right sensor. *Source:* Chart courtesy of Iris Power L.P.

discussed in Section 8.7.3. There is very high PD coming from the IPB beyond the right sensor (in Figure 8.5) (bottom plot in Figure 8.6), but very little PD from the generator (left sensor, top plot) or the IPB between the sensors (middle plot). The high PD activity is probably due to a broken IPB insulator close to the power transformer (Section 14.8.1). Without the time-of-flight method, the high IPB PD may have been attributed to the generator, and an unnecessary generator outage planned.

There are many practical problems with this method. The higher the speed of the electronics, the more accurate the system. With an LF PD system, the upper cutoff frequency is 1 MHz, and the risetime of the detected pulse will be 300 ns or longer. When a comparator switches to the high state with such a long risetime, i.e., when the pulse “arrives” is uncertain (i.e. does the comparator go high at the beginning of the pulse risetime, at the end of the pulse risetime, 300 ns later or some intermediate time). This creates uncertainty of at least 300 ns in the time of arrival at the comparator (when it goes high). Thus, with an LF system, there needs to be more than 300 ns of travel time between the sensors to locate the source of the PD or interference site to one of the three regions. With gas insulation, this implies the sensors must be more than 100 m apart, which is often not practical. In the VHF frequency range, the risetime of the detected PD or interference pulse may be as short as about 1 ns. Ideally, the uncertainty in pulse arrival time due to the risetime is 1 ns (in reality, it is longer since VHF electronics may take a few nanoseconds to change states). With a 1 ns uncertainty, the two sensors could be as short as 0.3 m apart to identify which of the three locations in Figure 8.5 the pulse is coming from.

Another practical problem associated with this method is when the pulse attenuates when traveling between the sensors, which is normal with power cables. This attenuation is significant in all dielectrics except gases, and the attenuation normally increases with frequency (i.e. attenuation in dB/m is higher at VHF than LF ranges). If significant attenuation occurs as a pulse travels, the one comparator may trigger, and the other does not, making decoding of the pulse origin impossible.

As presented in Section 12.14.1, for power cables it is sometimes possible to locate PD sites using just one PD sensor using a variation of the time-of-flight method based on time domain reflectometry.

8.4.3 Pulse Shape Analysis

With the advent of GHz bandwidth oscilloscopes in the late 1970s, it became apparent that individual PD current pulses may have different shapes [8, 9]. That is, laboratory experiments show risetimes may be shorter or longer, the pulse widths differ, and the oscillations after the first peak of the current pulse may have different frequencies and damping. As discussed in Chapters 2 and 3, part of the variation in shape may be due to the physics of PD (for example pulses tend to have a longer risetime at lower gas pressures, and large voids may have wider pulses as UV emission triggers PD in other parts of the same void). Some of the variation in shapes is also due to the local inductances and capacitances of the test object in the environment of the PD site. Pulse shape variation may also be caused by attenuation, distortion, reflections from impedance mismatches, and other traveling wave effects as the pulse travels from the PD site to the PD sensor. And of course, the PD detection system itself may change the pulse shape due to filtering or bandwidth limitations. It is also apparent that some types of PD-like interferences have pulse shapes that are different from most PD pulses.

The knowledge that PD current pulses and interference pulses may have different shapes lead to hardware that could be used to characterize the shapes of the detected pulses in the VHF and UHF ranges (clearly the LF and HF methods do not have the bandwidth to allow analysis of pulse shape characteristics since that information is filtered out). Hardware can measure in real time using FPGAs various characteristics of the shape such as

- the pulse risetime,
- the width of the initial pulse,
- the ratio of the magnitude of the initial peak of the pulse to the magnitude of the first opposite-polarity ring,
- the duration of the pulse including oscillations, etc.

Hardware-based pulse shape analysis is an important means of interference suppression for online monitoring of PD in rotating machines [10, 11] and is discussed further in Section 16.9. An older approach is to measure pulse width based on comparators and multivibrators is in [12]. The more modern approach is to digitize the input pulse and use FPGAs to decode the pulse shape information immediately after pulse acquisition, so that the entire pulse waveform does not have to be permanently stored, reducing post-processing analysis time and storage requirements.

TF maps (Section 8.9.2), where the stored individual PD waveforms are analyzed by software as a post-processing method, are alternative means of capturing wave shape information, as long as the ADC is operating at a high enough sampling rate.

8.5 PD Calibrator Hardware

For LF detectors compliant with IEC 60270, hardware is needed to calibrate the mV or mA signals from a PD sensor into apparent charge, i.e. pC or nC. Although the procedure could be applied for PD detection in the HF, VHF and UHF ranges, the apparent charge readings are not valid (see IEC 62478 and Section 6.6). However, special pulse generators with much shorter risetimes can be used in the HF to UHF ranges to perform a "normalization" or "sensitivity check" (Section 7.6).

The calibrator for LF detectors is simply an electronic pulse generator that creates a voltage pulse of magnitude *Vo* in series with a low-voltage or high-voltage calibration capacitor *Co* (Section 6.6.4). This will inject a charge of

$$Qo = Co \times Vo$$

For example, if the pulse generator outputs 10 V, and a *Co* of 100 pF is used, the injected charge is 1 nC. If the calibration is done with the test object energized, *Co* should be rated to the same voltage as the highest test voltage. Alternatively, *Co* can be a low-voltage capacitor if the calibration is done with the high-voltage supply set to 0 kV. In this case, *Co* cannot be left connected when the PD is being measured, which does slightly change the test circuit. For an accurate apparent charge calibration in the LF range, *Co* should be much less (<1%) than the capacitance of the test object, and in any case, <200 pF. With practical HV equipment, this is rarely a problem. However, it can be an issue with measurements on small test objects often used for research. In this case, a correction factor is required as described in IEC 60270 or in [1, 2].

Since calibration is only performed in the LF range, IEC 60270 requires the characteristics of the pulse generator to be

- pulse risetime <60 ns
- pulse duration >5 μs
- slow decay of pulse (1/lower cutoff frequency), or wide pulses with the same rise and fall times that do not interact
- under- and over-shoot <10% of steady state pulse level

The pulse repetition rate is not too critical, but it should be lower than the maximum input pulse frequency of the PD measurement system. Typically, it is a few kHz.

Although the pulse risetime should be 60 ns or less for LF systems, actual PD pulse rise times in atmospheric air are usually in the few nanosecond range (Figure 1.2). For HF and VHF performance and sensitivity checks, the pulse risetime should be about 1 ns. For UHF systems, the risetime should be <0.1 ns, which is difficult to achieve. It is good practice to use a pulse generator with a very low-source impedance, usually about 1 Ω, especially for VHF and UHF applications. This will ensure that most of the voltage is placed across the calibration capacitor and the test object.

Most pulse generators used for calibration and sensitivity checks are battery operated, since it is not only convenient, but it prevents ground loops that may change the pulse shape.

8.6 Special Hardware Requirements for Continuous Monitors

Until the late 1990s almost all PD measurement systems used in the field (as opposed to a laboratory or factory) were portable instruments that were used on a variety of test objects, and which were taken from site to site mainly for offline testing. The PD sensors and instrument are connected to the HV equipment to be tested for only a few hours, and usually the testing is done in a comfortable environment. If the PD instrument fails, it could be sent back to the vendor relatively easily, or sometimes the firmware and software could be updated from a user location, with little impact on HV equipment reliability. Since the PD system was often shipped to many different locations, the system had to be physically robust.

Continuous online PD monitoring creates additional constraints that the vendor needs to accommodate:

- Capacitive PD sensors are connected permanently to the power system high voltage, thus experience high-voltage AC continuously, plus any voltage transients that may occur. The failure of these sensors will trigger a ground fault.
- The environment for both the sensors and instrumentation can be extreme and affect PD system reliability.
- The system must work continuously for many years, often with no user intervention. If the system fails, the system must often be dismantled and returned to the vendor for repairs, which means monitoring stops for weeks or months.
- Many continuous monitoring systems are connected to digital networks. If these networks are connected to the internet or cell/mobile phone networks, there is a cybersecurity risk.

If these issues are not addressed, the reliability of the continuous monitoring system may be poor, which reduces user confidence that important insulation deterioration information may be identified before HV equipment failure.

8.6.1 Sensor Reliability

Antennae used for PD sensing rarely cause problems for power system reliability. However, HFCTs installed around power cables or installed at stator winding neutrals may cause issues. For machines, if the nearby power system experiences a ground fault in operation, the neutral voltage may raise to the normal line-to-ground voltage. Thus, HFCTs at the winding neutral must be insulated for the rated machine voltage. Rogowski coils are sometimes placed around the high-voltage insulated center conductor of power cables for PD sensing. At this location, there must be enough clearance between the Rogowski coil and the power cable to prevent PD or breakdown under all environmental conditions, since the Rogowski coil is exposed to the normal operating voltage of the power cable.

Capacitive PD sensors are generally permanently connected to the HV equipment in periodic or continuous online PD monitoring. If the capacitor dielectric breaks down, or the capacitor experiences a flash over the sensor surface, power system protective relaying will operate and take the HV equipment being monitored offline. The asset manager tends to be very upset if a condition monitoring tool meant to prevent insulation failures, in fact causes a failure. Hence, most capacitors that are permanently installed will have an increased creepage path to minimize the risk of flashover if the surface becomes contaminated (Figure 8.7) or have a high-voltage bushing with a long creepage path.

Figure 8.7 Photo of an 80 pF PD detection capacitor intended for permanent installation in rotating machines and air-insulated switchgear. The creepage distance between the high-voltage and low-voltage electrodes has been increased by using ridges molded into the surface. *Source:* Courtesy of Iris Power L.P.

Permanently installed capacitive PD sensors require much greater reliability than similar sensors used in offline testing. Every capacitive PD sensor must be given a routine withstand voltage test at least as high as the HV equipment is required to withstand. Each sensor should also have a PDEV much higher than the operating voltage. The capacitance (and dissipation factor) of each sensor should be measured and be within a reasonable tolerance, and be stable over the expected temperatures experienced in service.

In addition, many vendors will subject the sensor to a variety of "type" tests on a sampling basis:

- A voltage endurance test (i.e. exposing the sensor to a very high voltage for a fixed period of time, without failure) if the sensor uses a liquid or solid dielectric.
- A thermal cycling test to ensure all the components will work together at a range of temperatures without cracking or delaminating.
- A tracking resistance test to ensure that the sensor surface will not have discharge and electrical tracking if the surface becomes contaminated by dust, dirt, oil, or moisture.

The specifics of the routine and type tests depend on the type of HV equipment being monitored, the environment, and of course the ratings of the HV equipment. A good example of PD sensor requirements for rotating machines is given in IEEE 1434. A less stringent set of requirements for rotating machine applications is presented in IEC 60034-27-2.

8.6.2 Instrument Robustness

For continuous monitoring, it is important that the PD system be supplied from an uninterruptable power supply. This is especially key in electric power utility generating stations and substations since it is common practice to routinely shift the station electrical supply between sources, which involves a very short outage with each transfer.

In industrial and utility plants, circuit breakers and disconnect switches are opening and closing all the time. This causes huge voltage and current transients in the plant, which in turn may give rise to transient voltage rises in the plant's grounding system. The power supplies within the PD system must use surge protective devices and filters to minimize the risk of electronics failure.

Of particular importance are the PD signal inputs from the PD sensors, which are usually located remote from the instrumentation. Voltage transients can occur during circuit breaker operations, since the ground at the instrument may temporarily be at a voltage that is different from that at the PD sensor. If the voltage transients are more than a few tens of volts, the input electronics may fail. Surge protective devices, such as metal-oxide surge arrestors, can be useful here. However, such devices may change the shape and magnitude of the PD pulses, especially in the HF, VHF, and UHF frequency ranges, due to their relatively high effective capacitance. To avoid this, nontraditional protective devices such as high-speed, hot carrier diodes tied to the DC rails within the instrument or optical isolation may be needed.

Continuous PD monitoring systems must be able to operate under the environmental conditions in the plant:

- Temperature range
- Humidity
- Vibration
- Chemicals
- Radiation

If commercial electronic parts are used, they tend to stop functioning above 50 °C, which is often likely in systems exposed to the sun in a hot environment. Even most military-grade electronics cannot function above 70 °C. If the environment cannot be tolerated, the hardware may need to be installed in a suitable environmental chamber or located where other systems and computers are located. Unfortunately, this may involve the laying of hundreds of meters of wiring – which may cost more than the monitoring system. Long cable runs between the sensors and the instrumentation may also not be possible due to signal attenuation and distortion, especially with VHF and UHF detection systems.

8.6.3 Cybersecurity

Cybersecurity is concerned with ensuring that systems within a plant cannot be accessed, read, or modified by unauthorized users. This is an important issue for end users who install continuous PD monitoring systems. One concern is that the data from the PD system can be passed to their competitors, who may be able to make use of information, such as if a transformer or generator has a high risk of failure. Also, people not familiar with the functioning of continuous PD monitoring systems may assume that if it is compromised, the system can upset plant operation by opening/closing circuit breakers, etc.

Most PD system vendors offer (and encourage in the case of HV equipment OEMs) online support services, including data downloading and analysis services. Such remote support requires an internet or mobile telephone connection. In spite of best efforts, it seems dark forces are sometimes able to overcome internet or mobile phone security. The result is that many end users will not connect the continuous PD monitoring system to the internet or mobile phone network, thereby preventing third parties from accessing the data by any means. Instead, large utilities and industrial companies seem to prefer an in-house centralized condition monitoring system for all their assets and monitoring systems. Communications with each plant and system is by means of an intranet that is not connected to the outside world, thus avoiding most cybersecurity issues.

PD cannot reliably indicate exactly when HV equipment will fail, and in spite of efforts to suppress interference, false alarms of high PD can never be completely eliminated. Thus, continuous PD monitoring systems should not be connected to protective relaying that can operate circuits and/or shut down a process. As a result, the consequences to the power system of a cyberattack on a PD monitoring system via the internet or mobile phone connection is minimal.

8.7 PD System Output Charts

PD systems made before about 2000 typically only had a cathode-ray tube (CRT) display of the PD pulses vs. the AC cycle phase angle, and sometimes a peak or quasi-peak PD reading on a meter or digital display. All modern digital PD systems use software to produce graphs, as well as various numerical indicators representative of the PD activity. The following describes the most useful charts produced by the display software. Section 8.8 discusses numerical indicators of PD activity.

8.7.1 Pulse Magnitude Analysis (PMA) Plot

This chart, sometimes also called a pulse height analysis plot, displays the PD pulse count rate on the vertical axis vs. the PD pulse magnitude on the horizontal axis. An example is shown in Figure 8.8.

The vertical scale is the total number of pulses over the data acquisition period, or it may be normalized per unit of time, for example to the number of pulses per second (pps). The vertical pulse repetition rate scale is usually logarithmic, but some vendors may offer a linear option. The plot in Figure 8.8 is a "density" plot since the vertical scale is really the pulses per second per magnitude window, even though most vendors do not show the correct units. For example, the PD

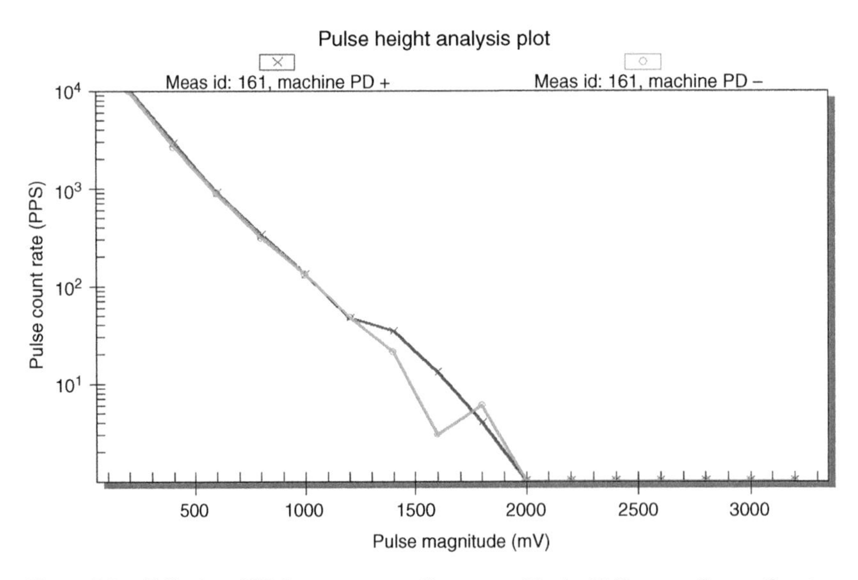

Figure 8.8 PMA plot of PD from a stator coil measured in the VHF range. *Source:* Courtesy of Iris Power L.P.

instrument used in Figure 8.8 had 16 positive magnitude windows between 200 and 3400 mV, with each window 200 mV wide (the count between 0 and 200 mV is not plotted). Thus, the data point at 1000 mV corresponds to 10^2 or 100 pulses per second measured between 800 and 1000 mV. If one changes the magnitude window width, or there is a pulse amplifier where the amplifier gain can be adjusted, the effective window width changes, and the PMA plot will have a different slope. That is, if an amplifier with a gain of 2 is inserted before the D/A converter in Figures 8.1–8.3 on the same 200–3400 mV range, the effective magnitude width is now only 100 mV. All other things being equal, the number of pulses per second per magnitude window will decrease by 50%. Therefore, changing the magnitude range or amplifier gain will change the appearance of the PMA plot. This occurs whether there are 16 magnitude windows or 1024. A cumulative PMA plot (where the pulse count rate is the total number of pulses below a particular magnitude) could be created that does not have this difficulty, but it is rarely used since the few higher magnitude pulses are harder to see among the thousands of smaller pulses. When comparing PMA plots over time, it is important to make sure the same gain and magnitude range are used.

The horizontal scale in Figure 8.8 is the pulse magnitude in pC, mV, or mA depending on the instrument. This horizontal scale is normally linear. With capacitive test objects and measurements in the LF range below 1 MHz, the horizontal scale is normally in terms of apparent charge (pC, nC). According to IEC 60270 and IEC 62478, the charge scale should not be used for partly inductive test objects or frequency ranges above 1 MHz, although it is common for many vendors to default to a charge scale, in spite of the IEC and IEEE standards.

Figure 8.8 shows the positive and negative pulses displayed as separate lines. This is common for HF and VHF detectors. For LF detectors in the narrow band mode (Section 6.5.2) and many UHF detectors that use demodulators (Figure 8.4), the pulse polarity information is lost since either the initial PD pulse initiates oscillations in Zm (or the quadripole), or the ADC is unipolar. Therefore, only a single line is plotted, which represent the total of positive and negative polarity. Some instruments may overcome this limitation by inferring the polarity of the pulse from the AC phase position (Section 3.10.1). This fix may lead to interpretation errors if there are interference pulses present.

Many PD systems enable the PMA lines from different tests (over time on the same phase, or from different phases) to be plotted on the same chart to enable easy comparison.

8.7.2 Phase-Magnitude-Number (Ø-*q*-*n*) Plot

This is a three-dimensional chart with the axes:

- AC cycle phase position (Ø) in degrees
- Pulse magnitude (q) in pC, mV, mA, etc.
- Number of pulses or number of pulses per second (n).

Figure 8.9 shows an example. Many find the three-dimensional plot hard to understand, so it is not as widely used today. The vertical pulse count axis is different from the vertical scale in the PMA chart. This axis is the number of pulses per second, per magnitude window, per phase window. Thus, if the phase window is 10° wide (i.e. there are 36 phase windows), there will be a higher pulse count rate than if the phase window is 1° wide (i.e. 360 phase windows). So, as with changes in gain or magnitude range, the character of the Ø-*q*-*n* chart changes when the number of phase windows is adjusted.

Many people find this type of plot hard to visualize, even though the software usually allows the plot to be rotated – so one can look at it from various positions. As a result, the PRPD plot, which presents the same information in a two-dimensional format, tends to be more popular.

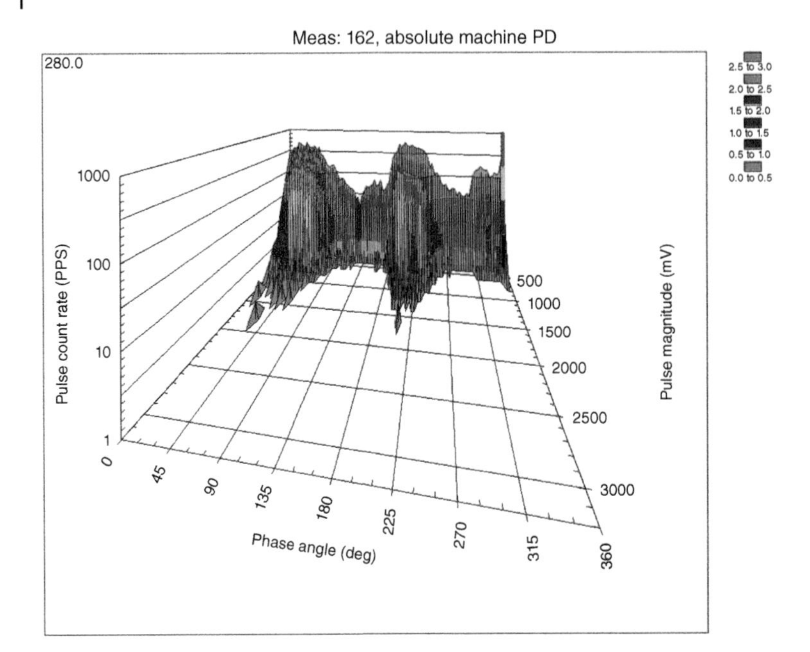

Figure 8.9 *Ø-q-n* chart of the same data shown in Figure 8.8. The pulse repetition rate is indicated by both the vertical axis and the color of the lines. The pulse polarity information is suppressed. *Source:* Courtesy of Iris Power L.P.

8.7.3 Phase-Resolved PD (PRPD) Plot

This is the most common chart for visualizing stored PD data in a two-dimensional format that is reasonably easy to understand. Figure 8.10 shows an example where there are 16 positive and 16 negative magnitude windows, together with 100 phase windows covering the 360° AC cycle. The vertical axis is the pulse magnitude in pC, mV, etc., and the horizontal scale is the phase position of the AC cycle. The pulse count rate (in pulses per second per magnitude window per phase window) is shown as different colored dots. The color key for the count rate is shown above the chart in Figure 8.10. Regrettably, there has been no standardization of the color scheme for the pulse count rate used by different vendors. The PRPD plot is actually a digital version of the CRT display used in older analog PD detectors, and widely used to identify PD types and interference in the 1960s [13].

The chart in Figure 8.10 is somewhat pixelated. This is due to there being only 16 positive (and 16 negative) magnitude windows, and only 100 phase windows. Systems with higher numbers of magnitude and phase windows will be less pixilated, as shown in Figure 8.11. Note that this higher resolution plot came from an LF detector that was not able to determine the pulse polarity. All PD is shown as positive PD, when in reality the PD pulses from about 0° to 90° are actually negative polarity PD. This detector has 1024 magnitude and 1024 phase windows.

PRPD charts are widely used for PD interpretation because the pulse pattern with respect to the AC cycle can give important insight into the possible causes of PD and aid in making sure the pulses are due to test object PD, rather than interferences (Section 10.5).

8.7.4 Trend Plot

For insulation condition monitoring using either offline/onsite testing or continuous online monitoring (Section 8.6), the most important chart is the PD trend over time, on the same test object. As discussed in Section 10.4.1, in general if the PD magnitude or other indicators of PD activity are

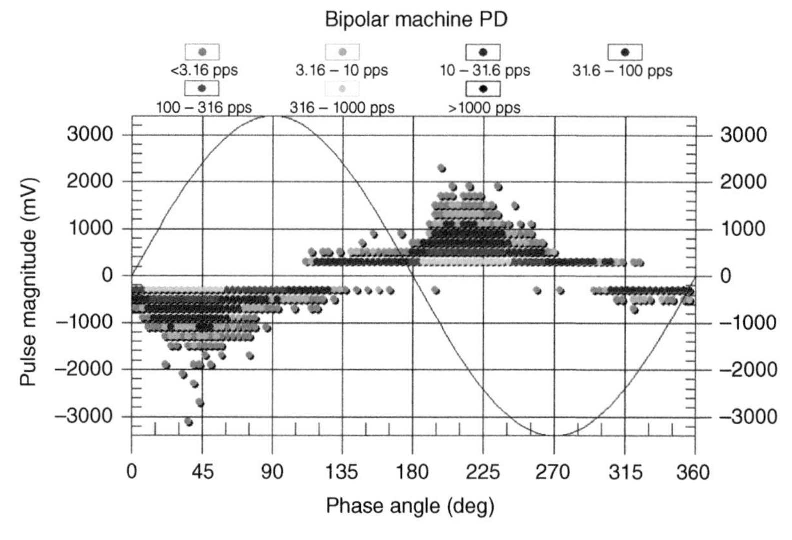

Figure 8.10 PRPD chart of the same data shown in Figure 8.9. *Source:* Courtesy of Iris Power L.P.

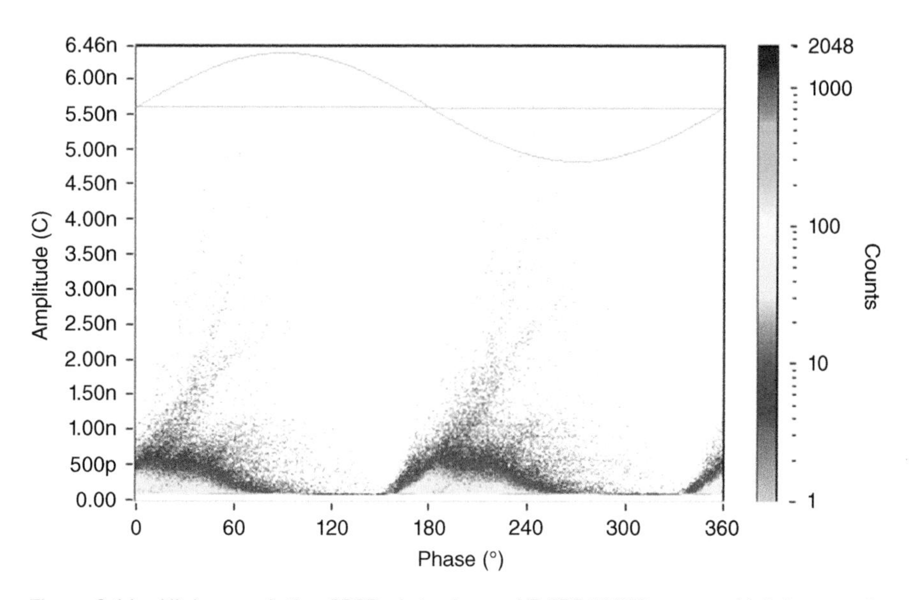

Figure 8.11 Higher resolution PRPD plot using an LF (IEC 60270-compatable) detector. *Source:* Courtesy of Iris Power L.P.

increasing over time, then it suggests that the volumes of the individual defects are increasing, more PD sites are being created, or both, and thus insulation aging is occurring. This may indicate to HV equipment owners that maintenance or replacement is needed. Section 8.8 presents some of the PD quantities that are used in trend plots. As discussed in Chapters 12–16, in many insulation systems PD activity does not monotonically increase until failure occurs. Instead, the PD may stabilize, occasionally reduce magnitude or even disappear, or otherwise be erratic, even though aging is continuing. Thus, in general, the PD trend may not be a good predictor when equipment failure will occur.

Trend plots are only valid on the same test object, with the same type of PD sensor and the same model of PD instrumentation operating in the same frequency range (Section 10.4.1). Since PD activity can be affected by test object temperature, loading (if the test object is operating in service)

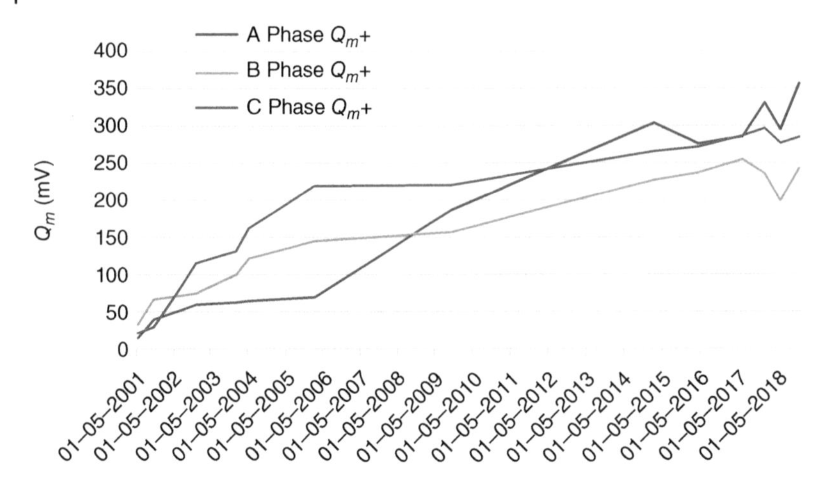

Figure 8.12 PD trend plot of the peak positive PD magnitude collected at the same machine operating temperature and load. *Source:* Courtesy of Iris Power L.P.

and ambient humidity, trend PD data must be collected over the years under about the same operating conditions (Section 10.6). Alternatively, the analysis software needs a feature to trend only the data that satisfies a set of data filter conditions based on operating conditions, ambient conditions, and instrumentation settings.

Figure 8.12 shows an example of the trend plot from online testing of a stator winding over 17 years, with only data displayed that is at the same winding temperature and load.

8.7.5 PDIV/PDEV Plot

For offline factory PD testing or onsite/offline testing, an important chart is the plot of applied AC voltage versus PD magnitude. The partial discharge inception voltage (PDIV) is the voltage at which PD is observable above a specified sensitivity as the voltage increases from zero. The partial discharge extinction voltage (PDEV) is the voltage at which PD is no longer observable at a specified sensitivity as the voltage is lowered from a voltage where PD is clearly occurring. For most types of test objects, usually those that employ only organic insulation, the PDIV and PDEV must be above the operating voltage in service, otherwise the PD will degrade the insulation and cause equipment failure. When the high-voltage equipment uses either inorganic materials (for example ceramic insulators for overhead transmission lines) or organic/inorganic composites (for example epoxy-mica stator winding insulation), then PD itself is usually not a direct cause of insulation aging, and the PDIV/PDEV may be below operating voltage.

Figure 8.13 shows a typical PDIV/PDEV chart. Such charts are normally automatically produced where a high-voltage supply slowly increases the AC voltage in steps from 0 to a specified maximum test voltage. At the same time, the PD system measures the AC voltage on the test object as well as Q_{IEC} or Q_m (Sections 8.8.1 and 8.8.2). Unlike the chart in Figure 8.13, as the voltage rises from 0 kV, the Q_{IEC} or Q_m should be zero (assuming the interference is successfully suppressed (Section 9.3). As the voltage continues to increase the PD magnitude will increase from 0. Once the peak magnitude is above the specified PD sensitivity (usually in terms of pC for IEC 60270 compliant detectors), this is defined to be the PDIV. The voltage continues to increase until a maximum specified voltage. Often the Q_{IEC} or Q_m is measured at the maximum test voltage. After a specified time at the maximum test voltage, the voltage is lowered until the PD is below the specified sensitivity level, measuring the PDEV.

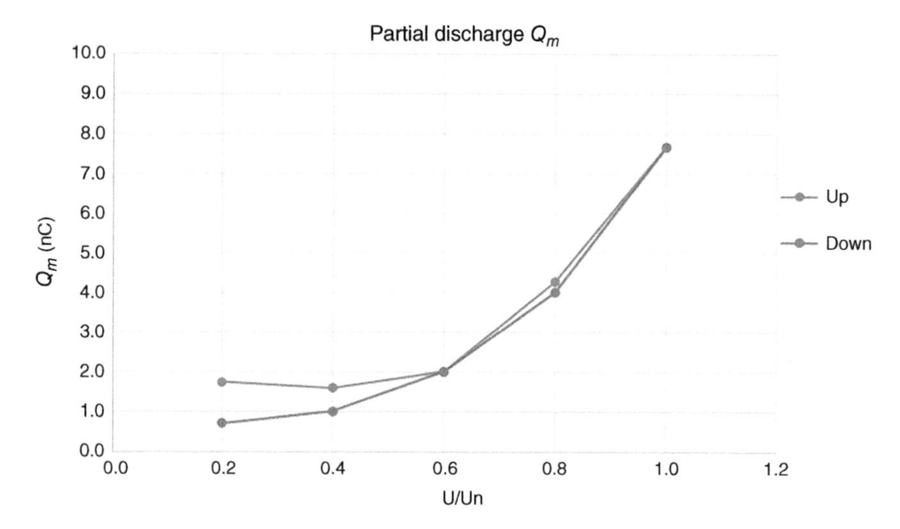

Figure 8.13 Typical PDIV/PDEV chart using a LF PD detector on a stator coil. The vertical scale measures Q_m at 10 pps, while the applied AC voltage is expressed as a percentage of the rated phase-to-phase voltage. If PDIV/PDEV is defined at a Q_m sensitivity of 1 nC, then the PDIV is below 0.2 Un, and the PDEV is a 0.4 Un. Un is the rated phase-to-phase voltage. This example may be acceptable for hybrid organic/inorganic insulation systems such as epoxy-mica, but not for equipment using purely organic insulation. *Source:* Courtesy of Iris Power L.P.

The PDIV and the PDEV are strongly dependent on the specified sensitivity of the PD. With the exception of rotating machines, all other common HV equipment specifications will define the Q_{IEC} or Q_m sensitivity for determining the inception and extinction voltages. Since the physics of PD is stochastic in nature, the Q_{IEC} or Q_m tends to jump around somewhat, which can make exact determination of the PDIV and PDEV somewhat difficult.

8.7.6 Scatter Plot

For offline PD tests, it is useful to plot a PD indicator vs. applied test voltage, as discussed in Section 8.7.5. Similarly, for online tests, it is sometimes useful to plot PD activity vs. test object temperature, humidity, test object current or load, etc. These are sometimes referred to as scatter charts, from their original use in statistical regression. How the PD activity changes with operating conditions can sometimes be used to determine the root cause of the PD (Section 10.6). Figure 8.14 shows an example of a scatter plot.

8.8 PD Activity Indicators

To facilitate trending over time and enable comparisons of measurements on different test objects, various "activity indicators" can be calculated by software from the PD data stored in memory. These are single numbers that characterize the PD activity in a measurement. There are two types of indicators:

- Indicators that are representative of the peak or quasi-peak PD activity, i.e. that are related to the highest PD magnitude,
- Indicators that are representative of the total PD activity including both the PD magnitude and the PD repetition rate information.

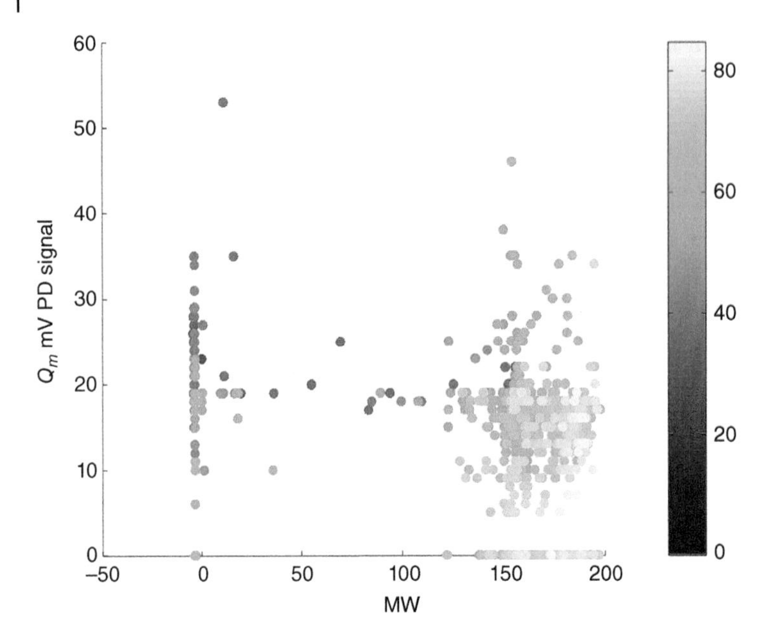

Figure 8.14 Plot of PD magnitude Q_m vs generator load for a 13.8 kV, 225 MVA generator. The color indicates the winding temperature (°C). In this example, as the load increases there is no obvious change in the PD activity. *Source:* Courtesy of Eric See-Toh of BC Hydro.

These indicators are automatically calculated by the software, but they do assume that there is no significant interference or noise (Section 9.2).

8.8.1 Quasi-Peak PD Magnitude (Q_{IEC})

For LF PD systems compliant with IEC 60270, by far the most common indicator for PD magnitude is the "largest repeatedly occurring" PD magnitude, sometimes called the quasi-peak magnitude, and often referred to by PD system vendors as Q_{IEC} (Sections 6.5, 6.6 and 8.5). This indicator dates back to the analog days of PD systems and is still only defined based on analog detection. Table 8.1, based on information in IEC 60270, shows how Q_{IEC} is measured based on a stream of equal-magnitude pulses that are equally spaced apart and have a pulse repetition rate of n pulses per second. R_{min} and R_{max} are the minimum and maximum responses (in percent of the true peak magnitude) from the calibrator. For example, at a pulse repetition rate of 5 pps, the calibrator will produce an output in pC that is between 76 and 86% of the actual injected charge. When there are 50 pps, the output is between 94 and 104% of the true charge produced by the calibrator. Thus, Q_{IEC} is essentially an indicator of the charge magnitude at a repetition rate of 50 pps, or one pulse per AC cycle in 50 Hz power systems.

The Q_{IEC} estimation method above assumes an analog circuit is used. This is implemented by a peak detector where the output of the amplifier or Zm is connected to a diode feeding a capacitor

Table 8.1 Q_{IEC} response of IEC 60270 systems to a series of pulses.

Pulse rep. rate, n (pps)	1	2	5	10	50	>100
R_{min} (%)	35	55	76	85	94	95
R_{max} (%)	45	65	86	95	104	105

to ground (see Figure 6.19). This circuit is required to have a discharge time constant of 0.44 s. There are many ways of simulating this peak detection method with digital PD systems in software. However, different software implementations may not produce the same results. This may result in different brands of PD systems estimating different Q_{IEC}'s, even with the same pulse train. Thus, if the user has access to different brands and models of PD detectors, it is incumbent on the user to make a comparison between the systems on the same test object to ensure that the quasi-peak PD readings are close.

Since Q_{IEC} is only defined for narrow band and wideband PD detectors in the LF (<1 MHz) range, other phenomena may lead to measurement errors. This can include superposition errors if the PD is occurring in relatively short lengths of power cable, due to reflections at the cable ends. Also, when there are many PD sites, the multiple PD pulses occurring within the processing time of the detector may be sensed as one large PD pulse. See Section 6.5.2.

Formally, Q_{IEC} is not defined for the HF, VHF or UHF ranges, although some vendors may still calculate it.

8.8.2 Peak PD Magnitude (Q_m)

Another indicator of the highest PD magnitude is Q_m, sometimes called Q_{max}. This quantity is easily derived from the PMA chart described in Section 8.7.1. It is defined as the magnitude of the PD at a specified pulse repetition rate. For rotating machine stator windings, the calculation is done at a pulse repetition rate of 10 pps (see IEEE 1434 or IEC 60034-27-1), and many PD system vendors now use this calculation for all test objects. A repetition rate of 10 pps (per magnitude window) may have become the most common basis to calculate Q_m because older analog instruments had a CRT for display, and with the common CRT phosphors used at that time, short-duration PD pulses with a repetition rate of 10 pps are easily seen on the display. Thus, the CRT display of the highest magnitude pulses and a digital calculation at 10 pps tended to agree. The use of 10 pps may also have become popular since one of the first widely used digital detectors (the corona or TVA probe, see Section 16.13.1) used the 10 pps criteria.

The repetition rate may also be 1 pps, or 50 or 60 pps, rather than 10 pps. However, if one uses a criteria of 1 pps, then the Q_m tends to be very erratic from measurement to measurement, compared to Q_m's using the 10, 50, or 60 pps criteria. If other repetition rate criteria are used, then the Q_m will be different.

Another way to quantify the peak PD magnitude that vendors use is based on statistical methods. All the pulses during the measurement interval are analyzed on a cumulative probability scale. The peak PD magnitude (Q_{95}) could be defined as the pulse magnitude that is higher than (say) 95% of pulses. A problem with this approach is that the peak magnitude will be dependent on the trigger threshold for acquiring the pulses (8.3.1), since a lower threshold will inevitably have more pulses.

The Q_m must be calculated with the same amplifier gain or the same magnitude and phase window width if comparisons or trending of Q_m are performed. See Section 8.7.1 for the rationale.

8.8.3 Integrated PD Indicators

The peak PD magnitude is only one indicator of PD activity. There are several other indicators that look at both pulse magnitude and the number of pulses per second – i.e. they are an indicator of the total PD activity. In general, if the pulse repetition rate is increasing there are more active PD sites within the HV equipment. Since digital integration is common with such indicators, they are sometimes referred to as integrated indicators. A consequence of all these integrated quantities is that the result may be the same for a few PD pulses of high magnitude as for thousands of pulses at multiple sites of low magnitude. Thus, the integrated indicators cannot distinguish between large pulses at only one or two locations in the HV equipment, and thousands of low magnitude PD pulses at many locations. Yet the insulation is more likely to fail at the sites with large discharges. This makes integrated quantities less useful for deciding if insulation maintenance is needed or not.

The following are the most popular integrated quantities.

The *average PD current* (I) can be calculated for IEC 60270-compliant detectors in the LF range on capacitive test objects using the formula:

$$I = \left(q_1 + q_2 + q_3 + \ldots q_i\right) / T$$

where q_i are the individual apparent charges associated with each PD and T is the time in seconds for the measurement. All of the PD pulses above a threshold must be measured. The current is in Amperes if the charges are in Coulombs. If the PD is measured in mV or any other unit except charge, then the result is NOT current, and it is sometimes referred to as the "nqs" or "NQS" indicator by some vendors.

The *average discharge power* P is an indicator of the average power in Watts supplied to the test object for PD activity. This calculation requires both the charge (q_i) in each PD pulse, plus the applied voltage across the test object at the corresponding PD event (V_i):

$$P = \left(q_1 V_1 + q_2 V_2 + \ldots\ldots . q_i V_i\right) / T$$

This is only valid for IEC 60270 detectors in the LF range, assumes a capacitive test object, and the units are Coulombs and Volts. An accurate measurement of the voltage is needed when each PD pulse occurs. Thus, the average PD power is usually only calculated in LF measurements done offline, when there is a resistive or capacitive, voltage divider to measure the test object voltage. It is difficult to understand the rationale of the average discharge power since the voltage across the voids at the time of the discharge is approximately constant, no matter what the applied voltage is (see Figure 3.2a).

The *quadratic rate* D is an integrated quantity that was widely used in France for decades. It assumes that the destructive power of the discharges is proportional to the square of the apparent charge in each discharge. This implies that large PD pulses are more important than small pulses, compared to I and P. For IEC 60270-compliant instruments in the LF range, it is calculated from

$$D = \left(q_1^2 + q_2^2 + \ldots . q_i^2\right) / T$$

where D has the units of Coulombs squared per second if q_i is in Coulombs and T is in seconds. If the PD is measured in mV or another non-apparent charge quantity, then D will have different units.

The *NQN* is an integrated quantity that is valid for both IEC 60270- and IEC 62478-compliant PD systems: that is for LF, HF, VHF, and UHF detectors that do not measure the PD in terms of charge. The *NQN* has been applied to both capacitive and inductive test objects and was originally defined in IEEE 1434:2000 for PD measurement in stator windings. As conceived by Smith and Lyles in the late 1980s, it is the normalized area under the PMA curve described in Section 8.7.1 and numerically calculated from:

$$NQN = FS\left(\log P_1 / 2 + \log P_2 + \log P_3 + \ldots . . \log P_{N-1} + \log P_N / 2\right) / N \times G$$

where the log is to the base 10, P_i is the number of pulses per second in magnitude window i, N is the number of magnitude windows between the lower threshold and the highest magnitude window, G is the amplifier gain (usually assumed to be 1), and FS is the highest magnitude the measurement was done at. As with other integrated PD indicators, the *NQN* changes if the gain or magnitude window width changes.

The *NQN* equation is a true digital integration, where there are no assumptions about the number of magnitude windows or the window width, as occurs in the other quantities described above. Taking the logarithm of the pulse count rate deemphasizes the contribution of high repetition rate pulses, which usually have lower magnitude. This is similar to the quadratic rate where higher magnitude pulses are emphasized.

8.9 Post-Processing Software for Interference Suppression and PD Analysis

Modern commercial PD instruments (Section 8.3) store either just the Ø-*q*-*n* data in a memory array, or capture the waveform of each pulse (above a threshold). In either case, software applications can do much more than produce the simple charts and PD activity indicators presented in Sections 8.7 and 8.8. These apps can help suppress the presence of interference in PRPD plots, create PRPD plots for each type of PD that might be occurring in a test object, or both. Additional information on the application of post-processing to suppress interference is in Section 9.3. The methods that have seen some commercial application include:

- Statistical processing
- Time-frequency mapping
- Three-phase synchronous processing (which one vendor calls 3PARD)
- Software-based data censoring
- Pattern recognition using artificial intelligence or expert systems

This is a rapidly evolving field. Every year researchers devise many new approaches for post-processing which are often presented at research-oriented conferences such as the IEEE Conference on Electrical Insulation and Dielectric Phenomena, the Condition Monitoring and Diagnostic Conference and the International Symposium on High Voltage, all of which are easily found by online searches.

8.9.1 Statistical Post-Processing

Applying statistical analysis based on the stored Ø-*q*-*n* data was the first commercial post-processing method (Section 1.5.3.3). This approach probably started since, when looking at just one PD polarity, for example from −45° to +135° of the AC cycle, the PRPD patterns in Figures 8.10 and 8.11 look similar to a probability distribution. On this basis, one can calculate the mean phase position and its standard deviation for either the negative or positive pulses. Note that only one polarity of pulses can be analyzed at a time.

The mean phase position indicates the phase-to-ground AC phase angle on which the PD is centered. As will be discussed in Section 10.5, void PD tends to be centered around 45° for one pulse polarity and 225° (for the other polarity) of the phase-to-ground AC cycle (Figure 8.10), while corona is usually centered around 90° and 270° [14]. In three-phase HV equipment where the three phases are co-mingled (e.g. metalclad AIS and stator windings), void PD may be phase-shifted +/−30° (see Section 10.5.4). Thus, by calculating the mean phase angle of each polarity of pulses, one can have an objective indication of the root cause of the PD.

The meaning of the standard deviation in a PRPD plot is more difficult to understand than the mean. If the phase angle standard deviation is small, then it implies that that the PD are clustered in a narrow range of phase angles, and perhaps there are only a few PD sites. If the standard deviation is large (say 90°), then perhaps there are many PD sites of different volumes and shapes, each triggering PD at different angles of the AC cycle.

Researchers noticed that PRPD patterns are sometimes skewed, with most of the PD activity occurring close to the AC voltage zero crossings. Figure 8.11 is a possible (but not perfect) example of a PRPD pattern that does not have a classic "bell-curve" of normal distribution shape. Thus, in addition to the mean and standard deviation, the skew (an indicator of symmetry) and kurtosis (if there is a low probability tail) of the *Ø-q-n* data of each polarity can be calculated. The formulae are in any introductory statistical analysis book.

Gulski, among others, published tables of the mean, standard deviation, skew and kurtosis of typical defects in HV equipment that cause PD [15]. Such tables seem to be of use when a single predominant cause of PD, with few PD sites, are present in the test object. If there are a large number of defects, and/or two or more simultaneously occurring types of PD, then the method seems to be less useful, or perhaps requires applying expert system or AI tools (Section 8.9.5). Today, few end users seem to be using this post-processing method.

8.9.2 Time-Frequency Maps

This has become one of the most popular post-processing methods and was first introduced commercially by TechImp. The *TF* map was conceived at the early 2000s at the University of Bologna as device to highlight in a simple way the differences in the waveforms of the pulses received by a PD instrument. The method requires an ADC with a sampling rate of 40 MHz or more. Each digitized pulse (above a threshold) is then stored in memory (Figure 8.3). The *TF* map uses the concept of an equivalent time length (*T*) and equivalent bandwidth (*F*) for each pulse, the definitions for which can be found in any standard book of signal theory (for example [16]). The *TF* map is calculated and represented as a dot in a time-frequency (*TF*) plane [17]. By inspecting the *TF* map (a task that can be performed visually, as the map is in two dimensions), clusters of pulses with different shapes can be highlighted and treated separately. In many cases, the pulses from a cluster will have its own unique PRPD pattern. For each cluster, the associated PRPD pattern be can be used for (Section 10.5):

- interference rejection if the PRPD pattern suggests that the pulses are not correlated with the voltage or show a correlation structure that is not related to the physics of PD, or
- PD source identification as only one source at a time is analyzed.

Before explaining how the *TF* map is calculated, Figure 8.15 presents an example of its application to an air-cooled turbogenerator. The complete pattern is very complicated as it is the superposition of four different phenomena. The detected pulses can be separated into four sub-groups based on their features in the *TF* map. Accordingly, four sub-patterns can be extracted, each of them pointing at a different type of phenomenon: exciter interference, crosstalk from another phase, slot discharges, and end-winding discharges.

The *TF* map calculation is relatively straightforward. Readers familiar with statistics will recognize a lot of resemblance of the *TF* parameter with the concept of standard deviation. The first operation is the normalization of the recorded pulse $s(t)$:

$$\tilde{s}(t) = s(t) / \sqrt{\int_0^L s(\tau)^2 \, d\tau}$$

In practice, normalization maps pulses with the same shape but differing by a multiplying constant into the same point in the *TF* plane. This is very important as PD pulses coming from the same location (same transfer function thus same waveform) but having different magnitudes must be treated as the same entity.

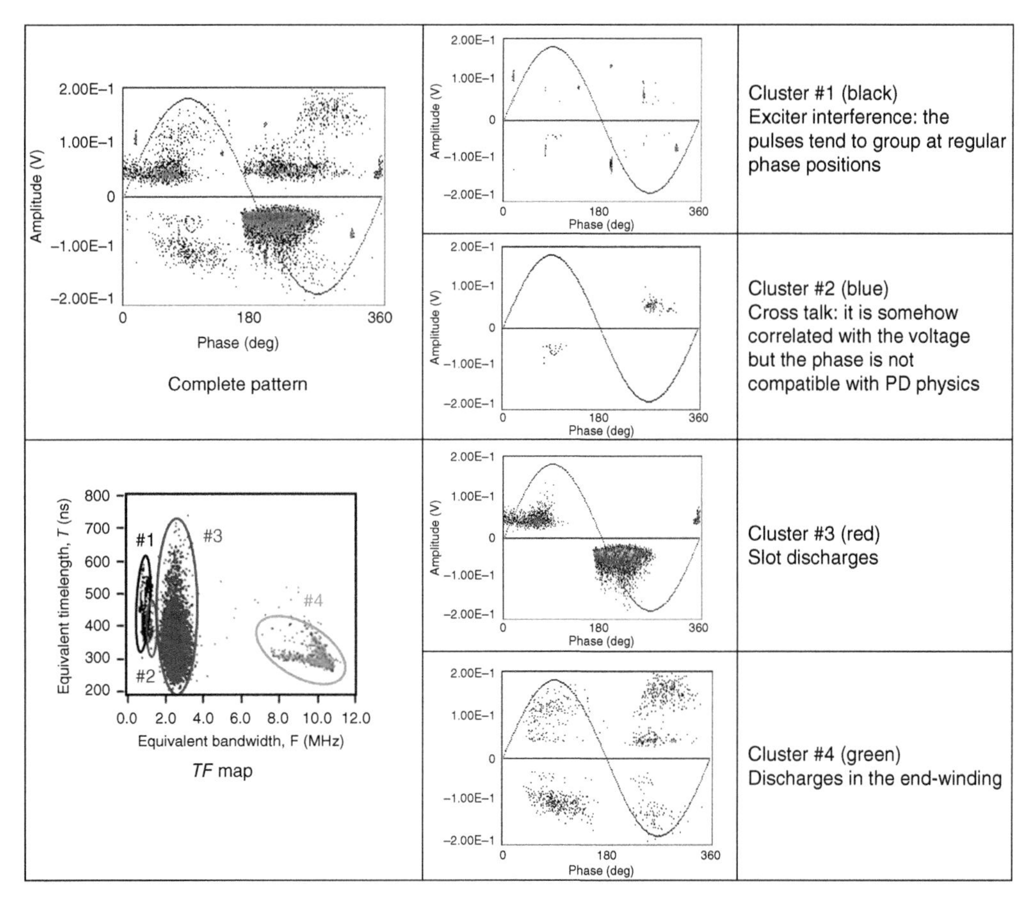

Figure 8.15 Example of application of the *TF* map on an air-cooled turbogenerator.

The following steps are performed in the time domain, dealing with the squared normalized pulse $\tilde{s}(t)^2$. First the "gravity center" of $\tilde{s}(t)^2$ is calculated:

$$t_0 = \int_0^L t\tilde{s}(t)^2\, dt$$

where t is time, L is the duration of the pulse. The next step is the calculation of the T parameter:

$$T = \sqrt{\int_0^L (t - t_0)^2\, \tilde{s}(t)^2\, dt}$$

Stressing again the analogy with probability and statistics, the T parameter is like the standard deviation of a probability function equal to $\tilde{s}(t)^2$, whereas t_0 bears analogy with the expected (or mean) value. The same operation is performed in the frequency domain dealing with the Fourier transform $\tilde{S}(f)$ of $\tilde{s}(t)$:

$$F = \sqrt{\int_0^\infty f^2 \left|\tilde{S}(f)\right|^2 df}$$

The "gravity center" (or "expected value") in the frequency domain need not be calculated as, in virtue of the symmetry of the spectrum, it is always at $f = 0$. It is important to observe that, thanks to the Parseval's theorem, the last equation can be worked out in the time domain, simplifying a lot the implementation on platforms such as digital signal processors (DSPs) or FPGAs.

As with any technical tool, the *TF* map has drawbacks. As an example, the superposition of noise with PD pulses can blur the clusters, leading to a sub-optimal separation of the sources or interference rejection (for this reason, it is always advisable to limit the acquisition window per pulse to the minimum). Furthermore, the *TF* map (as any other tool) is unable to separate pulses from different sources when attenuation phenomena tend to make them very similar in both the time and frequency domain.

8.9.3 Three-Phase Synchronous Pattern Analysis

This post-processing method requires the simultaneous capture of PD signals from all three phases of a test object, and is primarily used for online PD monitoring or offline transformer testing, where all three phases are energized at the same time. The technique was first investigated for instrument transformers and stator windings [18, 19]. It was first introduced commercially by Omicron and seems to be most useful for PD measurement in power transformers and stator windings [9, 20]. Most types of commercial online PD systems cannot use this post-processing method since they use a multiplexor to sequentially record the PD from each phase using a single channel.

Each of the three channels consists of a separate PD sensor and ADC for the hardware approaches shown in Figures 8.2 and 8.3. Each recorded pulse (above a threshold) is also time-stamped. When the pulses are recorded with the same time stamp (within a few hundred nanoseconds), it is assumed the pulse on all three phases came from the same discharge or interference event. The absolute magnitude of the pulse (in pC or mV) measured in each phase is then plotted on a three-dimensional plot with the pulse magnitude of each of the three phases being an axis. An example is shown in Figure 8.16. This plot is sometimes called a three-phase amplitude relation diagram – 3PARD. The red dot on the right chart in Figure 8.16 identifies a point in three-dimensional space with the address corresponding to the signal magnitude detected in each phase from one event (PD or interference). If there was no cross-coupling to other phases (as is typical in GIS and power cables), the red dots would always be on one of the axes, rather than in the volume between the axes.

When thousands of events are recorded for a few minutes, the red dots gather in a cluster in the three-dimensional diagram. The result is a chart shown in Figure 8.17 for a particular test object. Apparently, when there are different types of PD sources, or there are interferences from outside the test object, there will be a unique cluster of points on the three-dimensional chart. If multiple types of PD or interference are occurring at the same time, there will be multiple

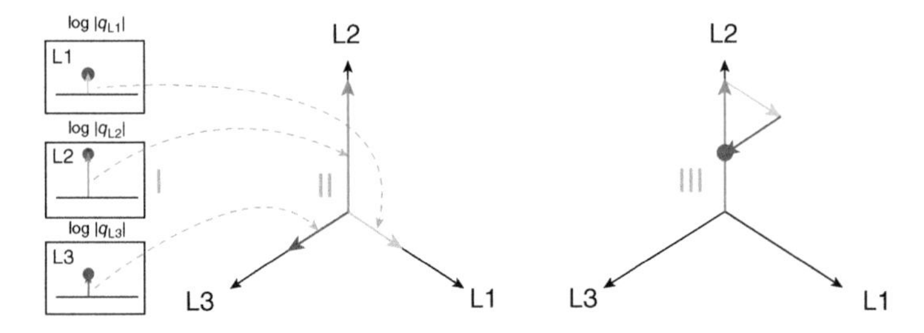

Figure 8.16 Example of the recording of a single event on Phase L2 (left plot), and the cross-coupled signals into phases L1 and L3. This enables a single point to be plotted on the right-hand "3PARD" chart (note that the red dot in the right chart is not on the phase L2 axis). *Source:* Koltunowicz and Plath [9].

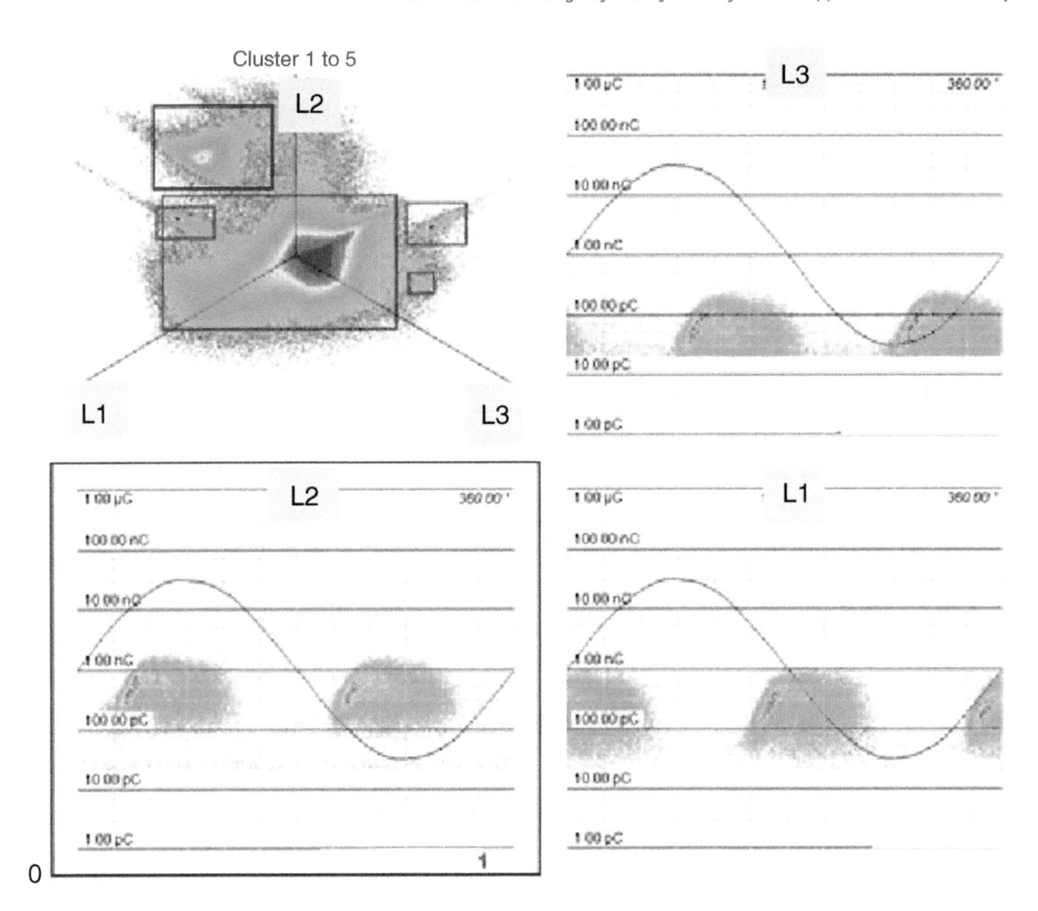

Figure 8.17 Three-dimensional amplitude comparison chart (upper left) for a three-phase test object. The rectangles in the upper left chart are clusters that have been manually identified, with PRPD patterns shown for some of the rectangles. *Source:* From Koltunowicz et al. [20].

clusters, as shown in the upper left of Figure 8.17. Each cluster can be circled (for example the red square in the upper right of Figure 8.17), and all the points within a circle can be used to create its own PRPD plot. The developers of this post-processing method indicate that each cluster tends to have its own PRPD pattern that is different from the PRPD patterns in other encircled clusters. PD specialists can then sometimes identify the source of the PD or interference for each cluster PRPD pattern. In this way, the method is similar to the *TF* map approach described in Section 8.9.2.

Users report that this method of separating different sources is useful, although considerable expertise is needed in identifying clusters and associating the PRPD patterns with specific types of PD or interference.

8.9.4 Software-Based Censoring

It has long been possible to make inconvenient data on PRPD plots "disappear" using editing tools like Photoshop™. For example in Figure 8.18, the complete PRPD chart is shown in Figure 8.18a, i.e. all pulse data is included. But the source of the pulses in clusters of data (the red squares in the

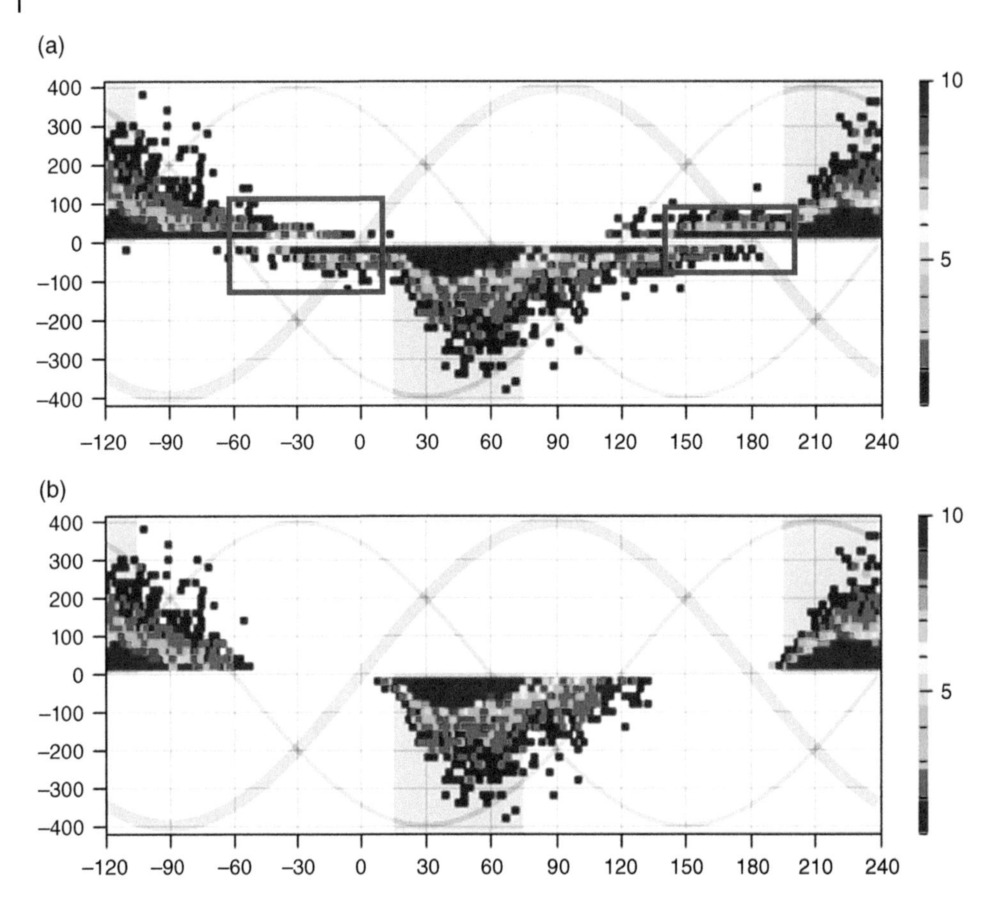

Figure 8.18 PRPD plots from an online VHF measuring system. The vertical scale is in mV. All the data is included in (a). Certain "nonideal" data is excluded in (b), leaving a classic PRPD pattern from void discharges. The censoring windows are outlined by the red rectangles in (a).

left of Figure 8.18a) may not be clear. That is, it is not due to phase-to-ground or phase-to-phase PD, so perhaps the clusters are due to cross-coupling of PD from another phase, and/or it may be due to interference. To present a clearer idea of the more classical PD pattern, the clusters of non-classical PD can be deleted from the file using a suitable software app, to produce the "cleaned-up" PRPD plot in Figure 8.18b.

This type of data censoring can be misused, especially if the complete PRPD plot is not shown, or it is not made clear that the data has been censored. Unlike the tools presented above, there is no physical basis for removing the "inconvenient" data. This method of post-processing should be avoided.

8.9.5 Artificial Intelligence (AI) and Expert Systems

Research into the application of AI and expert system tools for PD data analysis has been ongoing for more than 30 years. The purpose of such tools is usually to:

- identify different types of PD (corona, surface PD, void PD, etc.) within the HV equipment;
- identify where the PD may be occurring (for example the locations of PD sites in a length of power cable or GIS); and/or
- separate test object PD sources from interferences.

The idea is to do this automatically, so that general users of PD systems (as most users of PD systems are) can be as proficient in identifying actual PD sources and sites as well as the "experts" who use the PD system daily. Although commercial systems do not yet make normal PD system users as good as the experts, the rate of research has been increasing almost exponentially, both in the development of the AI tools and the specific application of the technology to PD data interpretation.

Expert systems were an earlier set of technologies where domain experts (in our case PD experts) created rules for analyzing features of the PD data such as polarity magnitude predominance, pulse repetition rate, phase angle of occurrence, and so on [21]. Unlike conventional algorithms, expert systems produce results even when data is missing. Most expert systems algorithms can also produce an indicator of the reliability of the analysis.

Probably the first commercial application of expert systems for PD data analysis is a program called MICAA, introduced by Iris Power in 1994 [21]. MICAA determined the most likely aging processes in rotating machine stator windings, based primarily on PD data. MICAA used Mycin reasoning to estimate the most likely cause of the PD using features such as the ratio of positive and negative PD activity as well as the effect of load and temperature on PD activity [19]. This software is no longer commercially available, partly because the software required more information than many users were willing or able to provide.

AI refers to a very wide range of tools that can create its own diagnosis based on the selected features, using either supervised learning or unsupervised learning. Supervised learning is where the AI system develops its diagnosis based on training data sets where the diagnosis is provided to it. Once trained, it produces a diagnosis on any new data. Generally, such systems do not learn after the initial training. Since about 2015, unsupervised learning using "deep learning" methods seems to have gained ascendency. Such systems continuously learn. However, they require huge sets of data to become effective, and when a diagnosis is produced, it is often not apparent what lead to the diagnosis (i.e. they are black boxes).

For both types of AI methods, the designer needs to develop a set of features for the system to analyze. Such features can include (but are not limited to):

- Raw $Ø$-q-n data
- $Ø$-q plus time of occurrence in a stream of pulses
- q, pulse risetime, pulse fall time, pulse width, $Ø$
- Moments of the statistical distribution (mean, standard deviation, skew, kurtosis) of the positive and negative PRPD pattern (Section 8.9.1)
- Time T and frequency F in a TF map (Section 8.9.2)
- The pulse magnitudes from the same event on each of the three phases (Section 8.9.3)
- Parameters from wavelet transformations

Hundreds of papers have been published from 2015 to 2021 on possible features and the AI methods for PD analysis. A comprehensive review of the most recent work, including an extensive analysis of the published work on deep learning and conventional neural network methods, has been published by Lu et al [22]. Some earlier and later research is in references [23–26].

To date, expert systems and AI are not widely used by owners of PD systems. Although many PD system vendors have incorporated some of these tools, it seems few general users have actually found them useful. The reasons for the lack of use are not clear, but it may be due to the lack of confidence in the diagnoses. This has been a common finding about the use of AI in many fields [27]. However, given the amount of research currently being invested in AI applications for PD analysis, hopefully, such tools will one day prove to be useful.

References

1 Kreuger, F.H. (1989). *Partial Discharge Detection in High-Voltage Equipment*. Butterworths.
2 Lemke, E., Berlijn, S., Gulski, E. et al. (2008). *Guide for Partial Discharge Measurements in Compliance to IEC 60270*, 366. Cigre Technical Brochure.
3 Lachance, M. and Oettl, F. (2020). A study of the pulse propagation behavior in a large turbo generator 2020 IEEE Electrical Insulation Conference (EIC), Knoxville, TN (22 June 2020 – 3 July 2020), pp. 434–439. IEEE.
4 Bartnikas, R. and Levi, J.E. (1969). A simple pulse-height analyzer for partial discharge rate measurements. *IEEE Transactions on Electrical Insulation* 18: 341–345.
5 Kurtz, M. and Stone, G.C. (1979). In-service partial discharge testing of generator insulation. *IEEE Transactions on Electrical Insulation* EI-14: 94–100.
6 Austin, J. and James, R.E. (1976). On-line digital computer system for measurement of partial discharges in insulation structures. *IEEE Transactions on Electrical Insulation* EI-11: 129–139.
7 Kelen, A. (1976). The functional testing of HV generator stator insulation. Cigre, Paper 15-03.
8 Holboll, J.T., Braun, J.M., Fujimoto, N. et al. (1991). Temporal and spatial development of PD in spherical voids in epoxy related to the detected electrical signal. *IEEE Conference on Electrical Insulation and Dielectric Phenomena*, Knoxville, TN (20–23 October 1991), pp 581–586. IEEE.
9 Koltunowicz, W. and Plath, R. (2008). Synchronous multi-channel PD measurements. *IEEE Transactions on Dielectrics and Electrical Insulation* 15 (6): 1715–1723.
10 Campbell, S.R., Stone, G.C., and Sedding, H.G. (1992). Application of pulse width analysis to partial discharge detection. *IEEE International Symposium on Electrical Insulation*, Baltimore, MD (7–10 June 1992), pp 345–348. IEEE.
11 Stone, G.C. and Warren, V. (2006). Objective methods to interpret PD data on rotating machine stator windings. *IEEE Transactions on Electrical Insulation* 42: 195–200.
12 Campbell, S.R. and Sedding, H.G. (1995). Method and device for distinguishing between partial discharge and noise. US Patent No. 5,475,312.
13 CIGRE (1969). Recognition of Discharges. WG 21.03 Report. Electra No. 11, pp 61–98.
14 Bartnikas, R. and McMahon, E.J. (1979). *Engineering Dielectrics Volume 1: Corona Measurement and Interpretation-STP 669*. ASTM.
15 Gulski, E. (1993). Computer-aided measurement of partial discharges in HV equipment. *IEEE Transactions on Electrical Insulation* 28 (6): 969–983.
16 Franks, L.E. (1969). *Signal Theory*. Prentice-Hall.
17 Cavallini, A., Conti, M., Contin, A. et al. (2004). A new algorithm for the identification of defects generating partial discharges in rotating machines. IEEE ISEI, Indianapolis, IN (19–22 September 2004), pp 204–207. IEEE.
18 Moreau, C. and Charpentier, X. (1996). On-line partial discharge detection system for instrument transformers. *Conference Record of the 1996 IEEE International Symposium on Electrical Insulation* 1: 65–68.
19 Sedding, H.G., Campbell, S.R., and Stone, G.C. (1992). Evaluation of coupling devices for in-service PD detection on thermal alternators. *Canadian Electrical Association Research Report 738G631*.
20 Koltunowicz, W., Gorgan, B., Broniecki, U. et al. (2020). Evaluation of stator winding insulation using a synchronous multi-channel technique. *IEEE Transactions on Dielectrics and Electrical Insulation* 27: 1889–1897.
21 Lloyd, B.A., Stone, G.C., and Stein, J. (1994). Motor insulation condition assessment using expert systems software. Proceedings of IEEE Pulp and Paper Industry Conference, Nashville, TN (20–24 June 1994), pp. 60–67. IEEE.

22 Lu, S., Chai, H., Sahoo, A., and Phung, B.T. (2020). Condition monitoring based on partial discharge diagnostics using machine learning methods: a comprehensive state-of-the-art review. *IEEE Transactions on Dielectrics and Electrical Insulation* 27 (6): 1861–1888.

23 Salama, M.M.A. and Bartnikas, R. (2000). Fuzzy logic applied to PD pattern classification. *IEEE Transactions on Dielectrics and Electrical Insulation* 7 (1): 118–123.

24 Raymond, W.J.K. et al. (2015). Partial discharge classifications: review of recent progress. *Measurement* 68: 164–181.

25 Yao, R., Li, J., Hui, M. et al. (2021). Pattern recognition for partial discharge using multi-feature combination adaptive boost classification model. *IEEE Access* 9: 48873–48883.

26 Govindarajan, S., Ardila-Rey, J.A., Krithivasan, K. et al. (2021). Development of hypergraph based improved random forest algorithm for partial discharge pattern classification. *IEEE Access* 9: 96–109.

27 Thompson, N.C., Greenewald, K., Lee, K. et al. (2021). Deep learnings diminishing returns. *IEEE Spectrum* 58: 50–55.

9

Suppression of External Electrical Interference

9.1 Impact of External Electrical Interference

External electrical interference (also known as electromagnetic interference – "EMI," or radio-frequency interference – "RFI") is encountered with most offline and all online PD measurements. In this book, noise refers to random, noncausal, nondeterministic electrical signals (i.e. "white" noise) arising from thermal noise in the electrical/electronic components within the measurements instruments themselves, whereas external electrical interference comprises any signals that are detected which are not PD pulses originating within the test object PD. Also often referred to historically as "disturbances," these signals are defined as not being test object PD, but nevertheless are often pulse-shaped signals, and thus may easily be confused with test object PD. Examples of both are presented below. These definitions are consistent with those in IEC 60270, but differ from those in some other PD standards.

Knowledge of the typical types of external interference encountered when performing electrical PD measurements is essential, since it affects the quality of any PD measurement and the probability of a false positive indication of high PD. The interference environment and its impact on the PD results tend to be different between offline tests performed in a factory or laboratory, and either offline or online tests performed in the substation or plant. Sections 9.1.1 and 9.1.2 discuss these differences in general.

This chapter also discusses methods that have been used to suppress the influence of interference. Many of these methods can be used with offline tests in the factory/laboratory, or with off- and online tests at the substation/plant site. Which methods are most effective depends on the nature of the interference and the type of test object. So Chapters 12–16 also present the most common interference-suppression methods for each type of equipment.

Sections 8.4 and 8.9 have already outlined how the interference suppression methods are implemented in the PD measurement system hardware or software, respectively.

9.1.1 Factory Testing

In offline tests on prototype or new HV equipment tested in the laboratory or factory, the presence of pulse-shaped interference is one of the main factors in determining the sensitivity of the PD measurement. The test technician in a factory QA PD test may think that any high interference that is present are legitimate PD signals and decides the test object failed the PD test, when in fact the equipment may have been well made. This then leads to unnecessary rejection of the

Practical Partial Discharge Measurement on Electrical Equipment, First Edition. Greg C. Stone, Andrea Cavallini, Glenn Behrmann, and Claudio Angelo Serafino.

HV equipment because of the positive (failing) result. This is termed a false positive. False positives of high PD create unnecessary costs due to extra testing being required, the scrapping of good HV equipment and/or delays in equipment shipping.

9.1.2 Condition Assessment Testing

As for factory QA testing, false positives can occur with onsite/offline PD tests and online PD tests of HV equipment to assess if insulation aging is occurring and maintenance is prudent. False positives can occur when high levels of external interference are assumed to be legitimate test object PD, leading the test technician to believe the insulation system is degrading, when in fact the true PD is low and the insulation system is in good condition.

Alternatively, the test technician may assume all the detected signals are caused by external interference – when, in reality, some of the signals are due to test object PD. This leads to a false negative because legitimate test object PD was attributed to be interference, and the insulation is then assumed to be in good condition, when the insulation is actually deteriorated. In practice, external interference is more likely to cause false positives than false negatives.

False positives of high PD during onsite/offline and online condition assessment tests are bad because the high readings will usually trigger more tests or disassembly of the HV equipment to verify the initial high PD readings. Some end users have also been known to unnecessarily replace or repair the HV equipment based on apparently high PD activity, without resorting to other confirmation. This adds costs and extends maintenance outages. More importantly, asset managers lose confidence in PD testing when the false positive rate is too high. This can lead to canceling of future PD testing and monitoring, thereby eliminating the benefits of PD testing to anticipate and avoid equipment failure.

9.2 Typical Sources of Noise and External Electrical Interference

9.2.1 Electrical/Electronic Noise

Typical sources of electrical noise are thermal (Johnson-Nyquist) noise in resistors, $1/f$ (“flicker”) noise, shot (Poisson) noise, “popcorn” or burst noise (typically associated with semiconductor devices), created by the movement of charge carriers such as electrons in electronic components. The Johnson-Nyquist noise signal increases with the square root of temperature and the square root of bandwidth. Thus, ultrawide band PD detectors will encounter higher levels of this type of noise [1].

For very sensitive PD measurements in the few pC range or less, electronic noise may be a problem. Keeping such PD instrument internal noise to a minimum requires careful circuit design and component choices. Other than reducing the measurement bandwidth or for example cooling the electronic circuits to very low temperatures, there is no practical way to suppress such noise without redesigning the measurement instrument.[1]

In addition to these component-based sources of electronic noise, the switching noise produced by digital gates and clocks in modern PD detectors produce a continuous baseline noise across the AC cycle (Figure 9.1). This type of noise can be suppressed by a circuit board layout that minimizes the lengths of the copper traces, employs a good ground plane, and terminates high-frequency

1 For example see: https://www.allaboutcircuits.com/technical-articles/electrical-noise-what-causes-noise-in-electrical-circuits/.

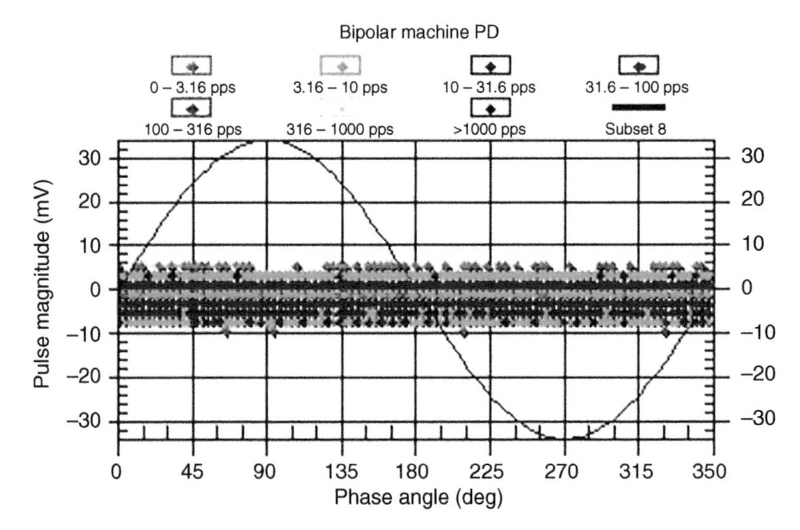

Figure 9.1 PRPD plot of noise across the AC cycle caused by electronic switching within a VHF PD instrument on a sensitive scale. *Source:* Courtesy of Iris Power.

signal traces in their characteristic impedance. The input signals, amplifiers, and ADCs can also be shielded from other circuits. This type of noise tends to be the most important type of noise in modern VHF and UHF PD measurement due to their wide bandwidth and usually determines the sensitivity of PD measurements.

9.2.2 External Electrical Interference ("Disturbances")

Probably the most common sources of external electrical interference are AM and FM radio broadcast stations, TV stations, and other radio-frequency communications transmissions (including mobile (cell) telephones, police and emergency VHF & UHF radio, radar, Wi-Fi, and Bluetooth). In fossil-fuel power-generating stations, many of these types of radio transmitters may be installed at the tops of chimneys (using their height to optimize coverage area), which can then cause very high levels of interference throughout the plant and associated substation. Power line carrier signals that allow communication between utility substations, generating stations, and system control centers can create substantial interference, especially with charge-based PD detectors such as those compliant with IEC 60270. All the above sources tend to occur at specific frequencies and are not synchronized to the 50/60 Hz power system. Such radio signals are easy to remove from PD sensors using analog or digital notch (band-stop) filters or by performing offline tests in electromagnetically shielded rooms.

Pulse-shaped interference is the most difficult to successfully suppress in all types of electrical PD testing, and especially in online PD tests. There are many possible causes of pulse-shaped interference, some of which are synchronized to the AC voltage, and some of which are not. Unsynchronized interferences occur across the entire AC cycle. Details on interference suppression techniques are in Sections 8.4, 8.9, 9.3, and 9.4.

9.2.2.1 PD and Corona from Connected Equipment

This type of interference is very common and has led to many false-positive indications of high PD in test objects. Such interference is synchronized to the AC cycle and are easily confused with test object PD. For offline PD tests, examples include corona from metallic high-voltage leads or suspended ground leads, which have too small a diameter. In online tests, corona may occur from bus work in substations or isolated phase bus (IPB) in generating stations. If it is raining or snowing in a substation, PD may occur on the surface of porcelain bushings. In these cases, the PD and corona are essentially harmless to the operation of the busses or bushings, and does not imply the test object itself has any PD. To prevent false positives in offline tests, these sources can be identified by energizing the leads, etc. without the test object connected. Then the sources must be eliminated by removing the source of the corona/PD (for example by increasing the radius of connecting leads, using corona shields around sharp points and/or partly conductive putty over any sharp points/edges). For online tests, where such interference can rarely be eliminated, other methods to separate test object PD from other PD/corona include time-of-flight separation, pulse shape analysis, and TF maps (Sections 8.4.2, 8.4.3 and 8.9.2, respectively).

9.2.2.2 Arcing from Poorly Bonded Metal and Connections

In offline tests, any metal objects that are "floating," i.e. not connected to ground or the high-voltage terminal, may acquire a high voltage due to capacitive coupling to nearby high-voltage components. If the floating metal is close to ground or high voltage, the potential difference may be such that the air breaks down between the metal object and, for example, ground. If nearby, this arcing can create very substantial interference that swamps the test object PD. Similarly, poor electrical connections in the high voltage or ground circuits can lead to contact arcing (Section 4.4). The solution is to make sure there are no floating metal objects in the HV test environment, and all connections are bolted (rather than using "alligator" clips).

In offline/onsite PD tests, the same issues can occur with floating metal. However, it is poor electrical connections elsewhere in the plant or substation that can be among the most serious disturbance sources in operating HV equipment that is being PD tested. Examples include:

- ungrounded fencing or barriers
- poor connections between parts of the ground grid
- poor bonding of isolated phase bus (IPB) enclosures in segmented busses or cable trays
- oxidized bolts or over/under torqued bolts in electrical connections
- ungrounded shields in current and potential transformers

These sources of interferences are synchronized to the AC voltage. Suppression in online tests, where it is assumed that the sources cannot be eliminated, involves the use of three-phase synchronous detection, TF maps, pulse shape analysis, or time-of-flight methods (Sections 8.4.2, 8.4.3, 8.9.2, and 8.9.3). In addition, experts looking at PRPD plots (Section 10.5) may ignore such signals based on their very high magnitude and the tendency for the pulses to occur near the AC voltage zero crossings of the phase they are occurring in.

9.2.2.3 Electronic Switching

This is an important source of interference pulses and is becoming more pervasive. The interference can be either synchronized or unsynchronized to the AC voltage. In generating stations, the generator rotor winding is often supplied with DC from electronic power supplies called "static exciters." Such supplies are fed from a three-phase AC source that rectifies the AC into DC using thyristors. The triggering of the thyristors creates switching noise resulting in six pulses per AC cycle (Figure 9.2). Snubber circuits and the relatively slow switching speed of thyristors usually limit the interference to less than about 100 kHz – but this is sometimes enough to swamp the PD measured with IEC 60270 PD detectors. Similar synchronized interference can come from industrial battery chargers common in generating stations and substations.

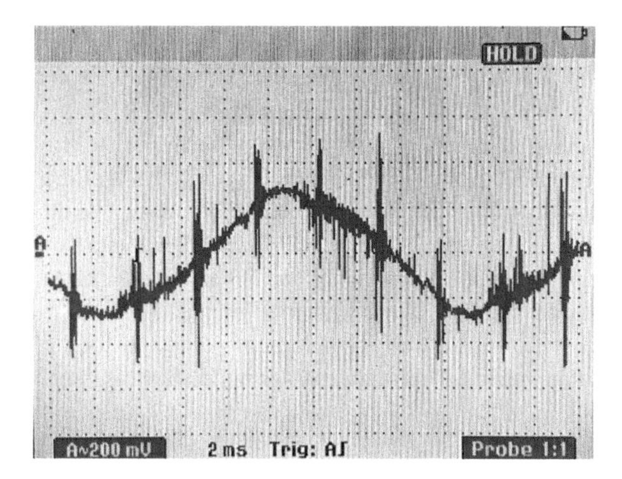

Figure 9.2 Oscilloscope image of static exciter pulses superimposed on the residual 60 AC voltage from a PD sensor. The static exciter pulses are equal-distant from each other and phase-locked to the 60 Hz AC supply. *Source:* Courtesy of Mladen Sasic, Iris Power L.P.

Perhaps more important is the unsynchronized sources of electronic switching in power electronics. Examples include:

- Switch-mode DC supplies commonly used for computers and many appliances. These supplies often switch at 100 kHz using fast MOSFET or IGBT transistors.
- Variable frequency invertors for motor speed control using voltage source-pulse width modulation technology. Such inverter fed drives are now widely used in industrial and generating plants. The switching rate is often at 2 kHz for medium-voltage motors and up to 50 kHz for low-voltage machines, with switching devices that change from on to off in 20–200 ns. Voltage spikes of several hundred volts with frequency content up to 20 MHz have been measured. As switching devices improve, the switching voltage has been increasing and the impulse risetime has been decreasing, which makes the interference from newer model invertors increase.

9.2.2.4 Slip Ring/Brush Arcing

Brushes and slip rings are used in motors and generators to get power to rotating elements. Such arcing also occurs in many motor-driven power tools such as electric drills. Even slip rings and brushes in good condition create some arcing. Similarly, shaft-grounding brushes in rotating machines often arc. Arcing is also common from moving overhead electric cranes in plants, which use brushes to obtain power from the rails. Such arcing can produce interference up to 10s of MHz and pervade a plant.

9.2.2.5 Lighting

Some types of illuminating devices can create significant electrical interference pulses. Fluorescent light is an example, since it involves glow-type gas discharges. Discharges tend to occur in older lamps due to intermittent operation and often the ballast creates its own interference. Mercury vapor and other types of arc lighting can also create significant interference. Interference from lighting is becoming less prevalent in high-voltage labs and plants since low-noise LEDs are rapidly replacing fluorescent and arc lighting.

9.3 Interference Suppression for Offline PD Testing

The most fundamental method to address interference is first to observe the signals in a PRPD plot (Section 8.7.3) with digital PD detectors, or even better, on an oscilloscope, with respect to the applied AC high voltage. As presented in Section 3.10, test object PD occurs at specific parts of the AC cycle. Especially with real-time observation on an oscilloscope (where PD pulses dance around the display), an experienced observer is often able to separate test object PD from all other types of interference, except PD from other connected apparatus. In the absence of a PD expert who has a "gut feel" for what is interference, various tools are available to suppress interference.

In offline PD tests there tends to be much better control of the high-voltage power supply for the test, as well as the environment, than with online testing. This often enables more effective interference suppression in offline tests.

9.3.1 Electromagnetic Shielded Rooms

Signals from radio/TV/mobile phones, as well as some types of interference, radiate RF signals into the environment. When very sensitive PD measurements are required (in the pC range and lower), these types of interference can be effectively eliminated using an electromagnetically shielded room, also called a Faraday cage. Most high-voltage labs include a solid metal (e.g. copper or mu-metal sheets) or metal mesh (e.g. copper screen) cladding on the walls, ceiling, and floor that suppresses most interference of this type originating from outside of the room.

There are also many commercial sources of shielded rooms that use bonded metal sheets, have no windows, use special metal door frame fixtures/fingers and feedthroughs to minimize leakage of external signals into the room. They also use LEDs for lighting from filtered DC supplies. Together with a filtered AC power supply, external radiated signals can be reduced by 60–100 dB, depending on the frequency. Although less relevant for PD measurement, if high permeability nickel-iron alloy (mu-metal) sheets are used, lower frequency magnetic fields can also be excluded.

Clearly, shielded rooms are used only in laboratory or factory offline PD testing. Due to the high sensitivity to PD required for some test objects, large, shielded rooms are often used by power cable, bushing, and transformer OEMs.

9.3.2 Good Practice for Test Set-Up

In offline tests, the set up the test equipment can lead to interference. Therefore, it is important to use good high-voltage practices to minimize the interference that may be introduced by the arrangement of the test equipment. These include:

- If possible, all ground loops should be avoided. A single-point ground is preferred to reduce the risk of signal pick-up in a ground loop.
- Use leads that are as short as possible, preferably with strips of copper sheets for ground connections to minimize lead inductance. If copper sheets are not available, use thick, braided copper leads. This becomes more important as the frequency of PD detection increase from the LF range to the UHF range.
- Both the ground leads and the high-voltage leads should not produce corona. Thus, they need to be of sufficiently large diameter to reduce electric stress. Where some sharp edges are unavoidable, use a partly conductive putty (Duxseal® is a common brand in some countries) to effectively increase the radius of the edges (and thus reduce the electric stress).
- Connections should be bolted to prevent contact arcing. Avoid using alligator clips for connections.

9.3.3 Power Supply Filtering

There are several possible sources of interference from the AC high-voltage supply used to energize test objects in offline PD tests:

- Noise or interferences may be conducted from the AC power system to the step-up transformer, and then capacitively coupled to its HV winding.
- The device that provides a variable AC voltage to the step-up transformer, a variable autotransformer (often called a Variac®, after a prominent supplier), can introduce interferences. These come from arcing at the carbon brushes that slide over the copper turns.
- The high-voltage test transformer itself may also be a source of PD and other interference, especially if it is of the resonant type.

The 50/60 Hz AC supply (in the 100–600 V range) available in laboratories, plants, and substations to energize the test transformer is often contaminated by power frequency harmonics and interference pulses from a variety of devices connected to the same supply. This interference can be suppressed by a low-pass filter that passes the 50/60 Hz, but blocks higher frequency signals. An alternative is to use isolation transformers that have a very low capacitance between the primary and secondary winding (typically <0.001 pF). This attenuates signals in the range of 100 kHz or higher from entering the test transformer. Commercial PD test transformers intended for PD measurement often incorporate the above measures to reduce these interference sources from the AC supply.

Another way to suppress all types of interference from the high-voltage test transformer is to use a filter/impedance between the test transformer and the test object/PD coupling capacitor (Z_B in Figure 6.1). The filter is also likely to improve detection sensitivity since it will partly block test object PD currents from flowing through the equivalent capacitance of the test transformer. The simplest filter is a resistor with low parasitic capacitance between the test transformer and the test object capacitance to create a low-pass, one-pole filter. The resistance must be low enough that there will not be a significant AC voltage dropped across it, but high enough in combination with the capacitance of the test object and PD detector capacitance, to attenuate frequencies in the PD measurement range. For example, if the combined capacitance of the test object and PD coupling capacitor is 5 nF, and a traditional IEC 60270 detector is used at 1 MHz, the resistor should be at least 30 Ω to suppress 1 MHz and above interference. This resister will have a negligible 60 Hz voltage drop across it.

The filter can also be an inductor with low inter-turn capacitance placed between the power supply and the test object to attenuate interference from the power supply. The inductor – usually more than the 1 mH range – should not be resonant with the capacitance of the test object and coupling capacitor in the PD detector frequency range.

Although costly, interference from the 50/60 AC Hz supply can also be avoided with low-capacitance test objects by using a pure 50/60 Hz oscillator signal fed to a linear amplifier. The output from the amplifier then supplies the step-up transformer to create the high voltage.

9.3.4 Signal Filtering

Some types of interference have most of their energy in a specific frequency range, whereas PD has a very broad frequency range from DC to hundreds of MHz (Section 7.1). For example, AM radio transmissions occur in the LF band of PD detectors. Such types of interference may be suppressed by filters. In some modern LF and HF PD instruments, the filters are incorporated within the instrument, and are of the band pass, digital type, with flexible lower and upper cutoff frequencies. The concept is to find a pass band where the PD signal-to-noise ratio is maximized.

With VHF and UHF detectors, either high pass, low pass, or notch filters can be inserted between the PD sensor detection impedance and the instrument. However, one needs to make sure the input impedance of the filter is the same as the detector, and the filter output is compatible with the detector input impedance.

9.3.5 PD Measurement Bridges

From the 1960s to the 1990s, low-frequency (i.e. IEC 60270 compliant) analog balanced-bridge measuring circuits were used to suppress severe noise (relative to the expected PD magnitude) from the AC supply. This balanced bridge is sometimes called a Kreuger bridge, in honor of the PD pioneer who created one [2]. Kreuger found a balanced bridge was often effective in reducing the interference from the power supply by as much as 60 dB, if the test object and the PD coupling capacitor have about the same capacitance. Figure 9.3 shows one version. This is typically only applied in HV laboratory testing since the test object (EUT) cannot be directly connected to ground. Instead, the low voltage side of the test object is connected to a variable measuring impedance S_A – usually a variable resister in parallel with a variable capacitance. The low voltage side of the coupling capacitor (C_B) is connected to ground via a different variable measuring impedance (S_B).

The signals from each leg are fed to a differential amplifier (or passive RF differential transformer) input, and the differential amplifier output is connected to a conventional detector for measurement. When a simulated PD or interference pulse is injected into the high-voltage side of the step-up transformer, S_A and S_B are adjusted to null out the response from the differential amplifier. That is, the "common-mode" signal from the power supply is subtracted from one another in the differential amplifier, yielding a net zero output when balanced. In contrast, the PD pulses in the test object may be as much as doubled in magnitude in terms of mV, after the differential amplifier. This is because a positive PD pulse voltage from the test object (measured across the measuring impedance of the test object S_A) is detected as a negative pulse in the leg with the coupling capacitor (across S_B), according to Kirchhoff's current law. This negative pulse from S_B is inverted in the differential amplifier and added to the pulse from the test object to yield a double magnitude output.

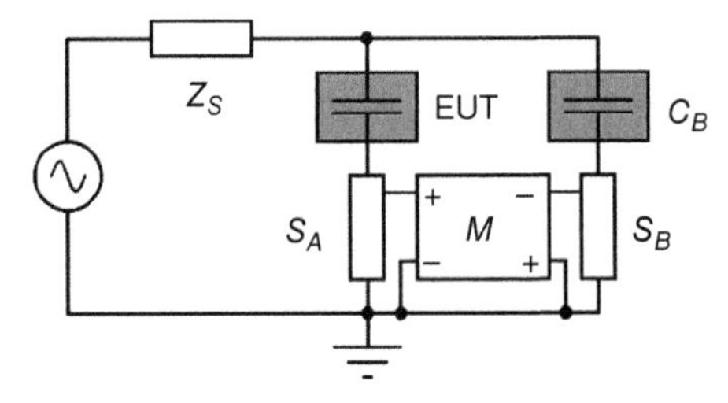

S_A variable PD detection impedance
S_B variable coupling capacitor detection impedance
EUT test object
C_B coupling capacitor
DA a differential amplifier
Z_S power supply filter

Figure 9.3 Example of an analog balanced bridge to suppress interference from the power supply. S_A is the detection impedance in series with the test object (EUT) and S_B is in series with the coupling capacitor.

As discussed in [2, 3], there are some limitations of this analog method to suppress power system noise:

- The test object cannot be directly grounded, which means it cannot be applied for most offline/onsite tests.
- The method is only likely to work for LF detectors since pulse travel times are ignored.
- An unusual amount of skill is often needed to achieve a balance.
- It tends to work for large test object capacitance only (for example long lengths of power cable).
- It works best when the coupling capacitor has about the same capacitance as the test object. The coupling capacitor must also be PD free at the test voltage (which is the same condition for all systems using coupling capacitors).
- If test objects have different capacitances, different high-voltage capacitors are needed. This can be true also when testing different parts of the same equipment (e.g. different windings in a transformer).
- Most importantly, it seems that instruments using the Kreuger bridge method have not been made for 20 or more years.

Some of these limitations were overcome when it was realized by Hashimoto and others that a balanced bridge is not needed; rather, only detection of the polarity of the pulses in each leg is required [4]. That is, in Figure 9.3 where S_A and S_B are now fixed detection impedances, a positive interference pulse from the HV power supply will be positive across both the detection impedances S_A and S_B. As mentioned above, a single PD pulse within the test object EUT will appear as (for example) a positive pulse across S_A and as a negative pulse across S_B, by Kirchhoff's current law. The relative magnitudes of the two pulses are unimportant, as long as the comparators (within M) on each detection impedance correctly detects the polarity of each pulse, and a logic circuit determines if the two detected pulses are of different or the same polarity. If the detected pulses have the same polarity, then the logic circuit declares the signal is due to noise, and the pulse is blocked from being counted in the PD instrument. If the pulses are of reverse polarity, it is PD in either the test object or coupling capacitor. The advantage of this method is that the requirement is removed for balancing the detection impedances for each of the test object capacitance and the coupling capacitor. Ian Black and his colleagues made a practical LF instrument based on this principle [5], which was commercially available for some time from Robinson Instruments. Laboratory PD detectors made in the past 20 years do not seem to have this capability.

The method was also applied to online transmission-class substation PD detection in the HF and VHF ranges and used HFCTs on equipment ground leads [6]. It was never commercialized.

9.3.6 Time-of-Flight

The time-of-flight method (Section 8.4.2) has also been applied to suppress interference pulses from the power supply, based on the difference of travel time between two sensors along a power cable or bus, between the power supply and the test object (Figure 8.5). This method is not effective in the LF range, since the two sensors must then be at least 300 ns apart (corresponding to 100 m of air-insulated bus or 67 m of cross-linked polyethylene shielded power cable).

In practical offline testing, the time-of-flight method is usually confined to the VHF or UHF detection range, since the distance between the pair of sensors can be as short as 2 or 3 m. The method is widely used for factory PD testing of stator windings and coils, where the two sensors are at either end of a shielded power cable connecting the high-voltage test transformer to the winding or coils.

9.3.7 PRPD Pattern Recognition

PD and corona have distinct characteristics when observed on a PRPD plot. When void PD is measured using a indirect detection circuit (Figure 6.2), and only one phase is energized, negative polarity PD pulses tend to occur between −15° and 105° of the AC cycle, and positive PD pulses tend to occur between 165° and 285° of the AC cycle (Figure 8.10). Note that in a direct detection circuit (Figure 6.1), the PD pulse polarity is reversed, i.e. the positive polarity PD pulses occur between −15° and 105° of the AC cycle, and vice versa (Figure 3.24). Figure 8.10 shows a textbook pattern associated with PD within voids. Similarly, negative corona pulses tend to be centered around the 90° position, and positive corona around the 180° part of the AC cycle (Figures 3.14 and 13.47). Pulses not in these positions of the AC cycle are not likely due to test object PD or corona. See Section 10.5 for exceptions to this.

Many types of disturbances do not have these polarity/phase position features. For example, if positive pulses are appearing at 140° of the AC cycle (assuming indirect detection), then the signal is likely caused by interference, or cross-coupling of the PD from another phase. An experienced observer of the PRPD patterns can thus discard such pulses "by eye" and focus attention on the pulses most likely to be test object PD or corona. Clearly, this assessment of what is PD and what is interference is somewhat subjective. PRPD pattern recognition is also an essential component to post-processing methods such as the TF map and three-phase synchronous separation (Sections 8.9.2 and 8.9.3, respectively), where different clusters are identified using known PRPD plots.

Artificial intelligence (AI) pattern recognition tools may permit inexperienced observers to distinguish between likely PD and likely interferences (Section 8.9.5), although to date their commercial success is limited.

9.3.8 Time-Frequency Map

The TF map method of post-processing signals (Section 8.9.2) is also an effective method of suppressing many types of interference from the HV transformer, especially in the HF or higher range. It depends on the pulse characteristics (duration and frequency) of the interference differing from the test object PD characteristics. Thus, it may not be effective for PD occurring close to the test object, because the pulses are similar to those of the PD within the test object. As an example, corona occurring on the HV step-up transformer output or the leads connecting to the test object may still be classified as test object PD. The TF map is less effective in the LF range, since the pulse characteristics are mainly determined by the detection impedance Z_m (or the quadrupole).

9.3.9 Gating

As discussed in Section 8.4.1, interference can be suppressed by detecting when an interference pulse occurs, and then preventing the PD detector from counting that pulse. This method is only effective when there is a means for detecting *only* the interference. For example, an auxiliary sensor (often an antenna) is placed near to the interference source and produces a much higher signal than test object PD, this auxiliary signal can be used to open a gate between the PD sensor and the pulse counter. Alternatively, if the interference is caused by electronic switching, the signal from the PD sensors is often at a different frequency. Thus using an appropriate band-pass filter, a gate signal can be triggered reliably. If the number of interference pulses is high, the gate may be "open" so much that legitimate test object PD may not be registered.

9.4 Online Interference Suppression

The interference levels in online PD testing are usually much higher than those in offline testing, and therefore cause much more trouble. The interference levels tend to be high in online tests since the test object is directly connected to the power system and any signals present there. Also, there is no ability to electromagnetically shield the test object in substations and plants. Practical online PD testing only became widely used with the move to the VHF and UHF detection frequencies and/or the development in the 1990s and 2000s of proven interference suppression methods.

PD measurement in the VHF and UHF frequency ranges tend to automatically attenuate interference remote from the test object since, in general, signal attenuation increases with increasing frequency [7].[2] The Faraday shielding from external interferences by equipment enclosures also tends to be more effective at higher frequencies for test objects like GIS, transformers, and rotating machines, with signals leaking into the test object only at windows, flanges, hatches, bushings, etc. Detection in the VHF and UHF frequency ranges also facilitates the use of noise suppression methods such as time-of-flight and pulse-shape analysis (Sections 8.4.2 and 8.4.3) [7].

The post-processing methods discussed in Section 8.9 also have enabled better noise suppression for online PD tests. In particular, the TF map (Section 8.9.2) and three-phase synchronous (Section 8.9.3) methods have seen widespread use for power cable, transformer, and rotating machine online testing.

Although all the noise suppression methods discussed in Section 9.3 (except for shielding, Section 9.3.1, and power supply filtering, Section 9.3.3) can also be used for online testing, the methods that are most effective tend to depend on the nature of the test object. Thus, Chapters 12–16 discuss the types of interference suppression used for each of the various types of HV equipment.

References

1 Boggs, S.A. and Stone, G.C. (1982). Fundamental limitations in the measurement of corona and partial discharge. *IEEE Transactions on Electrical Insulation* 2: 143–150.
2 Kreuger, F.H. (1965). *Discharge Detection in High Voltage Equipment*. American Elsevier.
3 Bartnikas, R. and McMahon, E.J. (1979). *Engineering Dielectrics Volume 1: Corona Measurement and Interpretation*. ASTM Publication 669.
4 Hashimoto, H. (1960). Problems of measurement of void discharge in high voltage cables of full reel length. *Electrotechnical Journal of Japan* 5 (3/4): 85.
5 Black, I.A. (1975). *A Pulse Discrimination System for Discharge Detection in Electrically Noisy Environments*, 239–243. ISH.
6 Stone, G.C. and Kurtz, M. (1987). Fault anticipator for substation equipment. *IEEE Transactions on Power Delivery* 2: 722–724.
7 IEC TS 62478 (2016). *High Voltage Test Techniques – Measurement of Partial Discharges by Electromagnetic and Acoustic Methods*. IEC.

2 Of course this same effect may reduce sensitivity to PD within a large test object that is more remote from the PD sensor.

10

Performing PD Tests and Basic Interpretation

10.1 Introduction

This chapter describes the basic procedures common for all types of electrical equipment to collect PD data in any frequency range using the electrical and electromagnetic methods in Chapters 6 and 7. This chapter also gives an overview of the test result interpretation process for test objects energized with power frequency AC voltage that is based on the PD activity indicators and charts presented in Chapter 8.

As described in Section 1.3, there are two main reasons PD tests are done:

- Factory acceptance tests (FATs) or onsite commissioning tests to determine if the HV apparatus meets specifications and is ready to be placed into operation.
- Condition assessment testing/monitoring of HV equipment already in service to determine if the electrical insulation is degrading or has a high risk of imminent failure. In this case, one is trying to determine the need for maintenance and/or avoid in-service failure.

In addition, PD tests may be performed to evaluate new insulation system designs or new insulating materials, or for fundamental research.

In both FAT and onsite commissioning tests, the PD inception and extinction voltages (PDIV/PDEV) are the main outcome for most types of HV equipment. A charge-based PD measurement method in the LF range (Section 6.5.1) using IEC 60270-compliant instrumentation is employed most often. PD detection in most types of equipment is via a high-voltage coupling capacitor/detection impedance, or bushing taps in the case of transformers (Section 15.7). The outcome is the PDIV or PDEV at a specified apparent charge sensitivity in pC. Less often, the PDEV may be measured in the HF, VHF, or UHF ranges, with a sensitivity usually specified in mV or in terms of RF signal strength (mV, mW, dBm, dBμ, etc.).

For condition monitoring, the PD magnitude at operating voltage and PRPD plots are the main outcomes, although the PDIV/PDEV may also be measured in onsite/offline tests. For condition monitoring, the tests may be charge-based in the LF range producing a pC or nC outcome, or use a measuring system in the HF, VHF, or UHF ranges where the PD magnitude is again most often reported in mV or in terms of RF signal strength.

In both cases, the actual measurement method and interpretation process is very dependent on the type of HV apparatus, e.g. HV transformers, GIS, AIS, stator windings, and power cable. Therefore, this chapter only gives an overview of the interpretation process and introduces some

Practical Partial Discharge Measurement on Electrical Equipment, First Edition. Greg C. Stone, Andrea Cavallini, Glenn Behrmann, and Claudio Angelo Serafino.

basic principles. For specific interpretation guidance, it is important to also read the interpretation sections for each specific equipment type covered in Chapters 11–16.

When high PD is detected in a factory test, or in a condition assessment test, then the interpretation may become more involved. Specifically:

- One may want to know where the PD is occurring within the test object, and/or
- The root cause of the PD may be investigated.

PD site location helps guide engineers where to look within the test object to find the source of the PD. This can speed any repairs that may be needed. Sections 5.3 and 5.4 discuss acoustic and optical methods for locating PD sites. In rotating machine applications, "corona probes" may also be used (Section 16.13). UV imaging and acoustic imaging devices have been particularly useful where PD and corona are not obscured by equipment enclosures, such as in offline tests on stator windings and AIS; as well as within outdoor substations and overhead transmission lines. If multiple PD sensors are installed, time-of-flight methods (Section 8.4.2) can also be used for PD site location. Chapters 12–16 discuss the location methods for each type of electrical equipment.

Identifying the root cause of any detected PD may also be useful in identifying the insulation system repair options (if indeed there are any options other than replacement). As far back as 1969, the PD patterns with respect to the AC cycle have been used to identify the root causes of PD [1]. With the introduction of digital PD technology and the widespread use of PRPD patterns, many papers and standards have been published that associate specific types of insulation defects with specific patterns. Post-processing using statistical analysis of the patterns (Section 8.9.1), T-F maps (Section 8.9.2), and three-phase synchronous PD analysis (Section 8.9.3) may all be helpful in recognizing patterns associated with specific insulation degradation processes when multiple types of PD are active.

10.2 PDIV/PDEV Measurement

The general procedure for determining the PDEV and PDIV is presented below, as well as some of the important factors that can influence their measurement.

10.2.1 Test Procedure

The PD inception and extinction voltages can only be measured in an offline test. With the exception of transformers, the test object is energized by a single-phase high-voltage 50/60 Hz power supply, which is normally a variable autotransformer (Variac™ is a well-known brand) connected to a partial-discharge free high-voltage transformer. The power supply must be rated to have sufficient output power capability (volt-amps) to enable it to supply the reactive power required to energize the capacitance of the test object (EUT) to the desired maximum test voltage. For high-capacitance test objects tested at high voltages, this volt-amp capability may be significant. Because of the large VA required at 50 or 60 Hz, sometimes the EUT is excited at a lower frequency, typically 0.1 Hz, to significantly reduce the reactive power (VA) required. For long power cables, AC supplies may be used, which are resonant between about 25 and 75 Hz. The high-voltage windings of power transformers are normally energized by a three-phase supply via the transformer's low-voltage winding (called an induced voltage test). Power transformers may use AC frequencies up to 400 Hz, to prevent the laminated steel core from magnetic field saturation during induced overvoltage tests. The typical AC supplies used for each type of power equipment (cables, transformers, etc.) are presented in the relevant equipment Chapters 12–16. Of course, the PD measurement system must also be able to synchronize to the AC voltage frequency to create the correct PRPD pattern.

The PDIV/PDEV test is done by slowly increasing the AC voltage to a predetermined maximum voltage that depends on the type of the test object. During the voltage increase, the presence of any PD is determined while looking at an oscilloscope trace with respect to the power frequency AC cycle or a PRPD plot from a digital PD instrument. The PDIV is the voltage at which PD is occurring above the required apparent charge sensitivity (often 1, 3, or 100 pC) and is determined from the oscilloscope or PRPD plot. The AC voltage continues to be increased to the predetermined maximum test voltage. At this point, the PRPD plots and PD magnitudes may be permanently recorded. Then the voltage is lowered until the classic PD pattern (again on an oscilloscope-type display or a PRPD plot) disappears. This is the PDEV. The PDEV is often lower than the PDIV due to the presence of electrons and ions from previous discharges.

When using a PRPD plot to determine if PD is occurring or not, the AC voltage must be raised and lowered in voltage steps to ensure the AC voltage is constant for the duration of the PD acquisition time of the instrument. These steps may be from a few seconds to a few minutes long, depending on the acquisition time of the PD instrument. Voltage steps are not needed when an oscilloscope or real-time digital instrument with an oscilloscope-like display is used. Usually, PDIV/PDEV tests take longer when the display is a digital PRPD plot, as compared to when an oscilloscope or a digital PD detector with a real-time simulated oscilloscope display is used, since the AC voltage can be smoothly increased without holding at each voltage step. The longer duration of the test when steps are used can affect the PDIV and PDEV values (Section 3.6).

10.2.2 Sensitivity

A critical factor in determining the PDIV and PDEV is the sensitivity of the measurement system to PD. In general, the higher the PD sensitivity, the lower will be the PDIV/PDEV (and thus the greater the risk that the test will fail). In the LF range, the sensitivity is determined primarily by the capacitance of the test object and other connected equipment. As discussed in Section 6.4.1.3, the larger ratio of test object capacitance to C_k (the detection or coupling capacitance), the lower will be the sensitivity (and the higher the PDEV will be). It is for this reason that when a factory or commissioning test is required, the sensitivity of the measurement must be specified, as well as the minimum PDIV or PDEV. IEC, IEEE, or national standards usually specify the required sensitivity (QIEC in pC for charge-based tests or suitable units for HF, VHF, and UHF tests) for each type of test object.

Sensitivity is also affected by the presence of noise and interference (Chapter 9) and the PD sensor/instrument's design. If a particular test arrangement is not sensitive enough, then alternative PD sensors or test instruments may be needed; or the noise and interference needs to be suppressed using one or more of the several methods described in Section 9.3.

10.2.3 Interpretation

Establishing the PDIV or PDEV depends greatly on the expertise of the person doing the test. Before the days of digital storage, the PD test operator was the only person to evaluate when the PDIV/PDEV occurred. Some operators may think they see pulse pattern on the oscilloscope screen that they assume is PD, when in fact it may be some sort of interference. Thus lots of experience was needed, and others that were not present at the test could not "second guess" the results. The development of digital PD detectors that store PRPD patterns at each AC voltage step, or have a "video playback" mode of a simulated-oscilloscope display, has allowed other experts not present at the actual test to retroactively confirm the PDIV and PDEV.

The stochastic (statistical) nature of PD means pulses may be occurring at one time and then disappear for a period of time (Section 3.6). As well, the PD pulses "bounce around" with respect to the AC cycle, further complicating the measurement. If the defect volume is small, the PD maybe quite intermittent, since as described in Section 3.6.1.1, exactly when PD occurs depends on having free electrons intersecting with the volume of a cavity while the electrical stress is high. For small voids, the probability of PD occurrence is low at any instant of time, and thus the PD is very intermittent, making the establishment of the exact PDIV/PDEV difficult. This has led to the use of X-rays or other ionization sources to provide lots of initiatory electrons in small voids, yielding more stable PRPD patterns and estimates of the PDIV/PDEV [2].

For almost all types of high-voltage equipment, the PDIV/PDEV is required to be well above the expected nominal operating voltage of the HV equipment. This is because any PD is expected to lead to failure of the organic insulation used in most types of electrical equipment via the electrical treeing or electrical tracking processes (Sections 2.4.1 and 3.7.2). However, experience indicates that in the specific case of rotating machine stator windings, the PDIV and PDEV are often below the operating voltage. This is acceptable for stator windings rated 3 kV and above since the main insulation is partly made with mica [3]. Mica can withstand PD without breakdown for many years or even decades.

Where the purpose of PD testing is to assess the condition of the insulation system, one looks for a decrease in PDIV or PDEV over time (under the same test conditions – i.e. test object temperature, ambient humidity, PD detection system). If the PDIV/PDEV are decreasing, it is an indication that aging is occurring. If the PDIV/PDEV is below the operating voltage of most types of test objects (excluding stator windings), then the PD source must be located and maintenance/replacement performed.

For some types of test objects, the PDIV/PDEV level may be an indication of what the root cause of the PD may be. For example, loose coils in the stator slot in rotating machines, if present, seem to have a lower PDIV and PDEV than for most other insulation aging mechanisms that occur in machines.

10.3 PD Magnitude and PRPD Test Procedure

The PD magnitude and PRPD plots can be acquired in both offline and online tests. The offline test may be part of the FAT or commissioning tests, in the factory or onsite, at any test voltage specified. Online tests are performed onsite, and are only at the normal operating voltage of the test object. Usually the PD magnitude and the PRPD in each phase of the test object is measured.

10.3.1 Offline Testing

In offline testing, the EUT must be isolated from the power system. Any embedded sensors (for example temperature sensors) must be grounded and connected equipment such as PTs, CTs, surge arrestors are removed or disconnected, with their outputs grounded or terminated in purpose-built impedances specifically designed for offline testing. The test object is then energized by a high-voltage AC power supply operating at the nominal power line frequency (50/60 Hz) or at other frequencies as stated in the corresponding equipment specification, as described in Section 10.2.1. Section 9.3 outlines measures that may be needed to reduce the interference.

For offline PD testing of most types of test objects, it is common to use a charge-based measuring system working in the LF range (IEC 60270 compliant). If the test objects can be modeled as a lumped capacitance at measurement frequencies <1 MHz, the calibration procedure in IEC 60270 (Section 6.6) enables the PD pulses to be measured in terms of apparent charge (pC or nC). Charge-based PD measuring systems will also calculate the quasi-peak apparent PD magnitude Q_{IEC} (Section 8.8.1) which is approximately the PD magnitude when about 1 pulse per AC cycle is occurring (i.e. 60 pps in a 60 Hz system). For distributed test objects like power cable or GIS/GITL, the PD can still be measured with a charge-based system in the LF range, but it is important to use the wideband mode to avoid PD pulse superposition effects (Section 6.6.2). For test objects like stator windings or transformer windings, there are usually natural frequencies in the LF range due to the inductive-capacitive nature of the test objects in combination with the coupling capacitor (Section 6.6.3). These natural frequencies may lead to amplification of the PD signals due to resonance. Thus, the calibration procedure in IEC 60270 is not formally valid, so that quoting Q_{IEC} in terms of apparent charge (pC or nC) is not accurate or valid, in spite of its widespread use.

An advantage of offline tests performed in the LF range is that there is likely to be less attenuation of PD signals between the PD site and the PD sensor, when compared to HF, VHF, and UHF measurements. Such attenuation can be significant at higher frequencies in long power cables as well as transformer and stator windings.

Sometimes offline tests are done with HF, VHF, or UHF PD measuring systems. The main reasons for this include:

- More sensitive measurement due to better interference suppression, which can be important in factory testing or tests done onsite where there may be a lot of arcing/sparking types of interference sources, or signals from electronic power supplies (especially from variable speed drives).
- Ability to locate the PD sites using so-called time of flight measurement methods (Section 9.3.6).

In this case, the PD is usually measured in terms of received signal level (mV, dBm, etc.) and cannot be directly related to apparent charge unless side-by-side tests are done with the charge-based measurement method in IEC 60270.

An important consideration in offline tests is to record the ambient conditions. The test object temperature may affect the PD. For example, the PD within voids tends to decrease as the temperature increases, whereas surface PD tends to increase from PD suppression coatings in power cables and stator windings as the temperature increases. Similarly ambient humidity may affect surface PD and corona. In some types of HV apparatus, the PD may also be affected by the current being carried through the conductors.

10.3.2 Online Testing

Online PD testing is used to assess insulation system condition in HV equipment. With online testing, the test object is connected to the power system and thus operates at the power system voltage. As a result, it is not possible to measure the PDEV or PDIV. Rather, the aim is to collect data during normal equipment operation over the years to detect any changes in the PD activity. Online PD measurement normally requires the permanent installation of sensors on the test object, although sometimes HFCTs or antennae-type sensors (Section 7.4) may be portable.

In almost all practical cases, online PD is measured in the HF, VHF, or UHF frequency ranges. The main reason for this is that interference levels in power systems tends to be lower at higher frequencies, and improved interference suppression methods are available, especially in the VHF and UHF ranges (Section 9.4). Thus, UHF methods tend to be used for GIS, power transformers, and hydrogen-cooled stator windings, while HF and VHF methods are used for power cables, AIS, and air-cooled stator windings (see Chapters 12–16). The disadvantage of measuring online PD in the HF and higher frequencies is that the PD signal is more likely to be attenuated more as it travels to the PD sensor, thus losing sensitivity. Also calibrating pulses into apparent PD charge magnitudes will not give an absolute indication of PD magnitudes, as it does for detection frequencies below 1 MHz (Section 6.6).

The data may be collected periodically or continuously. For periodic measurements, a portable test instrument is temporarily connected to the PD sensors on the test object, and once the data is collected, the instrument is moved to a different test object. This can be cost-effective since only one test instrument is needed for all the test objects in a company. However, if the test objects have purely organic insulation systems (cables, transformers, GIS), then after a test where no PD was measured, PD may start up and cause insulation failure before the next scheduled test. This reduces the usefulness of online PD testing for condition assessment (and of course produces false negative indications). This limitation has led to more and more installations of continuous PD monitoring systems. The capital cost of continuous PD monitoring is higher since more instruments are required.

With online testing of three-phase equipment, both phase-ground and phase-phase voltages are present. In most test objects like power cable systems, transformers, and single-phase GIS, this makes little difference since the insulation systems in each phase is physically (and often electromagnetically) isolated. However, in AIS, three-phase GIS and stator windings, the three phases tend to be physically close, if not intermingled. Sometimes PD may occur between phases rather than just between phase and ground. As discussed in Section 10.5.4, this will lead to changes in the PRPD plot. Also, when the phases are close to one another, there is usually significant cross-coupling between the phases. That is, PD in one phase may be coupled into another phase. This also makes the PRPD patterns more complicated (Section 10.5.5).

10.3.3 Differences Between Offline and Online Tests

Offline PD tests tend to be more sensitive than online tests since interference tends to be less severe. The main advantage of online testing is that the equipment does not have to be taken offline and disconnected from the power system to do the test. This also implies that online tests can be done more frequently and usually at a lower marginal cost than offline tests. However, even when both online and offline results are available, it is sometimes difficult to compare the results on the same test object, even when tested at the same voltage and using the same PD instrument.

In offline tests, the test object is usually at a lower temperature than occurs when the HV equipment is in service. Also, humidity and propensity for moisture to condense on the high-voltage components will usually be different between offline and online tests. Temperature and humidity can affect the PD activity (Section 3.8.2). In some test objects, the load current may create vibration due to magnetic forces, or temperature gradients, that may lead to additional PD (Section 10.6). Clearly offline testing may not find such vibration- or thermal gradient-induced PD.

The voltage across the insulation system may also be different between an offline test with a single-phase power supply, compared to the three-phase excitation when the test object is connected to the power system. In stator windings, AIS and three-phase GIS, there is both phase-to-ground and phase-to-phase insulation. In these types of equipment, PD can occur between the phases that is driven by the phase-to-phase (as opposed to phase to ground) voltage. Examples are shown in Figures 10.1 and 10.2.

When tested offline with a normal single-phase AC supply, to simulate normal operating voltage, the rated phase-to-ground voltage is usually applied to a single phase, with the other two phases

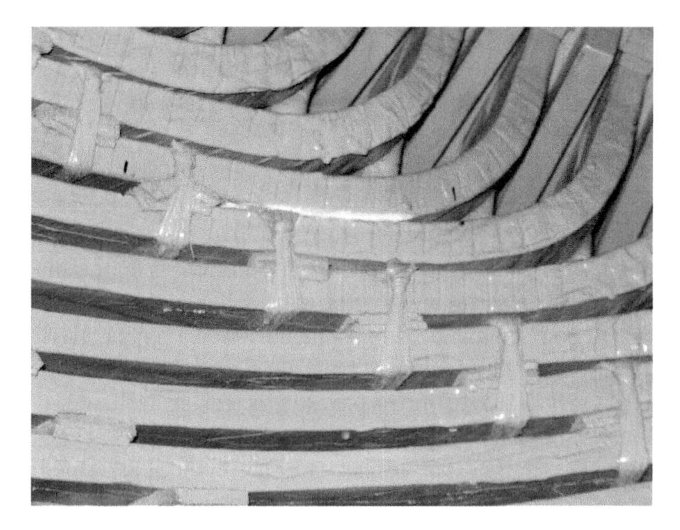

Figure 10.1 Interphasal PD (white powder) occurring on a stator winding between two adjacent phase-end coils in different phases that are too close together. *Source:* Courtesy of Iris Power L.P.

Figure 10.2 Photo of PD between the phase buses in a 13.8 kV metalclad air-insulated switchgear. The interphasal PD in the air gap between each bus and the red insulating support board lead to ozone and white powder. *Source:* Courtesy of Howard Sedding, Iris Power L.P.

grounded. This means the phase to ground voltage on the energized phase sees the same voltage as in online testing. However, the phase-to-phase insulation interfaces see only the phase-to-ground voltage across them, and not the phase-to-phase voltage. This means the phase-to-phase insulation is tested with only $1/\sqrt{3}$ or 58% of the in-service voltage. This much lower voltage is unlikely to produce PD or will produce much lower PD than that would occur during an online test. If the online PD is primarily occurring in the phase-to-phase insulation, as in Figures 10.1 and 10.2, this problem is unlikely to be detected in an offline test.

Another cause for dissimilar results between online and offline tests occurs in stator windings. In wye-connected windings, the AC voltage to ground decreases linearly from the phase-end terminal to the neutral or ground-end of the winding. Thus, PD is most likely to occur only in the winding sections where the voltage (electric field) is highest – near the HV terminals. PD will not occur at the neutral or ground-end, since the voltage (and electric field) is low. In an offline test, usually the entire stator winding is energized to the test voltage, including coils at the neutral end of the winding. This will create many more potential sites for PD activity to occur, possibly resulting in higher PD activity being observed compared to an online test.

10.3.4 Conditioning in Offline Tests

When performing offline tests to determine the PD magnitude and PRPD pattern, it is important to "condition" the test object to obtain stable PD measurements. Normally, when voltage is first applied to the EUT, the PD magnitudes will be higher, and there also tends to be a higher pulse rate. As the time the voltage is applied lengthens, the PD activity typically slowly decreases. For cavities within some types of insulation, this may be due to the gas pressure increasing due to gases produced by the previous PD, which tends to suppress further PD as a corollary of Paschen's Law (Section 2.3.1). For other types of insulation, for example polyethylene, the gas in the void created by the PD may have a higher breakdown strength. Another reason may be that conductive by-products from previous discharges within the cavity makes the cavity surface more conductive, reducing the electric stress within the cavity and thus suppressing PD.

Conditioning is done by:

- applying the AC test voltage for a period of time (typically 15–30 minutes, but the longer the better) before the PD results are captured.
- Alternatively, apply a higher test voltage than required for the PD test for a few minutes. The voltage is then reduced to the required test voltage and the PD can immediately be captured.

Both procedures will stabilize the PD.

The PD in operating equipment is expected to be in a steady-state situation since presumably the test object has been in operation long enough to stabilize. Thus, the concept of conditioning does not apply in online tests.

10.4 Interpretation of PD Magnitude

The most common result of an offline or online test at a set test voltage is the peak or quasi-peak PD magnitude as defined in Section 8.8. The quasi-peak magnitude (Q_{IEC}) has the units of apparent charge (pC or nC) and is measured by an IEC 60270-compliant LF PD measurement system. Q_m tends to be used with digital PD measuring systems in the HF, VHF, and UHF ranges. The peak PD magnitude (Q_m) is determined from a pulse magnitude analysis plot (Section 8.7.1) and corresponds to the PD magnitude associated with a specified number of PD pulses per second (pps), with a pulse repetition rate of 10 pps being common. Q_m (at 10 pps) tends to be higher than Q_{IEC}, everything else being equal, since Q_{IEC} corresponds to a pulse repetition rate of about 1 pulse per cycle or 50 (or 60) pps. Regrettably, for digital PD systems, each PD system vendor seems to calculate Q_{IEC} and Q_m differently, so that the Q_{IEC} (or Q_m) on the same test object at the same time may sometimes differ between vendors. This makes the comparison of PD magnitudes between different PD instrument vendors difficult.

In addition to peak PD magnitude, there are other "integrated" quantities that are produced by modern PD systems that are some combination of pulse magnitude and pulse repetition rate. As discussed in Section 8.8.3, common activity indicators include average discharge current, discharge power, NQN, and the closely related NQs.

10.4.1 Trend Over Time

For insulation system condition assessment, it is the trend of Q_{IEC}, Q_m, or any integrated quantity that is most powerful for establishing if failure is imminent or that the insulation is gradually aging. Under the same test conditions (voltage, humidity and test object temperature), if the PD activity stays the same over time, it is an indication suggesting that an aging mechanism involving PD is not accelerating. For most test objects, there will be no PD above the interference level, and then there is usually an abrupt jump in the activity indicating that something has changed which enables PD activity to start. Figure 10.3 shows an example on a power cable. In test objects using only organic insulation, it is prudent to take action as soon as possible, since the PD itself can lead to rapid insulation degradation. Due to the presence of mica in stator winding insulation, PD activity is usually present from initial manufacture; but after significant aging occurs (years or decades), the trend may gradually increase over the years, yet failure may still not be imminent (Figure 10.4). Thus, how quickly one takes action when PD increases depends on the nature of the test object. This is discussed in more detail in Chapters 12–16.

Note that for most test objects, if no PD is occurring in the EUT (which is usually the case), the activity measured is the level of the interference, which is not completely suppressed by the methods in Chapter 9. This level may change dramatically up or down over time, depending on which sources of interference are active at any one time. For example, the PD measured in a power transformer may actually be caused by transmission line or substation corona that depends on humidity and rain/snow conditions. Thus determining when PD has started can be a challenge and depends on the effectiveness of the interference suppression methods. Many false positive indications of PD problems in HV equipment are actually due to changes in interference over time.

PD magnitudes on the same test object, even within a short period of time, are often observed to vary by about 25%. Partly this is inherent in the stochastic nature of PD, but this variation can also be caused my small changes in voltage, temperature, load, and humidity. Thus, changes in Q_{IEC} or Q_m of less than 25% should not precipitate immediate action.

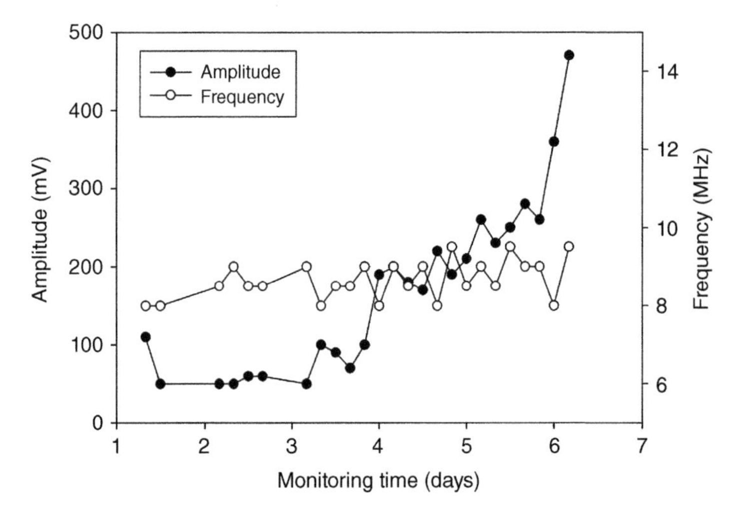

Figure 10.3 Trend in PD magnitude in a power cable where the joint failed after six days. *Source:* Cavallini et al. [4].

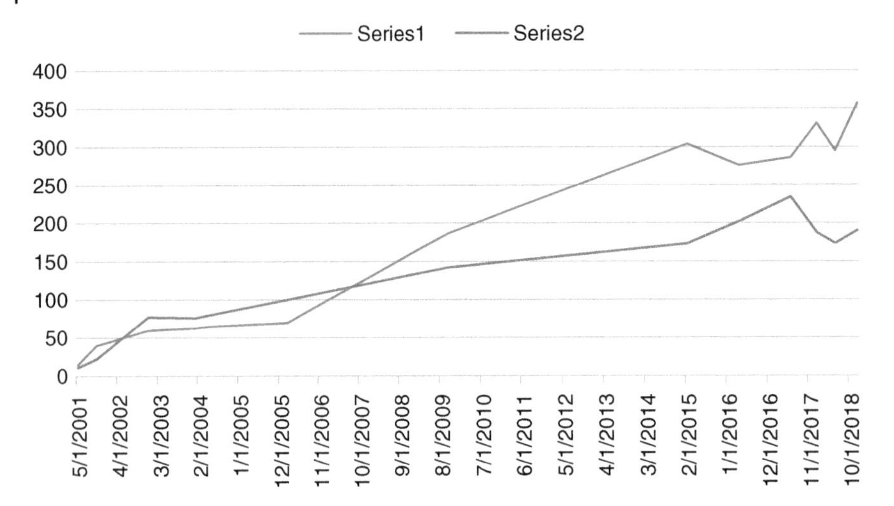

Figure 10.4 Trend in $+Q_m$ (blue) and $-Q_m$ (orange) measured online over an 18-year period on a 50 MW, 13.8 kV generator stator. The vertical scale Q_m is in mV. The data plotted is only that collected when the machine was operating at about 50 MW and at winding temperature of about 100 °C. *Source:* Courtesy of Iris Power L.P.

As stated earlier, the trend is only helpful if the data is recorded under the same operating conditions. Continuous online PD monitors make effective trending much easier since test results recorded under the same operating conditions can be easily extracted from stored results.

Trending is also only valid if the measurements are done with the same type of instruments – which normally means the same brand and model of PD instrument. This is especially true with HF, VHF, and UHF systems, which all tend to use slightly different approaches that the sensitivity check cannot completely correct for. When using an LF IEC60270 detector, the trends are only valid if the results are taken in the same frequency range.

10.4.2 High PD Level

Most types of new HV test objects have requirements to have a Q_{IEC} that is below a certain level of apparent charge when measured by IEC 60270-compliant systems. Any Q_{IEC} above the requirement will trigger re-measurement and probably an investigation as to the cause of the high PD. Thus, the absolute value of Q_{IEC} is useful, and commercially important. As outlined in Chapters 12–15, the maximum allowable Q_{IEC} tends to differ over a very wide range for different types of HV equipment.

A single measurement of Q_m measured with a HF, VHF, or UHF PD measurement system is much less meaningful. Each brand and model of PD measuring systems works differently and operates in different frequency ranges. Also, according to IEC TS 62478, there is no way to meaningfully "calibrate" the results so they can be compared between different brands/models of PD systems. Thus, the PD magnitude in mV, dBm, dBμV, etc. in one system will not necessarily be the same for the same level of PD activity from another system. This makes it impossible to scientifically set "high" Q_m levels valid for all measuring systems.

Interpretation of a single Q_m result using an HF and above measuring system for condition assessment is only possible if the result can be compared to similar test objects measured with the same PD system. HF, VHF, and UHF PD system vendors often have access to a large number of test results, and sometimes information on the insulation condition associated with the results. This allows them to build a statistical distribution of observed results for "similar" test objects. When compared to visual examinations of the test objects, it, then, is sometimes possible to establish what constitutes a high level of Q_m indicating that significant insulation deterioration is occurring. As discussed in Section 10.3.3, what is "high" for an offline test may differ from what is considered "high" in an online test. Q_m will also change if the magnitude and phase resolution of the PD measuring system changes (Sections 8.7.1 and 8.8.2).

Comparing PD levels acquired with the PD system from several identical test objects can also be useful. The test object with the highest Q_{IEC} or Q_m is likely to be closest to failure and probably needs maintenance first.

10.4.3 PD Polarity Effect

Since the 1960s, it has been known that the relative predominance of positive and negative polarity of the PD can give an indication of the location of the PD or corona within the test object [1]. Sections 3.5.1 and 3.10.1 present the physics of the "polarity effect." When using an indirect measuring system (Figure 6.2), which is the most common arrangement for off- and online tests, if the positive pulse Q_m or Q_{IEC} (on the negative half of the AC cycle) is greater than the negative Q_m or Q_{IEC}, the PD is likely to be occurring near the low-voltage conductors (Figure 10.5). If the negative PD (positive half AC cycle) is higher than the positive PD, the PD is likely occurring close to a high-voltage conductor. If the positive and negative PD activity is about the same, the PD is likely occurring in cavities within the insulation system, relatively distant from either high voltage or ground conductors. These polarity effects become clearer when observing PRPD plots (Section 10.5). If a direct PD measurement system is used, then the polarity effect will be reversed from that shown in Figure 10.5.

There are two limitations to this general observation. If multiple PD sites and/or multiple different aging processes are present in the EUT, then it is possible for PD to be occurring near the HV conductor and near ground – resulting in no apparent polarity predominance. This of course leads to a misleading conclusion that the PD is only occurring within the insulation, and not near both HV and ground conductors. The second limitation is that if the PD is occurring between phases in an online test, then the polarity effect does not occur, or rather – it will be misleading. The reason for this is presented in Section 10.5.4.

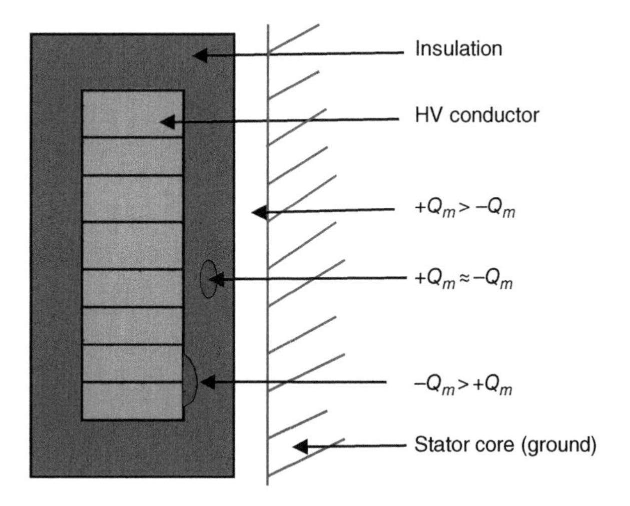

Figure 10.5 Cross-section of a stator coil in slot with PD occurring at three different locations and the resulting polarity predominance. Q_m+ refers to positive PD pulses measured with an indirect PD measurement.

10.4.4 Other PD Activity Indicators

Section 8.8.3 present other indicators of PD activity that are defined in various IEEE and IEC standards. In addition, many PD system vendors have their own proprietary PD activity indicators. All of these can be used for trending purposes to identify when the number of PD sites are increasing, or the total or average activity is increasing. However, unlike Q_m and Q_{IEC}, they are relatively unhelpful when used on their own, since there are no good/bad levels for such indicators in IEEE or IEC standards. Thus, they are only useful in comparison to other identical test objects measured with the same PD measurement system. These types of activity indicators are critically dependent on the threshold level above which the instrument counts the pulses (Section 8.3.1)

10.5 PRPD Pattern Interpretation

The phase-resolved PD (PRPD) pattern with respect to the AC cycle (Section 8.7.3) is perhaps the most powerful tool available to ensure that the measured signals are caused by test object PD rather than interference and to help identify the root cause of any PD detected. PRPD plots are applicable to offline and online PD tests, as well as for factory acceptance tests (FAT) and condition assessment. All reputable modern PD measurement systems produce PRPD plots, although some may have less amplitude or phase resolution than others.

Figure 10.6 shows a PRPD pattern from a 4.1 kV rated stator coil measured with a VHF measuring system using an indirect measurement circuit (Figure 6.2). The plot shows there are both negative and positive pulses. Note that pulse polarity information may be lost with narrow band LF (IEC 60270-compliant) measuring systems, or when HF/VHF/UHF measurement systems are used that "downshift" the PD signal to the frequency range of a conventional LF system. For example Figure 8.11 indicates that all pulses across the AC cycle are positive due to the downshifting. In these cases, the pulse polarity has to be inferred from the AC phase position. A direct measurement method (where the PD is measured across the detection impedance on the ground side of the test object (Figure 6.1)) will show the PD pulses with a reverse polarity from that shown in Figure 10.6 at the same phase position.

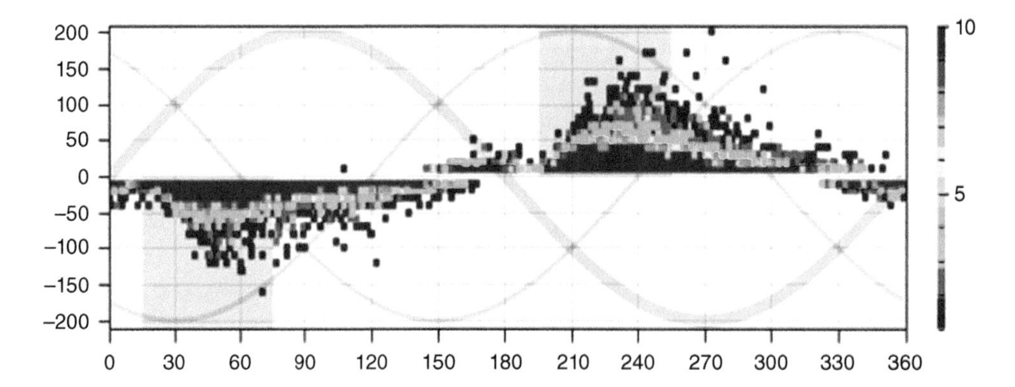

Figure 10.6 PRPD plot from a lab test on a 4.1 kV motor coil. The vertical scale is in mV, the horizontal scale shows the PD with respect to the line-ground AC voltage cycle and the color represents the number of pulses per second at that magnitude window width and phase window width (legend on the right). *Source:* Courtesy of Iris Power L.P.

10.5.1 Phase-to-Ground Patterns

Figure 3.24 indicated that phase-to-ground PD in HV equipment when displayed with respect to the AC cycle tends to occur between −15° and 105° or 165° and 285°. Space charge (in surface discharges and corona) and trapped charge (cavity and surface PD from previous discharges) may modify the electric field so that the PD may sometimes occur before the AC cycle voltage zero crossing or after the peak of the AC cycle, especially in cases of severe overvoltage. Unfortunately, the behavior of trapped charge (i.e. how much charged is trapped and the time it takes to decay) depends on many factors including:

- The insulation material itself, which influences how many and how deep the traps are,
- The overvoltage across the void at the time of previous discharges, which affects the energy of the charges,
- The temperature that affects decay time of trapped charge, as well as the humidity and/or contamination for surface discharges, and
- Cavity geometry that affects the local electric field.

Thus, the range of phase positions where even simple void discharge occurs may vary widely between different types of test objects. As presented in Section 10.5.4, in online tests, the phase position in a PRPD plot also depends on if the PD is occurring between phases or between phase and ground.

To many observers, the outline of PRPD plots for each polarity looks like a probability distribution function, with an associated mean and a standard deviation. Figure 10.7 shows an idealized sketch where the outline of the positive and negative PRPD plots are shown as normal (or Gaussian) probability distribution functions. With an indirect measuring system, the negative PD pulse occurs at a mean of 45° of the AC cycle and the positive PD occurs at a mean of 225° of the AC cycle. Describing the PRPD pattern as a statistical distribution enables statistical methods to be applied to the analysis of the patterns, and properties such as the mean, mode, and standard deviation can be calculated [5]. The standard deviation of each pattern gives an indication of how much

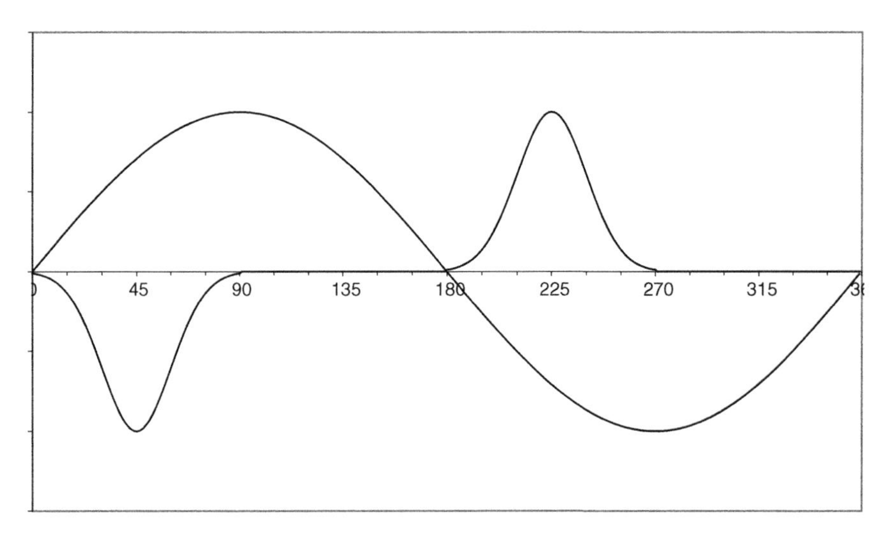

Figure 10.7 Sketch of the positive and negative PRPD pattern outlines as probability distributions. This and the following sketches assume an indirect PD measurement is made.

of the AC cycle over which the PD is occurring over. Normal-shaped distributions of PD are common for an insulation system containing a number of cavities of varying size.

Often the PRPD pattern does not look like a symmetrical normal distribution. Sometimes the PD seems to have a similar magnitude across up to half the AC cycle, i.e. it looks like a uniform probability distribution. This may indicate there is a fixed size of air gap between within the insulation system. If this uniform distribution is centered about 90° and 270° of the AC cycle, then it may be corona from a sharp-edged metallic protrusions. Alternatively, the distribution may be skewed, with higher and more PD near the AC voltage zero crossings and fewer and smaller pulses occurring near the peaks of the AC cycle, looking more like a right-angle triangle.

Gulski [6] and others have created catalogs of the primary statistical characteristics (mean, standard deviation, skew, and kurtosis) of the PD patterns associated with different root causes of PD (Section 8.9.1).

10.5.2 Generalized Phase-to-Ground PRPD Patterns

Given the wide variety of PD source mechanisms and locations within an insulation system, together with different shapes of cavities or air gaps, different PRPD patterns occur. Figure 10.8 shows sketches of different types of patterns associated with the different locations of PD sites

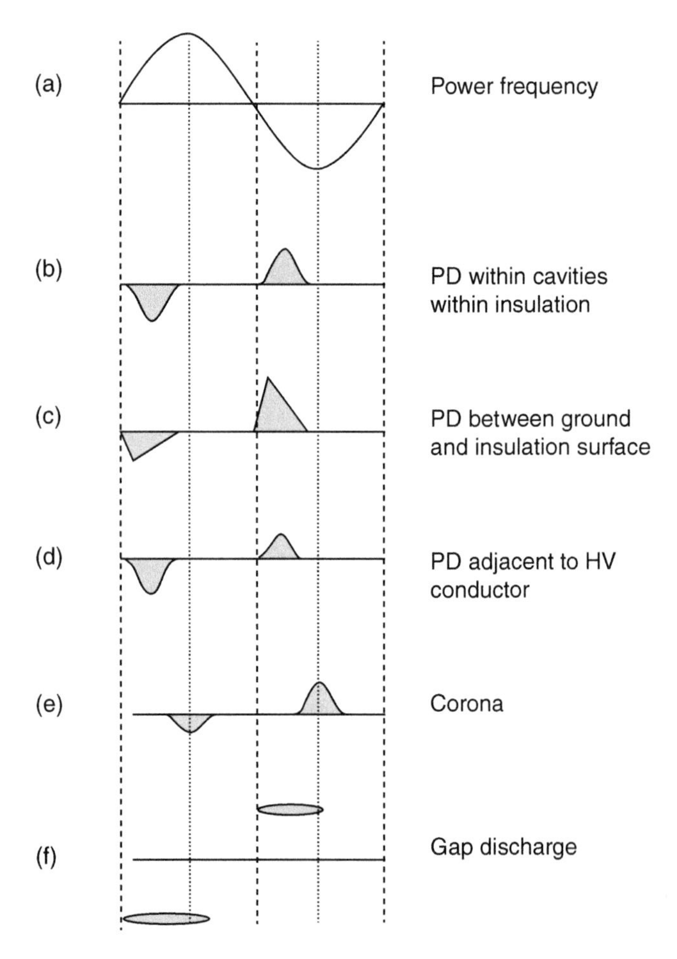

Figure 10.8 Idealized PRPD patterns from various types of defects in an insulation system where only one phase is energized using an indirect measurement system.

within an insulation system when energized by phase-to-ground voltage. Chapters 12–16 present some actual PRPD patterns for each type of equipment, some of which may vary greatly from the generalizations shown in Figure 10.8. Based on a literature survey, Shahsavarian et al. present published PRPD patterns from a wide variety of types of equipment and defects [7]. Indeed this paper shows little consistency in the PRPD patterns associated with similar defects in different types of HV equipment. This demonstrates the paramount importance of having a deep understanding of both the overall design and specific material properties of the insulation system in question when undertaking PRPD pattern analysis.

10.5.3 "Rabbit-Ear" Pattern

Figure 10.9 shows an idealized PRPD pattern that seems to be much loved by researchers. For each polarity, one can imagine a rabbit with a body and associated rabbit ears. The "ear" tends to follow the outline of the AC voltage. Although the origins of the name rabbit ear pattern are not clear, hundreds of papers have been written since 2008 or so about finding this pattern in test objects like power cables, transformers, GIS, and stator windings. In most cases, the authors have attributed the pattern to PD in voids. The rabbit ear pattern has been reproduced in analytical simulations for a single spherical void within a homogeneous solid insulation material, with the predominance of the "ear" depending on the conductivity of the of the cavity's inner wall [8].

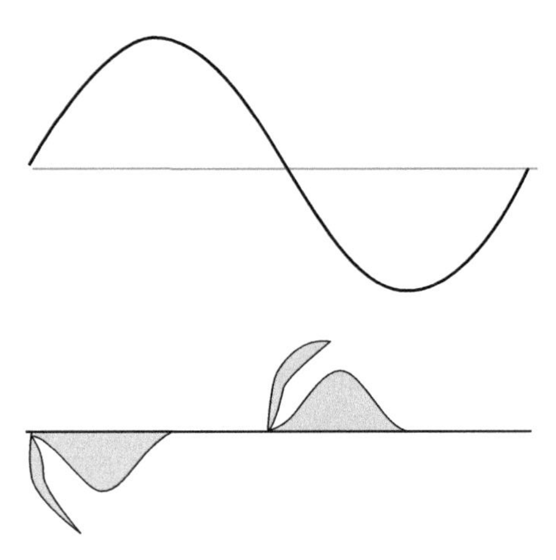

Figure 10.9 Rabbit ear PRPD pattern.

10.5.4 Phase-to-Phase PD in Online Tests

In addition to the effect of space and trapped charge, the apparent phase position of the PD may also shift if PD is occurring between phases in a three-phase power system. Such interphasal PD can occur in stator windings, AIS, and three-phase GIS (Figures 10.1 and 10.2). Figure 10.10 shows that the phase-to-phase voltage is phase-shifted from the phase-to-ground voltage (which is always used for PRPD plots) by + or −30°. Thus, if the interphasal PD is plotted on a PRPD plot referenced to the phase-to-ground voltage, the PRPD plots will appear to be shifted 30° to the right or the left, depending on which phase (to ground) voltage is referenced.

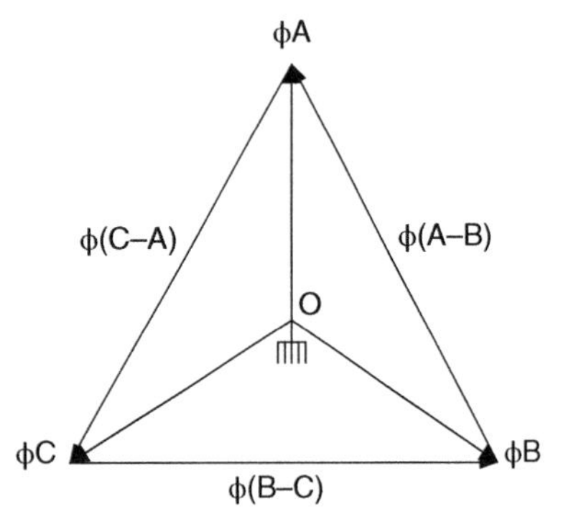

Figure 10.10 Vector diagram showing the relationship between the phase to neutral voltage in A, B, and C phases, with respect to the phase-to-phase voltages between phases.

Figure 10.11 shows an idealized sketch of the PRPD patterns caused by PD between phases A and B, with the top plot synchronized to the A phase-to-ground voltage, starting at 0°; while B phase is synchronized to the B phase to ground voltage, starting at −120° with respect to the A phase voltage. By shifting B phase 120°, a vertical line between the two plots indicates PD occurring at the same time [9]. Note that the PD pattern is shifted to the left by 30°

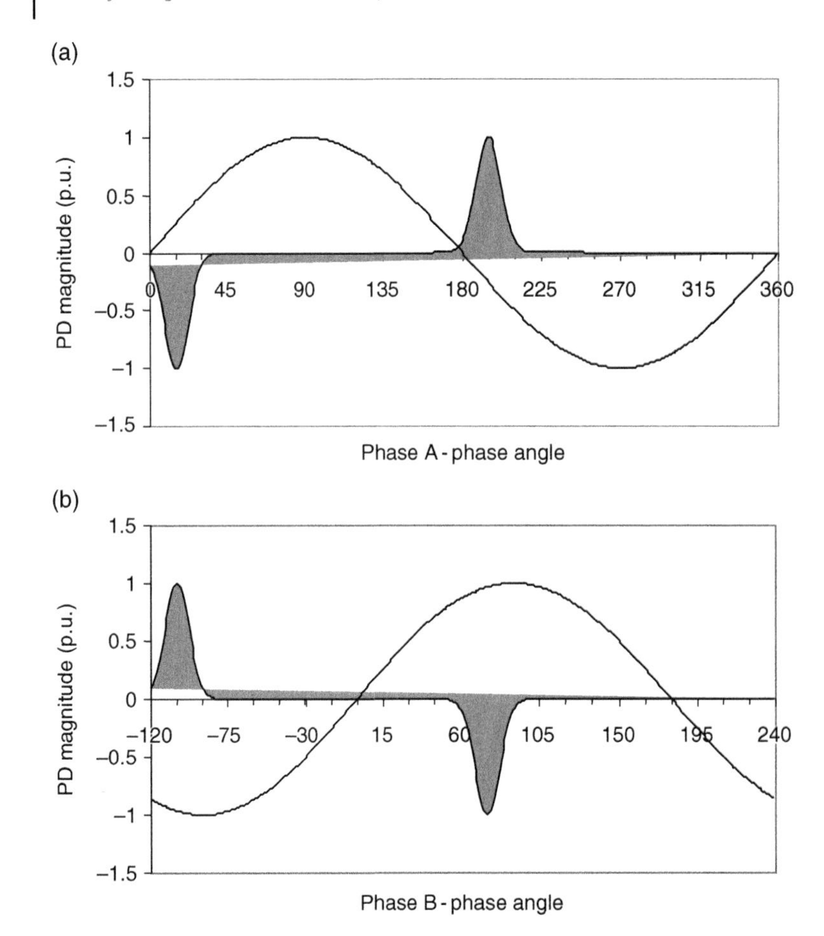

Figure 10.11 PD patterns for interphasal PD between A and B phases, when displayed with respect the (a) A phase-to-ground voltage or (b) B phase-to-ground voltage. Note that in (b) the AC phase angle starts 120° from that in (a), which aligns the two plots in time. *Source:* Courtesy of Iris Power L.P.

in the top plot, whereas it is shifted 30° to the right in the bottom plot. This shows the main characteristics of interphasal PD as displayed on phase-shifted PRPD plots:

- The PD occurs at the same time (i.e. a vertical line between the plots shows PD activity at the same instant of time).
- The PD in one phase is shifted to the right
- The PD in the other phase is shifted to the left
- For the vertically aligned PD, they have approximately the same magnitude, but opposite polarity.

These characteristics follow from how interphasal PD is created and Kirchhoff's current law. A PD in air between the end-windings (far from the stator core) of two adjacent coils in different phases in a stator (or two HV leads or bus-bars in AIS) is shown in Figure 10.12. The PD in air creates a current that flows through the capacitance of the coil in Phase B, which flows to the PD sensor at the B phase terminal, creating a current to ground. This current then flows up through the ground to the A phase PD sensor and then to the A phase terminal and completes the current loop via the capacitance of the A phase coil. Thus, the same PD pulse will create a positive

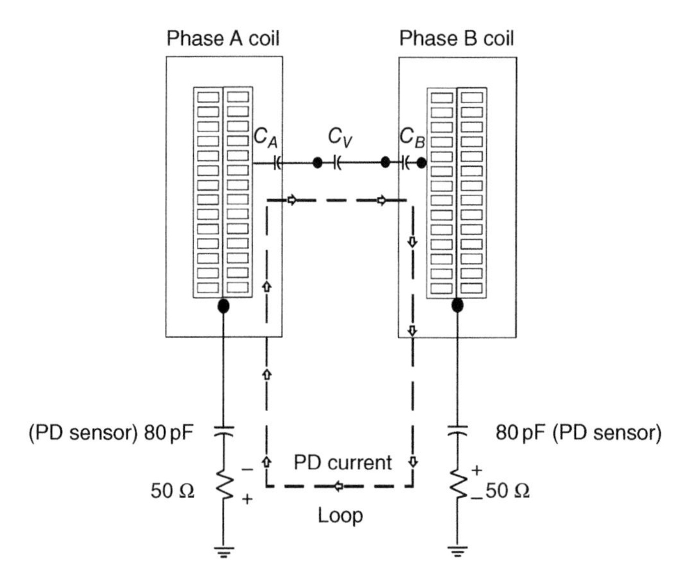

Figure 10.12 Equivalent circuit of a single PD pulse occurring in the air gap between two coils in a stator end-winding, which results in a polarity reversal of the detected pulse in each phase.

voltage drop across the detection impedance in Phase B while it creates a negative voltage across the detection impedance in Phase A. Even if there is positive predominance PD (Section 10.4.3) in Phase A, it will appear as negative predominance in Phase B. This is why the polarity predominance principle (Section 10.4.3) is misleading with interphasal PD.

10.5.5 Cross-Coupled Signals Between Phases in Online Tests

In online PD tests of three phase apparatus, it is possible in some types of test objects for there to be strong coupling of high-frequency signals such as PD pulses between the phases. This can lead to pulses appearing in the PRPD plots that are not interference, but occur in the "wrong" part of the AC cycle and/or with the "wrong" polarity for that part of the AC cycle. These "wrong" pulses are test object PD in one phase that is coupled into the other phases, making the PRPD plot in the other phases much more complicated.

Figure 10.13 shows a sketch of the PRPD in three phases, when PD is actually occurring only in Phase A. Note that the three plots are phase-shifted from each other, so that a vertical line indicates pulses occurring at the same instant of time. When PD in Phase A is a positive pulse in the negative half of the AC cycle, due to coupling between phases, it shows as a positive pulse near the peak of the positive half AC cycle in B phase and a positive pulse near the AC zero crossing in C phase. The signals in B and C phases due to cross-coupling from A phase must be ignored. Of course, if legitimate PD is occurring in all three phases, and there is cross-coupling between phases, the PRPD patterns can become very complicated, with pulses occurring across the entire AC cycle. This is why observing all three phases at the same time, but shifted in phase angle; or using the 3PARD type of analysis (Section 8.9.3) can be useful.

Cross-coupling is most likely due to capacitive coupling between the phases when the distance between components in different phases is small, and where there is no ground plane between the phases. Cross-coupling is unlikely in single-phase GIS, isolated phase bus, and single-phase power cables since the three phases have grounded conductors between them. Cross-coupling is also less likely in liquid-cooled and dry-type transformers since there is usually a relatively large

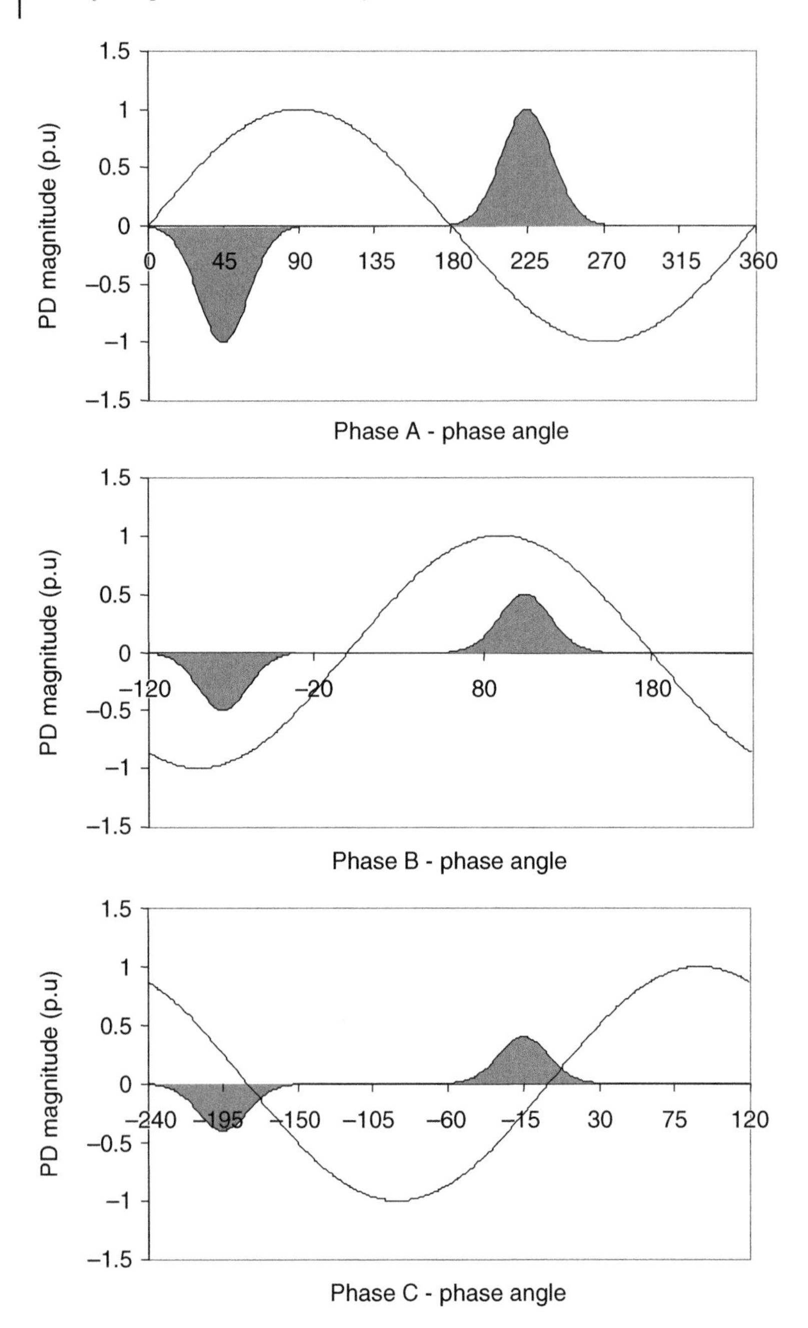

Figure 10.13 Sketches of PRPD plots where PD is occurring in Phase A, that is cross-coupled to B and C phases.

distance between the windings in each phase, reducing the capacitance between phases, and thus the opportunity for cross-coupling (although Table 15.4 shows as much as 35% cross-coupling between phases). This is also the case for AIS. The three phases in stator windings are in close proximity to each other, and there is no ground plane separating them in the end-winding region. Thus, cross-coupling of up to 50% is possible in stator windings, making PRPD interpretation much more difficult.

10.5.6 Simultaneous Occurrence of Multiple Aging processes

The idealized PRPD plots of phase-to-ground PD in Figures 10.8 and 10.9 assume that only one mechanism is producing the pattern, for example corona or PD in voids adjacent to the HV conductors. PD caused by a single process is very common in new equipment. However, as HV equipment ages in service, multiple failure processes can occur at the same time, much as humans may have several diseases at the same time as they grow older. When two or more failure processes are simultaneously occurring, the PRPD patterns can become very complicated, and identifying which mechanisms are occurring may be difficult if not impossible. Misidentifying the root cause of the PD may cause asset managers to plan inappropriate maintenance or postpone critical repairs.

Figure 10.14 shows a simple example involving just two aging processes occurring at the same time, in this case interphasal PD (only likely in AIS, three-phase GIS, and stators windings) and PD

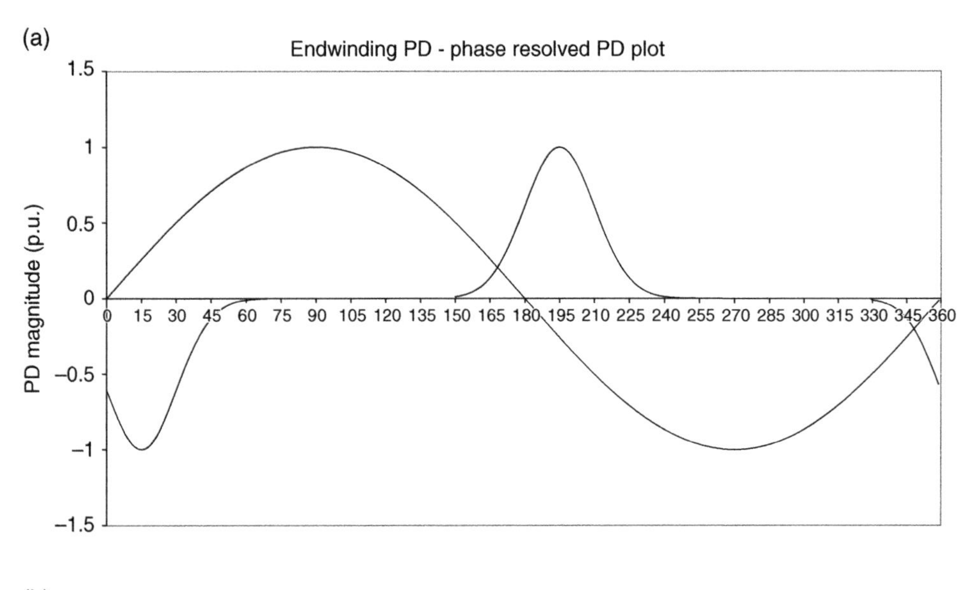

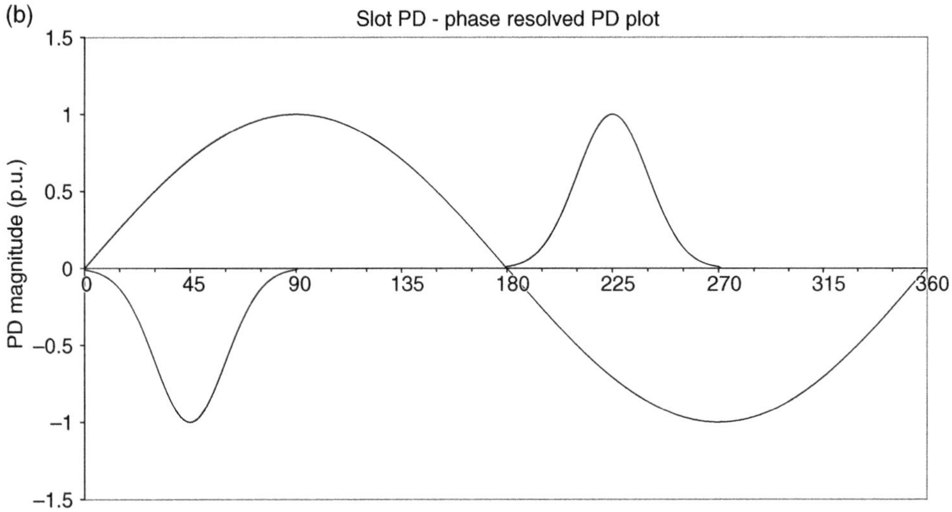

Figure 10.14 (c) Shows the idealized PRPD pattern when two separate failure processes in (a) and (b) are occurring at same time in a test object.

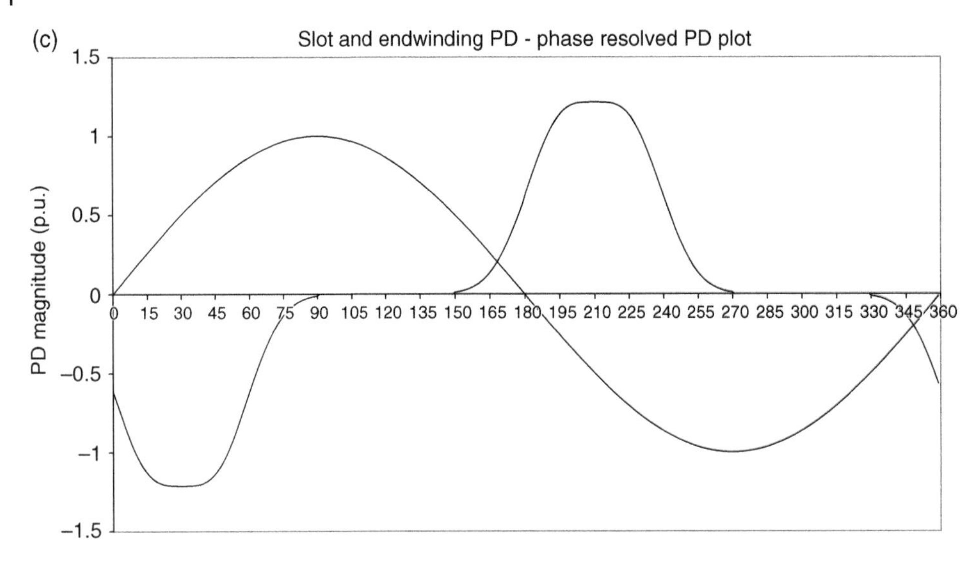

Figure 10.14 (Continued)

within voids in solid insulation. If one assumes each mechanism produces a classic Normal PRPD pattern, then the combined pattern has a shape that is flat-topped, perhaps indicative of a different aging process.

With online testing, the PRPD patterns can become very complicated since:

- Interference may contaminate the plots
- Cross-coupling causes PD in one phase to create pulses in the other phases
- Multiple processes that create PD may occur simultaneously.

The result is that even "experts" may have difficulty in deconstructing complex PRPD patterns to each of their actual causes. Certainly, in blind testing, experience to date indicates that expert systems and neural networks do not have a high probability of success in these situations.

The application of tools such as pulse shape analysis (Section 8.4.3), T-F maps (Section 8.9.2), and three-phase synchronous analysis (Section 8.9.3) can help to deconvolute different PD root cause and interference sources. However, these methods depend on the single-mechanism PRPD patterns (such as those in Figure 10.8) are always valid. As seen in [7], this is not always true.

10.6 PD Root Cause Identification Using Changes in Ambient and Operating Conditions

Since identifying the root cause of PD using PRPD patterns is not infallible, it is useful to be able to confirm the root cause of the PD in other ways. Of course, this can be done using techniques discussed in Chapter 5 or by a visual inspection after dismantling the EUT. However, for some types if HV equipment, how the PD (Q_{IEC}, Q_m or the PRPD pattern) changes with ambient conditions or, in the case of online tests, with equipment operating conditions, can help to identify the root cause of the PD.

For example, if PD measurements are performed on an outdoor substation in dry conditions, PD may be measured from some of the HV equipment, perhaps internal voids PD in a transformer.

However, when it is raining, both PD and corona may be occurring simultaneously. By performing tests under two different atmospheric conditions, the two different sources can be separated and identified.

For cavity-type PD in many types of test objects, the PD may be affected by temperature. Usually as the test object temperature increases, the PD level (Q_{IEC} or Q_m) decreases and the PDIV/PDEV increases. It is speculated that the decrease in PD with increasing temperature may be due to the volume of the voids decreasing as the insulation surrounding the cavity increases in volume due to the coefficient of thermal expansion. Alternatively, it may be caused by the shorter time to detrap charge and decrease void wall conductivity at higher temperatures. In any case, PD on insulation surfaces does not show a strong negative correlation between PD activity and temperature. Thus, if such a temperature effect is noted, it points to void-type discharges within the high-voltage insulation. In online tests, the influence of temperature is much easier to see if the PD is captured at different insulation temperatures.

In addition, in online tests, sometimes electromagnetic forces act on the insulation system as the current changes in the test object. If PD is occurring because of movement of the insulation, such as loose coils in stator winding slots (Section 16.4.3), then the PD may be affected by the load. Since most root causes of PD are not affected by load changes, this will help to confirm the root cause of the PD. This is discussed further in the equipment Chapters 12–16.

References

1 Cigre WG 21.03 (1969). Recognition of discharges. Electra No. 11, pp 61–98.

2 Rizzetto, S., Stone, G.C., and Boggs, S.A. (1987). The influence of X-rays on partial discharges in voids. *Conference on Electrical Insulation & Dielectric Phenomena – Annual Report* 1987, Gaithersburg, MD (18–22 October 1987), pp. 89–94. IEEE. https://doi.org/10.1109/CEIDP.1987.7736539.

3 Stone, G.C., Culbert, I., Boulter, E.A., and Dhirani, H. (2014). *Electrical Insulation for Rotating Machines*, 2e. Wiley/IEEE Press.

4 Cavallini, A., Montanari, G.C., and Puletti, F. (2006). Partial discharge analysis and asset management: experiences on monitoring of power apparatus. 2006 IEEE/PES Transmission & Distribution Conference and Exposition: Latin America, Caracas (15–18 August 2006), pp. 1–6. IEEE. https://doi.org/10.1109/TDCLA.2006.311544.

5 Snedecor, G.W. and Cochran, W.G. (1989). *Statistical Methods*, 8e. Wiley-Blackwell.

6 Gulski, E. (1993). Computer-aided measurement of partial discharges in HV equipment. *IEEE Transactions on Electrical Insulation* 28 (6): 969–983.

7 Shahsavarian, T., Pan, Y., Zhang, Z. et al. (2021). A review of knowledge-based defect identification via PRPD patterns in high voltage apparatus. *IEEE Access* 9: 77705–77728. https://doi.org/10.1109/ACCESS.2021.3082858.

8 Illias, H.A., Chen, G., and Lewin, P.L. (2011). Effect of surface charge distribution on the electric field in a void due to partial discharges. *Proceedings of 2011 International Symposium on Electrical Insulating Materials*, Kyoto (6–10 September 2011), pp. 245–248. IEEE. https://doi.org/10.1109/ISEIM.2011.6826278.

9 Warren, V., Stone, G.C., and Fenger, M. (2000). Advancements in partial discharge analysis to diagnose stator winding problems. *Conference Record of the 2000 IEEE International Symposium on Electrical Insulation (Cat. No.00CH37075)*, Anaheim, CA (5–5 April 2000), pp. 497–500. IEEE. https://doi.org/10.1109/ELINSL.2000.845557.

11

PD Testing of Lumped Capacitive Test Objects

11.1 Lumped Capacitive Objects

Most of the following chapters deal with offline and online testing of major types of equipment such as machines, transformers, switchgear, and cable systems. In this chapter, the focus is on testing of the many types of test objects that are not one of these types of major equipment but are rather are components of a larger system. Such test objects tend to be smaller in physical size and sometimes do not have a specific PD test standard associated with it. Due to their small size, the test objects behave as lumped capacitors.

There are many devices that can be treated as lumped capacitors. A (non-exhaustive) list of such objects includes:

- Cable splices, tested after manufacturing, but before being deployed in the field
- Cable joints and terminations
- Individual coils or bars used to manufacturer motors or generators
- Capacitors such as those used for power factor correction or surge suppression
- Spacers or insulators for switchgear (both AIS or GIS)
- Instrument transformers
- Surge arresters
- Bushings, capacitive dividers, connectors
- Power electronic modules and gate drivers
- Insulation models used for qualification purposes such as twisted pairs of magnet (winding) wire, motorettes, formettes [1]

Testing of lumped capacitive test objects is very common for the design verification and quality assessment of new electrical components. The PD tests are mostly done with charge-based PD detectors complying with IEC 60270 [2], since this standard is primarily used for capacitive test objects. The theory behind these measurements is described in detail in Chapter 6. Here, the focus is on practical details and recommendations regarding the aim of the tests (e.g. what is measured and what could be prescribed as integral part of the test), the way the test circuit is assembled, and how disturbances are suppressed. The tests can be performed either after manufacturing or after stressing the component (as an example, using high temperatures or high electric fields) to infer their reliability.

PD on these test objects may also be measured offline at site to help determine if maintenance or replacement is needed.

Practical Partial Discharge Measurement on Electrical Equipment, First Edition. Greg C. Stone, Andrea Cavallini, Glenn Behrmann, and Claudio Angelo Serafino.

11.2 Test Procedures

PD tests on lumped capacitive test objects to be used in substations, generating stations, and industrial plants are generally aimed at determining the apparent charge at a specified test voltage, although some prescription on PDIV and PDEV might also exist. The general procedures for measuring the Q_{IEC} and the PDIV/PDEV are presented in Sections 10.2 and 10.3. The tests are usually made following the IEC 60270 standard and should have a reproducible result independent of the detector used to carry out the tests (see [2] and Sections 6.5 and 6.6 in this book). Since these tests are sometimes done in the presence of the customer, it is of critical importance that the apparent charge of the PD is determined with high accuracy (even though, for some equipment such as coils, doubts exist about the validity of such an approach). If the measurement frequency range is changed, the system must be recalibrated for valid apparent charge (pC) measurements. The sensitivity level (minimum detectable PD) should be reported. It is important to notice that, if PD are not detected, one must specify that the EUT is PD free within the sensitivity of the detection circuit (e.g. 1 pC).

For wide-band detectors (as defined in IEC 60270), where PD pulse polarity can be recorded, the PD pulse polarity can be used to distinguish whether PD are incepted in the EUT or in other parts of the circuit (see Figure 11.1):

- If a direct measurement circuit is used (Section 6.2), it is possible to discriminate whether PD comes from the EUT or from the other components of the HV circuit (source, coupling capacitor, voltage divider, etc.). If PD are incepted in the EUT, positive and negative PD will be generally recorded when the derivative of the voltage is positive and negative, respectively. The opposite occurs if PD are originated in the rest of the HV circuit.
- For an indirect circuit (which is the most common test circuit in practice), separation of PD coming from the coupling capacitor, from those coming from the EUT or other parts of the circuit is also possible. In this case, if PD are incepted in the coupling capacitor, positive and negative PD will be recorded when the derivative of the voltage is positive and negative, respectively. The opposite occurs if PD are originated in the EUT or other parts of the HV circuit.

It is always useful to do a PD test with the EUT disconnected to ascertain if PD are incepted due to, e.g., protrusions in the high-voltage connection or on the ground, or if the coupling capacitor or the HV supply is not PD free. See also Sections 9.3.2, 10.2, and 10.3 in this book.

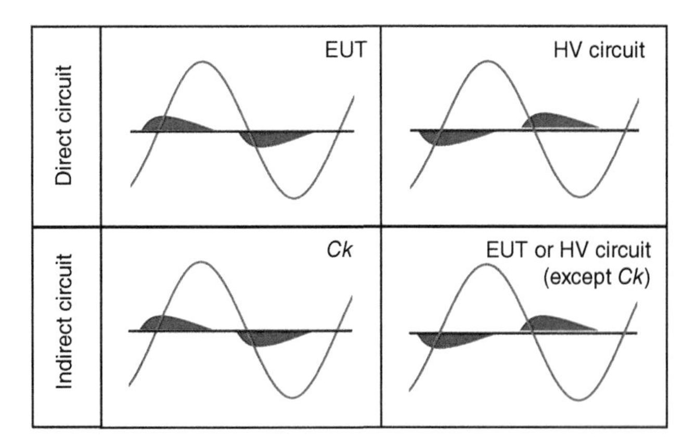

Figure 11.1 PD polarity analysis for direct and indirect circuits. *Ck* is the coupling capacitor (Section 6.2).

Good practice for PD measurements includes the following:

1) Using a coupling capacitor with sufficiently large capacitance compared with the test object. Alternatively, if the capacitance of the coupling capacitor is small compared to the test objects, the measurement can be carried out using a HFCT connected to the ground lead of the test object. In this way, the stray capacitances in parallel with the coupling capacitor will help to achieve a higher signal.
2) Using a certified calibrator and PD detector. The procedures for maintaining and ensuring the proper operation of both calibrator and detector are specified in [2]. It is customary to send calibrators to third parties at regular intervals to get an updated calibration certificate.
3) Use a calibrated voltage measurement chain (instrument transformer or capacitive voltage divider with a voltmeter).
4) Use a voltage source with an appropriate form factor for the voltage waveform.[1]
5) Reducing interference down to, at least, 50% of the maximum acceptable PD apparent charge (see suggestions in Sections 9.3 and 11.3).
6) Using a proper full scale for PD detection (typically, twice the limit for the maximum PD apparent charge).

During the tests, the voltage stress should reproduce the stress experienced by the device during service. Many devices can be regarded as a two-terminal capacitor, thus, their connection to the high voltage and the ground is straightforward. These devices include, among others, capacitors, capacitive dividers, surge arresters, and spacers. However, the connections of other devices might be less straightforward. As an example, the high-voltage winding of the instrument voltage transformer must be connected as in service (the high-voltage terminal to the high-voltage circuit, the other one grounded), whereas both the HV terminals of current transformers should be connected to the high-voltage circuit (see Figures 11.2 and 11.3). In case of doubts, it is always better to refer to the equipment-specific standard.

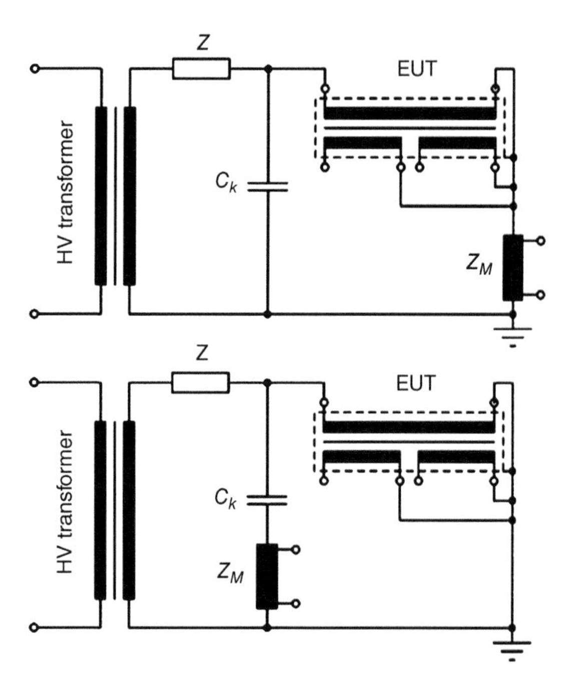

Figure 11.2 Direct and indirect detection circuit for instrument voltage transformers.

1 The form factor is defined for alternating waveforms as the ratio of the RMS value to the mean value of the rectified signal. For a perfectly sinusoidal voltage waveform, the form factor is equal to 1.11.

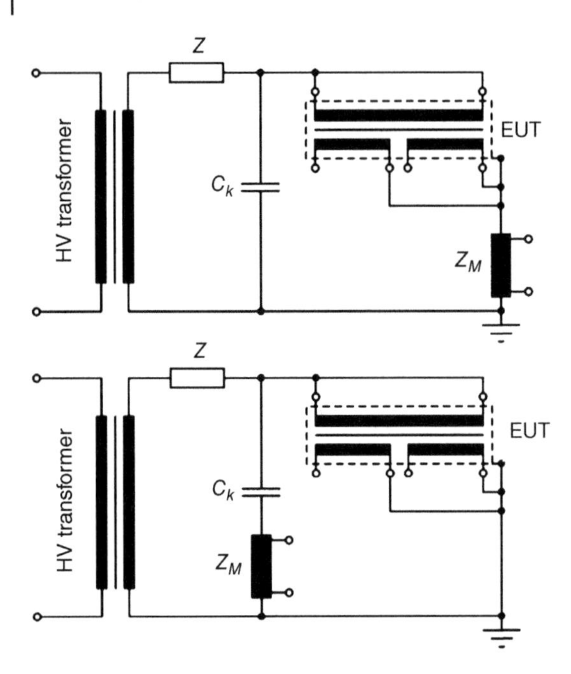

Figure 11.3 Direct and indirect detection circuit for instrument current transformers.

Another important point is that, very often, the way the test voltage is applied is standardized. This has an influence on the PDIV/PDEV. Equipment standards can include specifications on:

- The way the AC test voltage is raised:
  - In steps: the voltage increases at each step and the duration of each step.
  - The rate of rise of the voltage if a ramped voltage is used.
- The maximum voltage reached during the tests and the time this voltage must be applied to the EUT.
- Pre-conditioning of the EUT, including performing a withstand test prior to PD testing (see Sections 3.8.1 and 10.3.4).

11.3 Measures to Suppress Electrical Interference

As discussed in Sections 9.2.2 and 9.3, reducing the interference and noise can be achieved through different approaches, including:

a) Avoid PD inception in the high-voltage connections using appropriate shielding electrodes around the high-voltage connections and placing the conductors inside corrugated metallic pipes. To smooth out edges, partly conductive putties such as Duxseal® can be used. The walls of the test room should be far enough away. Be careful that tools and other metallic parts in proximity of the high-voltage circuit could act as floating metallic masses and become the site of PD activity. It is always recommended to keep the test area as clean as possible.
b) Use voltage sources, coupling capacitors, and voltage dividers that are PD free up to the maximum test voltage. If one is unsure, it is important to measure their PDIV.

c) Shielding the high-voltage circuit from electromagnetic interference. The most common approach (although very costly) is to use a shielded room (Section 9.3.1). A careful design of the shielded room entails using materials that can shield effectively both the electric and the magnetic field (the loop source-connections-EUT can couple with magnetic fields that can induce electromotive forces in the frequency range used for PD detection). The size of the apertures in the shielded room (for example ventilation and observation windows) must be designed in a way that, any electromagnetic wave radiated from the apertures, does not interfere with PD measurements.
d) Use high-voltage resistors (or inductors) in series with the high voltage to reduce disturbance from the source and the effect of electromotive forces induced by magnetic fields.
e) Shield signal connections from interference. This entails using coaxial cables with single or, if necessary, double shielding (RG-223/U coaxial cables). Four-layer shielding using RG-6/UQ coaxial cables is also possible. It is worth observing that coaxial BNC connectors can get damaged after repetitive use and pick up disturbances. This is a recurring issue during PD measurements. It is a good idea to replace the coaxial cables when noise issues are apparently impossible to correct.
f) Low-inductance ground connectors must be used. In general, round wires are not the best option at high frequencies. Braided flat wires or flat copper strips have much lower inductance.
g) Ground loops should be avoided. Ideally, the shielded (or testing) room should be provided with a separate grounding system. This is achieved (at relatively low frequencies) by covering the ground poles with a polymeric insulation foil. The ground loops should be carefully avoided in the test circuit. This is done by selecting a single connection point for all high-voltage equipment. Sometimes, it could be necessary to filter the AC 50/60 Hz supply used to power the detection equipment. Otherwise, battery-operated PD detectors can be a way to prevent disturbances from the 50/60 Hz AC supply coupling with PD measurements.
h) Selection of the bandwidth where PD signal exceeds noise can also be pursued provided that the bandwidth is below 1 MHz (in accordance with [2]).
i) Additional filtering techniques should be agreed upon with the customer, as they might mask PD activity if improperly used.

Prior to calibration, it could be useful to use the calibrator to check that the PD detector operates correctly. This is done by injecting calibrator pulses directly at the input of the PD detector.

11.4 Sensitivity Check

The sensitivity check serves to assess the minimum PD magnitude that the measurement system can detect with an acceptable signal-to-noise ratio (SNR). For measurements on capacitive objects compliant with [2], calibration is not only a means to calculate the apparent charge but also used to evaluate the sensitivity of the measurement in terms of pC. As a matter of fact, by progressively reducing the charge injected by the calibrator the operator can determine the minimum amount of charge that can be detected with a sufficient SNR.

The apparent charge calibration process is not valid for measurement frequencies above 1 MHz (Section 6.5). Above 1 MHz, IEC TS 62478 specifically indicates that calibration into apparent charge has no technical meaning (Section 7.6). Since many commercial IEC 60270 detectors may also measure in the high-frequency (HF) range (Section 7.2), a sensitivity check is typically performed. A sensitivity check is also normally performed for inductive-capacitive test objects, where calibration into apparent charge has no meaning as described in Section 6.6.3 (even though many commercial detectors produce an apparent charge result).

In practice, a calibrator such as in Section 6.6 or a pulse generator is used not to calibrate into apparent charge, but to (i) carry on a functional check of the system and (ii) determine the minimum PD magnitude that can be detected during the test. The sensitivity check is thus defined in a conventional way and the results are expressed in mV or pC (if using an IEC 60270 calibrator for this purpose). In practice, one must check the minimum signal (mV or pC) the measurement system is able to detect by injecting impulses of decreasing magnitude into a capacitive object having capacitance agreed upon by manufacturer and purchaser of the EUT.

The steps of sensitivity check are the following:

1) Agree on a pulse generator. For inductive-capacitive test objects, the rise time is important for the winding sensitivity check as resonances can be excited or not.
2) Establish a dummy load, generally a capacitor with capacitance close to that of the EUT (at 50 Hz).
3) Inject a pulse in the dummy load. The pulse must have an agreed charge/dummy load capacitance ratio, e.g. 1 mV = 1 pC/1 nF.
4) Check if the detection system can detect the calibrator pulse starting from large pulses and progressively decreasing their magnitudes. The minimum pulse that can be detect determines the sensitivity of the measurement (Figure 11.4).

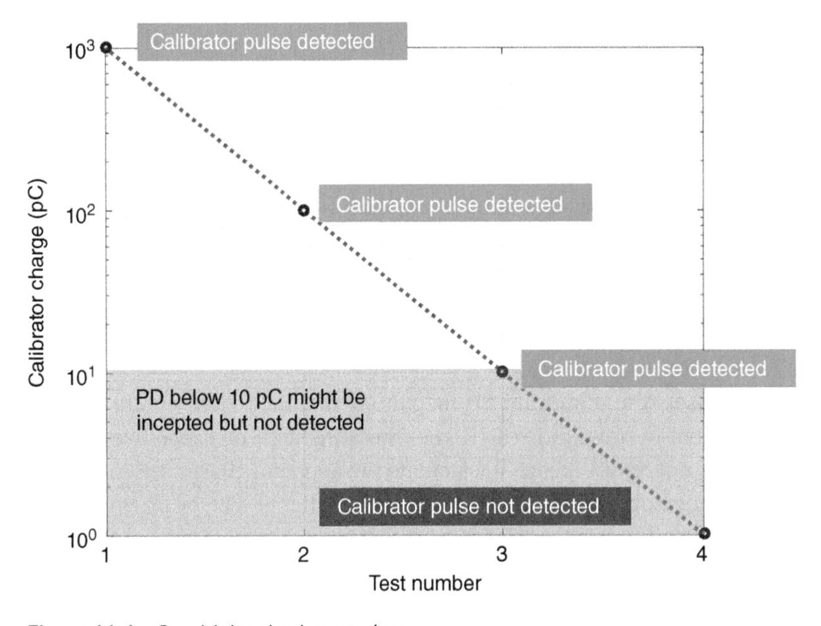

Figure 11.4 Sensitivity check procedure.

References

1 IEC Std. 60034-18-41 (2019). Rotating electrical machines – part 18–41: partial discharge free electrical insulation systems (type I) used in rotating electrical machines fed from voltage converters – qualification and quality control tests. https://webstore.iec.ch/publication/65356 (accessed 3 July 2020).

2 IEC Std. 60270 (2015). High-voltage test techniques – partial discharge measurements. https://webstore.iec.ch/publication/23841 (accessed 3 July 2020).

12

PD in Power Cables

12.1 Introduction

Power cables are essential to supply loads in distribution-class and transmission-class utility networks, where overhead lines are not possible due to space, esthetic, reliability, or safety considerations. Power cables are also widely used in industrial plants to connect power transformers to loads. Several books have been written that provide in-depth information on the application, design, aging, and repair of power cables [1, 2]. This chapter is concerned with partial discharge (PD) testing of medium voltage AC cables as well as high-voltage AC power cables. The chapter also provides a brief overview of medium voltage and high-voltage cable design and aging, concentrating on shielded polymer-insulated and mass-impregnated power cables rated 5 kV and above, as well as their accessories such as terminations and joints. Some comments on PD testing of high-voltage direct current (HVDC) apparatus such as high-voltage DC power cables are in Chapter 17.

12.2 Cable System Structure

Cable insulation can be polymeric, mass-impregnated (MI) cables, or oil-filled (OF) cables, which can be further subdivided into pipe-type and self-contained fluid-filled (SCFF) cables. Currently, polymeric cables are the preferred choice for HVAC and MVAC voltages. HVAC OF cable systems are still in use but are being phased out due to the environmental issues. Also, for OF cables, diagnosis is based mostly on dissolved gas analysis (DGA) rather than partial discharge detection [3]. In medium-voltage systems, the MI technology is still in use, but replacements are made using polymeric cables. MI cables, however, are still a popular choice for HVDC submarine transmission. The focus in this chapter is mostly on polymeric cables, with some notes about MI insulation.

The structure of accessories such as joints and terminations will depend on the insulation type and on the number of conductors in the cable system. Terminations will also depend on the cable insulation, type of equipment the cable is connected to (AIS, GIS, oil-filled transformers), and the number of conductors and their geometry (the latter being more typical of MV cable systems). This chapter concentrates on polymeric cable accessories.

Practical Partial Discharge Measurement on Electrical Equipment, First Edition. Greg C. Stone, Andrea Cavallini, Glenn Behrmann, and Claudio Angelo Serafino.

12.2.1 Cable Insulation

For high-voltage transmission lines, the simple geometry of a coaxial cylinder capacitor is the most efficient one to achieve an optimum use of the insulation (overhead lines are different but the insulation [air] is so abundant that one can afford to use it without restrictions). The coaxial cylinder capacitor was presented in Section 2.1.3.2. Using AC voltage waveforms, one can calculate the electric field at all points of the insulation using Equation (2.9). For HVDC cables, the temperature gradient within the cable can result in notable deviations from Equation (2.9), so the reader is referred to more specific books on the topic, such as [4].

If a power cable were to be manufactured using the insulation and conductors only, it would be unfeasible to treat it as a coaxial cylinder capacitor. To bend the cable, it is very difficult to use solid round or cylindrical conductors. Instead, it is necessary to separate the high-voltage round conductor and the return (grounded) conductor into strands (or subconductors). The result is sketched in Figure 12.1 (left) where the neutral strands do not uniformly cover the insulation (assumed as a hollow cylinder), giving rise to electric field concentrations at the conductor strands in contact with the insulation. Thus, the maximum electric field applied to the dielectric near the strands would be higher than that calculated using Equation (2.9). Without some other approach, the insulation would have to be thicker to compensate for the stranding. Besides, the sketched circular section of the strands in Figure 12.1 is ideal. Real conductor strands have protrusions or other random defects that will further increase the electric field.

To overcome these limitations, partly conductive "semiconductive" layers, also called shields or screens, are used. These layers have the function of smoothing the interface between the strands and the dielectric (see Figure 12.1 [right]) so that the coaxial cylinder capacitor assumption becomes valid. For mass-impregnated and oil-filled cables, the strands are wrapped with tapes loaded with graphite (Figure 12.2). The insulation consists of paper layers wrapped along the high-voltage conductor leaving butt gaps that allow the oil or the mass to penetrate within the insulation, achieving a complete impregnation.

For polymeric cables, tapes loaded with graphite were used up to the early 1980s. Since this technique had problems (Section 12.3.3), modern polymeric cables are manufactured using the triple extrusion process. The triple extrusion is performed by (i) extruding a semiconductive layer on the high-voltage conductor (the conductor shield), thus shaping it as a cylindrical electrode,

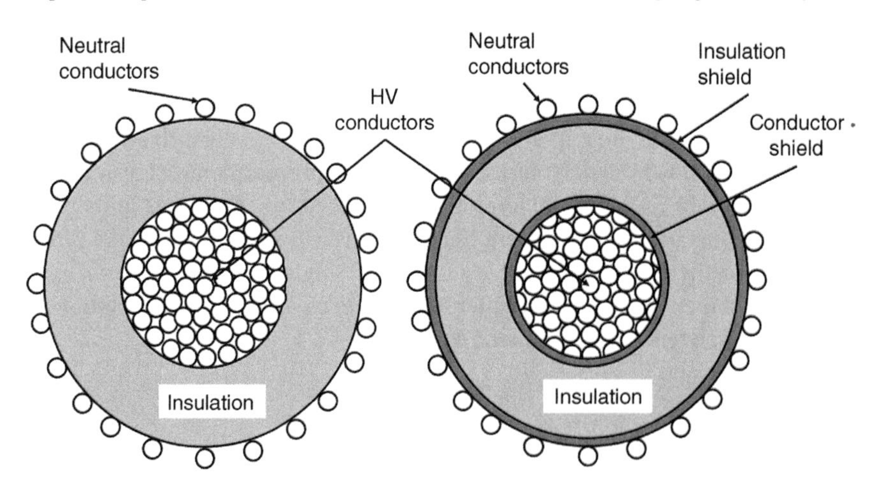

Figure 12.1 Structure of power cable (left) without semiconductive layers; (right) with semiconductive layers.

Figure 12.2 Cross section of an old submarine 30 kV mass-impregnated cable. The semiconductive tape is clearly visible around the high-voltage copper conductors.

Figure 12.3 Cross section of 400 kV polymeric cable. *Source:* Wdwd/Wikimedia Commons/CC BY-SA 3.0.

(ii) extruding the dielectric on top of the cylindrical high-voltage electrode, and (iii) extruding an outer semiconductive layer on the dielectric (the insulation shield), thus achieving a coaxial cylinder electrode configuration (Figure 12.3). The semiconductive layers are often realized using the same base polymer of the insulation but loaded with carbon black to increase their conductivity. The extrusion process is normally performed in a high-pressure nitrogen atmosphere to avoid contamination and bubble formation. The materials used in the extrusion process are also highly purified (originally these measures were for HV cables only, but in time they were adopted also for MV cables [5]).

Polymeric cables can be realized using ethylene-propylene rubber (EPR), polyethylene (most commonly, cross-linked polyethylene, XLPE), or polypropylene (PP). The choice between these materials depends mostly on the voltage level.

For medium or medium-high voltage cables (normally ≤35 kV), where design electric fields are generally not too large, EPR competes with XLPE. Compared to XLPE, EPR can withstand partial discharges for longer times, as EPR cables contain large fractions (up to 40%) of silica that, being inorganic, is relatively insensitive to partial discharges. EPR is also more flexible, has reduced thermal expansion, and is claimed to be less affected by water treeing (Section 12.3.2) compared with XLPE. However, the dissipation factor of EPR is 20 times larger than that of XLPE [6], which limits its application to cables with a maximum rated voltage of 150 kV. In contrast, XLPE cables are manufactured for voltages up to 500 kV.

Above 150 kV, and thanks to its better thermal performance, XLPE is more widely used than low-density polyethylene (LDPE) and high-density polyethylene (HDPE). However, XLPE's working temperature remains relatively low (90 °C, above which the mechanical properties decrease dramatically), which limits ampacity and thus power density. Looking for a suitable alternative, efforts were made to use PP. Despite being thermoplastic (XLPE is thermosetting), PP has a higher operating temperature (130 °C) but is considerably stiffer than XLPE and cannot be used alone. PP-based compounds (blends) suitable for manufacturing power cables were developed first by Prysmian in 2010 [7]. Compared with XLPE, the thermoplastic technology developed by Prysmian is free from ambers (translucent resins), can be filtered more effectively providing cleaner insulation, and does not need degassing (the last feature being the most appealing, as degassing can be a very time-consuming process). Also, PP is more easily recyclable and has a lower carbon footprint. However, compared with XLPE, PP is a relatively new technology that has not yet found widespread acceptance. Being both purely organic materials, XLPE and PP are vulnerable to partial discharges. Moderate partial discharge activity can lead to breakdown in hours or days. Thus, partial discharge detection in HV and EHV polymeric cables is vital. PD of even low magnitudes (the tens of pC range) always triggers preventive maintenance actions.

For mass-impregnated cables, the solid insulation is made using paper tapes wrapped around the high-voltage conductor. Thus, this type of insulation is generally indicated as taped insulation. Since water has a detrimental effect on cellulose, the paper should be dried to a moisture content of 0.1% in weight. Taping is realized leaving nonoverlapping butt gaps in different layers, so the mass can impregnate the insulation thoroughly. Since the mass is partially regenerative, this technology is less vulnerable to partial discharges than polymeric cables.[1]

12.2.2 Accessories

Accessories are either joints or terminations. Joints are necessary to connect two or more sections (lengths) of cable together to create a long run of cable. Terminations are used to connect the cable to other parts of the high-voltage systems such as transformers, motors, generators, switchgear, and overhead lines. Accessories are applied where the cable conductor and insulation shields are interrupted. Their main dielectric function is to control the electric field at the interface between the high-voltage conductor and the insulation shield at ground potential, to prevent PD.

As will be shown in the following, most problems in cable systems arise at the accessories. This is due to different issues:

- Cleanliness: joints need to be installed in the field, where it can be difficult to control contamination.
- Human errors: assembly is human-made, thus prone to errors.
- Electric field lines parallel to interfaces: interfaces (see Section 2.1.3.2) are a weak point as micro-cavities always exist due to imperfect mating [8, 9]. It is always preferable to have the field lines orthogonal to interfaces, but this is not possible in accessories.
- Possible ingress of contaminants from the points where the cable exits the accessory.

The most critical function within an accessory is that of electric stress control (also called stress grading or stress relief). This point is explained with a simple example. Assume that one must energize a coaxial cable. The high-voltage conductor at the cable end will be very close (a few mm) from the grounded insulation shield. When high voltage is applied, this short distance in air would have a high

---

1 For the interested reader, the high viscosity "mass" is normally a compound of poly-isobutylene (C_4H_8, a polyolefin derived from petroleum), resins (used for non-draining masses), paraffin, and bitumen.

risk of flashover (a full discharge connecting the high-voltage conductor to the grounded insulation shield), since the breakdown strength of air is much lower than the breakdown strength of the XLPE or EPR (Section 2.6). This can be partly improved by stripping back the insulation shield to increase the creepage distance between the HV conductor and the insulation shield, increasing the flashover voltage. This is enough for low-voltage cables, but not for high-voltage cables. As a matter of fact, the designed electric stress of the solid insulation can be large (10–15 kV/mm). When the insulation shield is removed, part of this field is transferred to the air where it can cause partial discharges. The amount of shield removed from the cable is irrelevant in this respect, as the insulation shield "senses" the high voltage conductor below the insulation. This situation is apparent in Figure 12.4. To prevent partial discharges at the end of the insulation shield, two main solutions have been devised:

- capacitive grading, and
- nonlinear resistive grading

Capacitive grading consists of gradually increasing the distance between the high-voltage conductor and the insulation shield (at ground potential). When the distance is sufficient, the field lines shown in Figure 12.4 will still exist, but the corresponding electric field values will be much lower, insufficient to cause partial discharges. This solution implies the creation of a partly conductive "stress cone" in the body of a termination (two stress cones for joints). A stress cone molded into a medium-voltage cable termination is shown in Figure 12.5.

Nonlinear resistive grading is based on the use of a heat-shrink or cold-shrink tube manufactured using materials whose resistivity decreases with the current density (typically, wide bandgap

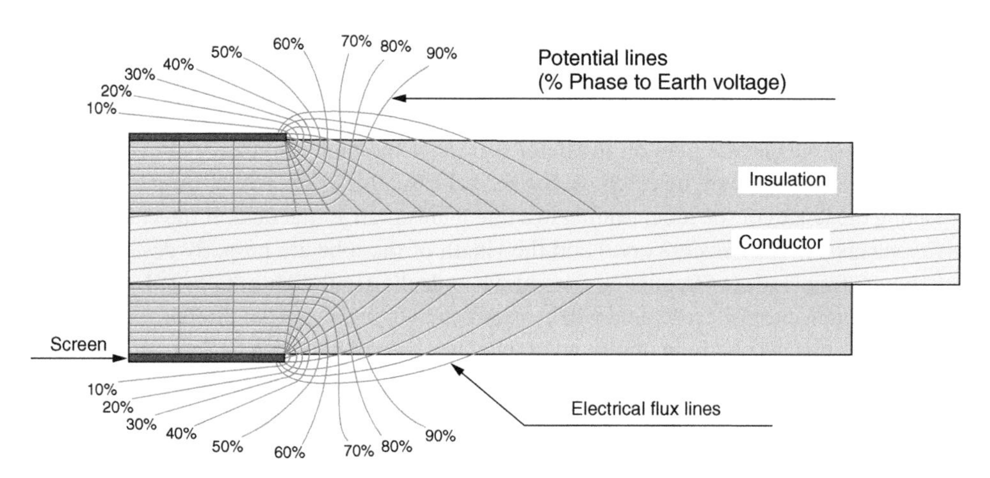

Figure 12.4 Electric field and equipotential lines in a coaxial cable with interrupted insulation shield or screen. The concentration of the equipotential lines near the end of the insulation shield indicates that the highest electric stress is there. *Source:* [10].

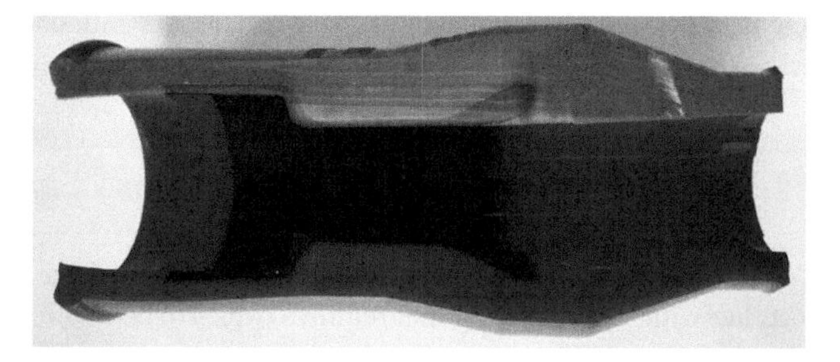

Figure 12.5 Stress cone (black, partly conductive material) embedded in a medium voltage cable termination.

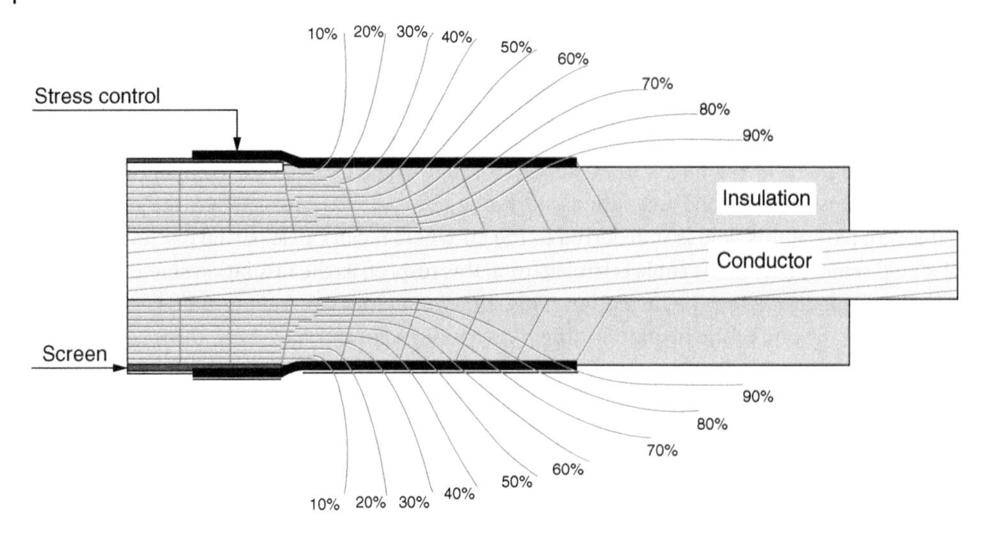

Figure 12.6 Effect of nonlinear stress control tubes in a coaxial cable with interrupted shield. *Source:* [10].

semiconductors such as silicon carbide or zinc oxide) embedded in a polymeric matrix loaded with carbon black that confers elasticity. During operation, the high voltage will cause a capacitive current to flow from the inner conductor to the tube. The density of the capacitive current flowing in the tube increases going from the end of the tube away from the shield to the shield itself. Since the resistivity of the tube decreases as the current density increases, the product of the resistivity times the current density, that is the electric field, will tend to stay constant. This can be observed comparing Figure 12.4 with Figure 12.6. In the first case, the equipotential lines are concentrated in proximity of the shield, thus the electric field on the outer surface of the cable can be very large, up to a point where partial discharges are incepted. In the second case, the equipotential lines are evenly distributed along the length of the tube, leading to a much lower tangential field. If the characteristic (resistivity versus current density curve) and the length of the tube are properly designed and matched to the AC frequency (or frequency content when dealing with lightning impulses) of the applied voltage, the tangential electric field will be insufficient to incept partial discharges.

As a final remark, it is necessary to observe that what was described above has general validity, as the need for stress control exists every time a cable needs to be terminated (or two cable lengths joined). Terminations as presented in Figure 12.5 are relevant for polymeric cables that are connected to air-insulated switchgear or overhead lines (often referred to as a pothead). Power cables, however, can be directly connected to power transformers or GIS systems. In this case, the part of the termination with the polymeric cable is very different compared to the other part (oil insulated or gas insulated). Furthermore, cable joints can also be asymmetric if, for instance, an MI cable is joined with a polymeric cable, or an XLPE cable is joined with an EPR cable, or if cables with different sizes need to be joined. All these differences influence the technology and aspect of the accessories (see as an example Figure 12.7) as well as the problems they might experience.

12.3 Cable System Failure Mechanisms

Considering the many technological solutions devised for manufacturing high-quality cables, together with factory testing that requires cables to be PD-free at very high voltages, the failure of extruded power cables themselves due to PD is rare. However, some aging may lead to PD during

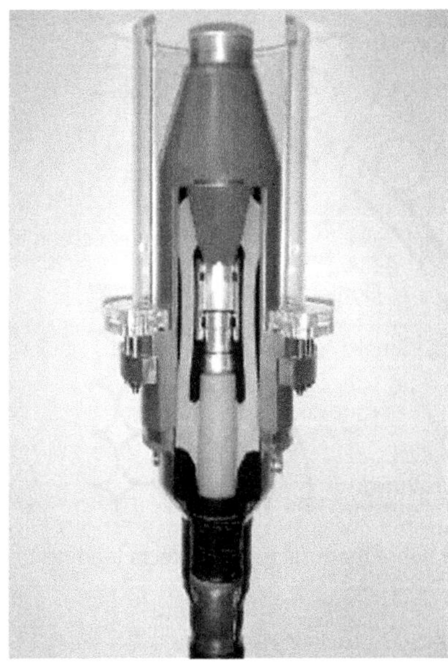

Figure 12.7 Comparison between terminations: (left) cable to air-insulated substation, (right) cable to SF_6-insulated substation termination. *Source:* Prysmian Group.

operation, especially in transmission-class polymer cables. Terminations and joints are more likely causes of power cable system failure. The following is a brief summary of failure processes that may occur and where PD may be expected.

12.3.1 Extruded Cable Manufacturing Defects

If PD occurs in the cable dielectric, they are often indicative of serious lapses in the production line and are usually screened out by factory quality control tests (Section 12.7). The most common reason for partial discharge inception in the cable itself is associated with mechanical damage to the jacket and insulation shield during pulling of the cable into its final position. Because the electric field is lower here than that at the high-voltage conductor, any PD tends to erode the insulation at a relatively lower rate. Medium-voltage cables, especially EPR cables, can withstand this type of PD activity for some time; utilities can therefore decide to postpone repair for a while. For high-voltage cables, this type of defect can lead to breakdown in much shorter times and is not acceptable. Other possible manufacturing defects are shown in Figure 12.8 but, as mentioned, in modern cables most are unlikely since they should be eliminated by factory PD testing.

There are sometimes concerns about the possible presence of metallic contaminants or protrusions from semiconductive layers. Such defects will not be immediately detected by the factory PD test since the defect is in intimate contact with the polymer (and not air). However, these imperfections can act as stress concentration regions leading to the onset of the space-charge limited field (SCLF) phenomenon, i.e. electric tree initiation [11]. The space charge injected at these protrusions limits the field to values around 250 kV/mm for XLPE, but induces mechanical strain in the insulation due to Coulombic forces. In addition, charge injection and extraction will lead to broken polymer chemical bonds. Given enough time, these two processes create a cavity at the defect. The cavity can enlarge up to a point where partial discharges can be incepted and, from there, an electrical tree can develop leading to breakdown (Section 2.4.1). This type of defect only becomes apparent after some time from both quality control tests in the factory or after laying tests in the field [11].

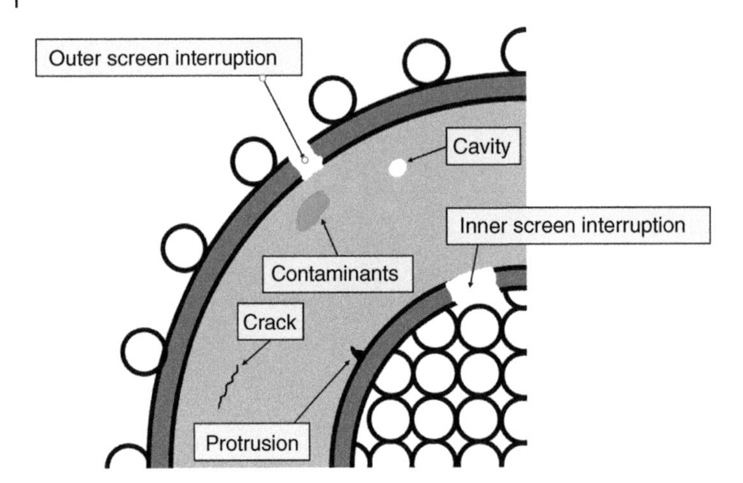

Figure 12.8 Power cable manufacturing defects leading to PD activity.

12.3.2 Aging of Polymeric Cables in Service

Extruded power cable insulation can age due to thermal and thermomechanical stress. Both types of stress might eventually create voids that produce PD prior to cable failure. Thermal aging occurs when the XPLE or EPR is operated well above rated temperatures for extended periods of time, for example due to incorrect cable sizing, during system overloads, undetected system faults or when the soil surrounding the underground cables become dry, increasing the thermal resistance of the soil and hence its operating temperature. The latter is an important issue in underground transmission-class power cables. Operation at high temperature leads to void creation and PD, especially near the conductor shield.

Load cycling is also an issue for transmission-class extruded cables. Load cycles lead to expansion and contraction of the high-voltage conductors, creating a mechanical shear force between the conductors and the conductor shield and insulation. After many cycles, fatigue cracks can appear, especially in aged cables and at interfaces, leading to PD.

By far the most likely failure process of older distribution class-extruded power cables is water treeing [1, 2]. However, PD does **not** accompany the growth of a water tree.

12.3.3 Water Trees

Water trees occur in extruded power cables operating in a wet ambient, and they eventually lead to cable failure. They are caused by salts (carboxylate salts and/or inorganic salts) transported by water due to dielectrophoretic forces. Morphologically, the water trees are tree-like structures that consist of rows of voids or *strings of pearls*, especially when they approach the opposite conductor (Figure 12.9). The diameter of these voids may vary from about 1 up to 5 μm, and relatively large separations between them may occur. The formation of these voids is an irreversible process: if a water tree is dried it becomes temporarily invisible, but as soon as water fills its voids again, the degradation returns to be evident. Nanometric (10–100 nm) tracks containing hydrophilic salt groups interconnect the voids. These tracks run through the amorphous phase of polyethylene and allow water to move within the insulation to fill up the voids.

Water trees are generally nonconducting although they are more conducting than the healthy insulation and have a higher permittivity due to the presence of polar species (water and ionic

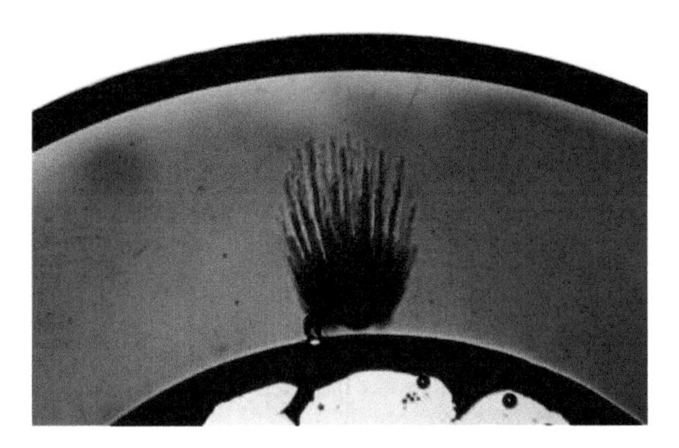

Figure 12.9 Example of water tree. *Source:* [12]/with permission from IEEE.

contaminants) and their conductivity tends to increase with time. Since cavities are normally filled with water, partial discharges cannot be incepted within water trees (even in the case they dry, the small diameter of the cavities, <5 μm, makes partial discharge inception a very unlikely phenomenon). However, the (semi) conductive tips of the water tree act as protrusions within the dielectric and tend to create regions where the electrical field is enhanced. The typical failure mechanism associated with water trees is rather complex as it generally involves a temporary enhancement of the field due to a lightning event or transient overvoltage. The sequence of events is thus [13]:

1) Space charge builds up at the water tree tips limiting the field to the SCLF (around 250 kV/mm for XLPE). Space charge and electric field induce mechanical strain in the region facing the water tree tips.
2) Due to the large current associated with the overvoltage, the temperature increases at the water tree tips. Above 90 °C the XLPE melts or softens. Above 100 °C, the water within the tree boils leading to overpressure.
3) Due to these effects, a cavity is formed. If the cavity is large enough, it can become the site of partial discharge activity, develop into an electrical tree, and lead to breakdown.

According to this, water trees cannot be diagnosed by partial discharge detection, but partial discharge detection can be used to prevent the failure if the water tree has created a defect (cavity, electrical tree) large enough to support partial discharges.

Water trees are generally a problem for XLPE cables manufactured before 1985, more frequently cable manufactured in the late seventies with semiconducting tapes and paint (the water could enter the insulation from the outer part of the cable, especially if the protective sheath was damaged). With the introduction of extruded (strippable or fully bonded) insulation shields, the likelihood of observing water trees was reduced greatly. Also, water trees were often originated by microvoids filled with water within the insulation. These microvoids were a by-product of the steam-curing of the insulation, a process used to achieve the full cross-linking of the XLPE. With the introduction of cross-linking processes resorting to peroxides, this type of defect has become almost irrelevant. Eventually, water could leak into the cable from the terminations, infiltrating through the conductor strands, eventually ending up in the insulation. This process is still possible today, but greatly alleviated using inter-strand fillers based on super-absorbing polymers. Furthermore, water tree-retardant XLPE (TR-XLPE) has helped to mitigate the problem. Yet, a large fraction of XLPE cables that could be potentially affected by water trees still exists. This is true for large industrial plants, such as refineries, steel plants, and cement mills, for example.

12.3.4 Aging of Mass-Impregnated Cable

For mass-impregnated cables, the main aging issue is exposure to water. Cellulose fibers in the paper tapes are strings of cellulose monomers, with fibers held together by hydrogen bonds that confer mechanical stability. Water can break the hydrogen bonds sticking to the monomers (hydrolysis). This leads to the embrittlement of the paper and the increase of the dissipation factor. An increase in dissipation factor leads to higher temperatures in the insulation that favor oxidation. A by-product of the oxidation of cellulose is water. Thus, a self-perpetuating process can be established that leads to the embrittlement of the paper insulation. Mechanical stresses (as an example, the strain due to the expansion and contraction of the conductor due to heating and cooling processes associated with load cycling, or the mechanical stress imposed by heavy vehicles passing above the cable if a road crosses the cable route) can then induce cracks in the insulation that can become the site of partial discharges if the viscous mass is unable to fill them.

Partial discharge inception is also favored by the water penetrating from a break in the sheath that acts as an electrode within the insulation enhancing the electrical stress. Thus, one of the most critical failure initiation mechanisms is the cracking or corrosion of the outer sheath and jacket. During heating, when the pressure of the impregnant can reach 8 bar,[2] the mass can be expelled from the cable (creating voids). During cooling, the cable pressure is below ambient pressure (0.8 bar), and moisture can be sucked in. Through the process mentioned above, PD can start to propagate within paper layers spreading between layers at each butt gap, as shown in Figure 12.10. Often, the carbonization tracks found in MI cables are several meters long.

In MI cables, rapid changes in cable loading forces the migration of the mass from the butt gaps into the paper layers.[3] This causes an intense partial discharge activity, which is generally short-lived. However, partial discharges tend to form waxes that fill the space liberated by the mass, quenching the partial discharge activity. The gasses liberated by the discharge process are absorbed quickly within the insulation. If partial discharges are particularly intense; however, the energy dissipated results in elevated temperatures leading to the formation of conductive carbon tracks. Eventually, these tracks can spread transversely between the paper layers and cause tracking, or treeing, slow deterioration, and eventually complete breakdown of the insulation as mentioned above.

12.3.5 Joint and Termination Problems

Most defects in cable systems are in the accessories, such as a termination or a joint (Figure 12.11). Accessories are assembled in the field, in conditions that are much harder to control than those in

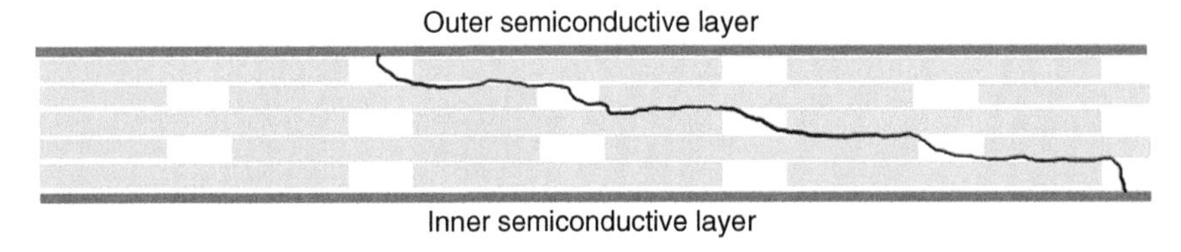

Figure 12.10 Propagation of PD activity in a taped insulation.

---

2 To withstand overpressures of 8 bar, the outer sheath is realized using lead. MI cables are, therefore, often indicated as Paper-Insulated Lead-Covered cables, or in short PILC cables.

3 Partial discharges are often found in MI cables due to thermal transients.

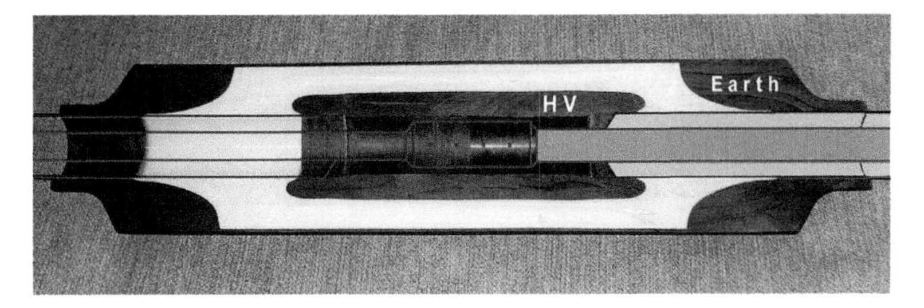

Figure 12.11 High-voltage cable joint for extruded cables. *Source:* Primasz/Wikimedia Commons/Public domain.

the cable factory. The assembly of terminations requires cleanliness to prevent contamination. Besides controlling the ambient conditions, the following details are very important:

- Stripping the outer semiconductive layer from the cable, cutting the insulation of the cable must be avoided. Cuts are like cavities where field concentration takes place, possibly leading to partial discharge inception. This type of defect is less probable for cables with fully bonded outer semiconductor shield, as this type of shield is removed by heating.
- The cut that separates the cable outer semiconductor shield from the part that is removed to expose the dielectric should be smooth to avoid forming protrusions on the interface between the cable and joint body.
- Mechanical alignment of the stress control system is very important: if the field control system of the accessory is not in contact with the outer semiconductive layer of the cable, the system is electrically floating, the electric field can exceed the partial discharge inception field.
- In joints, the high-voltage electrode should be placed correctly above the mechanical connector used to keep the conductors together.
- The correct amount of lubricant should be used to fill the microcavities at the cable/joint interface.

Imperfections at any of the above stages can create PD.

Accessories can also fail due to locked-in mechanical stress caused by cooling of the joint body too quickly during assembly. This type of process will not create PD because cracks are created when internal strains are released, normally at the very last moment and in a catastrophic way.

Operation at high temperatures can reduce the elasticity of the accessory body, leading to an imperfect mating between the joint/termination and the cable. Thus, the cable and the joint/termination can locally detach during load cycling, creating a delamination with the electric field tangential to the dielectric surface. This leads to tracking of the interface and, eventually, breakdown. Thermal stress and thermally induced migration can cause the lubricant to not perform its function of filling any air gaps at interfaces. In this case also, tracking and breakdown of the interface can occur. In addition, thermal stress can reduce the performance of the sealing systems that prevent contamination from entering the accessory from the aperture(s) where the cable enters the accessory. Typically, this will lead to the ingress of water and other conductive contaminants (salts) that can compromise the withstand capability of the insulation.

Thermomechanical stress is also possible, particularly in mass impregnated cables, with the introduction of renewable energy sources and their rapid load fluctuations. Aged MI cables (which often are a large fraction of the distribution grid) tend to be mechanically fragile and unable to withstand these increased stress levels that lead to the formation of cracks in the insulation. In some cases, due to the structure of the termination, a vessel containing the mass (for MI cables with draining mass, see Figure 12.12), the conductor can elongate up to a point where it becomes too close to the mass container leading to flashover.

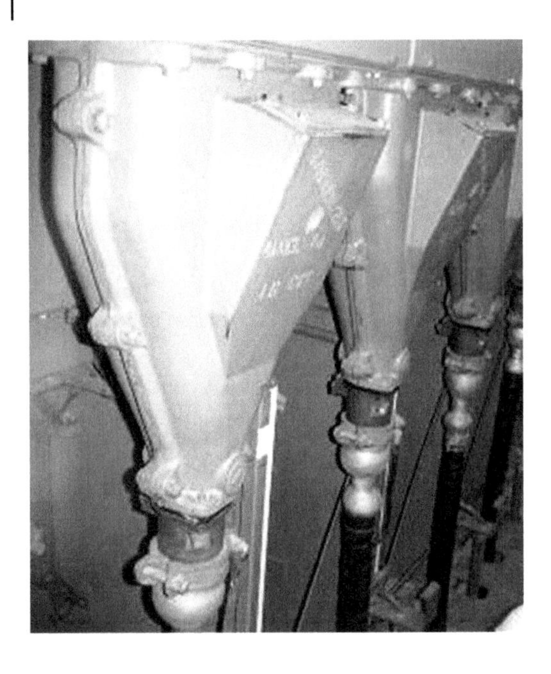

Figure 12.12 Wet-type terminations of MI cables with draining mass.

12.4 Cable PD Test Standards

Measuring PD in power cables can entail different types of measurement:

1) Quality control testing in the shop of
   - Cable lengths (require water terminations)
   - Accessories (require a cable mockup)
2) PD measurements of complete cable systems to ascertain that the circuit is PD free. These tests can be carried out for different purposes:
   - Commissioning (after laying), to ensure that cable pulling, accessory installation, and grounding were performed correctly.
   - Maintenance or recommissioning testing, to ensure that the repaired part of the cable is PD free.

For measurements (1) and (2), the general test method is IEC Standard 60270 [14]. However, as already mentioned in Section 12.4 and explained in Section 6.6.2, when dealing with cable lengths, propagation effects can affect the measurement of apparent charge due to superposition phenomena, particularly when using narrowband detectors. Thus, for power cables in particular, IEC Standard 60885-3 [15] is used. Indeed, for modern ultra-wide-band detectors that detect and record the PD pulse and perform apparent charge calculation through digital filters, superposition phenomena are not a concern.

For power cables, the standards do not specify the limits of the apparent charge recorded during the tests, but rather the minimum sensitivity of the measurements. As an example, the minimum sensitivity for power cable lengths is 10 pC in IEC standards and 5 pC in IEEE standards. For accessories, the IEC prescribes 5 pC, whereas the IEEE supports 3 pC. IEEE and IEC standards avoid defining the maximum allowable apparent charge in operation. Instead they specify:

a) a test voltage level that must be maintained for a given amount of time (as an example, >120% of the rated voltage for at least 3 seconds but no more than 60 seconds);
b) the conditions allowed to accept the cable or accessory if discharges above the prescribed threshold have been incepted during the previous step.

If step (b) must be performed, the voltage is reduced to prescribed levels and the presence of PD above the minimum sensitivity threshold is checked. In case PDs are not detected, the cable or accessory can be accepted. In practice, one can recognize that step (a) is a requirement on PDIV, while step (b) is on PDEV.

Regarding tests in the field (offline/onsite), standards are more a reference to the procedures used in the field than a document aiming at standardizing the measurements, see for example [16]. Alternatively, CIGRE brochures provide a lot of information and case studies, but they are not consensus standards. The reasons behind this lack of standardization are multiple:

- Measurements are mostly performed using ultra-wide-band detectors that are not compatible with the range of frequencies specified in [14]. The choice of the detector bandwidth depends on the noise spectral distribution, and it is difficult to standardize.
- Calibration is hard to perform as the calibration pulse injected between the high-voltage conductor and ground at a termination will be attenuated quickly if one uses ultra-wide-band detectors, as is normally the case.
- To prevent failures or damage to the insulation, the maximum test voltage levels should be applied to the cable for 60 seconds at maximum. Carrying out PD measurements on a multiplicity of accessories in such a short-time interval can be difficult.
- Tests performed after maintenance are on aged cables. The test levels prescribed in [17] are high, and there is a concrete risk that the cable can fail during the test.

As a final remark, IEEE standards explicitly support testing with alternative voltage sources: VLF and damped AC (DAC or OWTS) voltage sources.

Table 12.1 is a summary of the standards and consensus guidelines for cable testing. It is important to note that, very often, the customer and the manufacturer will debate about the standard to be applied trying to achieve the more or less stringent pass/fail criteria.

12.5 PD Test Sensors

12.5.1 Capacitive Couplers

Most factory PD tests use capacitive couplers connected in an indirect circuit using an IEC 60270 instrument to measure the PD (Section 6.2.2). The coupler typically has a capacitance of at least 10% of the cable length capacitance to achieve an adequate sensitivity; since cable lengths have capacitances of hundreds of nF, the coupling capacitor is normally 10–30 nF.

12.5.2 HFCTs on Neutral Grounding Leads

Usually capacitive couplers cannot be used for onsite/offline PD testing of cable circuits. There may be insufficient space to deploy a high-voltage coupling capacitor. In addition, shipping large capacitive couplers to site can be risky. Also, attenuation during propagation of the PD signal along the cable to the coupling capacitor (Sections 6.6.2.2 and 12.6) may prevent PD detection if the coupler is far from the PD site. On aged taped shielded cables, the corrosion of the shield overlaps can also increase attenuation of the PD signals [25]. For very long cable runs (say greater than a few km), the attenuation can be so strong that the PD pulse is "buried" below the noise (interference) floor. Thus, to achieve a sufficient sensitivity, it is necessary to detect the PD pulse as close as possible to the defect location. Since defects are more likely to be found at accessories (Section 12.3.4), it is customary to try to couple the PD source there.

Table 12.1 Summary of standards for cable testing and their aim.

Code	Title	Purpose
IEC 60270 [14]	High-Voltage Test Techniques – Partial Discharge Measurements	Root standard for the measurement of apparent charge
IEC 60885-3 [15]	Electrical Test Methods for Electric Cables – Part 3: Test Methods for Partial Discharge Measurements on Lengths of Extruded Power Cables	Describes test methods for PD measurements on lengths (reels) of extruded power cable in the factory. Does not deal with cable system after installation
IEC 60840 [17]	Standard Power Cables with Extruded Insulation and Their Accessories for Rated Voltages Above 30 kV ($Um = 36$ kV) up to 150 kV ($Um = 170$ kV) – Test Methods and Requirements	These two standards prescribe minimum sensitivity to be achieved during the measurements and the voltage profile to be applied. Criteria for accepting or rejecting the cable length or the accessory as specified as a function of the PDIV/PDEV achieved during the measurements
IEC 62067 [18]	Standard Power Cables with Extruded Insulation and Their Accessories for Rated Voltages Above 150 kV ($Um = 170$ kV) up to 500 kV ($Um = 550$ kV) – Test Methods and Requirements	
ICEA S-94-649-2000	Standard for Concentric Neutral Cables Rated 5000–46,000 Volts	This standard is similar to [17] and [18] as it prescribes voltage levels for testing. However, differently, it specifies that maximum PD level measured during the test as pass/fail criterion
IEEE 400 [19]	IEEE Guide for Field Testing and Evaluation of the Insulation of Shielded Power Cable Systems Rated 5 kV and Above	Discusses cable testing providing pros and cons of the different techniques for cable energization. Describes acoustic PD detection
IEEE Std. 400.2 [20]	IEEE Guide for Field Testing of Shielded Power Cable Systems Using Very Low Frequency (VLF) (Less Than 1 Hz)	The standard describes PD testing using VLF voltages
IEEE Std. 400.3 [16]	IEEE Guide for Partial Discharge Testing of Shielded Power Cable Systems in a Field Environment	Informative document discussing various aspects of PD measurements in power cables. Does not provide pass/fail criteria
IEEE Std. 400.4 [21]	IEEE Guide for Field-Testing of Shielded Power Cable Systems Rated 5 kV and Above with Damped Alternating Current Voltage (DAC)	Describes the features of PD during DAC tests and provides suggestions for pass/fail criteria selection
IEEE 404 [22]	IEEE Standard for Extruded and Laminated Dielectric Shielded Cable Joints Rated 2.5 kV to 500 kV	Provides minimum sensitivity, test procedure, and pass/fail criteria for accessories
CIGRE TB 182 [23]	Partial Discharge Detection in Installed HV Extruded Cable Systems	This CIGRE brochure describes sensors, signal processing and denoising principles, together with practical examples. It dates back to 2001, so detection methods are partly outdated
CIGRE TB 728 [24]	Onsite Partial Discharge Assessment of HV and EHV Cable Systems	TB 728 (2018) is an update of TB 182. However, it provides test levels and durations for every voltage class for new and aged cables. For new cables, the test levels agree with [17] and [18]

For terminations, the simplest solution is to use the lead connecting the cable insulation shield/neutral wires to the ground. The PD pulses in the cable will circulate to ground through this lead. This connection will be able to provide PD from all the accessories and cable lengths, but the sensitivity will be much higher for PD taking place in the associated termination. Besides the PD pulses, a 50/60 Hz displacement current will flow in the ground lead. Therefore, the ground lead can be used for both PD detection and synchronization using high-frequency current transformer (HFCTs) and Rogowski coils, respectively, as shown in Figure 12.13.

To measure PD in or close to joints, a similar approach can be adopted for offline tests on single phase cables used in three-phase circuits. The method is applicable, however, only to sectionalized joints, i.e. joints where the insulation shield and neutral conductors are interrupted at the joint. Sectionalized joints are used to transpose the neutral conductors at regular distances as shown in Figure 12.14a. For PD detection during offline tests, the neutral conductors in the same phase are directly connected to ground, as shown in Figure 12.14b; the HFCTs can be installed, for instance, on the lead connecting the shields of the cable lengths in the same phase-to-ground. During the tests, the HFCT will measure the PD (if any) occurring in the energized joint, in the cable lengths connected to the joint, and to the two halves of the joints connected to the cable lengths. As the capacitive 50/60 Hz current from two cable lengths in a phase will also flow to ground in the link box, AC synchronization can be achieved using a Rogowski coil, as done for the terminations. Figure 12.15 shows a link box prepared for PD measurements. Note that non-sectionalized joints (or straight joints) are also used in cable systems, but their purpose is to extend the cable line. PD detection in straight joints can be difficult unless the joints are equipped with embedded sensors or capacitive sensors are used, as explained in Section 12.5.3.

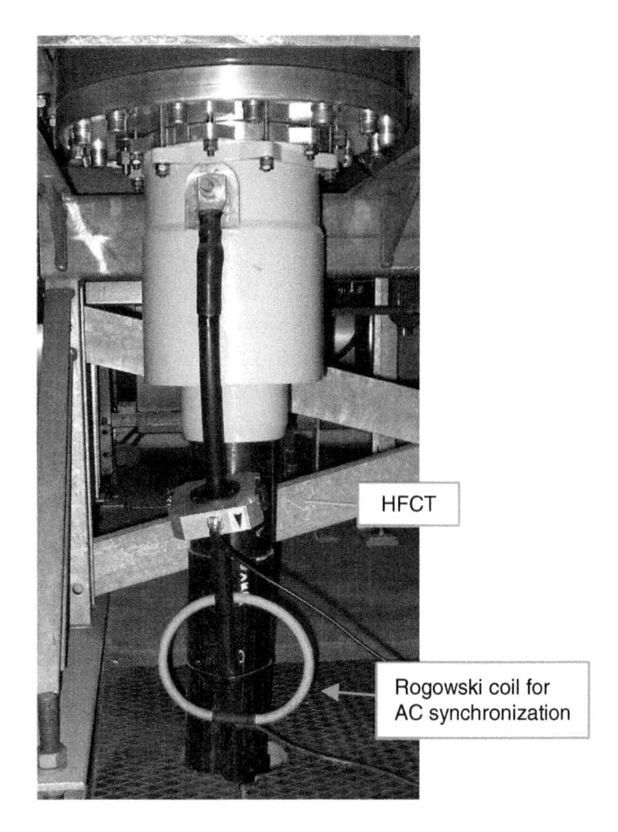

Figure 12.13 High-voltage cable termination with a HFCT and a Rogowski coil installed for PD detection and synchronization, respectively.

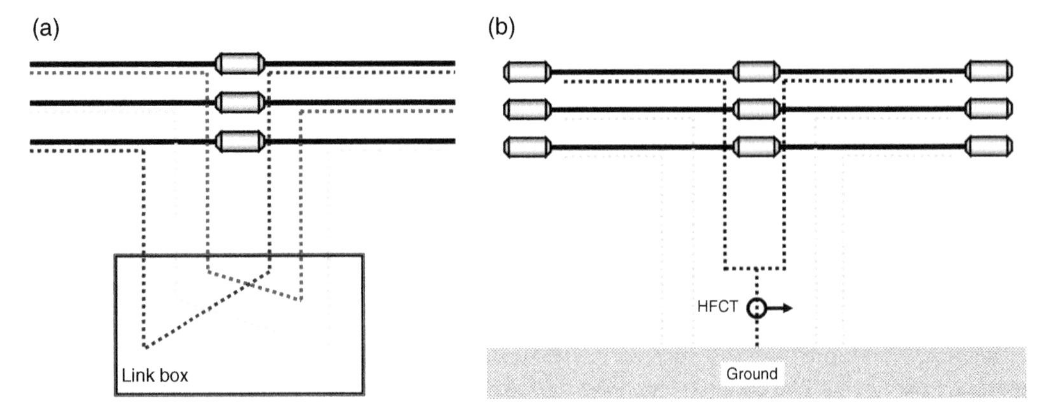

Figure 12.14 Connections of the shields in the link box (a) in operation, (b) during PD tests.

Figure 12.15 Link box prepared for PD measurement with a HFCT on a jumper cable.

12.5.3 Imbedded Capacitive Coupler

The most sensitive sensor for PD detection in installed cable runs is the imbedded capacitive sensor that can be installed in joints or cable terminations using stress cone technology. The capacitive sensors use the stress cone (Figure 12.5) as a capacitive sensor. Normally, the PD pulses are detected by monitoring the potential difference between the stress cone and an electrode placed on the cable insulation shield at some distance from the stress cone (Figure 12.16). Since the high-frequency resistance of the semiconductive insulation shield is high, the capacitive sensor is generally treated as a high-pass RC circuit with C in the order of 100–200 pF and $R = 50–200\ \Omega$. Typically, the cutoff frequency of the sensors is in the range of several MHz (i.e. the HF range), which generally prevents PD detection performed in the IEC 60270 frequency band (<1 MHz). Thus, readings from these sensors are expressed in mV. Only transmission-class power cables are likely to have these sensors.

12.5.4 Other Sensors

The sensors above are the most common and, in a sense, the most conventional ones. As for any technical solution, they have pros and cons. As an example, imbedded capacitive couplers at joints

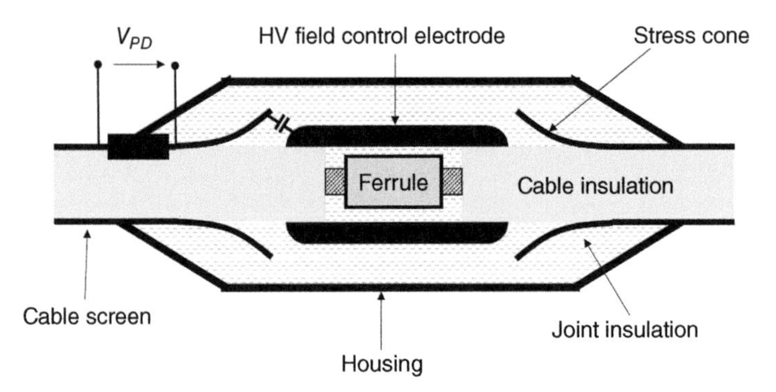

Figure 12.16 Principle of operation of joint capacitive couplers.

Figure 12.17 HFCTs clamped around the coaxial cables for shield transition.

are not suitable for retrofitting to existing cable systems (unless all joints are replaced, which is very expensive). HFCTs connected to the link boxes, as in Figure 12.14b, cannot be used in service since the configuration of the link boxes in service must be connected as in Figure 12.14a to fulfill their function. Indeed, as the link box is equipped with surge arresters (see Figure 12.15) to limit the shield voltage during transients, HFCTs can be installed around the coaxial cables for cable transposition as shown in Figure 12.17.

Other sensors have been developed to sense PD when the cable is in operation. In [26], the electric field in proximity of the cable body was simulated. It was found that, although attenuated, the electric field outside the cable is not negligible, as shown in Figure 12.18. Figure 12.18 also shows that the sensors must be placed at a relatively short distance from the PD source (<0.5 m) to collect a sufficiently large field. These simulations support the use of near-field (VHF/UHF) capacitive sensors located at a very short radial distance from the cable body and in proximity of the accessories.

In [27], a simple capacitive sensor was realized by removing the cable sheath and wrapping a copper tape around the cable on the outer semiconductive layer. The sensitivity was very good (3 pC), but the installation of this type of sensor requires that part of the sheath is removed, exposing the cable to the ingress of contaminants. Therefore, investigations were

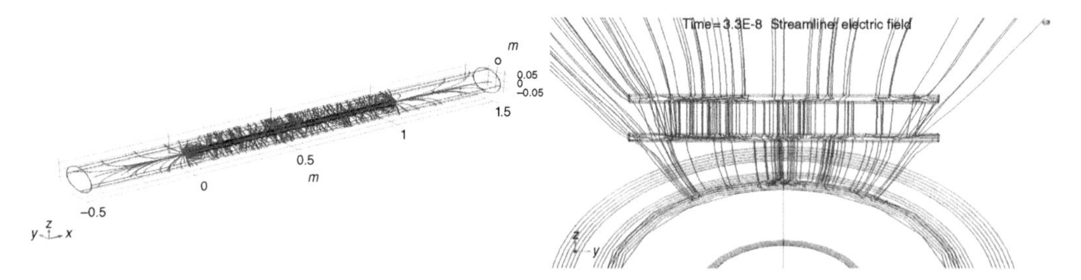

Figure 12.18 Simulation of the electric field created by the partial discharge pulse outside the cable. Right: axonometric projection showing the field lines along the cable, in proximity of the simulated PD source (x = 0.5 m). Left: cross-section of the cable and the sensor showing the electric field lines. *Source:* [26].

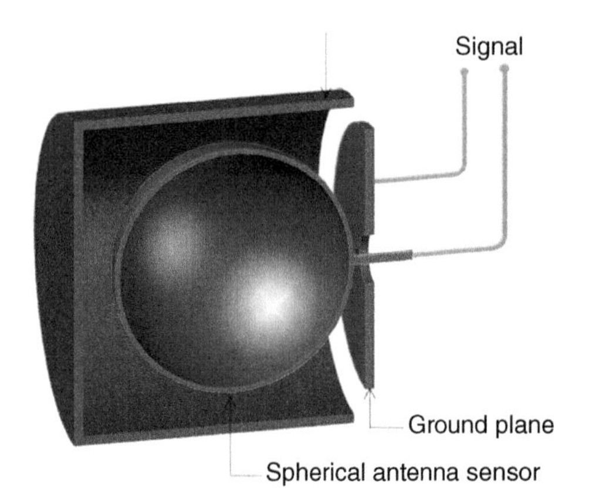

Figure 12.19 Spherical antenna (differential field probe) sensors. *Source:* [28]/MDPI/CC BY 4.0.

carried out to prove that PD detection can be carried out leaving the sheath intact (provided that the sheath is not a metal tape and not interrupted as in submarine cables). The results were positive. Figure 12.19 shows the structure of a spherical antenna (differential field probe) sensor that was successfully used to detect PD in a 138 kV XLPE cable circuit during a 24 hour soak test with the cable energized to the nominal line to ground of 74 kV [24]. A similar sensor is used in the Prysmian Prycam detector. The experience gained with the differential field probe prompted the power cable manufacturer, Prysmian, to also develop the so-called Prysmian wings. Compared to the Prycam, the Prysmian wings are passive, therefore, more suitable for monitoring.

Pairs of capacitive couplers can also be installed at both ends of the joints. If simultaneous PD measurements are carried out with sufficient time resolution, these couplers can also provide an indication of the location of the PD source (Section 8.4.2). If the difference in the arrival times of the pulses is below the time needed to cross the joint, the PD comes from the joint itself. Taking a propagation speed of 160 m/μs and a joint length of 2 m, this indicates that time differences below 12.5 ns need to be resolved to be able to infer whether the PD source is internal to the joint or not.

Starting from different assumptions, i.e. working on the magnetic field induced by the PD pulse, Techimp Altanova developed the flexible magnetic coupler (FMC), a sensor suitable for retrofitting existing cable lines by placing it on the cable body, as shown in Figure 12.20.

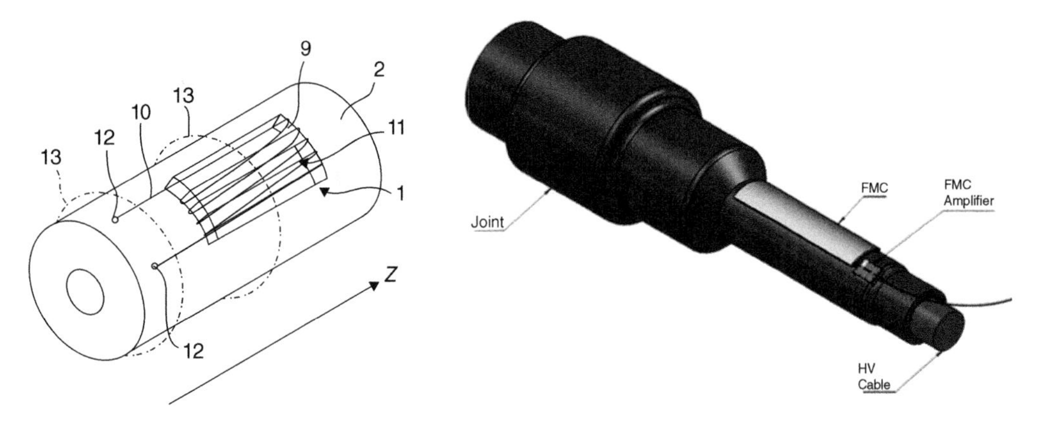

Figure 12.20 Flexible magnetic coupler (FMC): (left) sketch of the structure. *Source:* [29]/World Intellectual Property Organization. (right) FMC laid on the cable in proximity of a joint. *Source:* Image courtesy of Altanova.

12.6 PD Pulse Propagation and Detector Bandwidth

The topic of pulse propagation with a transmission line (including MV and HV power cables) is dealt with in Section 6.6.2.2. The most important effect of propagation is attenuation of the PD pulse along the cable line, which is synthesized in Figure 12.21. This attenuation is more severe at higher frequencies than lower frequencies. Thus, attenuation poses a problem of sensitivity, which in turn becomes an issue of what frequency band should be used for PD detection. Since noise and interference cannot be avoided, too large a bandwidth could lead to a sub-optimal detection due to too much thermal noise and possible interference from radio stations or other broadcasting systems. Too narrow a bandwidth could also lead to the overlapping of the signal with interference, preventing the separation of low-magnitude PD signals from interference. Since attenuation depends on the distance between the PD source and the detection point, so does the optimum bandwidth for PD detection. In [30], the concept of pulse equivalent bandwidth was used to characterize the frequency content of the PD pulses. In this context, the equivalent bandwidth (Section 8.9.2) is used to provide an indication of the optimum detection bandwidth. The equivalent bandwidth of a pulse is defined as the standard deviation of the pulse in the frequency domain[4] and corresponds approximately to a 6 dB attenuation. The following procedure can be adopted:

1) Derive attenuation and dispersion parameters from measurements on cable dielectric and semiconductive shield samples, or from a model cable.
2) Assume an ideal pulse (constant Fourier transform at all frequencies).
3) Calculate the pulse spectrum after a propagation at a distance x by multiplying the spectral components of the pulse times the attenuation (Equation (6.13)).
4) Calculate the corresponding equivalent bandwidth.

The results of such a procedure are reported in Figure 12.22.The optimum detection bandwidth of the detector declines rapidly after a few tens of meters from the source. At 100 m, it is approximately 20 MHz, a figure suggested already in [31]. At 10 km, it is in the range 100–200 kHz. These

4 The pulse needs be normalized to have unit RMS value, to avoid that different equivalent bandwidth values are assigned to pulses having different magnitude but same shape.

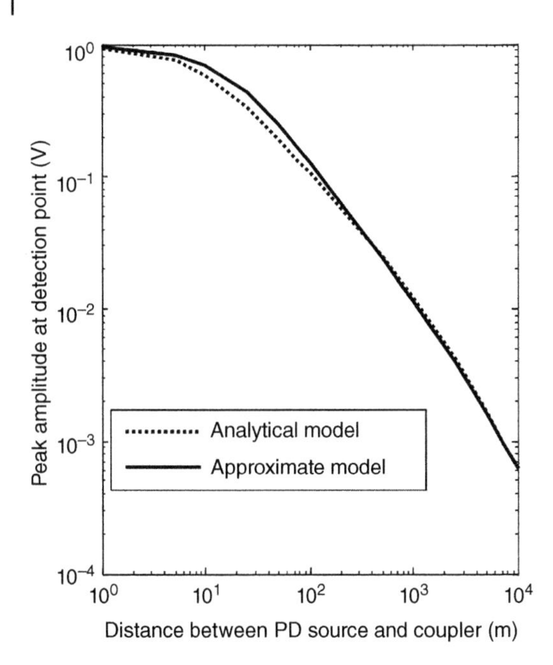

Figure 12.21 PD pulse peak voltage as a function of distance propagated (calculated using approximate and analytical model) [30]. Peak voltage at the source is 1 V. *Source:* [30]/ALMA MATER STUDIORUM.

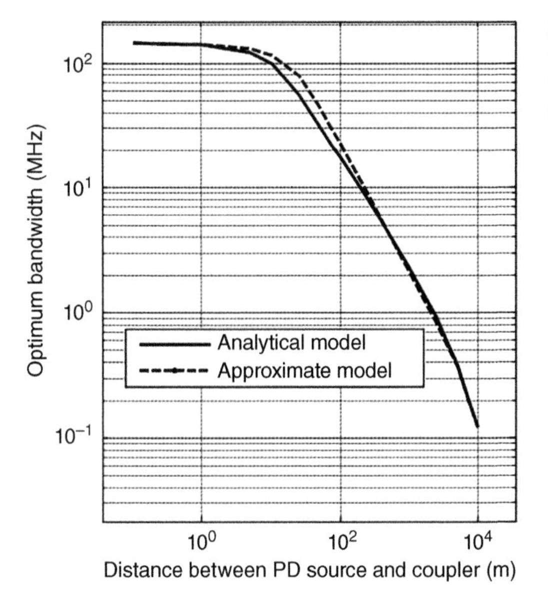

Figure 12.22 Optimum detection bandwidth as a function of the distance between the PD source and the coupler. *Source:* [30]/ALMA MATER STUDIORUM.

calculations, however, do not consider the nature of the interferences in the field. The cable having a length L can be regarded as an antenna with a high sensitivity to interference having frequency $f = c/L$, c being the speed of light. Theoretically, a cable 10 km long could act as an antenna for waves having a frequency of 30 kHz. The longer the cable, the lower will be the frequency content of the pulses, thus the optimum detection bandwidth. However, the longer the cable, the larger the disturbance at low frequencies (i.e. interference in cable circuits are likely higher at low frequencies than at higher frequencies). Summing up these two phenomena, it is not surprising to observe that PD detection can be impossible in long cables, unless very large PD are occurring. This problem can be solved by coupling the signal at the accessories (where the PD pulses have a large bandwidth and a large peak value and can thus be measured with a high signal-to-noise ratio).

Unfortunately, this solution is unapplicable to long submarine cables, due to logistic problems. PD in long submarine cable circuits remain a concern.

12.7 Factory Quality Assurance (QA) Testing of Power Cable

Factory QA tests are performed to ensure that each reel of power cable (and sometimes the associated accessories) are free from partial discharges at well above operating voltage. A minimum sensitivity should be achieved, and the acceptance/rejection criteria are based on PDIV or PDIV and PDEV, depending on the standard (Section 12.4). A high-voltage coupling capacitor with adequate capacitance (generally 10–30 nF) to achieve the desired sensitivity must be used. The measurement in terms of apparent charge (pC) requires that the upper limit of the detection bandwidth of the PD detector be below 1 MHz, as prescribed in the IEC 60270 [14]. This choice limits the possibility of using interference suppression methods presented in Section 12.12 and forces cable manufacturers to build shielded rooms (Section 9.3.1), install power supply filters (Section 9.3.3) and use good test set-up practices (Section 9.3.2) to limit interference during the measurements. During the tests, temporary terminations at each end of the cable must be used to prevent partial discharges at the cable ends. These terminations need to be removed after the tests. For high-voltage cables, temporary terminations typically use low-conductivity water to achieve a nearly uniform field distribution at the ends of the insulation shield to prevent the inception of spurious partial discharges (Figure 12.23). The water is circulated within the termination to maintain the temperature and the conductivity as constant as possible (typically, the dependence of conductivity on temperature is 5%/K [32]).

A problem with low-frequency PD measurements on cables is that PD pulse superposition can occur, as discussed in Section 6.6.2. According to [15], this problem can be neglected in three cases:

1) The cable terminal opposite to the detection system is connected to a high-voltage resistance matching the characteristic impedance of the cable.
2) The two ends of the cable length, if shorter than 50–100 m, are connected together to form a loop, and can be treated as capacitive test objects.
3) The PD detection system is equipped with a reflection suppressor, that is an electronic device that shuts off the detector upon receiving a PD pulse (thus preventing PD superposition). The shut off time should be designed based on the impulse response of the quasi-integration filter in the time domain.

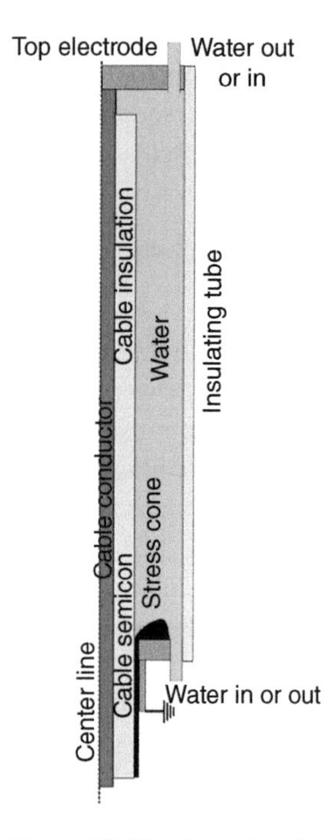

Figure 12.23 Example of structure of temporary HV water termination for PD testing. *Source:* [32].

In the other cases, one should derive a characteristic for the PD detector like that reported in Figure 6.22. However, predicting PD superposition is rather difficult, thus a double pulse calibrator is sometimes used [15]. This special calibrator produces two pulses with the same apparent charge where the second pulse follows the first with a time interval, which can be varied between 0.2 and 100 μs. By using the double pulse calibrator with different delays, one can reconstruct the PD detector characteristic as shown in Figure 6.22. Once the characteristic has been determined based on the propagation time of PD pulses through the length of the cable [15], this establishes the special cases where superposition issues can be neglected (not discussed here for the sake of brevity). Alternatively, measurements should be performed shifting the PD detector from one terminal to the other, to assess whether superposition phenomena are occurring or not.

There are different approaches aimed at performing the PDIV tests in conjunction with the standard AC withstand tests on the cable themselves. Let Uo be the rated line-ground voltage. Some manufacturers perform PD detection during the hipot test at of 3.5 Uo for 15 minutes for cables rated < 30 kV, or at 2.5 Uo for 60 minutes for cables rated > 30 kV. The PD sensitivity typically is 2-5 pC. The hipot tests are standardized in the IEC 62067, IEC 60840, and IEC 60502-2.

12.8 Energizing Cables in Offline/Onsite PD Tests

Factories will normally have a specialized 50/60 Hz supply to provide the required reactive power for energizing each reel of cable for both the PD tests and the withstand voltage tests. Offline tests performed after the cable has been laid are much more challenging in this regard, as they require a power supply that is portable and able to energize the capacitance of the entire power cable circuit. Suitable power supplies are expensive, are difficult to transport, and some of the options are controversial among experts. Most onsite high-voltage power cable PD tests cannot be performed at 50/60 Hz since power cables have large capacitances (×100 nF/km) that require large apparent power at 50/60 Hz. For example, in a 400 kV cable system, the current per kilometer of a single phase is about 6.9 A/km corresponding to 1500 kVAR/km). These numbers are huge, thus alternative ways of energizing the cable circuits are needed.[5]

Various methods have been used to energize such long cable circuits with a portable supply:

- Resonant test sets (RTS) at 50 or 60 Hz.
- RTS at an AC frequency that depends on the length of the cable circuit and the available inductive reactors to produce AC near 50/60 Hz.
- Very low frequency (VLF) test sets operating a 0.1 Hz or similar AC frequencies.
- Damped oscillating wave tests sets (OWTS).

12.8.1 Resonant Test Set

The resonant test set (RTS) normally features a series HV reactor. If the AC frequency is required to be at 50 or 60 Hz, the reactor needs to be variable, so that it can achieve resonance with the capacitance of the power cable. In this case, 50/60 Hz AC power can come from local power sources, via a variable autotransformer that is used to adjust the test voltage. Such systems are suitable for PD tests on medium-voltage power cable. Hipotronics is one supplier of fixed-frequency RTS.

If some variation of AC frequency is permitted in the RTS, then fixed-inductance reactors can be used (Figure 12.24). The reactor and power cable circuit (plus a voltage divider) is supplied by a high-voltage power electronics system that outputs a sinusoidal voltage waveform with controllable frequency (Figure 12.25). By matching the supply output frequency with the resonant frequency of

5 The Malta-Sicily interconnector system is a good example of current that needs to be supplied in an offline test how demanding can be supplying a long HVAC cable. The cable is 118 km long and operates at 50 Hz at a voltage level of 245 kV. It is one of the longest HVAC links in the world. To make the cable able to transmit the required power, high voltage reactors were installed at both ends of the cable, for a total of 340 MVAR [33]. Actually, few HVAC lines are so long. 10–12 km is more common, and would require a 29–35 MVAR test set. A device able to supply tens of MVAR that can be deployed in the field does not exist.

Figure 12.24 450 kV, 700 MVAR subsea cable testing resonant test set that includes several fixed inductance reactors. *Source:* [34]/with permission from IEEE.

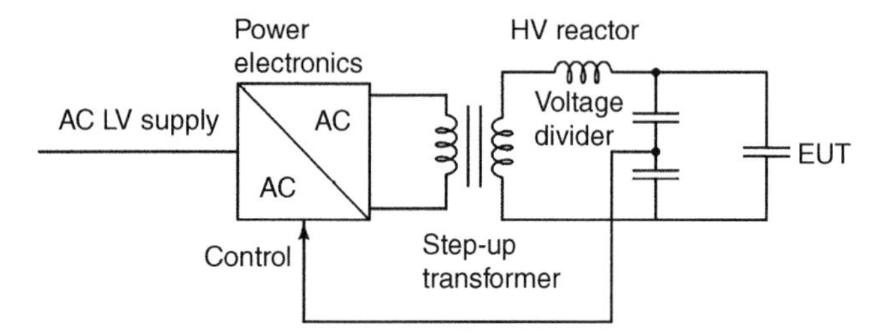

Figure 12.25 Scheme of a variable AC frequency RTS.

the reactor-cable series, high voltages can be easily reached providing only the active power required by the system (Joule losses in the conductors, dielectric losses in the insulation). For HVAC cables, the resonant frequency is often in the range of tens of Hz. Therefore, the voltages appearing across the insulation are like those experienced in service, and there is no need to investigate whether the supply frequency has a role in determining the PDIV/PDEV or the PD magnitudes. Often several reactors are used in series/parallel arrangements to achieve resonance at an acceptable frequency. The power electronics can create interference, which must be largely filtered out by the inductors or high-voltage filters. Variable frequency RTS are available from Highvolt, among others. Note that the PD detector needs to be able to synchronize to AC voltages different from 50 or 60 Hz.

Besides being used for offline PD tests, the same RTS is used for "soak" tests, i.e. tests where the cable system is energized at the rated voltage with no load for 24 hours. These tests are often performed to ensure that maintenance performed on the cable circuit has not introduced problems. Often soak tests can be coupled with PD tests, gaining important information on the cable reliability.

12.8.2 Very Low Frequency (VLF) Systems

VLF systems are electronic systems designed to operate in the AC frequency range 0.01–0.1 Hz, 0.1 Hz being the most typical choice. VLF systems can be manufactured to output a cosine-rectangular voltage waveform, which limits their applicability to withstand tests only, or sinusoidal voltages that can be used for voltage withstand tests, dissipation factor (tan-delta), and partial discharge measurements. The main advantage of VLF supplies is that the current needed to energize a cable circuit will be (say) 0.1/60 or 600 times lower than needed at or near power frequency. For medium-voltage cable testing, the supply can be plugged into a normal 120/240 V wall outlet.

VLF systems up to 75–200 kV are manufactured by High Voltage Inc. and Baur [35]. This voltage limits the applicability of VLF systems to cable circuits rated up to 140 kV for commissioning or recommissioning, or 250 kV for cable lines older than 15 years, where lower voltages are used. For testing partial discharges, VLF systems normally need some high-voltage signal filters to suppress the interference from the electronics. One of the problems of VLF is that the period of the voltages ranges from 10 to 100 seconds, and very long times could be needed to build up a suitable PRPD pattern and interpret the results in a meaningful way. Relatively few commercial PD instruments can synchronize to VLF frequencies.

Issues may arise when testing cables with nonlinear resistive stress control systems (see Figure 12.6), as opposed to the stress cone approach. Since the capacitive current at VLF is much lower than that at 50/60 Hz, the resistivity of the stress control tube is higher than in service, leading to the incorrect operation of the electric field control system. If the resistivity of the stress control tube is too high, the most likely outcome is that the electric field is such that the stress control system is effectively not present (see Figure 12.4), and partial discharges occur at the end of the insulation shield. This situation is recognized as an open issue in [36].

For VLF test sets operating at frequencies of 0.01 Hz (and for those providing a cosine-rectangular voltage waveform in particular), it is generally recommended that the field is kept below 10 kV/mm to prevent space charge injection and the risk of damaging the cable upon polarity inversion.

12.8.3 Oscillating Waves Test Set

An alternative to RTS and VLF systems is the oscillating waves test set (OWTS), also called a damped AC or DAC test set [37]. OWTS also features advantages and disadvantages. The circuit diagram of an OWTS is shown in Figure 12.26. A DC source charges the cables to the desired test voltage. The cable is in series with an external inductor, which is integral to the OWTS system. When the desired test voltage is reached, the switch disconnects the DC supply and short circuits the inductor-cable series circuit. The voltage across the cable thus follows an exponentially damped oscillation. Clearly, the better the quality factor of the inductor, the longer will be the transient, thus the slower will be the decay of the voltage applied to the cable. OWTS is made by Seba/KMT (part of Megger) and others.

The advantages of the OWTS system are mostly due to its portability, since the inductor is the largest component, whereas the HV DC supply is small. However, if the DC supply current is not large enough, the time between "shots" could become very long. An issue related to the OWTS is that if the quality factor of the inductor is not large enough, the exponential damping of the voltage coupled with the statistical delay of starting electrons for PD inception can lead to overestimate the PDIV. Besides, PD interpretation is more complicated as, cycle after cycle, the voltage decreases, and PD patterns are recorded at decreasing voltages. The relationship between PDIV/PDEV results from 50 to 60 Hz as well as RTS tests, and VLF/OWTS tests, is still being debated.

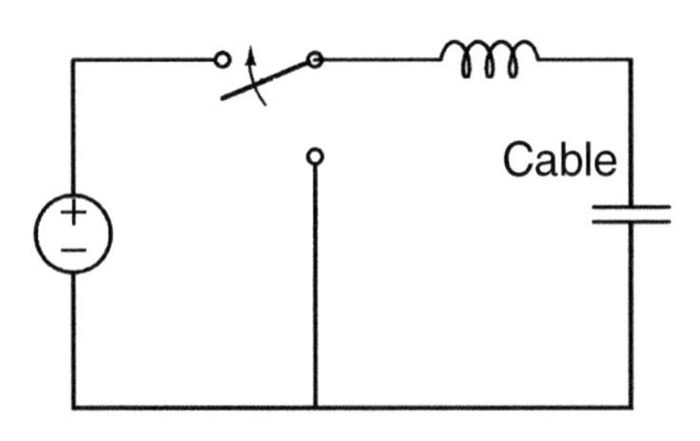

Figure 12.26 Schematic of an OWTS.

12.9 Offline/Onsite Testing

Offline/onsite tests are performed mainly for cable circuit commissioning tests to screen out defects related to an imperfect assembly of the joints and to find defects caused by pulling of the cable during the installation procedure. The main challenges with onsite/offline PD tests are:

- Energizing long lengths of cable circuits to the required voltages with portable power supplies (Section 12.8).
- Access points for performing the PD measurements. In the worst case, sensors may only be temporarily installed at the two ends of the cable circuit. This implies poor sensitivity to PD far from the terminals. In some cases (mainly transmission class circuits), additional sensors can be installed at cable joints.
- The PD signal in many cases can only be measured at frequencies above 1 MHz, thus the measured signal will be relative, not absolute (in terms of pC).

Onsite/offline PD tests are widely used for transmission-class circuits.

12.9.1 PD Detection

The main PD sensors are installed at the cable terminations at both ends of the circuit. Here, HFCTs are mainly installed on the neutral leads to ground (Section 12.5.2). For medium-voltage cables or high-voltage submarine cables, only measurements from HFCTs mounted on termination ground leads are generally possible. Onsite/offline PD tests on medium-voltage cables are less common. The reasons can be explained considering that the failure of an MV cable has a lower economic impact and that replacement circuits can be installed with lower costs. Often, automatic systems can switch the supply from a faulted MV circuit to a working cable in very short times.

For non-submarine, high-voltage transmission circuits, it is often possible to measure PD at the joints. Compared with online PD monitoring of power cables, the cable system can be operated in ways that enable temporary connection of the joints to the ground in the link boxes, using the coaxial cables used for shield transposition (Figure 12.13). Normally only transmission class power cables will be equipped with capacitive sensors in the accessories (Section 12.5.3), or be retrofitted with electric or magnetic field sensors placed in proximity of the accessories (Section 12.5.4). Measuring PD is straight forward with such sensors, although the measurement sensitivity remains to be demonstrated. For long cables, low-pass filters with different cutoff frequencies can be used to customize the bandwidth to the expected frequency content of the pulses based on the distance of the accessories one wants to probe (see Figure 12.22). When measuring from the terminations, PD from accessories a long distance from the line ends may be undetectable due to the similar frequency content of the PD pulse and noise/interference pulses.

12.9.2 Calibration vs Sensitivity Check

This type of measurement is not calibrated in the standard sense (Section 6.6). This is due to the fact that these measurements are not performed in a shielded room and, often, the sensors and/or the detection system need to operate above the IEC 60270 frequency limit (1 MHz) to suppress interference [14]. Thus, a sensitivity check should be performed.

In general, a sensitivity check (Section 11.4) can be performed only at the terminations where the high-voltage terminals are accessible. The check is done by injecting calibration pulses of decreasing amplitude into the high-voltage cable terminal. Indeed, the size of the terminations (several meters high for HV and EHV cable systems) forces the connection of the pulse generator between the HV terminal and ground to use very long cables. These connecting cables can have a large parasitic inductance that affects the calibrator pulse injected into the terminal and leads to an overestimate of the measurement sensitivity.

In the joints, two options are possible for a sensitivity check. For joints with capacitive sensors (internal or external), the scale factor (pC/mV) of the sensor can be established in the lab, using a dummy joint where the high-voltage conductor is accessible. Using the scale factor, the level of the noise in the field can be expressed in terms of pC. For PD measurements performed from link boxes, a similar approach can be pursued in the laboratory. Alternatively, a calibrator pulse can be injected directly into the HFCT site achieving a very optimistic estimate of the sensitivity (unfortunately, this is often the only feasible solution). Typically, the sensitivity is in the 10–20 pC range. In all cases, the sensitivity check is not performed in a standard way, and one must realize that what is provided is merely indicative. The use of filters matched to the noise in the field can lead to higher signal-to-noise ratios than expected, but sensitivity checks performed in the lab without these filters cannot be taken at face value.

12.9.3 Test Performance

As discussed in Section 12.4, most standards provide indications in terms of sensitivity, PDIV and PDEV for new cable systems. A recent CIGRE technical brochure outlines the levels in IEC standards and also provide suggestions for aged cable systems. These levels are summarized in Table 12.2 or Table 12.3, depending on the age of the cable [24]. As can be seen, for the commissioning of new cables or the recommissioning of cables whose age does not exceed five years, relatively high voltages are reached, and it may be acceptable for the PDIV to be less than the maximum test voltage. The pass/fail criterion is based on the PDEV, which should exceed 1.5 U_0 (U_0 being the rated line-to-ground voltage of the cable). For aged cables (>5 years), it is assumed that incepting PD activity during the tests (using a lower maximum test voltage) would lead to an unacceptable damage of the cable insulation. Therefore, the pass/fail criterion is that partial discharges are not incepted during the test (which means that the PDIV should exceed the test voltage).

If PD are detected and the cable does not pass the test, a key activity to be performed is the localization of the PD source. Localization must be performed to identify the part of the cable system that has to be replaced before recommissioning the cable. The techniques used to perform localization will be discussed in Section 12.14.

Table 12.2 Test conditions and pass/fail criteria for cables with an age <5 years.

Test frequency (Hz)	Test duration (min)	Voltage class (kV)	Test voltage	Pass/Fail criterion
10–300	60	66–72	2.0 U_0	PDEV > 1.5 U_0 No PD at 1.5 U_0
		110–115		
		132–138	1.7 U_0	
		150–160		
		220–230		
		275–285		
		345–400		
		500	1.5 U_0	

U_0 is the rated line-to-ground voltage.
Source: [24]/CIGRE.

Table 12.3 Test conditions and pass/fail criteria for aged cables.

			Age: 5–15 years		Age: >15 years	
Test frequency (Hz)	**Test duration (min)**	**Voltage class (kV)**	**Test voltage**	**Pass/Fail criterion**	**Test voltage**	**Pass/Fail criterion**
10–300	60	66–72	$1.5\,U_0$	No PD detected	1.1	No PD detected
		110–115				
		132–138	$1.4\,U_0$			
		150–160				
		220–230				
		275–285				
		345–400				
		500				

U_0 is the rated line-to-ground voltage.
Source: [24]/CIGRE.

12.10 Pros and Cons of Offline Versus Online PD Measurements for Condition Assessment

Onsite/offline PD testing can also be performed to evaluate the condition of the cable circuit insulation during its service life to determine if maintenance or replacement is needed. That is, PD testing can be used for condition monitoring, as well as for commissioning. Onsite PD testing for condition assessment of existing cable circuits can be performed either offline or online. The two types of tests have different advantages.

Offline tests are carried out by disconnecting the cable from the grid and energizing it using an external source. This choice enables:

- The possibility of raising the test voltage above the normal operating voltage level to check the PDIV, thus evaluating the safety margin (the upper limit for testing, unless agreed by the supplier, is often 27 kV/mm).
- Offline tests eliminate most of the interference from the grid. For example, poor electrical contacts, as well as surface or corona discharges from overhead lines, connections, and insulators can be prevented by using an external power supply with corona ring electrodes and other means to prevent PD occurring in the test circuit. This will reduce the risk of false positive indications of PD (and thus unnecessary investigations to find the phantom PD, or worse, replacing a good cable circuit).
- Using an HFCT PD sensor installed in link boxes, testing can be performed in cable circuits not equipped with permanent PD sensors.

The main disadvantage of offline testing is that at least a few days outage is required, during which the line cannot be used to transmit energy. Also, power system reliability decreases during the outage. This situation can be aggravated by delays in delivery of the test equipment.

The main advantages of continuous online monitoring compared to offline testing are:

- There is no need of an HV source; the cable is energized by the grid.
- It provides the greatest possible warning time that the insulation system is at risk. From the onset of detectable PD to when there is a high risk of cable system failure can be very short.

Depending on the failure process (treeing is relatively fast, whereas tracking along interfaces is relatively slow), the advance warning may be only hours to weeks. Clearly, offline tests would have to be very frequent to catch some incipient cable failures.

- PD phenomena can depend on the operating conditions (temperature, heating/cooling transients, 100–120 Hz vibrations due to electrodynamic forces between the conductors in the cable system, other activities external to the cable system). Monitoring PD activity can thus help to discriminate the influence of all these factors and identify the possible root cause of the PD (Section 10.6).
- Events such as lightning or transient overvoltages can damage the insulation leading to partial discharge inception. Online PD monitoring can detect the onset of this PD activity.

Besides the higher initial cost to set up a monitoring system, the main problem with online testing is that interference from other sources is measured along with PD from the cable. Therefore, denoising and PD source identification must be performed carefully. Otherwise, corona and surface PD from overhead line connectors and insulators may be detected, leading to false positives (PD outside the EUT are marked as PD in the EUT) or false negatives (when PD outside the EUT are recognized correctly, but they mask PD from the EUT). These risks are particularly high at the terminations, where the sources outside of the EUT are in proximity to the detection circuit (surface PD on a porcelain pothead bushing during rain, fog, and snow is harmless – but it is only 2 cm away from the polymer insulation in the cable termination). Typically, away from the terminals, interference is less of a problem since PD and other signals from the grid lose the high-frequency content and have lower magnitude the further they travel. Thus, online monitoring needs appropriate techniques to separate interference from PD in the power cable system (Section 12.12). Furthermore, the SCADA software used to collect the data should also be very effective to raise an alarm flag in a timely way. Failure to do so could result, on one hand, in a cable failure happening despite huge investments, and on the other hand, *crying wolf* too often because of false positive indications that lead asset managers to disregard the monitoring system information.

In summary, online measurements are probably the best solution for transmission-class cables despite a high initial cost for setting up the sensors, instrumentation, and the SCADA system to allow a large amount of information to be rapidly evaluated even by operators lacking a strong expertise in PD analysis. The reason is that assuming the false indication rate is low, no cable circuit outage is required.

12.11 Online Monitoring

As mentioned in Section 12.10, the main advantage of PD monitoring is that due to the fast rate of deterioration of organic material in the presence of PD, the time between PD inception and failure can be very short. As an example, Figure 12.27 shows burning marks due to PD activity in a new 400 kV XLPE cable joint subjected to 1.1 U_0 (U_0 is the rated line-to-ground voltage) for one hour [38]. The rapid evolution of the aging does not offer sufficient lead time to carry out PD measurements at regular times (e.g. once a year for AIS and rotating machines). This has prompted the development of continuous PD monitoring systems for cable circuits. These systems are costly (at least one PD detector is required in a joint bay), require an adequate infrastructure like a SCADA center, fiber optics for data transmission, and power for the PD instrumentation at joints and terminations. Sometimes these infrastructures are hard to realize – even installing 120/240 VAC circuits to power the instrumentation can be daunting. In some special cases, energy harvesting

Figure 12.27 Effect of PD activity in one defective joint after only one hour-long commissioning test at 1.1 U_0 of a 12 km-long 400 kV XLPE cable system. *Source:* [38]/with permission from IEEE.

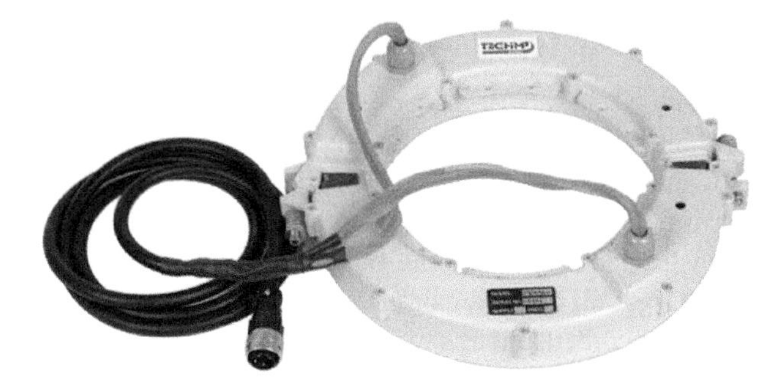

Figure 12.28 Permanent power supply (PPS) that can be used to power instruments at cable joints. This is clamped around the power cable, over the neutral wires. *Source:* Techimp-Altanova.

systems can be used, for example a permanent power supply can employ a large current transformer (see Figure 12.28) that exploits the 50/60 Hz magnetic field created by the load current in the cable if a 120/240 VAC supply is not available. Alternatively, solar panels can be used.

Another advantage offered by permanent monitoring system is that the PD source sometimes can be located as soon as it becomes active using the amplitude–frequency (AF) technique presented in Section 12.14

In general, a good monitoring system should have:

1) High sensitivity to PD.
2) The capability to separate different types of PD while suppressing interference without suppressing PD signals.
3) The capability to correlate PD activity with other phenomena such as load current, insulation temperature, transient overvoltages, ambient conditions(temperature and humidity) (Section 10.6).
4) The capability to discriminate harmful PD sources.
5) A diagnostic and notification system able to combine and synthesize all the information acquired minimizing the risk of false positives and false negatives.
6) The ability not to fail in the presence of high-voltage and high-current transients in the cable circuits.

The biggest challenge with continuous online PD monitoring systems for cables is achieving an acceptable false positive and false negative indication rate. Usually, an expert is needed to review

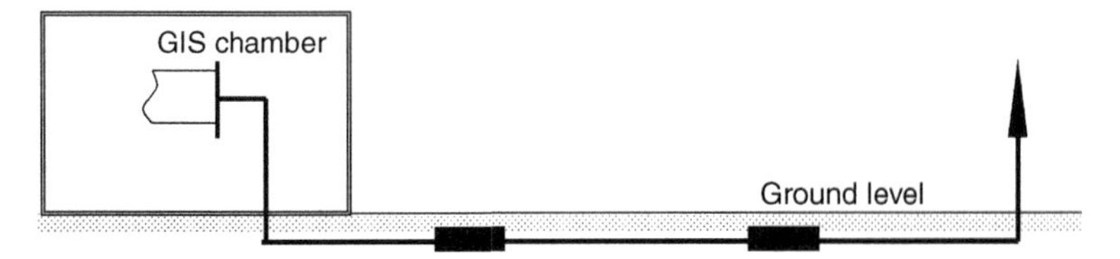

Figure 12.29 Cable system used for the demonstration of PD monitoring effectiveness. *Source:* Adapted from [39].

the PRPD patterns from an online PD monitoring system (PDM). Also, some companies find it difficult to decide in a timely fashion on whether or not to remove a cable from operation when a high PD alert occurs. Omicron, Power Diagnostix (Megger), and TechImp (Altanova), among others make such systems. However, online PD monitoring of cables is not as widespread as it is for machines, GIS, and possibly transformers.

Figure 12.29 shows the layout of a short 400 kV XLPE cable system used for testing the feasibility of online PD monitoring [39]. During preliminary tests, two phenomena, A and B, were detected and separated due to the different T–F map features of the PD pulses (Section 8.9.2) and recognized as discharges on the outer surface of the terminations (A), discharges internal to the joint (B), see Figure 12.30. The monitoring system was instructed to monitor the two phenomena separately. Figure 12.31 shows the evolution of the discharges on the outer surface of the termination. Very large PD (exceeding the detector full scale) were recorded, but this type of PD activity was likely on the porcelain surface of the termination and is not dangerous. Sometimes, very low values were observed, in conjunction with the drying of the external surface due to solar radiation in sunny days (measurements were carried out in the winter). In contrast, Figure 12.32 shows the evolution of the PD internal to the joint. As the damage progresses, the PD amplitude increases since interfacial tracking phenomena can develop on longer tracks. The monitoring was stopped after six days due to the failure of the joint.

These results emphasize the necessity of separating phenomena that occur outside the cable (often very large or with large repetition rates) from those that occur within the cable. As well, they demonstrate the point that failure can occur within days (six), so that periodic monitoring (for example every year) will be totally ineffective.

12.12 Interference Separation

Interferences in a cable can be ascribed to different phenomena:

- Coupling of radiated signals coming mostly from radio stations and power line carrier. An example of such interference is shown in Figure 12.33 for a 223 m long cable laid above the ground (this is a worst case as generally cables are below the ground where the conductivity of the soil can help to provide a partial shielding from interference),
- Pulses due to partial discharges and corona in the high-voltage system (that is from equipment connected to the cable, but not from the cable itself). For offline tests, this type of interference also includes the PD generated by the supply system (VLF or RTS), if any. Arcing and sparking nearby due to poorly bonded connections, slip rings, arc welding, etc. also can lead to high levels of interference in factory tests, onsite/offline tests and online monitoring.
- For offline tests, pulses due to the electronic circuitry of the supply system (VLF or RTS).

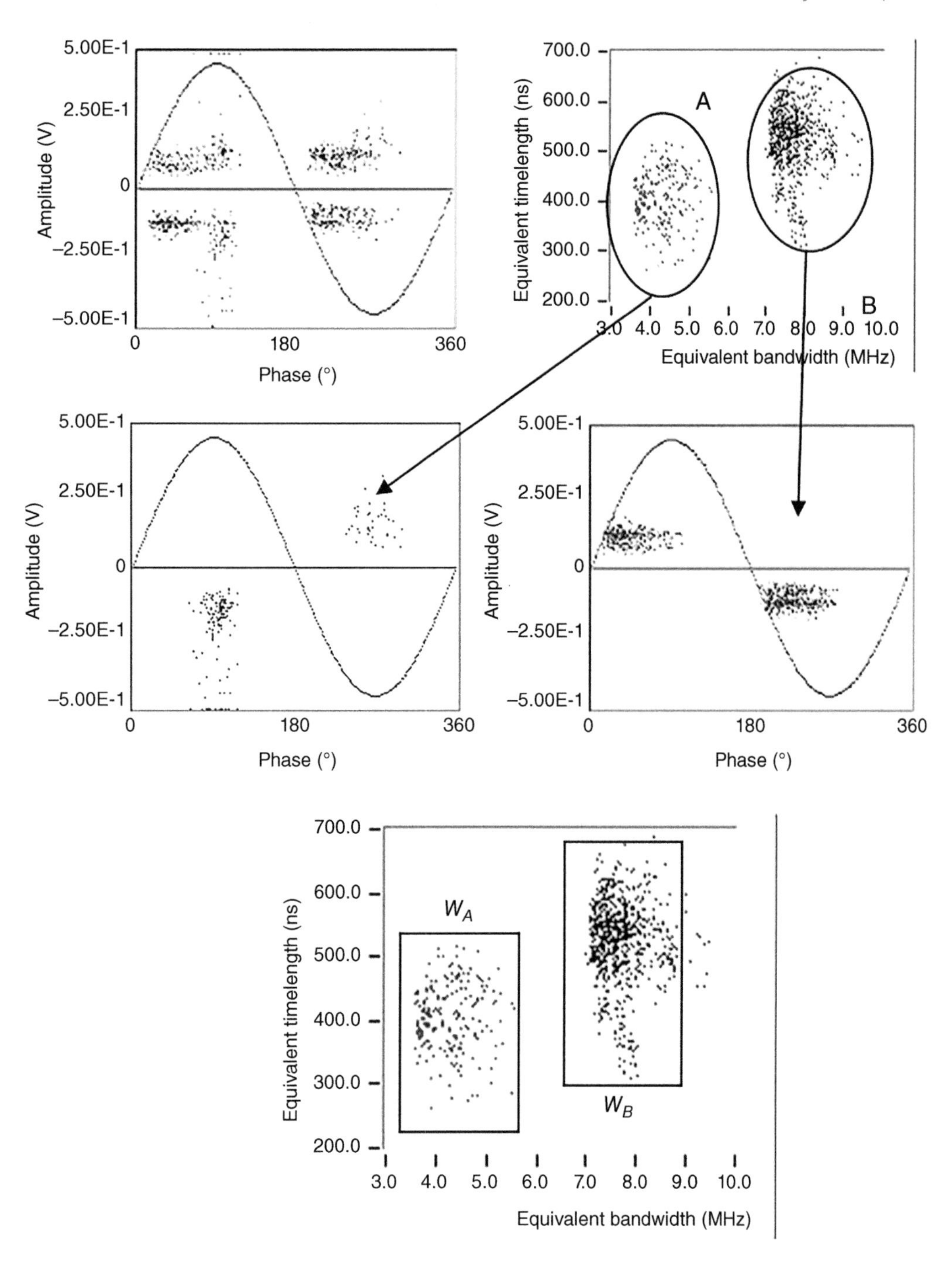

Figure 12.30 TF map analysis of the phenomena recorded in the cable system reported in Figure 12.29. *Source:* Adapted from [39].

It should be observed that while radiated signals are not greatly affected by the point at which they are measured along the cable, pulsed interference has a frequency content that decreases while traveling along the cable [41]. A self-explanatory representation of this is reported in Figure 12.34 [42]. This indicates that if PD detection is carried out at some distance from the terminations, pulsed interference will be less concerning and easier to suppress.

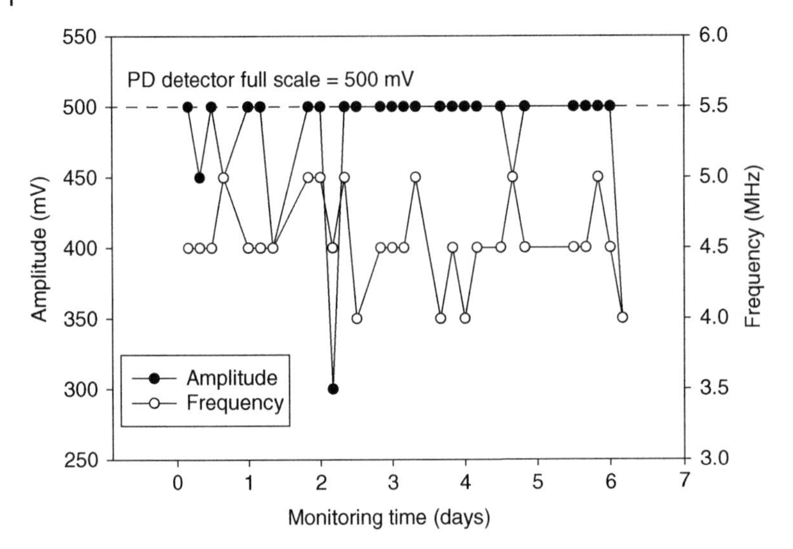

Figure 12.31 Monitoring of phenomenon A in the cable system of Figure 12.29, see also Figure 12.30. The right *y*-axis (Frequency) indicates the frequency content of the PD pulses. *Source:* [39].

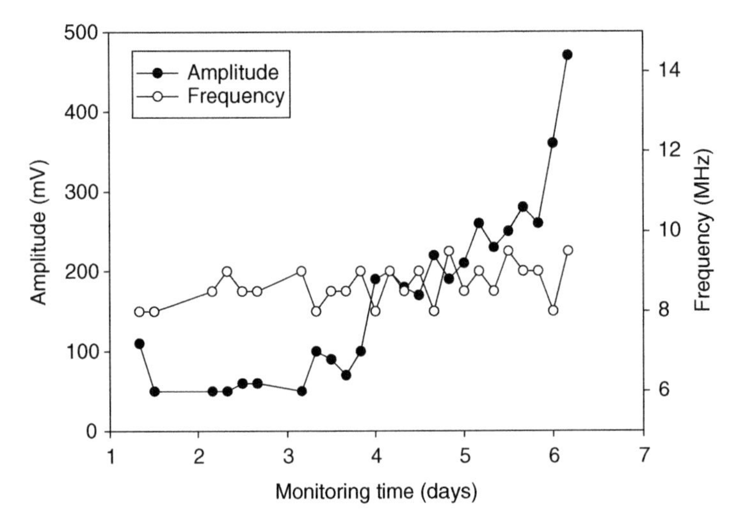

Figure 12.32 Monitoring of phenomenon B in the cable system of Figure 12.29, see also Figure 12.30. The right *y*-axis (Frequency) indicated the frequency content of the PD pulses. *Source:* [39].

Disturbance suppression must thus be carried out with different strategies depending on the type of measurement (time domain or frequency domain – Section 7.4), supply (online or offline), type of disturbance that one wants to reject, and distance between the PD coupler and the closest termination. A high-pass filter with a lower cutoff frequency of 1 or 2 MHz will generally reject pulse interference for measurements performed in the time domain. If measurements are performed in the frequency domain (Section 7.3.3), the frequencies where PD detection will be carried out will be above 1 or 2 MHz. For measurements performed at the terminations of a cable connected to an overhead line, the simultaneous acquisitions from the 3-phase amplitude relation diagram (3PARD) can help to discriminate corona or surface discharges taking place in the overhead line (Section 8.9.3). The 3-phase time relation diagram (3PTRD), which is based on the arrival time of signals [42], helps to discriminate whether signals originate in one phase or in another and

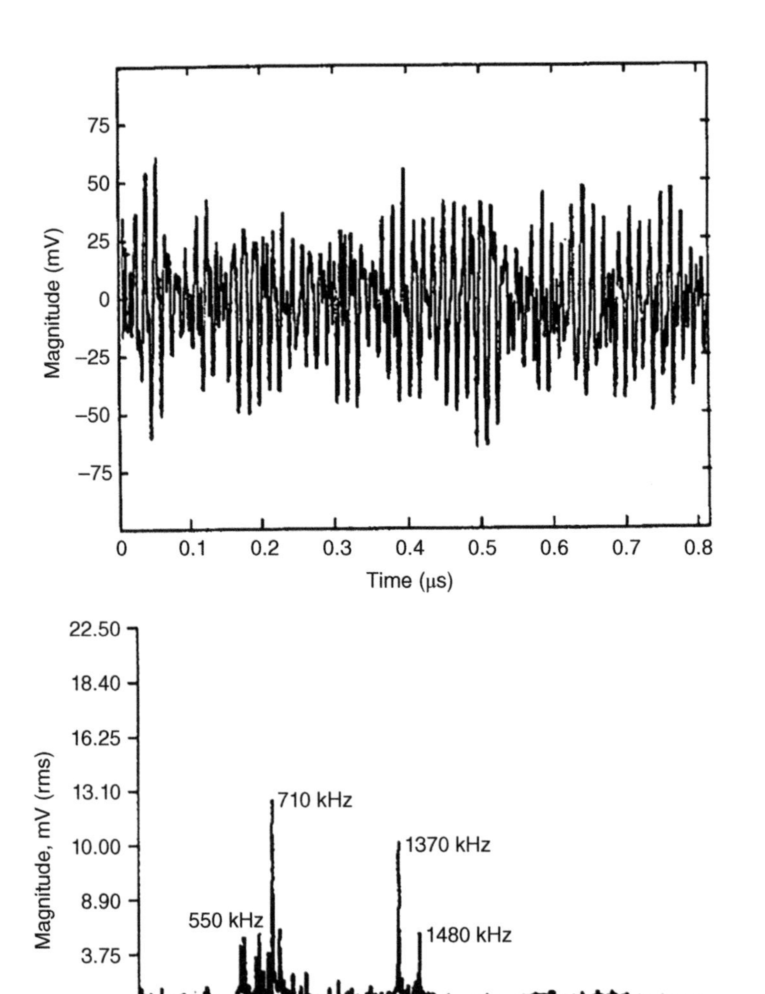

Figure 12.33 Noise in a 223 m-long cable laid above the ground. *Source:* [40].

can help to separate interferences particularly at a distance from the terminations, where pulses tend to be very similar.

The 3PTRD in Figure 12.35 is divided into six sections. The six possible order combinations of triplets of arrival times are also displayed in Figure 12.35. An ordered triplet of arrival times (t_1, t_2, t_3) is summarized by a vector originating from the center and having a length that increases with the delay between the first and the last pulse:

$$A = t_3 - t_1 \tag{12.1}$$

The phase of the vector is evaluated as

$$\alpha = \varphi_{Lx} + \frac{t_3 - t_2}{t_3 - t_1} \cdot 60° \tag{12.2}$$

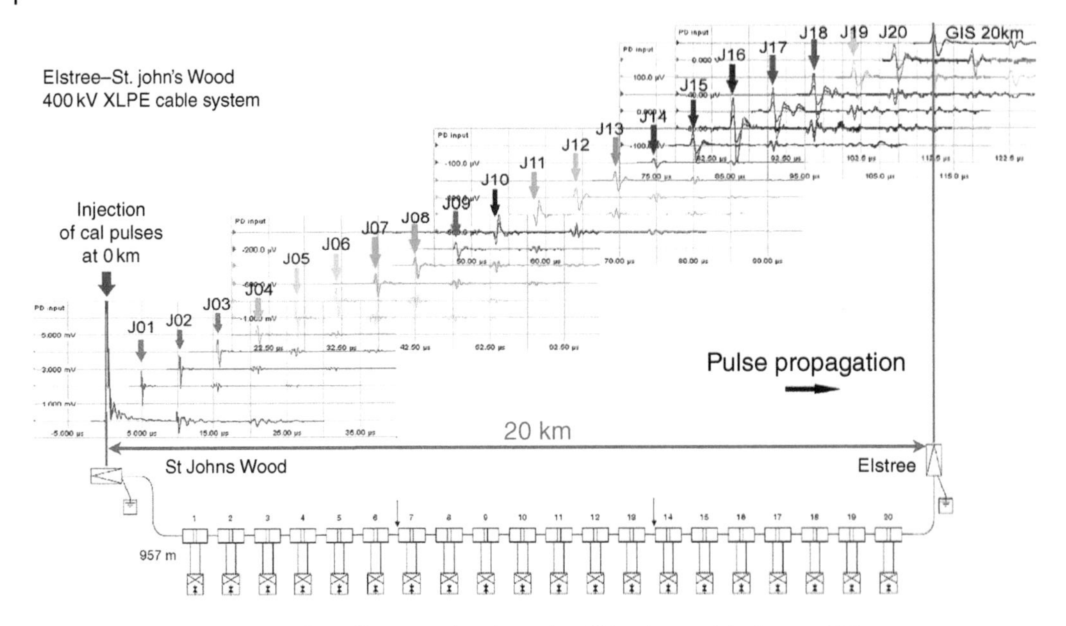

Figure 12.34 Propagation of a calibrator pulse through a 20 km-long cable. *Source:* [42].

where φ_{Lx} is an offset angle that depends on the phase where the first pulse of the PD triplet is detected ($\varphi_{L1} = 330°$, $\varphi_{L2} = 90°$, and $\varphi_{L3} = 210°$). The interested reader can find more information in [42].

To further suppress pulsed interference, post-processing of the measured data may be necessary. This would amount to collecting pulsed signals (PD or disturbance), extract information about the pulse waveshapes, and separate the pulses based on the features of the waveshapes. This can be done in the time domain using, for instance, the TF map (Section 8.9.2).

An example of signal propagation and separation of interference from PD pulses in a 400 kV XLPE cable system is shown in Figures 12.36 and 12.37. The test was an offline/onsite measurement. The cable system was prepared for the measurement by placing HFCTs in the link boxes, as shown in Figure 12.15. The supply was provided by an RTS, with the resonant frequency of 34 Hz. Figure 12.37 shows the readings typical of a healthy joint. The PRPD pattern is complex but using the features of the PD pulses (TF map), it can be decomposed in four subgroups. Cluster 1 is associated with a PD (note that the phase reference is leading by 90° since the PD detector was synchronized using the capacitive current in the cable). However, the PD pulse duration is more than 3 μs, evidence that suggests that this pulse has been attenuated and dispersed as it traveled along the cable to the PD sensor. As a matter of fact, this phenomenon was associated with PD from the power supply (an RTS). Cluster 2, giving rise to vertical bars and characterized by signals with a low-frequency content was associated with the operation of the power electronics in the RTS. Clusters 3 and 4 could raise suspicion that the pulses are PD, as they have high-frequency components. However, neither provides a meaningful PRPD pattern (i.e. pulses of the correct polarity and the normal positions in the AC cycle – Section 3.10) and are not a concern. Similar clusters were seen in a faulty joint (Figure 12.37). However, a new type of cluster (4) was observed. This cluster is characterized by a pulse-like signal having high-frequency content. Also, the PRPD pattern is typical of PD occurring in a cavity. The joint was replaced and dissected in the factory. Eventually, the root cause of PD activity was traced back to the incorrect alignment of the high-voltage stress relief electrode. It is apparent that some expertise is needed to do such analysis.

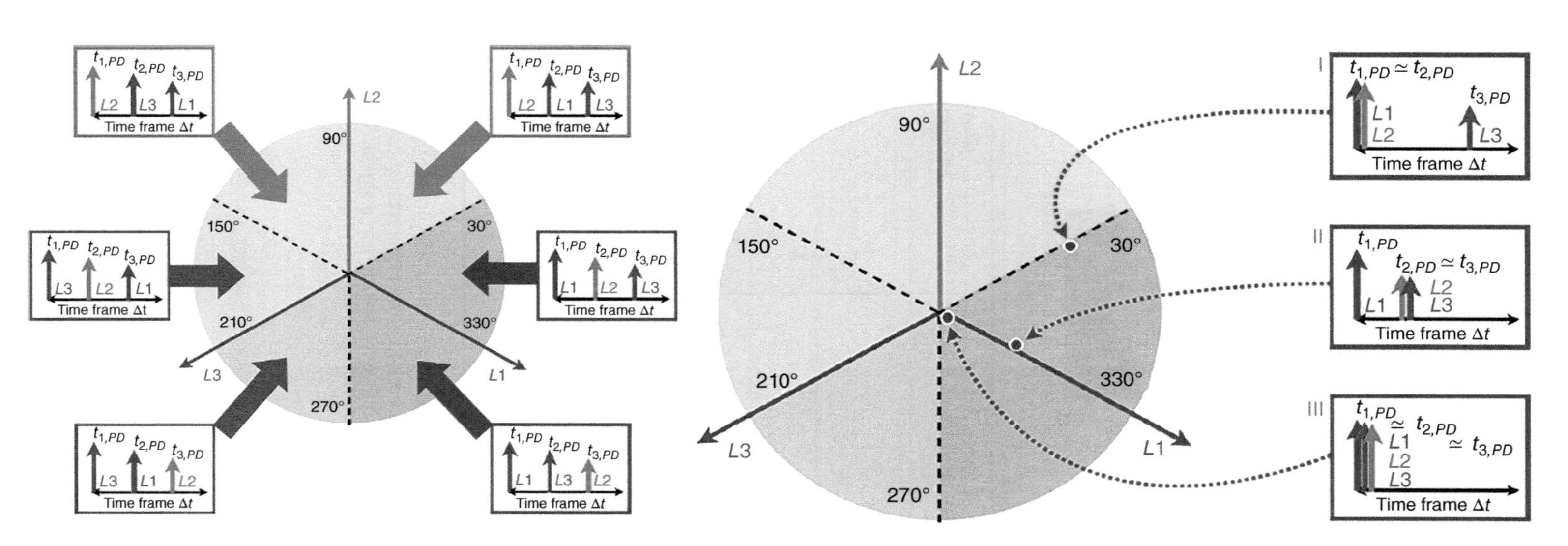

Figure 12.35 Derivation of the 3PTRD diagram. Left: Visualization segments of 3PTRD for six different arrival orders of the PD triple pulses. Right: Visualization of different time differences between the PD triple pulses within the *L*1 segment. *Source:* [42].

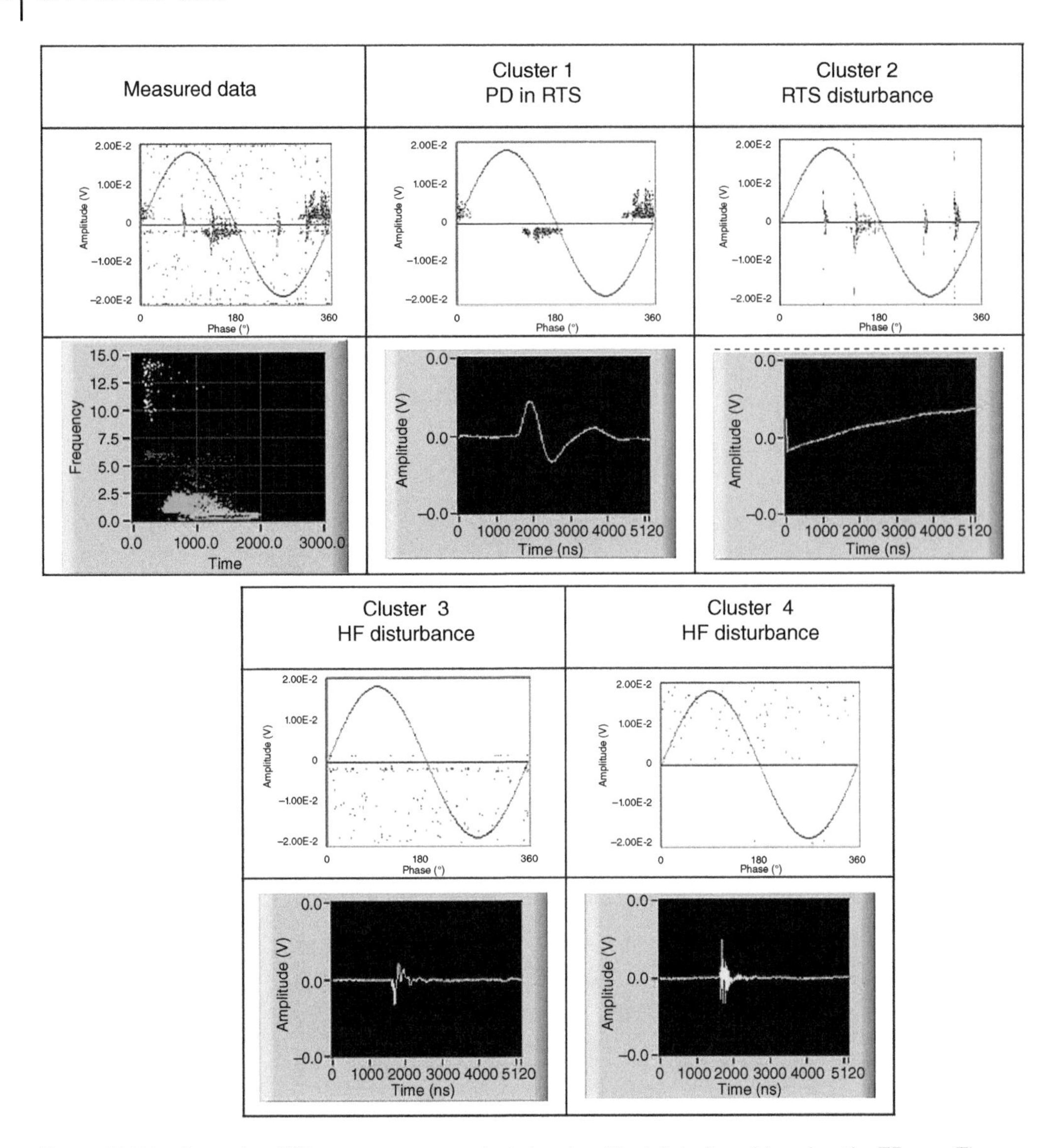

Figure 12.36 Example of PD measurement analysis in a healthy joint of a cable using the TF map. The measured data are separated in four clusters based on their characteristics in the TF map. For each cluster, the PRPD measurement is reported along with the typical detected signal. Note that in the PRPD plots the AC cycle is phase-shifted 90° from actual.

12.13 PRPD Patterns

Compared with, for example, rotating machines, PD in power cables have fewer types of PRPD patterns. In general, discharges occur in cavities, typically in the interfaces within accessories. When the cavity is in contact with the semiconductive layers at an interface, a certain degree of pulse polarity asymmetry can exist. Otherwise, they tend to be symmetrical. A few examples are shown in Figure 12.38, together with examples of interference.

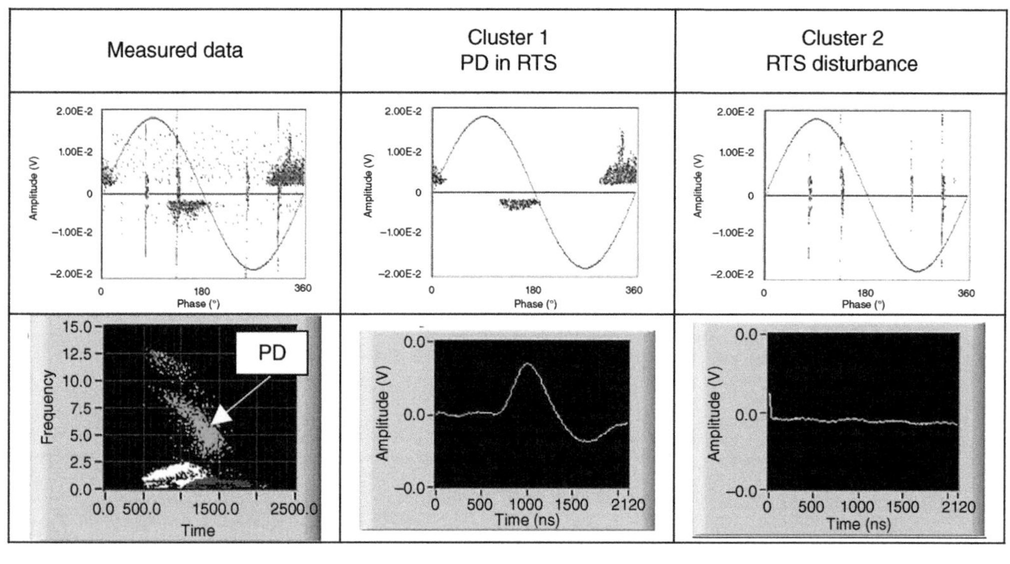

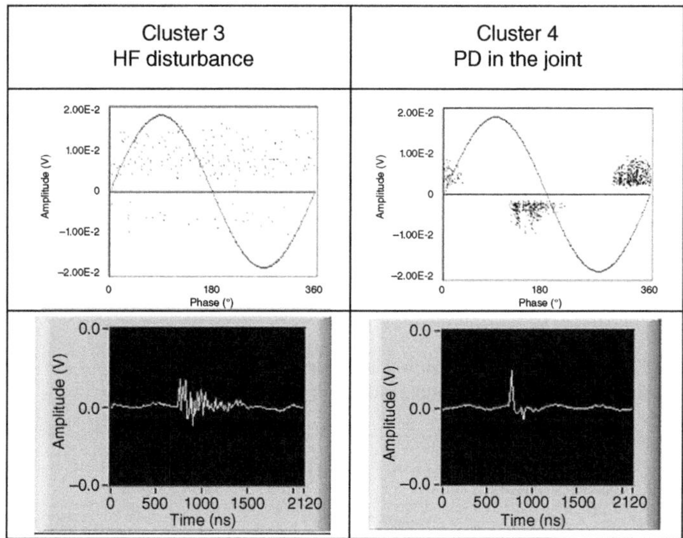

Figure 12.37 Example of PD measurement analysis in a faulty joint of a cable using the TF map. The measured data are separated in four clusters based on their characteristics in the TF map. For each cluster, the PRPD measurement is reported along with the typical detected signal. Note that in the PRPD plots the AC cycle is phase-shifted 90° from actual.

12.14 PD Source Localization

PD source localization is an extremely important part of PD measurements. If PD is detected, it is prudent to perform PD source location because:

- If PD is found to be occurring within the length of cable, it provides assurance that the signals are not due to interference, and
- Since the PD is most likely occurring at a termination or joint, identifying which accessory has the problem allows prompt repair or replacement, without disturbing other accessories.

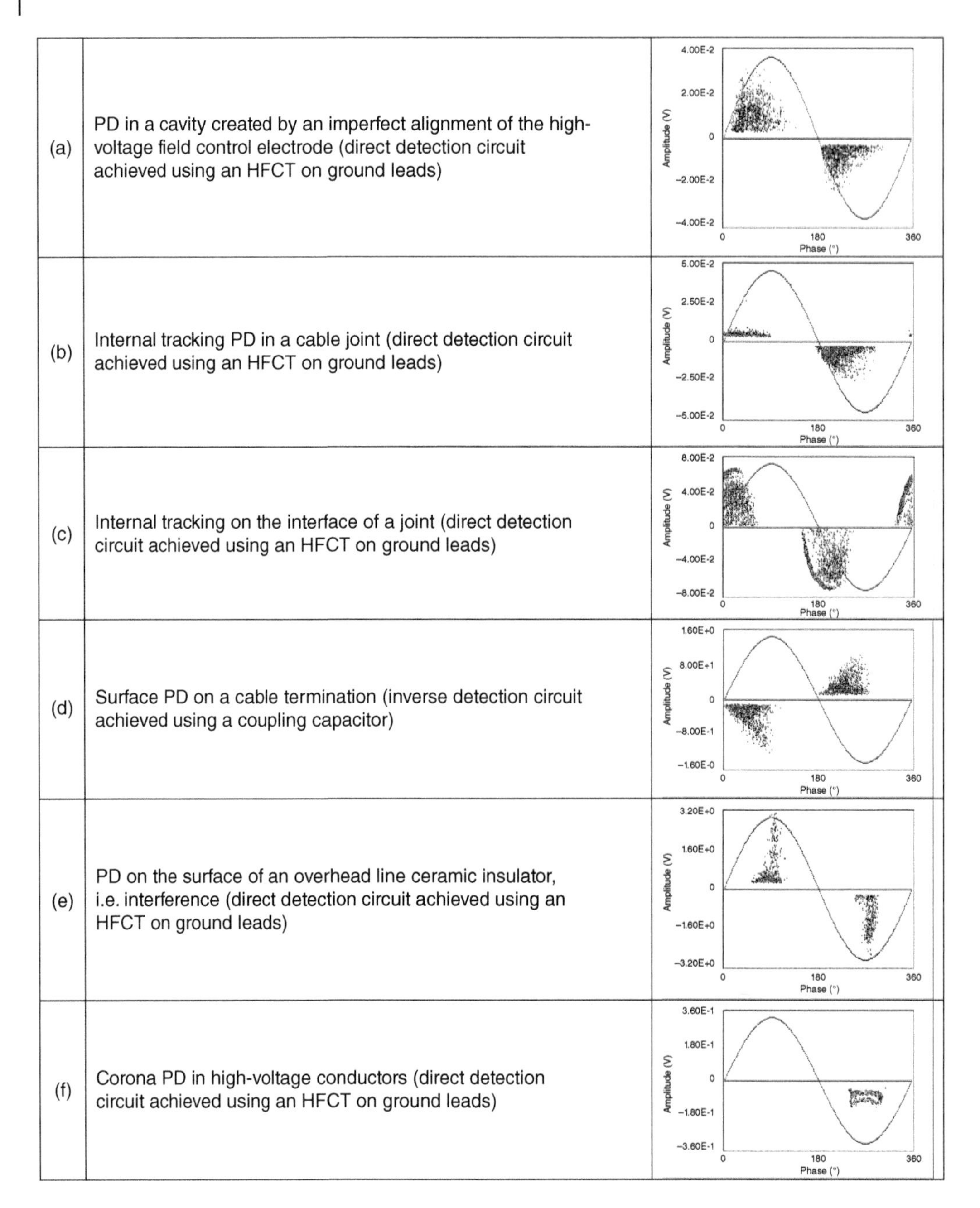

(a)	PD in a cavity created by an imperfect alignment of the high-voltage field control electrode (direct detection circuit achieved using an HFCT on ground leads)
(b)	Internal tracking PD in a cable joint (direct detection circuit achieved using an HFCT on ground leads)
(c)	Internal tracking on the interface of a joint (direct detection circuit achieved using an HFCT on ground leads)
(d)	Surface PD on a cable termination (inverse detection circuit achieved using a coupling capacitor)
(e)	PD on the surface of an overhead line ceramic insulator, i.e. interference (direct detection circuit achieved using an HFCT on ground leads)
(f)	Corona PD in high-voltage conductors (direct detection circuit achieved using an HFCT on ground leads)

Figure 12.38 Examples of PRPD pattern due to PD internal to cable systems and interferences.

Different techniques have been developed for PD site location, depending on whether tests are performed offline or online, on a short or long cable, or whether special hardware is available. Note in most situations only one PD site is likely to be active in a cable circuit, which simplifies the process of finding it. This is in contrast with, for example, rotating machines where hundreds of PD sites may be active at the same time. Although there may be only a few PD sites in a cable circuit, there may be many interference sites, especially with online monitoring.

12.14.1 Time-Domain Reflectometry

Time-domain reflectometry (TDR) is a technique that is applied on cables that are:

- Offline
- Open-ended
- Not too long

Note that "not too long" is a fuzzy definition since the actual length of the cable line that can be investigated through the TDR depends on several parameters such as:

- attenuation characteristics of the cable (Section 6.6.2.2),
- PD magnitude, and
- interference level.

It would be unwise to indicate the exact maximum length of cable that can be investigated using the TDR method. In spite of this, in many situations it is accepted that it can work for cable lengths up to 2 km.

The TDR principle for PD location in cables was devised by F. H. Kreuger in 1965 [43] and is described in Section 6.6.2.1. The PD sensor is connected to a digital oscilloscope, which must have a bandwidth of at least 100 MHz if the method is to locate the PD site within ± 1 m. It also assumes the coupling capacitors, HFCTs, or other sensors, also have about the same upper cutoff frequency (which corresponds to a detected pulse risetime of about 3 ns).[6] In a cable having length L, a PD pulse at a distance x from the coupler splits into two pulses each having magnitude equal to 50% of the original pulse. One is directed to the coupler and arrives there at a time t_1 after traveling a distance x (the distance of the defect to the near end, where the detector is located). The other one is directed in the opposite direction, arrives at the far end of the cable after traveling a distance $L-x$, and there is reflected. Eventually, the second pulse arrives at the PD coupler after traveling the whole cable length. Thus, the second pulse arrives at a coupler at a time t_2 after traveling a total length of $2L-x$ before reaching the coupler. Since the propagation speed of an electrical pulse in the dielectric is the same, and the observer at the detector does not know the time the PD takes place, the only quantity that can be measured is the difference Δt between the arrival times:

$$\Delta t = t_1 - t_2 = 2\frac{L-x}{v_p} \tag{12.3}$$

TDR can only be applied if the direct and reflected pulses are high enough to be seen above the noise floor. From Equation (12.3), it is clear that the worst case for detection is PD pulses occurring close to the PD coupler, since the reflected pulse arrives at the coupler after traveling ~$2L$. Thus, the 2 km limit mentioned above indicates that PD pulses have a magnitude (say) two times the noise/disturbance floor after traveling 4 km.

An example of a PD pulse and its reflections in a cable having a length L = 137 m are shown in Figure 12.39. Before the tests, the propagation speed of the wave was determined by injecting a pulse in the cable and measuring the time T_r between the injection of the pulse and the detection of its reflection at the same termination. Knowing the length of the cable L, the propagation speed

6 The sensors must not be connected to a quadripole used for IEC 60270 PD detection (Section 6.3), since this lengthens the PD pulse risetime to 300 ns or more, which implies a greater uncertainty zone for the location of the PD site.

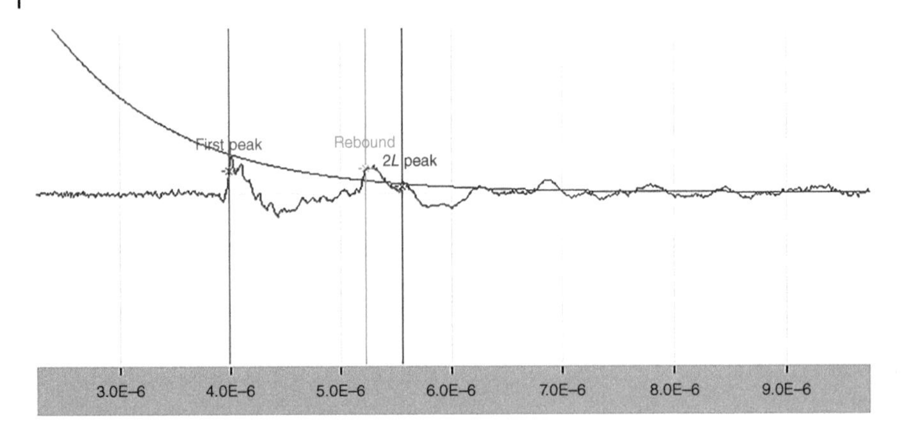

Figure 12.39 Example of TDR measurements on a digital oscilloscope (Rebound is the reflected pulse). *Source:* [44].

was estimated as $v_p = 2 \times L/T_r = 160$ m/μs. As can be seen, after energization, PD pulses tended to create a sequence of three detectable pulses.

- First Peak: corresponds to the PD pulse traveling directly to the near end (where the detector is located). It is received after traveling the distance x.
- Rebound: corresponds to the PD pulse directed to the far end and reflected to the near end. The rebound peak must travel $2L - x$ to be received at the near end.
- $2L$ Peak: corresponds to the first pulse that has been reflected at the near end and travels to the far end where it is reflected to the near end. The $2L$ peak is received after traveling $2L + x$ meters.

Thus, delay between the Rebound peak and the $2L$ peak is

$$\frac{(2L+x)-(2L-x)}{v_p} = \frac{2x}{v_p} \tag{12.4}$$

In the case presented here, this delay is 330 ns, corresponding to a distance x between the defect and the near end of ~27 m. Figure 12.40 shows the typical output of TDR measurements, the PD location map. When location is performed correctly, the locations tend to cluster at specific points corresponding to the defect positions. In this case, because of the short length of the cable, all PD detected is located at a defective joint, 27 m away from the coupler.

If time between the direct and the reflected pulses is twice the length of the cable, then the signal is either PD at the termination close to the PD sensors, or due to interference. The advantage of the TDR method is that only a single PD sensor is needed. Of course, the limitation is that the method has a lower probability of working in longer cable circuits.

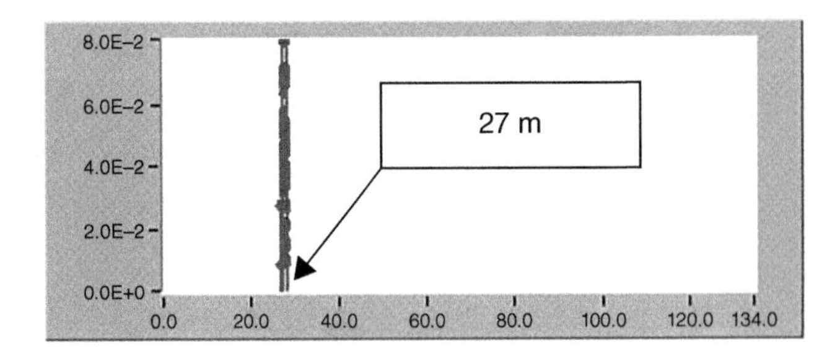

Figure 12.40 Example of TDR-based PD location mapping. *Source:* [44].

12.14.2 Time of Arrival

The time of arrival (ToA), also called the time of flight method is normally applied where PD signals traveling in cables can be coupled at two or more accessories [44]. Section 8.4.2 also discusses this method in general terms. Typically, this method is possible for

- high-voltage cables with imbedded capacitive couplers in joints and/or terminations (Section 12.5.3)
- high-voltage cables tested offline where access with the neutral connections at transpositions are accessible (Figure 12.41).

Each joint sensor is equipped with an analog-to-optical converter that converts the PD signal into an optical signal. The optical signal is sent via a fiber-optic link to a centralized detector where it is converted back to analog. The arrival time of the signals are recorded to identify the point of the cable system where PD originate. To do so, the fiber-optic links should have the same length. Alternatively, the differences in fiber-optic lengths should be compensated. Based on the arrival times of the signals at the different channels of the detector, the location of the defect can be assessed with good accuracy. Of course, this method only works if optical links to all sensors can be realistically temporarily installed, which is dubious in long cable circuits.

As an example, Figure 12.42 shows the readings achieved from the system represented in Figure 12.41. The pulse arrives last at joint 6, which is then the farthest away from the defect. It arrives at about the same time in joints 3 and 5, indicating that the defect is halfway between these two joints. In this case, the defect was located in joint 4, a straight joint where PD detection could not be carried out since PD couplers could not be installed there.

A different flavor of ToA analysis has recently been developed using HFCTs installed within sectionalized joints. VHF magnetic or capacitive couplers are located outside the accessories, at both sides of each joint. The arrival times at these sensors can be used to indicate whether the defect is outside the joint (and on which side of the joint) or within the joint. For this purpose, one should appreciate relatively short times since the propagation time in a joint is of the order of 10–20 ns.

Around 2010, a method was devised that uses a global-positioning (GPS)-based solution for online systems or for long cables when one is not able to deploy fiber optics along the length of the cable circuit. The idea is to add additional hardware to the PD detector that can time stamp the arrival time of PD pulses using the atomic clock installed aboard the satellites of the GPS system (Figure 12.43) [45]. Omicron and TechImp, among others, make commercial instruments using this approach [46].

12.14.3 Amplitude–Frequency (AF) Mapping

As discussed in Section 12.5, attenuation affects the PD pulse magnitude and frequency content. Dispersion affects the risetime and duration of the pulse (Section 8.4.3). The effect of attenuation and dispersion can be quantified using the peak value of the pulse as well as the equivalent

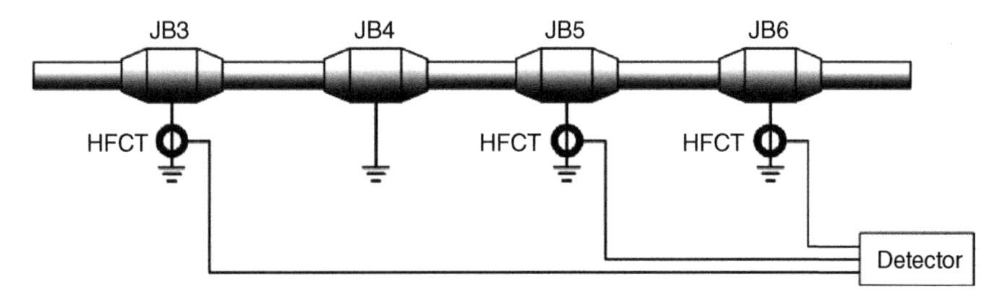

Figure 12.41 Example of ToA location localization. *Source:* [44].

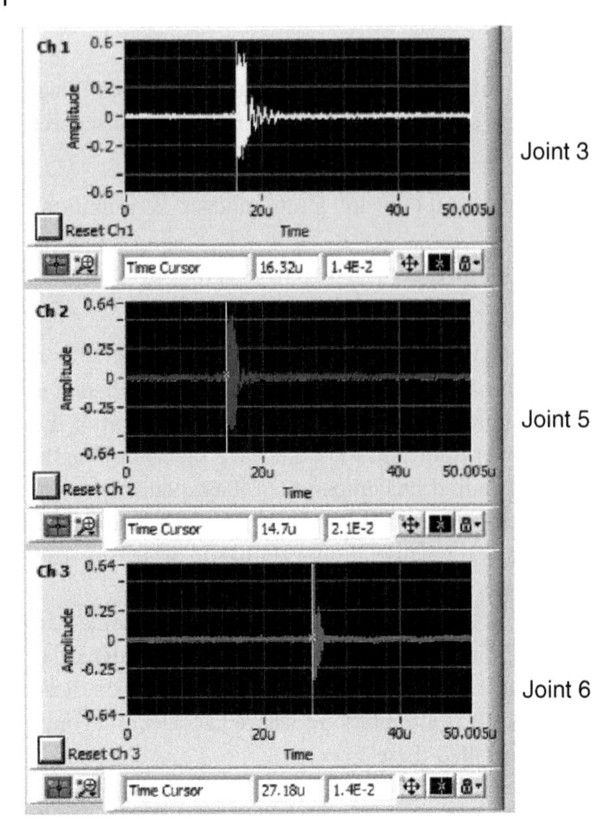

Figure 12.42 Example of readings from a ToA localization used in the cable system sketched in Figure 12.41. The cable circuit was 10 km long. *Source:* [44]/with permission from IEEE.

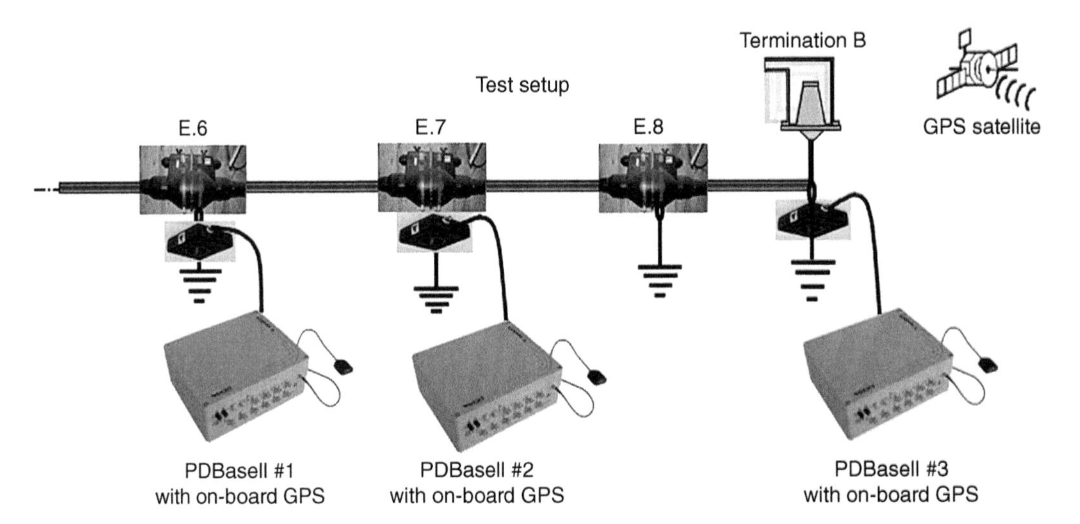

Figure 12.43 Sketch of GPS-synchronized ATA localization system. *Source:* Courtesy of TechImp/Altanova.

bandwidth and time lengths, as defined in Section 8.9.2. Based on the data presented in [47], these quantities are influenced by propagation as shown in Figure 12.44.

Figures 12.45 and 12.46 show and explain how AF mapping was used to rule out the effects of a PD source outside the EUT (Figure 12.45) and to locate the joint where PD was taking place (Figure 12.46):

- Figure 12.45 shows that the amplitude and the frequency content (bandwidth) of one PD event displayed a maximum at *T*2 (far end of the line, where the termination is installed); both

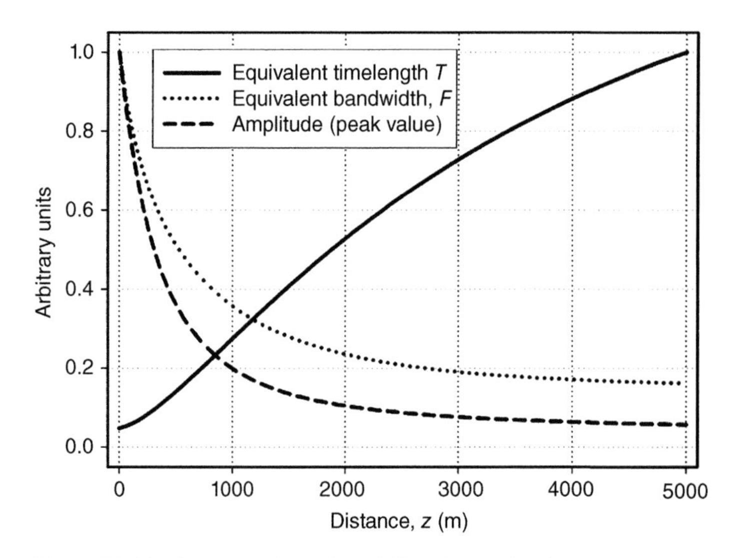

Figure 12.44 Estimated behavior of PD pulse peak value, equivalent time, T, and bandwidth, F, as a function of the distance, z, between PD source and detection point. The characteristics of the pulse at the source ($z = 0$) are used to normalize the data. The curves were achieved using the parameters of a distribution-class EPR cable. *Source:* [44].

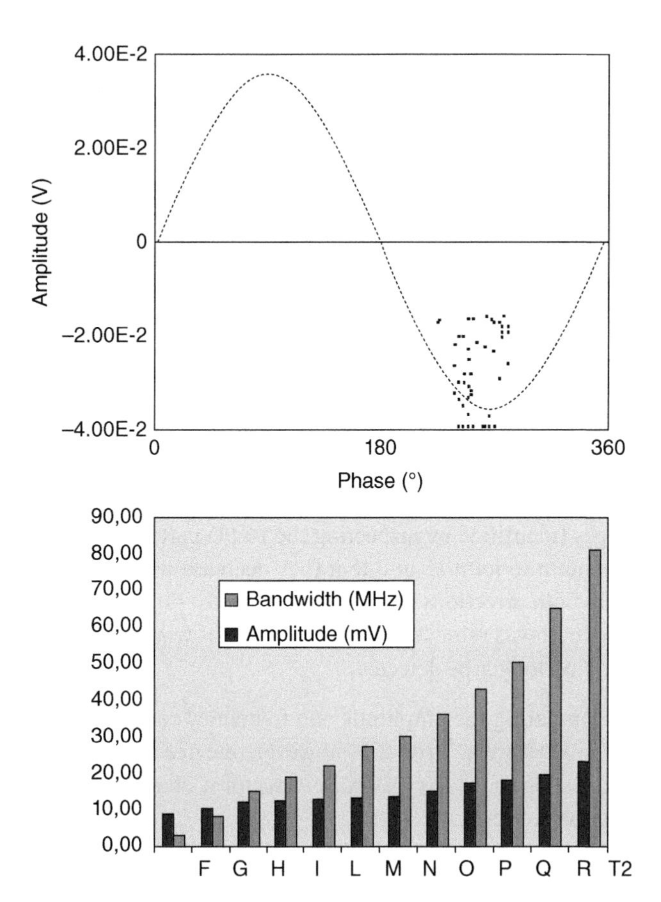

Figure 12.45 AF mapping of discharges on an outdoor insulator located at terminal T2: the amplitude and equivalent bandwidth both decrease as the pulses propagate toward the opposite terminal. *Source:* [44].

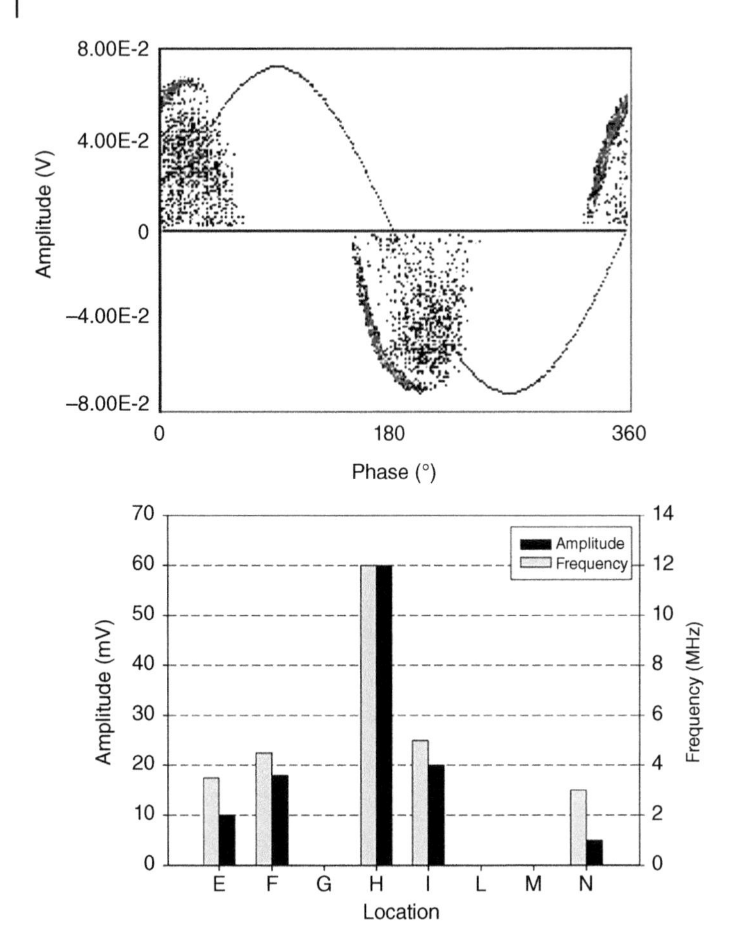

Figure 12.46 AF mapping of discharges within a section of a cable. The amplitude and bandwidth peak at joint bay H. However, due to the presence of the straight joints G (G, L, and M are straight joints) further tests were necessary to identify the PD source. The PD source was eventually located in joint H. *Source:* [44].

quantities decrease while the pulse moves along the line and is detected at the joints R, Q, P, … . Since the PRPD pattern is typical of discharges on the surface of a termination, this phenomenon is not a concern.

- Figure 12.46 shows that internal discharges (identified by inspecting the PRPD pattern) show a maximum of amplitude and frequency content in joint H, and that they decrease while the PD pulses are detected at different positions in both directions of the cable. The most probable location is, therefore, joint H. In this case, the joint was removed and burning traces were observed. After replacement, the phenomenon could no longer be detected.

Summarizing, for a given discharge event, comparing the amplitude and frequency content along the length of the cable can help to localize the PD source. To do this, however, one needs to be able to separate the effects of the different sources based on the PD pulse waveform characteristics. Also, multiple sensors are needed along the cable circuit.

References

1 Bartnikas, R. and Srivastava, K.D. (2000). *Power and Communication Cables: Theory and Applications*. IEEE. https://ieeexplore.ieee.org/book/5263919 (accessed 7 February 2022)
2 W. A. Thue. (2017). *Electrical Power Cable Engineering*, 3 CRC Press, 2017. https://www.oreilly.com/library/view/electrical-power-cable/9781439856451/ (accessed 7 September 2022).
3 CIGRE Technical Brochure 627. (2015). *Condition Assessment for Fluid-Filled Insulation in AC Cables.*
4 Mazzanti, G. and Marzinotto, M. (2013). *Extruded Cables for High-Voltage Direct-Current Transmission: Advances in Research and Development*. Wiley.
5 Wald, D. and Smedberg, A. (2008). Evolution of medium voltage cable technology in Europe. *IEEE Electrical Insulation Magazine* 24 (5): 31–35. https://doi.org/10.1109/MEI.2008.4635659.
6 J. C. Chan, M. D. Hartley, and L. J. Hiivala. (1993). Performance characteristics of XLPE versus EPR as insulation for high voltage cables, *IEEE Electrical Insulation Magazine*, vol. 9, no. 3, pp. 8–12, May 1993, doi: https://doi.org/10.1109/57.216782.
7 Belli, S., Perego, G., Bareggi, A. et al. (2010). P-Laser: breakthrough in power cable systems. *2010 IEEE International Symposium on Electrical Insulation*, San Diego, CA (6–9 June 2010), pp. 1–5. https://doi.org/10.1109/ELINSL.2010.5549826.
8 Hasheminezhad, M. and Ildstad, E. (2012). Application of contact analysis on evaluation of breakdown strength and PD inception field strength of solid-solid interfaces. *IEEE Transactions on Dielectrics and Electrical Insulation* 19 (1): 1–7. https://doi.org/10.1109/TDEI.2012.6148496.
9 Kantar, E., Panagiotopoulos, D., and Ildstad, E. (2016). Factors influencing the tangential AC breakdown strength of solid-solid interfaces. *IEEE Transactions on Dielectrics and Electrical Insulation* 23 (3): 1778–1788. https://doi.org/10.1109/TDEI.2016.005744.
10 Letvenuk, M., Beninca, M., Noglik, H., and Sheridan, P. (2017). The use of shear responsive stress control mastics as void fillers in medium voltage cable accessories. *IEEE Electrical Insulation Magazine* 33 (2): 16–23. https://doi.org/10.1109/MEI.2017.7866675.
11 Zheng, Z. and Boggs, S. (2005). Defect tolerance of solid dielectric transmission class cable. *IEEE Electrical Insulation Magazine* 21 (1): 34–41. https://doi.org/10.1109/MEI.2005.1389268.
12 Hvidsten, S., Holmgren, B., Adeen, L., and Wetterstrom, J. (2005). Condition assessment of 12- and 24-kV XLPE cables installed during the 80s. Results from a joint Norwegian/Swedish research project. *IEEE Electrical Insulation Magazine* 21 (6): 17–23. https://doi.org/10.1109/MEI.2005.1541485.
13 Boggs, S., Densley, J., and Kuang, J. (1998). Mechanism for impulse conversion of water trees to electrical trees in XLPE. *IEEE Transactions on Power Delivery* 13 (2): 310–315. https://doi.org/10.1109/61.660895.
14 IEC Std. 60270 (2015). High-voltage test techniques – partial discharge measurements.
15 IEC 60885-3 (2015). Electrical test methods for electric cables – Part 3: test methods for partial discharge measurements on lengths of extruded power cables.
16 IEEE Std. 400.3 (2007). IEEE Guide for Partial Discharge Testing of Shielded Power Cable Systems in a Field Environment. *IEEE Std. 400.3-2006*, pp. 1–44. https://doi.org/10.1109/IEEESTD.2007.305045.
17 IEC 60840 (2020). Power cables with extruded insulation and their accessories for rated voltages above 30 kV (Um = 36 kV) up to 150 kV (Um = 170 kV) – test methods and requirements.
18 IEC 62067 ED3 (2021). Power cables with extruded insulation and their accessories for rated voltages above 150 kV (Um = 170 kV) up to 500 kV (Um = 550 kV) – test methods and requirements.

19 IEEE Std. 400 (2012). IEEE Guide for Field Testing and Evaluation of the Insulation of Shielded Power Cable Systems Rated 5 kV and Above. *IEEE Std. 400-2012 Revis. IEEE Std. 400-2001*, pp. 1–54. https://doi.org/10.1109/IEEESTD.2012.6213052.

20 IEEE Std. 400.2 (2013). IEEE Guide for Field Testing of Shielded Power Cable Systems Using Very Low Frequency (VLF) (less than 1 Hz). *IEEE Std. 400.2-2013*, pp. 1–60. https://doi.org/10.1109/IEEESTD.2013.6517854.

21 IEEE Std. 400.4 (2016). IEEE Guide for Field Testing of Shielded Power Cable Systems Rated 5 kV and Above with Damped Alternating Current (DAC) Voltage. *IEEE Std. 400.4-2015*, pp. 1–62. https://doi.org/10.1109/IEEESTD.2016.7395998.

22 IEEE Std. 400.4 (2006). IEEE Standard for Extruded and Laminated Dielectric Shielded Cable Joints Rated 2.5 kV to 500 kV. *IEEE Std. 404-2012 Revis. IEEE Std 400.4-2006*, pp. 1–46. https://doi.org/10.1109/IEEESTD.2012.6220225.

23 CIGRE (2001). Partial Discharge Detection in Installed HV Extruded Cable Systems. TB 182. 2001.

24 CIGRE (2018). On-site Partial Discharge Assessment of HV and EHV Cable Systems. TB 728. 2018.

25 Guo, J.J. and Boggs, S.A. (2011). High frequency signal propagation in solid dielectric tape shielded power cables. *IEEE Transactions on Power Delivery* 26 (3): 1793–1802. https://doi.org/10.1109/TPWRD.2010.2099134.

26 Giglia, G., Ala, G., Castiglia, V. et al. (2019). Electromagnetic full-wave simulation of partial discharge detection in high voltage AC cables. *2019 IEEE 5th International forum on Research and Technology for Society and Industry (RTSI)*, Florence, Italy (9–12 September 2019), pp. 166–171. https://doi.org/10.1109/RTSI.2019.8895549.

27 Tian, Y., Lewin, P.L., and Davies, A.E. (2002). Comparison of on-line partial discharge detection methods for HV cable joints. *IEEE Transactions on Dielectrics and Electrical Insulation* 9 (4): 604–615. https://doi.org/10.1109/TDEI.2002.1024439.

28 Romano, P., Imburgia, A., and Ala, G. (2019). Partial discharge detection using a spherical electromagnetic sensor. *Sensors* 19 (5): 5. https://doi.org/10.3390/s19051014.

29 Testa, L. and Montanari, G.C. (2010). A sensor and process for detecting an electric pulse caused by a partial discharge. WO2010095086A1, 26 August 2010.

30 Tozzi, M. (2010). Partial discharges in power distribution electrical systems: pulse propagation models and detection optimization. Doctoral thesis. University of Bologna, Bologna. https://amsdottorato.unibo.it/2308/1/Tozzi_Marco_Tesi.pdf (accessed 22 September 2022).

31 Oussalah, N., Zebboudj, Y., and Boggs, S.A. (2007). Partial Discharge Pulse Propagation in Shielded Power Cable and Implications for Detection Sensitivity. *IEEE Electrical Insulation Magazine* 23 (6): 5–10. https://doi.org/10.1109/MEI.2007.4389974.

32 Sayah, A., Oussalah, N., and Boggs, S.A. (2016). Optimization of water terminations for testing of solid dielectric cable. *IEEE Transactions on Dielectrics and Electrical Insulation* 23 (1): 61–69. https://doi.org/10.1109/TDEI.2015.005296.

33 Lauria, S. and Palone, F. (2014). Optimal operation of long inhomogeneous AC cable lines: the Malta–Sicily interconnector. *IEEE Transactions on Power Delivery* 29 (3): 1036–1044. https://doi.org/10.1109/TPWRD.2013.2293054.

34 Belinski, E., Kaltenborn, U., Winter, A., and Melle, T. (2020). Testing of long AC and DC cables with resonant test circuits. *2020 IEEE/PES Transmission and Distribution Conference and Exposition (T D)*, Chicago, IL (12–15 October 2020), pp. 1–5. https://doi.org/10.1109/TD39804.2020.9300032.

35 High Voltage Inc. VLF Series. https://hvinc.com/products/vlf-ac-technology/vlf-series/ (accessed 14 February 2022).

36 IEEE 400.2 (2013). IEEE Guide for Field Testing of Shielded Power Cable Systems Using Very Low Frequency (VLF) (Less Than 1 Hz). *IEEE Std. 400.2-2013*, pp. 1–60. https://doi.org/10.1109/IEEESTD.2013.6517854.

37 Gulski, E., Wester, F.J., Smit, J.J. et al. (2000). Advanced partial discharge diagnostic of MV power cable system using oscillating wave test system. *IEEE Electrical Insulation Magazine* 16 (2): 17–25. https://doi.org/10.1109/57.833657.

38 Montanari, G.C. (2006). Partial discharge measurements: becoming a fundamental tool for quality control and risk assessment of electrical systems? *Conference Record of the 2006 IEEE International Symposium on Electrical Insulation*, Toronto, ON (11–14 June 2006), pp. 281–285. https://doi.org/10.1109/ELINSL.2006.1665312.

39 Cavallini, A., Montanari, G.C., and Puletti, F. (2006). Partial discharge analysis and asset management: experiences on monitoring of power apparatus. *2006 IEEE/PES Transmission Distribution Conference and Exposition: Latin America*, Caracas, Venezuela (15–18 August 2006), pp. 1–6. https://doi.org/10.1109/TDCLA.2006.311544.

40 Mashikian, M.S., Palmieri, F., Bansal, R., and Northrop, R.B. (1992). Location of partial discharges in shielded cables in the presence of high noise. *IEEE Transactions on Electrical Insulation* 27 (1): 37–43. https://doi.org/10.1109/14.123439.

41 Pommerenke, D., Strehl, T., Heinrich, R. et al. (1999). Discrimination between internal PD and other pulses using directional coupling sensors on HV cable systems. *IEEE Transactions on Dielectrics and Electrical Insulation* 6 (6): 814–824. https://doi.org/10.1109/94.822021.

42 Koltunowicz, W. and Plath, R. (2008). Synchronous multi-channel PD measurements. *IEEE Transactions on Dielectrics and Electrical Insulation* 15 (6): 1715–1723. https://doi.org/10.1109/TDEI.2008.4712676.

43 Kreuger, F.H. (1965). *Detection and Location of Discharges*. Chapter 5. American Elsevier Publishing.

44 Cavallini, A., Montanari, G.C., and Puletti, F. (2007). A novel method to locate PD in polymeric cable systems based on amplitude-frequency (AF) map. *IEEE Transactions on Dielectrics and Electrical Insulation* 14 (3): 726–734. https://doi.org/10.1109/TDEI.2007.369537.

45 Serra, S., Cavallini, A., Montanari, G.C., and Pasini, G. (2013). Device and method for locating partial discharges. US Patent US8467982B2, 18 June 2013. https://patents.google.com/patent/US8467982B2/en (accessed 8 September 2022).

46 Mohamed, F.P., Siew, W.H., Soraghan, J.J. et al. (2013). Partial discharge location in power cables using a double ended method based on time triggering with GPS. *IEEE Transactions on Dielectrics and Electrical Insulation* 20 (6): 2212–2221. https://doi.org/10.1109/TDEI.2013.6678872.

47 Steiner, J.P., Reynolds, P.H., and Weeks, W.L. (1992). Estimating the location of partial discharges in cables. *IEEE Transactions on Electrical Insulation* 27 (1): 44–59. https://doi.org/10.1109/14.123440.

13

Gas-Insulated Switchgear (GIS)

13.1 Introduction

Gas-insulated switchgear (GIS) is effectively used in dense urban areas or in other situations (e.g. the interior of hydropower dams, offshore platforms) where high energy-density is required but space for conventional outdoor AIS substations is simply not available. It is also favored for application in areas presenting atmospheric contamination, e.g. windblown sand, salt (ocean spray), or in the vicinity of mining or smelting operations. GIS has been used at AC voltages from 69 to 1200 kV. The high-voltage conductors are housed within grounded tubes containing pressurized gas, most commonly SF_6, as the main high-voltage insulation. As discussed in Section 2.3, the higher the gas pressure, the higher the design electrical stress can be while avoiding breakdown. The use of SF_6 allows a further increase in the design electrical stress, being both highly electronegative and exhibiting strong electron affinity (Section 2.3). The live (high voltage) parts within the GIS enclosure are supported by epoxy insulators, also called spacers. Figure 13.1 shows two examples of GIS substations, and Figure 13.2 shows the main components of a GIS substation.

This chapter will concentrate on the particulars of partial discharge phenomena in GIS and GIL (gas-insulated lines, a subset of GIS), with emphasis on the typical PD sources/defects found in GIS, the risk they present, and their PD signal behavior.

Since a large portion of PD detection and especially PD monitoring is today done using radio-frequency (RF) techniques, commonly and historically [1] referred to as "the UHF method," the RF signal environment presented by GIS will be covered along with typical signal processing techniques employed when using these methods. This chapter also touches on acoustic techniques, mainly used for locating PD sources, online PD monitoring, and the special considerations associated with HVDC GIS (see also Chapter 17).

13.2 Relevant Standards and Technical Guidance

The following international standards and technical specifications are relevant to GIS and GIL:

- IEC 60270 [2] is the main PD test method used for factory testing of GIS components.
- IEC TS 62478 [3], first published in 2016 as a Technical Specification, is widely used for offline/onsite and online PD testing in GIS and GIL.
- IEC 62271-203 [4] is the defining IEC standard for, as stated in its title, "gas-insulated metal-enclosed switchgear" and as such, is indispensable reading. It defines and specifies all aspects of

Practical Partial Discharge Measurement on Electrical Equipment, First Edition. Greg C. Stone, Andrea Cavallini, Glenn Behrmann, and Claudio Angelo Serafino.

Figure 13.1 Some examples of GIS. (with permission: Hitachi Energy (left), Siemens Energy (right)).

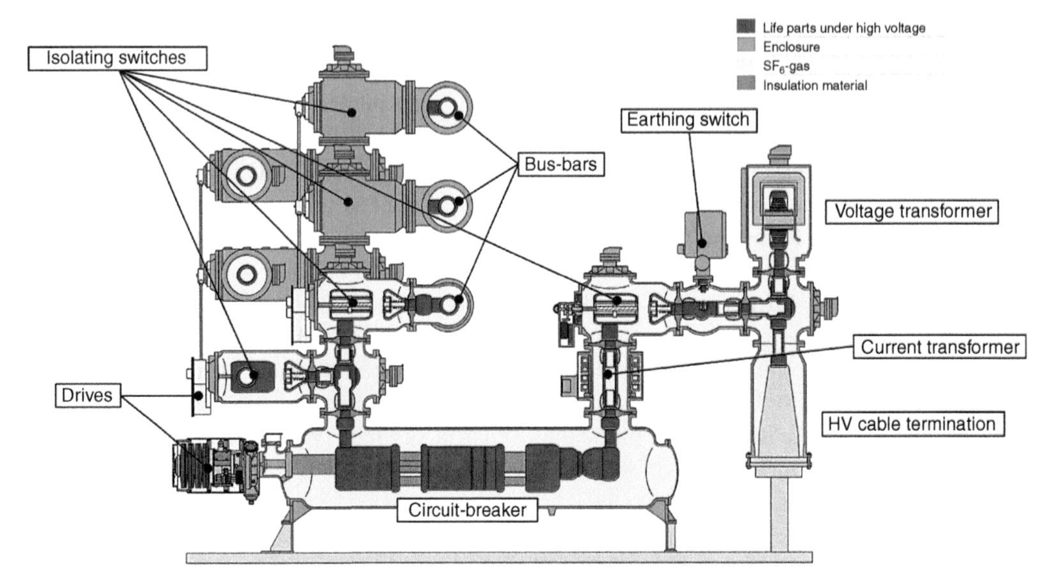

Figure 13.2 Typical components of GIS.

GIS. In terms of PD measurement technique, it references IEC 60270, but it specifically defines the PD test voltages for GIS along with pass/fail criteria. Section 10.2.101, "Tests after installation on site," and Annex C, "Technical and practical considerations of site testing" are both of paramount importance for GIS. IEC 62271-1 is the common specifications for high-voltage switchgear and control gear (of which GIS is a sub-category). As with IEC 62271-203, it is indispensable for anyone working with GIS, although it has relatively little to do directly with PD.

- IEC 60060-1 [5], IEC 60060-2 [6], and IEC 60060-3 [7], the three parts of the standard's title "High-voltage test techniques" govern all aspects of high-voltage testing of devices whose nominal operating voltage is >1 kV. As with IEC 62271-1, there is relatively little to do specifically with PD, but they all reference the horizontal standard IEC 60270.
- IEEE Std C37.122-2010 [8] is essentially a cursory summarization of IEC 62271-203, and in terms of PD testing of GIS, relies on referencing IEEE C37.301. Basic familiarity with it may be helpful in cases where it is specifically referenced (i.e. for projects in North and South America and occasionally parts of the Middle East).

- IEEE Std C37.301-2009 [9], as stated in its table of contents (p. viii) is the "IEEE adoption of IEC 60270:2000," together with two additional annexes. The second of these, Annex J – "PD pattern recognition" – shows a few of what appear to be screen-shots of time-domain plots of PD activity over very short acquisition times, several of which are misleading. This document can only be considered as interesting historical background.
- IEEE Std 1291-1993 [10] is now obsolete (superseded by IEEE Std. C37.122.1-2014) but is highly recommended for appreciating the historical background of PD testing.
- IEEE Std C37.122.1-2014 [11] can be seen as an alternative interpretation of and companion to IEC 62271-203. In section 4.12.2.2, "Electrical methods," onsite testing of GIS using the UHF method is briefly discussed, and reference is made to CIGRE TBs 444 and 525 for further information.
- IEEE SA P454, "Guide for the Detection, Measurement and Interpretation of Partial Discharges," is in preparation by the working group at the time of this writing (chaired by one of the authors). It is foreseen to essentially be an expansion and update to CIGRE TB 336 [12].
- CIGRE TB 654 [13], the 2016 update and expansion of CIGRE TF 15/33.03.05 [1], both of which were formulated to deal with the "calibration problem," which became clear in the early 1990s as RF techniques began being applied to detection of PD, especially on site. Both documents are very informative, explaining the background and procedure for carrying out the so-called CIGRE Sensitivity Verification, the main point being that there is no direct relationship between the received signal amplitude and charge (pC) when applying RF techniques (i.e. the "UHF method"). Both documents were published as "recommendations," but over the past two decades, they have assumed the status of consensus quasi-standards as a result of being required in countless specifications for GIS and GIS PD monitoring systems.
- CIGRE TB 444 [14] covers RF PD detection, re-stating that charge calibration of these methods is physically impossible (and explains why), and also that acoustic and UHF PD methodology continues to evolve (see comments regarding IEC 62478 above). It covers the state-of-the art regarding UHF PD sensors, signal propagation, measurement techniques, and includes sections on UHF PD measurement on HV transformers and VHF measurement on cables. Each chapter ends with a long list of references.
- CIGRE TB 502 [15] is a wide-ranging summary of state-of-the-art of onsite HV test methods including test procedures when accompanied by PD measurement (coping with electromagnetic interference [EMI], etc.), HV sources for onsite measurement (resonant test-sets), connection to the EUT, filtering, grounding, etc. It covers EMC for conventional PD test methods, conventional vs nonconventional methods for onsite PD detection, including acoustic, RF techniques, noise and interference reduction, PD type identification, and localization. There are separate sections dealing with GIS, HV cable, HV transformers, and rotating machines along with an excellent conclusion and an extensive reference list. Strongly recommended.
- CIGRE TB 662 [16] is sort of a follow-on to TB 444 and covers GIS, HV power cable, and HV transformers. The GIS chapter includes typical PD sources in GIS, conventional (IEC 60270) PD measurement vs nonconventional PD measurement using acoustic and RF methods. Several case studies outline onsite diagnostics of signals, which produced alarms on PD monitoring systems. Recommended as a companion to this book.
- CIGRE TB 525 [17] is an excellent, comprehensive discussion of GIS defect properties, detection, location, and identification of PD, risk assessment of typical GIS PD defects, and guidance for assessing failure probability. Highly recommended, essential reading, especially for those dealing with onsite PD measurements or online PD monitoring systems. The work of D1.03 Task Force 09 was summarized in [18].

13.3 The GIS Insulation System

Only very cursory coverage will be devoted to the details of GIS, just enough to present the basic concepts of the technology. Note that most of the material in this chapter is mainly based on single-conductor – single-phase – GIS, unless otherwise specifically mentioned. However, three-phase encapsulated GIS is made and is widely employed at the so-called sub-transmission level (usually meaning below 200 kV). Electrically, three-phase encapsulated GIS is not much different from single-phase GIS, but it is far more complex mechanically (think of an isolating switch or circuit breaker operating on three phases simultaneously and enclosed within the same compartment). It is also more complex from the standpoint of PD, because there can also be defects between phases. Figure 13.3 shows a photo of a three-phase insulator. The set of typical defects in three-phase GIS is the same as shown in Figure 13.6, but the PRPD patterns, depending on precise phase synchronization of the local electric field, are shifted in phase. Much more detail of the GIS is in references [20, 21].

13.3.1 Insulation System Components

GIS, defined by IEC as metal-enclosed switchgear, is generally in the form of a cylindrical outer enclosure at ground/earth potential surrounding inner conductor(s) – the live parts – at nominal system voltage (tens to hundreds of kilovolts) carrying currents up to several kA. A gas-insulated substation (also called "GIS") is built up of many such cylindrical modules containing the usual GIS building block components (Figure 13.2) such as:

- circuit breakers
- disconnector switches (isolating switches not suitable for switching under load)

Figure 13.3 Examples of epoxy insulating spacers in GIS; one for three-phase GIS is shown at lower right. *Source:* top [71], bottom left [34], center, Hitachi Energy, right, Blue GIS Siemens Energy, with permission.

- grounding (or earthing) switches (which can be slow-acting or fast-acting)
- voltage and current transformers (utilized for both metering plus control and protection)
- various bends (90° elbows), Tee- and X-junctions to allow interconnections
- HV cable-end terminations (containing a gas-cable adaptor bushing)
- modules containing flanges for, e.g., earthing switches
- surge arrestors (for insulation coordination and protection of e.g. transformers, HV cable)
- simple straight sections for bus-bars or feeder connections

In general the live, high-voltage, current-carrying inner conductors are supported by cast epoxy insulators almost always containing fillers (e.g. Al_2O_3, SiO_2, and so on) at each end of the compartment or module. The force or torque required to operate switching mechanisms must be transmitted from outside the enclosure (where electrical or manual drives are located) to the interior via insulating operating rods or shafts. Since the insulating gas is supposedly delivered humidity- and contaminant-free, in terms of dielectric integrity, it is the intrinsic quality of these solid insulating materials that essentially determines the operating lifetime of the GIS. A typical value for the withstand strength of SF_6 gas at 0.1 MPa (i.e. normal atmospheric pressure) would be ~88 kV/cm (compared to ~24.4 kV/cm for normal atmospheric air at 0.1 MPa), while a typical withstand value for the epoxy insulators would be about 150–200 kV/cm. These values are only for giving an impression of the relative dielectric withstand strengths; actual design values depend on a complex combination of physical material and design parameters. PD activity is one cause of deterioration and eventual failure for the epoxy insulators and operating rods, but at the same time, the signals produced by PD defects in solid insulation are often of relatively low magnitude. This necessitates that subsequent onsite PD measurements to assess the condition of the GIS must be made with high sensitivity to enable detection of insulation defects. This is especially true for particles/contaminants lying on insulator surfaces, both of which produce very low-level signal emissions.

Each cylindrical module has thick flanges at each end containing polished metal surfaces for O-ring seals that assure gas tightness, fixed with many bolts around their circumference. GIS compartments typically have an insulating disk at least at one end. The insulator disks are either a closed "barrier" disk insulator (thus defining the boundary of a single gas-compartment) or an open "support" insulator (which allows circulation of the gas within a larger gas compartment section). See Figure 13.3. These between-module insulators (also called spacers) are today almost always made of cast epoxy combined with various fillers (which are OEM proprietary) and also include an inner conductor (or conductors, in the case of three-phase encapsulated GIS), which allows attaching contact assemblies on both sides, which in turn interconnect the inner conductors of the two compartments being connected to one another. Each manufacturer has proprietary designs for coupling together the combination of compartment-end insulators and inner conductor.

On site, the GIS modules are hoisted or lifted carefully into position, undergo a thorough (but quick) final inspection and cleaning, and then bolted together. This onsite assembly stage presents the most opportunity for introducing contaminants (conducting/metallic particles, dust, dirt, excess silicone sealing grease, excess adhesive applied to threaded fasteners such as Loctite™, cleaning cloths, even tools have been found in GIS), which can then lead to failure during the onsite HV acceptance tests or cause a flashover later in operation. While it looks at first glance to be simple assembly work, the individual GIS modules must be thoroughly cleaned and close to perfectly aligned before being coupled together. The people doing the work (either on new GIS or during repairs) must be well-trained, experienced, and very conscientious. It is very definitely highly skilled labor and should never be rushed.

For much more detail and fundamental information on the design of GIS and GIL, the interested reader is referred to [20, 21].

13.3.2 PD Suppression Coatings

As in all cases of high-voltage (high electric field) insulation systems between metallic surfaces, it is also critical in GIS to control surface roughness and irregularities. Attempting this by polishing the opposing metallic surfaces – the external surfaces of the inner conductors and the inner surface of the enclosure – to a "mirror finish" is inappropriate, both from the standpoints of high-voltage physics (because such a surface magnifies the presence of any remaining irregularities or contaminants) and what is practicable in industrial production (imagine achieving such surfaces inside a 10 m long bus-duct enclosure). Instead, each manufacturer settles on a realistically achievable surface quality (roughness) and then adjusts their design criteria to match it (including a safety margin). One way to achieve higher surface flatness and homogeneity without resorting to excessive mechanical polishing is to use various coatings; these can be designed to mitigate problems with moving particles (e.g. by preventing them charging up so as to lift off and become mobile) and prevent field enhancement by the occasional small surface protrusion. As with rotating machines and HV power cable, the GIS OEM may sometimes apply insulating or slightly conductive ("semi-conductive") coatings to certain metallic parts of the GIS (conductors, shields) to favorably enhance or "steer" the dielectric properties, depending on local factors (Figure 13.4). Such specially designed coating solutions have lately attained high importance, especially in DC GIS owing to the special conditions and stresses imposed on insulation surfaces under HVDC (see Section 13.13).

13.3.3 Insulating Gas (SF_6 and Alternative Gases)

Sulfur hexafluoride (SF_6) was discovered by Henri Moissan and Paul Lebeau in 1901 and first used as gas insulation in circuit breakers in the late 1950s, though it was patented for such use in 1938. It is both highly electronegative and has very high electron affinity, meaning its electrons are very tightly bound to its quite symmetrical molecule; these properties make it an excellent electrical

Figure 13.4 Photo showing insulating paint on GIS enclosure. *Source:* Hitachi.

insulating medium, with a dielectric strength approximately 2.5 times that of air, and this increases linearly with pressures above 1 bar. It also exhibits quite high thermal conductivity, higher than normal atmospheric air, closer to hydrogen or helium, and also has a high specific heat (and thus it is good for cooling). It is also very dense, among the most dense gases, readily flowing (streaming) downhill to rapidly displace normal atmospheric air (and thus oxygen) in lower-lying spaces, thus posing the safety hazard of drowning in case persons are working in such an enclosed area. For this reason, factories and substations are usually equipped with SF_6 gas alarms to indicate large concentrations of the gas and warn personnel to stay out of low-lying rooms.

These properties all combine to make SF_6 an outstanding insulation medium, especially excelling in the extraordinarily demanding conditions that occur when a high-voltage circuit breaker operating under short-circuit (tens of kA) loads; its outstanding thermal and electrical properties allow it to quench the resulting arc reliably and efficiently and without leading to follow-on damage to the contacts. It is generally nonpoisonous except in the case of heavy electrical arcing and in the presence of moisture, in which case breakdown products can be produced, which are both hazardous to people and materials. Strict adherence to the proper protocols and procedures following such breakdowns eliminates most of the risk to service personnel, but great care and attention must be taken before opening switchgear following a major fault.

The major drawback of SF_6, which has (rightfully) gained very high notoriety in recent years, is that it is about 23,500 times more potent as a greenhouse gas than CO_2 by weight, with an atmospheric lifetime over 3000 years. This has led to very strict regulations regarding its use, especially regarding limits on leakage, which are fast-approaching absolute physical limits. Because of its fundamental importance in the electrical power industry, this has set off a flurry of R&D work over the past decade to come up with so-called alternative gases of which now there are several, e.g. GE Grid's g^3 ("gee-cubed") and Siemens' "Blue GIS."

Much more detailed information regarding both SF_6 and the new alternative gases is available in the technical literature [22–24] and from the GIS OEMs, but one characteristic in particular will be noted here. Because of its very high electron affinity, when electrons are broken free from the SF_6 molecule, the resulting risetimes of the PD pulse are extremely short. In 1992, the risetime of a PD pulse in SF_6 was considered to be around 1 ns, but in [25] from 2012 it is estimated to be around 24 ps, and the authors admitted that they may still have been measuring the limits of their equipment rather than actually measuring the true risetime. Recently, further work has been done [26], which indicates the risetime in SF_6 may be even shorter.

In terms of PD detection using RF techniques (Section 13.8), plugging these values into the standard equations means these risetimes transform to frequency content well in excess of 15 GHz, which is well past the 3 GHz upper limit of the UHF band. In the 1980s, researchers such as Steve Boggs and Brian Hampton [27–29] (see Section 1.5.3) began using RF spectrum analyzers and the fastest available oscilloscopes of the day to observe the RF signals produced by PD in SF_6 and immediately realized its strong RF emission efficiency at frequencies above 1 GHz. This led to the development of the so-called UHF method for detection, measurement, and monitoring of PD in GIS, which is explored in further detail below.

13.4 Typical PD Sources in GIS and their Failure Modes

Figure 13.5 shows that over 50% of dielectric failures in service have a high likelihood of producing PD signals – the rationale behind both sensitive PD measurements during GIS factory and onsite acceptance tests and later for PD monitoring. Figure 13.6 shows a diagram illustrating typical

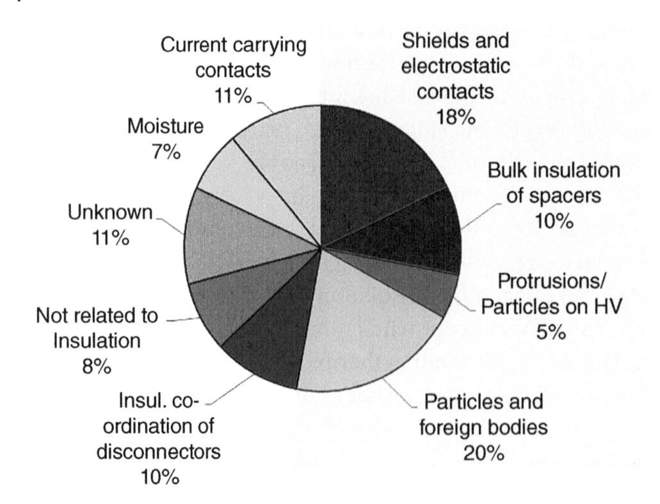

Figure 13.5 Proportion of failures in GIS/GIL due to dielectric defects (blue shaded). *Source:* [18].

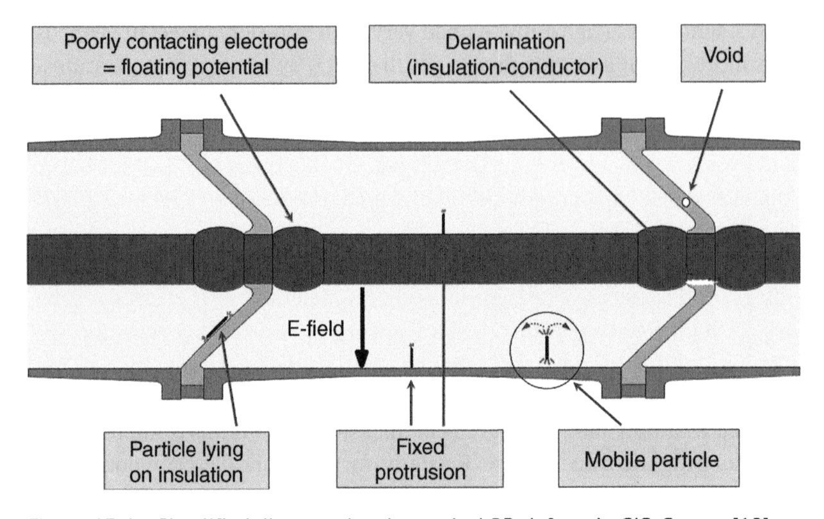

Figure 13.6 Simplified diagram showing typical PD defects in GIS. *Source:* [19].

examples of PD defects in GIS. This section will summarize their characteristics in terms of signal amplitude and behavior along with the risk they pose to operation.

13.4.1 Mobile Particles

Mobile particle, also known as "hopping," "dancing," or "moving" particles, consists of small (<5 mm) pieces of metallic debris, a typical source being shards or slivers of soft aluminum chiseled by the much harder threads of the steel bolts used to bolt the housings together at the housing flanges, although there are many other sources of such metallic/conducting particles. Photos of a couple of typical particles are shown in Figure 13.7. Every effort is made by GIS OEMs to achieve the highest possible level of cleanliness when assembling the GIS on site. Even so, small metallic particles can find their way into the interior of the GIS, either introduced during the onsite work or having remained hidden (i.e. in a low-field region or on the surface of an HV electrode) during

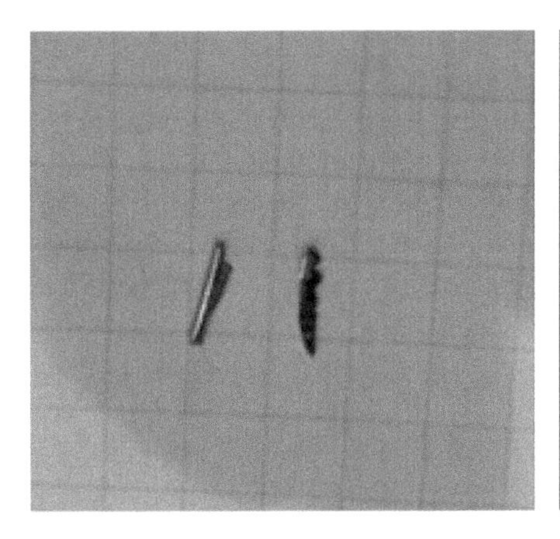

Figure 13.7 Photos showing some examples of particles found in GIS.

the factory acceptance tests, but which get shaken loose during transport and installation. Another source of conductive particles appearing on site is when pieces of defective coating (e.g. on switching contact surfaces or even insulating paint flakes containing traces of the aluminum surface to which they were applied) fall off and begin producing PD.

Mobile particles are considered one of the defects that pose the highest risk of flashover in GIS because they are mobile and thus unpredictable; as a result, there is an extensive body of technical literature devoted to them [30–36]. Larger particles (e.g. 10 mm or more) will have higher momentum, and assuming they achieve lift-off, may "fly" upwards into the gas volume between the enclosure and the high-voltage center conductor, bridge the gas insulation at just the opportune moment, leading to a so-called crossing flashover (also known as a "cleaning" flashover if it takes place during a high-voltage test and no other damage occurs). A more insidious risk is that they bounce onto and attach themselves to the surface of an insulator, where over time the sharp edges of the metallic shard can lead to surface discharges and carbonized tracking on the insulator surface. In this scenario, the formerly mobile particle has now become a "particle lying on insulation," which is a special case of its own covered below.

They are called hopping particles because they typically appear to bounce up and down on the bottom of the inside of the GIS housing under the influence of the AC electric field. They can also lie on the enclosure and slide or tumble about irregularly, as is the case if the particle has a high enough mass and/or finds itself in a low-field region where the field strength is insufficient to achieve lifting force to get the particle up and dancing. As the particle charges up in the electric field and bounces up and down, each time it makes and breaks contact with the enclosure, sparking takes place at that contact point – this is the PD pulse.

The movement metallic particles undergo in AC GIS typically falls into three modes, either:

- lying still ("dormant") on the enclosure,
- sporadically tumbling or sliding about, also termed "shuffling,"
- or freely "hopping"/"dancing" – moving about.

The signals produced by particles in these three modes are very different (and also depend on whether the enclosure inner diameter is coated or not) and thus their associated PRPD patterns (Section 13.12.1) are as well. In an actual operating GIS (e.g. a running substation), a particle will usually exhibit one of these PRPD patterns; although it is not unknown that it can exhibit the other modes, of motion, it is rare. In the laboratory it is possible to build a set-up and with the right particle, obtain all three PRPD patterns, although this requires very precise voltage control and also the particle has to have the right shape in order to remain stable in the three different modes. Although it is possible to have a particle "levitate" in AC GIS, this is very rare in real-world practice, normally requiring very carefully designed laboratory set-ups and conditions.

Depending on their location, the magnitude of the imposed electric field, their weight, material, geometric shape, and even the sharpness of their edges, particles in GIS typically produce anywhere from 2 to 20 pC apparent (IEC 60270) charge. In the RF (UHF) spectrum, moving particles are typically quite efficient RF radiators, so that even "sub 5 pC" particles will be visible above the noise floor of the measurement instruments (spectrum analyzer or broadband systems, Section 13.8).

Particle motion and the signal they produce are significantly different in the case that a dielectric coating is applied to the enclosure's surface (see Figure 13.4 and Section 13.12.1).

Finally, particle behavior in HVDC GIS is more complex than in the AC case, less predictable, and strongly polarity dependent as well (i.e. whether the inner conductor is at positive or negative polarity with respect to the grounded enclosure); see Section 13.13.

13.4.2 Floating Potential Discharge

Floating potential PD occurs when a conducting/metallic object is in a high AC electric field, but is (usually mistakenly or unintentionally) not galvanically connected (bonded) to either the grounded enclosure or (one of the) high-voltage inner conductor(s). One typical example is when a metallic object is laying on the inner surface of the GIS enclosure but is slightly insulated from its potential, due to oxidation, surface coating, insulating contaminants, etc. Another example occurs when too much adhesive (e.g. Loctite) has been applied to the threaded fasteners used to fix a shield-electrode to the high-voltage conductor – the shield electrode ends up insulated from the high-voltage conductor. Many variations on this theme are possible. In all cases, the isolated floating metallic object charges up capacitively in the AC electric field (see Figure 13.8). When the potential difference between the floating object and the neighboring conductor (either the grounded GIS enclosure or a high-voltage conductor) is high enough to exceed the withstand strength of the insulation medium (the Loctite in the latter example above, the oxide or coating in the previous example, or the insulating gas itself), a sparkover occurs. For a brief moment, the floating object and its nearby conductor are at the same potential – and then the floating object

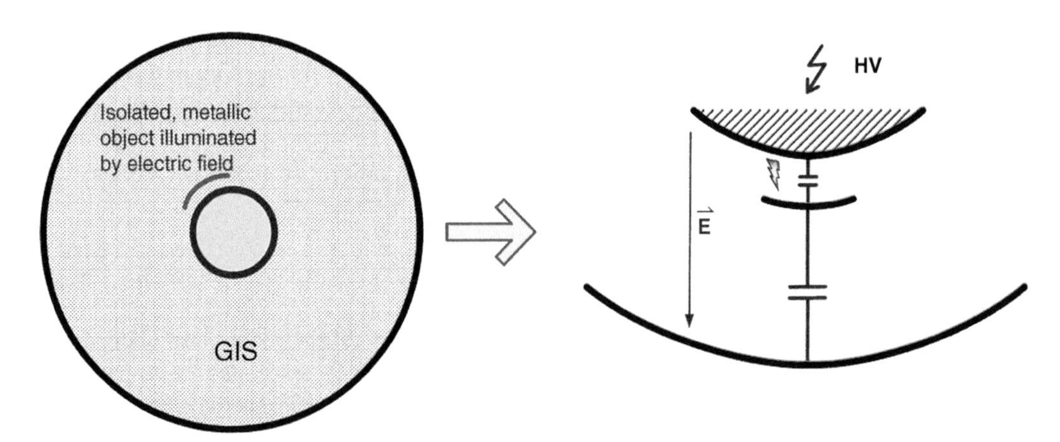

Figure 13.8 Conceptual diagram of a floating potential discharge.

begins charging up in the AC field again. The actual sparks – "micro-flashovers" or "micro-sparking" – are often over a distance of tens or perhaps hundreds of micrometers, but because the floating object is metallic, and often macroscopically large, the PD signal amplitude is quite high. In terms of apparent (IEC 60270) charge, floating potential PD can easily reach hundreds of pC, and perhaps a nano Coulomb (nC) or more. In the RF/UHF domain, it can produce very high-amplitude RF (UHF) signals (so high in fact as to sometimes overload the measurement systems), since the metallic objects involved often function very well as "transmitting antennas." Pulse rates for floating-potential PD are determined by the ("microscopic") conditions of the surfaces where the PD is taking place, but are often in the kHz range (e.g. very high intensity).

Floating potential PD is usually considered harmless for operation in the short term (as opposed to the random risk presented by moving particles as described above), but over the long term, the constant micro-sparking produces (typically) aluminum "nano-dust," consisting of nanometer-sized spherical particles. This is because each "micro-spark" essentially blows a crater into the opposite metallic surface; like a meteorite striking the surface of a planet, debris is thrown off in the form of tiny particles of molten aluminum. Because the SF_6 gas is so dense and the "nano-dust" particles are so small, they can remain suspended for long periods and drift around inside the GIS. They often end up settling on otherwise pristine insulator surfaces, where they form a dull-gray metallic coating. . .obviously not a good thing for the surface of a dielectric. With time, the surface area of the insulator in question sees a high enough concentration of this conductive contamination, and a flashover can occur – usually triggered either by a nearby lightning strike or the high-voltage transients, which routinely occur during switching operations.

13.4.3 Protrusions (Inner Conductor or Enclosure)

Conductive protrusions, either on the high-voltage inner conductor or on the grounded enclosure, may initiate a flashover. Conductive protrusions can be modeled as a point-plane (or needle-plane) geometry (Section 2.1.3.4 and Figure 13.9). The protrusions usually occur during manufacturing or maintenance, where the aluminum conductors are impacted by an object such as a tool, or there has been inadequate surface preparation. Because the electric field is higher near the inner conductor(s), protrusions there tend to produce PD more readily than those on the enclosure, although they may also remain undetected during the one-minute AC test voltage because of "corona stabilization" [37, 38]. Such a protrusion can flash over at 30% of the lightning impulse withstand voltage (LIWV) [17]. Protrusions on the enclosure may only reveal themselves during the 60 second AC withstand voltage test during which time it is often the case that the PD instrument output is not being recorded. This is a common mistake of false economy – if evidence of a protrusion on the GIS enclosure shows up during the routine HV test in the factory, it should be located and removed/eliminated.

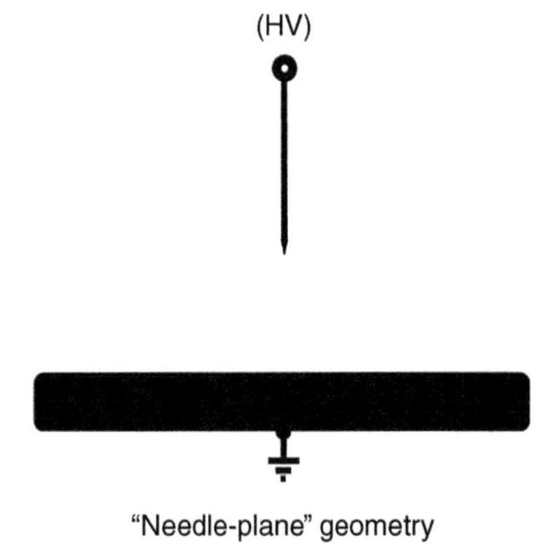

Figure 13.9 Conceptual diagram of a protrusion in GIS as a "needle-plane" geometry.

When the onsite HV acceptance tests are carried out, the voltages are slightly lower than during the factory test, and typically there is even more pressure to complete the test quickly. Defects like hopping particles or floating potential PD are normally quite easy to detect at these voltages, while particles lying on an insulator or small protrusions on the HV inner conductor (i.e. a live shield electrode) may only be detected with very high PD sensitivity at higher test voltage. The scenario posed by protrusions is that large ones (several millimeters long) are usually found immediately in the factory and never make it to site, while smaller ones (<2 mm) produce only very low-level signals. For example, a 2-mm long protrusion in 400 kV GIS may only produce 2–3 pC apparent charge and very low signal levels in the RF/UHF spectrum. PD pulse repetition rates for protrusions are generally 1–2 pulses per cycle of the AC power frequency. These low signal levels make them much less likely to be detected, but therein lies their inherent problem: the risk posed by protrusions is when high-level transient fields occur in the GIS, either through switching operations or lightning strikes. Since there is no time for the corona stabilization effect, which occurs during the relatively slow-changing AC field, they can lead to an instantaneous flashover without prior warning [37, 38]. Because of this combination of low signal level output but with a risk of flashover with little or no warning, it has always been emphasized that onsite UHF-method measurements be made with very high sensitivity, especially during onsite acceptance tests.

Protrusions should be prevented by careful manufacture of GIS components, and as stated above, protrusions are hopefully found using sensitive PD testing carried out in parallel with the routine factory test. They should never make it out into the substation. However, it is possible, though relatively rare, that during site assembly or maintenance either a metallic shard somehow attaches itself to a conductor for example by collision with another hard object such as a wrench or other tool, which raises a burr. Again, very careful attention during assembly should prevent such occurrences happening, but accidents happen, once more underlining the importance of a very high sensitivity PD measurement during the onsite acceptance tests.

13.4.4 Partial Discharges on Insulator Surfaces

Surface discharge in GIS typically occurs at the interface between solid insulation and the insulating gas as a result of dirt or contamination deposited on the surface of the insulating component (e.g. GIS disk spacers or an insulated drive rod), typically during onsite assembly of the GIS. Other examples include (but are not limited to) interfaces between the layers of the coaxial insulators comprising field-grading cable termination bushings (Figure 13.10), faulty composite insulators where the bonding between the fabric (e.g. polyester, Dacron, and so on), and filling resin (typically epoxy) is failing or contains contamination (e.g. metal filings from the fabric weaving machines), or similarly, when windings (or guard electrodes) in cast-resin voltage transformers separate from the insulation (the resulting PRPD patterns (Section 13.12) will often show variations ranging between “pure surface PD” and “delamination”).

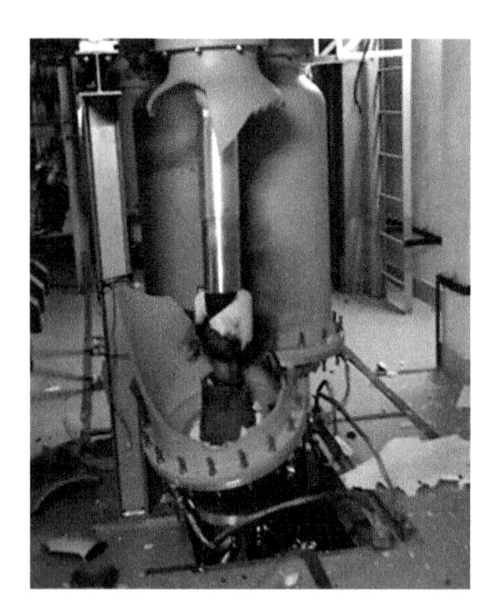

Figure 13.10 Photo showing damage following cable-end termination failures. *Source:* Dr. N. Zeller.

The key point regarding surface discharge PD is that the HV designer normally tries to avoid geometries, which result in a large component of the electric field running tangentially along an insulation–insulation interface, the so-called triple-point problem (Section 2.1.4.3). However, it is often unavoidable, and for this reason the tangential field component is held to the lowest possible minimum during the design process. Any contamination of these

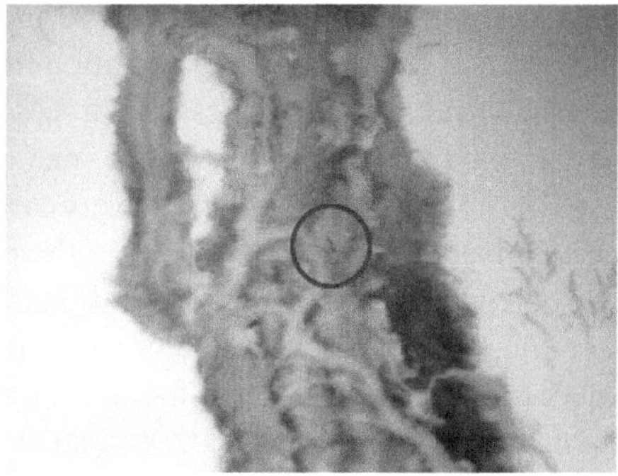

Figure 13.11 Photo showing surface tracking caused by a particle lying on the insulator (indicated by the red circle).

interfaces usually compromises their integrity since the designer assumes they are homogeneous, smooth, and clean. If the surface contamination is conductive (metal filings, "nano-dust" produced by a large floating-potential source) or partly conductive (carbon dust, other nonmetallic pollutants) and covers a large area, surface discharges usually produce quite high signal amplitudes, up to hundreds of pC apparent charge. Somewhat similarly to floating-potential PD, RF signal amplitudes can overload measurement instruments. Surface discharges also typically produce quite high pulse rates (kHz range), the exception being under HVDC where their pulse rates may be very low (see Section 13.13). Some PRPD patterns corresponding to such surface discharges are in Section 13.12.4.

A special case of surface discharge is single, isolated particles lying on an insulation surface, which shares some similarities with small protrusions (Figure 13.11). They typically produce only very low-level signals (1–2 pC apparent charge). Using either charge-based or RF (UHF) PD detection methods, signals from such particles on insulation may only be visible briefly during the 60-second AC high-voltage withstand part of the test. Yet if present, a reduction in the lightning impulse withstand voltage (LIWV) of up to 50% has been reported [39, 40]. Again, similar to the case of protrusions, particles lying on insulation might lead to a flashover in service in the presence of either lightning strikes or very fast transient overvoltages (VFTOs) produced by routine switching operations [17]. Again, this underlines the need for high signal-to-noise ratios (SNRs) when using UHF methods onsite, and, despite ambiguous guidance in the relevant testing standards, it is never a bad idea to observe the PD instruments during the 60-second AC HV withstand part of the test.

13.4.5 Voids, Cracks, Delamination of Solid Insulators

Voids, cracks, and delamination are all defects associated with the solid insulation components used throughout the GIS to support the high-voltage live parts (including the operating rods that connect external drives to switching devices) within the gas-insulated enclosure. The workhorse material from which GIS spacers – the disk or cone-shaped insulators between compartments that hold the current-carrying center conductor in place (Figures 13.3 and 13.6) – are usually cast epoxy combined with various fillers such as AlO_2. The risk of voids occurring is inherent to the casting process (because the epoxy resins employed undergo a reduction in volume as they cure and harden). Much effort and care is put into the manufacturing processes to prevent void formation, because, of course, once present, they remain within the insulator for their lifetime. While GIS is often touted to be extremely reliable, a major part of that reliability is based on very tight quality control of the solid insulation components contained within it.

Cracks or delamination either can be present from the time of manufacture of the insulating component or can form during operation from mechanical or thermal stress. Similarly to preventing voids forming during the casting process, very careful attention must be paid to material purity and handling and every step of the manufacturing process to prevent these defect types from occurring during manufacture, and very careful mechanical and thermal design (and rigorous type testing) must be carried out to prevent them occurring in service.

A somewhat different type of void is caused by delamination. One delamination scenario is when the epoxy in a GIS insulator separates from the metal inner conductor(s), a result of poor surface adhesion coupled with the well-known property that cast epoxy shrinks in volume during the curing process. If the void thus created is filled with atmospheric air and/or epoxy outgassing products (typically at partial pressures), PD will almost always initiate, although even if the delamination is open to the SF_6 gas, PD will also likely initiate because of the conductor electrode's surface being intentionally made rougher to aid adhesion. Once PD starts, it will exhibit both very high magnitude (tens to hundreds pC apparent charge and very high RF/UHF levels, comparable to floating-potential PD signal levels) and pulse rates (low kHz is possible) due to the high availability of electrons from the metallic surface. This behavior is totally different than typical signals from small (~1 mm diameter) voids located somewhere within the bulk of the insulator. Also, because of its high intensity, PD due to insulator–conductor delamination is often audible in the immediate vicinity of the insulator, even without using acoustic instruments. As with a high-intensity floating-potential PD defect, it may be possible to detect breakdown products in the gas over time, and insulators that have undergone such delamination and subsequent PD over time will show obvious damage around the region(s) where the PD was active. However, depending on the insulator design, this defect may not necessarily present substantial risk for operation and may well go undetected for years if no one hears it from outside or makes other onsite PD checks.

A second delamination scenario in GIS is much more dangerous in terms of risk. This is when the coaxial cone-shaped insulation and field-grading layers within cable-end termination bushings become separated from one another. This usually occurs during installation as a result of material or processing problems, poor installation work, contamination, or some combination of those contributing factors (often exacerbated by scheduling pressure). Being a key element in so-called insulation coordination, cable-end termination bushings play a critical role in interfacing the higher electric field design of the HV cable (whose main insulating medium is solid insulation, typically cross-linked polyethylene – Section 12.2.1) with the lower electric stress levels used in the GIS's gas insulation. These termination bushings are extremely complex and dependent on very high levels of material quality and homogeneity, close adherence to manufacturing tolerances, and close attention to cleanliness and detail in installation. If separation between any of the solid layers occurs, this will often result in unexpected high electric fields parallel to the surfaces, leading to the initiation of surface discharges, which in turn usually develop and expand very rapidly, leading to failure within hours or days (Section 12.3.5). The results are both catastrophic (it is not unusual for the cable-end bushing to explode, see Figure 13.10 and very costly to fix.

For this reason, it is important for utilities to carefully monitor for any PD activity appearing in GIS/cable terminations during the week or so following installation of new cables to make sure the bushings are discharge-free. It is typical that such PD activity starts at a low level (about 10–50 pC) but then rapidly increases in both amplitude and pulse repetition rate as the PD source(s) begin producing surface tracking on the interfaces where the PD occurs. If any sign of PD in such a cable-end termination bushing develops, it is much cheaper to de-energize the GIS, perhaps perform some tests, then replace the faulty termination, rather than take the chance of a catastrophic failure.

In all these cases, the void is filled with some mixture of gas, depending on how and when the defect formed and also its location within the insulator. In a narrow crack or void open to the high-pressure insulating gas, PD may never incept. However, if the void is filled with atmospheric air, or some mixture of outgassing products from the early manufacturing process, PD is more likely to initiate, assuming it is located in a region exposed to high electric fields during operation. This is because the dielectric constant of the surrounding insulation material is usually much higher – 4–6 or even higher. As discussed in Section 3.6, since the dielectric constant of the gas is essentially 1, the electric field focuses into the void.

For voids within the bulk solid insulation, the local insulation system consists of a gas-filled void surrounded by an assumed homogeneous insulator. With no conductor nearby, there is no obvious source of charge carriers that could initiate PD. As is frequently observed, PD often does not immediately initiate (Section 3.6.1), especially for small (<2 mm diameter) voids in solid insulation during the relatively short HV test intervals that are typically on the order of only 60 seconds. When PD does occur, the initiating electrons are produced when either cosmic rays or gamma rays produced by terrestrial background radiation ionize gas molecules within the void [41]. This situation in turn leads to insulators containing voids being installed in GIS, which subsequently go into operation, a fact verified when GIS equipped with UHF PD monitoring systems begin to register PD coming from those insulators.

Following research by organizations such as Ontario Hydro [21] and ABB, it was determined that the most effective way to reveal the presence of such small voids in production was to illuminate the insulator with a series of short, high-intensity X-ray pulses. The X-rays are not used to image the voids within the insulator (the required resolution would be prohibitively expensive) but rather to provide "start electrons" by ionizing the gas within any voids present, as explained above. If any voids are present, a "textbook" PRPD pattern corresponding to a void appears immediately following the X-ray pulse (see Section 13.12.5). The technique, known as "PXIPD" – "pulsed X-ray-induced partial discharge" – was implemented on an industrial scale and used in the insulator production facility of some companies [27].

13.5 Detection of PD in GIS

Different methods are used for detection and measurement of PD in GIS depending on its state of assembly. Charge-based methods using IEC 60270 (also called conventional LF or "baseband" methods) are usually used in routine factory testing because they allow calibration in terms of units of apparent charge (Chapter 6). This, in turn, allows testing against acceptance criteria set in terms of a maximum charge threshold (typically 5 pC). IEC 60270 [2] techniques work well for GIS because the individual GIS modules are lumped elements at frequencies less than 1 MHz and are essentially high-Q capacitors.

Following assembly on site, for onsite/offline and online PD testing/monitoring, VHF and UHF methods (Chapter 7) are more typically employed, especially the UHF method that has gained wide acceptance since being pioneered in the 1980s (Section 1.5.3.2). The reason for this is that the very short risetime of PD pulses in SF_6 gas (likewise for other alternative insulating gases under high pressure) transforms to a strong frequency domain signal in the UHF (and higher) spectrum. The metal enclosure and coaxial topology of GIS allow these UHF/RF signals to propagate for long distances with relatively little attenuation and dispersion compared to other types of test objects such as power cables and stator windings (Sections 12.6 and 16.5). The grounded GIS metal enclosure also acts as a fairly effective Faraday cage to shield the PD measurements from external interference. Detection in the UHF range implies the PD sensors are compact and relatively inexpensive (i.e. compared to traditional coupling capacitors) so that they can be installed throughout the substation, increasing the likelihood of detecting harmful PD defects.

13.6 Charge-Based PD Measurement in GIS

As is normal for most high-voltage equipment, GIS is subject to high-voltage AC testing before it leaves the factory, almost always including a PD measurement. Seen from an electrical perspective, individual GIS modules (aka shipping units) are lumped elements, essentially low-loss capacitors. This means they are very well suited to conventional PD testing based on IEC 60270 [2] in the LF range (Chapter 6).

The detection circuit most often used for GIS factory PD testing is the indirect method, using a coupling capacitor in series with a quadripole (Section 6.2.2). Since the capacitance of typical GIS sections is relatively low (hundreds of pF), high PD sensitivities can be achieved with coupling capacitors in the 500–1000 pF range. However, today there is an increasing trend among GIS manufacturers to assemble several GIS modules together, test them, and ship them in one piece; this even includes complete bays. On the one hand, this considerably simplifies onsite assembly of an entire GIS and also reduces the chances that contamination will be introduced (or other damage done inadvertently) as each individual module is installed on site. On the other hand, because the capacitance of such pre-assembled sections is higher, the sensitivity of the PD measurement falls because the ratio of the coupling capacitor to the capacitance of the EUT determines the absolute detection sensitivity (Section 6.4.1.3). Doing the work necessary to reduce the background noise level assumes even more importance for testing such integrated assemblies. Dealing with a low-level PD defect on site that remained undetected in the factory routine test, i.e. locating the PD site, then opening the GIS, removing the defect, and re-closing, will prove to be much more complicated and costly, especially if the completed assembly also has the secondary systems and cabling installed.

High PD sensitivity is of critical importance for factory routine testing of GIS for two main reasons:

- once the GIS module has been assembled and filled with clean, dry gas, it has become a high-value-added object. Plus, re-opening a GIS module always presents the risk of introducing contaminants (especially metallic particles) that may lead to subsequent failure in operation.
- two of the most critical defects – protrusions and particles lying on a spacer – usually produce very low signal amplitude (typically ≤3 pC). As previously explained, both of these defect types are difficult to detect yet present definite risk of flashover in the presence of switching transients or lightning strikes (i.e. to an overhead line or to the GIS bushing itself) [39].

The real determining factor in PD detection is not absolute sensitivity but rather signal-to-noise ratio (SNR) – see Section 9.1. SNR is of fundamental importance in PD detection, especially in light of the fact that PD signals, in the context of other man-made signaling functions, essentially resemble random noise. While we can definitely have a "noise problem" in PD measurement instruments, if, for example, their analog front ends are not well-designed (and therefore present high electronic noise levels), the much more likely limiting factor to achieving high SNR in PD detection is due to background interference, to be precise, EMI arising from outside of the PD test circuit.

With careful attention paid to preventing conducted EMI entering into the test circuit, very high PD sensitivities can be achieved, typically <0.5 pC. Such low background levels are required to assure that the outgoing GIS modules contain no small protrusions or free particles before being delivered to site. Achieving such low background levels is neither simple nor cheap, but it is important to keep in mind that IEC 60270 specifies the minimum reliable PD threshold level as being twice the background level. It is possible and not very difficult to quickly throw together a GIS test bay and achieve a background level equivalent to 1 or 2 pC. However, this typically results in the manufacturer issuing factory test reports saying the delivered GIS module has passed the routine factory test because no PD was observed above 5 pC. There are two problems with this:

- small protrusions and particles, both of which pose a real risk for operation, may be missed.
- it is possible that once the GIS module with a low output-level PD defect is installed on site, the PD signal may be picked up by a nearby UHF sensor, either during onsite acceptance testing or monitoring systems employing the UHF techniques.

It bears repeating that it is impossible to assess PD activity in terms of charge magnitude when employing UHF techniques (Section 7.5). On the other hand, a low-output defect remaining in a shipping unit when it left the factory may show up very clearly in the field when UHF techniques are applied. The conclusion is that routine factory PD testing of GIS should be carried out at the highest sensitivity possible to prevent defects leaving the factory. In the IEC 60270 frequency range (10 kHz to 1 MHz), typical EMI sources are AM radio broadcast signals and electrical noise from machines such as relay contacts in motor drives and brakes on overhead crane systems, variable speed motor drives, brush-noise on motors, switching-mode power supplies, etc. Depending on the exact and particular device and the coupling path, the electrical interference they produce can enter the IEC 60270 base-band test circuit as radiated emissions (through the air) or as conducted interference via the power supply for the high-voltage test-set, and/or even through the factory grounding system.

Given the wavelength at 1 MHz (the upper end of the IEC 60270 spectrum) is 300 m, only test objects of large physical dimensions will enable connections to act as efficient antenna. On the other hand, being in close range of a powerful AM broadcasting station will pose interference problems even with HV connections much shorter than a wavelength.

Although one can install the GIS factory test bay inside of a (very large, very expensive) shielded room (Section 9.3.1), the typical and very effective way to eliminate radiated EMI is to use an encapsulated HV transformer and encapsulated coupling capacitor for the test circuit components. This usually prevents radiated EMI getting into the test circuit because of the inherent shielding effect of the GIS enclosure (Figure 13.12).

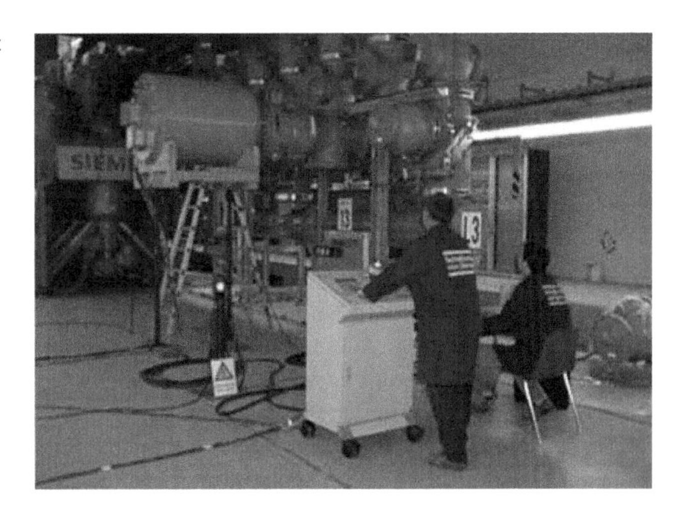

Figure 13.12 Photo showing resonant high-voltage test-set for onsite testing of GIS.

Typically, 90% of interference problems in GIS factory PD tests using the IEC 60270 method are due to conducted interference coming in from the AC test supply and facility grounding systems. Especially at the IEC 60270 frequencies, such conducted EMI can enter the test circuit at surprisingly high levels and can be extremely difficult to eliminate. Moreover, besides the "deterministic" interference sources mentioned above, PD or corona on other nearby electrical systems, while not posing any risk to those systems, can also enter via conductors.

Therefore, once the GIS HV supply, coupling capacitor, blocking impedance, and EUT itself are implemented in an entirely GIS-encapsulated set-up, it is usually necessary to totally isolate the incoming mains power supply with purpose-designed powerline filters (Section 9.3.3). These are typically EMI filters with very low-pass frequencies (ideally low enough to even block the third harmonic at 150/180 Hz), very steep roll-offs, and very high rejection. These filters must, of course, be able to carry the maximum current required for the highest test voltage levels without saturating their internal (high-Q) components. In a typical GIS plant, manufacturing for example 400 kV GIS, these filters will be housed in shielded metal enclosures and weigh several hundred kilograms because of the physical size of the high-current capability inductances (high-Q but exhibiting no self-resonance in the IEC frequency band) and capacitances (voltage and current handling capability) required to achieve sufficient rejection of the unwanted EMI.

Lastly, it may be necessary to somehow separate the IEC 60270 grounding system from the facility (factory) grounding system. This is because all of the systems and devices mentioned above often dump very high levels of EMI into the factory grounding networks, with the result that these EMI signals are often omnipresent throughout the facility. If this is the case, the high-level interference "riding" on the grounding network is conducted into the IEC 60270 measurement circuit in all sorts of insidious ways, making it hard to troubleshoot and get rid of. It is almost always impossible to "filter" the grounding system; separating it – i.e. constructing a totally separate grounding system for the HV test system – is the only way to eliminate the high levels of EMI found in most industrial facilities today. It is beyond the scope of this book to describe in detail what such a separated grounding system entails, but it is paramount that expert electrical engineers, electrical contractors, and the local electrical safety authorities are involved in the design and implementation of such systems to ensure total safety of the operating personnel.

In light of the previous paragraph, a final note regarding EMI filters is that, especially at IEC 60270 frequencies, to achieve their maximum performance (highest rejection of conducted EMI) and effectiveness, the filters must themselves "see" a very low-impedance, low-resistance connection to ground. In addition, the EMI filters are purposely designed to dump the interference they're installed to remove, onto the grounding system. . .and there can be other paths of ingress into the measurement circuits from the grounding system. If the interference suppression methods are insufficient, it may result in GIS modules containing defects getting shipped to site, which later result in expensive in-service failures.

Finally, even with the best available EMI filters and careful design, the PD measurement instruments may still exhibit unacceptable levels of background interference. Eliminating the last vestiges of background interference demands a high level of measurement expertise (both practical and hands-on), appropriate instruments and coupling devices, and last but not least, a very disciplined approach to troubleshooting. In fact, it is not unusual that the most well-engineered and installed systems will present the most intractable EMI troubleshooting problems.

For the reasons mentioned, performing onsite PD testing, e.g. during the onsite HV acceptance tests, is all the more difficult because the interference sources are much more difficult – even impossible – to control, and often more numerous and of higher intensity.

13.6.1 Charge Calibration

Calibration into apparent charge can be done by one of two methods. The first is to connect a typical IEC 60270 calibrator (Section 6.6.4) directly between the GIS inner conductor and the enclosure, for example by means of inserting a moveable conductor to contact the inner conductor to carry the calibrator signal. The second approach, described in IEC 60270, is by capacitively coupling a calibration pulse to the inner conductor (Figure 13.13). In this latter method, the calibration pulse generator is typically connected directly to one plate of the calibration capacitor C_0, often a UHF PD sensor or other similar small metallic plate, electrically isolated from the GIS enclosure and connected to the outside via an isolated RF coaxial feed-through connector. In such cases, the value of C_0 is usually very small ($<<1$ pF) and must be accurately determined, usually by comparison methods (i.e. checking the divider ratio under high voltage), in order for the PD calibration to be reasonably precise. It must also be noted that, due to the much smaller capacitance of C_0 so implemented (as the parasitic capacitance to the GIS inner conductor), this necessitates that the risetime of the charge calibrator used must be much shorter (i.e. $\sim<1$ ns).

It is normal practice for the PD calibrators to be themselves calibrated once per year, together with the test system high-voltage dividers. It is also good practice to check their output periodically with a suitable oscilloscope to confirm they are working properly and delivering their expected output levels. This is especially true in the case that calibration levels or the PD levels appearing during a test are much higher than the expected range.

The other important point, which is often overlooked, is to be sure that the AC phase synchronization input to the PD measurement instrument is correct (Section 8.3.2). That is, the positive and negative going zero crossings are in phase with the variable high-voltage test supply at all voltage levels. This stable, known phase relationship is, of course, critically important for the correct interpretation of phase-resolved partial discharge (PRPD) patterns (Section 8.7.3). At lower voltages, this can be checked with an oscilloscope via a suitable high-voltage probe, then with a suitable high-voltage divider (whose phase behavior is known). At higher voltages, say >100 kV, the most reliable, direct, and easy way to check this is to attach a protrusion to a high-voltage conductor somewhere in the measurement circuit. The high voltage is raised slowly until PD inception is just reached; this will result in a few pulses occurring exactly at 270° on the AC sine-wave (see the top left-hand PRPD pattern shown in Figure 13.47).

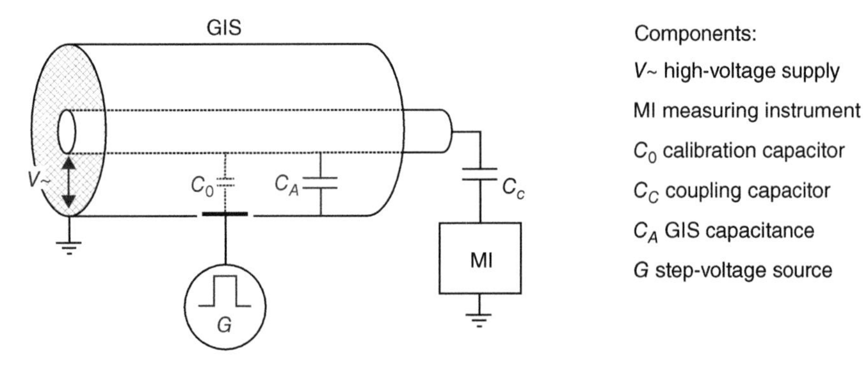

Figure 13.13 Diagram showing the "indirect" IEC 60270 calibration of GIS.

13.7 Application of Acoustic Techniques for PD Measurement on GIS

The electron avalanche making up a PD pulse consists of a super-heated (and thus partially evacuated) channel in the gas; when that channel collapses, an acoustic shockwave results when the surrounding gas rushes back into the channel, a process analogous to how terrestrial lightning produces a thunder clap (Section 5.4). This acoustic shockwave then propagates throughout the apparatus and can be picked up using suitable acoustic detection equipment. The audible crackling caused by corona on overhead high-voltage lines or in the high-voltage laboratory are commonly observed examples. Although there was an awareness of audible sounds accompanying partial discharge for decades, during the 1980s and 1990s, a particularly large concentration of research efforts led to significant advances in the area of acoustic detection of PD; a few of the most significant technical references are listed in the bibliography [42–45].

Today, acoustic PD detection is almost always based on using sensitive "acoustic energy" transducers made of piezoelectric crystals, which resonate at ultrasonic frequencies between 40 and 150 kHz (Section 5.4). Acoustic PD techniques divide themselves into manual methods (Figure 13.14) and instrument-based methods (see also Section 5.4.2). The manual instruments typically take the form of a hand-held acoustic "gun" or "pistol" housing the power supply and electronics with a metal contact rod on the front mechanically coupled to a piezoelectric crystal transducer inside the unit. These devices are very intuitive to use – the operator puts on the headphones, turns the unit on, and places the contact rod firmly against the surface of the GIS. However, "intuitive" does not mean "trivial," it is important to gain experience using these devices on GIS, hopefully in a high-voltage laboratory on an actual GIS with defects intentionally built into them. Depending on the model, quite sophisticated signal-conditioning is available, along with various output modes.

Instrument-based techniques typically consist of an electronic unit to which one or several piezoelectric sensors (Figure 13.15) are connected via coaxial cables and then mechanically attached to the GIS at selected locations. Suitable instruments can perform a wide variety of analysis including phase-resolved displays of the incoming acoustic signals or other displays of the pulses vs time.

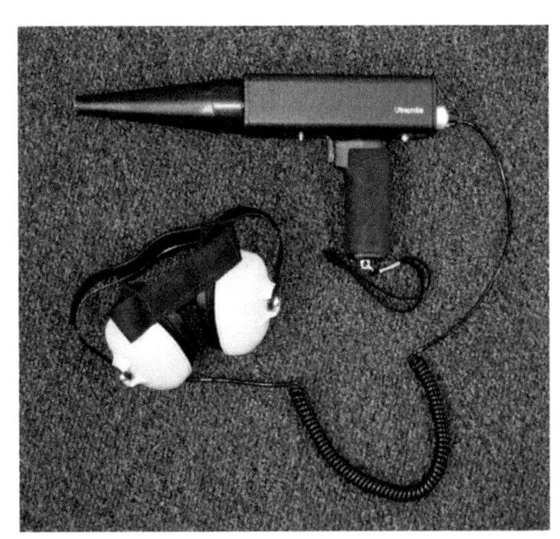

Figure 13.14 Manual acoustic methods: the acoustic "pistol."

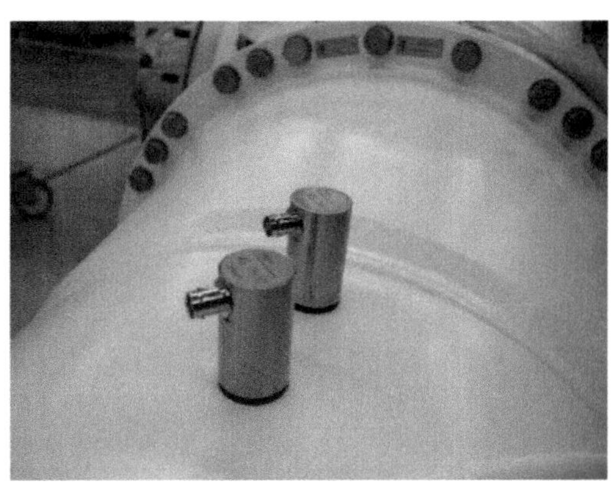

Figure 13.15 Acoustic sensors mounted on GIS.

The typical problem with either manual- or instrument-based acoustic PD detection is that the received signal is highly dependent on the type of PD and its exact location within the GIS.

The latter dependence on location is significant for two reasons, the first being that the acoustic output of the PD activity differs drastically dependent on where it is located, and the second reason is that the received signal level is extremely dependent on the exact transmission path between the PD site and the location of the acoustic sensor.

Table 13.1 lists the velocity of propagation and intrinsic damping (attenuation) of acoustic signals in some of the usual materials found in GIS. The values are not exact and may, of course, vary depending on the exact material composition used. The point is to observe how differently the materials behave. In GIS, on one hand, moving particles lend themselves especially well to measurement with acoustic methods. Indeed, acoustic methods are usually very effective for locating "hopping" particles – because typically the particles are mechanically impacting the enclosure as they fall back down onto it (under the influence of gravity, before being again lifted off under the influence of the AC electric field). In contrast, a small void located near to the center of an epoxy insulator in a 500 kV GIS, or to use an even more extreme example, deep within the field-grading bushing of a cable-end termination, will often prove difficult or impossible to detect due to the high intrinsic damping of these materials. Similarly, low-level acoustic signal emitters such as small protrusions on the inner conductor or conducting particles/contaminants lying on insulator surfaces will also prove difficult or impossible to detect using acoustic methods.

The extreme dependence of the received acoustic signal level on the propagation path between the PD site and the sensor is not hard to imagine given the acoustic signal transmission properties (Table 13.1) together with the complex topology of GIS. While the GIS enclosure, being made of aluminum, presents the acoustic signals with an excellent, low attenuation, high propagation speed medium, the ends of each enclosure, bolted together with an insulator flange, present an

Table 13.1 Acoustic properties of some of the materials used in GIS.

Material	Speed of sound (m/s)	Damping
Air	343	Mild. . .
SF_6	136	Strongly!
Aluminum	6420	Very low
Epoxy	1230	Strongly

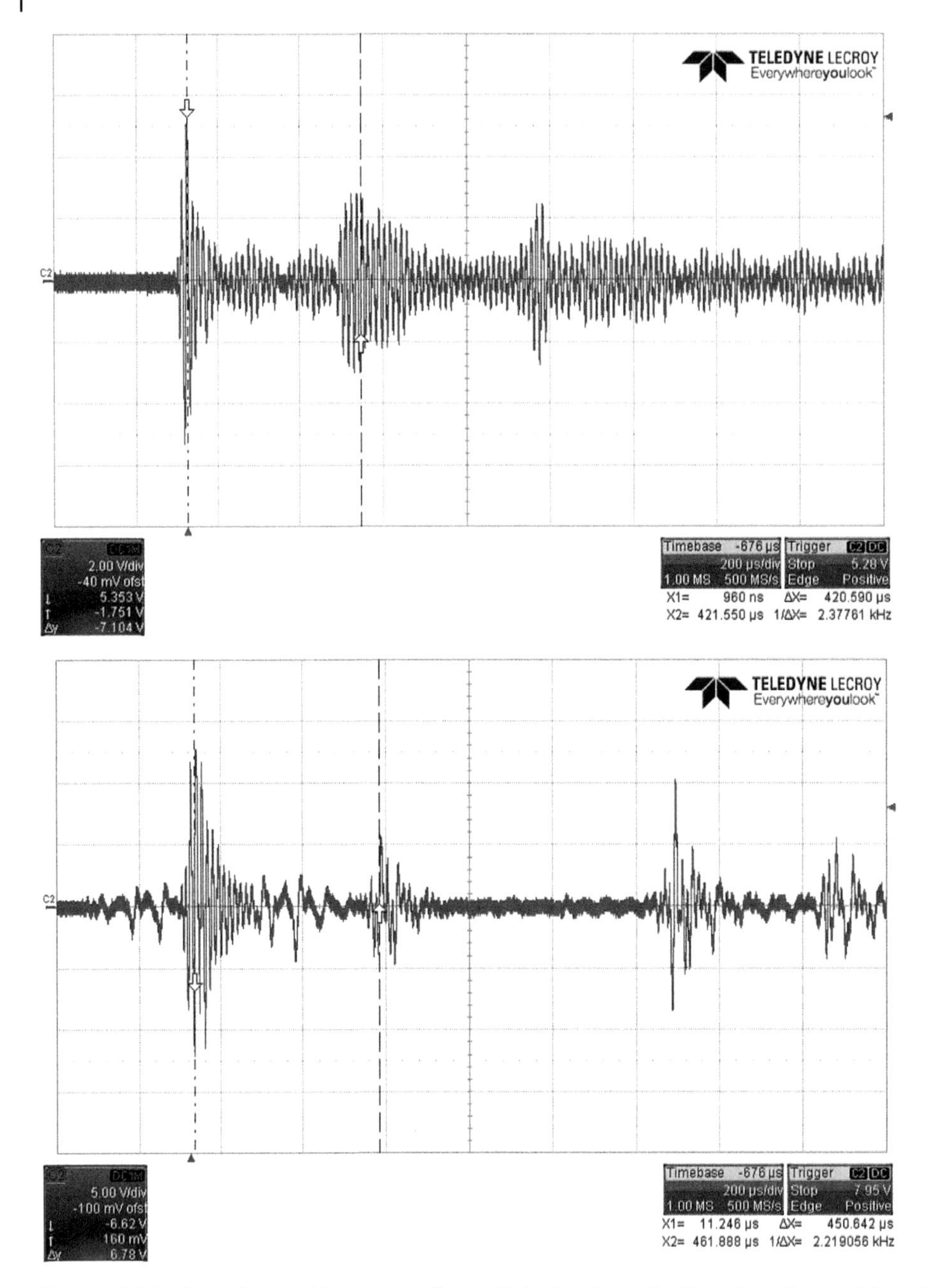

Figure 13.16 Acoustic signal from a hopping particle showing reflections from the module ends (top), signal from tapping the enclosure with a pen (bottom).

abrupt discontinuity at which a large proportion of the acoustic signal will be reflected (Figure 13.16). Besides a particle hopping on the inside surface of a GIS enclosure, the shockwave generated by any other PD signal will end up traversing all of the individual mechanical interfaces between it and the acoustic sensor. Also of note is that one of the fundamental characteristics of high pressure SF_6 gas is that it is very dense (in fact one of its past industrial uses has been as a sound-deadening material in multiple-pane windows).

The factors affecting acoustic detection of PD are very easy to observe. If a particle is set to hopping in a section of GIS, the acoustic signal heard in the headphones of the acoustic "pistol" shown in Figure 13.14 will be clearly audible and identifiable, even for particles on the order of 1 mm or even less. Moving past just one end-flange, to the next enclosure, the drop in amplitude is very pronounced. Even the relatively large particles used for the CIGRE Sensitivity Verification in a 400 kV GIS – aluminum shards ~3–4 mm long – can barely be heard past 3 or 4 enclosure flanges. Staying on the enclosure containing the hopping particle, it is astonishing to observe the wide variation in acoustic signal heard when moving the contact point around the length and circumference of the enclosure. This location dependence is even more pronounced at the neighboring enclosures. This is the crux of acoustic PD detection: *if you can hear* the PD signal somewhere, then the manual acoustic method is usually excellent for narrowing down the location of the defect site. However, often searching for PD types other than hopping particles (or floating potential discharges, which can also produce strong acoustic signatures) can be very frustrating and time-consuming.

Another obvious problem with acoustic PD detection is that HV equipment including GIS is typically resonating at twice the powerline frequency (i.e. 100 or 120 Hz). This fundamental electromechanical resonance will "pump" higher-order harmonics and can excite any loose parts or parts with their own natural acoustic resonance attached to or close to the GIS. Reports based on acoustic detection often claim the presence of PD signals, but careful examination reveals that the measurements are actually of parasitic resonances in the GIS structure. Borrowing a technique first used for transformer diagnosis (Section 15.11.5), one way to be sure that the acoustic signal being received is actually produced by PD is to use a signal from a UHF PD sensor (if available, or it could also be an external sensor temporarily installed) as a trigger for the acoustic PD measurement as described in [46–48]. Although acoustic PD measurement looks easy at first glance, especially employing temporarily attached piezoelectric sensors (Figure 13.15), it requires great care and extensive experience to use in practice.

Because of the strong location and propagation path dependency of acoustic PD detection and similarly to the case with RF methods, it is impossible to draw conclusions about the charge magnitude based on the received acoustic signal amplitude, outside of compact, purpose-built laboratory set-ups. However, because of the ongoing research and field application of acoustic PD detection on GIS in the 1980s and 1990s, a sensitivity verification for the acoustic method was proposed by the CIGRE Joint Task Force 17/33.03.05 [1]. Sometimes known as the "ball-drop method" (although the reference impulse can also be generated using a suitable piezoelectric coupler), it was decided not to include it in the revised and updated publication [13].

However, some experimentation and laboratory practice is recommended with whatever acoustic detection tools are intended for diagnostic use, especially on site. The best place to do this is in the high-voltage laboratory by intentionally placing defects inside a real GIS test section. Alternatively, whenever a defect is discovered in a shipping unit, it is good practice to spend some time seeing whether it can be heard acoustically (with the pistol in Figure 13.14) and to allow any other available field personnel who may use acoustics in the field to get practice on a real defect.

13.8 Radio-Frequency PD Measurement on GIS: The UHF Method

Owing to the problems inherent in attempting to apply conventional, charge-based PD measurement techniques (i.e. LF or "baseband" detection per IEC 60270) to multi-bay GIS on site (detailed below), along with the fact that PD in SF_6 generates signals whose RF spectra extend to several

GHz and beyond, Professors Steve Boggs [28, 29] and Brian Hampton [49, 50] began applying what became known as the "UHF method" for onsite PD measurements (see Section 1.5.3). Constructing simple sensors in the form of round metallic plates installed on the inside of maintenance flange covers and connected to the outside via RF feedthrough connectors, they quickly discovered PD signals could be measured at 1 GHz and above using RF spectrum analyzers along with the fastest oscilloscopes then available (e.g. the revolutionary Tektronix 7104, (Section 1.5.3)). The most significant advantages were relatively lower levels of interference (especially compared to the charge-based IEC 60270 frequency range), no need for a heavy, bulky-but-delicate coupling capacitor, and the possibility to locate the PD source using what they referred to as the "time-of-flight" or "time-of-arrival" method (Section 8.4.2).

Since the introduction of the ultrahigh-frequency (UHF) partial discharge (PD) measurement method in the late 1980s [28, 29, 49–55], it has become widely accepted as the de-facto standard for onsite PD measurement during HV acceptance testing, periodic online assessment for diagnostic investigation, and online monitoring, due to the high detection sensitivity and high signal-to-noise ratio (SNR), which can be achieved.

This chapter will discuss the UHF PD sensors, RF propagation environment inside the GIS, narrowband vs wideband techniques, practical aspects of spectrum analyzer-based PD measurement, and other practical aspects of applying the UHF method to PD in GIS.

Based on its historical association beginning with Brian Hampton and others in the 1980s, the "UHF technique" or "UHF method" has come to signify application of radio-frequency (RF) signal processing techniques for detection of PD in GIS. In fact, the term "UHF" is an acronym for "ultra-high frequency," which refers to a specific region of the RF spectrum (300–3000 MHz), but in many cases it can also be useful to look for PD activity below 300 MHz, in the VHF (very-high frequency) spectrum, i.e. 30–300 MHz.

For the context of this chapter and in keeping with its historic association in the field, we will use "UHF" as the generic term for PD measurement on GIS, but will sometimes use "RF" when the more general term is called for (e.g. "RF impedance").

Similarly, note that the terms "interference" – meaning external "EMI" (also known as "radio-frequency interference" – "RFI") and "noise" – have a long history of being used interchangeably in the PD field (Section 9.2.1). In fact, this is a fundamental technical error. These two terms mean very different things, but this confusion is easily remedied and understood by means of a simple thought (or actual) experiment. If we take either a wideband oscilloscope or an RF spectrum analyzer, connect a standard 50 Ω termination to their input and turn up the gain, the signal we see will be the noise "floor" of the instrument. There is no way to reduce that internal, intrinsic noise level without turning off the instrument or changing its internal architecture and components. Now, if we remove the 50 Ω termination and connect an antenna to the instrument's input – this can just be a short length of wire – we will see a huge increase in signal level, a host of random-looking signals will appear (unless the test is done inside a shielded room). This is interference – EMI or RFI – being "broadcast" by external devices which we are picking up with our little input "antenna." The difference between EMI/RFI and "noise" will be strictly adhered to here.

13.8.1 UHF Sensors

Starting back in the 1980s, UHF sensors were designed specifically for GIS. These typically consisted of circular metal plates, isolated from the GIS housing and mounted in convenient locations inside the GIS, such as the inside surface of maintenance hatch covers [49, 56]. Connection to the outside world was made via gas-tight feedthrough. The essential thing was that the metallic sensor was housed inside the GIS enclosure, thus profiting from the effect of its shielding against external interference (EMI).

As the UHF method became more well-known and more widely used, GIS OEMs began creating their own RF PD sensor designs and installing them in their GIS; today most manufacturers install UHF PD sensors in their transmission GIS (>200 kV) on a routine basis. Today there is a wide variety of designs, but many still resemble the first sensors developed by Boggs and Hampton (lower right-hand photo in Figure 13.17). Sometimes these sensors are called "antenna," but this is a technically incorrect use of the term. The rule of thumb in RF engineering is that conducting objects used for transmission or reception of electromagnetic waves begin acting as "antenna" at a distance of at least 10 wavelengths at their mid-band frequency [57, 58]. The UHF mid-band is 1650 MHz, corresponding to a wavelength of 18 cm. Therefore, this means that GIS UHF PD sensors are acting in their extreme near field, even in the largest diameter GIS; they are technically *not* "antennas." A primary consequence of this is that one of the main RF resonances appearing in the RF spectrum is at the frequencies at which standing waves occur between the PD sensor and the GIS center conductor. Therefore, GIS PD sensors are strongly influenced by all of their surroundings, especially metallic surfaces, within their installation location inside the GIS.

Engineers interested in picking up the weak RF signals generated by small PD defects desire the highest possible sensitivity, which usually means pushing the sensor element further inward into the GIS volume, i.e. toward the inner conductor. In contrast, the high-voltage GIS designer wants nothing more than to keep grounded (or near-ground) metallic PD sensors back behind the

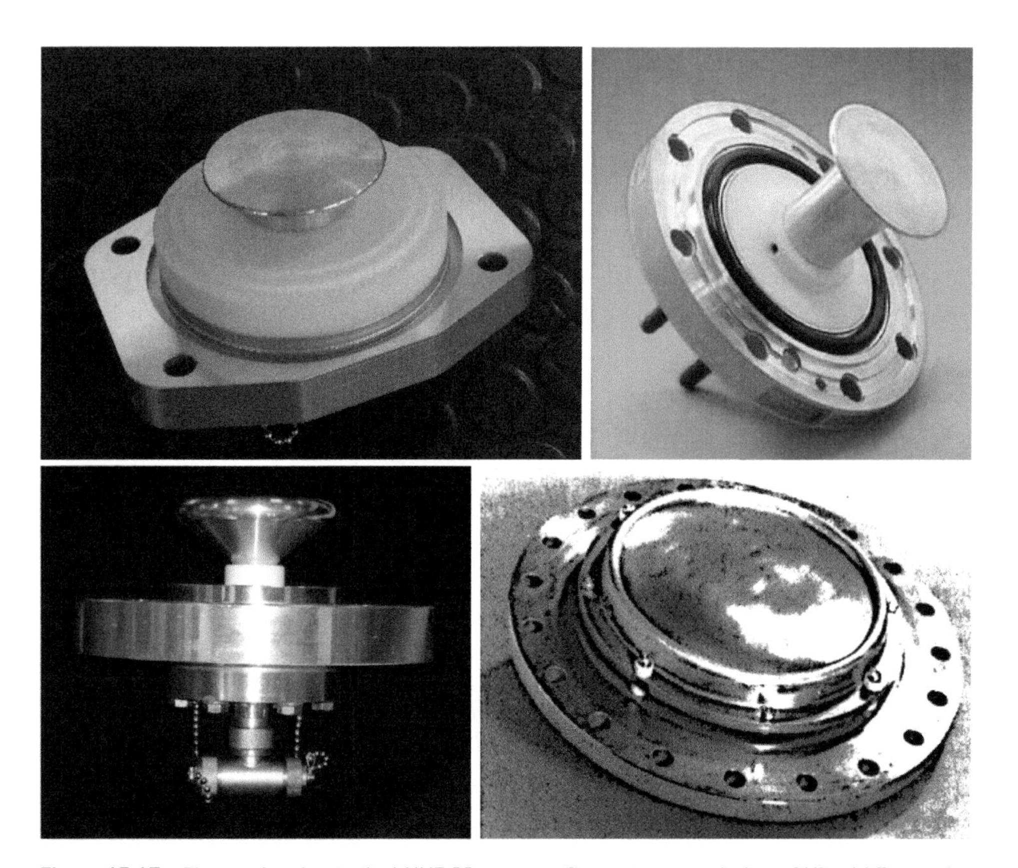

Figure 13.17 Photos showing typical UHF PD sensors. *Source:* top permission of Hitachi Energy, bottom right from [50].

enclosure's circumference. It is also intuitive to try to make the surface area of the sensor as large as possible in order to pick up as much of the radial electric field of the passing PD pulse as possible. However, this leads to the sensor having a lower self-resonance, being "slow" in the time domain [59]. This is important for accurate time-of-flight measurements (Section 8.4.2) and making it more efficient at coupling the lower frequencies where the major part of the energy of very fast transient overvoltage (VFTOs – very high amplitude, very fast risetime transients caused by GIS switching operations) energy is concentrated, thus making things more difficult for the designer of the monitoring system protection circuit. Lastly, with the strong cost pressure under which all GIS OEMs operate, PD sensors usually end up being designed for installation in unused flange positions. Some manufacturers have begun taking advantage of field-control electrodes in already existing earthing switches or other devices, and turning them into "synergistic" sensors by isolating them from ground and connecting them to a gas-tight RF coaxial connector to the outside. The sensor thus described would, therefore, comprise a floating element; one would observe sparking inside the RF coaxial connector under normal operation, as the sensor element charges up in the high-voltage AC field it "sees" inside the GIS. To address this problem, most manufacturers either equip the sensor with a shunt resistance or inductance (to short the 50/60 Hz power frequency current to ground), or equip the RF connector with a short-circuiting dust-cap (which is then removed during measurements or e.g. if a permanent PD monitoring system is connected to the sensor).

In the past, PD sensors were simply drawn and designed by hand, but it can take considerable trial and error to get high sensitivity across a multi-octave bandwidth while avoiding parasitic resonances from objects in the sensor's immediate vicinity. Today, the wide availability of sophisticated RF modeling tools enables "tuning" designs and avoiding problems with resonances or compromising dielectric design rules [60]. A few examples of some typical internal PD sensors are shown in Figure 13.17.

Another interesting innovation is to embed a PD sensor inside the GIS insulators as shown in Figure 13.18. This sensor takes the form of a metallic "guard" electrode cast into the GIS insulator disk between the outer flange and one of the three inner conductors (this is three-phase GIS). Contact between the sensor and the outside world is made via special adapters that are screwed into the sensor body from outside; the adapter facing out is a typical type-N connector.

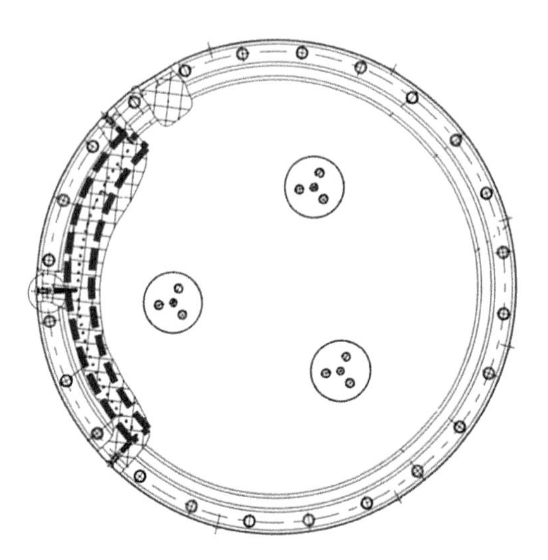

Figure 13.18 VHF/UHF PD sensor embedded in three-phase GIS barrier insulator (ABB).

This is an elegant solution because every barrier insulator contains three PD sensors. Barrier insulators are scattered at many locations throughout the GIS core, so there is no need for installing other PD sensors.

At the same time, UHF sensors began being placed inside the GIS, it became clear that RF signals from PD can leak out of the GIS at various locations, such as at unenclosed insulators, inspection windows (which allow observation from outside of contacts and verification of contact position in isolation and grounding switches), current transformers (which require a break in the GIS's metallic enclosure in order to operate), and so on. Thus, leakage enabled the use of external UHF sensors to detect GIS PD where internal UHF sensors were never installed. If well designed and matched to the opening at which they are applied, surprisingly high sensitivity can be achieved with external UHF PD sensors [61]. However, their sensitivity is lower since their usual mounting location is further outward from the inner volume of the GIS, plus it may be hard to get as much sensor surface area or physical size owing to the typically small size of the openings available for their application. Their inherently smaller size may also act like a high-pass filter, cutting off part of the UHF spectrum and thus reducing the amount of signal energy available. If not carefully designed, they may also be more susceptible to external EMI. This external UHF sensors are usually only used for carrying out measurements on GIS, which were never equipped with built-in UHF PD sensors, or for hunting down PD defects producing very weak RF signal output (e.g. a particle lying on an insulator in an isolation switch equipped with an observation window). Another example of their use is to inject a signal (using the same type of fast pulse generator used for the CIGRE sensitivity verification, Section 13.8.3) for diagnostic purposes during onsite PD measurements. Some examples of commercial external GIS PD sensors are shown in Figure 13.19.

Some vendors specify the bandwidth of UHF PD sensors and also "calibrate" them as if they were antennas (i.e. expressing their equivalent/effective height in terms of an ideal monopole), sometimes by using specialized test fixtures such as GTEM cells used for electromagnetically compatibility testing. However, note that UHF PD sensors are acting in their extreme near-field, which means their RF behavior is directly and fundamentally affected by the surrounding surfaces where they are installed in the GIS, requiring different test fixtures to simulate the actual GIS (this applies to both internal and external installation). Alternatively a UHF PD sensor's bandwidth and sensitivity can be estimated by plotting its frequency spectrum in the actual GIS using an RF spectrum analyzer on MAX HOLD for a few minutes' duration during the CIGRE sensitivity verification Step 1 (Section 13.8.3 [13]).

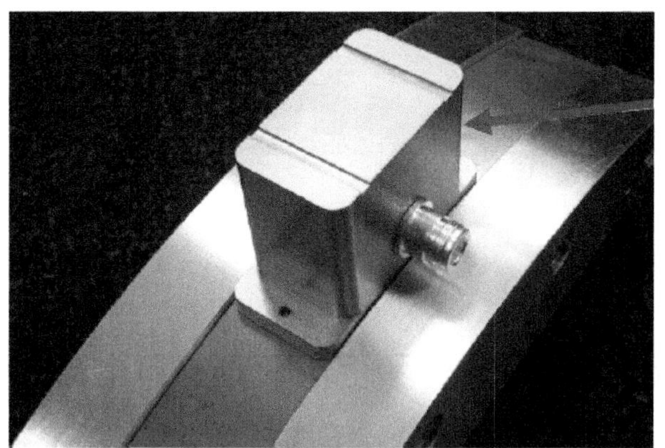
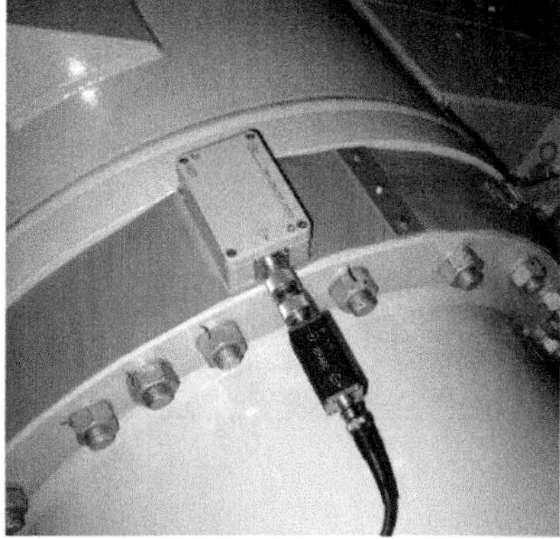

Figure 13.19 Examples of external RF sensors made by Qualitrol-DMS (left) and Power Diagnostix (right).

13.8.2 The RF Signal Propagation Environment in GIS

As explained in Section 13.3.3, the risetimes of PD pulses in SF_6 gas are now known to be ≤30 ps [25, 26]. From fundamental signal theory we know that the risetime of a pulse determines the upper frequency roll-off of its frequency spectrum.

The Fourier transform of a unit (Dirac) impulse (theoretical zero risetime) transforms to infinite bandwidth as shown in Figure 13.20. For the case of a real-world impulse with a finite risetime, the high frequency roll-off of its frequency spectrum can be determined by the following well-known "rule-of-thumb" relation:

$$\text{freq}_{-3\,\text{dB}} = 0.35 / t_r$$

If we assume a risetime of 24 ps [25], this results in a theoretical −3 dB upper roll off at about 14.6 GHz.

Figure 13.21 shows fast oscilloscope plots of some PD pulses in SF_6, which exhibit pulse widths around 0.5–1 ns or so [25]. Again, from signal and systems theory, we know that one way to determine the transfer function of a linear system – in this case, the transfer function (or frequency spectrum) of a given signal propagation path through the GIS – can be determined by using such an ideal step waveform, or a Dirac impulse as the stimulating signal.

If we set up a PD source in GIS, e.g. a hopping particle and take measurements at a UHF PD sensor, instead of seeing a single sharp impulse as shown in Figure 13.21 at the sensor output, or a perfectly flat RF spectrum as theory would predict, rather we see the transfer function of the

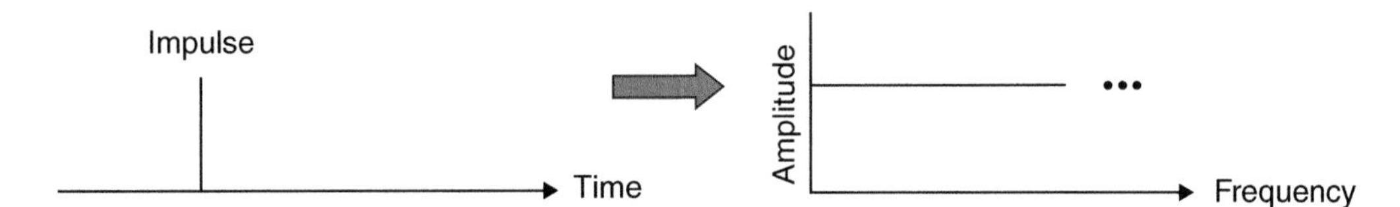

Figure 13.20 How the fast risetime of the GIS PD signal transforms to GHz frequency content.

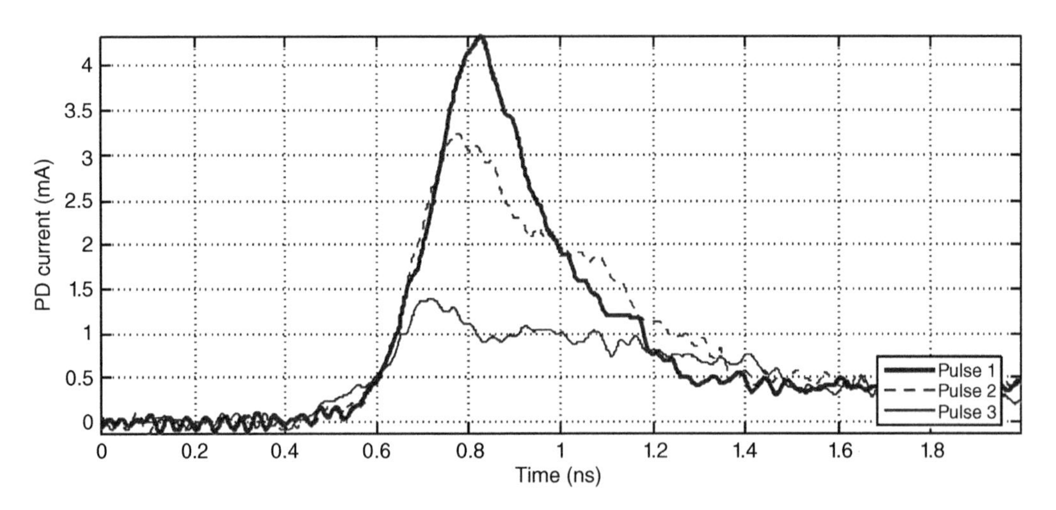

Figure 13.21 Actual PD pulse in SF_6. The horizontal time scale is 200 ps per division. *Source:* [25].

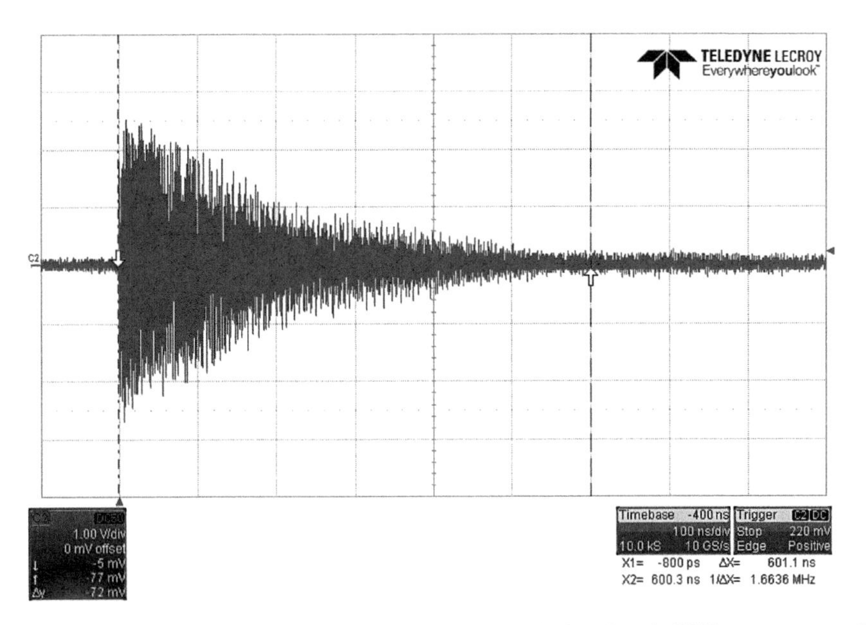

Figure 13.22 Time-domain measurement of an actual PD pulse via UHF sensor output. The horizontal scale is 100 ns/div.

propagation path in the form of the frequency-response spectrum. When observed in the time domain on a fast oscilloscope, the original, sharp, Dirac-like PD pulse has been transformed into a complex decaying transient, a sample of which is shown in Figure 13.22. What has happened to our fast risetime, Dirac-like PD pulse? And why isn't the spectrum a flat line, starting to decay at 14 GHz?

When looking at a simplified conceptual drawing of GIS or an actual section of long, straight busbar, GIS seems like a well-behaved coaxial waveguide, which should have excellent RF transmission characteristics. Indeed, in the early days of applying the UHF method, signal loss figures on the order of "2 dB/m" were often claimed [54]. However, based on much recent work [62–75], it turns out that the RF propagation characteristics of GIS are significantly more complicated.

Briefly, the nonideal RF propagation of GIS is due to three main factors:

- The interior dimensions of even smaller diameter (lower-voltage) GIS are on the same order of or larger than the wavelengths at which UHF PD measurements are being taken.
- GIS components such as circuit breakers, disconnector and grounding switches, 90° "elbow" bends, "tee" and "X"-junctions, voltage and current transformers, enclosure diameter changes all affect RF signal propagation. Instead of the uninterrupted symmetry of a long straight coaxial waveguide, all of these components present RF impedance discontinuities.
- A third and related factor is that GIS is of course designed to carry electric power; therefore, it contains field control electrodes, switching devices, and their associated contact assemblies, and many other metallic parts and structures.
- The combination of these factors means that RF standing waves occur throughout the GIS whose wavelength (and thus frequency) corresponds to the distance between any two metallic surfaces.

As soon as the electromagnetic wavefront produced by the PD pulse begins to spread out from the PD site, it begins to interact with all of the interior surfaces present within the GIS structure. Each

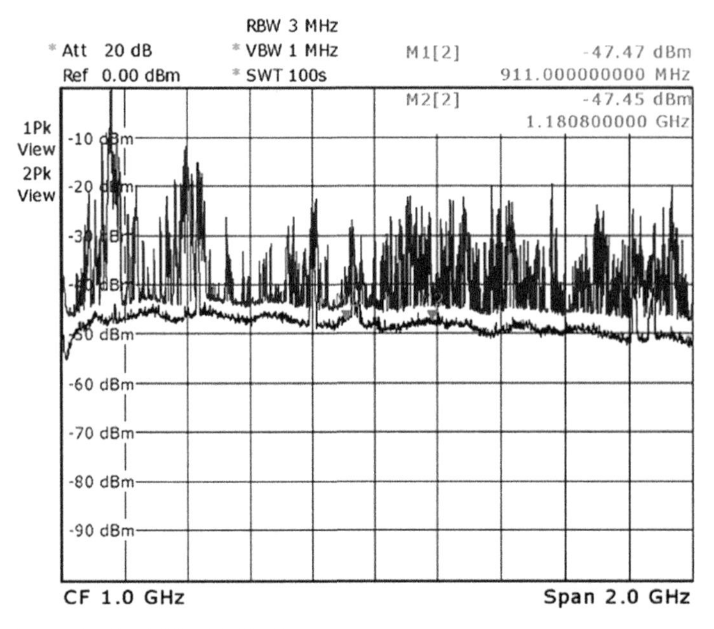

Figure 13.23 The RF spectrum of a typical PD pulse from a metal particle in GIS.

of the standing waves results in constructive interference (producing sharp peaks in the RF spectrum) or destructive interference (producing sharp notches in the RF spectrum). Again, because the internal dimensions of the GIS are of the same dimension as UHF wavelengths, it is clear that if we move the location of the PD defect and/or of the PD sensor, the combination of RF standing waves will change and therefore the position of the peaks and notches in the frequency-response spectrum of the transmission path will also change. In other words: the exact transfer function between a PD defect and the PD sensor at which we measure the signal is both extremely complex and unknown. In fact, if we measure the RF spectrum between PD sensors at the same location but compare the measurements between all three (single) phases, the three different spectra have discernable differences. (In fact, the CIGRE sensitivity check – covered in the next section – can pass on one or two phases but fail on the others.)

Instead of a close-to-ideal coaxial waveguide, these factors combine to make GIS resemble something more like a series of interconnected frequency-dependent RF resonators and notch filters. This results in an RF spectrum (equivalent to the transfer function) with the expected extension well into the GHz region, but which is filled with sharp peaks and valleys, indicating strongly resonant behavior, as can be seen in the example shown in Figure 13.23.

A further typical RF characteristic of GIS is attenuation of higher frequency components as shown in Figure 13.24, the top plot of which shows the signal received from an artificial pulse being injected in the adjacent bay while the bottom plot shows the obvious roll-off at the higher frequencies 6 bays distant. Note that this effect becomes more pronounced with higher voltage ratings because the diameter of the GIS increases, which in turn increases the opportunity for resonances to occur at longer wavelengths/lower frequencies; hence the RF transmission spectrum in larger-diameter GIS rolls off at an even lower cutoff frequency. Such effects are common to all RF transmission waveguides. In addition, the topology of the GIS exerts a powerful influence on its RF transmission response as shown in Figures 13.25 and 13.26.

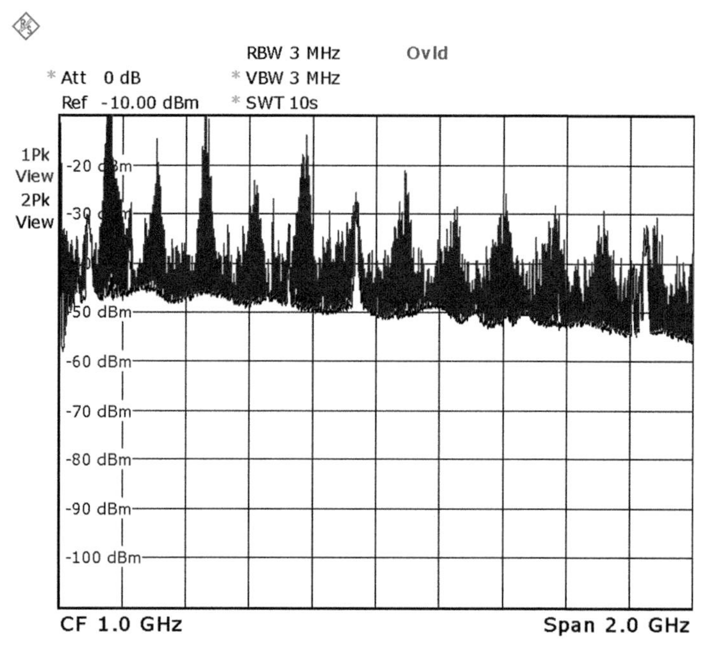

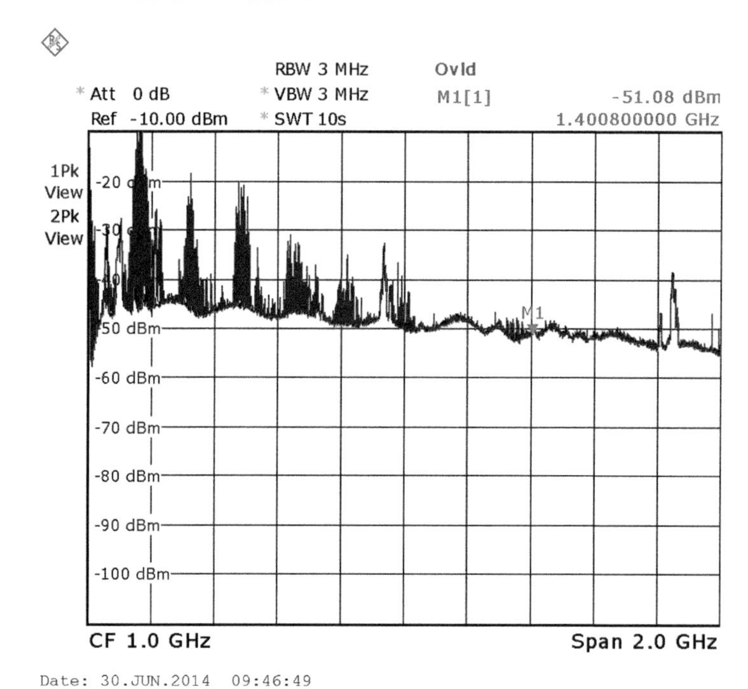

Figure 13.24 Decay of GIS RF PD signal as a function of distance (top is near source, bottom is 6 GIS bays away).

13.8.3 The CIGRE Sensitivity Verification

Because the PD signals die off with distance in GIS (Figure 13.24), this means that PD sensors need to be distributed at strategic locations throughout the GIS to be sure of picking up small PD signals from a hopping particle, as an example. The transfer function of the signal propagation path in GIS

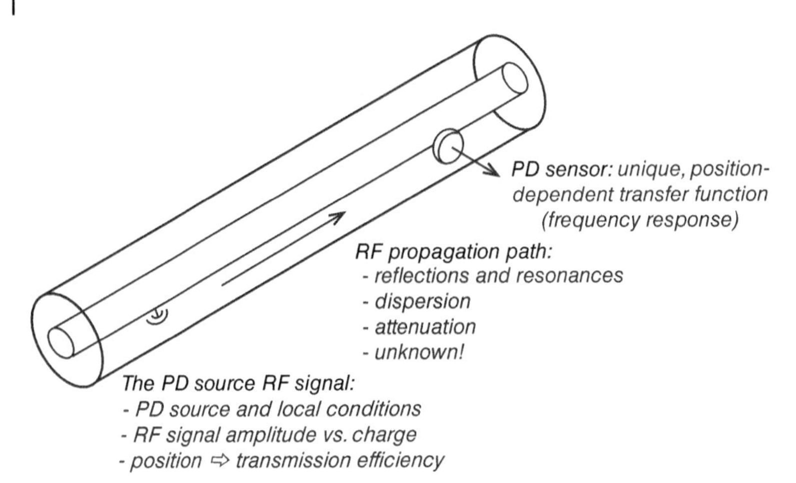

Figure 13.25 Some of the factors affecting the PD signal from origin to sensor output. *Source:* [76].

between a PD site and the next available PD sensor is highly variable and frequency dependent [55, 75, 76]. This underscores the fact that it is physically impossible to assess the charge value (pC) of a PD defect based on the received RF signal amplitude at the PD sensor, because we cannot know how much of the signal was lost before it got there. It also means that we need some way to verify that our PD sensors are not spaced so far apart as to miss weak (but important) PD defects anywhere in the GIS. As a result, a CIGRE task force was convened to address these issues and is the subject of the next section.

Already by the early 1990s, it was clear that it is impossible to quantify the charge value of a PD signal based on the received UHF signal amplitude due to the variation in propagation loss between the PD site from where the signal is emanating and the PD sensor location. On the other hand, GIS operators want some assurance that the UHF method is achieving sufficiently high sensitivity so as to detect PD defects posing real risk to the GIS. This led a CIGRE Task Force to publish an analysis of the problem in 1999 [1]. Although published as a recommendation, it has assumed the significance of a de-facto standard, being often written into specifications and contracts for GIS PD measurements during onsite acceptance tests or for permanently installed PD monitoring systems. It was extensively revised and expanded with the publication of CIGRE TB 654 [13]. Some Important aspects of the sensitivity verification (or sensitivity check) procedure are presented below. Step 1 is usually referred to as the factory or laboratory step because it typically takes place at the GIS OEM facilities. A test section of GIS as shown in Figure 13.27 is set up in the high-voltage laboratory or factory final test bay, connected into the PD measurement circuit and calibrated as specified in IEC 60270 (see Section 6.6). One GIS enclosure is equipped with an RF PD sensor, "C1," and a metallic particle is also placed in the same enclosure. A second enclosure some distance away (e.g. 2 m) is equipped with a UHF PD sensor "C2." The high voltage is then raised until the particle begins hopping regularly so as to provide a reasonably steady pulse rate. The sensitivity check specifies that the particle should generate about 5 pC apparent charge measured by the IEC 60270 test circuit (some small variation is allowed). The apparent charge given off by a particle is dependent on several variables including its weight, length, the sharpness of its edges, and of course the local electric field strength in the GIS. It is not unusual that several particles must be tried before one produces the requisite 5 pC. Once the particle is hopping steadily, the RF signal it generates is recorded by the UHF measurement device at sensor C2. The CIGRE TB 654 procedure only specifies a "UHF measuring device"; but typically an RF spectrum analyzer is placed on

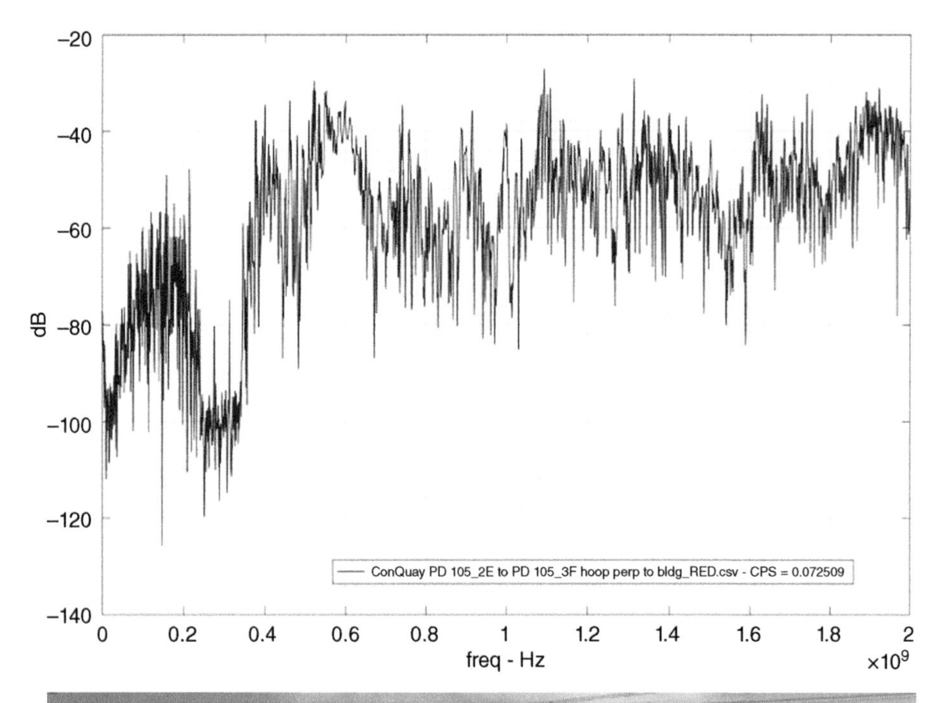

Figure 13.26 Example of RF frequency response through a U-shaped section of GIS. *Source:* [75].

MAX HOLD mode and the RF spectrum is accumulated for 10–20 minutes until it is totally filled in (corresponding to the black trace in Figure 13.28). Alternatively, the "UHF measurement device" can also be the broadband front-end of a PDM monitoring system (again: it must be the same type of PDM system hardware as will be subject to the Step 2 onsite verification). It should be noted that, although it is nowhere expressly stated, it is strongly recommended against use of a wideband

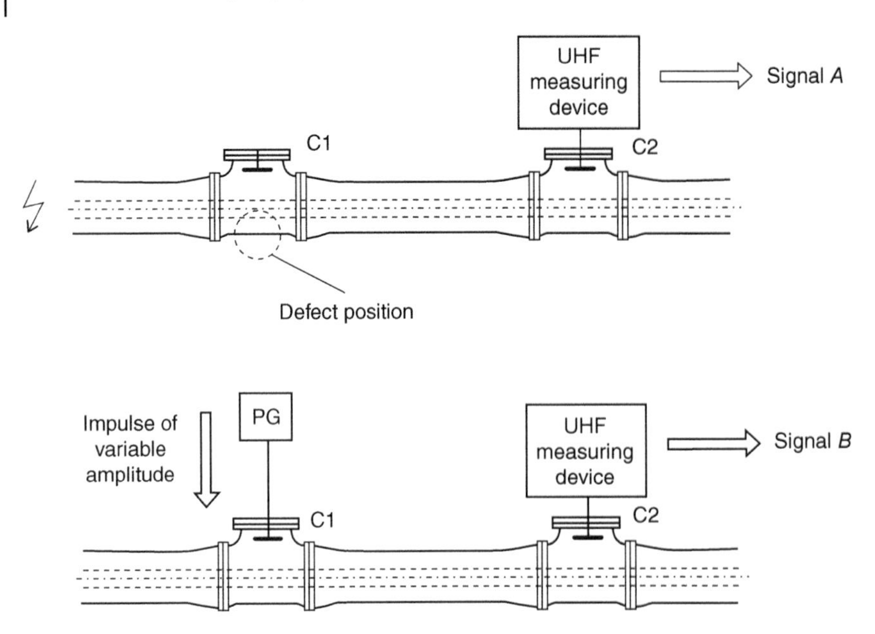

Figure 13.27 Diagram of the CIGRE TB 654 sensitivity verification for the UHF method. *Source:* [77].

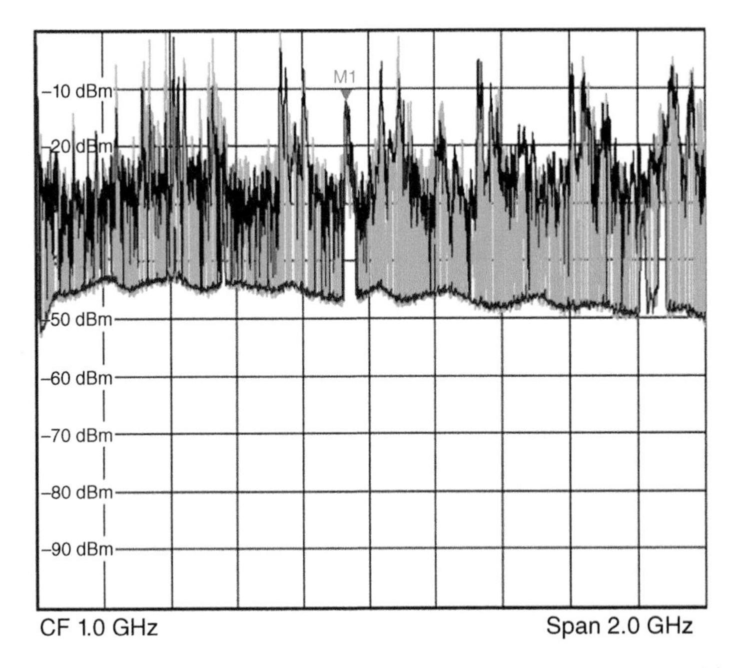

Figure 13.28 Showing a close match between 5 pC particle spectrum (black trace) and pulse generator.

oscilloscope as the "UHF measurement device." Even in e.g. "MAX HOLD" or "persistence" mode, time-domain-based measurements can often produce anomalous peak readings, which are not accurate representations of the actual signal-level relationships, due to strongly varying local resonances along the signal transmission paths in GIS.

In the second part of Step 1, the high voltage is then turned off, and a suitable pulse generator, one having a risetime sufficient that its output spectrum exceeds the measurement spectrum of the

UHF measurement device (<<1 ns), is connected to the PD sensor C1 located in the same GIS enclosure as the hopping particle. Its output level should be adjusted so that the signal received at C2 matches the signal recorded from the hopping particle as closely as possible. An example of a good match between the UHF signal from the actual hopping particle (black trace) and the pulse generator (green trace) is shown in Figure 13.28.

In Step 2 of the sensitivity verification, known as the "onsite" part, the same type of pulse generator as was used in Step 1 is connected to each PD sensor and the output observed (recorded) at the next adjacent sensors. If the pulse generator signal is visible, that sensor to sensor path is considered passing. This means that, should a PD source equivalent to a hopping particle producing 5 pC apparent charge be present anywhere along that sensor-sensor path, it will be picked up by the measurement (or monitoring) system. The pulse generator is turned off and connected to the next sensor, its signal is observed on the next adjacent sensors, and so on, until all of the sensor–sensor paths in the GIS have been checked.

Several important points need to be observed or kept in mind:

- This sensitivity verification is not a "calibration" in the sense of IEC 60270. It is a performance check based on an equivalency measurement.
- Exciting a PD sensor with a pulse generator is a fundamentally different signal generation process than that of a hopping particle. They are also not spatially co-located, an important factor at these wavelengths. Careful inspection of the spectra recorded at C2 of the particle and the pulse generator will show obvious differences; this is in part due to this fact. This is also the reasoning behind the previous statement recommending against using time-domain-based instruments (oscilloscopes) for the sensitivity verification.
- It may not be possible to achieve as close a match between the particle's UHF signal and that of the pulse generator in Step 1 as shown in Figure 13.28. The CIGRE recommendation makes allowances for this. . .but this is again another reason that this procedure cannot be considered a "calibration."
- In all cases and as stated in both CIGRE publications, it is imperative that the same GIS type, the same sensor type, the same pulse generator type, and the same type of "UHF measurement device" (e.g. PDM system, spectrum analyzer, pre-amplifier type, etc.) are used in both Step 1 and Step 2. (In the case of a commercial spectrum analyzer being used for the "UHF measurement device" – essentially a standard laboratory instrument – the actual analyzers used for Step 1 and Step 2 do not have to come from the same manufacturer, but the settings used must be the same).
- Neither the original CIGRE recommendation [1] nor the update of TB 654 [13] give any guidance as to how far above the background noise floor the received signal in Step 2 needs to be, only that it is visible. However, as we will see in the section covering the 'time-of-flight' technique (used for locating the PD source), if the SNR of the incoming signals is poor, this may make the TOF measurement inaccurate or even unusable. Furthermore, if the incoming PD signal from a real defect is too close to the background noise floor, there will be insufficient dynamic range to create a meaningful PD pattern, making defect type interpretation difficult or impossible. In other words, sufficient SNR to obtain clear PRPD patterns (to enable accurate recognition of the defect type) and accurate time-of-flight measurements (to locate the defect) are key to enable reliable defect detection, recognition, location, and removal, in order to assure risk-free operation.
- The real PD defect used in Step 1 is typically specified to be a moving particle producing 5 pC apparent charge according to IEC 60270. However, other PD defects may be used if agreed upon

between the GIS OEM, the PD measurement or monitoring vendor, and the GIS operator. The typical alternative is a small protrusion on the inner conductor producing an apparent charge of say 2 pC. In that case, obviously, this will have the effect of forcing the number of PD sensors to increase significantly, because they will need to be closer together to pick up the much weaker signal.

The CIGRE sensitivity check has a direct impact on PD sensor placement. As a GIS OEM accumulates experience over the years running sensitivity checks on their GIS, they will usually settle on a set of fixed locations where sensors are almost always installed, especially in the "core" of the GIS (i.e. the switching bays) as well as understanding how far apart they can be on outdoor feeders, such that the CIGRE Step 2 has a high probability to pass on each sensor-sensor path. Also they begin to discover problematic sensor locations and geometries, which tend to fail the Step 2 onsite test.

Some further remarks regarding placement of UHF PD sensors are in order here. On one hand, GIS OEMs are under strong financial pressure to install as few PD sensors as possible (also reducing the number of nodes of a PD monitoring system), which means pressure on the GIS designer to push the distance between sensors to the limits at which the CIGRE Step 2 can be passed on site ("artificial pulse just visible"). This is false economy not only due to the factors mentioned above (negative impact on defect recognition and location) but also because, if the PD sensors are too far apart, certain critical defects may be missed during onsite acceptance tests. CIGRE statistics clearly show that a thorough, high-sensitivity PD measurement carried out at the time of the onsite acceptance tests provides the best guarantee of operational reliability at energization [15]. However, experience gained through onsite PD measurement has shown that certain critical defects such as small protrusions and particles lying on insulator surfaces – defects whose RF (UHF) signal output energy is relatively low – drastically lower the insulation withstand strength in the presence of lightning or switching impulses. These defects emit proportionally much less RF signal than the 5 pC hopping particle defect normally used for the CIGRE TB 654 Sensitivity Verification Step 1 [76, 78, 79]. This means that, if the distance between UHF PD sensors is increased to the limits where the CIGRE Step 2 test "just passes," there is a chance such critical defects may be missed during the acceptance tests. And echoing what was mentioned above, locating PD defects whose signal levels are low will become difficult or impossible if the PD sensors are too far apart. This can result in the uncomfortable scenario of seeing a weak signal during the PD measurement, but not being able to make the TOF measurement work; this would then necessitate "exploratory opening" of more GIS compartments in the search for a defect. The problem is that each time a GIS is opened, it introduces more risk of contamination upon re-closing. It is well known for decades that the risk of a flashover following such intervention is as high as that from the original PD defect itself, sometimes higher.

A more detailed discussion of these issues can be found in [78].

13.8.4 UHF PD Signal Acquisition: Narrowband vs Wide Band

From the earliest days of applying the "UHF method" to the detection of PD signals in GIS, both narrowband and wide-band[1] techniques have been employed with their respective advantages and disadvantages. A pictorial representation shown in Figure 13.29 below illustrates the difference between the two strategies, both of which were employed by Hampton and his colleagues.

As explained in Section 13.8.2, the sharp peaks and valleys typical of GIS PD spectra (e.g. those of hopping particles shown in Figures 13.23 and 13.28) are the result of the constructive and

1 Note that the terms "narrowband" and "wideband" as used here should not be confused with the definition of those terms in IEC 60270.

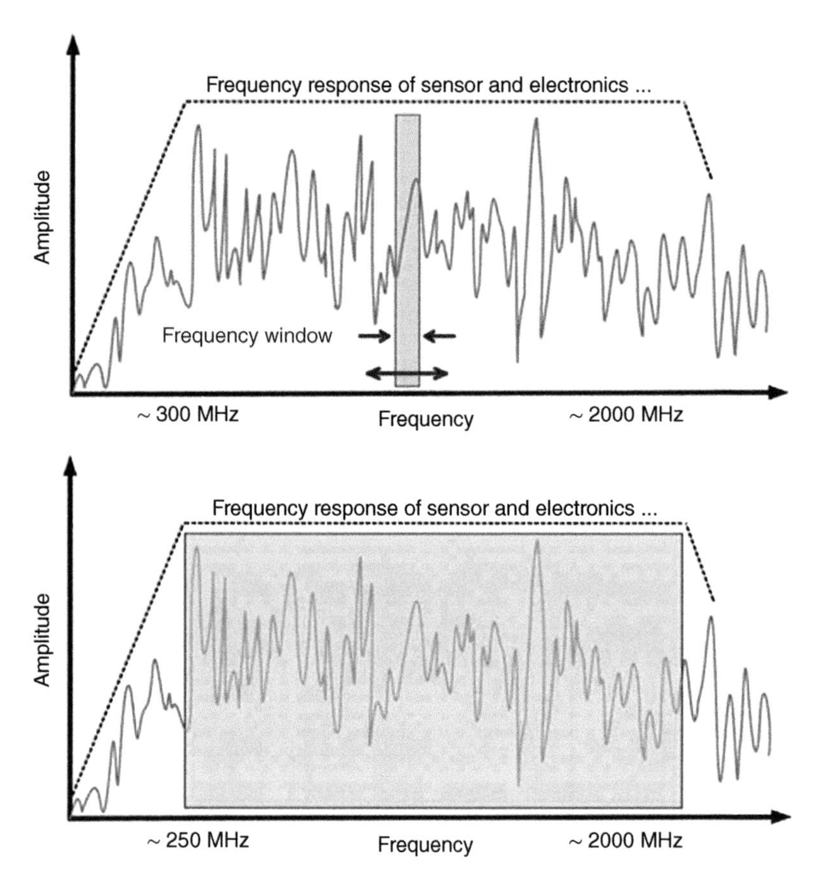

Figure 13.29 Narrowband vs. broadband acquisition of PD in GIS. *Source:* [19].

destructive interference caused by the very high-Q resonances of RF signals within the GIS. Using an RF spectrum analyzer, an experienced operator scans through the RF spectrum looking for peaks in the PD signal away from external interference in order to obtain a strong PD signal with a high signal-to-noise ratio (SNR).

However, because installing an RF spectrum analyzer at each PD sensor (along with a skilled operator) would be impractical, for online PD monitoring systems (both portable systems for temporary checks and permanently installed continuous systems), simpler wide-band techniques are employed. Typically, this consists of relatively simple broadband RF receivers being connected to each PD sensor. These broadband RF "front-ends" consist of high-gain RF receivers driving variations of AM (amplitude modulation) "envelope" detectors that are relatively cheap to build in large numbers required for large PD monitoring systems (Figure 13.33 right).

13.8.5 Narrowband UHF PD Detection with RF Spectrum Analyzers

If no PDM system is already installed, the signals coming from the UHF PD sensors (either permanently installed in the GIS or external sensors temporarily employed for diagnostics or on older GIS with no internal sensors installed) can be fed into a spectrum analyzer, allowing an experienced operator to search for regions of the RF spectrum where the PD signal can be easily observed (Figure 13.31). The user must be well versed with the spectrum analyzer and its internal

Figure 13.30 RF pre-amplifier and transient protection connected to UHF PD sensors. *Source:* Swiss Mains.

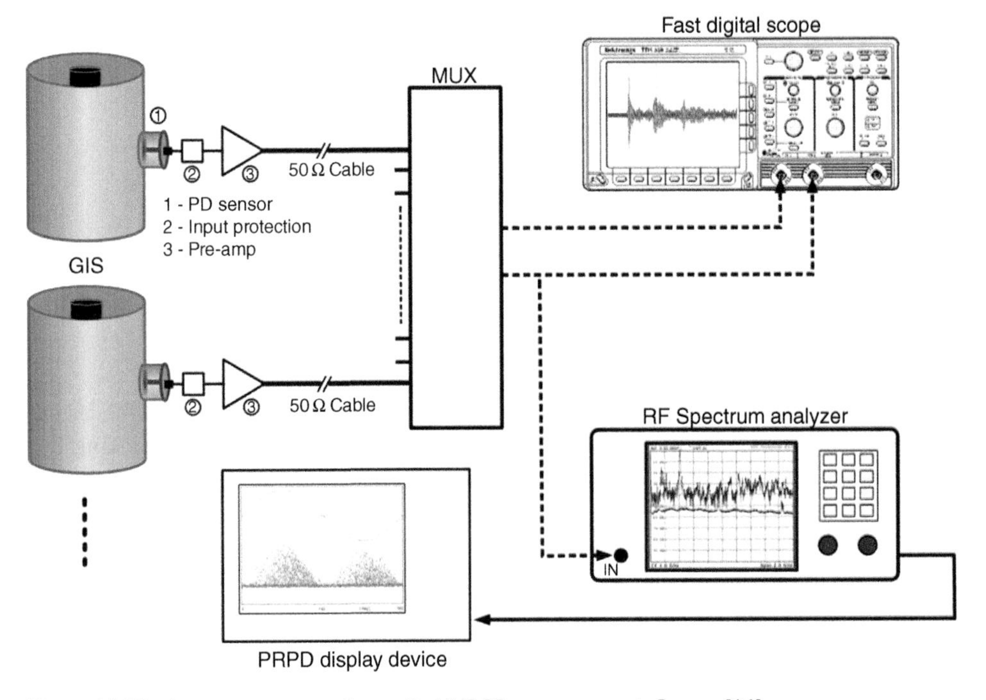

Figure 13.31 Instrument set-up for onsite UHF PD measurement. *Source:* [16].

sub-systems and its operating principles [80]. Once a PD source (defect) has been identified, a fast oscilloscope can be connected to sensors on either side of the defect (Figure 13.31), and the difference in the time-of-arrival of the signals from the two sensors is used to calculate its position; this is known as the "time-of-flight" (TOF) technique, discussed in detail in Section 8.4.2.

Starting in the early 1990s, GIS OEMs such as ABB Switzerland began deploying UHF PD measurement systems for use during the onsite high-voltage acceptance tests of GIS. However, some hurdles had to be overcome to make these early systems work, based as they were on RF spectrum analyzers. With a large number of UHF PD sensors to be covered in a test section, RF multiplexers had to be developed, which would allow the operator to switch rapidly between sensor inputs. Because of the physical size of the GIS test sections, this also meant that the cables connecting the test equipment to the UHF PD sensors could often be 20 m or more. Even with the best low-loss, broadband coaxial cable available at the time (10.5 mm dia. "10BASE5 thick Ethernet" cable), work at the ABB Corporate Research Center demonstrated the utility and need of broadband, high-gain preamplifiers connected directly at the PD sensor (Figure 13.30) to drive the UHF PD signal down the cable to the multiplexer and finally the input of the spectrum analyzer. Power for the remote pre-amplifiers was supplied by a "phantom" DC power supply inside the multiplexer unit, which would send +15 VDC up the center conductor of the coaxial cable. Figure 13.31 shows a concept drawing of a narrowband UHF PD set-up with some onsite examples shown in Figure 13.32.

As discussed in Section 8.6.2, if a flashover takes place within the GIS, the resulting very fast transient overvoltages (VFTO) in the GIS will also be picked up by the PD sensor, instantaneously

Figure 13.32 The equipment shown in Figure 13.31 set up for onsite diagnostic measurements.

destroying the sensitive pre-amplifier electronics. Switching transients from circuit breaker and disconnector-switch operation will also usually result in instantaneous destruction of the delicate pre-amplifiers. This necessitates connecting an input-protection unit (IPU) between the pre-amplifier and the UHF PD sensor whose job it is to shunt the VFTO energy to ground before it can enter and destroy the pre-amp. There are many different ways to design the IPU units, but in all cases their design is challenging because the VFTO energy is several orders of magnitude higher than what the pre-amp inputs can withstand and also in the same general frequency region of interest (VFTO energy in GIS is considered to be concentrated in the tens of MHz, maxing out at about 100 MHz). Achieving IPUs that are robust and could stand up to field use and the requisite exposure to many switching transients and flashovers (during HV testing) and still exhibit low attenuation in the desired PD sensing frequency range continues to be one of the toughest challenges in UHF PD measurement methods for GIS.

Some vendors use RF pre-amplifiers at the UHF sensors (Figure 13.30), rather than at the PD instrument. First, from the standpoint of a pure gain budget, it may appear that placing the pre-amplifier at the GIS PD sensor, between it and the coaxial cable back to the multiplexer, is the same as just placing it between the cable and the multiplexer. As is seen from looking at parabolic dishes for radio astronomy, radar, or satellite TV, RF designers often place the pre-amplifier directly at the antenna (equivalent to our UHF PD sensor) in order to obtain the best possible SNR (signal-to-noise ratio). Also, GIS PD sensors are not stable RF impedances – they are far from resembling stable 50 Ω source impedances. By placing the RF pre-amp at the PD sensor's output, it acts as a buffer to drive the long cable with a stable source impedance [81]. However, pre-amplifiers at the sensor will increase cost and are more likely to be damaged by impacts. RF pre-amplifiers often have a minimum of 30 dB gain (more is preferable, but difficult to achieve in a compact enclosure) over a frequency range of ~50 to ~2000 MHz. Beyond 2 GHz, the cost of gain (along with other components) goes up quickly without significant benefit. A flat frequency response out to 2 GHz is necessary and sufficient to preserve the high-frequency content of incoming PD signals; this allows their sharp leading edges to be clearly observed on a fast digital oscilloscope for making accurate time-of-flight measurements to locate the PD source (Section 13.10.5).

IRF pre-amps will amplify any EMI/fixed interference sources getting into the PD sensor as well. In most cases, the operator simply avoids these background EMI sources when hunting for PD signals with the spectrum analyzer. However, in rare cases, there may be a very high-level EMI source present (for whatever reason), which exceeds the linear input overload threshold of the RF pre-amp. Depending on the design of the transient protection, many will let such high-level EMI through (since, as already explained, they are designed to block VFTOs that are of much greater energy levels). The result can be that the pre-amps are driven into saturation, which can have various different effects on their output signal, depending on their individual design and the frequency and range of the EMI source. This scenario can be especially bothersome in the case that the high-level EMI source is highly sporadic.

With sufficient training and practice, the operator can learn to recognize the typical tell-tale flickering of PD signals in the RF spectrum (if used) and to search for regions in the RF spectrum where the PD signals have the highest SNR (where the PD signal is large and the background noise level is low). Once such a region is found, the spectrum analyzer technique can also be used as (very expensive) tuned filter to down-convert the UHF PD signal to the LF IEC 60270 frequency range, which can then be fed directly into the input of a PD analyzer to produce a clean PRPD pattern for signal interpretation. The combination of the spectrum analyzer's ability to allow the operator to inspect the entire available frequency spectrum, avoid external interference, and use it as a tuned filter in combination with a PRPD display device continue to make it the most sensitive method for detecting PD in GIS using the UHF method. However, it requires much more expertise to use the spectrum analyzer method and of course can only be used for short durations (rather than continuous monitoring) in this manner.

For large HV test sections, another disadvantage with the spectrum analyzer approach is the multiplexing problem. As explained in the Section 13.4, some defects can have very low pulse rates, which means the operator needs to dwell on each sensor for 30–60 seconds or more to be satisfied there is an absence of PD. Meanwhile, the signals from all other sensors are not being surveyed. There is some possibility that PD sources that can suddenly turn on and off (such as certain types of low-level floating-potential or surface discharges) along with erratically behaving particles can be missed. The experienced operator learns to scan through the sensors at different speeds and intervals and develops a sense of where a sporadic signal might be present. Besides this "blanking" problem inherent with using multiplexers, the operator also has to be conscious of the problem of the portable high-voltage source that can overheat if driving a large test section during the offline testing. Online continuous PD monitory does not suffer from these problems.

Somewhat similar to and in some cases related to the use of multiplexers, despite its advantages in processing power for PD in GIS, the usual superheterodyne RF spectrum analyzer is actually not the perfect instrument for searching for PD signals because it is inherently a swept frequency instrument. In the case of a very low-level PD source (e.g. particle lying on insulation) at some distance from a given sensor, the case may be that only a handful (and in the worst case, one) resonance peak in the RF spectrum is excited by the PD signal. Since the spectrum analyzer is continually sweeping its resolution bandwidth "window" through the frequency spectrum, and since the PD pulses only have a duration of about 1–2 ns at maximum, the PD source has to "fire" at the moment the analyzer swept window is at the resonance frequency where the signal is visible at that particular sensor. Again, experienced operators learn to play with different combinations of sweep speed and dwell time on each sensor to overcome these problems and not miss PD that is there, but it is an important point to keep in mind.

Low-amplitude, low pulse-rate PD sources may only be "visible" on a few peaks in the frequency spectrum. Thus, the most important goal is to achieve the highest possible SNR to make low-amplitude PD signals clearly visible above the background noise "floor"; sufficient dynamic range is required to generate a clear and distinct PRPD pattern to be able to correctly identify the defect type. This means expending the maximum practicable resources on each link of the signal-processing "chain" [16]:

- the best available low-noise transistors in the pre-amplifiers,
- solid RF design of the IPUs and pre-amps,
- very low-loss and low dispersion coaxial cable (at least RG-214, preferably better),
- low-loss switching matrices inside the multiplexer, and
- high-quality interconnect cables (e.g. RG-223) to the spectrum analyzer and scope.

Besides having good quality, low-noise components, another easily overlooked aspect of RF PD measurements needs to be emphasized. PD signals are inherently very weak (<<1 μW at the sensor output) and stochastic in character; in other words, PD signals are noise-like. Because we are searching for what essentially resembles an electrical noise signal, if all we see while carrying out a PD measurement is what we think is background noise, we will assume there is no PD. Therefore, it is important that the operator carefully observes the background noise environment while taking measurements, especially if there are any odd changes or long periods of seeing anything, including interference. It is important to constantly check all of the components in the signal processing chain to make sure component gains and losses are in the range they should be. Even relatively easy to find PD signals such as a small particle or a void can be missed due to a partly broken connector or an IPU or pre-amplifier that has been damaged by an overload or switching transient.

Moving from sensor to sensor and observing only the ubiquitous background noise "floor," the operator may assume there is no PD present when in fact an actual PD signal is not even getting to the spectrum analyzer.

The preferred type of spectrum analyzer for GIS PD detection is the traditional superheterodyne swept-frequency type (e.g. Rohde & Schwarz FSL/FSVR or HP/Agilent 8594/E4403B). Compact "handheld" units that have appeared during the past 20 years or so do not have the range of control and flexibility of the larger more traditional portable units that have wider resolution bandwidths and, therefore, provision for the down-converted "video" signal output to the PD display instrument. Neither the new generation of "real-time" spectrum analyzers nor the FFT-based "spectrum analysis" capability of many fast modern digital oscilloscopes are capable of replacing this "tuned-filter receiver" functionality of superheterodyne spectrum analyzers.

A final remark concerning use of RF spectrum analyzers for onsite PD diagnostics concerns very low pulse-rate PD sources. For example, aged voids or other solid insulation defects which have transitioned from producing pulses every half-cycle when new, but which after some time (months, years) begin sporadic, bursty behavior. These sources can sometimes turn off for hours or days at a time. Obtaining an envelope spectrum using a swept superheterodyne spectrum analyzer will require allowing the analyzer a very long acquisition period, perhaps even several days. Any EMI sources will also be included in the acquired "MAX HOLD" spectrum and will also make generating a PRPD pattern in "ZERO SPAN" mode via the "VIDEO" output very slow and tedious. In the case of such work, e.g. searching for an insulation defect recorded by an online PD monitoring system, it would be advantageous to use a PD instrument with a broad-band front-end to obtain a PRPD pattern, because in such cases it will be more efficient at capturing isolated bursts of PD pulses from such low pulse-rate defects.

13.8.6 Broadband UHF PD Detection and Measurement

For online PD monitoring of GIS by end users, several PD system suppliers such as DMS-Qualitrol and Power Diagnostix provide either portable instruments specifically for GIS periodic online testing or permanently installed systems for continuous monitoring. Such PD monitoring systems are much easier for GIS owners to implement for condition assessment purposes and require much less expertise to use than RF spectrum analyzer measurements in Section 13.8.5.

The systems typically consist of a broadband (~250 to ~2000 MHz) front-end unit connected at or very close to the PD sensors, again with transient protection and sometimes RF pre-amplifiers. A small control instrument is connected to a computer, which also displays the PD signals. Typically, the broadband front-ends contain a combination of amplifiers driving an envelope detector/AM demodulator whose output bandwidth is around ~1 MHz, three decades below the UHF band (Section 8.3.5). The main instrument sends DC voltage (typically +15 VDC) up the coaxial cable to the front-end to power its electronics (if present). The main instrument receives the down-converted pulses from the front end. In a portable system, the operator walks through the GIS, connecting the front-end module to each PD sensor and monitors for a few minutes the display for any sign of PD activity. It can be helpful to have several inputs and several front-ends to allow simultaneous measurements at different sensors or across different phases to facilitate trouble-shooting. Figure 13.33 shows a commercial portable unit and its front-end connected to a PD sensor.

Since these broadband "PD checking" systems dispense with the spectrum analyzer (and the need for the specialized know-how required to use it), they are easier and faster to use for a quick look just to see if PD activity is present. However, being broadband, they act as a sort of signal "vacuum-cleaner" – if there are strong interference signals present, especially pulsed EMI, these systems do not have the capability of a spectrum analyzer to go to a quieter region of the spectrum.

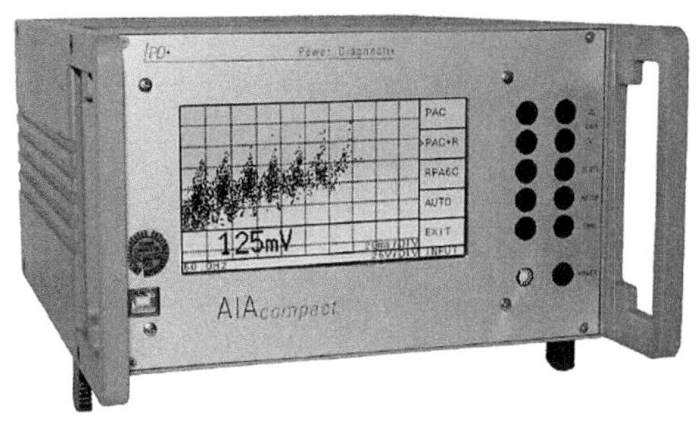

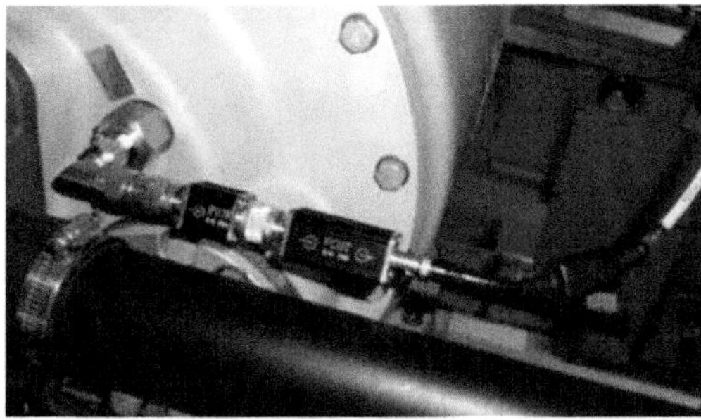

Figure 13.33 Portable broadband PD detection instrument. *Source:* Diagnostix Systems GmbH.

Still, in most cases, they can provide the GIS OEM or substation owner a useful snapshot to know whether there is PD in operating GIS. Once PD is detected, further investigations using a spectrum analyzer will provide more information. These portable and continuous online systems primarily depend on UHF detection and the GIS aluminum enclosure to reduce the risk of interference causing false positive indications.

Further information about online PD monitoring systems is contained in Section 13.11.

13.9 GIS Routine Factory Test

Routine factory testing of GIS is normally done in compliance with IEC 62271-1 (clause 8.2), IEC 62271-203 clause 7.1 [4], IEC 60060-1, IEC 60060-2, and IEC 60060-3 [5–7]. The equivalent IEEE documents are in [8–11]. The PD detector works in the IEC 60270 frequency range, i.e. below 1 MHz. The key outcome from such tests is the PD inception voltage (PDIV) and the PD extinction voltage (PDEV) (Section 10.2), with a recommended sensitivity of ≤1 pC. An example of the equipment used for factory PD tests is shown in Figure 13.34. The voltage steps for a factory test cycle are shown below in Figure 13.35.

An example routine GIS factory test is carried out in the following steps (corresponding to the numbering in Figure 13.35):

1) The test (high) voltage is raised from zero[2] to the specified power frequency withstand voltage V_{PFVW}. (see Tables 2 and 3 of IEC 62271-203 [5]) (If it is not possible to reach 0 V, then the test voltage must be raised from well below V_{PD}.)
2) The test voltage is then maintained at V_{PFVW} for a time $t_1 \geq 60$ seconds.
3) Thereafter, the test voltage is gradually reduced to the specified partial discharge test voltage V_{PD}. If PD activity is observed during t_1, the extinction voltage V_e (indicated by the red dot in Figure 13.35), the voltage level at which the PD signal disappears (i.e. falls below the background measurement noise level) is recorded.
4) At the specified partial discharge test voltage level V_{PD}, the voltage is maintained for a time $t_2 > 60$ seconds and the PRPD instrument is started. During this period no phase-resolved PD activity is permitted, i.e. no PD activity above the specified pass/fail threshold is allowed.
5) Following the interval at V_{PD}, the test voltage is decreased to zero.

2 If PD activity is observed and diagnostic measurements are carried out, the test voltage need not be returned to 0 V, but should be lowered well below the observed PD inception voltage V_i and then slowly raised again to capture the moment PD activity resumes; this defines the PD inception voltage V_i which should be recorded in order to help understand what type of defect is causing the PD signal activity.

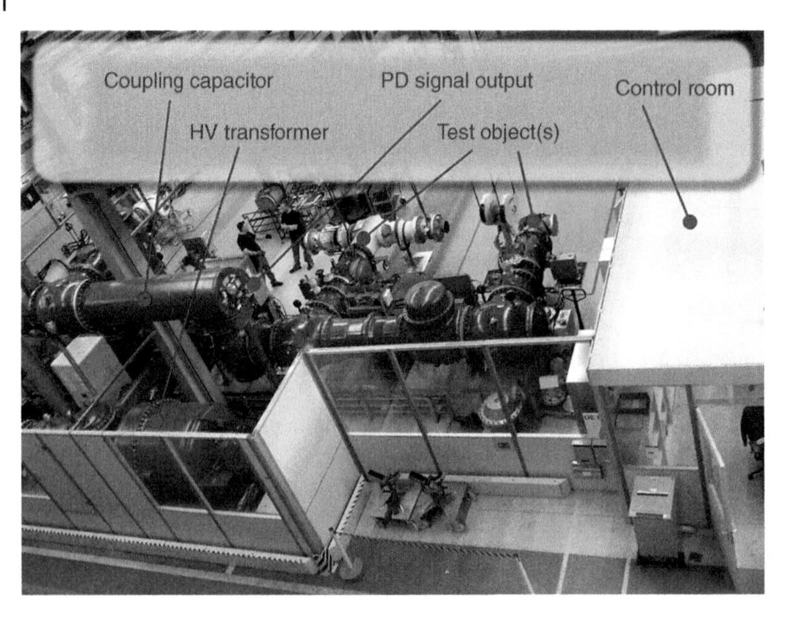

Figure 13.34 Factory test bay for conventional (IEC 60270) PD testing of GIS. *Source:* ABB.

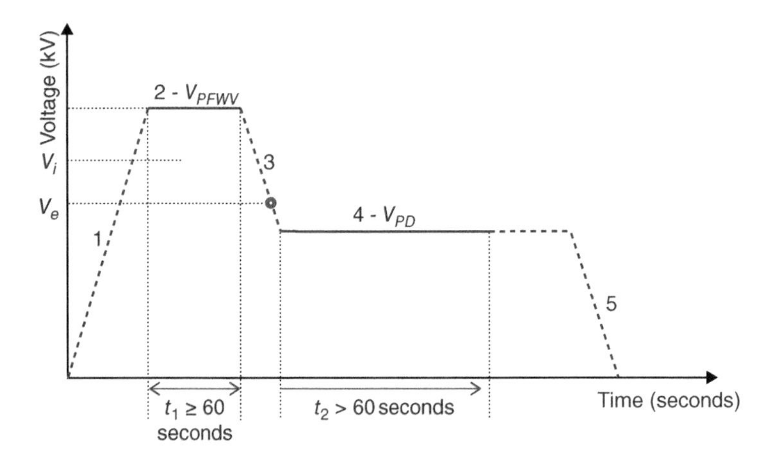

V_{PFVW} = power-frequency withstand voltage, the highest voltage the EUT is subjected to per IEC
V_i = PD inception voltage (PDIV)
V_e = PD extinction voltage (PDEV) – indicated by the red dot
V_{PD} = partial discharge test voltage, at which PD is recorded for an extended time, e.g. five minutes.

Figure 13.35 Example of voltage level steps for routine factory test of GIS modules.

In some cases, the GIS end user may specify a certain PD test voltage V_{PD}, in which case the above steps should be carried out and documented accordingly. If V_{PD} is not specified differently by the customer (as is the usual case), then V_{PD} is calculated according to the following equations from the applicable standard (IEC 62271-203) [4]:

- For single-phase encapsulated GIS: $V_{PD} = 1.2 * Ur/\sqrt{3}$ (units of kV)
- For three-phase encapsulated GIS: $V_{PD} = 1.2 * Ur$ (units of kV)

(here *Ur* refers to the nominal phase-to-phase operating voltage of the GIS)

Once a PD source has been detected in the factory routine test, the next task may be to locate the PD defect (in the case its location is not obvious). Since the upper frequency range of IEC 60270-based PD measurement is limited to 1 MHz, corresponding to a wavelength of 300 m, it is not possible to employ “time-of-flight” location algorithms that are used with UHF frequency detection. In such cases, it is normal to use manual acoustic (ultrasonic) instruments to quickly locate PD sources in such daily PD testing work. Additionally, when planning a GIS test bay (Figure 13.34), it is strongly recommended to install UHF PD sensors in strategic locations for use in locating PD signals either in the GIS being tested, or in the test bay itself, should they occur in the future.

13.10 Onsite PD Measurement of GIS

Nowadays PD measurements almost always accompany the HV factory acceptance tests of newly erected GIS on site, extensions to existing GIS, and in some cases following extensive highly intrusive repairs, typically at end-user request. The applied high voltage does not come from a stationary 50/60 Hz HV transformer as in the factory test bay, but rather from a portable HV test-set. Formerly site acceptance tests consisted only of a high-voltage test. However, decades of experience have shown that the most effective means to guarantee a high level of operational reliability for a new GIS (or following extensive repair work) is a carefully performed, high-sensitivity PD measurement during final onsite acceptance testing [15, 17, 82, 83]. If any PD activity is observed, the defect is located and removed, and the affected section re-tested. Despite the higher costs generated by the additional equipment and time required for the PD measurements, this has been shown to be the best way to deliver a defect-free GIS with a low probability of early “teething” failures following energization.

Many GIS OEMs continue to specify 5 pC as an acceptable pass criteria (based on IEC 62271-203 [4] (para. 6.2.9.102): “The maximum permissible partial discharge level shall not exceed 5 pC. . .”). This may be false economy. The problem is that, as already stated, it is impossible to determine the charge level of a PD defect using RF techniques (i.e. the UHF method). The situation can occur in which a GIS module contains a low-level PD defect whose apparent charge is below 5 pC, therefore allowing it to pass the routine factory test and be shipped to site. But then, if a UHF PD monitoring system is installed on the GIS and it picks up the signal from the defect, the end user most likely would not accept it. Rightly so, because, depending on the defect type, there is a chance that it will eventually lead to a failure. For the same reason, any PD defects that appear, i.e. any phase-resolved PD activity observed during the onsite acceptance test, should be located and removed if at all practicable.

When a low-level PD defect is picked up during onsite acceptance tests (or later, from a continuous PD monitoring system), often the GIS OEM attempts to demonstrate its charge level is below 5 pC by, e.g., comparing the received UHF signal level to the results of the CIGRE Step 2 onsite test. Besides the fact that it is impossible to assess a defect’s actual apparent charge from its UHF signal, dealing with such low-level signals in this way is even more risky for other reasons. Being close to the detection noise floor, it may be impossible to generate a PRPD pattern with high enough resolution to identify the defect type (and thus enable an accurate assessment of its risk). Also, certain PD defects such as small protrusions or particles lying on insulation surfaces exhibit low UHF signal level but pose a high risk of flashover under switching or lightning impulse conditions [39, 76]. Especially during onsite acceptance tests, it is better to take the time to locate and remove the defect to achieve the highest possible level of quality and reliability before the GIS is (re-) energized.

13.10.1 High-Voltage Resonant Test-Set for Onsite/Offline Testing

In some cases, for example following repair work on an older GIS, the re-worked section may be energized from the grid and observed for an agreed-upon time (e.g. one to several hours); this is commonly known as a "soak test." However, for onsite acceptance testing of new GIS, simply connecting it to the grid and monitoring for PD is unacceptable and also hazardous. It is unacceptable because nominal operating voltage does not sufficiently stress the dielectric integrity of the newly assembled GIS, hazardous because if a "teething" flashover occurs, the damage could be extensive, given the power available from the network.

For these reasons, portable AC test-sets are brought to site and used to generate the necessary test voltages to satisfy the relevant IEC or other standards. Today this is a relatively settled technology and most systems are designed to operate in a similar way: they set up a high Q-factor resonance circuit consisting of the capacitance of the GIS combined with either fixed inductors fed with a variable-frequency power supply (as is used for transformers – see Section 15.8), or a fixed-frequency power supply with variable inductors (Figures 13.12 and 13.36). Clearly such an arrangement will result in the applied AC test voltage not being synchronized with the AC mains, nor will it typically be equal to the nominal AC operating frequency (50/60 Hz). The standard organizations (IEC, IEEE) understand this and thus allow onsite testing of the GIS at frequencies in the range 10–300 Hz [15].

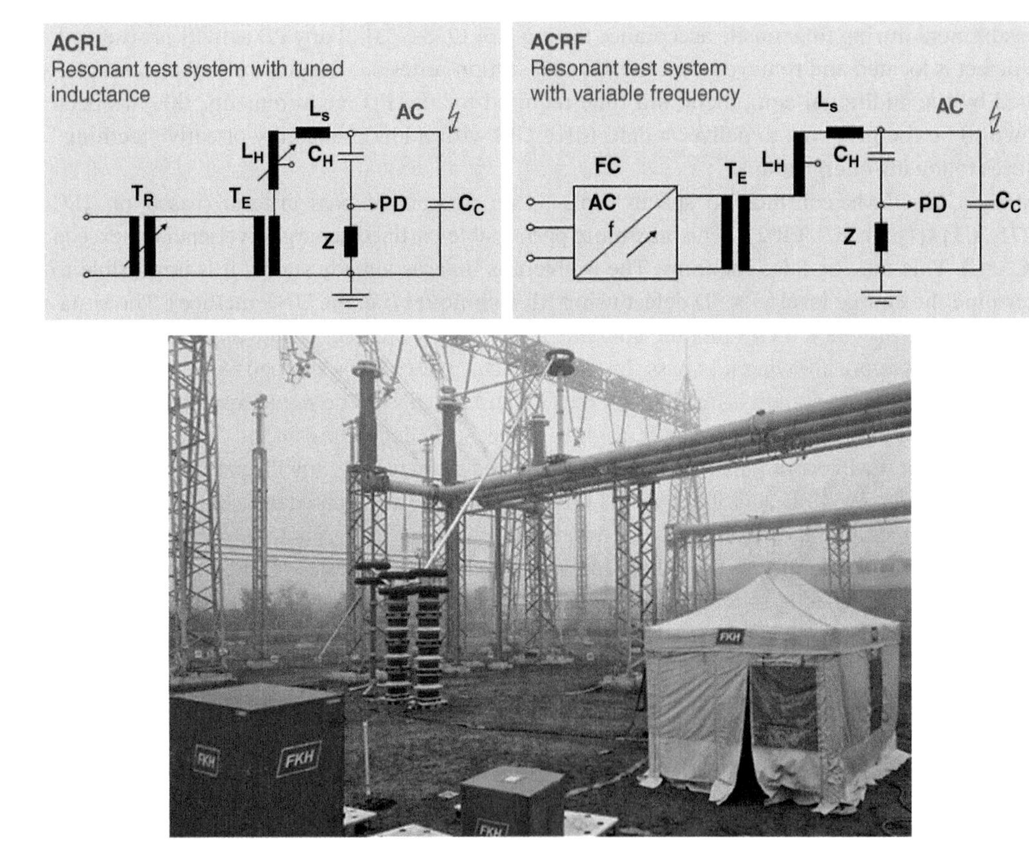

Figure 13.36 Types of portable resonant test-sets for onsite high-voltage test of GIS. *Source:* [15] top, FKH (photo).

Primary power comes from the usual three-phase mains power available on site, or if the local mains supply is inadequate, they can be energized from a portable generator.

It is outside of the scope of this book to delve deeper into the complexities of resonant test-sets, but a couple of points should be emphasized:

- In every case these devices pose hazard to both man and machine; extreme care and caution must be exercised setting them up with special attention being paid to grounding and personnel safety.
- The variable-frequency test-sets generally rely on high-power semiconductor devices – e.g. power MOSFETS, thyristors, or IGBTs – to generate the AC sine-wave output excitation (see also Section 15.8.2). These devices switch high currents with risetimes in the range of 10–30 ns; it is extremely challenging to prevent very high levels of EMI escaping them. This is especially problematic if attempting to apply conventional IEC 60270 techniques for PD measurement; the switching impulses will often obscure the PD measurement. Gating techniques (Section 9.3.9) can help somewhat, but the magnitude and jitter of the EMI will result in the PD input essentially being blocked for much of the time. UHF techniques will generally perform much better than IEC 60270 methods (their high pass "knee" frequency being far above the frequency range of the EMI generated by the power electronics).

13.10.2 Problems with Applying IEC 60270 PD Method Onsite

As explained in Section 13.6, conventional charge-based PD measurements according to IEC 60270 are well suited for routine factory tests because they enable setting pass/fail criteria based on a predetermined or agreed-upon PD apparent charge level. This functions very well for individual GIS modules, which present/manifest themselves as compact lumped capacitors. In addition, it is important to recall that the PD measurement sensitivity in the IEC 60270 measurement circuit is fundamentally determined by the ratio between the capacitance of the EUT – the GIS module – and the coupling capacitor (Section 6.5). However, once the individual GIS shipping units arrive on site and are assembled into a complete substation, the total capacitance can add up to several nanofarad, thus requiring a very large coupling capacitor to achieve the requisite high PD sensitivity. This leads to the following problems:

- the large coupling capacitor presents a further high load to the portable HV power supply
- high voltage coupling capacitors are heavy and delicate; shipping costs are high
- to achieve good EMI rejection, the coupling capacitor needs to be flanged onto the GIS
- connecting the coupling capacitor, therefore, requires opening the GIS (adding further risk)
- floor space, supports, ingress, etc. are required for the coupling capacitor in the building
- mitigating the EMI susceptibility of IEC 60270 measurements is even more difficult on site.

Besides all of the physical problems with the coupling capacitor itself, the last point is at least as important if not more so. The background interference (EMI) problems encountered when applying IEC 60270-based measurements in the factory (described in Section 13.6.1) are far worse in the onsite environment. Conducted EMI "riding" on onsite grounding systems is often an order of magnitude (or more) greater than encountered in the factory. Lastly, if there are sources of external surface discharges or corona, i.e. from nearby overhead lines or an adjacent AIS substation, the external EMI levels can become prohibitively high, and mitigation techniques that might work under the controlled setting of the factory are insufficient or impossible.

IEC 62271-203 [4] quotes a maximum PD acceptance level for GIS as 5 pC, but that is assuming a charge-based PD detection is being implemented for the onsite test. Yet in the same document,

Annex C correctly states: "In actual site conditions, a noise level below 5 pC is hard to achieve." If we look to IEC 60270 concerning the minimum reliable PD level, it states: ". . .the background noise level should preferably be less than 50% of a specified permissible partial discharge magnitude (. . .)." This problem is exacerbated by the coupling capacitor problem and the much higher EMI disturbances encountered on site, cited above.

For these reasons, it is extremely difficult for onsite PD measurements to achieve the high PD sensitivity required to guarantee the assembled GIS is free of potentially harmful PD defects. Another basic disadvantage of applying IEC 60270-based methods for onsite PD measurement of GIS is that if a PD signal is detected, other methods are required to locate the source. Indeed, it was in response to all of these problems and to overcome them that provided the motivation in the 1980s to develop and apply the UHF method for onsite measurement of PD in GIS.

13.10.3 Applying the UHF Method for Onsite/Offline PD Measurement

The main goal of UHF PD detection and measurement for onsite/offline acceptance testing is to maximize the signal-to-noise ratio (SNR) of the PD signals to all forms of background noise and interference. This is especially true for onsite UHF measurements for the following reasons:

- it is the last chance to detect, locate, and remove any remaining PD defects in the GIS before energization,
- a high SNR enables clean, high-resolution PRPD patterns to be obtained, which are necessary to recognize the PD defect type and thus the corrective action needed,
- a high SNR enables clean TOF (time-of-flight measurements – Section 8.4.2), which are used to locate the PD defect(s). The SNR of the incoming PD signals is directly related to the errors in TOF spatial resolution [76]; see Figure 13.39 below.

If the GIS being tested is not equipped with an online PD monitoring system, it is recommended that the narrowband methods described previously in Section 13.8.5 be employed to achieve the highest possible sensitivity (SNR) to any PD sources present. The only additional recommendation is, when using a multiplexer connected to many PD sensors (i.e. on a large test section), it is important to resist the temptation to step the multiplexer rapidly through each PD sensor. Signal activity should be observed at each PD sensor long enough to be sure there are no signals, ideally 60 seconds, but for at least 30 seconds, and all of the sensors in a test section should be viewed at least twice. This is, of course, assuming the GIS *has* sensors. If this is not the case, e.g. testing is being carried out on an older GIS after repair or extension or as part of a general condition assessment, then external UHF sensors may be applied at appropriate locations (Section 13.8.1 and Figure 13.19). In such cases, it is also recommended to attempt to carry out a provisional CIGRE Sensitivity Verification check, by injecting pulses of a reasonable amplitude (e.g. 5–10 V) at each temporary sensor location and looking to see if the signal is visible at the next sensors. Such a provisional test, while not strictly in line with the CIGRE TB 654 recommendation [13], will at least give some indication of the overall sensitivity of the set-up being used to monitor for PD activity.

If the GIS being tested is equipped with a permanently installed online PD monitoring (PDM) system, and it has passed the CIGRE Sensitivity Check (Section 13.8.3), it is recommended to use the internal UHF sensors for the PD measurement during the acceptance test. This is assuming that the PDM system is equipped with special software for use during commissioning tests (which speeds up sensor polling to update a multi-sensor display in near real-time), plus the capability to synchronize to the phase of the AC waveform of the high-voltage resonant test-set power supply (10–500 Hz, per IEC 60060-3 [7]).

Assuming the PDM system fulfills the previous two requirements, using it to monitor for PD activity during the onsite/offline acceptance tests offers two important advantages:

- dispensing with cabling all of the PD sensors in the test section to the multiplexer, and
- the PDM system will allow simultaneous observation of all of the sensors on the HV test section and thus eliminate the "blind time" due to the sequential measurement of each sensor, inherent to multiplexed systems.

During the high-voltage part of the acceptance test, it is often the case that any defects present will make themselves obvious. Careful observation of the whole test section and remembering which sensors showed activity may help find the defect when it becomes less active as the voltage is reduced to the "PD checking" level.

Another advantage of using the installed PDM system for the onsite/offline HV acceptance test is that the system can store any signal activity observed – sometimes known as the GIS "fingerprint" – for future reference. The signals from the sensors should also be stored at the time of the test, just before energization, to record external interference signals (or e.g. signals coming from an existing GIS or other externally connected equipment) which will be useful for future reference.

If, during the HV test, a PD signal is observed, unless its location is obvious (from the displayed signal strength relationships) or the installed PDM system can perform localization (based on distribution of a high-resolution clock signal and signal time sampling at each PD sensor), it will be necessary to have additional diagnostic equipment on hand. Essentially this would consist of a diagnostic kit described in Section 13.8.5 and shown in Figures 13.31 and 13.32 [16]. If the PD signal cannot be localized as described above, the diagnostic instruments will then be used to locate it using either acoustic techniques (especially effective if the defect is a moving particle) or the UHF TOF method as explained below in Section 13.10.5.

In either case, whether using narrowband PD detection or an installed PDM system, the principal problem with all RF-based methods remains: it is impossible to assess charge based on the UHF signal amplitude [1, 3, 4, 13–17, 76]. Therefore, when employing the UHF method for onsite PD tests, it is strongly recommended to act on any phase-correlated PD activity observed, meaning locate the defect and remove it. If it proves impossible to locate, the signal's behavior should be carefully documented in the test report for later reference.

Another tip that can be helpful when carrying out UHF PD measurements with either narrowband or broadband (i.e. the PDM system) equipment: even though it may already have been carried out, it can sometimes be useful to inject pulses in the part of the GIS being examined using the CIGRE TB 654 Step 2 onsite test. Doing so may help understand unusual or "missing" signal behavior.

13.10.4 Interference Encountered During Onsite UHF PD Measurement

When Boggs, Hampton, and others demonstrated the powerful advantages of employing the UHF method for onsite PD measurement of GIS [14, 16, 29, 53], especially regarding its higher immunity to background EMI (compared to IEC 60270-based methods), the background EMI environment was rather quieter than it is today. Much has changed over the past 40 years and now the VHF/UHF frequency regions are filled with man-made interference, most typically various mobile communications and radar (e.g. L-band). One of the most obvious and ubiquitous signals that simply did not exist in those early days is mobile (cell-phone) telephony, specifically the GSM I and II bands, located at 900–1000 MHz and 1800–2000 MHz, respectively (the actual frequency allocations differ by country and region). As 3G, 4G, and 5G have been rolled out, operators in some countries and regions are already shutting down 2G networks,[3] though in others they will probably remain in operation for some time.

Besides the 2G mobile telephone bands, there is a plethora of communications signals operating in the VHF/UHF region, which were not there in the early days. Industrial VHF mobile multi-party radios built by companies such as Motorola and Yaesu are typically found operating around 400 MHz, being often observed at GIS in the Middle East, as well as at large thermal power generation and desalination plants. It can sometimes be the case that going lower in frequency is helpful in certain diagnostic situations, where one runs into the FM broadcast band (88–108 MHz). Besides these few examples, the reader is directed to the ITU (International Telecommunications Union) frequency allocation charts for each country, in which they will note a rich collection of emitters in the VHF and UHF bands. These may enter the GIS through the usual apertures (overhead line bushings, current-transformer gaps, compensators, cable terminations, etc.) and pop up in their PD measurements. Use of a spectrum analyzer usually rapidly reveals these signals as (more or less) fixed-frequency emitters that can then be avoided when searching for actual PD signals.

13.10.5 UHF PD Source Location: The Time-of-Flight (TOF) Technique

One of the most important advantages of employing UHF methods for PD assessment of GIS is the ability to locate the source of PD signals using the so-called "time-of-flight" (TOF) method (Section 8.4.2) [3, 16, 49]. This is based on connecting a fast digital storage oscilloscope (DSO) to two PD sensors via two equal-length coaxial cables (note: RF pre-amplifiers as described above in Section 13.8 are strongly recommended to enhance SNR). The operator adjusts the DSO to display the incoming PD pulses from the two sensors, and then uses the DSO to measure the difference in the time-of-arrival of the two signals. This time difference is then used in the equation shown in Figure 13.37 to calculate the location of the PD signal source. Screenshots from typical TOF measurements are shown in Figure 13.38.

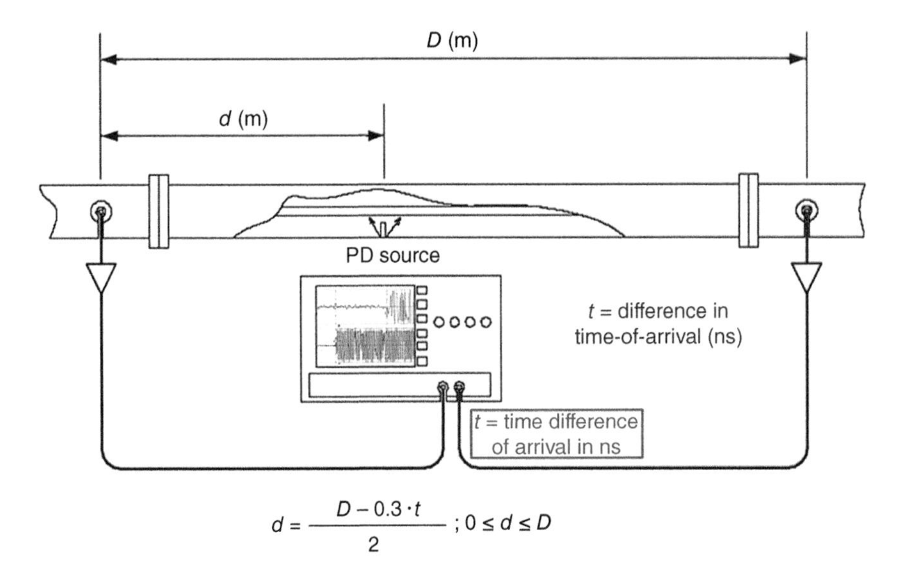

Figure 13.37 Illustration of time-of-flight method used for location of PD sources in GIS. *Source:* Adapted from [3, 16, 19].

3 This is a pity because checking for the presence and healthy signal level of 2G signals in the 100–2000 MHz PD measurement band functions very well as a check that the PD measuring system is working.

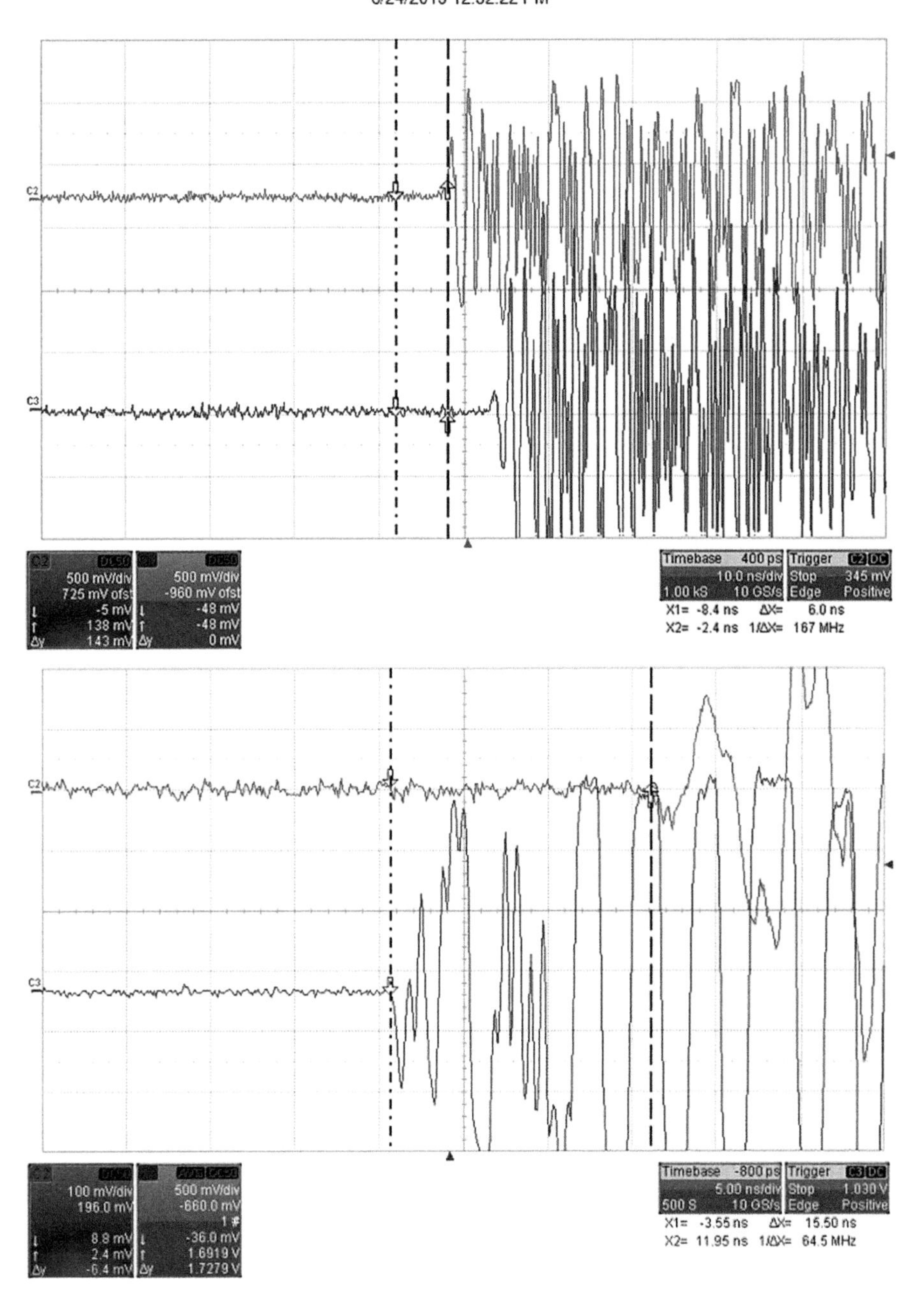

Figure 13.38 Actual examples of clean onsite TOF measurements.

An important condition for successfully locating a PD source using the TOF technique is that the SNR of both incoming time-domain signals is sufficient to get the most accurate fix possible on the moment the signals climb out of the background noise "floor." The combination of a weak PD defect signal, PD sensors that are a bit too far apart, and/or the presence of strong pulsed interference leaking into the GIS (e.g. airport radar coming in through overhead line bushings while

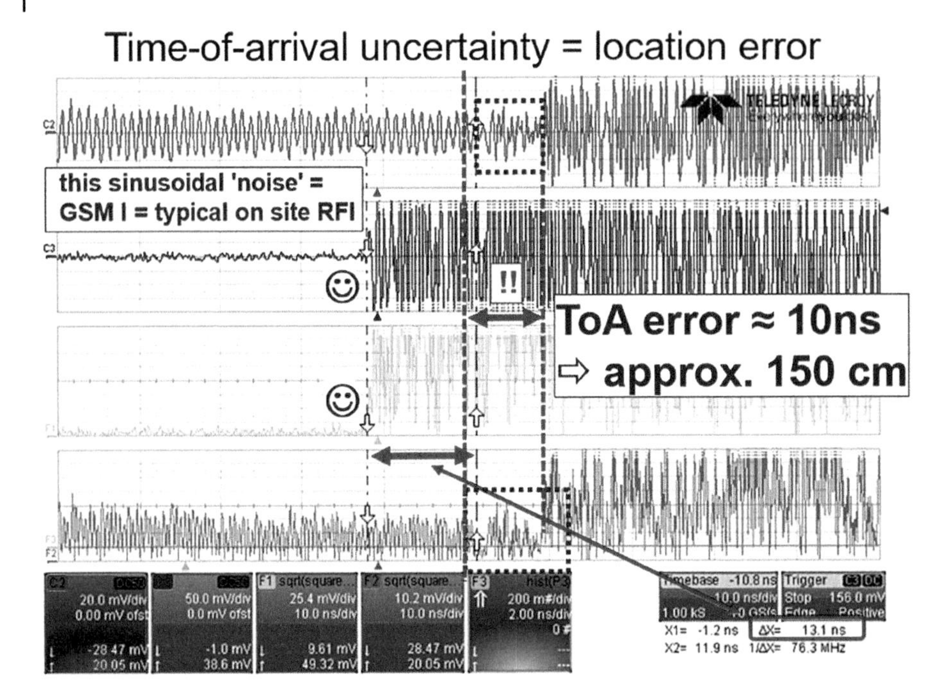

Figure 13.39 How poor SNR effects the precision of TOF defect location.

hunting for a small particle in a feeder) can prove very frustrating and time-consuming. If the PD sensors being used to pick up the signal to perform the TOF measurement are far apart, poor SNR may make it difficult to locate the exact start of one or both of the incoming signals. It is important to understand that an uncertainty in the signal start time will lead to an error in the difference of time-of-arrival, which will in turn lead to a spatial error in locating the PD source; as shown in Figure 13.39, this can be quite substantial. Whether a PD source is on one side or the other of a spacer insulator can lead to substantial differences in the time required to access the compartment of interest. Modern DSOs with sophisticated triggering features can help in such situations. Swapping which incoming signal is the primary trigger can often lead to unexpected success. Setting the oscilloscope in single-shot mode, carefully setting the cursors, taking the time difference, then repeating this several times, each time trying to forget the last reading or the expected reading can help. After taking 10–20 readings, they will often converge. In a similar vein, very sophisticated DSOs may allow setting up an automated delay-time measurement, dumping the values into a histogram after each shot, which can be evaluated after acquiring the signal for long periods (e.g. overnight or over a weekend, in the case of defects exhibiting very low pulse rates).

In principle, TOF measurements are straightforward and often work out quite quickly; the PD source is located within half an hour or so, and the determined location makes sense. Other times, problems crop up. Strange values that lie outside the expected time-delay window usually indicate something obvious such as a strong signal reflection owing to an RF impedance discontinuity in the measurement path; change settings swap the triggering around as mentioned above, and try again. Also, as noted above in Section 13.8.5, anomalous values or other odd behavior may indicate component problems along the signal chain. Reduce the high voltage to zero (if testing offline), check all of the components for proper functioning, and repeat the TOF measurement.

If faced with odd timing values, another point to consider is the measurement of the physical distance D [m] in Figure 13.37. There are philosophical opinions as to how to take the measurement, running the measurement tape strictly down the center-line of the GIS (the "follow-the-inner-conductor" method) vs cutting corners slightly (the "RF signals will snake their way along the shortest possible path" method). The best way is to take both values and write them down, then take some delay time values from the DSO, and determine what makes sense. In most cases the most probable location of the PD defect can be found with a few such iterations.

13.11 Online Continuous PD Monitoring (PDM) of GIS

Soon after the UHF method began being applied to detect and locate PD in GIS [29], continuous online monitoring of GIS began in the late 1980s by connecting sensitive broadband UHF receivers to PD sensors distributed throughout the GIS [50]. The goal was to detect PD defects posing operational risk and shut down the affected section of the GIS before a flashover could occur (thus leading to both an outage and serious damage to the GIS). Since those early days, there has been a steadily increasing number of GIS substations equipped with PDM systems along with a large body of technical publications covering all aspects of the subject [19, 84–87].

Assuming the PDM system is well designed, and the CIGRE Sensitivity Verification (Section 13.8.3) has been passed, any signals appearing in the GIS whose RF signal strength exceeds the CIGRE Step 1 level (usually, a 5 pC hopping particle) will trigger an alarm (event) and will also be stored in the PDM system memory. The capability to observe and analyze the PRPD patterns and their temporal behavior is the most powerful advantage of these systems. Also, besides actual PD, this capability is often useful to identify and reject external interference signals (either EMI, or PD and corona external to the GIS) as well.

Typically these systems trigger an alarm to notify the end user that signals indicating PD activity have been registered and action should be taken. For the foreseeable future, the systems have not reached the level of sophistication and reliability necessary to be directly integrated into the substation control and protection system. That is they are not being used to shut down automatically a section of the GIS based on PD activity. They remain stand-alone systems with their alarm outputs connected to the utility's network control system. When a PDM system issues an alarm, a PD expert needs to look at the system's output display (typically PRPD patterns and a record of signal activity) and then go to site to carry out further investigations.

Despite the ongoing trend to install PDM on GIS – even down to 66 kV "distribution" equipment – there are several general problems with these systems:

- Although prices have come down in recent years, they remain relatively expensive.
- The typical outputs – PRPD patterns and signal behavior – require an expert to interpret.
- It is well known that flashovers can occur with no or only very short prior warning from the PDM.
- The systems continue to be plagued with high rates of false positive alarms [87].
- End users often (mistakenly) believe that installing PDM improves the reliability of the GIS itself.

At the time of writing, CIGRE working group D1.66 is approaching completion of a new Technical Brochure: "Requirements for UHF PD monitoring systems for gas insulated systems." Publication is expected toward the end of 2023. This document should be reviewed when specifying a GIS PDM. PDM systems are complex, and substation and end-user-specific. End users looking to install and operate PDM systems need to learn as much as they can about what the systems' actual capabilities are before deciding whether to invest in them. More importantly, it needs to be determined how the PDM systems will be integrated in the overall GIS operational concept and asset management program.

13.11.1 Typical GIS PDM System Components

UHF PDM systems generally consist of UHF input protection, amplifiers, filtering, and detection circuits – typically some variation of AM demodulators or envelope detectors (Figure 8.4) to convert the broadband UHF signals to a much lower bandwidth of around 1 MHz (i.e. equivalent to the IEC 60270 frequency range). This is different from the approach of VHF/UHF systems for equipment such as rotating machines where the actual PD pulse shapes are captured, allowing for instantaneous recognition of the pulses as due to PD or interference (Sections 8.4.3, 16.9.3.1, and 16.9.3.3).

The second part of the PDM system digitizes (Section 8.3) the incoming analog signals, applies AC line-frequency phase synchronization to them, and aggregates them together to forward them (usually via a fiber-optic network) to the third part of the system, the main control and display rack in the GIS control room. Note that the boundaries between the front-end components, A/D conversion and synchronization, and the way sets of sensors are combined, vary between system manufacturers, but the basic functionality is very similar. In the control room, the central "head end" rack contains the control and communications for the PD sensor network along with an industrial PC, data storage, and communications interfaces. The latter connect the PDM system control rack to the GIS control system alarm panel (typically based on IEC 61850) along with links to the end user's IT network and/or to the outside internet to allow for remote access by the PDM or GIS OEM. Figure 13.40 shows simplified component views of typical PDM systems. These systems continue to be developed into the present day, but their basic concept and building blocks remain about the same.

13.11.2 GIS PDM System Alarm Triggering

Because of the combination of the highly stochastic nature of PD signals together with the problem of false positive alarms due to interference, one of the biggest challenges facing both PDM system designers and end users has been trying to determine when an incoming signal indicates actual GIS PD and thus warrants an alarm.

Assuming the GIS OEM is using high-quality UHF PD sensors and installed them throughout the GIS, and that the combination of the PDM system and the sensors has passed the CIGRE Sensitivity Verification (Section 13.8.3), the end user will expect the system to notify them (via an alarm or alert) if PD equivalent to a 5 pC particle (the typical reference defect for the CIGRE test) appears anywhere within their GIS. In other words: any pulses received that exceed the UHF amplitude (recorded during the CIGRE Sensitivity Verification Step 1 test, Section 13.8.3) should trigger a system alarm. If we think of a GIS PDM system as being analogous to the fire alarm and smoke detector systems in a modern office building, this would be the reasonable assumption and expectation – we want to know if something is "burning" in the GIS.

However, with real GIS PDM systems, setting the "alarm" thresholds at such a low amplitude level (e.g. equivalent to a highly sensitive smoke alarm in the analogy used above) would result in most of the GIS PDM systems in the world being in a constant state of alarm. There are simply too many sources of interference to run the alarms triggering thresholds this close to the noise floor [87].

Therefore, rather than relying on only the incoming signal crossing a simple amplitude threshold, PDM vendors typically set up algorithms, which take into account the dynamics of both amplitude and temporal behavior of the incoming signals. The typical set-up is to count how many incoming pulses exceed a set amplitude threshold in a given time interval. This initial low-level

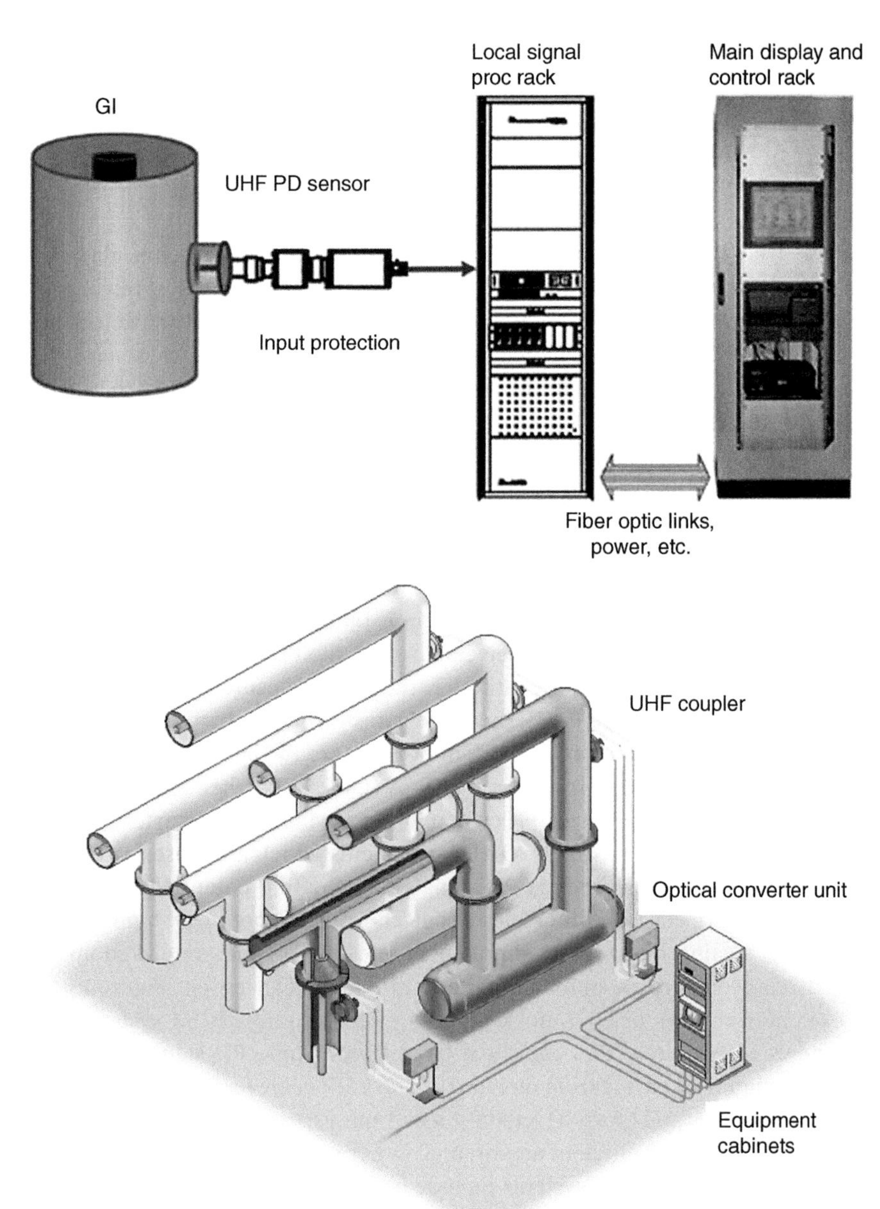

Figure 13.40 Basic building blocks of typical contemporary GIS PDM systems. *Source:* Hitachi (top), DMS/Qualitrol (bottom).

threshold might be a few dB above the level recorded by an equivalent 5 pC particle during the Step 1 test, as an example. A low-level alarm (sometimes called "an event") will be triggered if say 5–10 pulses exceed the low-level threshold in a 15-minute time window (seen from the perspective of pulses-per-halfwave, this is still a very low threshold in relation to the powerline frequency). A higher level warning or alarm will be triggered if many more pulses exceed the low-level threshold, if that signal remains active for several 15-minute time intervals, or if pulses start to regularly exceed a second, higher level, say 50–60% of full-scale display amplitude.

The above is only one simple example of a strategy for deciding which incoming pulses should trigger an alarm output on a GIS PDM system. It represents a compromise between missing what might be a critical defect (e.g. a large particle suddenly appearing in a disconnect switch) vs constant false alarms arising from external disturbances. Setting up the alarm algorithms and thresholds to try to capture low-level PD defects which pose a risk to operation vs trying to avoid a constant stream of false alarms continues to be a challenge for the GIS OEMs and PDM suppliers, as well as for the end users. It is an ongoing topic of lively discussion in the various technical working groups and as of this writing the problem remains a work in progress. Instead of just relying on algorithms that look at the combination of the amplitudes and statistics of occurrence of the incoming pulses as described above, artificial intelligence is already being used by some vendors to first determine if the pulses are associated with known PD defect patterns. However, these PD interpretation engines are still in the early stages of development and some way to go to achieve the required level of accuracy and dependability.

13.11.3 The GIS PDM System False Alarm Problem

Among online PDM systems for electrical equipment, those for GIS probably suffer the highest level of false alarms (false positive indications) The problem is inherent to their design, which is based on installing high-gain, broadband RF receivers on all of the PD sensors throughout a GIS. As discussed previously in Section 13.8.1, although the GIS enclosure is quite good at shielding against external RF interference (EMI), there are paths through which external EMI can leak into a GIS. The most obvious and most efficient of these are the bushings connecting the GIS to outdoor overhead high-voltage transmission lines (OHL). Together with the OHL conductors themselves, the bushings act as very efficient antennas, which pick up external EMI and funnel it into the GIS. Besides the bushings, OHL components which are loose, corroded, or contain sharp edges and themselves produce PD signals (e.g. floating potential or surface discharges) or corona (from sharp edges or other surface defects), often at both very high amplitude levels and pulse rates. Again, these signals simply flow into the GIS and are picked up by the PDM system, which has been purpose-built to monitor for PD levels of a few pC. Figure 13.41 shows a couple of examples of outdoor OHL noise coupling into an online PDM system (the right-hand graphic shows PRPD plots from the three phases with same PD signal offset by 120° – a sure sign that the signal is coming from outside the GIS). HV cable terminations, which often have open insulation flanges and (partial) separation of grounds, can also act as ingress for EMI, although they are usually not as efficient an antenna. Other interference sources that are known to trigger false alarms in GIS PDM systems include high-pressure gas-discharge lamps (which, however, are disappearing as LED lighting takes over), relay sparking (e.g. on cranes), and other switching transients, even from other nearby high-voltage equipment, aircraft and shipboard radars, DECT cordless telephone systems (often used for the wireless remote control of overhead cranes), etc.

Various combinations of filtering (Section 9.3.4) and gating (Section 9.3.9), as well as software-based interference suppression methods (Section 8.9) can often help to reduce the impact of external interference. However, all of these measures affect the PDM systems' sensitivity to actual PD signals. GIS end-user specifications often demand that the PDM systems suppress false alarms, but the overall physical context of PDM monitoring on real operating GIS essentially makes this impossible, because many of the disturbance signals responsible for triggering false alarms themselves closely resemble actual PD signals, thus blocking them will likely compromise the PDM system sensitivity.

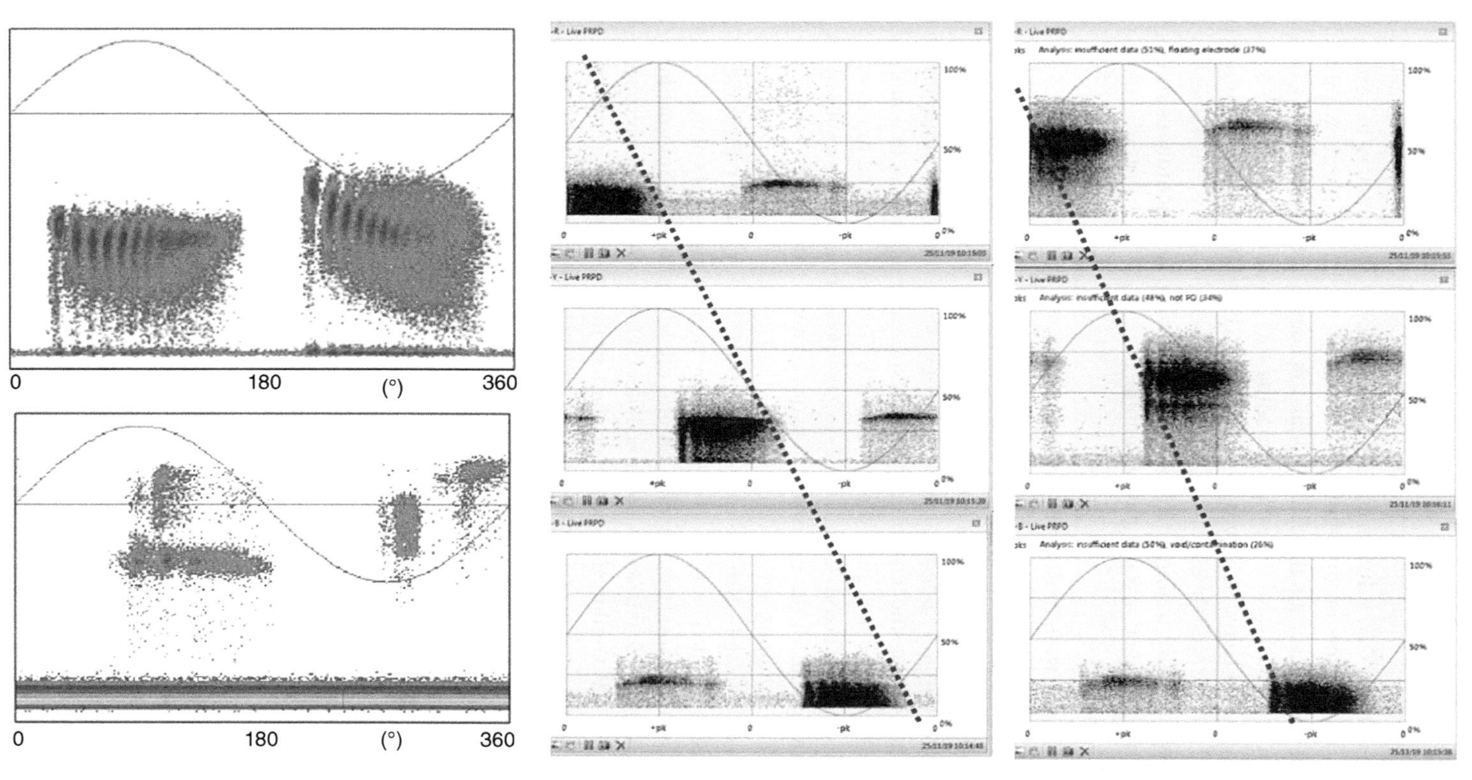

The PRPD patterns at the left show typical overhead line (OHL) interference entering the GIS (thus also the UHF PDM system) via the outdoor bushings. The patterns are typical of floating-potentials and/or contact noise. The PRPD patterns at the right were taken at two OHL exits showing identical activity across the three phases (red-yellow-blue, from top to bottom); the dotted line indicates the tell-tale 120° phase offset.
(The red lines have slightly different slopes because of slight amplitude differences between phases.)

Figure 13.41 Typical EMI entering through outdoor OHL bushings showing tell-tale 120° phase shift.

Short of a fully reliable and operational automated PD pattern recognition algorithm, another possibility to help prevent false alarms – especially those arriving from strong outside sources such as OHLs – would be for the PDM systems to sift out signals that arrive simultaneously at adjacent sensors on different phases (assuming single-phase GIS) or that arrive simultaneously at two adjacent exits connected to outdoor OHLs. In this case, the time window does not have to have ns resolution; if signals appear at sensors on different phases on adjacent building exits within say the same μs window, it can be fairly certain that these signals are coming from outside the GIS. This is even more the case if the signals (and their corresponding PRPD patterns) are offset by 120° as already mentioned and shown below in Figure 13.41. Modern GIS exhibit relatively few defects; it is exceedingly rare that defects will appear on adjacent phases, and virtually impossible that they will share exactly the same pulse timing or PRPD patterns. Thus, the synchronous three-phase analysis methods described in Section 8.9.3 may be helpful.

13.11.4 Real-World Application of GIS PDM Systems

Returning to the analogy of a fire-alarm system in a large commercial building, suddenly a buzzer sounds and a red LED starts blinking on an alarm panel. Before calling the fire brigade, most experienced building security guards or supervisors will send someone to the location indicated on the panel to see what has caused the smoke alarm to go off, acknowledging the fact that more knowledge is required before taking the next step. With online GIS PDM systems (and in fact any of the continuous online systems described in Chapters 12–16), the problem is more complex. First, it is likely that the PD alarm appears in the end user's network control center, which means someone from the maintenance department needs to actually go to the GIS site, which may be tens to hundreds of kilometer distant. Typically, that person will look at the PDM control rack monitor in the GIS control room, but is confronted with PRPD patterns – mysterious clouds of colored pixels, and strip-chart displays of signals and alarms turning on and off. What does the information mean? Should that section of GIS be taken out of service immediately? Or if not: when? The affected section may be connected to a 450-MW combined-cycle power plant, or the downtown of a large metropolitan area, with real significant consequences that will affect many people should an outage occur.

In addition, if the end user is already aware of the fact that some flashovers occur following PD activity that precedes it by only seconds or minutes or a few hours, and has no "PD expert" available to interpret the signal's patterns and temporal behavior, panic and frustration can set in when a signal appears. Those feelings may be further amplified in the case that the affected GIS has a history of lots of false alarms, or on the other hand, if it has usually been "silent." What to do?

The response of the GIS PDM community to the "no expert available" problem has been to try to develop sophisticated pattern-recognition algorithms to interpret the PRPD patterns (Section 13.12), possibly along with other signal characteristics. The goal of these systems is to indicate what the defect is (often in the form of several pattern types weighted in terms of probability) to give an indication of its actual operational risk and thus give some guidance about what mitigating steps should be taken and when (e.g. that part of the GIS needs to be taken out of service immediately or if intervention can wait for the next available maintenance window). However, at this time, none of the PDM vendors yet have such defect-identification software running, which is 100% reliable, despite decades of work in both academia and industrial R&D labs (Section 8.9.5). Developing these artificial intelligence systems to make automatic PD pattern recognition more reliable would require a huge R&D effort, first to develop the algorithms and second (the larger problem), an exhaustive effort to build up a "training database" of accurate and realistic examples of real PD signals. Until now, no one has been willing or able to commit the necessary resources to assure success.

The necessity of effective defect recognition algorithms was recognized as one of the most important requirements for future GIS PDM systems in CIGRE Working Group D1.66 (currently running), especially for addressing the high number of false alarms. Besides the technical hurdles involved in development of these systems, one of the group tasks is to recommend reliable, uniform methods potential end users of PDM systems can use to test the PD recognition algorithms. Again, the Technical Brochure will probably be published in 2023 or 2024. When published, it is recommended reading for those contemplating acquiring and installing PD monitoring on GIS.

At present, when an alarm occurs on a GIS PDM system – and the end user already recognizes it is not a false alarm, based on previous episodes – the best thing to do is to immediately call the PDM system vendor experts or the GIS PD experts – preferably both – send the data files (or at least some screenshots of the displayed PD patterns along with the signal's time behavior) to get an opinion as quickly as possible.

Another aspect is PD signal "intensity," that is, the combination of amplitude and pulse count or pulse repetition rate. Different types of PD in different parts of a GIS exhibit very different risk profiles. In the case that a large particle signal suddenly appears in a bus-bar isolating switch, this should be taken seriously and mitigation steps be decided on quickly. Similarly, signals indicating a void suddenly appearing after a new HV cable has been connected to an existing bay where there was previously no PD signal is cause for immediate shutdown of the cable, since such signals may indicate serious problems with the cable termination. On the other hand, in the case of a low-amplitude void signal localized to an insulator spacer, it may be decided to keep an eye on the signal for changes, until such time as the affected section can be taken out of service and the spacer exchanged. In all such cases of "watch and see," another point is important to emphasize. An increase in PD amplitude is not the sole determiner of risk. In fact, in some insulation systems, disappearance of PD signals implies the discharge activity has transitioned over to so-called pulseless or glow discharge, or that the defect has carbonized, reducing the electric field in the defect, which is typical of electrical treeing in solid insulation. In other words: once the "pulsed" PD signal disappears, real damage might be being done to the solid insulation. This has also been noted in rotating machine PDM (Section 16.11.1).

Once it has been established that an alarm triggered by the GIS PDM system really is due to GIS PD, the usual next step is for an expert to go on site, take some measurements with the type of equipment described in Section 13.8.5 to correlate with what the PDM system is showing, and then locate the signal source. Once the defect type is known and located, a risk assessment can be made to guide the next decision(s), i.e. take an outage and immediately remove the defect, wait for the next scheduled maintenance opportunity, or leave the defect in the GIS and use the PDM system to observe whether its behavior changes. In any case, it is paramount that the end user understands what the PDM system can and cannot do, and to have clearly defined processes and procedures in place to effectively deal with PDM system alarms (and signals) when they occur.

13.11.5 Do I Really Need a PDM System?

Over four decades of experience with GIS has shown that it is highly reliable [82, 83, 85], but at the same time, breakdowns do occur. Figure 13.5 showed a pie chart that indicated just over half of GIS failures in service are dielectric in origin, seemingly making PDM systems an attractive means to detect the signals from these failure mechanisms and take action before an actual flashover takes place.

However, as noted in the previous sections, GIS PDM systems are expensive, subject to high numbers of false positive indications, and require expert intervention should signals appear. In

2018, a CIGRE paper [86] discussing the economic viability and effectiveness of GIS PDM systems was published, based in part on a study of the total life-cycle costs of 70 420-kV GIS equipped with online PDM systems. It showed that the cost reduction of unscheduled maintenance actions along with power outages were actually lower than the cost of the implementation and operation of the PDM systems. The conclusion was that the overall failure rate for GIS would have to be much higher than what is observed in the industry (based on CIGRE statistics) to break even. Given the current sensitivity of PDM systems along with their "prevention rate" (the number of PD events that were detected early enough to take effective mitigation measures divided by the total number of events, including false alarm situations that needed to be addressed), it was found that PDM systems are only justified for GIS deemed as critical whose outage costs are extremely high [86].

The overall experience with GIS PD monitoring remains mixed and has not yet been definitively proven to significantly increase system (network) reliability. Before investing in a PDM system, the GIS end user needs to understand what the systems can and cannot do. They need to know what to do when alarms occur and what resources are available to call on to assure a timely and effective response to enable the system to do what it was designed to do: prevent in-service faults due to PD defects in the GIS. This is one of the most important and most overlooked aspects of implementing PDM systems on GIS. Clear and simple processes and procedures must be in place to define what to do when alarms occur:

- who should be contacted
- what outputs are expected from those personnel
- specific time limits for making decisions
- names of decision makers in the network organization, fallback procedures, etc.

At present, lacking reliable automated defect interpretation algorithms, every PD alarm needs to be taken seriously, examined by the requisite 'expert', assessed for risk and a decision made regarding next steps. In most cases, this will mean a lot of activity in the beginning phase of dealing with a new PDM system, but as the end user begins to know how the system (each PDM system has its own "technical personality") and GIS interact, things will settle down. They will know which signals are likely due to external interference and when to take a "new" signal seriously and act quickly to resolve it [87].

13.12 GIS PD Signal Examples and PRPD Patterns

This section shows some example PRPD patterns (Sections 8.7.3 and 10.5) for the typical PD defects found in GIS, following the same order as in Section 13.4. However, as anyone who has worked in the field of partial discharge knows, PD signals are highly stochastic and PRPD patterns can vary greatly from the near-ideal "textbook" examples presented here.

13.12.1 Moving Particles

Below are some PRPD patterns of moving or "hopping"/"dancing" particles in GIS, taken under quite ideal conditions (Figures 13.42–13.44). Typical characteristics are that the pixel-cloud is bounded by a sine-wave profile with the leading/left-hand part (lower values of phase-angle) having somewhat sparser pulse/pixel density and the trailing/right-hand part having very much higher pulse/pixel density, but again, this can vary widely depending on the microscopic characteristics of the discharge site. It is also not uncommon for the two halves of the pattern to exhibit

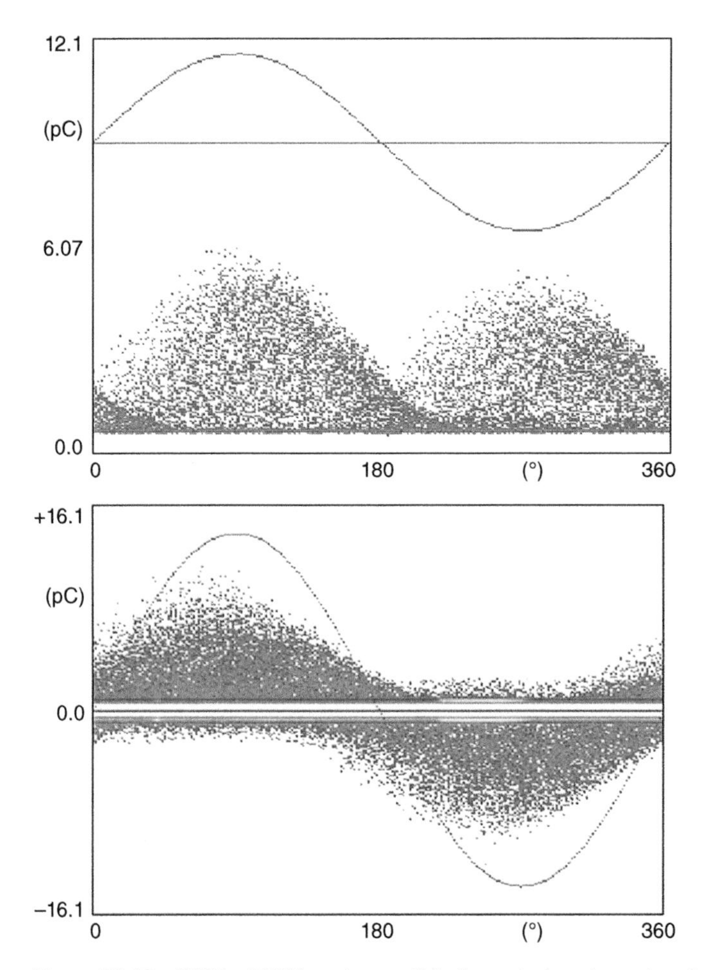

Figure 13.42 PRPD of GIS hopping particle (uncoated enclosure); unipolar (*l*), bipolar (*r*).

different relative amplitudes, for the same reason as with floating-potential discharges. If the particle(s) is (are) in a stable place, the signal(s) are usually constant. However, during the onsite HV acceptance tests, particles may move to a lower field region and drop into purpose-built traps in which case the signal disappears, essentially forever (the traps are designed such that the electric field at their bottom never increases to the level required to lift particles back out). Figure 13.44 shows a PRPD of a particle "shuffling" on a coated enclosure (i.e. the applied electric field is not strong enough to achieve lift-off/"hopping").

13.12.2 Floating Potential Discharges

Typically, PRPD patterns produced by floating-potential discharges consist of two horizontal bars of pulses ("floating clouds") centered just to the right of the respective zero-crossings of the HV AC sinusoid, as shown in Figure 13.45. It is typical that the magnitude in either charge-based or UHF measurement scenarios of the two "floating clouds" is slightly different, owing to the fact that the voltage at which the micro-sparking occurs is slightly polarity dependent. This is itself dependent on the micro-/nanoscopic structure of the surfaces opposing each other where the sparking is taking place.

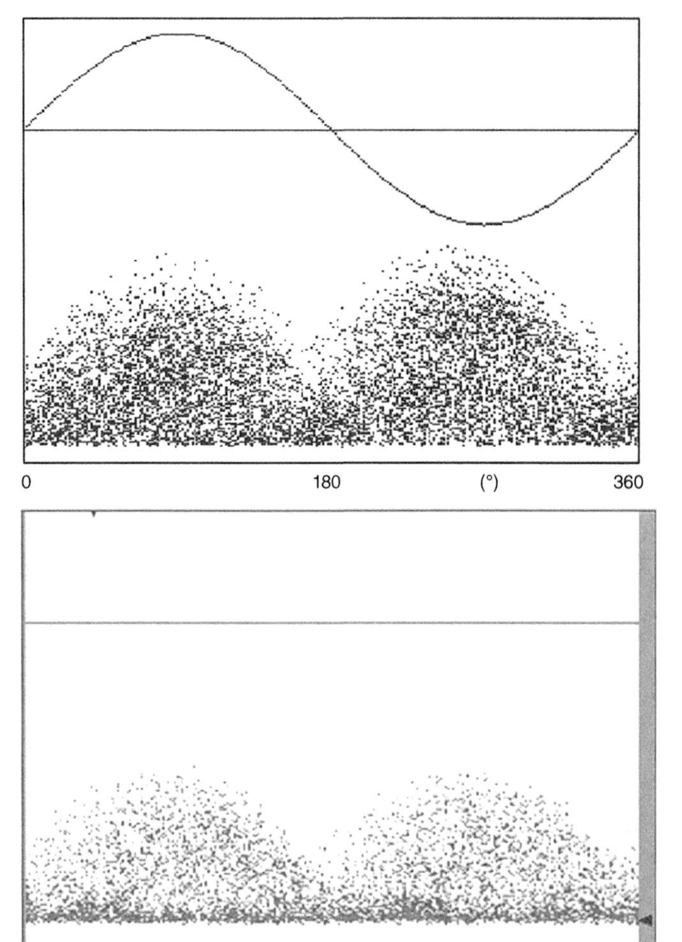

At top, via spectrum analyzer on "zero span" center frequency 1.786 MHz: at bottom, taken via a broad-band monitoring down-converter (PDIX FCU)

Figure 13.43 Unipolar PRPD patterns of hopping particles taken via UHF.

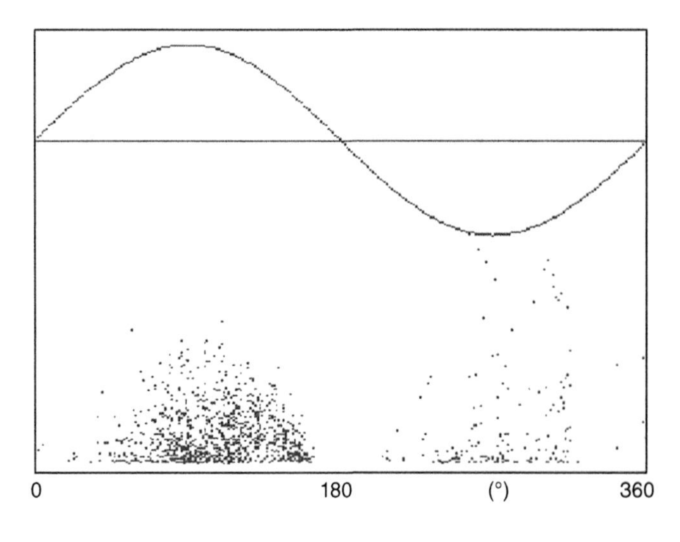

Figure 13.44 Example PRPD of a particle "shuffling" (i.e. not hopping) on the enclosure.

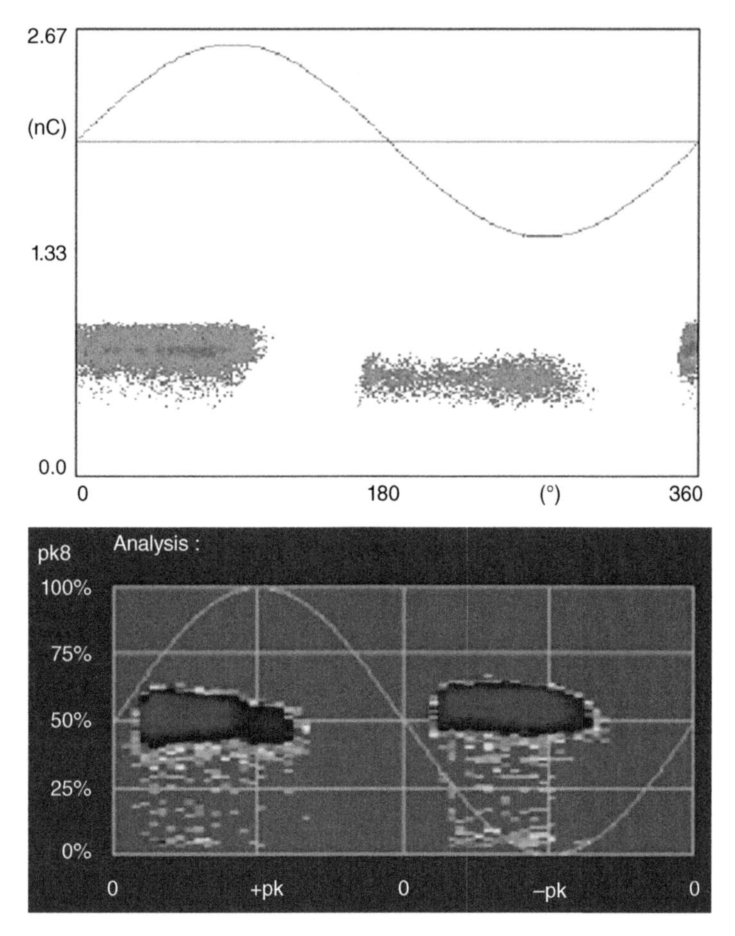

Figure 13.45 Typical PRPD patterns of floating potential discharge with an IEC 60270 detector (top) and an online continuous GIS PDM (bottom).

However, despite the relatively simple geometric explanation for the floating-potential sparking mechanism, this sparking occurs at microscopic distances. It is often the case that surface paints or coatings, localized geometries (which in turn control the electric field intensity at the sparking site), (semi-)insulating contamination, and vibration may be dynamically changing under the influence of the AC current, and many other factors can alter the apparent magnitude of the PD at each particular phase position. This can lead to floating-potential PD looking like extended horizontal clouds or very irregular shapes, which at first glance do not seem to indicate their origin, as in Figure 13.46.

In the case when other defect types have been (mostly) ruled out, one of the basic defining characteristics of floating-potential PD can help make diagnosis more certain: because it involves sparking across a gap between two metallic surfaces, its voltage dependence is extremely sensitive, plus the inception and extinction voltages are nearly equal. During high-voltage testing, either in the factory or on site, floating potential discharges switch on and off abruptly as the voltage is raised and lowered around the inception point.

The cause of less well-defined floating potential PD patterns, i.e. PRPD patterns that deviate from the neat dual horizontally aligned clouds (Figure 13.45, top) brings us back to the definition

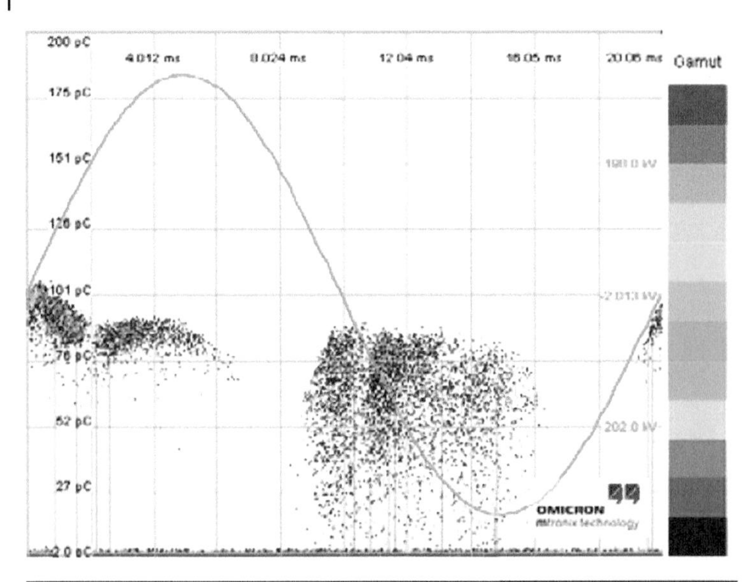

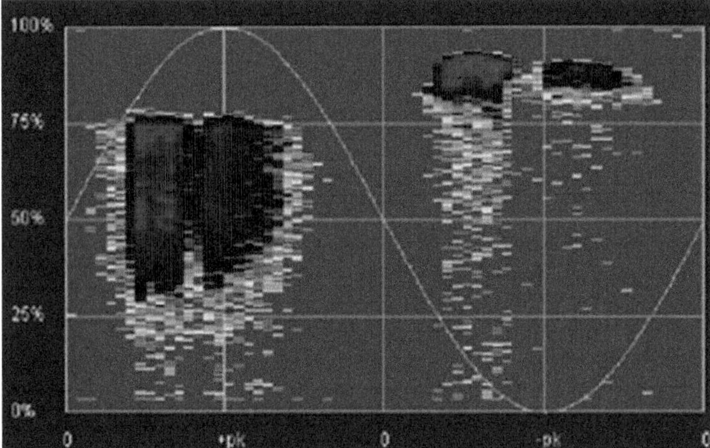

The patterns were taken with an Omicron MPD600 (top) and a Qualitrol PDM system (bottom). Note there is also a 180° phase difference. The "missing pixels" are most likely glow discharge.

Figure 13.46 PRPD patterns of "strange" floating potential PD.

of PD: all PD is local. Outside of a purpose-built set-up designed to produce floating PD in the lab, we can never be sure of exactly where the discharges between the floating "plates" is taking place, and any deviation from the "normal"/assumed conditions can produce variations in the PRPD pattern, including peculiarities in surface roughness or the presence of coatings (intentionally applied as in paints or unintentionally as in, i.e., leftover silicone grease) or any other contaminants can change the local discharge conditions and thus lead to deviations of the pattern from the ideal.

Lastly, floating-potential PD signals, although typically among the highest amplitude signals seen in GIS (they can get quite huge and even overload front-ends and pre-amplifiers), are likewise notorious for "turning on" and "turning off" abruptly. This is due to the physical nature of the local discharge conditions. It is not unknown that they "short-circuit" themselves for some time until the mechanical shock from circuit-breaker switching breaks the connection between the two metallic surfaces, and the PD starts up again. A similar mechanism is when the loose or floating part mostly has some galvanic connection, which sporadically breaks; the floating-potential PD switches on for a while (as the part vibrates under the influence of power-frequency vibration in the adjacent systems) and then dies out again.

13.12.3 Protrusions

Signals caused by protrusions – unless the protrusions are very large (>>1–2 mm), are among the smallest, both in apparent charge and in the RF (UHF). For this reason, they may be very difficult to locate, but it is essential to do so and to remove them prior to operation of the GIS because of the risk of flashover they pose in the presence of switching or lightning transients.

Figure 13.47 shows the PRPD pattern at different test voltages. At inception, PD pulses first appear in a tight cluster at 270° (which is useful for checking correct phase synchronization!). As the voltage increases, the pattern at 270° progressively broadens, and the first pulses appear centered at 90°, typically fewer in number but higher in magnitude. As the voltage is increased, the magnitude of the positive half-wave pulses forces a gain reduction, thus the much lower-magnitude pulses in the second half-wave may disappear (right-hand pattern).

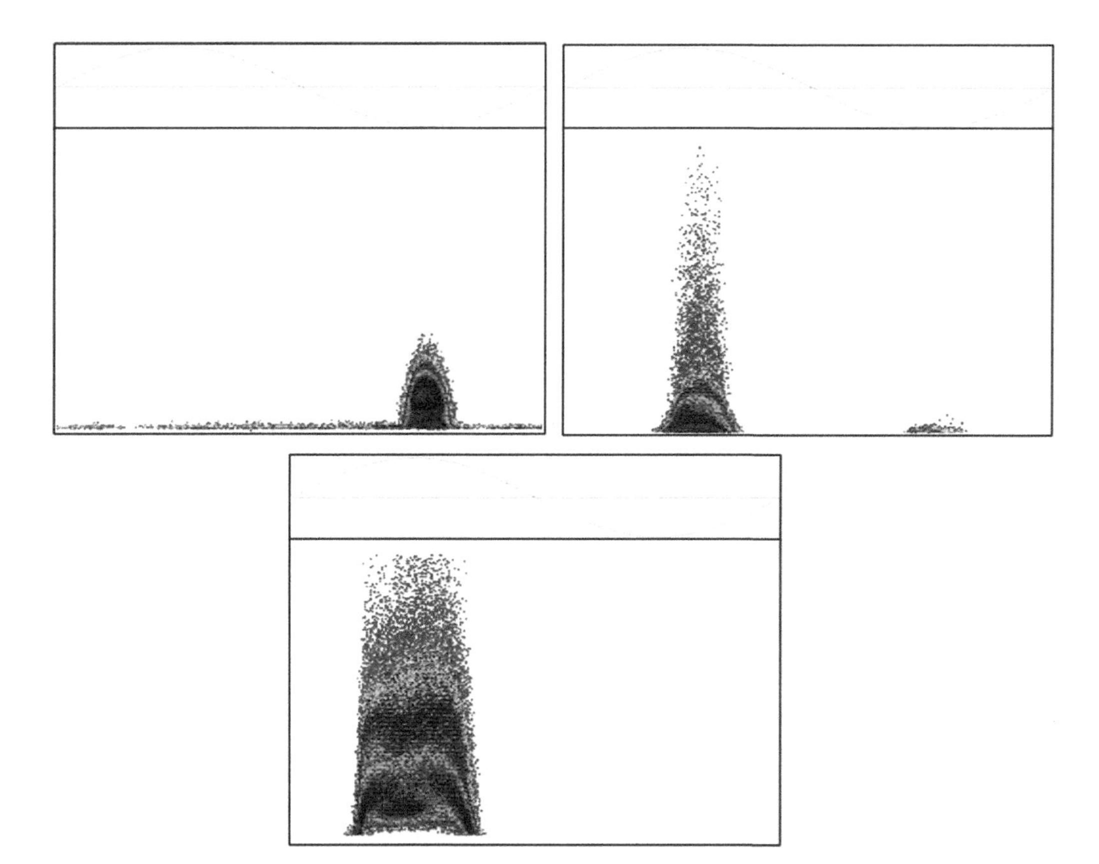

Figure 13.47 PRPD patterns of a 3-mm protrusion on the GIS inner conductor with increasing voltage (clockwise, from the top left).

13.12.4 Surface Discharges

Surface discharge signals are usually both high in amplitude and pulse-count (pulse repetition) rate. In GIS, they should be taken seriously because they typically result in damage (carbonization) of the affected insulation surface, thus potentially providing a conductive path between poles (HV-HV in three-phase GIS, or HV to ground in single-phase GIS) and thus a flashover (Figures 13.48 and 13.49).

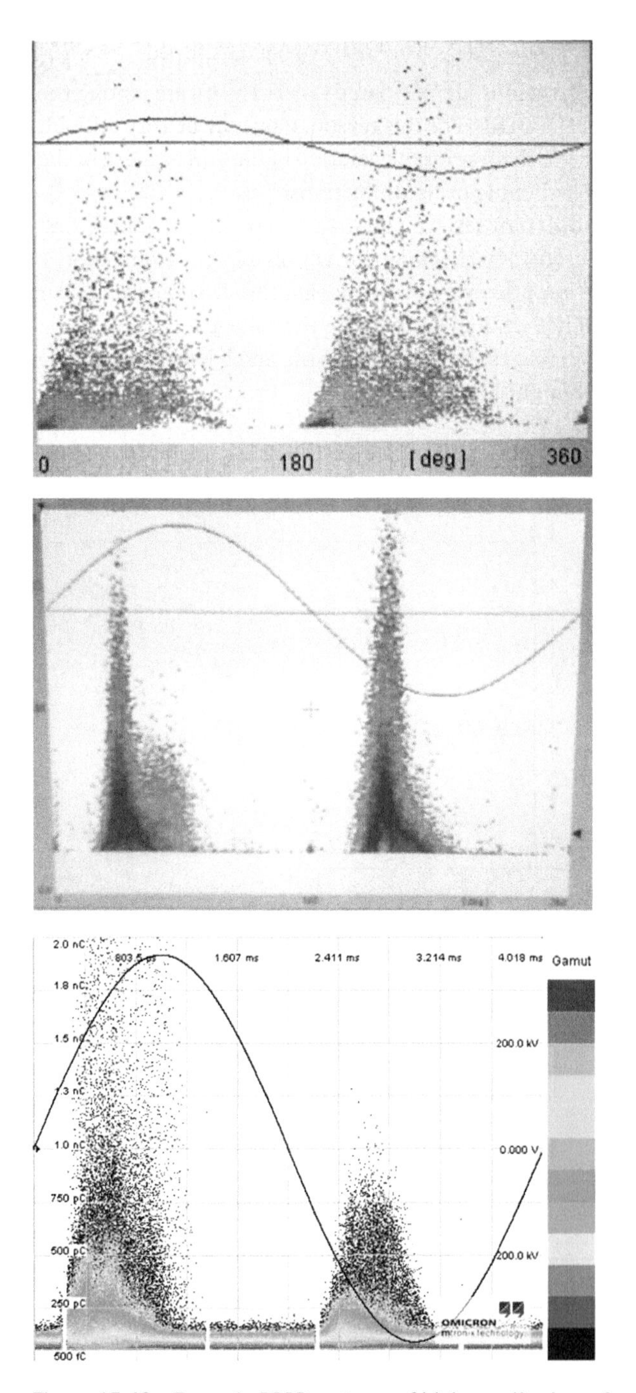

Figure 13.48 Example PRPD patterns of high-amplitude surface discharges (UHF measurements).

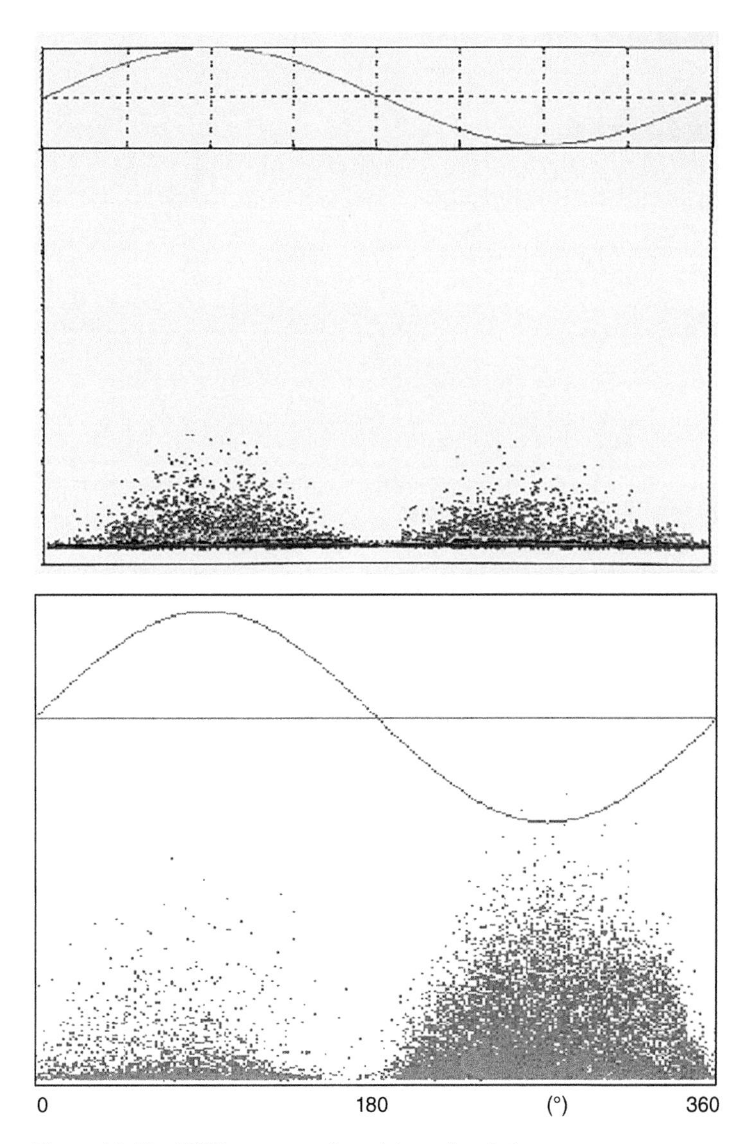

Figure 13.49 PRPD patterns of particle on insulation.

13.12.5 Voids, Delamination

Some typical void patterns are shown in Figure 13.50. Voids are among the most complex of PD phenomena because the internal environment of the void undergoes change from the moment the first PD pulse is incepted. Although tiny, the PD pulses are essentially plasma and thus act to physically damage the solid insulation materials in GIS (typically filled epoxies) over time (Section 3.4). This means the internal surfaces become eroded and carbonized. This can go so far as to make the walls conductive, essentially making the void into a conductive sphere/obloid. PD may then cease completely, but any protrusions exiting from the void surface may go on to initiate treeing in the solid insulation material, normally exhibiting much reduced (or no) PD signals. In other instances, the PD behavior becomes sporadic as shown in the plot of the discharge history of a void from a GIS PDM system shown in Figure 13.51. Also, note the trend plot shows that the

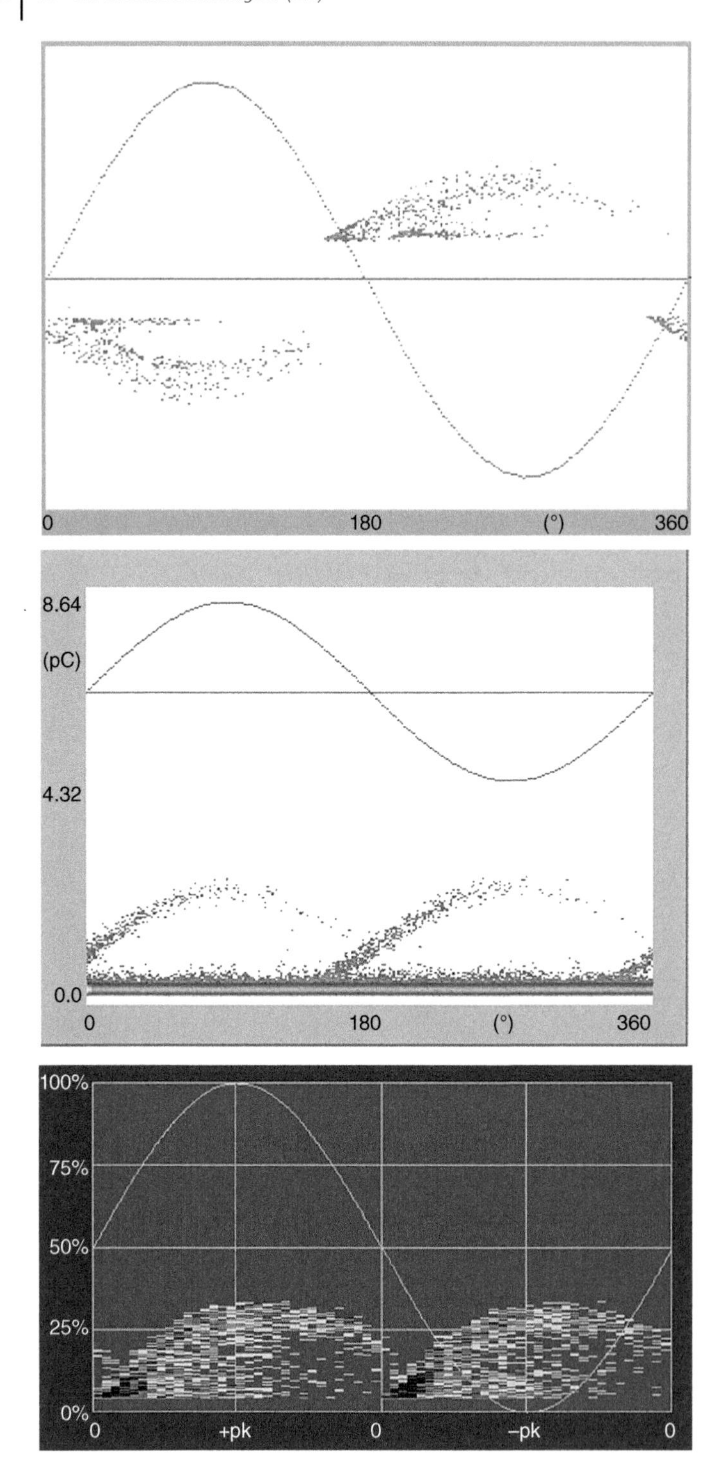

Figure 13.50 Some typical void patterns, some of which are like "rabbit ears" (Section 10.5.3).

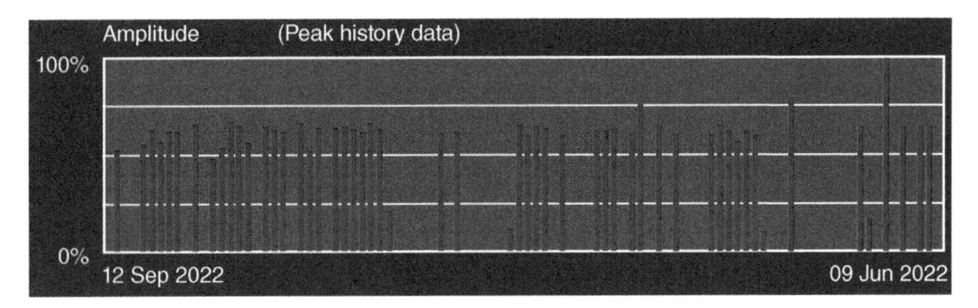

Figure 13.51 Trend plot from a GIS PDM system showing sporadic behavior of void PD.

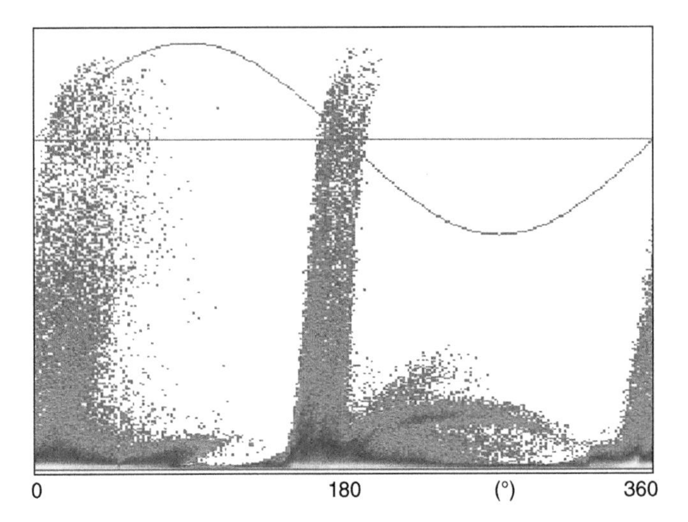

Figure 13.52 Example showing delamination between GIS insulator and inner conductor.

amplitude of the PD is very constant, typical of voids (the single large pulse at the right is likely due to another transient signal).

Lastly, depending on the precise details of construction, GIS insulators can sometimes exhibit PRPD patterns combining elements of pure void discharges (as in Figure 13.50) together with surface PD due to delamination at the inner conductor, an example of this is shown in Figure 13.52.

13.12.6 External Interference (EMI, RFI)

Finally, some examples of external interference are shown in Figure 13.53. After having looked at the example PRPD patterns of actual PD signals shown in this section, the highly ordered groupings of pixels shown in Figure 13.53 clearly differentiate themselves from the rather amorphous "clouds" produced by actual PD activity. Such strongly regular alignment – especially well-defined, vertical linear stacking – of pixels is typical of deterministic signals, i.e. signals generated for some purpose by other electrical equipment. It is important to note two points: such man-made signals are not always periodic; and they may also not be AC phase-synchronous. In either case, they will "walk through" the PRPD phase domain; depending on their repetition rate, the pixels will "smear" across the PRPD pattern, essentially increasing the noise-floor level.

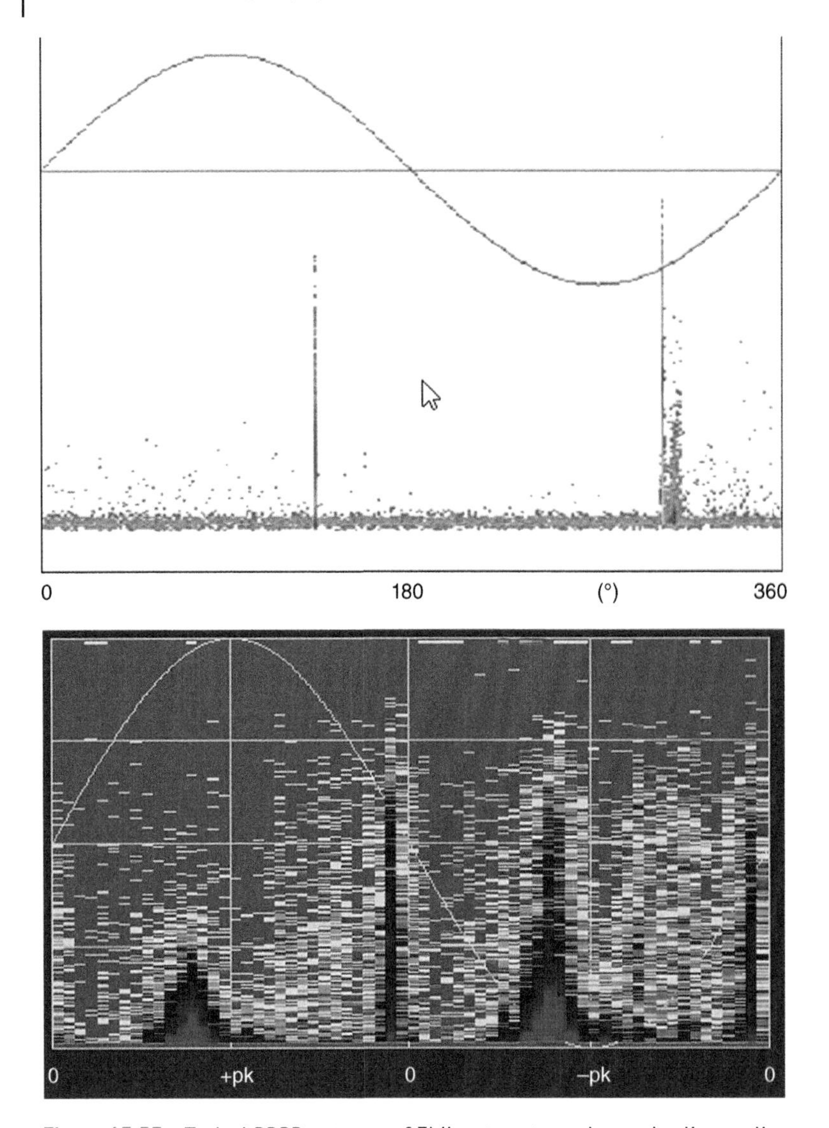

Figure 13.53 Typical PRPD patterns of EMI, note extremely regular, linear alignment of pixels.

13.13 HVDC GIS: Special Considerations

The great majority of the installed base of GIS and GIL in the world operates at 50 or 60 Hz AC. However, the growing importance of high-voltage DC (HVDC) in transmission systems has increased the need for GIS that operates under HVDC (direct voltage) conditions. The measurement of PD in DC systems (Chapter 17) is rapidly evolving. Thus, this section only briefly reviews the measurement of PD in HVDC GIS. Much of the following section is based on the work of CIGRE working group D1.63 [88] and CIGRE joint working group D1-B3.57, summarized and recently published as CIGRE Technical Brochure 842 [89]. D1.63 has not yet issued their Technical Brochure at the time of this writing. Although specific aspects of partial discharge form only a relatively small part of TB 842, it is important for anyone approaching dielectric testing (including PD measurement) of HVDC GIS.

As discussed in Section 17.2, the dielectric stresses on the insulation system in HVDC GIS are fundamentally different from that in AC GIS. The discharge pulse repetition rate may be very low for direct voltage applied to solid insulation, i.e. PD pulses usually occur much less frequently under DC conditions. The result is that the damage caused to the insulation by the PD activity may also be minimal [90]. Another characteristic of HVDC systems is that charges can build up and accumulate in and on the insulation components [90, 91]. Under the right conditions (especially during transient overvoltages), this can cause an abrupt flashover, without having revealed itself via any increased PD signal activity beforehand (as might be expected under AC operation). Therefore, although PD pulse rates may be very low under DC conditions, any PD activity should be taken seriously [90]. Such effects add to the urgency to develop and improve techniques to interpret PD activity under HVDC conditions. With far fewer pulses, and no PRPD pattern to assist in diagnosis (Section 17.5), determining which pulses are important becomes even more critical for HVDC.

Under HVDC, the electric stress distribution is dominated by the resistivities and volume of the dielectrics (Section 17.2), rather than the capacitances as is the case for AC stress. However, the effects of voltage changes are often more pronounced in HVDC GIS since the distribution of electric field stress is not simply a function of the volume or surface resistivity as would be the case if the voltage is held constant. PD activity may also be strongly influenced by any ripple voltage "riding" on the applied HVDC operational (or test) voltage. Such ripple is unwanted and usually suppressed as much as possible, but investigations have shown that if PD is present in HVDC insulation systems, it can become synchronized with the frequency of the ripple. In some cases, it may be possible to take advantage of this effect for diagnostic purposes [90]. In addition, pulseless partial discharge (also known as "glow discharge" – Section 4.7) is often observed with DC voltage.

An end user implementing HVDC GIS will likely expect the GIS to be equipped with a state-of-the-art UHF PD monitoring system, and they may expect the GIS OEM and/or the PDM system vendor to be able to interpret the PDM systems' outputs. However, implementing PDM systems for application to HVDC GIS (as well as other DC equipment) poses unique challenges that have not yet been fully addressed. The biggest challenge is to effectively differentiate HVDC PD from external interference in order to give sufficient advance warning of failures due to actual defects while simultaneously suppressing false alarms. PDM systems applied to HVDC GIS need to process and display the incoming signals in such a way as to enable unambiguous defect identification, just as in the case of PDM systems applied to AC GIS systems. But the special characteristics inherent to HVDC make this task far more difficult.

As discussed in Section 10.5, the PRPD display algorithm (along with other similar phase-synchronized display types) has become the dominant method for PD defect identification because of the clear and fundamental link between the patterns displayed and the physics of the local partial discharge process. There is neither periodic polarity reversal nor the ubiquitous zero-crossings present in HVDC GIS, and therefore, there is no periodic signal to synchronize to. Therefore, there is no PRPD pattern under HVDC conditions (Chapter 7).

At present, no consensus algorithm has yet emerged for displaying PD signals under HVDC conditions that allows defect identification with equivalent effectiveness as the ubiquitous PRPD display algorithm does for AC systems. Lacking such a capability, by far the biggest challenge will be trying to discriminate between pulses coming from real PD activity inside the GIS verses from external pulsed interference, or whether the incoming pulses belong to one or more defects. Lacking the regular field reversal and zero crossings under AC, the incoming signals from a PD sensor are essentially a random stream of pulses with no obvious relationship to one another, and at this time, there is no obvious way to display them so as to be able to differentiate them based on their origin.

For these reasons and until more effective algorithms are found and agreed to for displaying PD signals under HVDC conditions, PDM systems for HVDC GIS will probably require several different signal processing algorithms (or perhaps even other detection methods) operating in parallel to determine which incoming signals are significant indicators of actual PD and which can be ignored. An example of such augmented signal processing would be time-stamping of incoming signals with very high resolution (1 ns) as suggested in [90] to aid in identifying incoming pulses from different sensors as belonging to a common source (whether it be a real PD source or interference).

Because of this relative lack of experience with PD measurement and behavior under DC conditions, the recommendation of CIGRE JWG D1-B3.57 [89] is that factory routine testing of HVDC GIS modules (shipping units) be carried out using AC and very sensitive PD measurements to eliminate as many PD defects as possible. This includes particles, small protrusions, and contamination on insulator surfaces, all three of which pose risk for operation under HVDC conditions, and also demand a high SNR PD measurement. Assuming this test regime is successful, theoretically the only PD defect type likely to be introduced during assembly of the HVDC GIS on site is a mobile particle. Since this is the defect type on which most of the recent new work in HVDC GIS PD has focused, it should be relatively easy to diagnose – and hopefully locate – on site. In addition, HVDC GIS of the latest generations usually contain special countermeasures adapted to deal with the occurrence of the so-called "firefly mode" of moving particles (Section 13.13.1) such as special coatings and particle traps [89].

Owing to the same factors, CIGRE JWG D1/B3.57 also recommends carrying out the onsite acceptance test of HVDC GIS using the same test techniques as with AC GIS. The test should be carried out with a sensitive PD measurement using built-in UHF PD sensors, using the same procedures, test voltage levels, conditioning and measurement intervals, etc. [89].

As discussed in Section 17.5, much of the recent work in HVDC GIS PD research has focused on algorithms, which display statistical variations in the differences in time and amplitude relationships of the incoming PD pulse stream, re-visiting pulse-sequence concepts first developed in the 1990s [92–100]. To help illustrate the difference between these time-domain pulse-sequence analysis (PSA) algorithms and the well-known phase-domain PRPD algorithm for AC, Figure 13.54 shows a comparison of the parameters used for the different display types [101, 102].

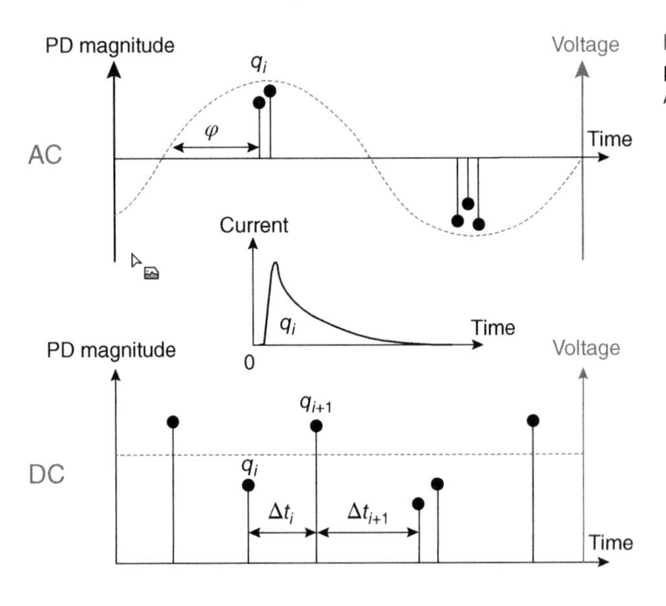

Figure 13.54 The difference in parameters evaluated to display PD in AC (PRPD) and DC. *Source:* [101].

One example shows different graphic presentations of HVDC PSA data is presented in [101, 102]. Figure 13.55 illustrates how the different so-called NODi (sic) plots are constructed using the parameters shown in Figure 13.54.

Figure 13.56 shows few samples of NODi patterns vs various PD defects. Again, the relatively straightforward correspondence between a particular PD defect and its PRPD pattern is not as clear

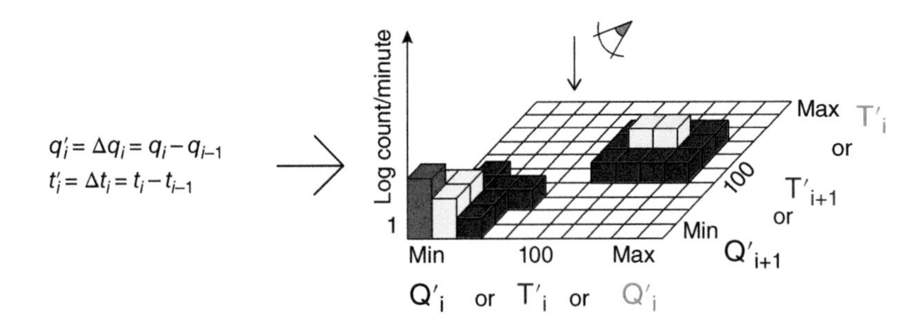

Figure 13.55 Diagram showing how a NoDi pattern is plotted from the parameters in Figure 13.54. *Source:* [102].

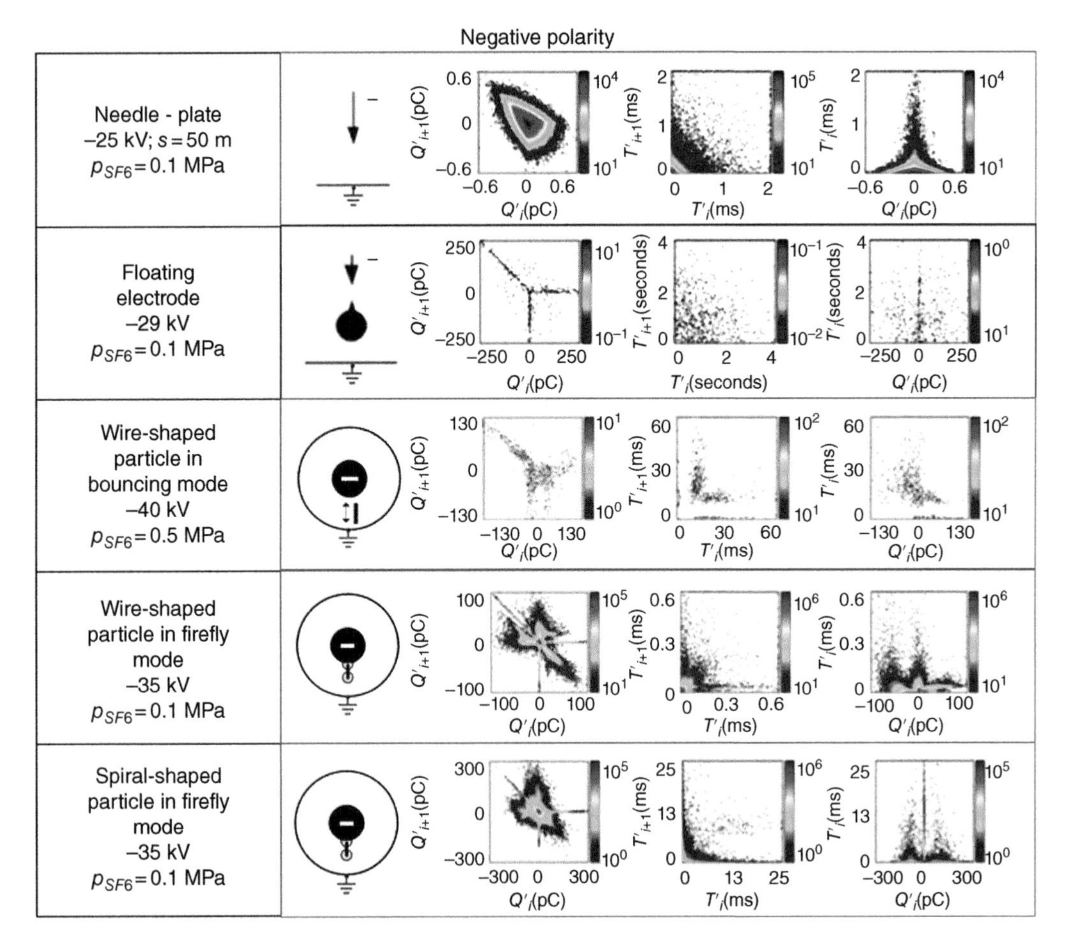

Figure 13.56 Sample HVDC GIS PD defects (negative polarity) with corresponding NoDi* patterns. *Source:* [89].

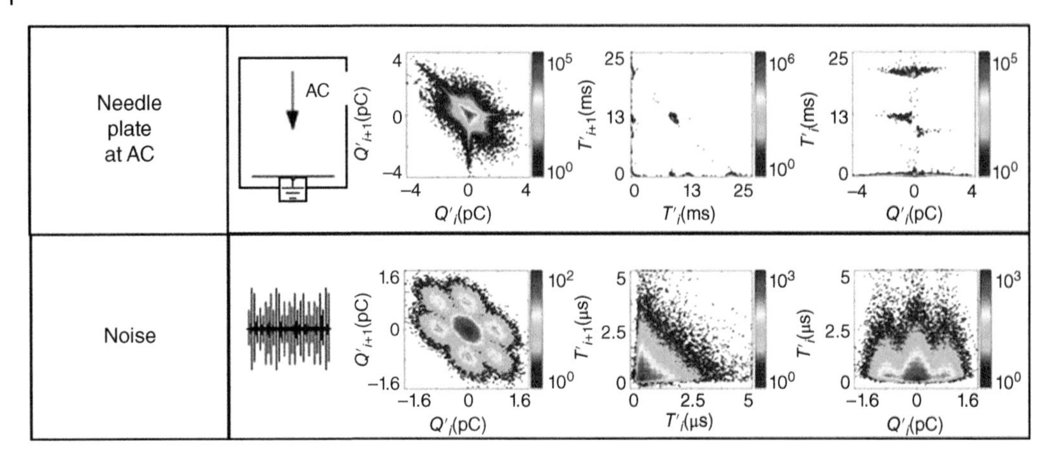

Figure 13.57 NoDi* patterns of needle-plane PD at AC along with noise. *Source:* [89].

in the DC case; still, it can be seen that the PSA (NoDi) patterns do seem to show some differences, which suggest they can be used for defect identification. For comparison, Figure 13.57 shows NoDi patterns for a simple needle-plane geometry (Figure 13.9), but under AC conditions, along with some plots of white noise. As already stated, much work remains to be done to identify effective display algorithms for HVDC PD [103].

13.13.1 PD Defect Characteristics Under HVDC

Section 13.4 provided a list of typical GIS PD defects in AC GIS. The following discusses their typical behavior under DC conditions:

1) Mobile Particles: These behave totally differently in HVDC GIS compared to under AC conditions. Instead of the ubiquitous hopping or "dancing" motion of a particle on the bottom of the GIS enclosure, particles in HVDC GIS have been observed to behave in two distinct modes of motion. In the first, the particle is bouncing, i.e. flying – between the HV electrode and ground electrode/enclosure, as shown in Figure 13.58. The lift-off electric field, the period/frequency of back-and-forth motion, and the position at pseudo-resonance [89] depend on the geometry and

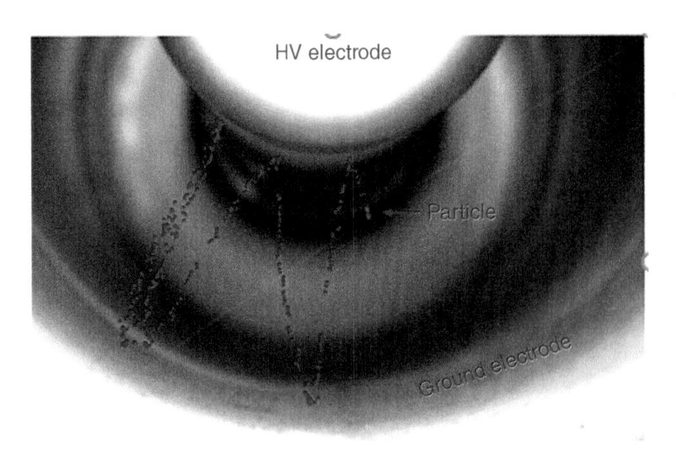

Figure 13.58 Video track of particle oscillating in HVDC GIS. *Source:* [104]/with permission from IEEE.

separation distance of the electrodes, the particle's size, shape and weight, the gas type and pressure, and in particular the voltage magnitude. The second distinct mode of particle motion in HVDC GIS, the so-called firefly effect, takes place if the particle is able to get rid of its absorbed charge before reaching the electrode. This is characterized by the particle being seen to hover just beneath the center conductor and may remain in one location or move laterally back and forth along it. Its PD signal behavior consists of (roughly) periodic superposition of glow discharge and the usual PD pulse discharge. This mode has mostly been observed with a combination of a high electric field and a negative center conductor polarity together with a particle having sharp edges [89].

Comprehensive investigations of the different modes of particle motion in HVDC GIS are documented in detail in [104] in which PD signal activity in both the UHF and IEC 60270 frequency ranges was recorded simultaneously with video of the particle's movements. Besides their significantly different modes of motions, particles in HVDC GIS can cause flashovers just as they do in AC GIS, but only when rapidly traversing the space between the inner conductor and the enclosure, what is termed "crossing"-flashover.

Because of the much wider variation in the modes of motion of particles in HVDC GIS, traditional particle "traps" are probably of little use, but work has been done on other geometric forms and positions of traps (e.g. at the inner conductor) specifically aimed at capturing particles under HVDC conditions with some demonstrable success. In addition, special coatings to prevent particle lift-off have also shown promise [89].

2) Floating-Potential Discharges: In the classic sense, this is two metallic surfaces separated by a gas-gap. As in the case of AC GIS, these defects usually produce a constant, continuous stream of PD pulses of similar and usually quite high amplitude and pulse rate, but which can, however, vary ranging from milliseconds to seconds. Similar to the case in AC GIS, floating potential may turn on and turn off for extended periods, depending on local conditions at the discharge site, but the PD activity will almost always resume after short-term voltage changes.
3) Protrusions: In HVDC GIS, protrusions actually pose two distinct risk scenarios. Just as in AC GIS, even quite short protrusions (2 mm) present a risk of instantaneous flashover in the presence of transient overvoltages due to e.g. lightning strike or switching transients. The other risk posed by protrusions is specific to the HVDC case, especially when the protrusion has a negative polarity. If and when PD incepts at the tip of the protrusion in HVDC GIS, there is nothing to stop the resulting stream of electrons. Instead, protrusions in HVDC GIS essentially become small current leaks. Depending on polarity, protrusions on either the HV (inner conductor) or ground (enclosure) side can produce so-called pulseless (glow) discharges close to the inception voltage. This pulseless discharge cannot be detected by electrical PD detection methods. For both reasons, it is essential to eliminate protrusions from HVDC GIS through very sensitive PD measurement during both routine AC factory testing and the onsite HVAC acceptance tests.
4) Surface Discharges: Discharges due to contaminants such as conductive particles or dust lying on the surface of insulators. In the case of an isolated particle lying on an insulator surface, if the electrical field is high enough, such that PD incepts, low magnitude PD pulses occur and build up a region of stable space charge at the particle's tips. This space charge reduces the field enhancement created by the particle's sharp tips with the result that the PD activity will disappear relatively rapidly. However, exactly the same as in AC GIS, particles lying on insulator surfaces are very difficult to detect but at the same time they strongly reduce the insulation withstand strength at the surface of the insulator. Similar to the AC case, the PD inception voltage level and the breakdown voltage are very close, so that these defects present a high risk of flashover in the presence of transient overvoltages due to lightning strikes or switching impulses.

5) Voids, Cracks, Delamination: In most cases, voids located within solid insulators and cracks open to the insulating gas produce one electron avalanche (PD pulse) once the applied DC voltage reaches a high enough value to produce sufficient electric field strength at the defect site and a starting electron is available. The discharge pulse repetition rate may be very low for direct voltage applied to solid insulation because the time interval between discharges at each discharge site is determined by the relaxation time constants of the space charge from previous discharges.

References

1 CIGRE TF 15/33.03.05 (1999). *Partial Discharge Detection System for GIS: Sensitivity Verification for the UHF Method and the Acoustic Method*. Paris: CIGRE.

2 IEC 60270:2000+AMD1:2015 (Consolidated Version). "High-Voltage Test Techniques – Partial Discharge Measurements" (Currently in Revision as of This Writing).

3 IEC TS 62478. High Voltage Test Techniques – Measurement of Partial Discharges by Electromagnetic and Acoustic Methods (Ed. 1, 2016).

4 IEC 62271-203. Gas-Insulated Metal-Enclosed Switchgear.

5 IEC 60060-1. High-Voltage Test Techniques – Part 1: General Definitions and Test Requirements.

6 IEC 60060-2. High-Voltage Test Techniques – Part 2: Measuring Systems.

7 IEC 60060-3. High-Voltage Test Techniques – Part 3: Definitions and Requirements for On-site Testing.

8 IEEE Std C37.122-2010. IEEE Standard for High Voltage Gas-Insulated Substations Rated Above 52 kV.

9 IEEE Std C37.301-2009. IEEE Standard for High-Voltage Switchgear (Above 1000 V) Test Techniques – Partial Discharge Measurements.

10 IEEE Std 1291-1993. IEEE Guide for Partial Discharge Measurement in Power Switchgear.

11 IEEE Std C37.122.1-2014. IEEE Guide for Gas-Insulated Substations Rated Above 52 kV.

12 CIGRE TB 366 (2008). Guide for Partial Discharge Measurements in Compliance to IEC 60270. December 2008.

13 CIGRE TB 654 (2016). UHF Partial Discharge Detection System for GIS: Application Guide for Sensitivity Verification. April 2016.

14 CIGRE TB 444 (2010). *Guidelines for Unconventional Partial Discharge Measurements*, CIGRE WG D1.33 (ed. E. Gulski et al.). Paris: CIGRE.

15 CIGRE TB 502 (2012). *Guidelines for On-site Partial Discharge Measurements*. Paris: CIGRE.

16 CIGRE TB 662 (WG 1.37). (2016). Guidelines for PD Detection Using Conventional and Unconventional Methods, August 2016.

17 CIGRE TB 525 (2013). RISK Assessment on Defects in GIS Based on PD Diagnostics.

18 Schichler, U., Koltunowicz, W., Endo, F. et al. (2013). Risk assessment on defects in GIS based on PD diagnostics. *IEEE Transactions on Dielectrics and Electrical Insulation* 20 (6): 2165–2172. (summary publication of CIGRE TB 525).

19 Behrmann, G., Koltunowicz, W., and Schichler, U. (2018). State of the art in GIS PD diagnostics. *2018 IEEE Conference on Condition Monitoring and Diagnsotics (CMD)*, Perth, Australia (23–26 September 2018).

20 Koch, H.J. (ed.) (2022). *Gas Insulated Substations*, 2e. Hoboken, NJ: Wiley/IEEE Press, **ISBN-13:** 978-1119623588 **ISBN-10:** 1119623588.

21 Boggs, S.A., Chu, F.Y., and Fujimoto, N. (1986). *Gas-Insulated Substations: Technology and Practice: Proceedings of the International Symposium on GIS*. Pergamon Press, ISBN-10: 0080318649, ISBN-13: 978-0080318646.

22 Seeger, M., Smeets, R., Yan, J. et al. (2017). Recent development of alternative gases to SF_6 for switching applications. Electra No. 291, 2017.

23 CIGRE TB 849. Electric Performance of New Non-SF6 Gases and Gas Mixtures for Gas-Insulated Systems, 2021.

24 CIGRE TB 802. Application of Non-SF_6 Gases or Gas-Mixtures in Medium and High-voltage Gas-insulated Switchgear Investigated Various Aspects of Non-SF_6 Gases or gas mixtures in medium and high voltage gas-insulated switchgear Use in Substations.

25 Reid, A.J. and Judd, M.D. (2012). Ultra-wide bandwidth measurement of partial discharge current pulses in SF_6. *Journal of Physics D: Applied Physics* 45: 165203.

26 Ohtsuka, S. and Kotsubo, T. (2021). Rise time properties of partial discharge current pulse waveforms in SF_6 gas with picosecond order and its formation mechanism. *IEEJ Transactions on Fundamentals and Materials* 141 (1): 40–47. (in Japanese).

27 Tehlar, D., Riechert, U., Behrmann, G. et al. (2013). Pulsed X-ray induced partial discharge diagnostics for routine testing of solid GIS insulators. *IEEE Transactions on Dielectrics and Electrical Insulation* 20 (6): 2173–2178.

28 Boggs, S.A., Ford, G.L., and Madge, R.C. (1981). Detection of partial discharges in gas-insulated switchgear. *IEEE Transactions on Power Apparatus and Systems* PAS-100 (8): 3969–3973.

29 Boggs, S.A. (1982). Electromagnetic techniques for fault and partial discharge location in gas-insulated cables and substations. *IEEE Transactions on Power Apparatus and Systems* PAS-101 (7): 1935–1941.

30 CIGRE WG 15.03 (1994). Effects of Particles on GIS and the Evaluation of Relevant Diagnostic Tools. CIGRE Session, Report 15-103, Paris, 1994.

31 Wohlmuth, M. (1995). Criticality of moving particles in GIS. *Ninth International Symposium on High Voltage Engineering*, Graz, Austria (28 August to 1 September 1995).

32 Schlemper, H.-D. (1996). Akustische und elektrische Teilentladungsmessung zur vor-ort Prüfung von SF6-isolierten Schaltanlagen. PhD dissertation. University of Stuttgart.

33 Cooke, C.M., Wooton, R.E., and Cookson, A.H. (1997). Influence of particles on AC and DC electrical performance of gas insulated systems at extra-high-voltage. *IEEE Transactions on Power Apparatus and Systems* PAS-96 (3): 768–777.

34 Wooton, R.F. (1997). Investigation of high-voltage particle-initiated breakdown in gas-insulated systems. EPRI Report EL-1007, March 1997. https://www.epri.com/research/products/3002002657.

35 Schlemper, H.-D. and Feser, K. (1997). Characterization of moving particles in GIS by acoustic and electrical discharge detection. *10th International Symposium on High Voltage Engineering*, Montreal (25–29 August 1997), p. 87.

36 Niemeyer, L. and Seeger, M. (2015). Universal features of particle motion in AC electric fields. *Journal of Physics D: Applied Physics* 48: 435501, 11 pp.

37 Hinterholzer, T. and Boeck, W. (2000). Breakdown in SF_6 influenced by corona-stabilization. *2000 Annual Report Conference on Electrical Insulation and Dielectric Phenomena (Cat. No.00CH37132)*, Victoria, British Columbia (15–18 October 2000), pp. 413–416.

38 Seeger, M., Niemeyer, L., and Bujotzek, M. (2008). Partial discharges and breakdown in SF_6 at protrusions in uniform background fields. *Journal of Physics D: Applied Physics* 41 (18): 185204.

39 Neuhold, S., Brügger, T., Bräunlich, R. et al. (2018). *Return of experience: The CIGRE UHF PD Sensitivity Verification and On-site Detection of Critical Defects*, D1-304. Paris: CIGRE.

40 Schurer, R., Feser, K. (1998). PD Behaviour and Surface Charging of a Particle Contaminated Spacer in SF_6 Gas Insulated Switchgear. *1998 Annual Report Conference on Electrical Insulation and Dielectric Phenomena (Cat. No.98CH36257)*, Atlanta, GA (25–28 October 1998).

41 Niemeyer, L. (1995). A generalized approach to partial discharge modelling. *IEEE Transactions on Dielectrics and Electrical Insulation* 2: 510–528.

42 Leijon, M. and Vlastos, A. (1988). Pattern recognition of free metallic particles motion modes in GIS. Conference Record of the 1988 IEEE International Symposium on Electrical Insulation, Boston, (5 June 1988).

43 Lundgaard, L. (1992). Partial discharge – part XIII: acoustic partial discharge detection – fundamental considerations. *IEEE Electrical Insulation Magazine* 8 (4): 25–31.

44 Lundgaard, L. (1992). Partial discharge – part XIV: practical application. *IEEE Electrical Insulation Magazine* 8 (5): 34–43.

45 Lundgaard, L. (2001). Particles in GIS characterization from acoustic signatures. *IEEE Transactions on Dielectrics and Electrical Insulation* 8 (6): 1064–1074.

46 Markalous, S.M. (2006). Detection and location of partial discharges in power transformers using acoustic and electromagnetic signals. PhD thesis. University of Stuttgart, Germany.

47 Tenbohlen, S., Hoek, S.M., Denissov, D. et al. (2006). Electromagnetic (UHF) PD Diagnosis of GIS, Cable Accessories and Oil-Paper Insulated Power Transformers for Improved PD Detection and Localization. CIGRE 2006, D1-104.

48 Gross, D. and Soeller, M. (2012). On-site transformer partial discharge diagnosis. *2012 IEEE International Symposium on Electrical Insulation*, San Juan, Puerto Rico (10–13 June 2012).

49 Hampton, B.F. and Meats, R.J. (1988). Diagnostic measurements at UHF in gas insulated substations. *IEE Proceedings C* 135 (2): 137–144.

50 Hampton, B., Pearson, J., Welch, I. et al. (1992). *Experience and Progress with UHF Diagnostics in GIS*. Paris: CIGRE, paper 15/23-03.

51 CIGRE WG 15.03 (1992). Diagnostic Methods for Gas Insulating Systems. CIGRE Session, Report 15/23-01, Paris.

52 Bargigia, A., Koltunowicz, W., and Pigini, A. (1992). Detection of partial discharges in gas insulated substations. *IEEE Transactions on Power Delivery* 7 (3): 1239–1249.

53 Pearson, J., Farish, O., Hampton, B. et al. (1995). Partial discharge diagnostics for gas insulated substations. *IEEE Transactions on Dielectrics and Electrical Insulation* 2 (5): 893–905.

54 Judd, M.D., Farish, O., and Hampton, B.F. (1996). The excitation of UHF signals by partial discharges in GIS. *IEEE Transactions on Dielectrics and Electrical Insulation* 3 (2): 213–228.

55 Behrmann, G.J., Neuhold, S.M., and Pietsch, R. (1997). Results of UHF measurements in a 220 kV GIS substation during on-site commissioning tests. *10th International Symposium on High Voltage Engineering (ISH)*, Quebec, pp. 451–455.

56 Welch, I.M. (1997). Capacitive couplers for UHF partial discharge monitoring. Technical Guidance Notev TGN (T) 121, Issue 1, January 1997, National Grid Company, Coventry, UK.

57 Kraus, J. and Carver, K. (1973). *Electromagnetics*. McGraw-Hill.

58 Volkais, J.L. (2007). *Antenna Engineering Handbook*, 4e. McGraw Hill.

59 Hans, G.S. (2015). *The Art and Science of Ultrawideband Antennas*, 2nd Revisede. Artech House, ISBN-13: 978-1608079551, ISBN-10: 1608079554.

60 Obrist, R., Smajic, J., and Behrmann, G. (2016). Simulation based design of GIS sensors for PD measurements. *IEEE Conf. on Electromagnetic Field Computation (CEFC)*, Miami, (13–16 November 2016).

61 Neuhold, S., Heizmann, T., Bräunlich, R. et al. (2007). Experience with UHF PD detection in GIS using external capacitive sensors in windows and disk-insulators. *XVth International Symposium on High Voltage Engineering (ISH)*, Ljubliana, T7-480.

62 Kurrer, R. and Feser, K. (1997). Attenuation measurements of ultra high frequency partial discharge signals in gas insulated substations. *Int(ernational)' Symp. High Voltage Eng. (ISH)*, Montreal (25–29 August 1997), Vol. 2, pp. 161–164.

63 Doi, M., Muto, H., Fujii, H., and Kamei, M. (1998). Frequency spectrum of various partial discharges in GIS. *Proceedings of 1998 International Symposium on Electrical Insulating Materials. 1998 Asian International Conference on Dielectrics and Electrical Insulation. 30th Symposium on Electrical Insulating Ma*, Toyohashi, Japan, (27–30 September 1998), paper C2-2.

64 Judd, M.D. and Farish, O. (1999). Transfer functions for UHF partial discharge signals in GIS. *1999 Eleventh International Symposium on High Voltage Engineering*, London (27–23 August 1999), pp. 74–77, Vol. 5, 1999.

65 Muto, H., Doi, M., Fujii, H., and Krunei, M. (1999). Resonance characteristics and identification of modes of electromagnetic wave excited by partial discharges in GIS. *Proceedings of the 1999 Eleventh International Symposium on High Voltage Engineering*, London (27–23 August 1999), pp. 70–73, Vol. 5.

66 Meijer, S., Smit, J.J., and Girodet, A. (2002). Estimation of UHF signal propagation through GIS for sensitive PD detection. *IEEE International Symposium on Electrical Insulation*, Boston (7–10 April 2002).

67 Okabe, S., Yuasa, S., Kaneko, S. et al. (2006). Simulation of propagation characteristics of higher order mode electro-magnetic waves in GIS. *IEEE Transactions on Dielectrics and Electrical Insulation* 13: 855–861.

68 Hikita, M., Ohtsuka, S., Teshima, T. et al. (2007). Examination of electromagnetic mode propagation characteristics in straight and L-section GIS model using FD-TD analysis. *IEEE Transactions on Dielectrics and Electrical Insulation* 14 (6): 1477–1483.

69 Hoshino, T., Maruyama, S., and Sakakibara, T. (2009). Simulation of propagating electromagnetic wave due to partial discharge in GIS using FDTD. *IEEE Transactions on Power Delivery* 24 (1): 153–159.

70 Hikita, M., Ohtsuka, S., Ueta, G. et al. (2010). Influence of insulating spacer type on propagation properties of PD-induced electromagnetic wave in GIS. *IEEE Transactions on Dielectrics and Electrical Insulation* 17 (5): 1643–1648.

71 Hoek, S.M., Kraetge, A., Koch, M. et al. Emission and propagation mechanisms of PD pulses for UHF and traditional electrical measurements. *2012 IEEE Int'l Conf on Cond. Monitoring & Diagnosis*, Bali (23–27 September 2012), paper A-146.

72 Okabe, S., Ueta, G., Hama, H. et al. (2014). New aspects of UHF PD diagnostics on gas-insulated systems. *IEEE Transactions on Dielectrics and Electrical Insulation* 21 (5): 2245–2258.

73 Behrmann, G. and Smajic, J. (2016). RF PD signal propagation in GIS: comparing S-parameter measurements with an RF transmission model for a short section of GIS. *IEEE Transactions on Dielectrics and Electrical Insulation* 23 (3): 1331–1337.

74 Gross, D. (2018). Partial discharge signal transmission of distributed power engineering equipment. *2018 International Conference on Diagnostics in Electrical Engineering (Diagnostika)*, Pilsen, Czech Republic (4–7 September 2018).

75 Behrmann, G., Franz, S., Smajic, J. et al. (2019). UHF PD signal transmission in GIS: effects of 90° bends and an L-shaped CIGRE Step 1 test section. *IEEE Transactions on Dielectrics and Electrical Insulation* 26 (4): 1293–1300.

76 Behrmann, G.J., Gross, D., and Neuhold, S. (2020). Limitations of attempting calibration of partial discharge measurements in VHF and UHF ranges. *2020 IEEE Conference on Electrical Insulation and Dielectric Phenomena (CEIDP)*, (21 October 2020).

77 Schichler, U., Koltunowicz, W., Gautschi, D. et al. (2016). (CIGRE D1.25), UHF partial discharge detection system for GIS: application guide for sensitivity verification. *IEEE Transactions on Dielectrics and Electrical Insulation* 23 (3): 1313–3121.

78 Stewart, B.G., Judd, M.D., Reid, A.J., and Fouracre, R.A. (2007). Suggestions to augment the IEC 60270 partial discharge standard in relation to radiated electromagnetic energy. *2007 Electrical Insulation Conference and Electrical Manufacturing Expo*, Nashville, TN (22–24 October 2007).

79 Hoek, S., Bornowski, M., Tenbohlen, S. et al. (2008). Partial discharge detection and localisation in gas-insulated switchgears. *Stuttgarter Hochspannungssymposium*, Stuttgart (21–22 June 2008).

80 Agilent (2004–2006). *Spectrum Analyzer Basics, Application Note 150*. Agilent Technologies Inc., 5952-0292, CA, USA.

81 P. Horowitz, W. Hill, (2015) *The Art of Electronics*, 3, Cambridge University Press, ISBN 978-0-521-80926-9

82 CIGRE Joint Working Group 33/23.12. Insulation Co-ordination of GIS: Return of Experience, On Site Tests and Diagnostic Technique. Electra No.176, pp. 67–97, February 1998, Paris.

83 CIGRE (2012). Reliability of High Voltage Equipment – Part 6, GIS Practices, October 2012.

84 Lee, H.H., Yoon, K.T., Behrmann, G. et al. (1998). A new partial discharge monitoring system at Labrador and Ayer Rajah 400kV GIS substations. *Proceedings of the 12th Conference of the Electric Power Supply Industry (CEPSI)*, Pattaya, Thailand (November 1998).

85 CIGRE (2012). Final Report of the 2004–2007 International Enquiry on Reliability of High Voltage Equipment. Part 5 – Gas Insulated Switchgear.

86 Schlemper, H.D., Riechert, U., Behrmann, G., and Manz, J. (2017). Application and Benefits of UHF On-line Partial Discharge Monitoring in GIS. CIGRE Colloquium Winnipeg, D1-086, Fall 2017.

87 Al Shaiba, Z., Bhatnagar, C., Dave, Y., and Karimbanackal, J.A. (2020). Partial discharge vs noise in online GIS partial discharge monitoring systems-experience of KAHRAMAA. *CIGRE GCC Power 2020 E-conference*, (25–26 November 2020), Session B-1.

88 Ronald Plath et al., CIGRE D1.63. Partial Discharge Detection Under DC Stress. Technical Brochure in Preparation, 2022.

89 CIGRE (2021). Technical Brochure 842. Dielectric Testing of HVDC Gas-Insulated Systems.

90 Judd, M., Siew, W.H., Hu, X. et al. (2015). Partial discharges under HVDC conditions. *Euro TechCon*, Stratford-upon-Avon, UK (1–3 December 2015).

91 Madhar, S.A., Mor, A.R., Mraz, P., and Ross, R. (2021). Study of DC partial discharge on dielectric surfaces: mechanism, patterns and similarities to AC. *International Journal of Electrical Power & Energy Systems* 126: 106600.

92 Patsch, R., Hoof, M., and Reuter, C. (1993). Pulse-sequence-analysis, a promising diagnostic tool. *Proc of 8th ISH '93, Vol IV*, Yokohama, pp. 157–160.

93 Hoof, M. and Patsch, R. (1995). Pulse-sequence analysis: a new method for investigating the physics of PD-induced ageing. *IEE Proceedings – Science, Measurement and Technology* 142 (1): 95–101.

94 Fromm, U. and Kreuger, F.H. (1994). Statistical behaviour of internal partial discharges at DC-voltage. *Japanese Journal of Applied Physics* 33: 6708–6715.

95 Fromm, U. (1995). Partial discharge and breakdown testing at high DC voltage. PhD thesis. Technical University of Delft (NL), September 1995.

96 Fromm, U. (1995). Interpretation of partial discharges at DC voltages. *IEEE Transactions on Dielectrics and Electrical Insulation* 2 (5): 761–770.

97 Morshuis, P., Jeroense, M., and Beyer, J. (1997). Partial discharge part XXIV: the analysis of PD in HVDC equipment. *IEEE Electrical Insulation Magazine* 13 (2): 7–16.

98 Morshuis, P. and Smit, J.J. (2005). Partial discharges at DC voltage: their mechanism, detection and analysis. *IEEE Transactions on Dielectrics and Electrical Insulation* 12 (2): 328–340.

99 Pirker, A. (2019). Measurement and representation of partial discharges at DC voltage for identification of defects of gas—insulated systems (in German). PhD thesis. Technical University of Graz.

100 Geske, M., Neumann, C., Berg, T., and Plath, R. (2020). Assessment of typical defects in gas-insulated DC systems by means of pulse sequence analysis based on UHF partial discharge measurements. *VDE Conference High Voltage Technique 2020*, Berlin (9–11 November 2020).

101 Pirker, A. and Schichler, U. (2016). Partial discharges at DC voltage – measurement and pattern recognition. *2016 International Conference on Condition Monitoring and Diagnosis*, Xi'an, China (25–28 September 2016).

102 Pirker, A. and Schichler, U. (2018). Partial discharge measurement at DC voltage – evaluation and characterization by NoDi* pattern. *IEEE Transactions on Dielectrics and Electrical Insulation* 25 (3): 883–891.

103 Madhar, S.A. (2021). Defect identification through partial discharge analysis on HVDC, partial discharge fingerprinting. PhD dissertation. TU Delft, September 2021.

104 Wenger, P., Beltle, M., Tenbohlen, S. et al. (2019). Combined characterization of free-moving particles in HVDC-GIS using UHF PD, high-speed imaging, and pulse-sequence analysis. *IEEE Transactions on Power Delivery* 34 (4): 1540–1548.

14

Air-Insulated Switchgear and Isolated Phase Bus

14.1 Introduction

Air-insulated switchgear (AIS) may be known by many other names. In North America, it is often called metalclad switchgear, medium-voltage (MV) switchgear, or distribution-class switchgear. In the UK and other countries, it may be referred to as MV switchboards. In any case, for the purposes of this book, AIS is used to describe equipment that is enclosed within grounded metal enclosures consisting of a series of circuit breaker cubicles connected to different loads and isolated from one another by metal walls, where air at atmospheric pressure is the main insulation medium between the phase busses and to ground. The cubicles may contain circuit breakers, fused disconnect switches, contactor starters, and capacitor banks. A grounding switch, protective relays, potential transformers (PTs) and current transformers (CTs) for metering, and protection purposes are also usually associated with each circuit breaker or cubicle. Usually, the main supply bus is also connected to potential (voltage) transformers (PTs/VTs), again for both metering and protection purposes. In summary, an AIS usually consists of a "line-up" of many circuit breaker cubicles each about 1–2 m wide and perhaps 3 m high. Depending on the number of circuit breakers (cubicles), the total length of the line-up may be from 5 to 20 m or more (Figure 14.1). The AIS is typically rated from 3 to 38 kV. There are many books that describe the design and function of metalclad switchgear, including a chapter in [1]. IEEE C37.20.2 and IEC 62271 Part 200 give the general technical requirements for medium-voltage AIS [2, 3]. This type of switchgear is ubiquitous, forming the backbone of electrical power distribution in towns and cities, industrial plants, generating stations, refineries, university and business campuses, high-rise buildings, etc.

AIS is supplied from one or more in-feeds (often via power cables or air-insulated bus duct from a nearby step-down power transformer) that energize a common set of three-phase busbars that extends the length of the line-up. There is no grounded metal barrier between the phases, as occurs in isolated phase bus (Section 14.8) and most GIS (Chapter 13). In industrial applications, the circuit breakers are connected to the common busbars to control the power going to individual motors or to step-down transformers to supply lighting, heating, or other industrial loads. In distribution utilities, the circuit breakers are connected to power cables that supply different areas of a city with residential, commercial, or industrial loads. In the past, the circuit breakers used contactors in air (air-magnetic breakers), or sometimes oil breakers. Today the most common type of circuit breaker is the vacuum breaker or SF_6 breaker.

Sometimes the AIS is installed outdoors (Figure 14.2). In other cases the AIS may be inside a building, as shown in Figure 14.1.

Practical Partial Discharge Measurement on Electrical Equipment, First Edition. Greg C. Stone, Andrea Cavallini, Glenn Behrmann, and Claudio Angelo Serafino.

Figure 14.1 Photo of a metalclad switchgear line-up. *Source:* Courtesy of Powell Industries.

Figure 14.2 View of an outdoor AIS with most rear panels open, showing the power cables going to the loads with the associated CTs (three cubicles on the left) and the PT cubicle on the lower right. The common three-phase horizontal bus visible on the right is behind metal barriers in the three left cubicles. The green capacitors in the upper right cubicle are used for online PD monitoring. *Source:* Courtesy of Iris Power L.P.

Although not formally switchgear, this chapter also briefly discusses the design, failure, and PD testing of isolated phase bus (IPB), which also uses air as the main insulation within a grounded metal enclosure.

14.2 AIS Insulation Systems

Prior to the 1960s, the common three-phase busbars running through the cubicles, as well as the busbar connections to the breakers and power cables in each cubicle, were usually supported with porcelain post insulators. Since porcelain is inorganic, it is very resistant to surface tracking due to pollution (Section 3.7.2). However, porcelain is expensive and prone to crack or shatter; so when tracking-resistant fiberglass-reinforced polyester (FRP) became available in the 1960s, this rapidly

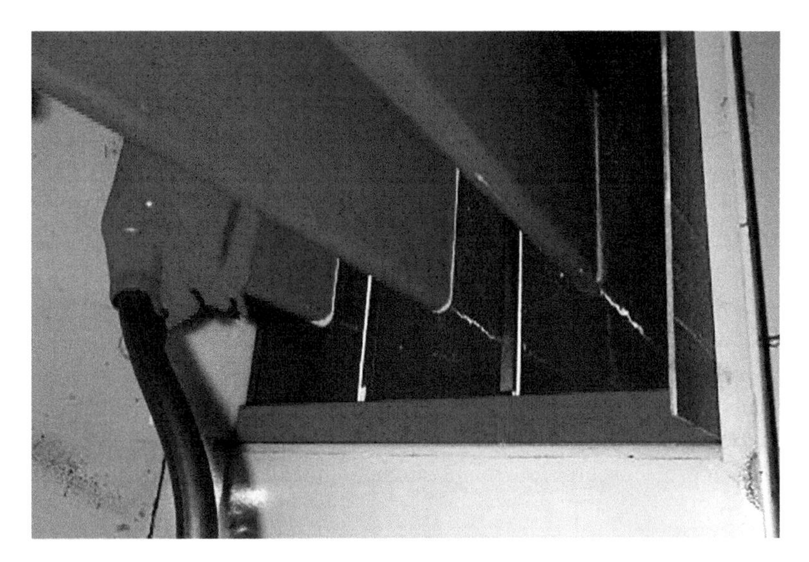

Figure 14.3 Photo of the three-phase busbars being supported by insulating boards between cubicles. Note that the copper busbars are coated with a thin epoxy coating to improve the BIL.

replaced porcelain post insulators to support the high-voltage busses. The FRP sheets (or boards) usually employed in AIS are normally referred to as NEMA grade GPO3, since it is relatively resistant to electrical tracking (although not nearly as good as porcelain) and somewhat flameproof. Glastic is a well-known manufacturer of such sheets. Figure 14.3 shows the vertical insulating boards supporting the common bus between cubicles. The GPO3 sheets may also be mounted horizontally (Figure 14.4).

Usually the copper busbars are coated with a thin epoxy film (Figures 14.2–14.4). The purpose of the coating is to improve impulse withstand voltage (BIL, basic insulation [or impulse] level) for both the phase-to-ground and the phase-to-phase components of the insulation system, in the event of lightning or switching impulse voltages. The epoxy bus coating may also restrict the movement of any arc along the bus if an arc is established.

Figure 14.4 Photo of the black traces from electrical tracking (before in-service failure) over two of the three phases. The tracking over the horizontal GPO3 support sheet that would have eventually led to a phase-to-ground or (even worse) a phase-to-phase fault is not visible. *Source:* Courtesy of Iris Power L.P.

Electrical insulation is also used in the PTs and CTs used for metering and protection. Usually such instrument transformers are made from a cast epoxy (Figure 14.2).

Often any bolted connections between busbars or other leads are covered in a flexible silicone rubber boot or a boxed dust cover. The box may also be compound filled to increase the BIL. Silicone rubber is very resistant to surface tracking (Section 3.7.2) compared to purely organic insulation materials.

14.3 Insulation Failure Processes

Most failures that occur in AIS are due to rodent entry causing shorts, or from multiple issues that may occur with the circuit breakers themselves (which are primarily due to mechanical or electrode aging problems, although electrical tracking over circuit breaker bushings or insulation cracking is possible). The following briefly describes the insulation aging processes that produce PD. Examples of many insulation problems are shown in References [4–11].

14.3.1 Surface Electrical Tracking

Figure 14.4 shows a metalclad switchgear cubicle where surface electrical tracking over the epoxy coating on two busbars is clearly visible (the black tracks). As discussed in Section 3.7.2, if insulating surfaces are coated with partly conductive contamination, electrical tracking can occur. Tracking is a PD phenomenon. In switchgear that is outdoors, the contamination is most likely, and it is accelerated if dust or other pollution has settled on the insulation, which further reduces the surface resistance. This process is less likely in indoor locations, particularly if the building is air conditioned to reduce humidity. When the bus is dry due to atmospheric conditions, the tracking process may stop, and the PD may not continue, even though significant tracking has already occurred. This makes this type of PD very erratic over time [4, 10].

Figure 14.5 Photo of damage to the busbar insulation caused by PD occurring in an air space between the busbar and the gray insulating support board. The busbar was pulled back from the support board to reveal the damage (yellow stripe on busbar).
Source: Courtesy of Powell Industries.

14.3.2 Air Gap PD

Often insulated high-voltage leads or busses go through cutouts in the FRP sheets. If there is an air space between the lead or busbar and the FRP, partial discharge can occur. Figure 14.5 shows an insulated high-voltage lead going through an FRP sheet where an air gap is present, leading to PD in the gap. Even though the ground and the other phases are distant from the lead, if partly conductive pollution is present, the surface of the FRP board may essentially be at ground potential, which will increase the electric stress across the air gap, leading to PD. This PD does not cause failure, but it may initiate tracking along the board surface to ground or to another phase, especially if the FRP sheets have a part line (i.e. the two half boards are mated). Tracking seems to be accelerated when such part lines exist. After years, a ground or phase-to-phase fault may result.

14.3.3 PD in Cast Epoxy Components

The instrument transformers in AIS are often made from epoxy using a vacuum casting process. If the casting process is not under control, or the epoxy formulation is not correct, voids may occur within the casting. Voids near a high electric field region may result in PD. One hopes that such manufacturing flaws are caught during factory PD testing of the component. Usually the epoxy resin contains inorganic fillers such as silica to try to match the coefficient of thermal expansion of the epoxy to that of the copper windings. If the fillers are not well mixed, mechanical strain is created on cooling after the epoxy is cured, or in operation due to thermal cycling that may lead to fatigue cracks. These voids and cracks create PD within the casting, leading to electric treeing (Section 2.4.1). Another failure mechanism in cast epoxy equipment is when the epoxy resin shrinks back from metallic surfaces during curing of the epoxy because of poor design, manufacturing, or adhesion; such defects are known as "delamination." Considerable effort is put into achieving good adhesion between the epoxy resin and metal surfaces, by roughing them up, chemical treatments, or a combination of such techniques. Due to the large difference in the thermal coefficient of expansion of metallic parts and the resin, surface adhesion is very important because once delamination starts (i.e. in manufacturing), the separation gap will often enlarge itself due to thermal cycling of the equipment. Since epoxy insulation is an organic material, once the PD initiates, failure will likely occur at some point, perhaps within only weeks or months.

14.3.4 PD in Cable Accessories

Switchgear usually includes many power cable joints and terminations, and perhaps "load break" elbows (plug and socket). Section 12.3.5 identifies several ways that PD can occur in such power cable accessories, often leading to failure. The accessories within the AIS would normally be detectable by any AIS PD testing.

14.4 PD Sensors

For offline factory or offline/onsite PD tests, conventional IEC 60270 detectors are normally used, with the 50/60 Hz voltage supplied by a test transformer. The PD sensor is usually a 1 nF or so capacitor and detection impedance, in an indirect measurement (Section 6.2.2).

For online testing, a wide variety of PD sensors have been developed and implemented commercially (see below). Most of these sensors work in the VHF range, mainly to reduce the impact of electrical disturbances.

14.4.1 TEV Sensor

One of the most popular sensors for AIS online PD testing is the electromagnetic TEV or "transient earth voltage" sensor, where "earth" is the British term for "ground." As described in Section 7.4.3, there are two versions of the TEV. The original version has a metal plate that forms a capacitor to the outside of the metal enclosure of a switchgear cubicle. A discharge occurring within the switchgear will create an RF signal that will radiate to the outside surface of the cubicle metal enclosure at gaskets on panels and doors, or through any ventilation openings, inducing voltage pulses ("transients") on the outside surface of the cubicle enclosure. The TEV was invented by Dr. John Reeves about 1979 at a British organization called the Electrical Council Research Centre (a predecessor of EA Technology) [12]. As discussed later, the TEV can be built into a portable instrument that is sequentially moved along to each cubicle and compartment of the switchgear line-up. These instruments may have a provision for a second TEV sensor, which may be placed on an adjacent cubicle. The relative time of arrival of signals from TEVs on different cubicles can then be used to locate the cubicle with the most significant PD activity using the time-of-flight method (Sections 8.4.2). The signals are usually detected in the lower end of the VHF range. More details on this sensor, which have resulted in a database of over 100,000 measurements, are presented in [5–7].

A second type of TEV, which measures the voltage between two capacitive electrodes a few centimeters apart, is installed across the main metal enclosure of a cubicle and its access door, i.e. straddling across the gap. The gasket between the door and the main frame is an impedance where transient voltages caused by PD within the enclosure will create a potential difference [8].

Both types of TEVs create a small voltage pulse in the millivolt range when placed outside of the cubicle enclosures. When connected to a suitable instrument, most suppliers measure the voltage magnitude in relative terms, i.e. dBm: dB referenced to 1 mV. The TEV sensor itself is relatively inexpensive compared to other sensors and can be either portable or permanently installed. The TEV sensor does not need to be installed within the AIS – it can be used on the exterior of the AIS, thus requiring no outage for installation. Some providers, including EA Technologies, embed the sensor within a handheld instrument. If the sensors are installed on the outside of switchgear cubicles, then the sensors may also measure disturbances outside of the AIS, including any sparking from poor electrical connections or corona from outdoor substations. Hence, only PD above the background interference can be measured. There are now many manufacturers of TEV sensors.

If the TEVs are connected to a PRPD type of instrument, the PRPD pattern will be more complex than normal since the PD signals from all three phases will be detected by the one sensor along with any other strong PD signals entering the AIS from outside sources.

14.4.2 Capacitive and HFCT Sensors

For online AIS PD monitoring, the second most common sensors are high-voltage coupling capacitors connected to the bus (Section 6.2.2) and HFCTs (Section 6.3.3) that are installed on the connection between power cable neutrals and ground. The capacitors supplied by one major AIS vendor (Eaton) and a third party (Iris Power) tend to be rated 80 pF, since they are also used for online PD measurement in rotating machines (Section 16.6.2). These sensors in combination with the associated instrumentation work in the HF or VHF range, depending on the vendor. The capacitors (one per phase) do not need to be installed in every cubicle. Suppliers normally recommend they be installed in every second cubicle [9–11]. Where capacitors are used, the vendor must perform extensive testing to ensure that the capacitors themselves do not lead to in-service failures (Section 16.6.2).

HFCTs are generally installed on each power cable, on the ground connection of the cable termination and operate in the HF and sometimes the low VHF range. Capacitors and HFCTs are more expensive than TEVs or other antenna-type sensors and require an outage to install since they must be placed within the AIS. However, such sensors tend to be inherently less sensitive to external disturbances since they are within the Faraday shield of the AIS enclosure and can use a variety of interference suppression methods including time-of-flight discrimination, pulse-shape analysis, T–F maps, and three-phase synchronous analysis (Sections 8.4.2, 8.4.3, 8.9.2, and 8.9.3, respectively).

14.4.3 RF Antenna

The original corona probe (also known as the TVA probe) was used to measure and locate stator winding and switchgear PD in offline testing since the 1950s (Section 16.13). As discussed in Section 16.13.1, the probes use a "loopstick" RF antenna (usually 10 turns of copper wire around a ferrite rod) to measure PD in the 0.5–5 MHz range. A variation of this RF antenna tuned to 300 kHz was first employed to measure PD online within switchgear cubicles in 1985 [4]. Research then showed that the center frequency of 300 kHz provided the highest PD signal-to-noise ratio [4]. More recently, other researchers have used other types of RF antennae to detect PD within AIS [13], although the detection frequency was not disclosed. The higher the measurement frequency, the easier it is to locate the cubicle with the PD, since the higher frequencies attenuate PD more quickly (Section 6.6.2.2). Outages are required to install the RF antennae within the cubicles. Research indicates that the sensors should be installed in every second or third cubicle to locate the cubicle with the highest activity, based on magnitude [13]. The signals are measured in units of mV or dBm.

These sensors can detect PD both on the insulation surfaces in air, as well as within voids, for example in instrument transformer (CT/PT) epoxy castings. Note that PD from all three phases is detected with one sensor, so the PRPD pattern is much more complex, often resulting in six clusters of pulses, rather than the normal two clusters per AC cycle.

14.4.4 Ultrasonic Sensors

Inexpensive ultrasonic microphones (Section 5.4.1) have been used as sensors to detect surface PD, if they are within the switchgear cubicles. Research indicates that acoustic signals from PD in switchgear tend to be highest about 40 kHz [4]. Alternatively, contact-type piezoelectric sensors temporarily or permanently mounted on the exterior surface of a cubicle can detect vibration from surface PD within the enclosure (Section 5.4.2). Such contact-type ultrasonic sensors are incorporated into some brands of TEV instruments.

Note that acoustic sensors will only detect surface tracking PD or PD in exposed air gaps. PD within voids in a casting will not generate an acoustic signal that can be measured in a remote microphone or contact vibration sensor because the acoustic impulse is strongly damped by the epoxy. Acoustic PD measurement is modulated at twice the power frequency. Thus, the acoustic sensors will detect PD from all three phases within the cubicle, so the PRPD patterns will be a mixture of the PD from three phases.

14.5 Commissioning and Offline/Onsite Testing

Offline PD tests for acceptance testing of new AIS, or condition assessment of operating AIS, most often use the charge-based LF method in IEC 60270 (Chapter 6). For testing, the power cables must be disconnected, or the circuit breakers are in the open position, which means some parts of the switchgear will not be energized and therefore defects causing PD will not be identified in these unenergized locations. Under these conditions, the capacitance of one phase of the AIS is modest and a 10 kVA test transformer may be adequate to energize the bus. Sometimes PTs can be back-fed from a variable autotransformer to excite each phase to the required test voltage. Note that in a single-phase test (with the other two phases grounded) at rated line-to-ground voltage, the voltage between phases will be only 58% (1/√3) of the normal voltage between phases. If a three-phase test transformer is available, this should be used since both the phase-to-ground and the phase-to-phase insulation undergo the same voltage stress as occurs in operation.

IEC 62271-200 provides details on using the IEC 60270 method for AIS PD testing [3]. It does not provide a recommendation on the bandwidth used; however, unless there is excessive interference, the wideband mode should be employed so that the pulse polarity can be measured and to minimize any effects from natural frequencies (Section 8.2). For single-phase tests on AIS, IEC 62271-200 suggests 1.3 times the rated line-to-line AC voltage be applied, following which the voltage is then gradually lowered to determine the PDEV (Section 10.2). The standard suggests that for new equipment the Q_{IEC} be below 100 pC at 110% rated line-to-line voltage (applied line-to-ground). 100 pC is also the suggested max PD level in a Canadian AIS standard created in 1989 [14, 15]. The 2022 version of IEEE C37.20.2 also suggests (but does not require) an offline PD test on new switchgear [2]. It proposes that each phase has a PDIV greater than 1.2 times the rated line-to-ground voltage, with a Q_{IEC} sensitivity of 100 pC.

If PD is detected in commissioning tests or offline/onsite tests, the PD can usually be more precisely located using corona probes (Section 16.13). If the PD is occurring on insulation surfaces and metal barriers can be removed to expose them, ultrasonic microphones or acoustic imaging devices (Section 5.4.1) can help to locate the PD sites. Alternatively, ultraviolet imaging cameras can also locate surface PD (Section 5.3). Such methods will help zero in the defect locations and thus will facilitate repairs.

It is important to note that the actual PD levels and PDEVs depend strongly on the humidity and temperature of the AIS during the tests. Even poorly made switchgear with airgaps around busbars will normally be PD-free when the AIS is clean and dry.

Any commissioning tests to measure the PDEV and PRPD patterns can be used as a baseline to determine if insulation aging is occurring over time, using the same test procedure and measurement frequency range. However, the strong impact of humidity makes it difficult to trend the PD activity over time for condition assessment purposes.

14.6 Online PD Monitoring

Online PD measurements, acquired by either periodic testing or continuous monitoring, rarely use LF test methods such as that described in IEC 60270, due to the many sources of interference. For RF or electrical methods, interference sources include (Section 9.2):

- Sparking from poor (e.g. corroded or intermittent) electrical connections, whether on the high-voltage circuits or the ground circuits, including poorly grounded metal structures, can create massive interference. This is usually the main cause of interference.
- Corona from any outdoor overhead busses, post insulators and the external surfaces of transformer bushings that supply the AIS. Such corona is almost always harmless (except from an EMI point of view), but may overwhelm important PD within the AIS.
- As with all other types of PD measurements, interference from power tool operation, invertor fed drives, communication systems, etc. can also lead to false positive indications.

When acoustic sensors are used, acoustic interference sources include:

- Metallic objects throughout the AIS vibrating at twice the powerline frequency (100/120 Hz), a result of being excited by the magnetic fields generated in current-carrying conductors.

- The movement of gravel due to people walking, wind, or vehicles. Gravel is often present surrounding outdoor AIS.
- Rain and sleet hitting outdoor AIS.
- Small leaks in nearby high-pressure steam or gas pipes.
- Impact noise due to vibrating components.

There are two basic types of PD systems used for online PD testing of AIS:

- Systems where the sensors are installed within the switchgear. This requires an outage to retrofit the system sensors and is usually more expensive, but are generally less likely to produce false positive alerts of AIS PD due to interference.
- Systems where the sensors are outside of the AIS enclosure. These can include systems where the sensors/instruments are portable (enabling "surveys" by test technicians of the PD activity in each cubicle); or systems where the sensors are mounted permanently on the outside surface of each cubicle and connected to a continuous monitor. This approach is generally much cheaper and does not require an AIS outage. However, it has a higher risk of false positive indications and is more likely to incorrectly indicate the cubicle with active PD sources.

14.6.1 Systems with Sensors Within the Switchgear

The most common sensors used for online electrical PD detection are capacitors (often 80 pF) connected to the bus, and/or HFCTs installed on the power cable neutral shield connections to ground. Figure 14.2 shows 80 pF PD sensors permanently installed in an outdoor metalclad switchgear. It is also possible to permanently install RF antennas, TEV sensors, or ultrasonic microphones to be located within each cubicle. The measuring systems usually work in the HF or VHF range. The sensors can be connected to a termination panel on the AIS enclosure for periodic survey testing or be directly connected to a continuous monitoring system.

The main advantage of installing sensors within all or a high percentage of cubicles is that the metal enclosure provides some measure of suppression of external interference, especially in the case of electrical detection where the enclosure serves as a partial Faraday shield.[1] Many suppliers of PD systems for AIS also augment the natural isolation from external interference with other interference-suppression techniques such as:

- Time-of-arrival analysis to separate interference such as PD from an overhead in-feed bus (Section 8.4.2). The same method can be used to identify which cubicle has PD, if there are sensors in say every other cubicle.
- Pulse shape analysis (Section 8.4.3) or T-F maps (Section 8.9.2) to separate interference from local PD, or PD far from the sensor in other cubicles.
- Three-phase synchronous detection (Section 8.9.3).

The conducted PD currents or RF signals will propagate from cubicle to cubicle along the line-up. One researcher suggests that the signal loss around 10 MHz using capacitive sensors or HFCTs is about 30% per cubicle along the line-up [11]. Experiments using an RF antenna (with an undisclosed detection frequency) show the attenuation to the adjacent cubicle can range from 3 to 10 dB [13]. These attenuation effects imply that, to some degree, one could find the

1 It is only "partial" since insulating boards supporting the main bus allows PD signals to radiate between cubicles to some extent. In addition there are discontinuities in the shield at vents, bushings and cable terminations.

cubicle with the highest PD activity based on signal magnitude alone. However, all providers suggest additional methods such as using acoustic sensors, time-of-flight, or visual inspection to confirm this.

The advantage of using traditional PD sensors such as capacitors and HFCTs (provided that three HFCTs are installed in each cubicle) is that the phase with the most PD activity can be determined, and PRPD patterns (Sections 8.7.3 and 10.5) may identify the root cause of any detected PD – such as surface tracking or void discharges. RF antennas and acoustic microphones will detect PD from all three phases at the same time, and thus the PRPD plots can be very complicated if PD is occurring on two or more phases or between phases, even if a reliable AC phase synchronization signal (Section 8.3.2) is provided.

14.6.2 Systems with Sensors Mounted Outside the Switchgear

TEVs employed outside the switchgear are by far the most popular method to detect PD within AIS cubicles in online PD tests, probably because they are much cheaper to implement and do not require an outage to install.

In a PD "survey," a handheld instrument contains the TEV sensor with associated electronics to measure the PD level in dB (Figure 14.6). It seems the natural background noise level is about 5–10 dB (above 1 mV), but this should be confirmed by placing the probe against some other (non-AIS) metal structure. The magnitude of the activity is measured by placing the TEV instrument against the metal door or panel of each cubicle. By repeating the measurement on all the other cubicles, one can sometimes locate the cubicle that is likely to contain the PD source, based on signal amplitude alone. However, devices like that shown in Figure 14.6 often have an external input that can be connected to an auxiliary TEV sensor placed on adjacent cubicles. The instrument then determines which TEV detected the signal first (this is a variation on the time-of-arrival method in Section 8.4.2). With repeated measurements on adjacent cubicles with the two TEVs, one can have additional confidence of the location of the PD sites. Some training and experience are needed to collect data effectively. When data is collected, it is important to record the humidity

Figure 14.6 Photo of a portable TEV instrument pressed against a metal cubicle enclosure to measure the PD activity within the cubicle. The instrument contains both a TEV capacitive sensor and a contact piezoelectric sensor to measure acoustic signals within the cubicle. *Source:* Courtesy of EA Technology.

Figure 14.7 Photo of TEVs (blue objects) installed on AIS cubicles as part of a continuous monitoring system. *Source:* Courtesy of EA Technology.

and ambient temperature (from which absolute humidity can then be calculated) in the AIS environment, since moisture conditions will have a major influence on whether PD activity is taking place or not. Surface PD and electrical tracking are relatively slowly evolving aging mechanisms, thus PD surveys can be conducted about once per year.

Some providers will augment the TEV sensors with a piezoelectric ultrasonic contact sensor to detect the ultrasonic signals within the AIS cubicle associated with PD (Section 5.4.2). The combination of the TEV and piezoelectric sensor gives greater confidence that surface PD is occurring within the switchgear.

Alternatively, one can permanently mount the TEVs on the external metal surface of each cubicle and connect the sensors to a continuous monitoring instrument (Figure 14.7). The advantage of such continuous monitoring is that it will be measuring the PD continuously. It is well known that surface PD in AIS is highly dependent on atmospheric conditions, with PD sometimes disappearing even though there is considerable damage to the insulation due to PD and tracking [4, 11]. Continuous monitoring ensures that the periods of high activity are captured. In some systems, the data from the TEVs can be augmented by HFCT sensors on the power cable [8].

14.7 PD Interpretation for AIS

The general principles for PD interpretation are presented in Chapter 10. As discussed in Section 14.5, new AIS is expected to be PD-free at a voltage substantially above the operating line-to-ground voltage, with a sensitivity (Q_{IEC}) of 100 pC, using IEC 60270 detectors. If any PD is detected, then its source can be partly identified by observing the PRPD patterns (Section 10.5). The location of the PD can be determined from RF corona probes (Section 16.13), UV cameras, or acoustic imaging cameras (Chapter 5).

As with other types of apparatus, the trend in PD activity over time is the key indicator for online monitoring systems. With AIS, the PD is much more variable than for other types of equipment, due to the significant impact of the atmospheric conditions on any surface PD activity (which is the main type of PD occurring in most AIS). Thus the trend needs to account for the daily and seasonal changes in atmospheric conditions – although there is no formula to correct for the impact of (say) humidity on PD. If the PD is due to surface PD or surface tracking, the aging processes are slow. Thus, trending the activity over a few years will provide warning that aging is occurring, and that maintenance would be prudent. An increase of 100% (or 6 dB) of the peak activity over a year is significant. Unfortunately, if the PD is occurring in voids or cracks within epoxy castings or power cables, the PD may lead to failure in days or weeks and thus the longer-term trends will not be useful. Readers are reminded that PD testing can only indicate that flaws are present and/or aging is occurring. PD trending cannot be used to predict when failure will occur.

Most of the published data that comes from online tests on AIS is measured with TEV sensors. The U.S.-based National Electrical Testing Association (NETA) has suggested that if the levels are above 20 dBm from a TEV mounted external to the AIS enclosure, then it is important to verify the source by other means and perform a visual inspection and/or repairs [16]. This level, of course, depends not only on the magnitude of the PD but also on the design of the TEV sensor and the propagation of the VHF signal to the outside of the panel where the sensor is placed, including the size of the opening at the panel or door gaskets. This means that the 20 dBm alert value should be used with considerable caution.

14.8 PD Measurement in Isolated Phase Bus

Isolated phase bus (IPB) is air-insulated, medium-voltage bus that is widely used to connect generators to the step-up (or GSU) transformer for connection to the transmission grid (Figure 14.8). For each phase, there is usually a center hollow aluminum tube that is supported coaxially within a larger grounded outer aluminum cylindrical enclosure by ceramic post insulators, as described in IEEE C37.23 [17]. The phases are usually placed adjacent to one another horizontally, although sometimes the three phases may be stacked on top of each other. Unlike AIS, there is a grounded metal barrier between the high-voltage inner conductors of each phase and it prevents phase-to-phase failure. To allow for axial expansion as the busbar is loaded (increasing the busbar temperature), there are usually flexible links (heavy braided tinned-copper wires) between busbar segments, between the bus and the generator terminals, as well as between the bus and the GSU transformer terminals (Figure 14.9). In large IPBs, there are usually dozens of bolted electrical connections on the flexible links. An excellent review article on IPB design and failure was written by Timperley [18]. The air in positive pressure IPBs is slightly above atmospheric pressure.

Figure 14.8 Photo of IPB connecting a generator to the GSU transformer. This IPB includes a circuit breaker, but most IPBs do not include a circuit breaker. *Source:* J. E. Timperley Consulting. LLC.

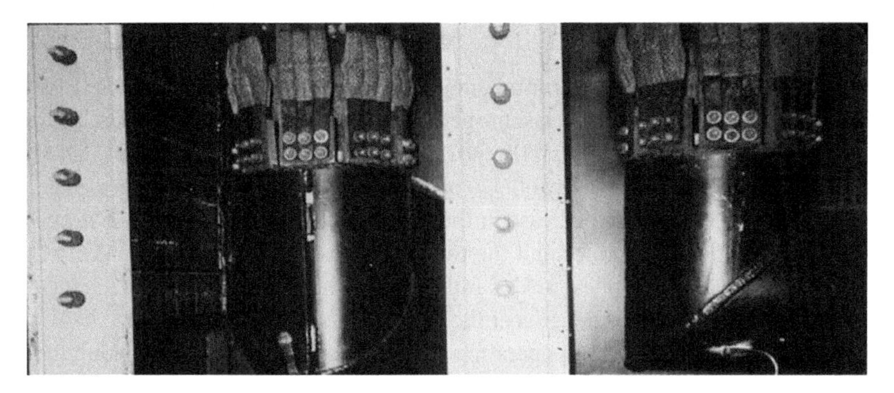

Figure 14.9 Photo of the heavy-braided flexible links on two phases between the HV IPB terminals and the generator bushings.

14.8.1 IPB Deterioration Processes

Unlike AIS, electrical tracking due to contamination rarely leads to IPB failure. The main insulation in IPB, other than air, is from post insulators that are almost always porcelain, and very resistant to electrical tracking (Section 3.7.2). Flashover due to contamination may only occur when liquid water or heavy carbon pollution is present within the IPB.

It is more common that the post insulators crack or break due to thermal expansion and contraction (caused by load cycling) when there is no mechanism to compensate for those forces. Often cracks in the porcelain may lead to PD, sometimes of very high intensity, but this usually does not result in electrical failure of the IPB since PD does not significantly deteriorate porcelain. If the crack is at the metal ends of the post insulator, then normal 100/120 Hz vibration can lead to contact sparking (Section 4.4). Usually PD from cracked post insulators or contact sparking does not affect the reliability of the IPB. However, such PD/contact sparking will be a substantial source of interference for online PD measurements on the transformer (Section 15.12.5) or the generator (Section 16.9.1). If there are several broken post insulators, the mechanical stability of the center conductor may be affected, especially when there are high fault currents.

Poor electrical contacts can occur at the flexible links within the bus and at the connections to the generator or transformer (Figure 14.9). It can be caused by corroded surfaces, incorrect tightening force, or small gaps between the insulator top and the busbar (sliding bus models). If even one of the bolts used to fasten the flexible link to the busbar is not properly torqued, contact arcing can occur (Section 4.4). If the contact arcing is confined to just one or two bolts, there is little risk of IPB failure. However, when many bolts are not properly torqued, hot spots may develop which can eventually lead to melting of the aluminum conductors and IPB failure. Such sparking can be easily detected by online PD measurements. Offline tests may not detect this problem, since there is no current flowing through the connections, and there may be no vibration to make and break the intermittent contact points.

Somewhat surprisingly, metal foreign materials such as nails, screws, and even tools are sometimes found within the IPB [18, 19]. As long as the IPB passes the AC voltage withstand test, foreign materials cause few future problems. But if the foreign materials move to the wrong location, for example near a post insulator, or water enters the IPB, it may trigger PD and eventual breakdown.

14.8.2 Offline PD Tests

Manufacturers of IPB may perform factory or commissioning PD tests, but it seems that the IEEE and IEC standards do not require them. However, some generating utilities specify such tests. They normally require the offline test be done using an IEC 60270 LF charge-based detector. The PDEV is measured, usually with a minimum sensitivity of Q_{IEC} of 100 pC. Offline tests may be performed on an IPB that has seen service to detect a decrease in the PDEV over time, or an increase in the Q_{IEC} or Q_m. The flexible links to the generator and the transformer need to be removed. The AC transformer to energize the bus can be modest in kVA rating due to the relatively low capacitance of one phase of bus and the relatively short lengths of the IPB (usually <100 m).

If PD is detected below the phase-to-ground operating voltage, the source of the PD should be investigated. PD sites can be located using capacitive sensors at each of the IPB and a VHF detection system that employs the time-of-flight method (Section 8.4.2). Alternatively, an RF corona probe or EMI sniffer may be able to locate the PD sites by finding the region with the highest signals (Section 16.13), provided that the IPB enclosure is not completely welded. Also contact-type piezoelectric sensors (Section 5.4.2) may locate PD sites based on the strongest acoustic signal that is measured from the IPB enclosure. Using RF, EMI, and contact acoustic probes may be somewhat difficult since the IPB is often suspended high above the floor, necessitating some means of hoisting a technician to be in proximity to the IPB along its length.

14.8.3 Online PD Monitoring

High PD is most likely associated with cracked porcelain insulators, although it may also be an indication of surface PD due to contamination on the porcelain insulators (which is unlikely to be a cause of failure). Poor electrical connections are a source of PD-like sparking that can be either due to poorly torqued bolts at connections or where the metal ends of post insulators are not solidly connected to the ground and high-voltage aluminum tubes. Formally, this is not PD, but most online PD measurement systems will detect such arcing/sparking during operation when high currents are flowing through the IPB and 100/120 Hz magnetic forces are present.

A variety of methods to detect PD online in IPB have been proposed, although there seems to be no preferred method. As with AIS, there seem to be few users of the LF method in IEC 60270, due to interference and its inability to locate where the pulse signals are originating. The online methods that have been most applied are:

- EMI monitoring using an HFCT on the generator neutral together with EMI sniffers (Sections 16.13) [18]. This method requires a skilled practitioner to separate IPB signals from all other interference from other equipment in the generating station.
- Capacitive sensors installed at two or more locations along the IPB. 80 pF capacitors are the most common. If the generator is already equipped with such sensors (one at the generator terminals and the other some distance along the IPB) for online generator PD detection (Section 16.9.3.1), then PD monitoring of the IPB is also possible. However, additional capacitive sensors may be needed at the transformer and along the IPB to allow better sectionalization of the IPB to ensure PD/sparking is detected along the entire length of the IPB. The PD/sparking sites are located using the time-of-arrival principle between a pair of sensors (Section 8.4.2). This approach has the lowest risk of false positive indications.
- TEV sensors mounted at different locations along the IPB, preferably near any interfaces between busbar segments, where the RF energy can radiate through gaskets. The advantage of this approach is that the IPB does not need to be opened to install the sensors. The time-of-arrival method can be

used to locate the most probable location of the PD/sparking sites. The disadvantage is that the TEV will also be sensitive to electrical interference external to the IPB, including other nearby sources of PD or sparking.

Both periodic and continuous monitoring systems are available.

14.8.4 Interpretation

Ideally there will be no PD or sparking sources within the IPB. If signals are measured, then it is prudent to trend the activity levels over time. Measurements should be taken at least once per year. Normally, when a new problem is encountered, the signal level will go up by orders of magnitude in just a few months. The root cause of the PD may be established from PRPD plots (Section 10.5.2). Note that sparking at poor electrical contacts tends to be centered on the zero crossings of the AC voltage cycle. Examples of the use of online PD testing using capacitive sensors and EMI to find PD and contact sparking problems are in [19].

References

1 Blair, T.H. (ed.) (2017). Switchgear. In: *Energy Production Systems Engineering*, 467–482. Wiley/IEEE Press https://doi.org/10.1002/9781119238041.ch19.

2 IEEE (2022). IEEE Std. C37.20.2. IEEE standard for metal-clad and station-type cubicle switchgear.

3 IEC (2021). IEC 62271-200*:2021*. High-voltage switchgear and control gear – Part 200: AC metal-enclosed switchgear and control gear for rated voltages above 1 kV and up to including 52 kV.

4 VanHaeren, R., Stone, G.C., Meehan, J., and Kurtz, M. (1985). Preventing failures in outdoor distribution-class metalclad switchgear. *IEEE Transactions on Power Apparatus and Systems* PAS-104 (10): 2706–2712. https://doi.org/10.1109/TPAS.1985.319111.

5 Mackinlay, R.R. (1990). Discharge measurements on high voltage distribution plant. *IEE Colloquium on Developments Towards Complete Monitoring and In-Service Testing of Transmission and Distribution Plant*, Chester, UK (31–31 October 1990), p9.

6 Brown, P.M. (1996). Non-intrusive partial discharge measurements on high voltage switchgear. *IEE Colloquium on Monitoring and Condition Assessment Equipment*, Leatherhead (5 December 1996).

7 Davies, N. and Jones, D. (2008). Testing distribution switchgear for partial discharge in the laboratory and the field. *Conference Record of the 2008 IEEE International Symposium on Electrical Insulation*, Vancouver, BC (9–12 June 2008), pp. 716–719. https://doi.org/10.1109/ELINSL.2008.4570430.

8 Caprara, A., Cavallini, A., Garagnani, L., and Guo, J. (2018). A novel approach for continuous monitoring of partial discharge phenomena on medium voltage equipment. *2018 IEEE Electrical Insulation Conference (EIC)*, San Antonio, TX (17–20 June 2018), pp. 495–498. https://doi.org/10.1109/EIC.2018.8481094.

9 Smith, J.E., Paoletti, G., and Blokhintsev, I. (2002). Experience with on-line partial discharge analysis as a tool for predictive maintenance for medium voltage (MV) switchgear systems. *2002 IEEE Petroleum and Chemical Industry Conference*, New Orleans, LA (23–25 September 2002), pp. 155–161. https://doi.org/10.1109/PCICON.2002.1044997.

10 Blokhintsev, I., Kozusko, J., Oberer, B., and Anzaldi, D. (2017). Continuous and remote monitoring of partial discharge in medium voltage switchgear. *2017 IEEE Electrical Insulation Conference (EIC)*, Baltimore, MD (11–14 June 2017), pp. 205–208. https://doi.org/10.1109/EIC.2017.8004669.

11 Benzing, J., Patterson, C.L., Cassidy, B.J. et al. (2012). Continuous on-line partial discharge monitoring of medium voltage substations. *IEEE-IAS/PCA 54th Cement Industry Technical Conference*, San Antonio, TX (14–17 May 2012), pp. 1–12. https://doi.org/10.1109/CITCON.2012.6215702.

12 Reeves, J. (1979). High voltage partial discharge measurements – improving their significance. PhD thesis. University of Manchester.

13 Pestell, C., Caves, D., and Bowen, J. (2017). Qualitative analysis of partial discharge radio frequency emission propagation in medium voltage metal-clad switchgear. *2017 IEEE PCIC*, Calgary, AB (18–20 September 2017), pp. 185–194. https://doi.org/10.1109/PCICON.2017.8188737.

14 CSA (2018). CAN/CSA-C22.2 NO. 31-M89 (R2000): switchgear assemblies.

15 Gillespie, M.T.G., Murchie, G.B., and Stone, G.C. (1989). Experience with AC hipot and partial discharge tests for commissioning generating station cables and switchgear. *IEEE Transactions on Energy Conversion* 4 (3): 392–396. https://doi.org/10.1109/60.43240.

16 ANSI/NETA (2019). ANSI/NETA MTS-2019, Standard for Maintenance Testing Specifications for Electrical Power Equipment and Systems, 2019 Edition. Section 11: On-Line partial discharge Survey for Switchgear.

17 IEEE (2015). IEEE 37.23-2015, Standard for metal-enclosed bus.

18 Timperley, J.E. (2022). Isolated-phase bus—design, operation, deterioration, and failure. *IEEE Electrical Insulation Magazine* 38 (3): 15–23. https://doi.org/10.1109/MEI.2022.9757854.

19 Singh, A. and Hughes, M. (2018). Partial Discharge Activity in Isolated Phase Bus (IPB) – Case Studies from UK Power Stations. CIGRE, Paper A1-203, 2018.

15

Power Transformers

15.1 Introduction

Power transformers and autotransformers are an integral and critical component of modern electric power transmission and distribution systems. Power transformers take the relatively low output voltage (usually 27 kV or less), high-output current power from generating stations and converts it to high-voltage (up to 1000 kV), low-current power suitable for transmission lines. Such transformers are often referred to as generator step-up transformers (GSU). Similarly, at the load centers, a power transformer converts the high-voltage, low-current power from the transmission lines to lower voltage, higher current power needed for local power distribution. In addition, power transformers are used in utility distribution systems. References [1–3], as well as many other books, provide information on the application and design of power transformers in modern power systems. Many power transformers also have online tap changers (OLTCs) that enable the output voltage from the high-voltage winding to be adjusted during normal operation by including more or fewer turns normally in the low-voltage winding.

Until the 1950s, verification of the dielectric strength of transformers was conducted by applying a withstand voltage (over voltage) in the factory for a relatively short time. If electrical breakdown did not occur within the test period, the insulation systems were considered good. Since the 1950s, a challenge was undertaken of analyzing the phenomena that occur in power transformer insulation prior to breakdown by measuring the partial discharge activity (known as corona at the time) [4, 5]. This involves measuring the low-energy discharges located in tiny defects within the insulation, which by eroding the insulating materials over time, lead to total failure of the insulation. The new frontier has led to a substantial change in transformer testing philosophy by moving to a more cautionary approach, which involves using reduced voltage tests, for much longer times, but still significantly greater than the operating voltage. During the test, the level of PD is monitored, immediately providing information on the behavior of the insulation in the face of the imposed dielectric stresses.

As is the case for all other types of electrical equipment in this book, it should be borne in mind that, although the aging phenomena of the insulation due to PD are known, it is difficult to create a reliable relationship between the level of PD measured and the state of insulation degradation. In particular, one cannot determine the remaining life of transformer insulation based on PD measurements alone.

Currently, the measurement of PD in transformers helps to confirm the correct design, the quality of the insulating materials used, as well as the adequacy of the treatment and drying

Practical Partial Discharge Measurement on Electrical Equipment, First Edition. Greg C. Stone, Andrea Cavallini, Glenn Behrmann, and Claudio Angelo Serafino.

processes. This information at the beginning of its life provides a limited guarantee on what the life of the transformer may be, which also depends on the operating conditions to which it will be subjected. This chapter examines most of the aspects related to the measurement of PD on transformers, focusing attention on practical aspects with the hope of guiding the reader to identify the best practice to be implemented according to their needs.

As with the other equipment chapters, it is first useful to have a basic knowledge of how the transformers are constructed and the insulation aspects needed to ensure their reliability. For this reason, the properties of the insulating materials used in transformers are first presented, including how they change as a function of time under thermal and electrical stress, contamination, and how they influence the likelihood of PD inception. Some basic transformer construction features are also described. These aspects can be extended to other types of electric equipment insulated with solid insulation materials impregnated in oil, more commonly defined as composite insulation such as reactors, phase-shifting transformers and autotransformers. After this material is presented, the factory, onsite/offline and online PD measurement methods are discussed. Since dissolved gas in oil analysis (DGA) is another common tool for transformer insulation assessment, case studies comparing DGA and PD results are given.

15.2 Transformer Insulation Systems

References [1–3] provide detailed information about all aspects of power transformers, including the insulation systems and manufacturing methods. Reference [6] presents information specific to the design of the solid insulation components used in transformers. The following presents an overview of the insulation systems in medium-voltage power transformers (dry-type transformers, which are air-cooled) and high-voltage, liquid-filled power transformers. These two types of transformers are the most likely to require PD testing in the factory, as well as onsite/offline PD testing or online PD monitoring. Most of this chapter concentrates on liquid-filled power transformers.

15.2.1 Dry-Type Transformer

Dry-type transformers do not use liquids such as transformer oil to insulate and cool the transformer. Instead, they are passively or forced-cooled by air and insulated with solid insulation such as polymer films, fiberglass, and fiberglass reinforced with epoxy (sometimes called prepreg). Compared to liquid-cooled transformers, air-cooling confines dry-type transformers to medium-voltage applications and relatively low power ratings. For the same power rating, dry-type transformers are bigger than liquid-cooled transformers. The high-voltage windings are encapsulated or cast with a polyester or epoxy resin. The low-voltage winding is usually rated less than 1000 V and is placed closest to the laminated steel core for each phase. The high-voltage winding surrounds the LV winding. The cooling system provides many places for air to circulate in the low- and high-voltage windings, as well as between the windings. There are many different designs of dry-type transformers, with open wound (impregnated using a "dip and bake" process – Figure 15.1) and cast resin types (using a vacuum casting process – Figure 15.2) being the most common. Vacuum cast epoxy windings are mechanically stronger, less prone to environmental contamination, and can operate at higher voltage ratings due to smaller internal voids. The windings are made from either copper or aluminum in the form of rectangular conductors or wide foils.

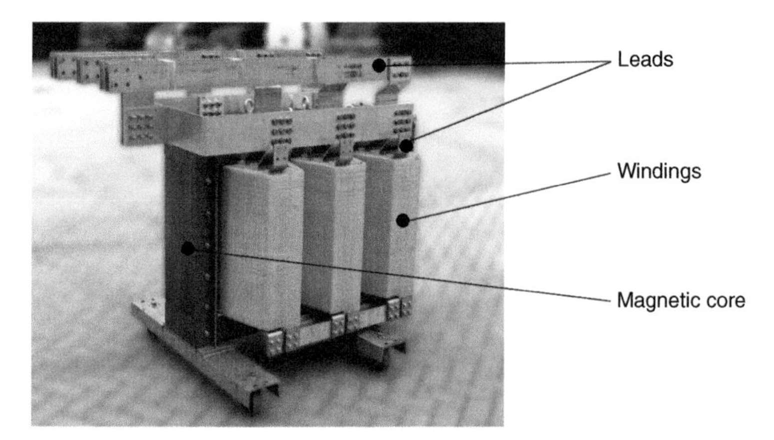

Figure 15.1 "Open-wound" three-phase dry type power transformer. *Source:* Courtesy of Ortea, Italy.

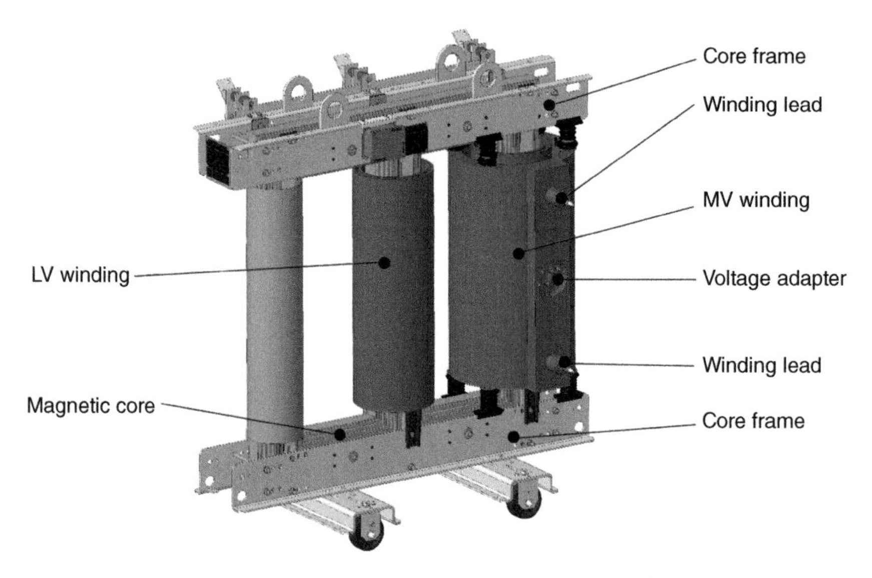

Figure 15.2 A three-phase vacuum cast epoxy power transformer. *Source:* Courtesy of SEA Transformers, Italy.

Transformers with vacuum cast windings in a polymer resin (epoxy is now the most common resin) using a mold typically operate at ratings ranging from <1 kV up to 36 kV and at rated powers that are usually less than 20 MVA. Cast resin windings are widely used as transformers for inverters in energy production plants from renewable sources, in utility distribution systems, low-temperature installations, installations on ships and railway trains, as well as where the risk of fire or environmental contamination needs to be minimized.

The casting of the resin is the most delicate phase in the manufacture of the windings. It must take place in a vacuum when the resin is fluid (around 50–60 °C) and cured at high temperature so that air pockets are not left within the resin volume. Cavities in epoxy become the site of PD and affect the mechanical integrity of the structure. The windings are normally made with aluminum foil strips previously covered with an insulating polyamide film for the turn insulation as shown in Figure 15.3. Aluminum also has the advantage of having a coefficient of thermal expansion similar to the epoxy, thus reducing internal mechanical strain.

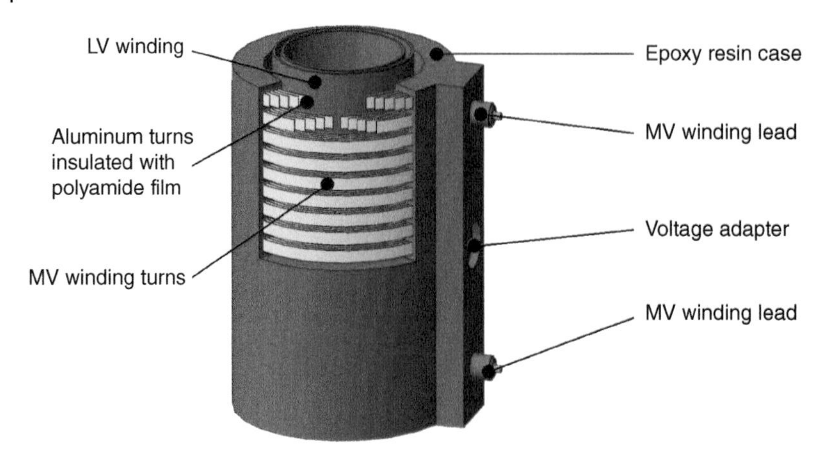

Figure 15.3 Typical resin incorporated medium-voltage winding.

15.2.2 Materials Used in Liquid-Filled Paper-Insulated Power Transformers

Liquid-filled power transformers are rated from about 15 to >1000 kV and can be either single phase or three phase. They have power ratings up to 1000 MVA. In large high-voltage transformers, the insulation system is made from organic solid-insulating materials and insulating fluids such as mineral oil (i.e. derived from petroleum), natural esters (made from vegetable oils such as canola or soybean), or synthetic esters. The use of insulating fluids is intended to insulate components from one another, as well as to conduct the heat caused by losses in the transformer windings and laminated steel core to the environment via heat exchangers. The oil acts as an insulator when it fills the space between two conductors at different voltages since the breakdown strength of oil is more than an order of magnitude higher than the breakdown strength of air (Section 2.6). The heat removal function occurs through forced or natural circulation of the oil, depending on the power of the transformer. In the case of very large power transformers, or special apparatus where the losses reach high values, special pathways for oil flow within the windings are created to better direct oil flow.

In addition to the oil, solid insulation materials are also used in liquid-filled power transformers. The solid insulation increases the breakdown voltage and helps to provide mechanical support to the windings. Cellulose-based solid insulation materials such as kraft paper and pressboard (also called transformer board) [6] are still used as the solid insulation in liquid-filled power transformers, more than 100 years after they were introduced.

The construction is very complex, and the control of the electric field distribution requires great attention. The use of solid organic materials is mainly to permit reducing the distance between live parts as much as possible and to build structures capable of conferring mechanical strength to the copper windings.

15.2.2.1 Mineral Insulating Oil

A guiding principle for the dimensioning of transformer insulation systems is the so-called "volume effect," that is the experimental confirmation that a liquid or solid insulation has a lower breakdown strength if the volume of the material subjected to the electric field is increasing [1, 2, 7]. This effect depends on the fact that the insulating medium is never 100% pure since metal particles, cellulose fibers, and gas-filled bubbles can be found in the oil of any transformer. Larger volumes increase the probability that these contaminants are found in the

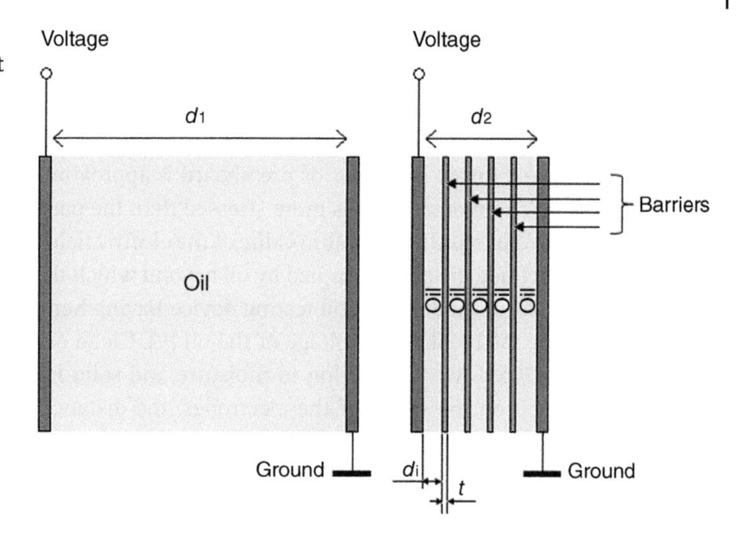

Figure 15.4 Reduced distance between the electrodes brought about by using solid insulation barriers in oil. The breakdown voltage is the same for both arrangements.

insulation, leading to statistically lower breakdown strength. Therefore, the insulation system contains several barriers of solid insulation (pressboard) that confine the oil to different regions, thus improving the breakdown strength. Figure 15.4 shows the theoretical advantages represented by barriers of solid organic material placed between two electrodes immersed in oil. The insulation thickness "d1," of the electrodes immersed in oil alone, is much greater than the distance between the electrodes immersed in oil with the addition of barriers of thickness "*t*," spaced "di" apart. The thickness of the barriers (t) must be as thin as possible. The use of organic solid insulating materials such as pressboard impregnated and immersed in oil has the advantage of decreasing the insulation thickness. This reduces the dimensions and mass of large transformers.

The first applications of mineral oil used as an insulating medium in transformers date back to the 1890s. Before this date, the transformers were mainly insulated in air or with paper soaked with bitumen. Mineral insulating oil is a mixture of hydrocarbons containing minimal quantities of nitrogen, oxygen, and sulfur, which are elements that reduce the quality of the oil. To limit the insulating liquid oxidation, antioxidant additives could be added. The most widely used additive is BHT (butylated-hydroxy-toluene), a phenolic antioxidant also called DBPC (di-ter-butyl-*para*-cresol). Some other additives for improving specific performances like pour-point or preventing unwanted effects like static electrification, or metal corrosion, are sometimes used. This category includes pour-point depressants, metal passivators, gas-absorbing, anti-bacterial, etc. [1, 2, 8].

The fundamental characteristics required of an insulating fluid for use in transformers are:

- High dielectric strength (Section 2.6)
- Low electrical losses (power factor)
- Low dielectric permittivity
- Arc quenching properties
- High cooling properties
- Low viscosity, even at low temperatures
- Lubricating properties

The relative permittivity (ε) of oils consisting of hydrocarbons assumes values of about 2.0–2.2 at room temperature and decreases with increasing temperature. It plays a primary role in the sizing of the insulation as ε affects the distribution of the electric field when different insulating materials are used in a system. The distribution of the electric field borne by the individual insulating materials is an inverse function of the relative permittivity (Section 2.1.4), and this allows for the dimensioning of the insulation components to avoid areas with high electric stress. The permittivity of insulating paper or pressboard is around 4 and, therefore, it can be deduced that the electric field inside the paper or pressboard is approximately half of what is obtained on average in the oil. Therefore, the oil is more stressed than the paper or pressboard.

The dielectric strength is the maximum value of the electric field in alternating voltage (50/60 Hz) applied across a defined distance occupied by oil beyond which the electric breakdown occurs. For oils in common use, a standardized oil testing device having hemisphere electrodes 2.5 mm apart is used to measure the breakdown voltage of the oil [9]. Clean oil has a breakdown voltage in the range from 30 to 70 kV AC. In addition to moisture and solid impurities, the dielectric strength depends very much on the shape of the electrodes, the distance between them, and the rate at which the voltage is increased. The presence of moisture in the oil is important, as it enhances the negative effect of impurities by significantly lowering the breakdown voltage (BDV). For these reasons, the measurement is conventionally referred to a specific standard and results obtained with different methods cannot be compared.

The behavior of the oil with respect to steep-fronted voltages from lightening (with a risetime of 1.2 μs and a fall time to half value of 50 μs) is different. The impulse BDV, usually in the range of 110–220 kV assuming a 2.5 mm spacing between metal hemispheres, is significantly influenced by the composition of the oil rather than by suspended particles and humidity [10, 11]. Since nitrogen and sulfur compounds have a negative influence on impulse breakdown, the breakdown voltage under lightning impulses depends a lot on the degree of oil refining.

In practice, the design of power transformer insulation depends mostly on the properties and the stress applied to the oil during events characterized by voltages having steep fronts, e.g. switching, lightning, and chopped-lightning impulses. These steep fronts create high-frequency components up to about 1 MHz that affect predominantly the oil due to its lower permittivity. It is the impulse strength that determines the thickness and design of the transformer insulation system.

The losses in the oil dielectric can be considered under two aspects. The first refers to the conduction current through the liquid: it increases with increasing temperature and as the electric field itself increases. It is usually expressed in the form of conductivity and depends on the presence of ionic species and little on the structure of the hydrocarbon. With AC excitation, the losses are due to polarization and are expressed by the dissipation factor (DF, or tan δ, which is very close to power factor). DF is very temperature-dependent and is affected by small traces of polarizable or colloidal substances. It is associated with the production of heat which, if not properly removed, can cause thermal instability in the insulation.

Also of importance is the oil viscosity. A lower oil viscosity significantly improves the impregnation of the solid organic insulation present in the transformer. Since the viscosity of the oil decreases with increasing temperature, the impregnation of the solid insulation is usually carried out at temperatures in the range of 50—60 °C.

Among the other characteristics required of the insulating oil for use in transformers is oxidation stability. The contact of the oil with oxygen present in the atmosphere, together with temperature and the presence of metals with catalytic action such as copper, leads to oxidation reactions whose by-products are acids, polar substances, and peroxides [12]. The progressive accumulation of these by-products causes a degradation of the dielectric characteristics of the oil and of the cellulose-based insulation as well as the corrosion of metals. The resistance to oxidation depends on the effectiveness of the refining process and the presence of natural inhibitors that are able to block the chemical reactions of oxidation. To improve oxidation stability, inhibitors can be added, thus giving rise to the inhibited oils [6, 12, 13].

When subjected to high electrical gradient, insulating oils tend to develop a substantial amount of hydrogen (a phenomenon known as "gassing") along with other hydrocarbon gases. This can lead to the presence of bubbles, and thus PD. It is, therefore, advisable to avoid the phenomenon of "gassing."

15.2.2.2 Natural and Synthetic Ester Liquids for Transformers

In recent decades, alternatives to mineral oil (which is derived from petroleum), such as synthetic and vegetable esters, have been increasingly used. The chemical structure is different from that of mineral oil which is a hydrocarbon mixture formed by thousands of different molecules (naphthenic, paraffinic, and aromatic).

Vegetable esters are triglycerides (similar to olive oil) coming from renewable sources like vegetable seeds (soybean, canola, sunflower and others). Vegetable esters have a low resistance to oxidation, so they come with oxidation inhibitors and are now produced on an industrial scale for power transformers. Unfortunately, plant-based esters have a higher pour point and worse thermal and dielectric characteristics. A key advantage is that they are almost totally biodegradable. They also reduce the aging rate of cellulose due the liquid's high water tolerance, which produces a sort of continuous dehydrating process in the solid insulation. Plant-based esters have a greater resistance to fire, which as far as transformers are concerned, is of fundamental importance. Due to their low oxidation stability, high viscosity, and high dielectric losses, they were, at first, only used for the construction of capacitors for electronic applications. In time, their oxidation stability was improved so that vegetable esters were introduced in power transformers. Initially, due to the lack of experience, they were used only in small distribution transformers (mostly pole-mounted units). In recent years, however, vegetable esters have been increasingly used in high-power, high-operating voltage transformers.

When subjected to a temperature of 120 °C or more, some vegetable ester produces a smaller amount of methane than mineral oil, but a greater amount of ethane which is the key gas for diagnosing anomalies of thermal origin. The production of carbon monoxide is similar for vegetable ester and mineral oil, while the presence of carbon dioxide is much higher in the vegetable ester.

Synthetic esters can be used as an alternative to natural esters. Synthetic esters are also of vegetable origin, but are obtained from a chemical reaction between a fatty acid and a polyalcohol. Synthetic esters have been used in transformers for wind farms and trains, but their use in power transformers for transmission and distribution is less widespread than the use of vegetable esters, since they are more costly and more prone to be hydrolyzed.

Table 15.1 summarizes the main chemical and physical characteristics for the three most common types of insulating fluids used in power transformers.

15.2.2.3 Solid Insulation Materials – Paper and Pressboard

The insulation system in liquid-filled transformers incorporates solid materials derived from cellulose and generically defined as organic materials. Cellulose is widely present in plant cells where it forms the membrane that covers them and represents at least half of their total weight.

The cellulose used to make insulating materials is obtained from the wood in which it is found, together with encrusting substances that bond the fibers together, giving it mechanical strength. Cellulose for electrotechnical use is obtained from pine or fir trees where it forms 40–50% of its mass, of which 20–30% is lignin and 10–30% is hemicelluloses mixed with a minimum quantity of other substances such as resins, waxes, fats, and mineral salts.

Table 15.1 Main properties of different transformer insulating liquids.

Property	Unit	Natural ester	Synthetic ester	Mineral oil
Biodegradability	—	Readily	Readily	Slowly
Oxidation stability	—	Susceptible	Excellent	Good
Density at 20 °C	g/cm^3	0.92	0.97	0.88
Viscosity at 40 °C	mm^2/s	42	30	9
Water saturation at 23 °C	ppm	1100	2600	55
Breakdown voltage	kV	>75	>75	>70
Tan δ at 90 °C	—	$<5\times10^{-3}$	$<6\times10^{-3}$	$<2\times10^{-3}$
Permittivity ε at 20 °C	—	3.2	3.2	2.2
Flash point	°C	>300	>250	160–170
Fire point	°C	>350	>300	170–180
Thermal conductivity at 20 °C	W/mK	0.17	0.14	0.13
Specific heat at 25°C	J/kgK	1880	1880	1860
Coefficient of expansion	$°C^{-1}$	0.00074	0.00075	0.00075
PD inception voltage (IEC 61294)	kV	34.0	28.2	38.2
PD inception voltage (suggested modified IEC 61294)	kV	25.6	22.3	23.2

Source: Adapted from [14, 15].

The paper is obtained through a series of processing steps. The first stage consists of pulping, which separates the fibers through hydrodynamic action. The next stage is refining, which uses water to remove hemicellulose, lignin, and other polar contaminants; next are mechanical actions such as crushing, cutting, and rubbing the fibers together. This step is used to give greater flexibility and an active surface. Very refined fibers are robust, have a greater density and hardness, and can be processed more easily.

The refined fibers are mixed with other substances such as adhesives, dyes, fillers, etc. in quantities depending on the type of paper needed. The diluted mixture is finally distributed evenly on a wire cloth on flat or roller machines and water is eliminated by dripping, suction, pressing and heating. The sheet is then subjected to a further process to obtain the desired density and thickness for a specific use. Further details are in [6, 16].

Inside transformers, cellulose is found in thin and thick sheets. The former consisting of paper (more specifically, kraft paper or thermally upgraded paper). The thick sheets are pressboard (also known as transformer board). Paper is used for turn-to-turn insulation, whereas pressboard is used for phase-to-ground and phase-to-phase insulation, as well as to give mechanical stability to parts of the winding (for example the high-voltage cables connecting the windings to the bushings).

Kraft Paper Paper intended for electrotechnical use is characterized by long and soft fibers well suited to manufacture products with excellent mechanical characteristics and good chemical and thermal stability. This is also known by the term "kraft" which, in German, means "resistant." Table 15.2 summarizes the main properties of paper for electrotechnical use in transformers.

A characteristic of the paper which is particularly critical in transformer manufacture is its hygroscopicity. Humidity represents one of the main causes of aging for paper. If exposed to the atmosphere, the paper absorbs water until it reaches a temperature-dependent equilibrium. The

Table 15.2 Main properties of insulating papers.

Property	Unit	Value	Note
Thickness	μm	80–130	
Density	g/m^2 (*)	30–60	Conductor insulation
Density	g/m^2 (*)	100	Main insulation
Mechanical strength	kg/mm^2	10	Machining direction
Mechanical strength	kg/mm^2	5	Orthogonal direction
Breakdown voltage	kV/cm	100–150	Dry paper
Tan δ	—	10–30 × 10^{-4}	Dry paper
Resistivity	MΩcm	100–600 × 10^6	Dry paper
Dielectric constant (ε)	—	~≤2.2	Dry paper

* the scientific way to report the density is "grams per centimetres cubed" (0.7–0.8 in the case of kraft paper) but, in the common practice, the density is reported in grams per square meter".

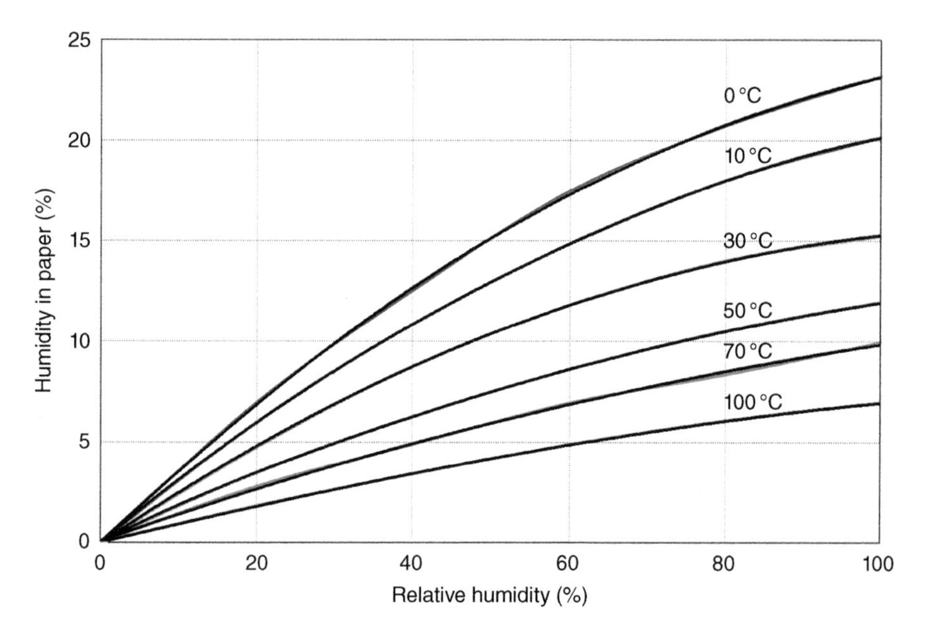

Figure 15.5 Equilibrium curves of humidity in paper. *Source:* [17].

equilibrium is reached after a few weeks if the paper is impregnated with insulating oil. Figure 15.5 shows the equilibrium curves of the humidity present in the paper for a given value of relative humidity of the environment at different temperatures [17].

The presence of humidity adversely affects some of the main properties of kraft paper, and in particular:

- reduces mechanical strength
- reduces the discharge inception voltage at 50/60 Hz
- increases the value of Tan δ and losses
- decreases resistivity
- increases the dielectric constant ε

The graph in Figure 15.6 shows the effect on Tan δ as a function of the humidity present in the paper at two different temperatures. The increase in dielectric loss is proportional to the humidity of the paper and is much more pronounced as the temperature increases.

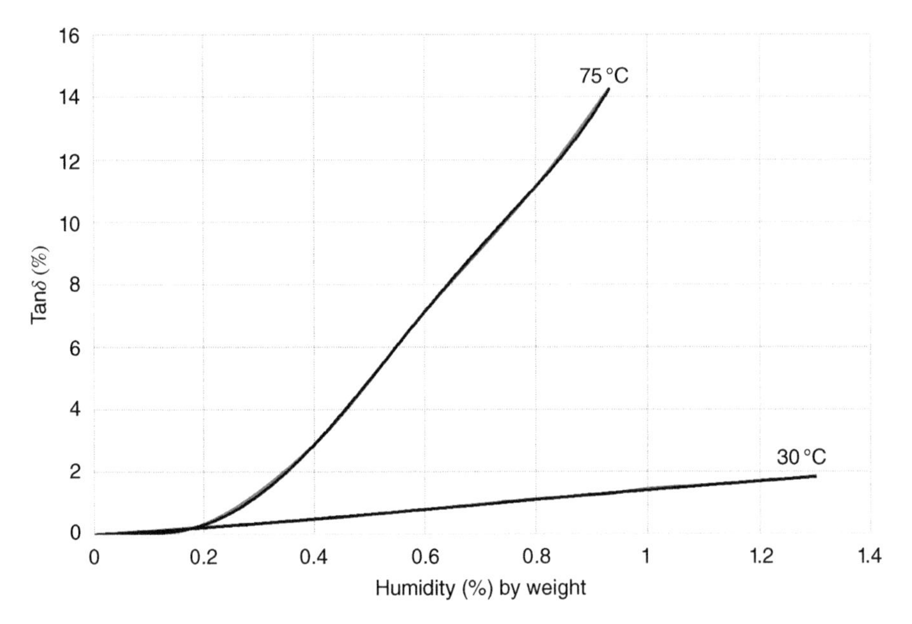

Figure 15.6 Dissipation factor vs humidity in Kraft paper.

The high temperature resistance of insulating materials is another characteristic to consider when designing a transformer. The maximum working temperature of the paper obtained from unmodified cellulose is 105 °C. However, as the temperature exceeds about 70 °C, the paper becomes brittle. Humidity and the presence of oxygen are elements that further accelerate the aging process due to oxidation and hydrolysis reactions [18, 19].

To obtain a better resistance to thermal aging, it is possible to treat the cellulose with stabilizing substances that slow down the aging process due to the temperature. This type of paper is known as thermally upgraded and is used for the insulation of parts exposed to higher temperature [17, 20].

Pressboard Pressboard sheets are used to create barriers interposed between the phases and the ground, or between the phases, to create cylinders, spacers, hoods, angular rings, etc. These elements must prevent the windings from movement and ensure a good circulation of the oil to achieve optimal cooling.

Pressboard is obtained by overlapping and pressing layers of thin paper together in the presence of moisture which is then extracted by drying. Glues are not used to bond the sheets together. The typical thickness of the pressboard is about 8 mm. Depending on the type of drying and pressing, three types of pressboards can be distinguished, which differ from each other in density, shape retention, ability to be shaped and ease of impregnation with oil:

- Standard pressboards are obtained from vacuum drying pressboard and have a density of 0.9–1 kg per decimeter cubed and humidity approximatively 5% which, subjected to further pressing with rollers, reaching a density of 1.15–1.30 kg per decimeter cubes.

Table 15.3 Main properties of transformer board.

Property	Unit	Value
Density	kg per decimeter cubed	1.25
Breakdown voltage	kV/cm	200–250
Tan δ at 20 °C	—	4×10^{-3}
Tanδ at 70°C	—	7×10^{-3}
Permittivity ε	—	4.4–4.5
Surface dielectric strength	kV/cm	30

- Preformed pressboard elements are obtained by molding without undergoing pressing which is followed by the drying process until a humidity equal to approximately 5% is reached; typical density is approximatively 0.9 kg per decimeter cubed.
- Transformer board is produced through a process of dehydration, pressing, and drying that takes place in special hot presses to improve the mechanical and structural characteristics. The pressboard thus obtained has a plastic behavior and mechanical characteristics that are superior to the previous standard types, and a density equal to 1.25 kg per decimeter cubed. Table 15.3 shows the main characteristics of transformer board impregnated with mineral oil.

15.2.3 Typical Construction Arrangement in Oil-Filled Transformers

Figure 15.7 schematically shows the positioning and function of the cylinders, barriers, and spacers used to hold the cylinders in the correct position allowing the circulation of the oil for cooling. All components are made with pressboard.

Cylinders and barriers are components used to improve the breakdown strength in the parts of the transformer where the electrical field is highest (Section 15.2.2.1). These high-field areas are usually located near the output leads of the high-voltage windings, in the spaces between the phases,

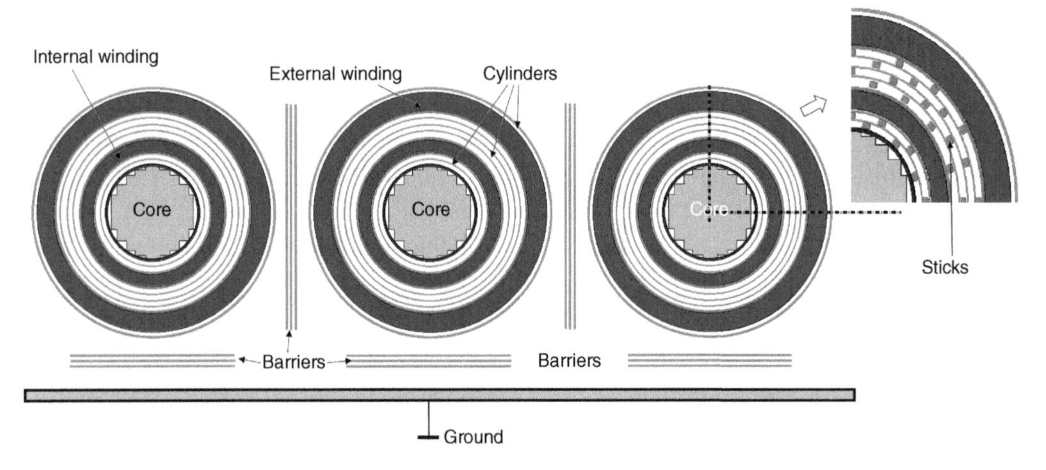

Figure 15.7 Schematic arrangement of insulating barriers, cylinders, and sticks in a three phase, 3-leg transformer. The external winding is usually the high-voltage one.

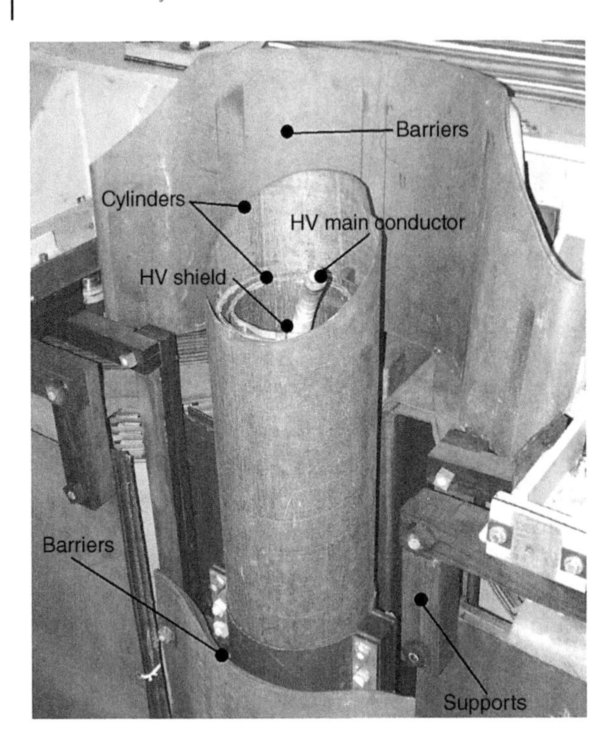

Figure 15.8 Typical arrangement of pressboard components at a high-voltage winding output lead.

and between the phases and the ground. Figure 15.8 shows the detail of a high-voltage output where the cylinders and barriers are indicated that reduce the overall insulation distances to ground.

Figure 15.9a shows the positions of the barriers placed between the phases which permits smaller distances between phases ensuring the necessary level of breakdown voltage (Figure 15.4). It is possible to identify the sticks that ensure, in this case, the positioning and mutual spacing of the upper and lower rings, and the supports made of multilayer insulating material, which support the paper-insulated conductors for the connection of the regulating winding sockets at the online tap changer (OLTC). All these parts are placed in areas where the electric field is particularly high; their sizing, the materials they are made from, their shape, and their correct positioning are essential for the proper functioning of the transformer, not only in relation to the dielectric strength but also and above all as regards the onset of PD.

There are also structures dedicated to conferring mechanical strength. As an example, the supports that keep the cables in specified positions on their way to the bushings (fixation elements), or to the sockets of the OLTC. The structures are made from overlapping sheets of treated beech wood, glued together, until the desired thickness and dimensions are obtained. This type of material is mainly used in parts of the transformer where the electric fields are not high. In parts of the transformer where the electric field takes on high intensity, multilayer structures obtained with overlapping, glued and pressed sheets are used.

As mentioned above, the copper conductors in the windings are covered with kraft paper tapes. The paper tapes are normally 25 mm wide with a thickness that varies depending on the use. To insulate both the winding and main lead conductors (see paper insulated conductors in Figure 15.9a and an insulated main output conductor in Figure 15.9b), the thickness of the paper tape is 120–150 μm. Figure 15.9b shows a high-voltage output lead consisting of a metal tube (copper, brass, or aluminum) with a diameter ranging from 80 to 120 mm and a few millimeters thick. This tube forms a shield for the main HV conductor within the tube, through which the winding current flows. The internal conductor is insulated with paper tape to avoid accidental contact with the

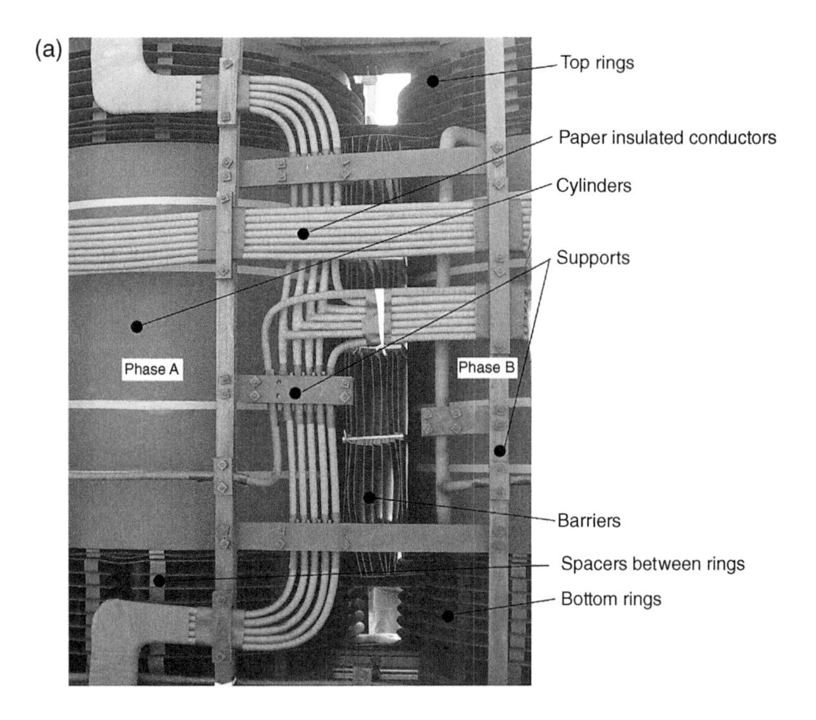

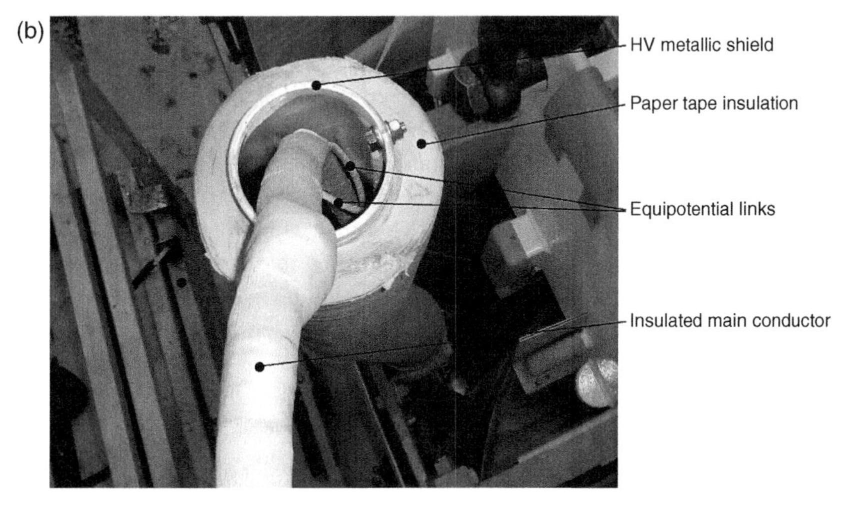

Figure 15.9 (a) Typical arrangement of the space between phases. Top and bottom rings are expressly molded diaphragms to guide the oil outside the winding. (Note: the picture is from a used transformer, the barriers were parallel after manufacturing, but were distorted due to thermal stress). (b) Detail of the high-voltage connection between the winding and the HV bushing connection outline.

shield. Such intermittent contacts could cause contact sparking (Section 4.4) and create gas byproducts dissolved in the oil. The external coating of the shield is critical; the thickness of this layer is usually of the order of a few centimeters and is made by wrapping paper tape continuously. The considerable thickness of the layer (a few cm of wrapped paper, as shown in Figure 15.9b) can hamper the oil impregnation process, thus affecting the reliability of the connection over time. The main conductor and the shield must be connected at a single point to prevent conductive parts of this assembly from floating.

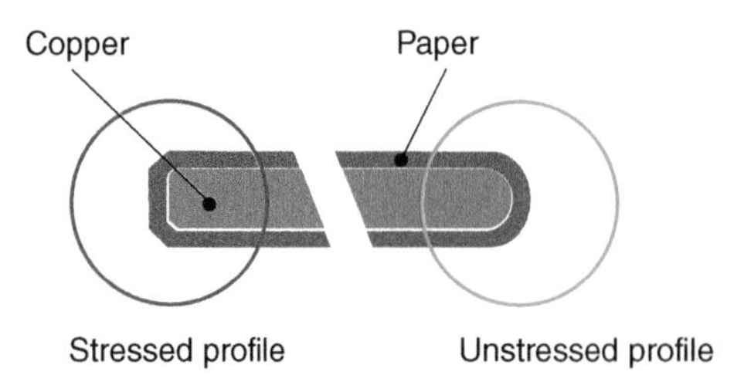

Figure 15.10 Stressed and unstressed profile of winding conductors.

The same paper tapes used to insulate the leads are employed as turn insulation within the windings. The windings consist of rectangular section copper wire (also known as winding wire). The wire is precoated with a polyamide-imide insulating enamel. The wire is then taped with several layers of paper and later impregnated with oil (Figure 15.10). The paper tape has a thickness of 60–80 μm and must endure the high temperatures of the winding conductors, maintaining a good mechanical strength to avoid compromising the insulation between the turns or between the sub conductors in the same turn.

The same considerations apply if the winding is made with a transposed cable instead of magnet wire. The cable consists of a number of sub conductors that are insulated from each other with the use of enamels and special resins. The cable is dried in an oven. This structure has considerable mechanical strength. The cable is then covered with layers of paper 120–150 μm thick to increase insulation thickness. This type of conductor is particularly suitable for making windings with high current ratings, ensuring mechanical strength needed to resist winding distortions from the magnetic forces that result from nearby short circuits.

The paper tapes in direct contact with the winding conductors must be able to withstand high temperatures without altering the mechanical properties. Paper aging due to operation at a high temperature causes the molecular bonds in the cellulose to break. This is manifest by the degree of polymerization (DP) of the paper, which is above 1200 in unaged paper tape, but gradually reduces with thermal aging, until the yield strength of the cellulose becomes unacceptable at a DP of about 200 [17]. For this reason, there is a tendency to use so-called Thermally Upgraded paper or, in even more critical cases, aramid (Nomex™) paper. These materials must be able to withstand the operating temperatures as well as the temperatures that occur during winding manufacture. The latter involves oven drying and polymerization (curing) of the resins of the transposed cable. A mechanical stress also occurs during manufacture caused by the pressure exerted on the windings to achieve the design dimensions; after the drying treatment in the oven, the windings are subjected to axial forces that can even reach several hundreds of tons.

Although the practical electrical breakdown strength of mineral oil is about 40 kV AC rms/mm, and the paper on its own is 12 kV/mm, the AC breakdown strength of the composite oil-impregnated paper is about 64 kV/mm. The surprising performance of the composite insulation compared to that of the individual components justifies the success of this material, which has remained substantially unchanged for a century.

The manufacture of a transformer with oil–paper insulation involves various processing steps that inevitably take place in the normal air environment in the factory. Usually manufacturing occurs in an environment with slight air overpressure to keep dust and insects outside. However, since humidity is likely present, the moisture is then extracted from the insulation by first drying in an oven followed by a preliminary pressing of the individual windings to bring them to the desired dimensions. The final assembly of the windings involves processes that can take a few days, during which all the organic cellulose insulation can again absorb moisture from the environment. Once the windings have been completed, the active components are subjected to a vacuum drying process where most of the moisture present in the solid insulation is extracted. The moisture-free components are then placed in the transformer tank, the tank is sealed and a vacuum is drawn down to ~0.5–1 mbar. After degassing under vacuum, the tank is then flooded with hot oil to facilitate the impregnation process whose duration depends on the mass of the organic insulation, typically on the order of a few days. The transformer is then equipped with all the other components necessary for its operation such as coolers and an oil conservator. At this point the transformer is ready for its final test.

15.3 Typical Causes of PD in Dry-Type (Cast Resin) Transformers

PD in transformers can have different origins. In dry, resin-cast transformers, they essentially depend on the quality of the resin insulation and in particular on the presence of air pockets (i.e. voids) that have remained trapped due to imperfect casting. In this case, the material is not homogeneous, and PD may occur in such cavities. Another cause can be found in the presence of foreign materials that contaminate the liquid resin before being poured into the appropriate molds.

In addition, cracks can form in the resin because of mechanical stresses, sometimes associated with rapid thermal cycling and/or inadequate mixing of the resin/hardener/fillers. The cracks create a severe dielectric discontinuity, triggering PD. The processing steps for casting are of fundamental importance to prevent PD. The resin curing phase must be carried out in such a way that the solidified material is homogeneous, free of contaminants, or small air bubbles. During operation of the dry-type transformer, the resin may be exposed to high operating temperatures as well has 100/120 Hz magnetic forces caused by the currents in the conductors. If short circuits occur in the power system, there may be very high fault currents which can greatly increase the magnetic forces acting on the windings. The short circuits may create cracks in the insulation that cause PD, especially if there is ambient humidity and pollution.

In this type of transformer, the presence of PD deteriorates the insulation due to electric treeing and/or surface electrical tracking (Sections 2.4.1 and 3.7.2), with consequent loss of dielectric strength. The measurement of PD constitutes an effective method of identifying mechanical and other defects in the insulation to avoid serious insulation failure.

Due to modern electric field simulation techniques, the presence of PD caused by design error is uncommon.

15.4 Typical Causes of PD in Oil-Filled Transformers

There are many reasons that PD develops in liquid-filled transformer insulation systems. Some causes are due to manufacturing of the transformer. Some are caused by insulation system aging.

15.4.1 Defects in Solid Insulating Materials

PD sometimes is caused by defects in the solid insulation materials used to manufacture the transformer. Solid insulation materials, consisting of cellulose, can have defects due to their complex manufacturing process. During the process phases, the materials can be contaminated by metal particles or other residues that can give rise to PD. Careful quality control, based on X-ray analysis, can be useful in verifying the absence of metal impurities within the insulating materials obtained from cellulose.

For voltages up to 170 kV components such as pressing rings, strips, and so on are usually made of wood. For higher voltages, they are made by gluing and pressing together sheets of paper or other insulating material (the glues must be suitable for the purpose). The pressing process must result in components as void-free as possible and not introduce impurities or humidity. The latter is removed by drying processes that must be carried out carefully without damaging the product due to temperatures that are not compatible with the materials.

Ultimately, the absence of PD in organic insulating materials depends both on the purity of the base materials used in the manufacture and on the care taken during manufacturing. Components such as pressing rings, spacers, cylinders, output connections from the windings, and other insulating components are normally found to be subjected to particularly high electric stress so that their quality plays a decisive role in the presence of PD.

It is also important that the copper conductors have rounded edges rather than sharp edges (Figure 15.10) to avoid electric field enhancements that may trigger PD (Section 2.1.3.4). Processing residues should be absent, and the paper must be able to withstand high temperature and be free of contaminants.

15.4.2 Defects in the Core Structure

The construction of a transformer laminated steel core can introduce imperfections leading to PD. The first significant manufacturing is the cutting of the magnetic steel sheets and the assembly of the laminated core. The cutting of the sheets often creates burrs with sharp edges that must be removed. Also, the stacking of the core laminations could damage the thin layer of insulation between individual sheets. Two or more magnetic sheets that are not perfectly insulated from each other can become the site of low-energy discharges and localized overheating.

The laminated magnetic core must be mechanically strengthened by a structure made of ferrous material which forms a cage around the magnetic core. There are two manufacturing techniques. One where all parts of the structure are electrically connected (closed core frame); another one where the two shells of the frame are kept isolated (open core frame). There is no real preference regarding this aspect, but some fundamental principles typical for each technique must be ensured to avoid PD or contact sparking (Section 4.4).

In the first case, good continuity must be ensured between the ferrous parts by using conductive (normally copper) links of adequate cross section as shown in Figure 15.11. In the connections visible in Figure 15.11, during operation at full load, currents reaching a few hundred amperes can circulate; if the copper links overheat at the bolts, this may lead to bubbles in the oil and thus PD.

In the second case, it is necessary to ensure good insulation between the two parts of the frame. In addition to ensuring sufficient insulation, the materials used must also be able to withstand the continuous mechanical stresses due to the 100/120 Hz magnetostriction forces in the core during operation. Figure 15.12 shows a detail of the insulation. During operation at full load or due to the stray flux that may occur in the frame, voltages may occur which, in the event of inadequate insulation, can give rise to high electric stress, and lead to PD.

15.4.3 Defects Arising During Factory Assembly

The assembly of the entire active part of the transformer (core, windings, connections, etc.) involves the use of tools that can create dust and small metal fragments. In addition, such contamination may be introduced from the factory environment. This debris may lead to PD. The windings for transformers rated 400 kV and above are usually made in controlled positive air pressure

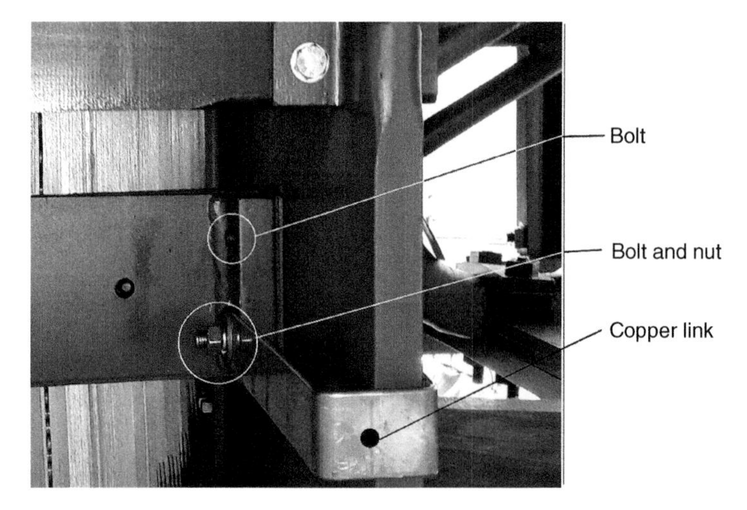

Figure 15.11 Bypass links between parts for a closed core frame.

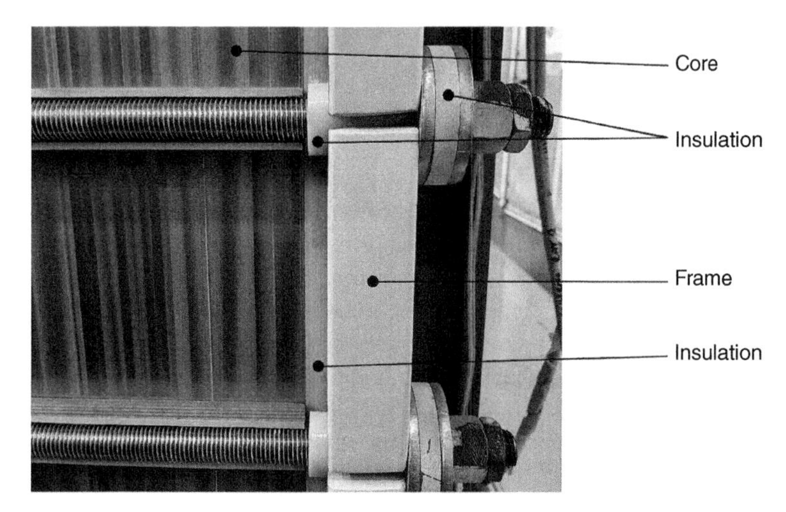

Figure 15.12 Insulation between parts of an open core frame. *Source:* Courtesy of Elettromeccanica Tironi - Italy.

environments to minimize contamination from dust and other elements that can compromise their dielectric characteristics.

Whatever the voltage ratings of the winding, once the construction phase is complete, the windings are treated in ovens for a preliminary moisture extraction from the insulation by heating only. Subsequently, the winding is subjected to axial pressing to obtain the required winding design height and the parallelism between the base and the winding head. The applied force can reach a few hundred tons and depends on the type of winding, the insulating materials and the type of conductor. The pressing force must be correctly selected to avoid tilting of the conductors (Figure 15.13). Conductor tilting may damage the paper covering the conductors, which may lead to PD.

While the active parts are being assembled, the organic insulation is exposed to the atmosphere. This exposes the winding to moisture and contamination such as dust or metal particles.

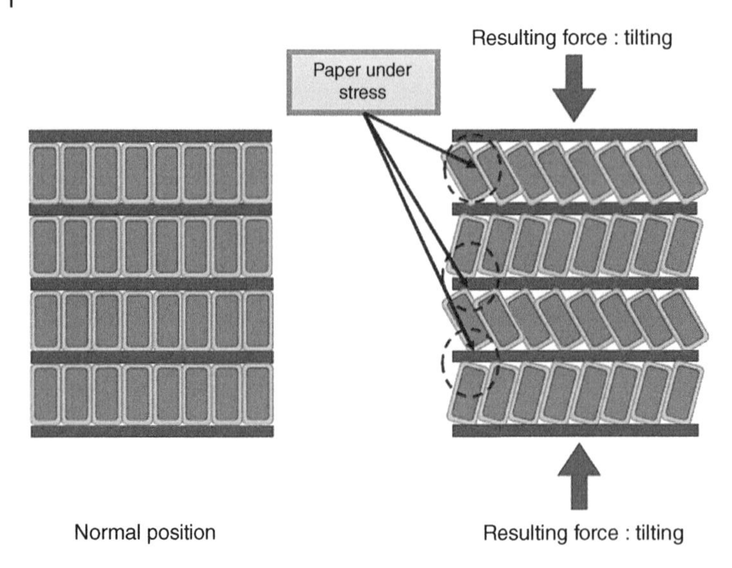

Figure 15.13 Effect of an excessive pressing force.

Such pollutants present on the insulating surfaces in contact with oil, barriers, cylinders, or slats can significantly compromise the surface dielectric strength and can cause PD, which if persisting for a long time, can lead to failure.

The fully assembled active part is treated by a "vapor phase" drying process (i.e. treatment in an atmosphere of kerosene vapors). The process is carried out in a heated autoclave, under vacuum, in which the kerosene vapor is injected and subsequently extracted from the autoclave. The extracted volume will be composed of kerosene vapor mixed with the moisture coming from the cellulose. (In a separate process the moisture is extracted from the kerosene and reused.) All the humidity that was absorbed by the cellulose insulation during assembly should be completely extracted. The vapor phase treatment takes only a few days compared to the traditional oven treatment, which takes more than a week. Once the humidity extraction operations have been completed, the core and windings are installed within the transformer tank, and the tank is closed to start the oil-impregnation phase. The time to insert the active components into the tank must be as short as possible so that the dried cellulose insulation has little time to reabsorb ambient humidity.

Another approach is to extract moisture by drying the active part directly in the transformer tank. In this case, the heating of the active part takes place by circulating low-frequency (few mHz) current in the windings to heat itself. A vacuum is continuously applied to the tank by vacuum pumps in order to extract the moisture released by the heated insulation material. Dry oil (the same used for filling) is sprayed on the active part by nozzles previously positioned on the top cover of transformer tank; the oil is circulated in an oil-treatment plant so to continuously heat it and extracting absorbed moisture during spraying process on the active part. Kerosene is not used in this onsite procedure to reduce the risk of explosion and handling difficulties of this fluid on site. The properties of the oil ensure an efficient extraction of moisture from the cellulosic parts. This solution is very helpful when the drying of the active parts must be performed on site (for example after an onsite major maintenance and when the transformer is of considerable weight and large, and therefore difficult to transport with oil).

Oil impregnation is a very important step in the manufacturing of a transformer and has a fundamental role to prevent the inception of PD. The impregnation must be such as to allow the oil to penetrate deeply into all the cellulose insulating parts to create a homogeneous, gas bubble-free insulation system to prevent PD. This is critical for the dielectric strength both at the operating frequency and with steep front surges. The impregnation takes place with heated oil, free of atmospheric gases and hydrocarbon gases, and takes about three days in the case of mineral oil. Impregnation with natural ester liquids takes a few days longer due to the higher viscosity of the fluid.

15.4.4 Defects Arising During Onsite Assembly

After successfully passing routine manufacturing tests, the transformer must reach the substation where it will be assembled for operation. Large power transformers with high operating voltages are transported without the oil and without all the accessories such as the bushings, the cooling system, the conservator, and so on. The oil used for the impregnation during the factory routine tests is removed from the transformer at the factory. To keep the solid insulation dry during transportation to the substation, the transformer tank is filled with dry air or nitrogen with low moisture content (typical dew point of the gas in the cylinder is −60 °C or lower) at an overpressure of about 20 kPa.

In the case of transformers that use natural esters, it is recommended not to use dry air or fluids that contain oxygen to avoid oxidation of insulation surfaces that are already impregnated with natural ester. In fact, these fluids tend to oxidize and therefore compromise the main dielectric characteristics. In any case, contact of the active parts with the atmosphere for long periods must be avoided to prevent exposing them to the absorption of moisture present in the atmosphere.

The oil used to fill the transformer on site is almost always a different oil from that used for tests at the factory. The oil to be used usually arrives on site directly from the refinery. As small quantities of oil intended for different uses such as lubrication, hydraulic systems, etc., can contaminate large volumes of dielectric oil, it is important that the transportation tanks have not previously been used for other types of liquids. The quality of the oil has a direct impact on the presence or absence of PD. Physical-chemical checks on an oil sample are therefore also recommended before filling the transformer tank at the substation.

Onsite final assembly operations must take place in a clean environment. The exposure of the transformer during installation to the humidity in the atmosphere, dust and other pollutants, processing residues, etc. may contribute to PD. In the following, some examples of issues introduced during final assembly are outlined.

The removal of nitrogen or dry air, used to protect the active parts during transport, and the subsequent filling with oil, must be done with extreme care. As a rule, the first filling with oil has the purpose of protecting the organic insulation from contact with the atmosphere which will be inevitable, even if very limited, during the assembly of the accessories. First, the nitrogen or dry air in the tank for shipping to site is extracted using vacuum pumps. Then heated oil is introduced into the tank until the active parts are covered so that insulation materials are not in contact with the atmosphere during the subsequent steps.

The assembly of accessories must be carried out with great care. The HV bushings are very delicate components that must be transported, stored, and handled with extreme care and protected from contact with moisture. Thus, the assembly of the bushings is a particularly critical process as it involves very detailed operations to be performed with extreme care, avoiding impacts that could damage these delicate components. The connection of the bushings to the cables coming from the windings often involves the positioning of shields that must be at the same voltage as the conductors coming from the winding. The parts must be mechanically secured in a reliable way.

At the end of the assembly of all the accessories, the transformer is filled with oil until the level required for operation is reached. This last operation must be completed by creating a vacuum to remove any residual air. Trapped air (typically in bushing turrets) can cause PD in operation or, even worse, discharges that can lead to transformer failure.

15.4.5 PD Caused by Aging During Operation

PD can occur due to aging of the insulation system during the operational life of a transformer. Recognized causes of PD initiation include severe operating conditions, failure events, premature aging of organic insulation, construction defects that eventually lead to PD, as well as contamination and humidity. A nice overview of the degradation of solid insulation used in liquid-filled transformers is given in [21].

One of the causes of PD during the operational life of a transformer is localized overheating in the windings operating at high voltage. This can occur because of prolonged operation at high load, deficiencies in the cooling system or due to accidental localized obstructions in the winding cooling channels. This overheating leads to thermal aging of the paper that surrounds the winding conductors in the high-temperature regions. Thermal aging leads to paper embrittlement and loss of mechanical and electrical strength. Cracks or tears in the tape lead to PD in high-stress regions, and electrical tracking over the paper surface and/or electrical treeing through the paper. When the turn insulation shorts, a very high circulating current flows in the shorted turn and will likely rapidly morph into destructive failure.

Localized PD and tracking phenomena have also been found in insulating support elements such as wooden blocks or plywood positioned in points of the transformer where the electric field is very high. These include the upper parts of the windings near the outputs, as highlighted in Figure 15.14. In this case the main culprit is to be found in the quality of the material from which the supports and barriers are made, and the carefulness with which it has been subjected to dehumidification treatment and subsequent oil impregnation. The conditions of maximum electric

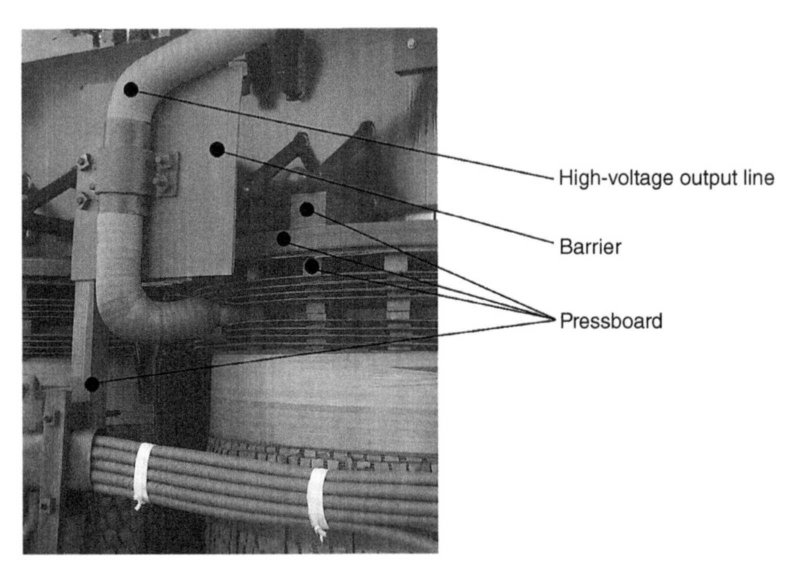

Figure 15.14 High-voltage stressed areas where PD is likely, leading to surface electrical tracking.

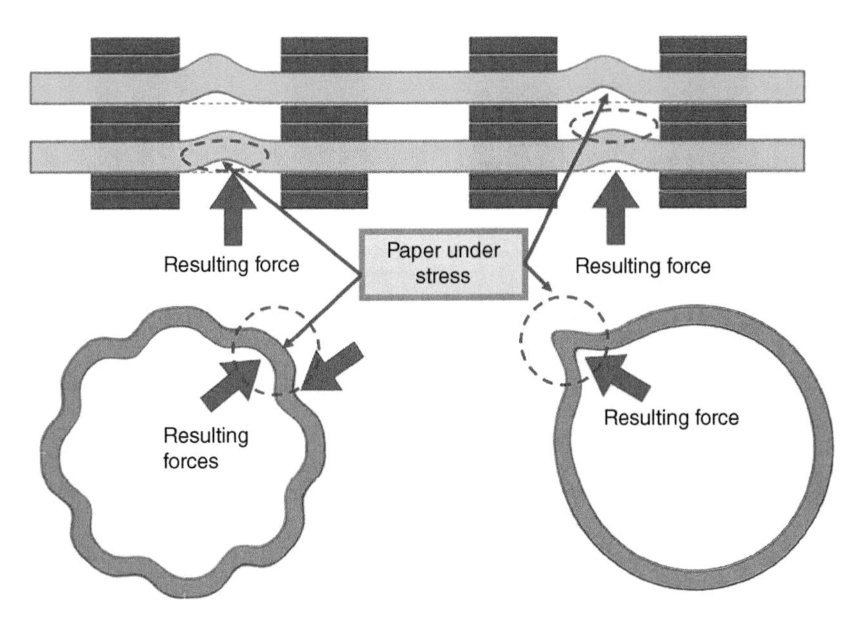

Figure 15.15 Effects of magnetic forces on paper insulated copper conductors.

field and maximum temperature are almost coincident; the resulting stress can be fatal in the long run even in the presence of an insulation system apparently in good condition at the time of manufacture. If these components are multilayer, the compatibility of the adhesives used to make the multilayer and its construction process in general play a fundamental role, as does the accidental presence of moisture accumulation.

Another possible cause of PD is the aging of the paper that covers the winding conductors. Aging occurs mainly due to prolonged operation at temperatures beyond the limit tolerated by the paper which, as a rule, is 105 °C. This condition leads to the weakening of the paper and, due to the 100/120 Hz winding vibrations, mechanical damage to the paper, and may give rise to deterioration of the insulation properties and the onset of PD. This damage is particularly severe in the case of short circuits in the power system when very high magnetic forces are created within the transformer. Some of the effects resulting from electrodynamic stresses are shown in Figure 15.15.

As observed above, PD can compromise the insulation causing a discharge between turns, or between sections made up of several turns (discs) of a single winding. If shorts develop, high circulating currents lead to conductor melting and failure. The parts most subject to this phenomenon are the ends of the windings, where stray magnetic fluxes near the coils induce local currents and thus localized overheating (hot spots). To limit the thermal effect caused by the stray flux, the turns that make up the ends of the windings are usually made with transposed conductors made up of several subconductors to reduce eddy current losses. In many cases, the transposed conductor used in these parts is taped with thermally upgraded paper, which is more resistant to high temperatures, allowing operation up to about 115 °C. The consequence of the phenomenon described is visible in Figure 15.16.

The compound action of the thermal and mechanical phenomena can give rise to PD, which can quickly evolve into electrical tracking and turn shorts.

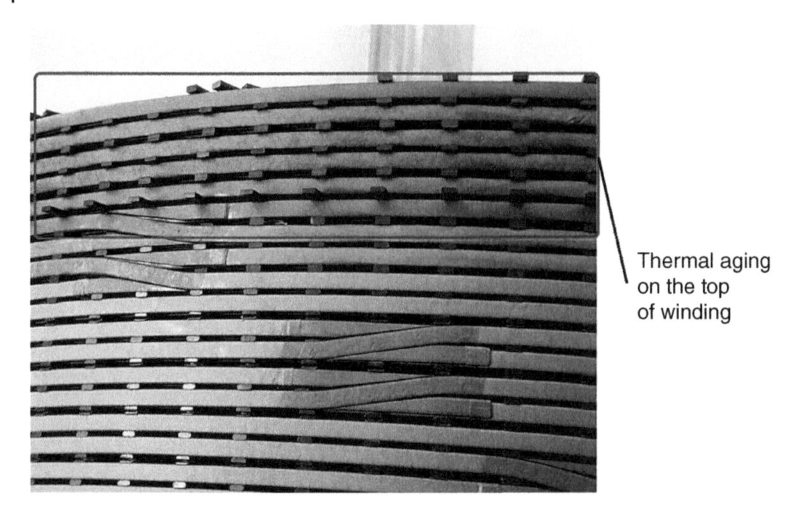

Figure 15.16 Thermal aging on the top of a high-voltage winding due to stray magnetic flux.

15.4.6 Partial Discharges Due to a Poor Electrical Design

For high and extra-high voltage power transformers, the electrical design must focus on the ends of the windings and on the connections between the windings and cables connecting them to the bushing. These are the regions where, if precautions are not taken, the electric field might be large enough to incept PD at minor flaws. In the end-windings, the electric field is often high at the round corners of the conductors (Figures 15.10 and 15.17). Figure 15.18 shows the equipotential lines and the electric field in the end-winding. The electric field in the end-winding can be almost twice the electric field within the winding cylinders. These regions can thus become site of PD or, worse, they can incept streamers (Section 2.3.2) that may lead to flashover in the oil during lightning or switching surge transients.

To prevent PD and streamers, the electric field in the end-windings is reduced by adding shields (shielding rings). These shielding rings are conductive and have the same potential as the last turn

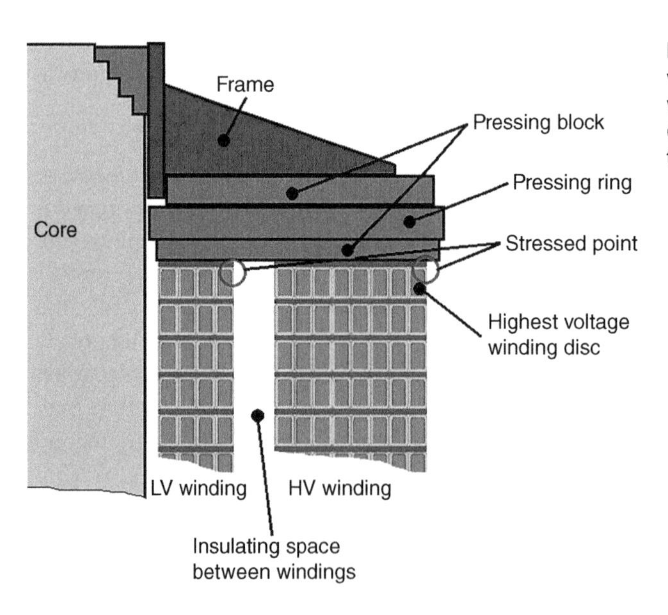

Figure 15.17 High- and low-voltage windings arrangement in the end-winding, showing that the highest electric stress occurs at the corner of the top-most copper turn.

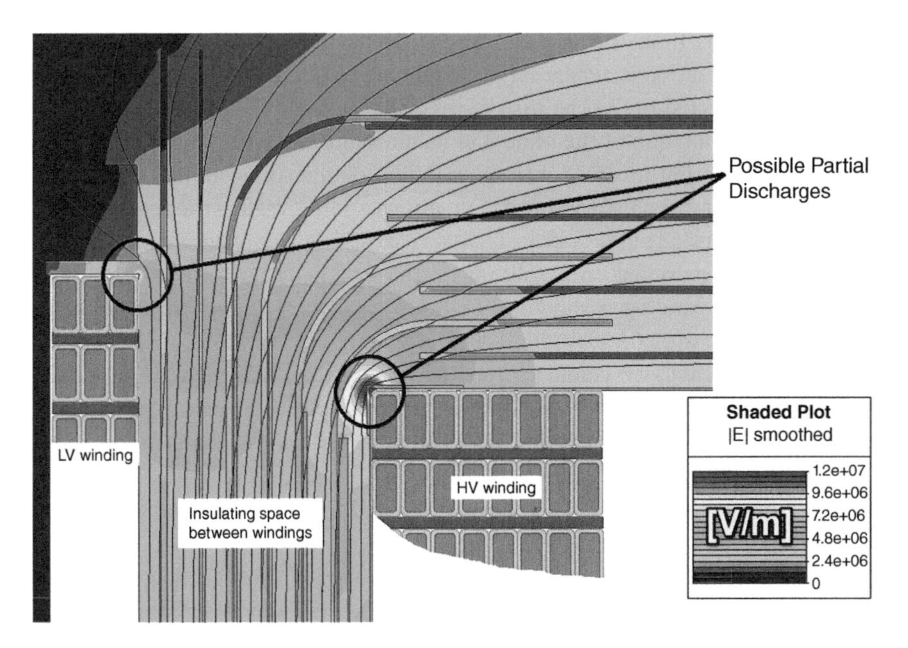

Figure 15.18 Dielectric stress at the top of the LV and HV windings using a finite element simulation.

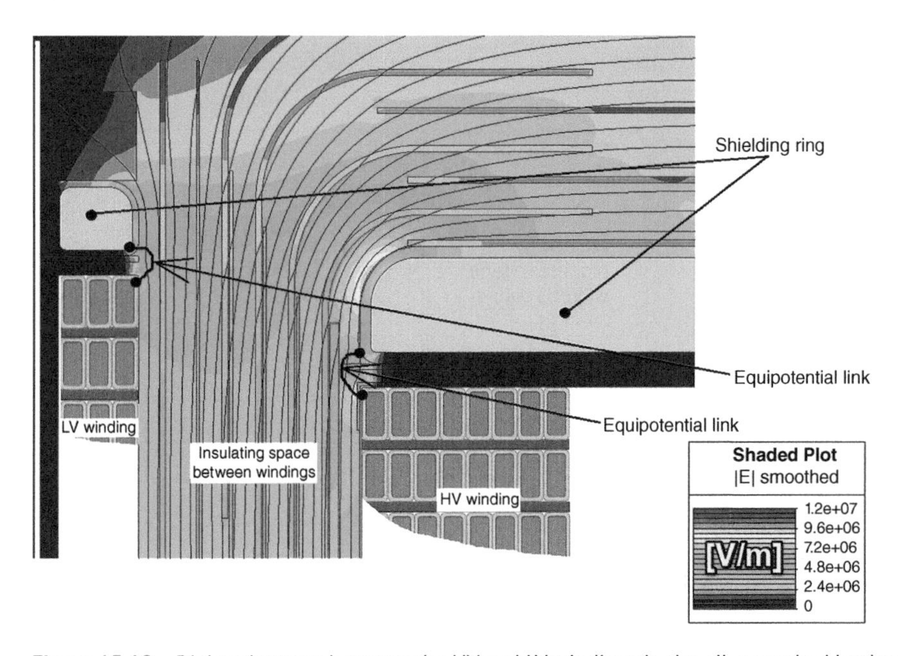

Figure 15.19 Dielectric stress between the HV and LV windings in the oil, smoothed by the shielding ring.

in the winding, reducing the electric field at the copper turn corners down to values similar to those within the cylinders, as shown in Figure 15.19. These values should be low enough to prevent the inception of PD in the oil. The reduced electric field should also be low enough during lightning and transient surges to prevent flashover.

The shielding rings are realized using solid insulation, normally multiple paper layers, covered with a metallic tape. The shield is eventually covered with some layers of kraft paper to minimize the risk of emission of electrons from the metallic tapes. The ends of the metallic tapes are separated to avoid the formation of loops, where large currents could circulate, giving rise to hot spots. The metallic tape is connected to the last turn of the winding with a braid that is covered with kraft paper. The proper installation of the braid and its mechanical connection to the shield are critical to ensure that PD are not incepted during operation. If the connection between the metal tape and the winding is lost, the shield assumes a floating potential, raising the risk of PD; if the braid floats in the oil, the inception of PD is almost certain. Avoiding these phenomena requires a careful choice of materials and manufacturing solutions.

The final structure of the end-winding is shown in Figure 15.20. The shielding ring is located between the winding and the pressing (or clamping) ring. Therefore, the mechanical strength of the shielding ring must be sufficient to withstand compressive forces without displacing the metallic tapes, especially during short circuits when axial forces can be as large as several tens of tons.

Similarly, large electric fields can occur at cylindrical conductors (such as connection cables between components). If the conductor diameter is too small, as shown in Figure 15.21a, higher electric stresses result (Section 2.1.3.2), possibly initiating PD. For these reasons, the cables

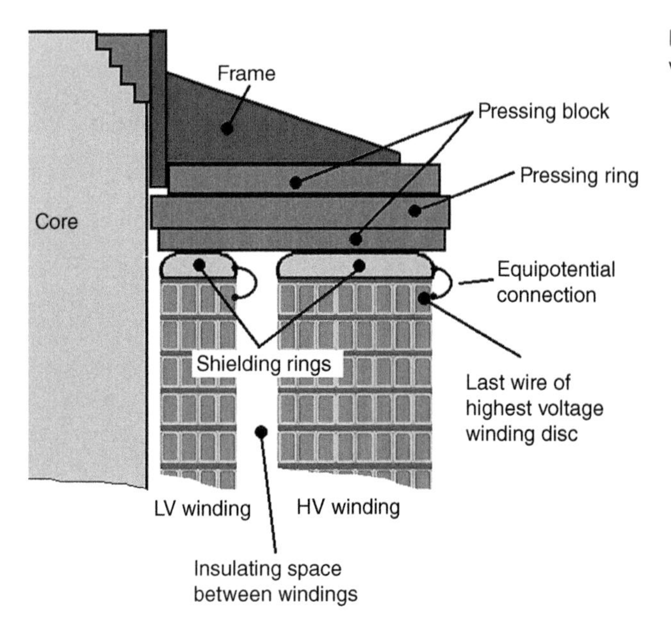

Figure 15.20 Final arrangement of a winding with the equipotential ring.

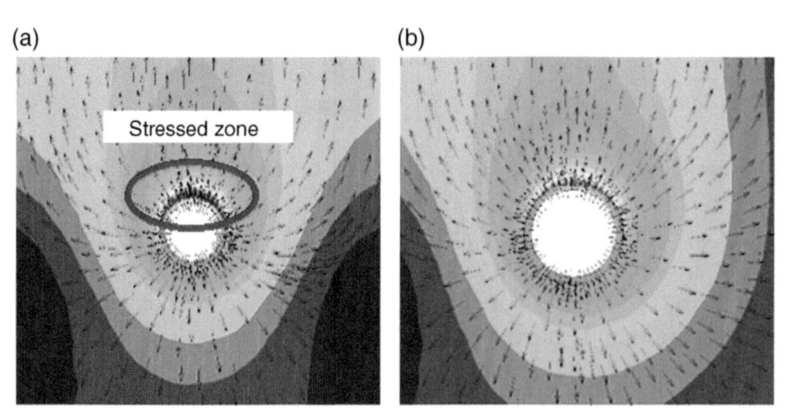

Figure 15.21 (a) Stressed zone in a small diameter conductor; (b) lower stress with a larger diameter conductor.

connecting the windings to the bushings are normally made of flexible conductors whose cross section is designed to carry the load current. These conductors are then inserted into metallic pipes (Figure 15.9b) whose diameter is designed to achieve a low electric field in the oil. The copper conductors and the shield pipe are electrically connected, and thus equipotential.

15.5 Relevant Standards

The measurement of PD on transformers is documented by IEC and IEEE standards. In addition, CIGRE has published a consensus document about transformer PD measurements [22]. The purpose of these documents is to establish the characteristics of the PD measuring system to perform a correct measurement ensuring repeatability and reproducibility and to define the procedures to be adopted when carrying out the measurement.

15.5.1 IEC 60270

IEC 60270 is the general purpose, horizontal standard specifying how to measure charge-based (i.e. pC and nC) PD on electrical equipment during tests with direct voltage and alternating voltage generally, with an AC frequency not exceeding 400 Hz (Chapter 6) [23]. This standard is often used in factory PD measurements for all types of equipment, describes the measurement of PD at frequencies <1 MHz, and outlines how to calibrate the PD in terms of apparent charge (Section 6.6). It does not give specific guidance for transformers.

15.5.2 IEC 62478

IEC 62478 discusses the advantages and technologies used in measuring PD using acoustic techniques and RF techniques in the VHF (30–300 MHz) and UHF (300–3000 MHz) bands [24]. It is a compliment to IEC 60270, which specifies direct, charge-based PD measurement at frequencies less than 1 MHz. IEC 62478 is normally applied for online PD measurements of GIS, transformers, and rotating machines during normal operation. Note that PD measurements above 1 MHz cannot be calibrated in terms of apparent charge (pC or nC). Instead PD magnitudes are normally measured in mV, dBm, dBμV, or other units of signal strength. The standard gives no guidance on acceptance criteria.

15.5.3 IEC 60076-3

IEC 60076-3 is the standard that is directly concerned with the factory electrical tests on (primarily) liquid-filled transformers [25]. The measurement of PD in transformers is usually performed simultaneously with an induced overvoltage[1] test at increased AC frequency – typically up to about 400 Hz (Section 15.8). The standard establishes the procedures for carrying out dielectric tests and the minimum applicable test values; it also establishes the minimum distances in the air between the phases and to ground if these are not specified by the buyer. The standard also contains the main test parameters that must be observed for the execution of all dielectric tests.

IEC 60076-3 also lists the most important aspects to be considered when measuring PD. The PD measurement is to be compliant with IEC 60270, i.e. made at <1 MHz. The measurement

1 An induced overvoltage test is one where the low voltage winding is energized, to excite a high voltage in the high voltage winding. By using an AC frequency higher than 50/60 Hz, a higher than rated voltage can be induced in the high voltage winding with a lower risk of saturating the magnetic core.

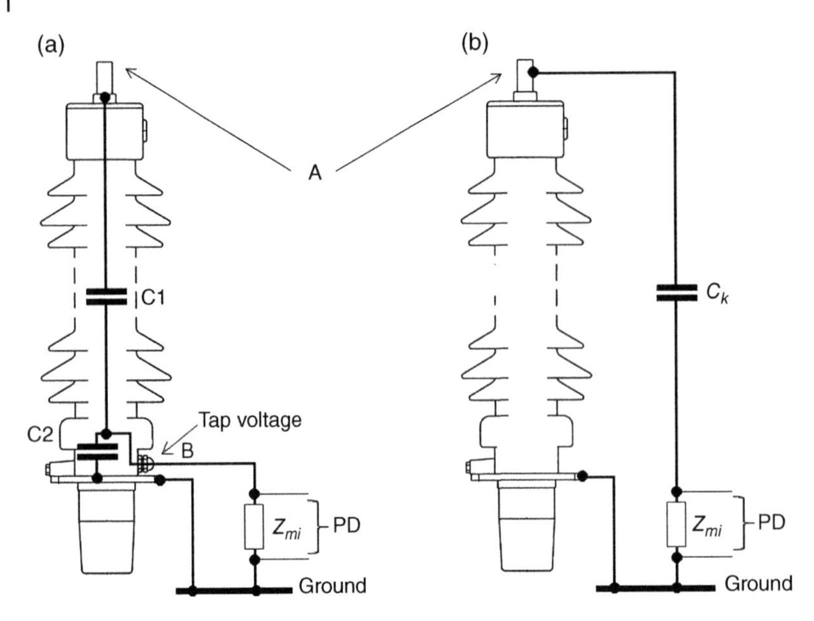

Figure 15.22 Alternative PD coupling arrangements for transformers. A is where the calibrator is connected. (a) Bushing with tap adapter (b) bushing without tap adapter.

procedures are outlined in IEC 60076-3 and indicate that the PD units of measurement is apparent charge, how the calibration must be performed (in spite of the cautions on calibration into apparent charge for inductive test objects in IEC 60270 and Section 6.6.3), and how many PD measurement points are needed. The standard also provides information relating to the measurement that is normally performed during the test with induced overvoltage at increased excitation frequency.

The PD is detected using either an external capacitor (C_k) or the power transformer bushing capacitance tap (Figure 15.22). In both cases, a measurement impedance is connected as shown in Figure 15.22. Point "A" indicates where the calibrator must be connected during the calibration procedure.

A very important part of IEC 60076-3 is the acceptance criteria of the test (see Section 15.10.3). IEC 60076-3 and IEC 60076-11 show the limit value of the background noise, the maximum level of PD, which must not be exceeded, at selected voltage levels (Section 15.10.2).

15.5.4 IEC 60076-11

The standard for the measurements of PD on dry-type transformers is the IEC 60076-11 [26]. Since dry-type transformers are different from transformers immersed in insulating liquids in terms of both rated power and rated voltage, and have profoundly different construction characteristics, the test procedures are inevitably different. The standard 60076-11, therefore, limits itself to establishing the voltage levels, the times of exposure to the test voltage, and the acceptance criteria by setting the maximum PD charge allowed for PD and refers to 60076-3 for any other consideration.

15.5.5 IEEE C57.12.90

The IEEE C57.12.90 standard is somewhat like IEC 60076-3 [27]. It indicates how to perform a factory test with an induced overvoltage. Since the test voltages required cause saturation of the core of the transformer, the standard indicates the criterion for establishing the minimum AC frequency that must be adopted. The guide establishes the test voltages that must be applied and the time for which each voltage level must be maintained. The PD test acceptance criterion is that if the PD limit values exceed the maximum expected, the standard indicates what actions to take between end user and manufacturer. Normally, the transformer is not immediately rejected but further investigations are suggested.

15.5.6 IEEE C57.113 and C57.124

Most of the relevant IEEE documents for PD measurement in power transformers are recommended practices as they analyze the technical aspects of transformer testing in much more detail than the IEC standards, at the expense of procedural details. For the electrical measurement of PD, the relevant guides are C57.113-2010 for liquid-filled transformers [28], while for dry transformers the document is C57.124-1991 [29]. In both IEEE documents, the "wideband" method (Section 6.5.2) in IEC 60270 is the preferred PD measurement method for factory testing. These documents provide technical details relating to the choice of the values of the various components that are used for a measurement of PD, such as optimal capacitance for the coupling capacitor (C1 or C_k in Figure 15.22), the specifications for the measurement impedance, Z_{mi}, and the calibrator. C57.124 has not been updated since 1991, so does not include advancements made in the past 30 years.

15.5.7 IEEE C57.127

A guide for using acoustic methods to detect and locate PD within liquid-filled transformers is given in IEEE C57.127 [30]. The guide provides information on the processing of acoustic signals from PD sources within power transformers and reactors. It also provides information on the features of the instrumentation to be used and on how to install the acoustic sensors. There are indications about the procedures to be adopted and criteria for interpreting the results to identify the position of the source of PD within the transformer tank.

15.6 PD Pulse Propagation and PD Detection in Transformers

The PD events within a transformer result in pulse currents, electromagnetic signals, and acoustic waves that propagate in different ways:

- Conducted currents: direct propagation through the transformer windings
- Displacement currents: propagation through capacitive coupling
- Electromagnetic waves: propagation by electromagnetic radiation
- Ultrasonic waves: propagation of acoustic waves in the oil and in the tank.

Some propagation modes dominate at low frequency and others at higher frequencies. Pulse propagation has a profound effect on the signal detected at the transformer terminals.

15.6.1 PD Current Pulse Propagation Through Stray Capacitance

Figure 15.23 shows a high-frequency equivalent circuit of a winding. This model is relevant for PD measurements below the 1 MHz range used in IEC 60270 PD detectors (Chapter 6). L_w represent the inductive parameters of the individual sections of a transformer disk winding

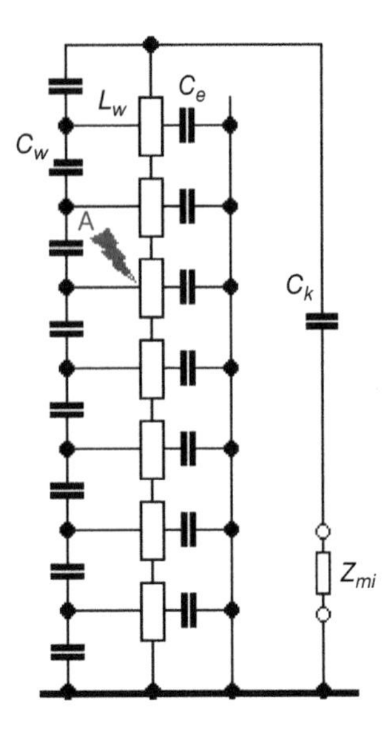

Figure 15.23 Equivalent circuit of a single phase in a transformer winding at frequencies below 1 MHz. PD at the indicated location results in the oscilloscope trace in mV across Z_{mi} is shown on the right.

(typical for high-voltage windings). The capacitance between winding elements accounts for the propagation of displacement currents: C_e represents the parasitic (or stray) capacitance to ground and C_w represents the parasitic capacitance between individual winding sections [31]. If a PD signal originates at the PD point indicated by A in Figure 15.23, the path that the signal must follow before reaching the coupling capacitor C_k affects the signal that will be available across the coupling impedance Z_{mi}.

The transfer of the PD signal to other windings in the transformer (cross coupling), such as between the low- and high-voltage windings or between windings in different phases, takes place through a purely capacitive coupling. Generally, due to the value of the capacitances, the coupling between windings belonging to the same phase is greater than the coupling between windings of different phases. Figure 15.24 shows a generic second winding not connected to ground and the capacitance C_{iw} between the two windings, as well as the capacitance to ground (C_e) of the second winding. In this case, the PD signal that occurs in the indicated point A (Figure 15.24) reaches the two measurement impedances on the second winding, Z_{mi2} and Z_{mi3}, via PD coupling capacitors (C_k). Since the coupling between the first and second winding is based on high-frequency displacement currents, the propagation within the second winding will mostly occur through the turn-to-turn stray capacitances and thus through a path essentially consisting of capacitances. In this case, the signals available on the three PD measurement impedances will have similar waveforms but very different magnitudes from each other. That is, the PD signals at measurement impedances, Z_{mi2} and Z_{mi3}, will have a lower magnitude than the signal available across the measurement impedance Z_{m1}, assuming the

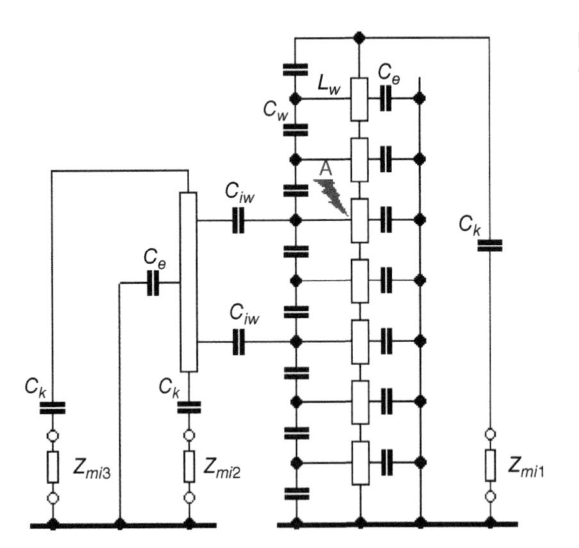

Figure 15.24 Equivalent high-frequency circuit between windings.

PD is in the winding associated with Z_{mi1}. This aspect can be of help, as discussed in detail in Section 15.10.1 to locate PD sites. The PD signal measured on the second winding will be smaller, have a longer risetime and a longer duration than the signal at the PD site.

15.6.2 UHF Propagation

The signal generated by the PD also propagates in the form of VHF and UHF electromagnetic (EM) waves [24, 32]. The propagation of the electromagnetic wave is affected by the nature of the materials it passes through, by the characteristics of the transformer tank (which has its own natural frequencies) and in general by the geometric shape of the elements that are along the path of the EM wave, which cause strong reflections, attenuation, and dispersion. The different permittivity of the solid and liquid insulation materials can introduce propagation delays of the order of 0.2–0.3 ns and have a limited influence on the measured signal spectrum. Metallic discontinuities result in a greater delay in the propagation of the signal, typically 3–5 ns, and show a significant alteration of the PD signal spectrum. For these reasons, the EM signals generated by PD undergo attenuation and propagation delays between PD sites and the detected signals. The received signal depends on the location of the PD source, the characteristics of the sensor (e.g., its resonance frequency, directionality, sensitivity, etc.), and most significantly, on the exact propagation properties (transfer function) of the complex RF signal path between them. The result is that the VHF/UHF PD signal available at a PD sensor is characterized by multiple resonance frequencies that depend both on the harmonic content of the original PD signal and on the typical natural frequencies of the transformer tank, which in turn depend on the size of the tank, and those of the sensor.

15.6.3 Acoustic Propagation

Some of the energy in a PD event also propagates in the form of acoustic waves in power transformers (Section 5.4) [33, 34]. Each PD event produces a short-lived pressure wave that generates an elastic wave in the oil that propagates inside the transformer until it reaches the walls of the tank where it can be picked up through special sensors with suitable characteristics. The process is like that of an earthquake with an epicenter at the point where the PD originates, but with microscopic dimensions. Since the PD signal is an impulse of very short duration, the harmonic content of the acoustic signal produced is characterized by frequencies that can vary in the range 10–300 kHz. In the case of propagation in oil, it can be imagined that the propagation occurs similarly to a series of concentric spheres with the acoustic source in the center; when the acoustic waves reach the tank, it starts to propagate within the tank walls as a series of concentric circles. Along their path, the acoustic waves undergo attenuation that mainly depends on the medium in which they propagate, on the materials, and on the geometry of the objects that occupy the space as well as on the geometry of the space itself. These considerations make the propagation of acoustic signals a complex problem. PD can originate in multiple points of the transformer, some of which are close to the external wall of the tank where an acoustic sensor can be positioned.

For PD signals originating between the turns of an internal winding, or on the surface of the magnetic core, the acoustic wave must cross different regions on its path to the tank: oil channels, barriers of organic insulating materials, and copper windings. Each region has its own geometry and propagation characteristics (such as characteristic impedance, speed, group delay, attenuation, dispersion). These features influence the propagation time as well as harmonic content of the signal; in this regard, it must be considered that the solid insulating materials behave like low-pass filters.

Much simpler is the case in which, between the source of PD and the sensor, there is a direct path consisting of a single medium (oil) with a limited series of barriers of solid insulating material. In this case, the signal undergoes minor alterations from its original shape and reaches the sensor with a shape and magnitude that is easier to interpret.

An aspect that deserves consideration is that PD often occurs with a relatively short time between pulses on the order of tens of microseconds. Consequently, the acoustic emissions from sequential PD pulses tend to superimpose at the sensor, making it difficult to interpret. Finally, the propagation speed in mineral oil of an ultrasonic pulse varies from about 1240 to about 1300 m/s for temperatures in the range from 50 to 80 °C and that the attenuation is proportional to the square of the signal frequency [35].

15.7 Sensors for PD Detection

Some PD sensors are used for testing in the low-frequency range required by IEC 60270 detectors. Others are intended for use in the VHF and UHF frequency ranges covered by IEC 62478.

15.7.1 Impedance Connected to Bushing Tap

There are at least three methods for detecting PD in the low-frequency range with an IEC 60270 detector. For transformers rated higher than 72 kV, the most common solution is to use HV transformer bushing capacitance (bushing tap).

Bushing taps are commonly provided to measure the dissipation factor of the main capacitance C1 (Figure 15.25) during the bushing life. The bushings used in high-voltage transformers consist of a graduated insulation, which is made by interposing multiple floating electrodes made from

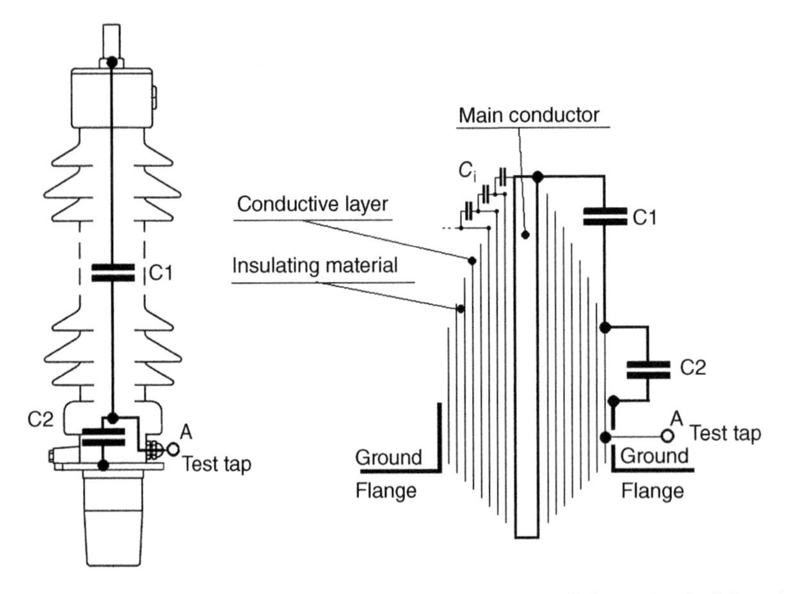

Figure 15.25 A transformer bushing composed of a set of electrodes (solid vertical lines in the right sketch) separated by solid insulation (oil paper), which makes the electric field between the HV main conductor and ground uniform.

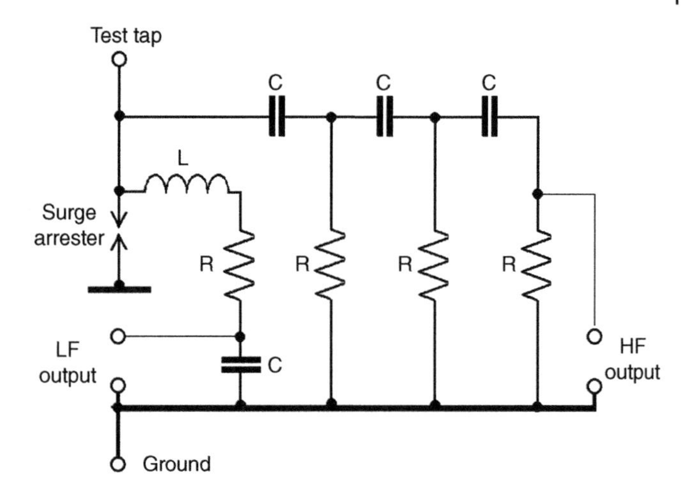

Figure 15.26 Typical circuit for a quadripole connected to a bushing tap. The LF output is an AC frequency phase reference, while the HF output goes to an IEC 60270 PD instrument.

sheets of metallized paper, separated by layers of solid insulating material (paper or polymer films). The bushing is then impregnated with insulating oil or epoxy resin [36]. The capacitance between pairs of electrodes is designed such that the electrical stress is made uniform between the HV conductor and the grounded transformer tank. In practice, the bushing is realized as a series of coaxial cylindrical capacitors each having capacitance Ci. Figure 15.26 shows the structure of a bushing; the capacitance C1 represents the series circuit of capacitors Ci. C2 is the capacitance of the last insulation layer to which the bushing tap is connected. C1 and C2 thus form a capacitive voltage divider. During normal operation of the transformer, terminal A in Figure 15.25 is connected to the ground and C1 is the capacitance of the bushing to ground, while C2 is short circuited to ground. To detect a PD signal in the transformer or bushing, it is necessary to remove the ground connection at A and connect a measuring impedance (or quadripole – Sections 6.3.1 and 6.3.2) between point A and ground to transfer the PD signal to an analyzer. WARNING – this should NOT be done under live conditions).

Different types of impedances can be connected between point A in the figure and ground depending on the type of analyzer to be used. The most common impedance is a quadripole (Section 6.3) whose input is connected between the bushing tap and the ground while the output is connected to the PD instrument input. This type of impedance is used when carrying out PD measurements in accordance with IEC 60270. Normally, the quadripole has two outputs: the high-frequency PD signal and a low-frequency voltage synchronized to the applied AC voltage. Figure 15.26 shows a basic electrical diagram of a quadripole equipped with a filter for the separation of the low-frequency AC voltage from high-frequency PD signals.

15.7.2 Coupling Capacitors

High-voltage transformers rated 72 kV and above are normally equipped with the bushing tap. Unfortunately, such taps are rarely found in the bushings for lower voltages. If the bushing tap is not available, high-voltage coupling capacitors (C_k in Figure 15.22b) suitable for factory PD testing compliant with IEC 60270 are commercially available for voltages up to 100 kV. The coupling capacitors measure the PD using an indirect detection method (Section 6.2.2). Normally these coupling capacitors are sold by the suppliers of PD instruments and are sized for correct operation with the detection impedance or quadripoles (Section 6.3) supplied by PD instrument manufacturers.

15.7.3 HFCTs

Another type of sensor commonly used for PD detection in power transformers is a high-frequency current transformer (HFCT), see Section 6.3.3. The HFCT is made with a ferrite core suitable for high frequency, with the primary circuit consisting of a single turn and typically 10 turns in the secondary winding with low inter-turn capacitance. The shape of the magnetic core is normally toroidal to ensure the best magnetic coupling. Due to the high-pass characteristics of the ferrite core, this type of sensor does not allow AC voltage synchronization of the measurement, thus an alternate method of synchronizing PD with the AC cycle for PRPD plots is needed. The connection diagram is very simple and is shown in Figure 15.27, which shows the connection of the HFCT to the grounded bushing tap or to ground side of an external coupling capacitor. The typical frequency response of an HFCT is shown in Figure 6.7. Upper frequency cutoffs as high as about 100 MHz are possible for this application. It is possible to increase the number of turns of the secondary winding to obtain a higher output signal, but this will reduce the upper cutoff frequency due to higher inter-turn capacitance and self-inductance.

The HFCT sensor is also suitable for performing PD measurements by inserting the HFCT in series with the transformer winding (direct circuit – Section 6.2.1). In this case, the HFCT is clamped around the neutral ground connection (Figure 15.28). In this case, the HFCT must be of the split core type (Figures 6.8 and 15.29) so that it is not necessary to disconnect the neutral connection to insert it into the opening of the HFCT. The measurement of PD on transformers using the direct coupling method is only possible using an HFCT as the quadripole, which is not normally designed for this purpose; the low-frequency current for which a quadripole is designed is a few hundred mA and is not compatible with the current flowing to ground during a test. That is, all the current flowing through the transformer passes through the primary winding of the HFCT and the high-pass characteristic should be selected in a way that the winding current does not saturate the magnetic core. One trick to avoid ferrite core saturation due to 50/60 Hz high current, when using HFCT as shown in Figure 15.29, consists of inserting a thin, non-magnetic gap (using paper or plastic foils about 1 mm thick) in the split space of the HFCT core (the "critical surfaces" in Figure 15.29). In this case, the low-frequency saturation is avoided

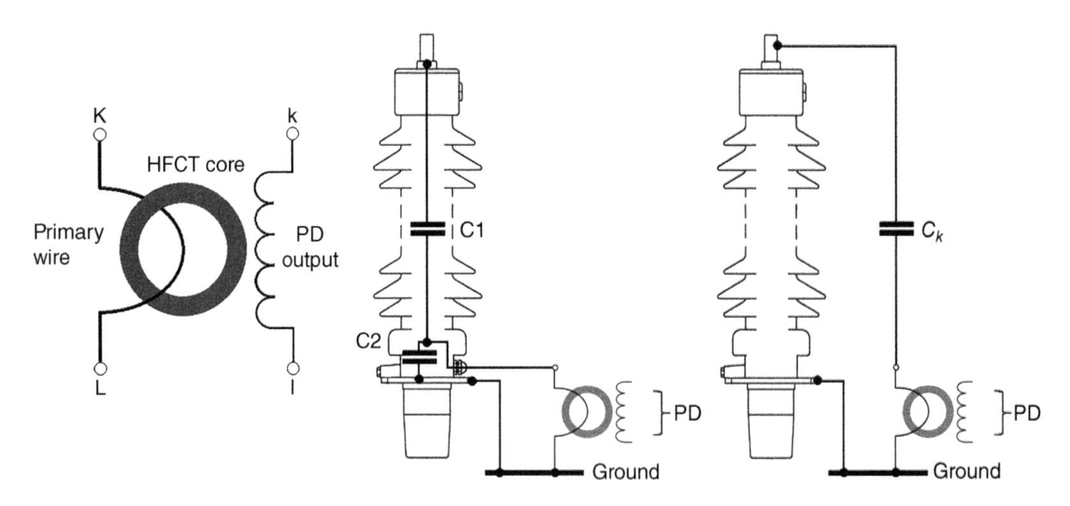

Figure 15.27 Typical HFCT diagram (left) and typical connections via a bushing tap or a coupling capacitor.

Figure 15.28 HFCT arrangement directly connected to the neutral.

without compromising the wide band of the HFCT. However, this gap will slightly reduce the HF signals magnitude. The HFCT is also particularly suitable for measurements of PD in the neutral or ground connection on single-phase power transformers.

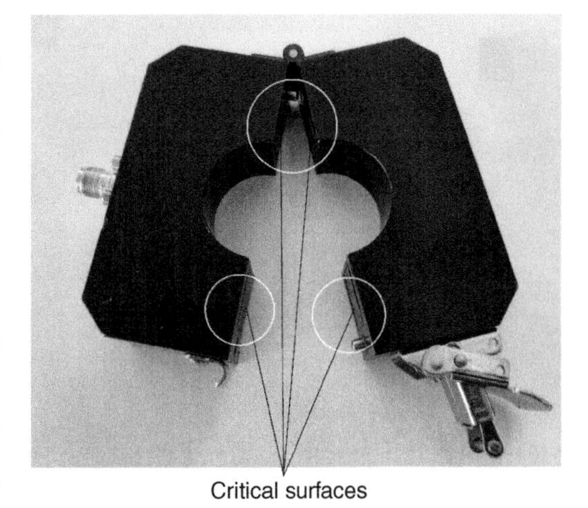

Figure 15.29 Split core HFCTs that can be installed on the neutral connections of power transformers.

15.7.4 VHF/UHF Sensors

VHF/UHF sensors (Sections 7.4.4 and 7.4.5) are also popular to capture PD signals, especially with online PD monitoring systems. These sensors are suitable for permanent installation. Historically, UHF sensors were first developed for GIS applications (Section 13.8.1), and then modified for application in power transformers. Most UHF sensors (see for example Figure 15.30) require special arrangements for mounting on and through the transformer tank, and therefore are usually only installed during transformer manufacture. Such tank-mounted UHF sensors are very similar to those used in GIS (Section 13.8.1). However, there are VHF/UHF sensors that can be installed via the oil drain valve

Figure 15.30 UHF PD sensors that are specially designed to be installed within the transformer and mounted on the transformer tank wall using a dielectric window. *Source:* BSS-HS.

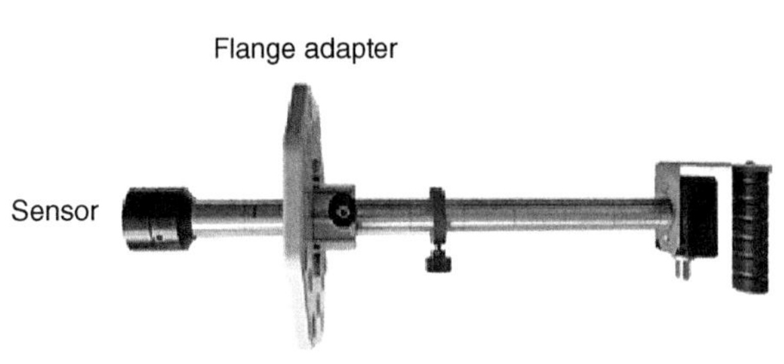

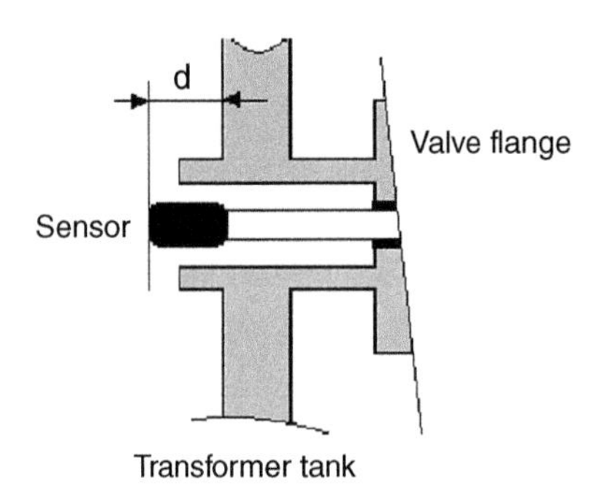

Figure 15.31 UHF sensor that can be retrofitted to an existing transformer via the oil drain valve. *Source:* BSS-HS.

(Figure 15.31). Such valve-mounted UHF sensors will prevent the proper use of the oil valve. Also, a valve-mounted antenna is often not close to the HV parts of the windings, which is where most PD activity is expected (Section 15.6.2). Therefore, if VHF/UHF sensors are to be used for the detection of PD during operation, it is best to install the tank-wall mounted type. The valve type, however, can be retrofitted to some types of drain valves.

The number and the optimal position of tank-mounted sensors depends on the dimensions of the transformer. The minimum number is four antennas, suitably positioned according to the most likely positions of PD sources. For large transformers, the number of sensors could rise to six, but it depends on how much precision and definition one wants to obtain from the measurement.

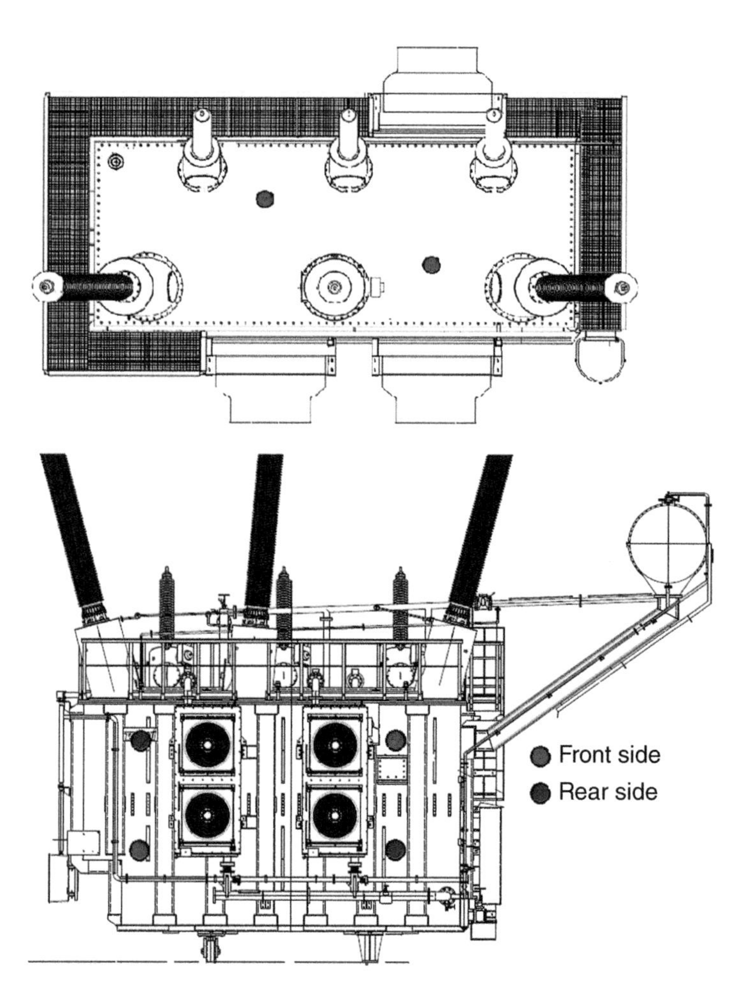

Figure 15.32 Typical arrangement of six tank-mounted UHF sensors (four on the sides and two on the top) of the tank. A typical four sensor arrangement would consist only of the sensors on the front and back side without sensors on the top plate.

Figure 15.32 shows the typical positions of the VHF/UHF sensors positioned on the transformer top cover and on the tank sides. CIGRE has recommended, for the rated voltages of 420 kV, 245 kV, and 123 kV, that the minimum distance between the sensors and the regions at the greatest potential (i.e., where PD is more likely to occur) be 1.5 m, 1 m, and 0.8 m, respectively. [37].

VHF/UHF sensors operate in the 200 MHz to 3 GHz range and cannot be used for measurements in the low-frequency (<1 MHz) range with IEC 60270 detectors. Thus, the output from VHF/UHF antennas cannot be calibrated in terms of charge (i.e. pC or nC) [24]. Such sensors are instead used for onsite/offline or online PD measurements and/or monitoring. They are suitable for a timely detection of the PD and to help in locating PD sources, especially for online monitoring systems. To verify functioning of the sensor, conventional calibration systems (Section 6.6) cannot be applied since they do not generate VHF/UHF electromagnetic waves. Instead, a sensitivity check is made using the procedure in IEC 62478 where short risetime (<1 ns) pulse generator is connected to one of the UHF sensors. That is, this sensor is used as a broadcasting antenna, and the signal received at another UHF sensor is measured to evaluate the sensitivity of the transmitter/receiver chain. However, although methods for the standardization of the response of these sensors have been proposed [38, 39], this process will not lead to a correlation between the magnitude of the PD in terms of apparent charge and what is detected by the antenna. Ultimately, the values detected by the different types of sensors can only be trended, rather than compared with an absolute value.

A sensitivity check used for comparability between UHF measurement systems and sensors is described in appropriate documents [37]. See also Section 13.8.3 for the related GIS sensitivity check. The sensitivity check method considers two factors:

- The influence of the UHF sensor's sensitivity and that of the UHF instrument characteristics, including accessories like cables, pre-amplifier, etc. The UHF instrument's influence is corrected by using a defined and invariable test signal as a reference for all recording devices (as for the calibration method used in IEC 60270 for electrical PD measurement).
- The sensitivity of the UHF sensor can be addressed by a characterization of UHF sensors using the antenna factor (AF) measured in a special reproducible setup, i.e. a GTEM cell.

When VHF/UHF sensors are used in transformers, it must not be overlooked that the antenna is placed in direct contact with insulating fluids. The fluids can reach high temperatures perhaps up to 120 °C, and maybe as low as −40 °C in a spare transformer stored outside in a cold climate. Also, the materials used in the antenna must be tested for compatibility with the fluid, just as gaskets, and other accessories in the transformer are. The aim is to avoid problems with oil leaks.

An aspect that concerns the phenomenon of magnetically induced mechanical vibrations at twice power frequency and its harmonics (up to the 4th or 5th). In addition, there may be noncontinuous mechanical phenomena at different frequencies but of greater intensity. These include:

- movements of the fluid as a result of high fault currents caused by power system short circuits.
- the sudden movement of the fluid due to earthquakes and near-by equipment vibration.

These can have negative repercussions on sensor sensitivity or on the reliability of associated electronic circuits (if they are installed aboard the sensor).

When positioning the VHF/UHF sensors through the oil fill/drain valves, it is necessary to ensure that the sensor does not interfere with the live parts of the transformer. For this reason, it is important to check the distance "*d*" shown in Figure 15.31 by carefully following the instructions provided by the transformer manufacturer. For a correct UHF sensitivity check, the insertion depth needs to be the same as for the sensitivity measurement. Higher insertion depths lead to only minor decreases or increases of the signal amplitude, but are usually not favorable in transformer installations due to the high electric field stresses.

15.7.5 Acoustic Sensor

Acoustic emission (AE) sensors are used to detect the acoustic signals emitted by PD sources within transformers (Section 5.4.2) [30]. Unlike VHF/UHF sensors, AE sensors can be placed on the outside of the transformer tank and are considerably smaller in size. Magnets are often used to fix the acoustic sensors to the transformer tank. Although such sensors are an independent method for detecting PD in liquid-filled transformers, AE sensors are mainly used as an alternative method to locate PD sites within the tank.

Figure 15.33 shows the structure of an acoustic sensor (in this case, an accelerometer) with an integrated signal preamplifier. The piezoelectric element converts acceleration to charge. It is contained within a metal case together with the signal conditioning circuit that sets the sensor operating frequency

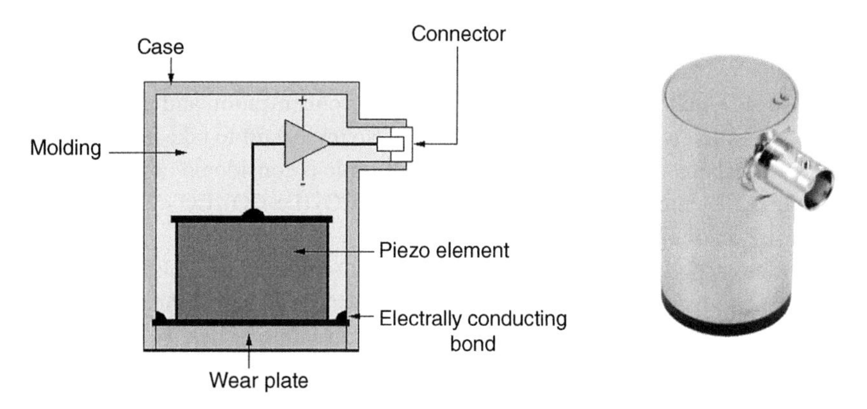

Figure 15.33 Piezoelectric accelerometer typically mounted on the exterior of a transformer tank to detect the acoustic signals accompanying PD. The sensors typically are a few centimeters in diameter.

Figure 15.34 Typical acoustic sensor bandwidth for PD measurement in transformers.

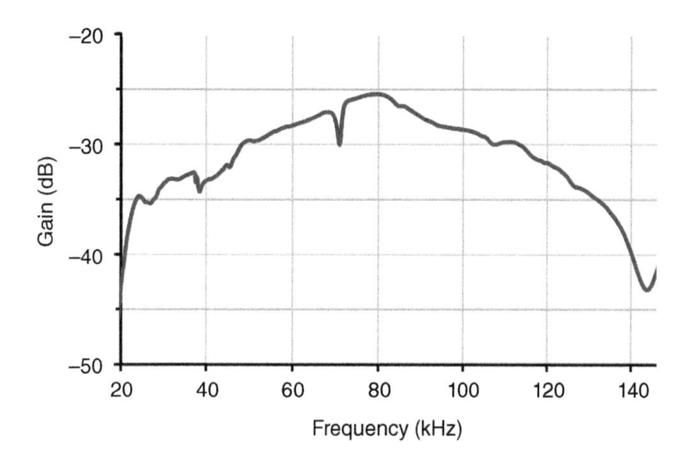

and produces an mV output. The charge amplifier requires power, which is typically delivered via a phantom power supply as low-voltage (10–20 V) DC carried on the center conductor of the coaxial signal cable from the measuring instrument (often a spectrum analyzer or oscilloscope). Figure 15.34 shows the typical passband of an acoustic sensor suitable for measurements on transformers.

In choosing the acoustic sensors to be used for locating PD sources in transformers, it is necessary to take into consideration some important aspects. When fixed on the tank of a transformer, the acoustic sensors are subjected to vibrations from the environment, as well as low-frequency electric and magnetic fields. As the sensor contact surface is usually made of an insulating material and the case is metallic and grounded via the output coaxial cable, attention must be paid to the potential difference between the sensor case and the transformer tank to avoid damaging the internal electronics and/or the piezoelectric element of the sensor. For these reasons, the sensor electronics must have good electromagnetic compatibility and an excellent rejection of high-frequency signals typical of the background noise.

15.7.6 Nonelectric Sensors: Laboratory DGA and Online DGA

Dissolved gas analysis (DGA) is a reliable and effective tool for identifying the presence of PD as well as other aging phenomenon in paper and oil-insulated equipment [40–43]. DGA is now the most widely used method to verify the reliability of oil–paper insulation systems. By measuring the concentration of some specific hydrocarbon gases in the oil, it is possible to establish whether low-, medium-, or high-energy electrical discharges are occurring inside a transformer. The typical gases indicative of the presence of PD are hydrogen and acetylene whose concentration and trend allow estimation of the energy associated with the phenomenon. It is also important to take into consideration the concentration of carbon oxides. For example, CO_2/CO ratio is considered normal if it is in the range between 3 and 11. The increase in CO in comparison to CO_2 (i.e. $CO_2/CO < 3$) indicates PD is occurring in organic insulation.

This type of analysis can be carried out by laboratory testing of 100–250 cc oil samples (stored in glass syringes) taken from the transformer. The sampling should always be taken at the same location of the transformer. The samples may have contamination and sealing issues. For example, hydrogen levels measured in the lab can be lower than those in the transformer since hydrogen tends to escape the test tubes if they are not properly sealed.

A second method to monitor the dissolved gas in the oil within the transformer consists of online monitoring systems. There are various types of commercial systems available that use different measurement principles. Compared to an analysis conducted in a laboratory, the online systems are less accurate, but are not affected by contamination and sealing issues. Online DGA provides a reliable evaluation of the trend of the concentration of individual gases, which helps in formulating hypotheses regarding the intensity of the phenomena. Interpretation of the meaning of the concentration and the nature of gases dissolved in oil is discussed in several documents [40–43].

In general, DGA is complementary to the electrical detection of PD. However, DGA cannot indicate the location or even the phase that is experiencing PD.

15.8 AC Supply for Offline Testing

The offline PD test on transformers is normally performed by energizing the LV winding to excite the HV winding to the specified test voltage. It is common to use a supply with an AC frequency higher than 50/60 Hz to avoid core saturation at high test voltages. For very large transformers, the test supply must be capable of delivering 600–1000 kVA. There are different methods of obtaining an adjustable voltage with increased frequency.

15.8.1 Motor-Generator Test Sets

The most common type of power supply uses a frequency converter made from rotating machines: a synchronous generator driven by an asynchronous (induction) motor having fewer magnetic poles than the generator. Thus, the AC frequency from the generator is higher by a factor equal to the ratio between the number of poles of the generator compared to the number of poles of the motor. The output voltage from the generator may be as high as about 400 Hz. This type of power supply creates relatively low interference, especially if a generator has a brushless exciter. The voltage is changed by varying the generator excitation current.

Figure 15.35 shows a basic diagram of the power system described. The "soft start" drive S gradually raises the motor M to the desired speed to minimize inrush currents. Once the rated speed is reached, the motor is connected directly to the power grid as S is excluded by means of the by-pass switch BPS. This eliminates interference generated by the soft start system electronics.

A variant of the circuit of Figure 15.35 is shown in Figure 15.36. In this solution, the generator G is an induction machine with a wound rotor. The motor mechanically drives the wound-rotor induction generator. This type of system is also called a doubly fed induction generator.

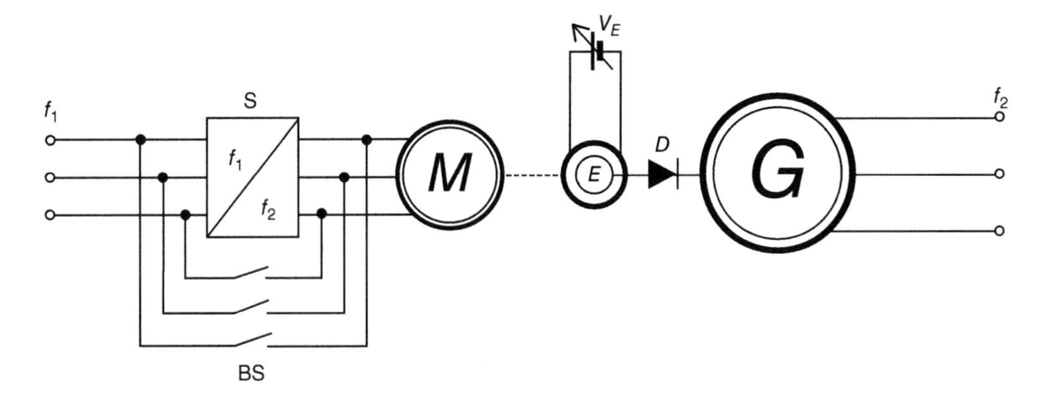

M : Motor
E : Exciter
V_E : A.C. regulated exciting voltage
G : Generator
D : Rotating diodes
V a.c. : Auxiliary A.C. supply
S : Soft motor starter
BS : By pass switch

$f_2 > f_1$

Figure 15.35 Transformer supply using an induction motor-synchronous generator arrangement, with the generator producing a higher AC frequency than 50 or 60 Hz.

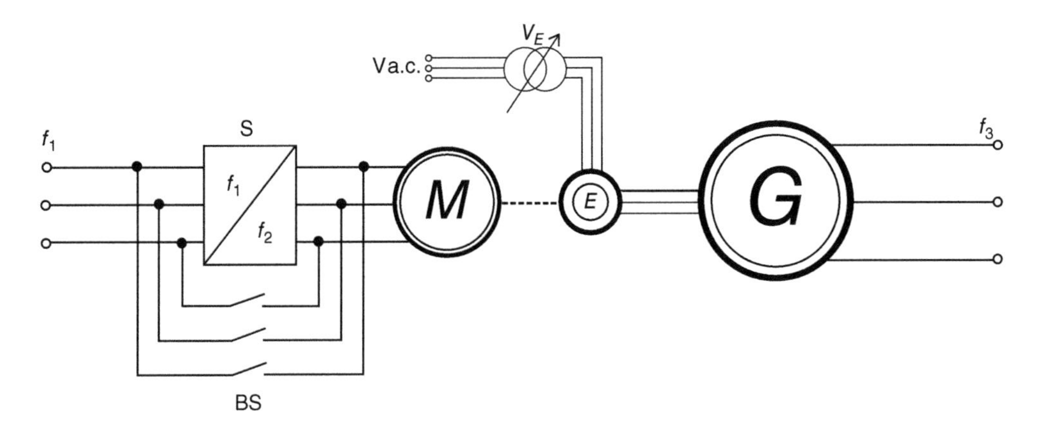

M : Motor
E : Exciter
V_E : A.C. regulated exciting voltage
G : Generator
D : Rotating diodes
V a.c. : Auxiliary A.C. supply
S : Soft motor starter
BS : By pass switch

$f_3 > f_1$

Figure 15.36 Frequency converter with AC exciting the generator rotor winding.

The three-phase generator rotor winding is supplied from an electronic frequency converter (also called a variable speed drive or inverter fed drive). The output AC frequency from generator (f_2) is a combination of the ratio of the number of magnetic poles on the generator to the number of poles on the motor, and the frequency of the AC voltage supply to the wound rotor winding.

The generator rotor winding is supplied from the frequency converter via slip rings; the latter are known to cause interference with the PD measurement due to contact sparking (Section 4.4).

15.8.2 Electronic Variable AC Supplies

In some special cases, such as mobile power systems for carrying out onsite/offline PD tests, solutions based on variable-frequency electronic power supplies are often adopted. Figure 15.37 shows a simplified diagram with the main elements that make up a suitable variable frequency power supply for power transformers. This method is convenient due to its lower weight, which favors its portability. However, such systems may not be as robust, and the presence of switching noise from the power electronics may cause interference.

The static frequency converter supplies a sinusoidal power supply voltage that can be controlled in voltage and in frequency by means of a suitable control unit. The converter supplies the active and reactive power necessary to supply the low voltage winding of a special step-up transformer (SUT). The step-up transformer adapts the low voltage supplied by the converter to a voltage level compatible with the voltage level of the winding of the transformer under test. The step-up transformer is equipped with various voltage taps to optimize the appropriate voltage value. The low-voltage filter helps to reduce the interference generated by the static frequency converter. Although the latter is designed to generate a sinusoidal voltage, due to the switching of the power electronic devices, a certain amount of noise generated by switching can leak into the PD measurement chain. The high-voltage filter decouples the transformer under test from the rest of the circuit for the purpose of measuring PD. During the test, to optimize the apparent power required, it is possible to search for the frequency that best suits the characteristics of the transformer, lowering supply current to the convertor.

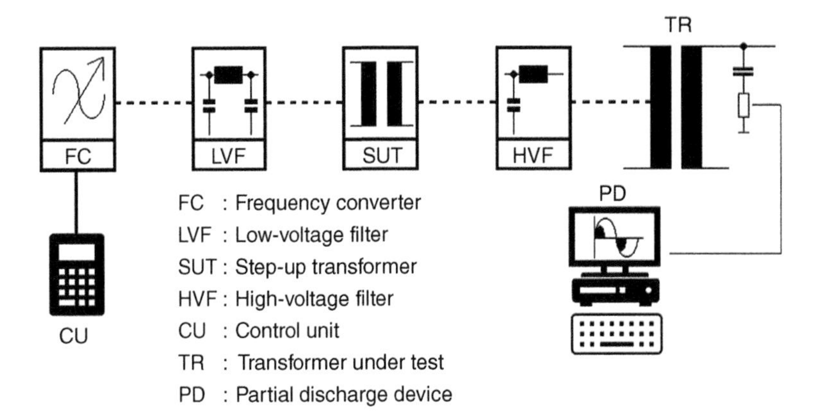

Figure 15.37 Typical diagram of a static frequency converter used to energize a power transformer (TR).

15.9 Precautions Against Background Noise and Interference in Electrical PD Testing

Among the most critical aspects when performing PD measurements on power transformers (or in fact any test object) is the separation of the transformer PD signals from unwanted signals that may dominate the PD and lead to false indications of insulation system problems (Chapter 9).

This electrical interference comes from countless sources, each with very different characteristics. Before addressing how to suppress interference, it is appropriate to reiterate the meaning of the term "interference" (see Section 9.1). Interference can be a repetitive signal with regular frequency that can be synchronous or asynchronous with respect to the AC supply voltage. In contrast, a continuous asynchronous signal with respect to the supply voltage is indicated here as noise (normally, electronic noise – Section 9.2.1). For convenience, we can define these two families of signals by the term "unwanted signals." It is best to actually eliminate the unwanted signals (for example by turning off high-pressure vapor lamps, or stopping cranes and other activities such as electric arc welding), before trying to separate the PD signals from the unwanted ones.

The factors that influence the presence of unwanted signals are many and depend mainly on

- the environment where the measurements are being made,
- the power and auxiliary power sources,
- the instrumentation used, and
- the care with which the measurement circuit is prepared.

15.9.1 Test Site Arrangement

The environment in which the transformer is located significantly affects the presence of unwanted signals on the measurement of PD. In particular, it must be taken into consideration whether a measurement is being carried out in a HV lab or at a substation (onsite). In each of the situations, interference signals of different natures and origin may be present.

In factory PD testing, the transformer is usually tested in an electromagnetically shielded room (Section 9.3.1). The shielded room is designed and built to exclude as much external interference as possible. Such shielded rooms are equipped with a metal floor, walls, and celling, all continuously bonded together. It is best if the floor and walls are made from conducting sheets rather than a metal mesh (the larger the metal mesh size, the less effective it is for blocking high-frequency components). The floor and walls should be grounded with specially designed and manufactured grounding rods (normally insulated in the first meters below the soil to avoid interference from stray currents). The size of large power transformers with their bushings implies a large shielded room is needed. This makes the room costly and technically complex to implement to eliminate external interference. Typically, even a well-made shielded room for transformer testing will still experience background interference of about 50–70 pC in an IEC 60270 measurement.

Background interference caused by the lighting system may not be negligible. It is a good idea to avoid lighting systems that use gas discharge lamps, or, alternatively, turn off the lighting system during the measurement of PD. As discussed in Section 9.3.2, the test environment should be tidy by minimizing the presence of unnecessary equipment, which could give rise to partial discharges external to the transformer. Metal parts, even small ones, not connected to ground, can give rise to floating-potential type discharges, which will directly interfere with the PD measurement; being synchronous with the test voltage, it can be extremely difficult to deal with them.

The environment when performing offline/onsite tests are quite different from factory tests since it is not possible to use shielded rooms. Thus, there is greater interference from RF broadcasts, powerline carrier signals, corona from nearby substation equipment, etc. Higher levels of interference are inevitable. The only measures that can be implemented are to be careful in setting up the measurement circuit and to use instruments and sensors more suitable for use onsite (Sections 15.9.4 and 9.3.4–9.3.9).

15.9.2 AC Supply Interference

As discussed in Section 15.8, the offline PD test on transformers is normally performed by energizing the LV winding to excite the HV winding to the specified test voltage. Motor-synchronous generator supplies (Figure 15.35) tend to produce the least amount of interference. Electronic variable frequency supplies (Figure 15.37) usually produce the greatest interference.

15.9.3 Measurement System Arrangement

From the point of view of limiting interference, the precautions that are adopted in setting up the measurement circuit are of fundamental importance (Section 9.3.2). As in all circuits intended for the measurement of very low magnitude, high-frequency signals, it is essential to ensure that the signal cables between the source and the detector are kept as short as possible. For signal transmission, it is important to use shielded cables with a characteristic impedance compatible with the input impedance of the detector. In general, minimizing the length of all connections (high-voltage cables, signal cables, ground connections) is difficult because of the physical size of the transformer. Recently many IEC 60270 PD system vendors offer optical fiber connections and battery-operated instruments, which reduces some of the difficulties in testing large power transformers.

The connection between the bushing tap and the quadripole input must be made with a shielded cable and must be very short to avoid altering the PD measurement bandwidth. The connections between the quadripole and the detector are made with shielded coaxial cables with characteristic impedance matching the input of the detector (usually 50 Ω).

The ground connections should **not** be made using standard stranded cable, but instead use wide copper strips. The former has higher inductance that can attenuate PD signals. Figure 15.38 shows an example of a flat ribbon conductor for the ground connections of the PD signals.

An important precaution is to power the PD instrument through an isolation transformer with low interwinding capacitance and filters to block electrical interference from the 120/240 V power supply, thus preventing them from being transferred to the measuring system (Section 9.3.3). Battery-operated instruments further improve the decoupling of the entire measurement system from main power supply interference.

Figure 15.38 Copper tape for ground plane.

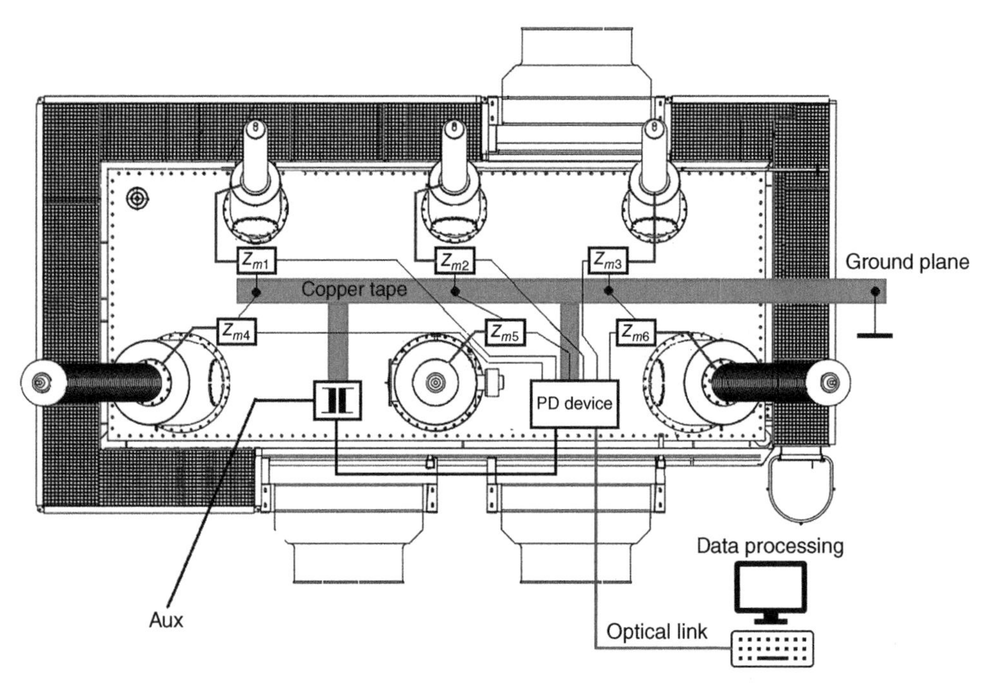

Figure 15.39 Typical layout for PD measurement on the top of large three-phase power transformer.

Figure 15.39 shows a measurement layout where all the equipment involved is placed on the top of a large autotransformer. A PD instrument compliant with IEC 60270 is powered via an insolation transformer equipped with a decoupling filter. The measurement layout is based on the use of a multi-channel PD detector connected by optical fiber to the control and processing device. The connections between the bushing taps and the quadripoles are made with unshielded insulated cables that is less than 50 cm, so that each quadripole is as close as possible to the bushing tap. The quadripoles are connected to the measuring instrument by coaxial cables with a characteristic impedance equal to 50 Ω, also with the shortest-possible length. A copper tape is used to create a ground plane for all the quadripoles.

To avoid the generation of corona from the HV bushings during the overvoltage tests, spherical-shaped conductive shields can be temporarily installed (Figure 15.40).

15.9.4 Instrument-Based Noise and Interference Separation

Even when all possible precautions are adopted, some interference is always present and cannot be eliminated completely using the methods above. It is, therefore, necessary to resort to hardware and software methods implemented in some commercial PD instrumentation to separate PD from interference (Sections 9.3.4–9.3.9). For example, some types of instruments are equipped with tunable band-pass filters. By properly selecting the center frequency and the passband of the filter, it is sometimes possible to minimize interference caused by radio stations, power line carrier signals, etc. Note that some vendors make instruments that are intended to be IEC 60270 compliant, but may actually operate up to 10 MHz or so. As discussed in Section 6.6, measurements above 1 MHz cannot be correctly calibrated in terms of apparent charge (pC, nC, etc.), and thus frequencies above 1 MHz will not be compliant with the pass/fail criteria for transformers. A frequency

Figure 15.40 Electric stress reduction at the tops of the bushings using temporarily installed spherical metallic shields.

spectrum analyzer (built into some commercial instruments), and used by a PD expert, is useful in selecting the best center frequency and passband for the PD measurement that minimizes interference. If the filter band pass is changed for the PD measurement, the calibration must be repeated using the same settings.

In transformer PD measurement, the AC supply frequency is always greater than 50/60 Hz, and this further helps in separating unwanted signals from e.g. corona from neighboring transmission lines in offline/onsite tests. All the signals that appear synchronous with the AC supply voltage on the PRPD pattern (Section 8.7.3) are certainly signals attributable to the AC supply voltage and can be transformer PD or PD/corona/interference from connected equipment. The latter appear on the PRPD pattern in positions and shapes that are easily recognizable by an expert. Note that the PD instrument must be able to synchronize to the frequency of the AC supply energizing the power transformer, and not just 50/60 Hz.

Time-frequency maps are another method to separate interference from transformer PD. The method is described in Sections 8.9.2 and 9.3.8. Each detected pulse is transformed into a plot of equivalent pulse duration vs frequency. In many cases, interference will show as a cluster of signals on this map, which are at a different location than clusters caused by PD pulses. If a cluster of pulses has a PRPD pattern (Section 8.7.3) that is not consistent with PD, then the pulses in the cluster are likely caused by interference [44]. These signals can then be deleted in a reconstructed PRPD plot. Some skill is needed to identify clusters, and to decide which PRPD patterns are consistent with PD and what patterns are consistent with interference.

For three-phase tests on three-phase transformers, the method in Section 8.9.3 (called 3PARD by the vendor) has also been useful in separating PD from interference [45].

A very useful technique to identify unwanted signals is to perform a preliminary acquisition at a voltage about 10% of the test voltage and to analyze the nature of the acquired signals. Normally, it is possible to exclude the presence of PD at this low voltage. Thus, any recorded signals are caused by interference. Figure 15.41 shows a PRPD plot of a transformer tested at low voltage where only unwanted signals are present. The signals caused by noise are not synchronized to the AC.

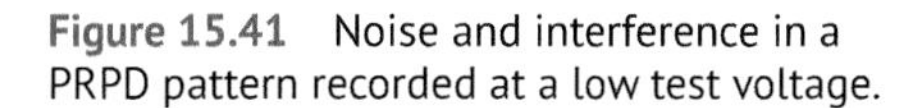

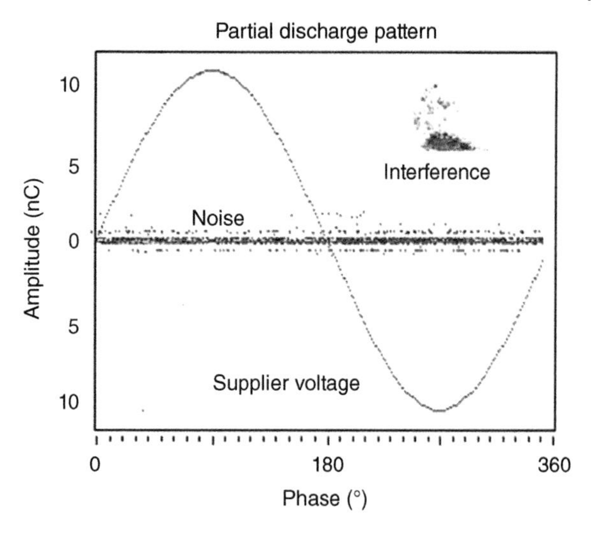

Figure 15.41 Noise and interference in a PRPD pattern recorded at a low test voltage.

The synchronous interference signals (at about 270°) cannot be attributable to transformer PD due to the low-test voltage. Such pulses must be ignored as the voltage is raised.

15.10 Factory Acceptance Testing of Transformers

PD tests are performed as routine tests prior to shipping the transformer from the factory to the substation or generating station. The PD test is to ensure that the transformer has been designed properly and manufactured correctly according to standards and any technical specifications provided by the customer. The PD measurement is performed at the same time as the induced AC overvoltage test.

15.10.1 Conventional IEC 60270 Charge-Based Measurement

The PD measurement performed during the factory acceptance test provides the apparent charge in pC, which is compared with a reference value to establish whether the transformer satisfies the contractual requirements. IEC 60270 establishes the test method and calibration process (Chapter 6). Most often the high-voltage bushing capacitance tap is used to couple the PD signal (Figure 15.27). The entire measuring circuit must be calibrated by injecting a known charge between the bushing HV terminal and its grounded flange while measuring the detected PD signal (Section 6.6). The injected charge must be proportionate to the value that represents the acceptance limit: for measurements on transformers this value is usually 500 or 1000 pC and depends on the characteristics of the calibrator itself. The calibration process can also be used to estimate cross-coupling between windings (for example the LV and HV windings in the same phase). This is determined by injecting charge into the terminal of one bushing and recording the response on all other bushings. The information about the inter-winding transfer characteristics can be useful for an initial localization of any source of PD. Table 15.4 shows the effect of the injection of 1000 pC at the six different bushings of a three-phase transformer. For example, 1000 pC injected into the primary winding of Phase U is detected as 250 pC (Phase V, primary winding) and 100 pC on the secondary winding in Phase w.

Table 15.4 Example of the cross-coupling between phases and between HV and LV windings of a 250 MVA, 400/155 kV autotransformer.

Primary side			Secondary side		
Phase U Channel 01	**Phase V Channel 02**	**Phase W Channel 03**	**Phase u Channel 04**	**Phase v Channel 05**	**Phase w Channel 06**
1000 pC	250 pC	100 pC	350 pC	100 pC	100 pC
250 pC	1000 pC	250 pC	150 pC	350 pC	150 pC
250 pC	100 pC	1000 pC	100 pC	100 pC	350 pC
350 pC	150 pC	150 pC	1000 pC	350 pC	150 pC
150 pC	350 pC	1500 pC	250 pC	1000 pC	250 pC
150 pC	150 pC	350 pC	150 pC	350 pC	1000 pC

The values shown in the Table 15.4 are representative for one type of transformer and cannot be generalized for all types of transformers. The cross-coupling depends on the values of the stray capacitances between turns, between winding sections, between windings and to ground. Besides, being a frequency-dependent characteristic, the transfer coefficient between terminals is highly influenced by the center frequency and the bandwidth of the filter used in compliance with the IEC 60270 standard. A wide band approach, according to IEC 60270, standard is normally preferred if the noise and interference levels are low and tolerable.

The attenuation observed in the cross-coupling between terminals can be high. PD pulses generated for example at terminal U could be confused with a 100 pC background noise at terminals W, v, or w. This is yet another reason to reduce interference as much as possible.

15.10.2 Test Procedure

The procedure for carrying out factory PD measurements is specified in standards such as IEC 60076-3 and IEEE C57.113 for liquid-filled transformers and IEC 60076-11 or IEEE C57.124 for dry-type transformers. These procedures are also often summarized in specific documents that the individual transformer OEMs incorporate into their quality assurance system. The operating procedures specify all the precautions that must be adopted in setting up the measurement circuit to minimize interference, the procedures that must be followed for calibration, the method of applying the test voltage (often specified in the IEC or IEEE standard), and the method of data collection and reporting.

Before proceeding with the test, it is necessary to evaluate the intensity of the background interference (Section 15.9). IEC 60076-3 suggests the interference must be less than 50 pC.

As discussed in Section 15.8, because of the high-voltage rating of large power transformers, the voltage on the top of the HV winding is often induced by exciting the low-voltage winding with a 3-phase supply. Usually, higher AC frequencies than 50/60 Hz are used since this prevents saturation of the laminated steel core at the test voltages that are higher than rated voltage (Section 15.8). The induced voltage test must have a duration of about an hour or so, since the statistical delay time for PD inception (Section 3.6) can be significant in liquid-filled transformers. Standards such as IEC 60076-3 set test voltage levels and durations for the various levels of rated voltage of the transformer under the test. This standard also suggests the induced overvoltage test at increased frequency follow the sequence:

- voltage value that must be applied at the beginning of the sequence
- rate of rise of the voltage to reach the final test voltage
- time during which a certain voltage value must be maintained
- voltage values at which the PD are to be measured.

The amount of time at increased voltage depends on the frequency of the test voltage. If the frequency is at least double the rated frequency, the time is reduced, proportionally, according to a given formula, but never below the minimum duration established in the standard. Figure 15.42 shows a typical test voltage application sequence as suggested in IEC 60076-3. The diagram shows the voltage values, and the relative duration at each voltage step, together with the voltages at which the PD must be measured.

15.10.3 Factory Test Pass/Fail Criteria

The PD measurement acceptance criterion is mainly based on verifying the total absence of PD at operating voltage. The test outcome is the Q_{IEC} at specific voltages (see Section 8.8.1, IEC 60076-3 and IEC 60076-11), as well as the *PDIV* and the *PDEV* (Section 10.2). The acceptance criteria can also be agreed between suppliers and customers.

If the *PDEV* is only slightly higher than the requirement, it is important to observe the evolution of PD over the duration of the test. It may be that at a given test voltage value, the measured PD magnitude (Q_{IEC}) increases, or the number of PD pulses per unit of time increases, or both. In this case, the test duration (Figure 15.42) should be increased. In the event of an uncertain result, the search for the sources and locations of the PD using acoustic or UHF methods should be pursued.

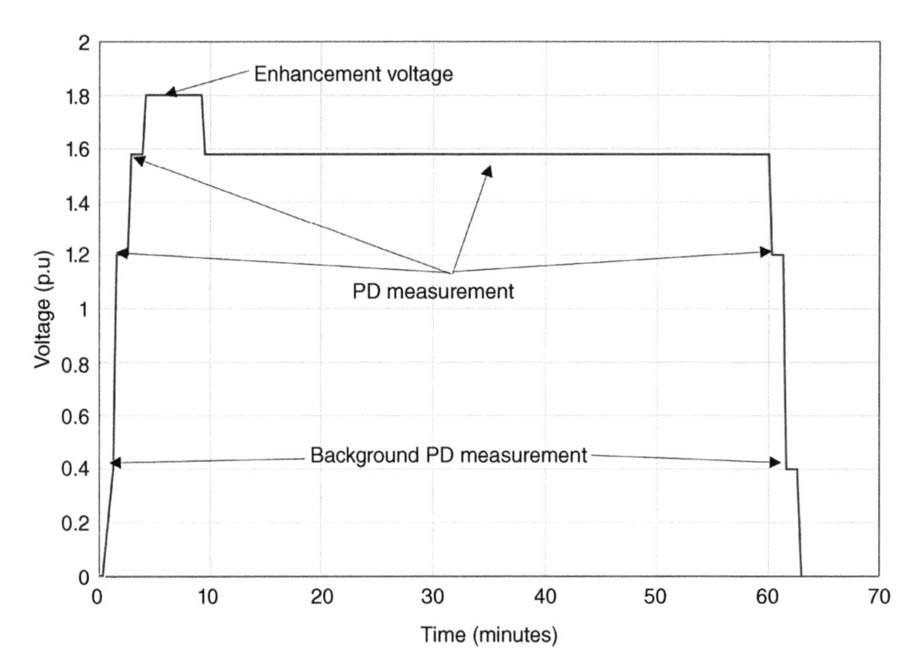

Figure 15.42 Typical sequence of rise and drop voltage where the vertical scale is based on rated voltage.

If the permissible PD levels have been exceeded at the test voltage, some of the signals may be due to interference (Section 15.9 and Chapter 9), rather than test object PD. To confirm this:

- Check if the PD is synchronized with the AC test voltage (i.e. analyze the PRPD pattern – Section 8.7.3) using a PD instrument or oscilloscope.
- Check if the signal comes from the power source. In this case, the use of a filter inserted between the power supply and the transformer under test can help (Section 9.3.3). This type of interference is frequent when using a variable-frequency power source that incorporates an electronic frequency converter. Such convertors produce interferences caused by electronic switching.
- Check if the PD comes from outside the transformer. It is very important to take care of the layout of the circuit to minimize ground loops. Also, metal objects should be grounded and/or far from the power supply and EUT, since floating objects may create their own PD (Section 9.3.2).
- Check the factory lighting system, especially if it is made up of gas discharge lamps, as they can create interferences that may be confused with power transformer PD. The lights should be turned off.

When performing measurements in environments not intended for high-voltage testing, unwanted signals are very frequent. The most common source is the presence of ungrounded metallic objects nearby which can lead to floating-potential discharges. Often these are sources that are difficult to identify as being external (since they be synchronous with the test voltage) except through repeated testing where the experience of those who perform the measurements plays a very important role. In these cases, tools capable of locating the sources can be of use such as directional ultrasonic microphones and acoustic imaging devices, UV cameras, and RF probes (Sections 5.3–5.5). As discussed in Sections 15.7.4 and 15.7.5, one can also verify that the PD is occurring within the transformer using investigation methods that are based on ultrasonic sensors or ultrahigh frequency (UHF) sensors. If it has been established that the PD is inside the transformer, even more investigation is needed.

It may also be helpful to measure PD with one phase at a time being energized. This will exclude PD originating between the coils in different phases. A particular example where this is useful is when PD is occurring in the HV bushing. In this case, we can still speak of phenomena internal to the transformer, as the bushings are an integral part of the transformer itself. Confirmation that the PD is within the bushing can be obtained through (a) a polarity analysis or (b) a calibration procedure, which involves injecting the calibration signal at the ends of the coupling capacitor C1 (point A in Figure 15.22a, b) as well as between the high-voltage terminal and ground. Procedure (b) is normally applied when using narrowband detectors (Section 6.5.2) that do not provide polarity information. The calibration signal applied between the high-voltage terminal of a given bushing and ground can be read on the remaining bushings attenuated by a constant that depends on the electrical parameters of the transformer, in particular the stray capacitance between phases (Section 15.10.1). If the calibration signal is applied to the ends of C1 (between point A and point B in Figure 15.22a) of a given bushing, the signal is detected on the remaining bushings with a much higher attenuation coefficient. Repeat the measurements by identifying the PDIV and PDEV (Section 10.2), the PRPD patterns at different test voltages and the varying PD intensity and repetition frequency.

If PD is detected during a factory test:

- The PD may be due to the imperfect execution of the drying process or too short an oil impregnation time. In this case, it may be advisable to repeat these manufacturing processes or simply wait some time to allow complete impregnation. The vacuum filling process, which ensures the absence of air inside the transformer, also plays a fundamental role to prevent PD.

- A brief burst of high PD can lead to discharges in the oil with consequent variation in the PDIV and PDEV. In this case, the original conditions can be restored by waiting a few hours. However, the phenomenon must be investigated.
- PD sources are difficult to identify even by removing the transformer from the tank and looking for tracking marks, etc., unless the PD amplitude is very high and occurring for long enough periods of time to cause damage.

If the transformer fails the PD test due to PD levels beyond the maximum limit allowed by the contract, IEC 60076-3, etc., then the acceptance/rejection of the transformer is subject to agreement between the parties in question.

15.11 Onsite Offline Testing

Offline/onsite PD tests are time consuming and costly (in terms of equipment and specialized operators) and are performed only rarely. They require a specific approach as the test conditions are very different from those in the factory.

15.11.1 When Onsite Tests Would Be Helpful

The measurement of PD performed on site is suggested after events that could have damaged the active parts of the transformer with particular reference to solid insulation. These events can be of multiple origins and types; they can be due to the transport and the assembly operations of the transformer but, above all, during the operational life of the transformer, they may be needed in the case of extreme transient events such as short circuits in the grid with high through-fault currents involving the transformer.

Sometimes the liquid is drained from the transformer after factory testing due to shipping considerations. During transport from the manufacturer to the destination plant, a period in which the windings are in the tank, but the oil has been replaced by an equivalent volume of dry air or nitrogen. All accessories such as bushings, heaters, and the oil conservator are not mounted to minimize the clearance for transportation. The sealing of the tank must use flanges with gaskets specially prepared for travel. A cylinder connected to the tank is needed to ensure the pressure inside the tank is slightly higher (0.2 bar) than atmospheric pressure. The integrity of the insulation is dependent on the presence of a dry gas inside the tank, which protects it from contact with the atmosphere, preventing the organic insulation from absorbing moisture that could cause PD. During transport, if the gas leaks out of the tank, the solid insulation will be exposed to the atmosphere, possibly for months. If, upon arrival at the substation, the pressure of the gas within the tank does not exceed the atmospheric pressure and the cylinder is empty, it is advisable to perform another PD test.

During transportation, the transformer may be subjected to shocks or vibrations beyond a tolerable limit. Also, the windings may suffer insulation damage or mechanical damage causing PD. This becomes even more critical if the transformer is shipped by water (oceans, lakes or rivers), where the transformer is not supervised by specialized personnel for a long duration. During such transport, the dry gas supply may be exhausted. Furthermore, the mechanical stresses on the windings may be high due to ocean waves. The effects of the time spent in the ports of departure and destination are also not negligible, where the transformer could be exposed to the atmosphere, extreme temperature variation, and not adequately supervised.

Another very delicate activity is onsite assembly and installation. The attention and care that is dedicated to the final set-up of the transformer manufacture is of fundamental importance, especially for the assembly of accessories, bushings and the final filling with oil. It is essential to perform an onsite/offline PD test to ensure the quality and reliability of the transformer.

An onsite PD test is recommended if the transformer is temporarily out of service due to the operation of the Buchholz (gas overpressure) relay, and the DGA (or the analysis of the gases collected in the Buchholz relay) points at a dielectric fault. In this case, the PD test takes on significance since the extent of the phenomenon that caused the Buchholz relay tripping is unknown. An onsite-induced overvoltage test allows the gradual increase in test voltage, minimizing the risk of incurring a destructive fault. Besides confirming what was highlighted by the gas analysis, PD measurement can help to identify the type of fault and its location (Section 15.11.5).

Once the transformer is in operation, onsite PD tests can be carried out periodically based on the analysis of DGA results and operational conditions. Using DGA (Section 15.7.6), the presence of PD is inferred based on the presence of gases such as hydrogen and acetylene in the oil at concentrations higher than normal in healthy transformers. Thus, the DGA results might trigger further analysis involving PD testing onsite. However, considering the large volume of oil in high-power transformers, the effect of abnormal operation may not be detectable in a short time through the analysis of the gases dissolved in oil. Thus, onsite PD tests can be requested if the transformer has been subjected for a reasonably long time to abnormal operating conditions such as, for example, overload conditions or voltages beyond the maximum permitted level.

Onsite PD tests are often carried out after transformer maintenance operations. There are cases in which bushings or on-load tap changers are replaced, insulating parts are restored, or maintenance operations are carried out on high-voltage connections. In all these cases, PD measurements can be useful to verify that the replaced/restored parts are in good conditions, that human errors did not occur during the assembly, or that metallic tools were not forgotten inside the tank.

In all these cases, having made the necessary cost/benefit assessments, the measurement of PD, simultaneously with the induced overvoltage test, can give operators some assurance on the reliability of the transformer.

15.11.2 Scope and Aim of an Onsite/Offline Test

When the PD measurement is performed following the operation of the Buchholz relay, the test is to verify if PD is the root cause of the high gas amount, and/or to identify the type and location of the PD. It is very difficult to predict how long it will take the PD to cause insulation breakdown. It depends on the location of the PD, and the associated energy of the discharges (which cannot be evaluated based only on the apparent charge measured at the bushing – see Sections 2.4 and 8.8). If the source of PD is in the vicinity of the winding insulation, near the connections between the windings and bushings, or, in general, where the electric field is high, it is worthwhile to start evaluating remedial actions. However, there are examples in which the discharges occur where there are no insulating parts, for example:

- discharges within or on the surface of the transformer core
- metal parts that have lost their connection to ground or high-voltage components, creating contact sparking (Section 4.4).

In these cases, the seriousness of the discharges must be carefully evaluated by identifying its location. Ultimately, the objective of the PD test is to establish whether the transformer is suitable to be put into service, even for a defined time and with deratings or precautions.

The approach is different when the measurement of PD is performed following a maintenance outage where pass/fail criteria are clearly established in a contract with the service provider. Under these circumstances, the purpose is to verify the absence of PD or, if PD activity (such as Q_{IEC}) does not exceed the levels specified in the contract or, by default to IEC 60076-3. Major maintenance operations can be of various kinds. They range from almost complete overhaul with replacement of the windings and reconstruction of the core to the simple replacement of the bushings or the on-load tap changer. In the first case, it would be advisable to completely repeat the factory insulation tests including the lightning impulse and switching impulse tests, especially if the voltage level of the transformer is higher than 245 kV. However, it is difficult and expensive to carry out these tests on site. PD tests with either applied or induced voltage at increased frequency is easier to carry out and can still provide important information regarding the quality of the maintenance work.

PD tests are also very useful in the case of less intrusive maintenance work such as the replacement of bushings and on-load tap-changers, especially if the latter also involves the replacement of the socket selector. In this case, in addition to ascertaining if the transformer is fit to return to service, the objective is to confirm the successful outcome of the maintenance and compliance with the work contract terms.

15.11.3 HV Supply Systems

For onsite/offline PD tests, the biggest problem is to optimize weights, dimensions, and rated power for the power supply system capable of an increased AC frequency and adjustable voltage. Section 15.8 dealt with some increased frequency systems based on motor-generators and static frequency converters that are used in factory tests. The question is which method is most suitable for a mobile test set.

The motor-synchronous generator method (Figure 15.35) is characterized by a non-variable AC frequency, considerable weight, but an excellent robustness both from the mechanical and electrical point of view. Such supply systems can withstand short duration overcurrent caused by dielectric failures of the transformer during the test.

The static converter method (Figure 15.37) has the possibility of adjusting the frequency over a wide range (reducing real power demand from the power source due to resonance) and is lighter. However, such systems are more delicate from a mechanical point of view, as the connections tend to wear out over time. Also, the electronics are not well suited to endure overcurrent transients due to winding failure or OLTC tap transitions.

In both methods, the voltage output must be adapted to the winding voltages of the transformers under test. In the case of motor-generator converters, the adjustable voltage can range a few kilovolts (typically 2–3 kV). Static converters voltage is only adjustable for a few hundred volts (typically 400–500 V); therefore, a step-up transformer (Figure 15.37) with multiple voltage taps is required. This step-up transformer must be suitable for operation at higher frequencies up to about 400 Hz and with adequate power.

It must be determined if a three-phase or a single-phase AC supply will be used (normally IEC 60076-3 requires three phase). Generator methods (Figure 15.35) are normally in a three-phase configuration. While both three-phase and single-phase static converters are available, the solutions differ in terms of weight and volume because of the different structure of the step-up transformer. Currently, IEC standard 60076-3 only permits a three-phase power supply for the induced overvoltage test. In contrast, IEEE C37.113 permits both three-phase and single-phase test methods.

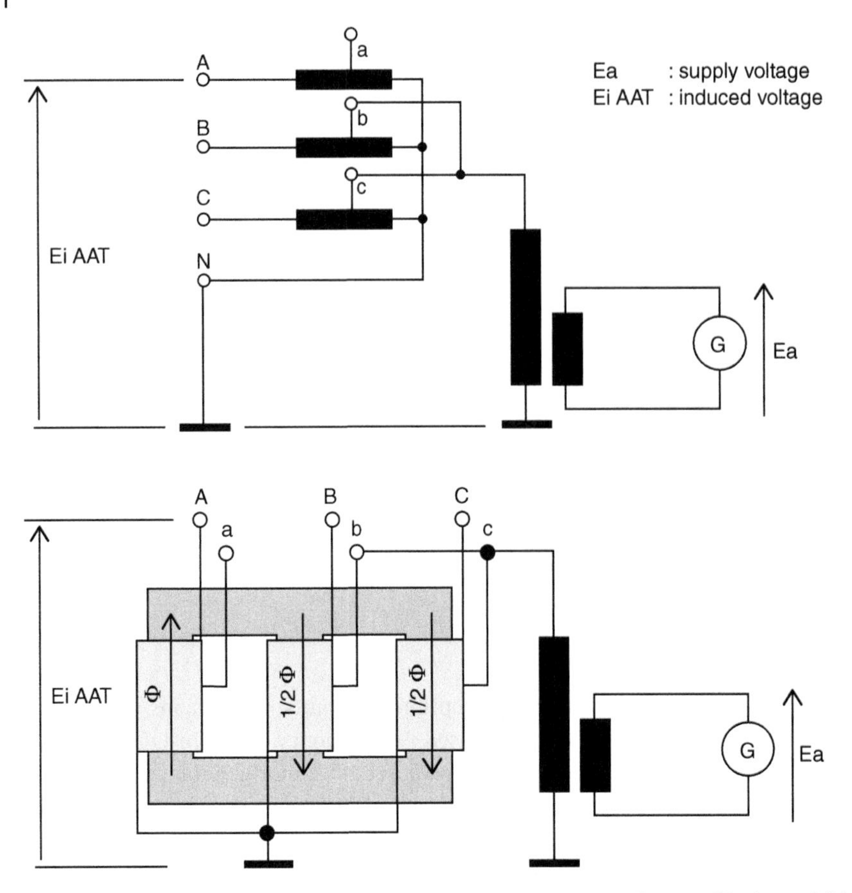

Figure 15.43 Single-phase supply for a three-leg core autotransformer. *G* is the variable frequency/variable voltage supply in Figure 15.37.

Having the possibility of adjusting the test frequency has the advantage of optimizing the current required by the power supply, by being able to balance the mutual contributions of inductive reactive power, due to the magnetization of the core, and the capacitive reactive power due to the capacitances typical of the transformer winding. It is, therefore, advisable for the static frequency converter to be equipped with fine frequency adjustment, typically 10 mHz, to easily find the operating frequency that draws the least current. To minimize the background noise, a fully static converter (Figure 15.37) must be equipped with appropriate output filters.

To test autotransformers, a single-phase power supply can be used by exploiting the structure of the autotransformer. This can reduce the cost of the test supply as well as its weight and dimensions. Autotransformers are characterized by two windings in series in each phase. Normally, the lowest voltage winding is higher than 100 kV and the neutral point is connected to ground. Figure 15.43 shows how to power a three-phase autotransformer using a single-phase step-up transformer. The phase under test is A, which is brought to the test voltage by the flux generated on phases B and C, which are in parallel and energized with a voltage equal to half the voltage required for the test. The fluxes generated in the columns of phases B and C sum in the column of phase A inducing the required test voltage.

The above scheme is applicable only on autotransformers with a three-leg core. Besides reducing the ratings of the supply system (only 50% of the rated voltage of the secondary is necessary

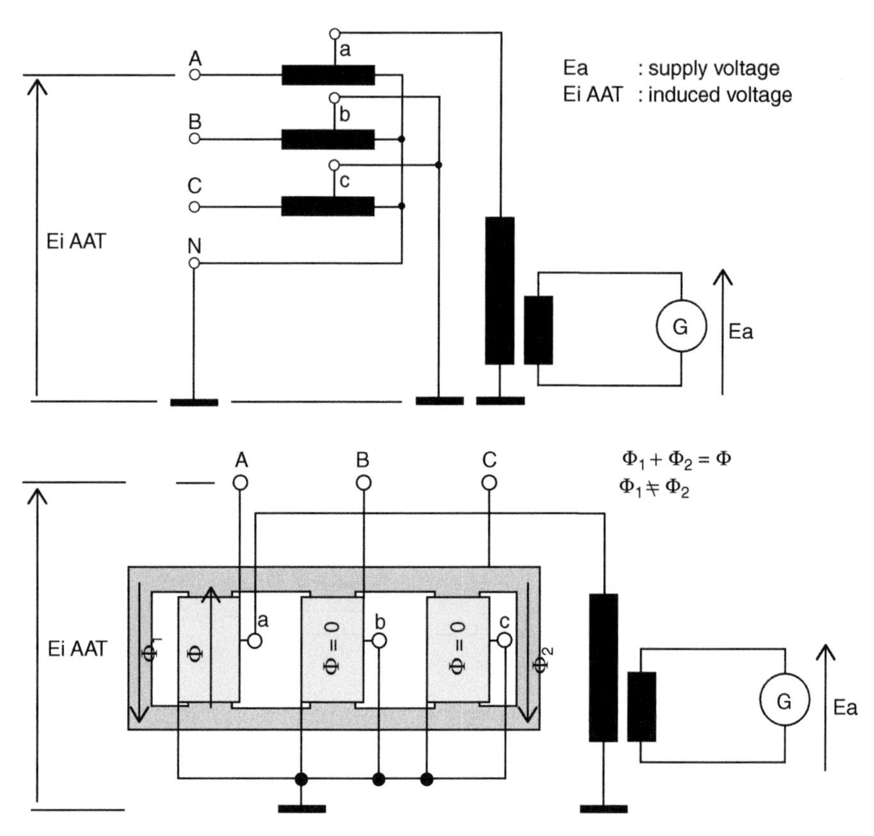

Figure 15.44 Single-phase supply for a five-leg core autotransformer.

for testing), the scheme has the advantage of limiting the voltage on the phases not under test. Thus, it is likely that PD will not occur in the phases B and C. This helps to locate in which phase PD is occurring.

In the case of five-leg autotransformers, the applicable scheme becomes the one represented in Figure 15.44. The phases not under test are connected to ground and, therefore, the voltage across them is zero. This method is possible since the flux generated on the phase under test finds its return path in the side legs. The scheme has the disadvantage of having a supply voltage equal to the test voltage, but it has the great advantage that the phases not under test are not subjected to dielectric stresses and, therefore, not subject to PD.

For power transformers with a delta-connected medium-voltage winding (typically, a generator step-up transformer), schemes that connect the power supply to the medium voltage winding can be used. If the variable frequency power supply is of the three-phase type, there are no precautions required. If a single-phase power source must be used, the scheme reported in Figure 15.45 can be applied to three-leg and five-leg transformers and has the same advantages as the scheme of Figure 15.43.

In transformers with five or three core legs with the delta connections within the tank, the power supply circuit shown in Figure 15.46 can be applied to the connections on the medium voltage bushings. This type of transformer offers the opportunity to short-circuit the winding of the phases not under test, obtaining the advantage of testing one phase at a time, as already found in the power supply diagram of Figure 15.44.

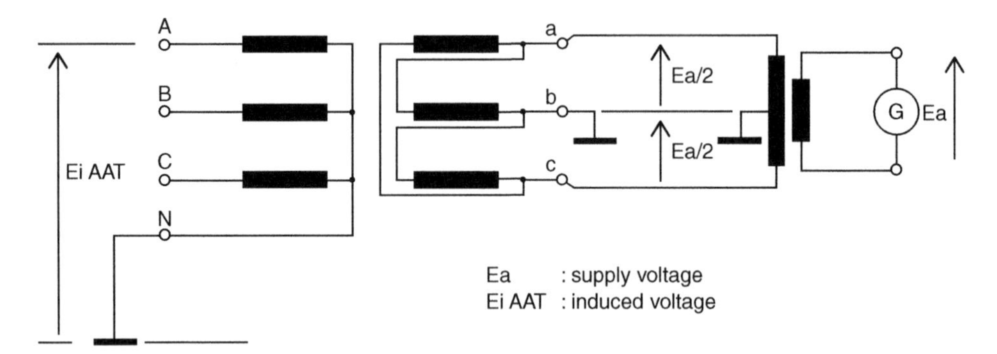

Figure 15.45 Single phase supply for a delta-connected generator step-up transformer.

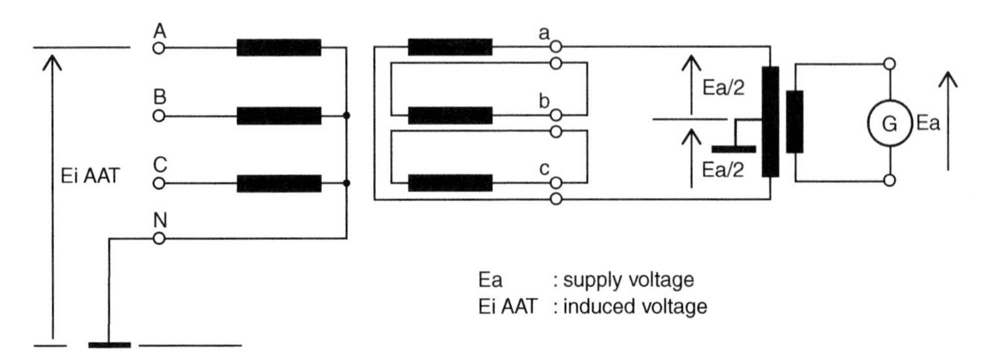

Figure 15.46 Single-phase supply for a step-up five-leg delta-connected transformer.

Single-phase power supply schemes have the advantage of obtaining the test voltage with relatively low supply voltages (about half). This aspect helps in discriminating the phase of any PD since only one phase at a time is fully energized, with the remaining phases subjected to half the test voltage.

15.11.4 Typical Onsite Test Procedures

The test procedure for onsite/offline PD measurement depends on the reason for such testing. When the test is performed to verify compliance with a contractual agreement, such as at the end of major maintenance, the purpose is to ascertain the absence of PD and/or verify that the measured Q_{IEC} (Section 8.8.1) is below the maximum allowed by the standard or contract. In this case, the test procedure given by the referenced standards is applied with any exceptions or limitations stated in the contract.

If the onsite/offline test is triggered by a Buchholz relay trip or a DGA result, then the focus of the testing is to investigate the root cause. In this case, it is important to measure the PDIV and the PDEV. In addition, it may be useful to apply the test voltages for longer durations to see how PD activity changes over time. If PD is confirmed, then acoustic methods (Section 15.11.5) and/or UHF methods (if UHF sensors have been previously installed for online monitoring – Section 15.12) can be used to locate the PD sites within the transformer.

If a Buchholz relay trip occurred, it is possible to establish if it was tripped because of a slow or fast developing issue. If the accumulation of gas has not been sudden, a medium-/low-energy discharge is likely occurring, and the PD measurement is useful to identify its presence and possibly its location inside the transformer. If there is a rapidly increasing amount of gas in the Buchholz relay, the PD measurement can indicate if it is safe to re-energize the transformer.

Following a Buchholz relay trip, if the DGA suggests that PD is active in the transformer, the offline/onsite PD test helps to evaluate whether the PDIV and/or PDEV is below operating voltage. If the PDIV and PDEV is significantly below operating voltage, then sufficient gas may accumulate at the test voltage to trip the Buchholz relay during the PD test. In this case, a PD measurement according to IEC 60270 wide-band mode is preferred. If UHF PD sensors (Sections 15.7.4 and 15.12) have been previously installed within the tank, or an UHF sensor can be inserted via the oil drain valve, then these sensors can be used to confirm the DGA results that PD may be occurring. If multiple internal UHF sensors are present, then time-of-flight and triangulation methods (Section 8.4.2) by an experienced technician may give an initial assessment of the PD locations. However, the correlation between the apparent PD charge and the quantity of gas produced is extremely weak.

Onsite measurement of PD is also useful when the DGA suggest the presence of PD, even if the Buchholz relay did not trip. In this case, it is probable that PD is already present at operating voltage, but not of sufficient intensity to cause gas bubbling due to oil saturation (and therefore the operation of the Buchholz relay). As in the previous case, the procedure must focus on the time spent at each test voltage level established for the test and on determining the PD inception and extinction voltages.

15.11.5 Acoustic Investigations

If PD has been detected in a transformer below the operating voltage, it is useful to identify the location of the PD sites inside the transformer. Depending on the location, it may be possible to return the transformer to service for a period with a low risk of failure despite the PD. Acoustic methods for PD site location within transformers has been investigated since the 1980s (Section 5.4.2 and [46]). As mentioned in Sections 15.6.3 and 15.7.4, a single discharge generates an acoustic wave that propagates within the oil in the transformer tank, reaching the walls, where they can be coupled by acoustic emission (AE) sensors sensitive to ultrasonic frequencies. The propagation speed of the acoustic signals in the oil is shown in Figure 15.47. By placing multiple sensors in

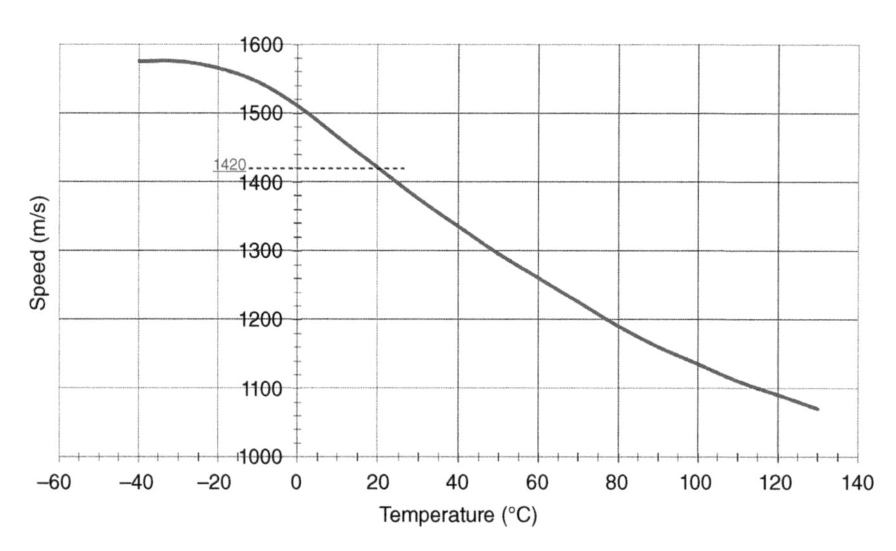

Figure 15.47 Speed of ultrasonic waves in mineral oil. *Source:* Graph courtesy Terna.

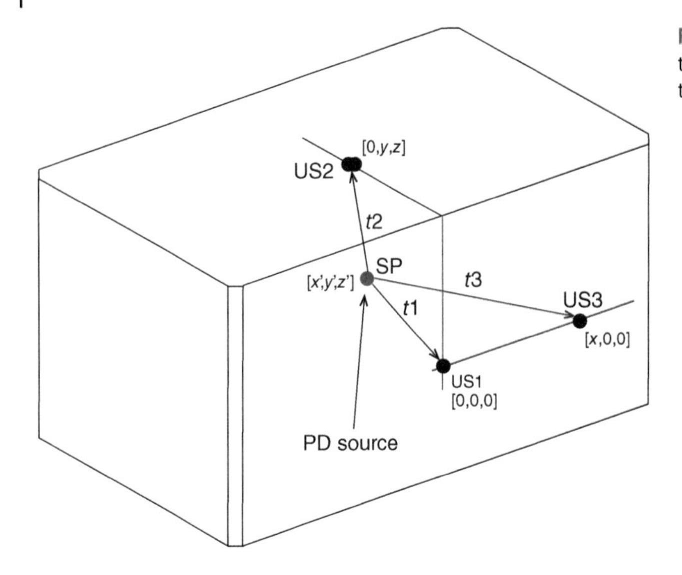

Figure 15.48 Typical arrangement of three acoustic sensors on a transformer tank.

different positions on the outside of the tank, at known distances, it is possible to identify with reasonable confidence the point where PD originates [30, 33, 34].

Figure 15.48 shows a rough sketch of the positioning of three ultrasonic sensors on the external walls of an idealized transformer tank. The coordinates relating to the position of each individual sensor must be such as to optimize and simplify the final calculation to establish the coordinates of the PD source. The times *t*1, *t*2, and *t*3 represent the times taken by the acoustic signal generated by a single PD pulse to reach each ultrasound sensor. Once the propagation speed of the acoustic wave in the oil is known, it is possible to establish the distance of each sensor from the source, assuming a direct propagation path. This distance represents the radius of a sphere on whose surface lies the source of PD, which will have the coordinates of the intersection point of the three spheres of radius *r*1, *r*2, and *r*3. Multichannel oscilloscopes are used to measure the relative time of arrival of the three signals. The oscilloscope is usually triggered by an electrical PD signal, either from the bushing tap or a UHF sensor within the tank. Since the travel time of the electrical pulse to the bushing tap is essentially zero compared to the propagation of the acoustic waves, the electrical signal can be taken as the time of the discharge. Figure 15.49 shows an acquisition of acoustic signals coming from three sensors and correlated with the PD signal acquired by bushing tap.

In this type of measurement, however, we must consider some difficulties that arise in practice [33, 47]. In Section 15.7.5, it was mentioned that the propagation of the acoustic wave inside the transformer is affected by the different types of materials and barriers encountered. Propagation through these different materials causes attenuation, reflections, refractions, and eventual propagation along the steel walls of the tank. These phenomena hinder the evaluation of the distance from the PD source calculated using speed and delay time with respect to the electrical origin of the signal (assuming a simple spherical wave).

The repetition rate of the PD signals may also cause the resulting acoustic signals to overlap each other, making it difficult, if not impossible, to reliably associate an acoustic signal with a given PD signal. In these cases, it is recommended to start the location assessment at a low test voltage where there is no PD, and then gradually rise the test voltage until only a few discharges occur. Each time an electrical pulse triggers the oscilloscope, the three acoustic responses should be stored in the oscilloscope memory. The desired results are not always obtained at the first attempt but, by repeating the procedure several times, a satisfactory result can often be achieved despite the countless variables that affect the measurement. The positioning of the sensors may require several attempts to maximize the intensity of the signal received and the delay times between the electrical and the acoustic signals. The oscillogram in Figure 15.49 shows a situation in which two PD pulses are sufficiently separated in time to avoid the overlapping of acoustic signals that are clearly identifiable; this situation is obtained if test voltage is just above the PDIV.

There are commercially available instruments capable of acquiring the acoustic signals of interest, processing them, and returning the coordinates of the identified source directly. They are equipped with graphical software interfaces reproducing the transformer under test highlighting the position of the acoustic sensors and the estimated position of the source. An example from the Omicron PDL 650 is shown in Figure 15.50.

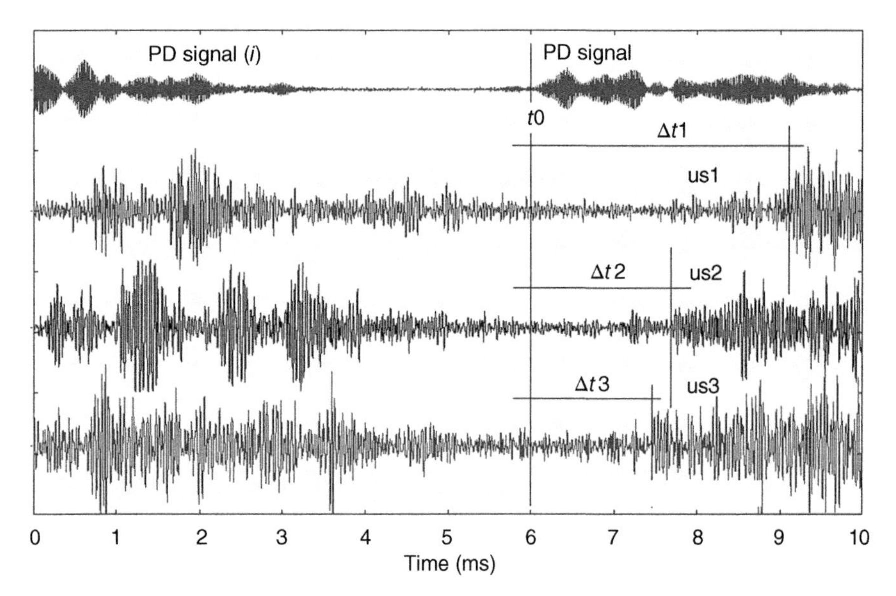

Figure 15.49 Example of an acquisition of acoustic signals from three sensors (bottom three traces) and an electrical PD signal (top trace).

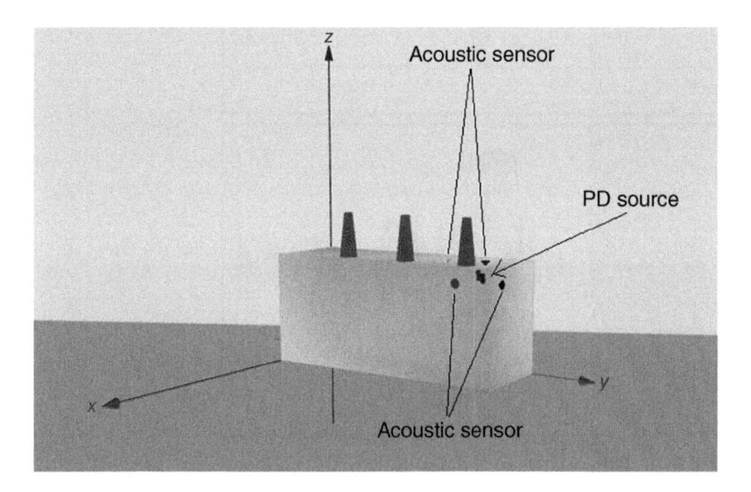

Figure 15.50 Example of a post processing analysis based on acoustic signals (40 MVA, 132/16.8 kV transformer).

The acoustic signals picked up by the sensors on a commercial instrument are shown in Figure 15.51. The software can establish the coordinates of the source based exclusively on the acoustic signals; however, the trigger can also come from a PD coupling capacitor, the bushing tap, or any other suitable sensor as HFCT and VHF/UHF antennas. This method of locating PD sites

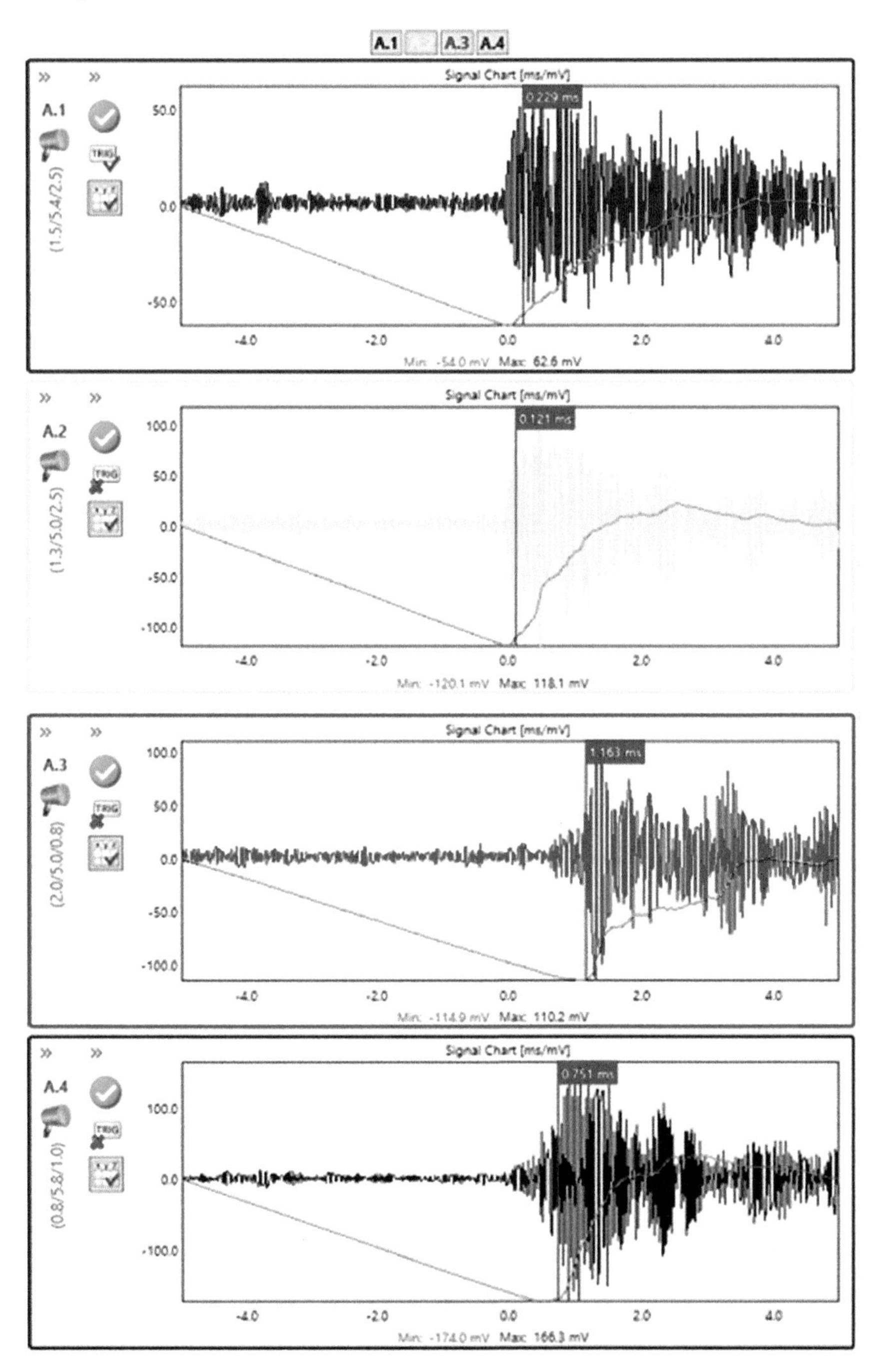

Figure 15.51 Typical output from Omicron PDL 650 to help locate PD sites based on the signals from four acoustic sensors (40 MVA, 132/16.8 kV transformer).

does require considerable experience. If there are multiple PD sites, the analysis may become very complex, and the result remain ambiguous.

In conclusion, acoustic methods to locate PD sites is of fundamental importance to planning if the transformer can be returned to service or if immediate repairs are needed. The final decision depends on a plurality of data from the PDIV/PDEV, PD magnitude and repetition rate as well as the estimated location of the PD site(s) and the type, concentration, and growth trend of the gases dissolved in oil. The risk of transformer failure increases if the PD site is near the highest voltage parts of the winding, the HV leads to the bushing and/or in the spaces between turns and between the windings. If this finding is accompanied by an analysis of the gases dissolved in oil indicating that carbon dioxide and carbon monoxide are present, this further increases the risk of transformer failure. In contrast, if the PD site(s) are not close to cellulose materials, and the DGA does not suggest the involvement of cellulosic materials, although one should still be cautious, the risk of imminent failure is lower.

In summary, the result of an acoustic survey is affected by many variables, most of which are difficult to manage. However, it is an essential tool for assessing the risk of failure.

15.12 Online PD Monitoring

Continuous online PD monitoring systems are now available from several vendors, including BSS, Camlin Power, DMS-Qualitrol, Omicron, Power Diagnostix, and TechImp/ISA-ALTANOVA. Compared with offline measurements, continuous online PD monitoring (PDM) in a transformer enables the earliest possible detection of some (but not all) aging processes that can lead to transformer failure. For aging mechanisms that PD can detect, the PDM will usually identify insulation system issues well before an online DGA system can detect them.[2] However, online PD monitoring systems are relatively expensive, and can be prone to false indications due to electrical interference. Using such systems can be challenging, since the output needs to be kept under continuous observation by personnel capable of correctly interpreting the results. The analysis of alarms from a PDM system usually needs to be undertaken by PD specialists.

Research into online PD monitoring started in earnest in the 1990s [48]. Although online PD tests can be done periodically using portable instrumentation, most commercial systems are intended to be continuous monitors of the sensors that have been permanently installed. Continuous online monitors are preferred because PD in transformers may be erratic over time, and the time between the detection of PD and transformer failure may be only weeks or months, since purely organic insulation is used (which may quickly degrade in the presence of PD). PDM systems may use bushing tap sensors, internal UHF sensors, or both. Continuous online monitoring of PD activity is currently possible due to the development and marketing of various types of instruments suitable for permanent installation in the field, either directly mounted on the transformer or in a position very close to it. Such systems must be reliable, suitable for outdoor use, and able to withstand continuous and transient mechanical vibrations, especially if installed directly on the transformer tank.

2 In fact the results from a PDM system and an online continuous DGA monitor are complementary and together help users to identify root causes with much more confidence. This is why many PDM system vendors also supply DGA monitors.

15.12.1 When Online PD Monitoring Would Be Helpful

Whether or not to install a PDM system mainly depends on the criticality of the transformer in question, i.e.:

- the impact of its failure on the power system and the public image of the utility
- is there a back-up transformer ready to replace the failed unit?
- if yes, how long does it take to replace the failed unit?

Special transformers, such as phase shifter transformers, or other types for which there are no spare units, may take years to replace. In these cases, it is important to detect PD activity and intervene well before failure. There are also particular network situations that can cause heavy and frequent stresses on the transformer (for example locations that see many high-current or high-voltage transients). In these cases, a continuous PD monitoring system can be very useful to determine if the insulation has been damaged by the recurring transients.

It is important to recall that PD monitoring systems needs to be properly polled on a regular basis, and the data collected must be continuously processed and analyzed by specialist personnel to translate them into useful information to take appropriate actions. The data analysis is often based on a statistical approach, analyzing PRPD patterns, and the evaluation of trends. To date, it is difficult to perform such analysis automatically using software tools such as artificial intelligence (AI), although they may be of benefit to those who have very little experience with PRPD patterns. Although alarms based on trend and PD activity thresholds can be set, a definitive evaluation is always the result of the analysis by personnel capable of integrating information from multiple sources. Thus, the owner of PDM systems must provide adequate technical resources to use the information from online PDM systems. Otherwise, the PDM will be of limited use.

15.12.2 Interactions with Other Online Monitoring Systems

The information from an online PDM should always be compared to the results from other types of online monitoring systems that may be installed on a transformer. One of the most common online systems measures the moisture content in the oil. These are found in almost all liquid-filled transformers. If the moisture content is high, then there is a greater likelihood of PD occurring in high electric stress regions [49].

Online DGA monitoring systems can now be found in most critical high-voltage transformers. Different DGA systems are available, ranging from analyzers providing the concentration of a single gas (usually hydrogen), to systems able to measure three or five gases (useful for analysis based on Duval's triangle or pentagon [43]. Lately, online DGAs to measure nine gases have become available. However, from the simplest to the most complex, online DGAs can provide indications that PD is occurring since PD creates certain gasses.

In addition to online PD monitoring, online moisture content, and online DGA monitors, there is an online monitor specifically for transformer bushings equipped with a bushing tap. Bushings are a critical component of the transformer and often their insulation system is stressed by transient overvoltages due to lightning and switching surges, and thus aging is possible. Systems have been developed to measure the bushing dissipation factor (DF) and capacitance (C) online, using the bushing capacitance tap. Such systems may give warning of the aging and failure of transformer bushings by detecting changes in C and DF over time [50].

The most complete monitoring configuration for liquid-filled transformers involves:

- Moisture in oil measurement
- Evaluation of the presence/absence of PD using DGA. The advantage of a DGA is that it does not produce false indications due to electromagnetic interference
- Timely presence of PD and information about their source (PD monitoring system)
- Presence of PD in the bushings (PD monitoring system and/or changes in the capacitance or dissipation factor of the bushing insulation).

These different types of monitoring systems provide important complementary information. Ideally, all these systems are present, since this will allow an insulation system analysis with the highest possible confidence. Vendors often now combine two or more of these systems onto a single hardware and software platform, making analysis easier.

15.12.3 Features of an Online PD Monitoring System

Like any other continuous online monitoring device, a PDM system has some fundamental requirements. The monitoring devices are mounted directly to or near the transformer, which is normally outdoors and exposed to atmospheric conditions. Rain, humidity, or snow must not compromise the functioning of the system. The mechanical stresses also can represent a critical aspect when the PD instrumentation is in contact with a transformer, or even more critically, with a reactor. The temperature at which the PDM is required to operate can reach up to 70 °C when exposed to direct sunlight in a hot environment such as in the Middle-East. Near the north or south poles, the temperatures may be as low as −50 °C. In addition to the ambient temperature, the operating temperature of the transformer must be taken into consideration. It is, therefore, important that the monitoring device is housed in an enclosure that guarantees mechanical strength, a degree of protection against atmospheric conditions and excess temperature extremes in order to ensure higher reliability.

Particular attention must also be paid to electromagnetic compatibility. That is, the instrument enclosure and all the electronic devices within it need to be sufficiently insensitive to interference that is inevitably present in the power system (e.g. corona from overhead lines, magnetic fields due to transformer operation, high-frequency interference due to transients from faults and circuit breaker operation). Figure 15.52 shows an example of an integrated system for monitoring PD, gases dissolved in oil and bushing DF and C.

Figure 15.52 Box including multiple continuous online monitoring systems. *Source:* Courtesy of Camlin Power.

When using the bushing tap as a PD sensor, it is also extremely important to ensure that the tap connector is mechanically stable and waterproof, otherwise there is the risk that bushing

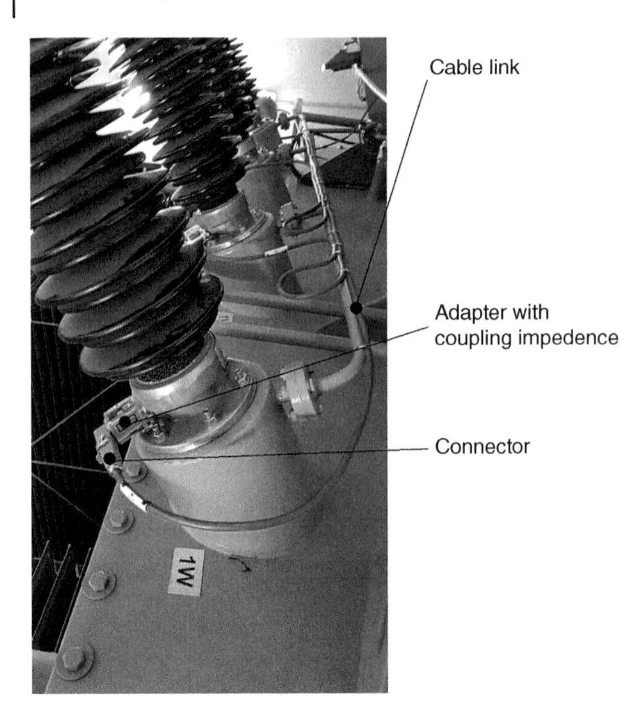

Figure 15.53 Typical arrangement of the connection of the coupling impedances (quadripoles) with the bushing tap. *Source:* Courtesy of Camlin Power.

insulation may be damaged by partial discharges incepted between the capacitive tap and the ground, or by moisture absorption. Figure 15.53 shows a typical connection to the capacitive socket of the bushing of a single adapter both for the measurement of PD and for monitoring the bushing status.

For PD signals derived from a bushing tap, it is important for the system to automatically short the terminal of the capacitive socket to ground when the detection impedance or quadripole fails (i.e. the impedance between the capacitor tap to ground becomes infinite). Figure 15.54 shows two examples of damage that can involve the adapter in the capacitive socket of the bushing and damage the bushing itself, as well as disabling proper functioning of the PDM. In Figure 15.54a, the

Figure 15.54 (a) Traces of oxidation caused by an imperfect seal, (b) Consequences of an electrical defect in the coupling impedance and arcing corrosion.

traces of oxidation caused by an imperfect seal of the adapter are evident while Figure 15.54b shows the consequences of an electrical defect in the coupling impedance, which was not followed by the automatic grounding of the capacitive socket.

15.12.4 Suppressing Interferences

As discussed in Section 9.1, electrical interference may lead to a high risk of false positive indications of transformer insulation problems. Compared to other equipment such as cables, GIS, and machines, the interference environment in large power transformers is probably the most severe for continuous online PD monitoring. There are several sources that can lead to very high interference activity:

- Corona from transmission lines, high-voltage busbars, and nearby post insulators in substations, which tends to be most active in rain, snow, and fog conditions.
- PD on the outside surface of the transformer bushings under rain, snow, and fog conditions. The amplitude of such discharging interference can be very high, being so close to the transformer and transformer bushings, yet it is essentially harmless to the safe operation of the substation.
- Radio station and cell phone interference coupled in via the transmission lines.
- Power line carrier, which is used by utilities to communicate between generating stations and substations.

The surface discharges on transformer bushings are particularly pernicious, since they are only a few cm away from the oil–paper insulation within the bushing and have many of the characteristics of PD within the transformer or bushing.

The main decision about installing a continuous PD monitoring on a transformer is the frequency range at which the system will work. Currently, there are three main commercial options:

- Systems working in the UHF range, which require sensors inside the transformer tank to be effective.
- Systems workings in the 1–30 MHz range (the HF range), which use the bushing taps on each phase to detect the PD signals.
- Systems that use both HF and UHF sensors.

UHF systems are inherently less susceptible to interference, largely because the transformer tank forms an effective Faraday shield, which blocks most signals external to the transformer from being detected (although some interference signals may still enter the transformer tank via the bushing apertures). With multiple sensors (four or more) within the transformer tank, it is also possible to confirm that the signals detected are transformer winding PD using time-of-flight methods (Sections 8.4.2 and 15.11.5). If there is only one UHF sensor within the tank, perhaps just a UHF antenna retrofitted via an oil drain valve, it should be noted that if PD is occurring on all three phases, the UHF sensor will detect a very complex PRPD pattern, since there may be six clusters of PD activity across the 50/60 Hz AC cycle.

Alternatively, one can use the bushing tap on each phase to detect the PD signals. Commercial systems offering this option work up to 20 MHz or so. They are much cheaper to retrofit; however, they are more likely to give false positive indications due to bushing surface PD and interference from the rest of the substation/plant. Interference suppression may be further enhanced using notch filters, the T–F map (Section 8.9.2), or 3-phase synchronous pattern analysis (Section 8.9.3), all of which require some skill to implement. Such methods may have to be continually tweaked since the interference sources often change over time. Any indications from a bushing tap system should be followed up with a DGA test or by locating the PD source (Section 15.11.5), to ensure it is within the transformer.

Some vendors have developed a hybrid system whereby both one or more UHF sensors within the tank, and the three bushing taps are used. The main sensors are from the bushing taps. However, digital gating is performed such that the bushing tap signals are only recorded if the UHF sensor(s) also sense a PD pulse. If the UHF sensors do not detect a signal, then the bushing tap sensors are probably responding to external interference, and the signals are ignored.

15.12.5 PDM Implementation

Many commercial PDM systems use UHF sensors (Section 15.7.4) due to their lower false indication rate. It is only possible to install most types of UHF sensors (which are similar to round metal plates – Figure 15.30 and Section 13.8.1) during transformer manufacture, or when major maintenance is being performed (e.g. removal of the windings from the tank). One type of UHF sensor that can be retrofitted without a major outage, can be installed via the oil fill/drain valve (see Figure 15.31). However, such oil valve sensors measure the PD signal in one location only, as opposed to many locations with pre-installed UHF sensor. If multiple UHF sensors are installed within the tank, comparisons of signals from different sensors increases confidence that the signal is from within the tank, and it also may enable PD sites to be located, using the time-of-flight approach used for acoustic location (Section 15.11.5).

To reduce the cost of electronics, the UHF signal from the sensors is not directly processed by UHF electronics. Instead, a demodulator detection method is used which effectively down-converts the UHF signal to a lower frequency (Figure 8.4) – often the LF or HF frequency ranges. This is also the approach taken for UHF PD detection in GIS (Section 13.8.6) and is in contrast with VHF/UHF systems in rotating machines where the electronics are capable of directly of processing VHF and UHF frequencies (Section 16.9.3). Unfortunately, the demodulation method loses the pulse polarity information, making PD interpretation more difficult. It also means the time-frequency map method (Section 8.9.2) cannot be used to distinguish between different sources of pulse signals, since the pulse characteristics are dominated by the demodulator.

The synchronous three-phase analysis method of signals from bushings tap can be implemented (Section 8.9.3) to further reduce noise and eliminate cross-coupled signals from other phases (Figure 15.55). This technique does not rely on the pulse wave shape and pulse polarity being preserved.

15.12.6 Basic Interpretation of PDM Results

General comments on interpretation are found in Chapter 10. For *offline* transformer PD tests in the factory, commissioning tests at site, or after major maintenance, the interpretation process is to make sure that the PDIV and PDEV are above operating voltage or some other voltage specified in a purchase contract. If PD is found below operating voltage, the PRPD patterns are examined (Section 15.13) and the PD sites are located (Section 15.11.5) to determine the root cause of the PD. Before the transformer is put into service, the PD sites are almost always repaired.

Interpretation of the PD from continuous online monitors is primarily about trending any signals over time. Normally there should not be any measurable PD from the system during operation. If any transformer winding PD is detected – it is potentially hazardous. Alerts (i.e. alarms) of the presence of PD need to be carefully investigated using other online monitors (if available), as described in Section 15.12.2, to determine if the alert can be correlated with signals from the DGA, etc. If not, the alert may be due to electrical interference. An expert also needs to review the PRPD patterns to determine if the patterns are

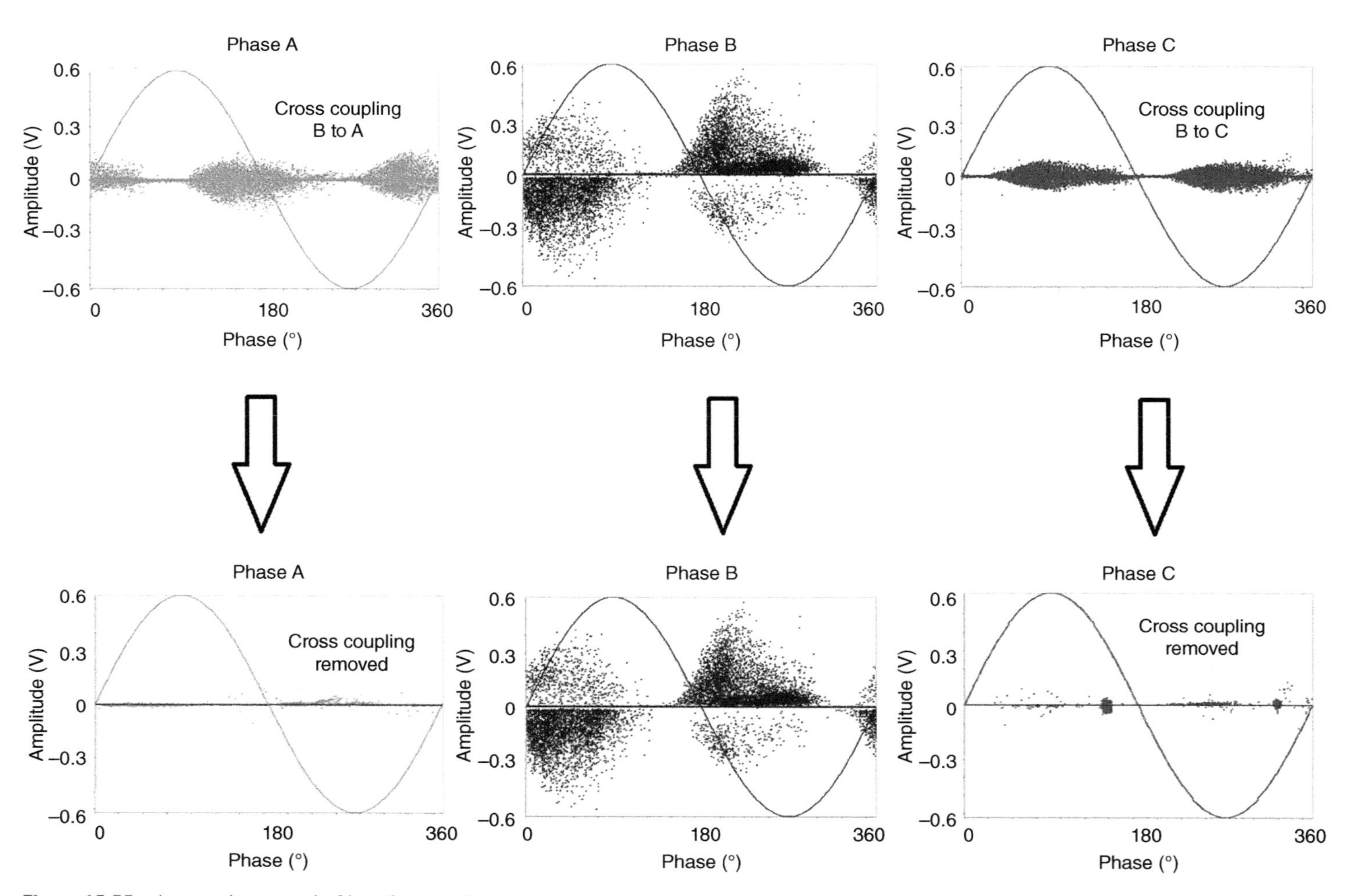

Figure 15.55 Automatic removal of interference. *Source:* Courtesy of Camlin Power.

consistent with transformer PD (Section 15.13). In addition, it is useful to try to locate the PD sites to make sure they are within the bushing or transformer. The acoustic technique in Section 15.11.5 using temporarily applied ultrasonic sensors on the outside of the transformer tank is one way to locate the PD sites. Another is to use the signals from multiple UHF sensors mounted on the inner tank walls (if present) to locate the PD based on pulse travel time. If PD is confirmed within the transformer, the PD activity (either the peak magnitude, the pulse repetition rate, and/or an integrated indicator of both such as NQS [Section 8.8.3]) is trended over time. The rate of increase, together with information from the DGA or moisture content monitor, are used to determine if the transformer needs to be immediately removed from service or if maintenance can wait until a more convenient time.

A good example of correlation between a continuous DGA system and a PD online monitoring system is shown in Figure 15.56. The PDM is connected to the voltage tap of the HV bushing by means of a quadripole and continuously monitors PD activity. The values of PD are in nC since the system has been previous calibrated.

Clearly, this example shows the advantage to having the DGA and PD trends available simultaneously. In the diagrams in Figure 15.56, it is possible to observe every time the PD activity increases, hydrogen and acetylene concentration (which are the key gases when PD is occurring) increase too. The PD activity is not persistent but occurs sporadically.

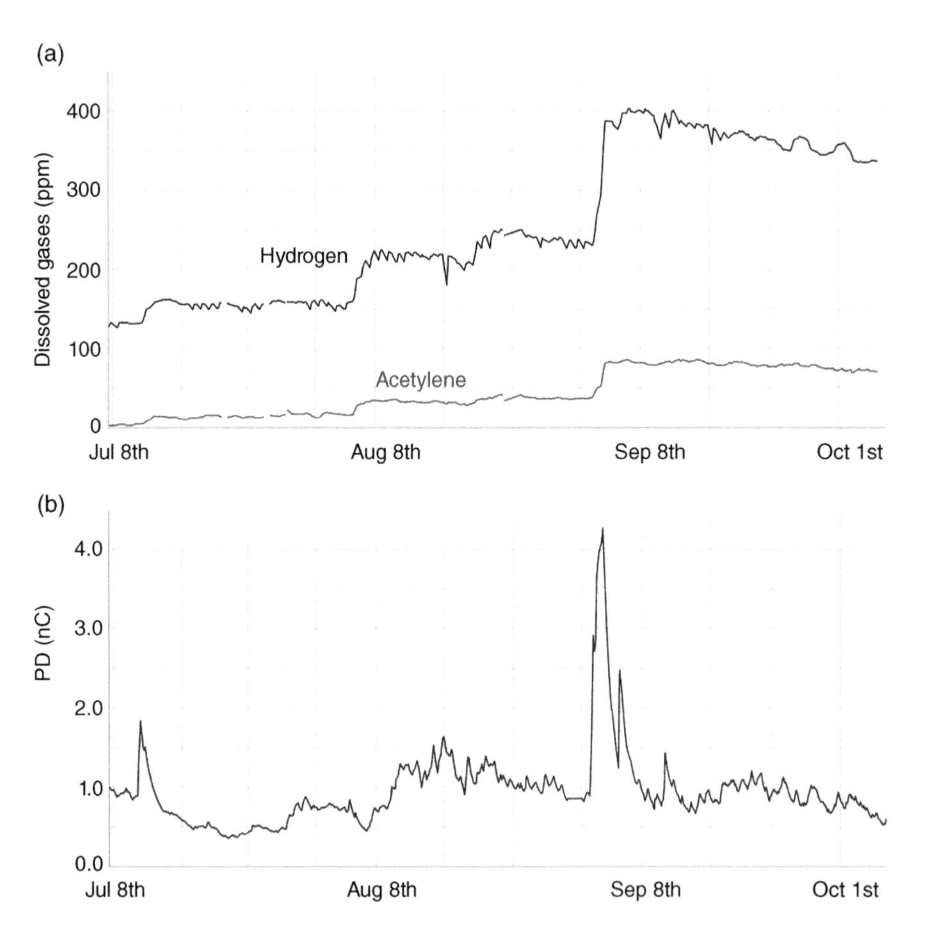

Figure 15.56 86.6 MVA, 400 kV single-phase (one of a three phase arrangement) shunt reactor. (a) hydrogen and acetylene trend; (b) PD trend.

15.13 Typical PRPD Patterns

When performing a PD measurement on transformers, it is important to analyze the information that the PRPD patterns provide. A complete analysis of the PD phenomenon should also include DGA (Section 15.7.6), localization by the acoustic method (Section 15.7.5) or by means of UHF sensors (Section 15.7.4), and the information they can provide via PRPD patterns. The PRPD pattern depends on the nature of the source of the PDs and where they originate. A fundamental contribution in the knowledge of the PRPD comes from the comparison between the shape of these and what was found during the inspection of the transformer in search of the defect. Thanks to the experience gained, it is sometimes possible to associate some typical PRPD shapes with specific defects found by visual inspections of the transformer. The PRPD patterns reported below refer to defects that gave rise to PD and are mainly related to the possible causes listed in Section 15.4.

15.13.1 PRPD Pattern Related to Defects in Solid Insulating Materials

Surface tracking on barriers or cylinders, beginning of carbonization inside wooden supports, pressboards, or paper wrapped connections are one source of PD (Section 15.4.1). These defects are often due to an imperfect drying process or impregnation procedure (too short a time or the oil is not hot enough) or contamination. In this case, the cause of PD may be recognized in defects arising from factory manufacture (Section 15.4.3). Figure 15.57 shows the tracking defects found during a visual inspection. Figure 15.58 shows the PRPD patterns found.

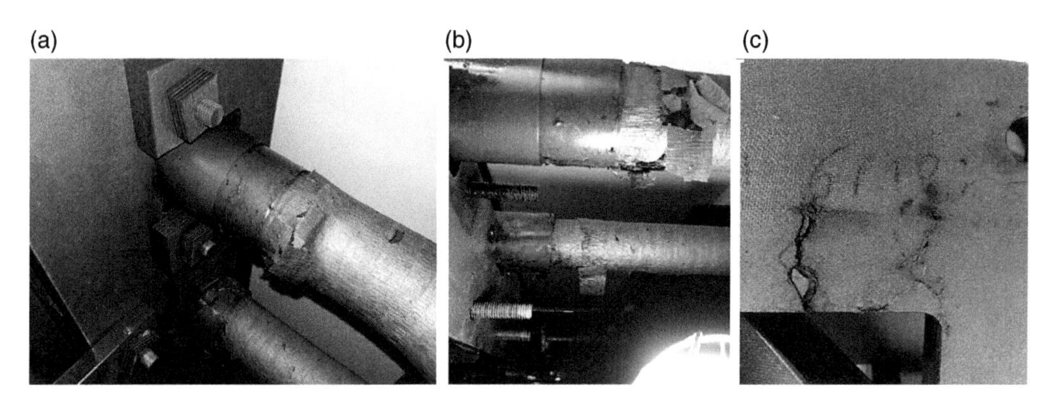

Figure 15.57 Examples of surface tracking due to PD on the connection insulation (a) and (b), and on a wood support (c).

15.13.2 PRPD Pattern Related to Defects in the Core Structure

Defects in the core may have various origins (Section 15.4.2). They can originate from improper tightening of the core frame since assembly in the factory or be the consequence of thermal and electromechanical stresses that occur during the operational life of the transformer. Thermal or electromechanical stresses can damage the magnetic core and the magnetic flux may cause discharges between laminations and/or localized overheating that lead to burning of adjacent pressboard near the joints in the core (i.e. between a leg and the yoke). Figure 15.59 shows a defect found in the magnetic core of a 250 MVA, 400/150 kV autotransformer. An offline/onsite PD measurement had the PRPD pattern in Figure 15.60. PD signals had been processed by the frequency/time map.

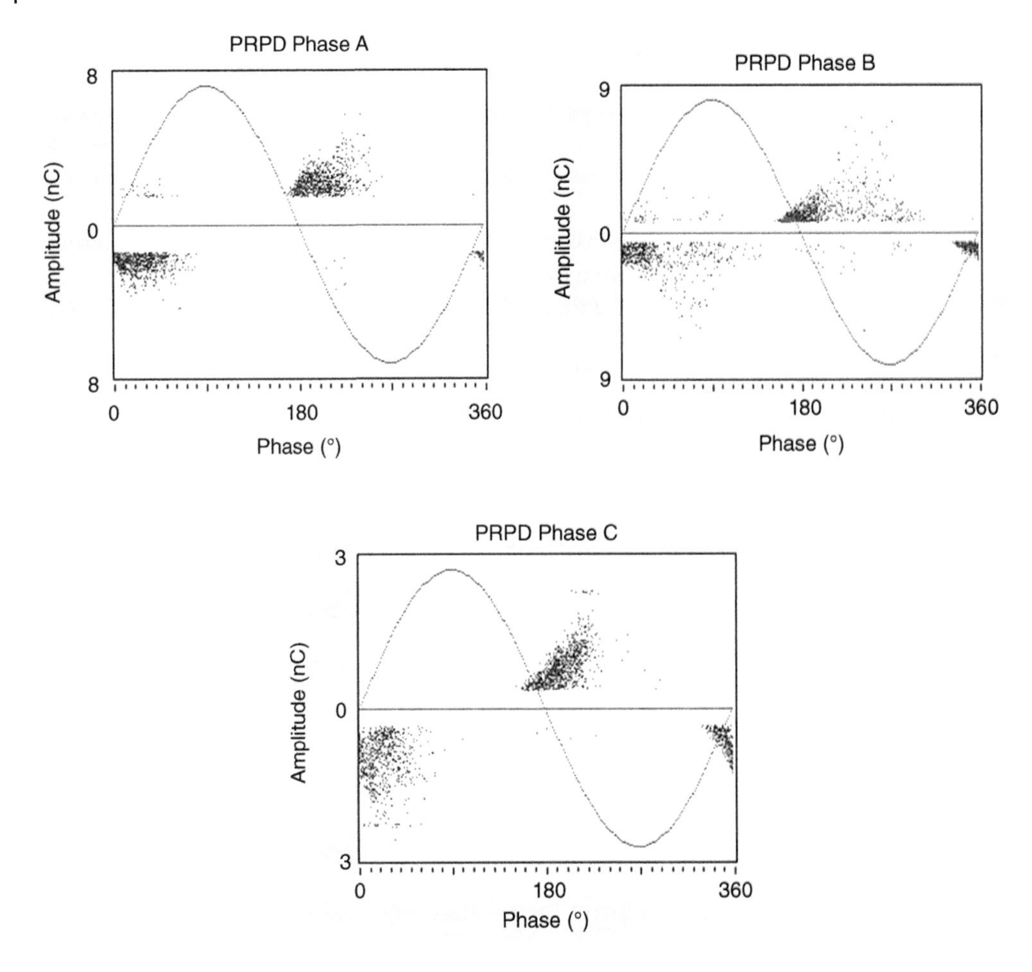

Figure 15.58 PRPD patterns measured in an onsite/offline test on the transformer in Figure 15.57, presumably due to surface tracking.

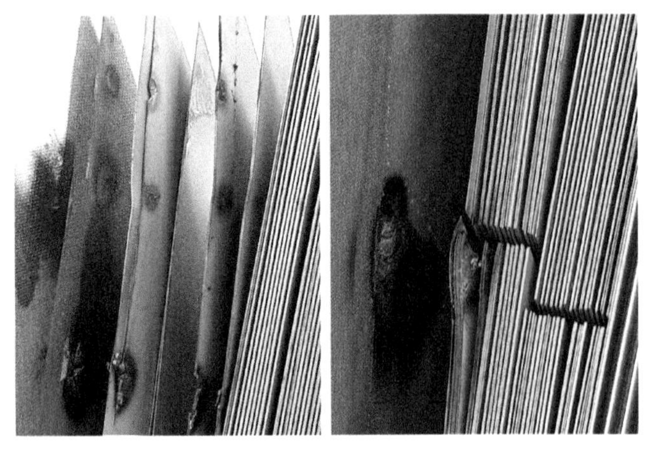

Figure 15.59 Discharge damage involving the magnetic core laminations (right) and the pressboard (left) at two locations.

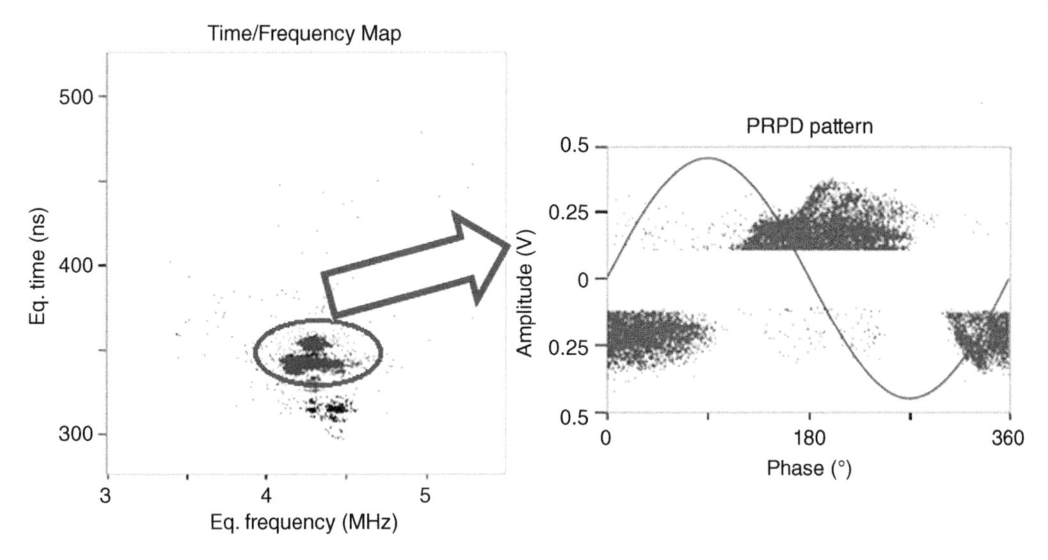

Figure 15.60 Time/frequency map (left) and PRPD pattern of a specific cluster.

15.13.3 Floating Discharge Due to Unbonded Bushing

Bushings undergo intense quality control tests at the factory. Therefore, the presence of PD in new bushings is uncommon. Discharges, however, can be caused by faulty installation of the bushing onto the transformer. A very critical aspect of installing the bushing is the coupling between the cable from the HV winding and the inner conductor of the bushing. During assembly, care must be taken to ensure that the HV cable and the metallic pipe inside the bushing are connected properly. Figure 15.61 shows an example of improper onsite assembly of an OIP (Oil Impregnated Paper) bushing leading to insulation burning in a 420 kV bushing from a 250 MVA power autotransformer. The connection, between the metal pipe of the lead, the main conductor and the metal pipe inside the bushing, was improperly arranged so PD occurred in that zone with burning on the paper tape (Figure 15.61). An offline PD measurement was performed before operating the autotransformer and PD was measured (Figure 15.62).

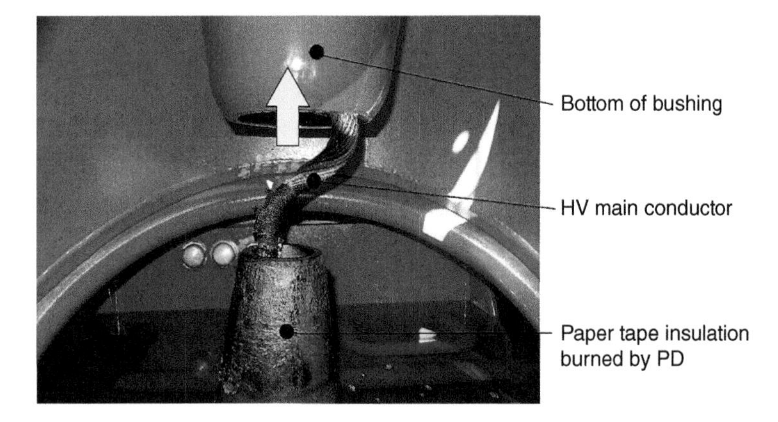

Figure 15.61 Burning on the paper wrapped lead from winding to bushing due to high-intensity PD (improper equipotential coupling).

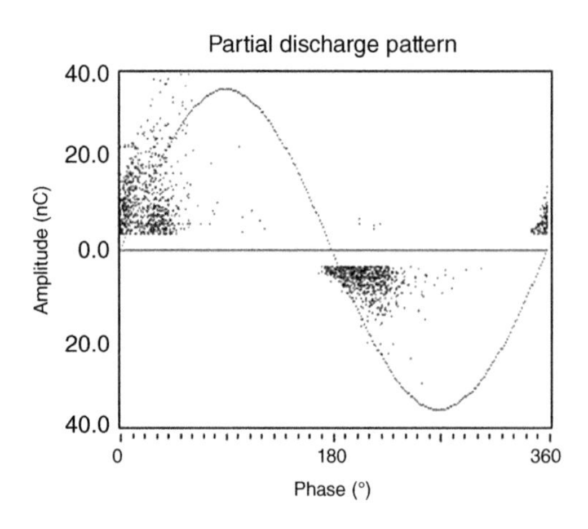

Figure 15.62 High-intensity PD in the bushing. The PD was measured offline using the bushing tap. The PD is within the bushing since the PD polarity is reversed from PD occurring within the transformer itself.

15.13.4 PRPD Pattern Due to Trapped Air Within the Transformer After Oil Filling

Another example of PD due to an improper onsite assembly is presence of air within the transformer. This is usually caused by incorrect onsite oil filling. The final oil filling must be done under vacuum with the transformer completely assembled and injecting hot oil slowly from the bottom of the tank. If the process fails, PD normally occurs immediately. Figure 15.63 shows the PRPD pattern measured when air is within a 250 MVA, 230/135 kV autotransformer.

To solve the problem, it was required to repeat the final oil filling under vacuum. The oil level in the transformer tank was first lowered and then refilled under vacuum.

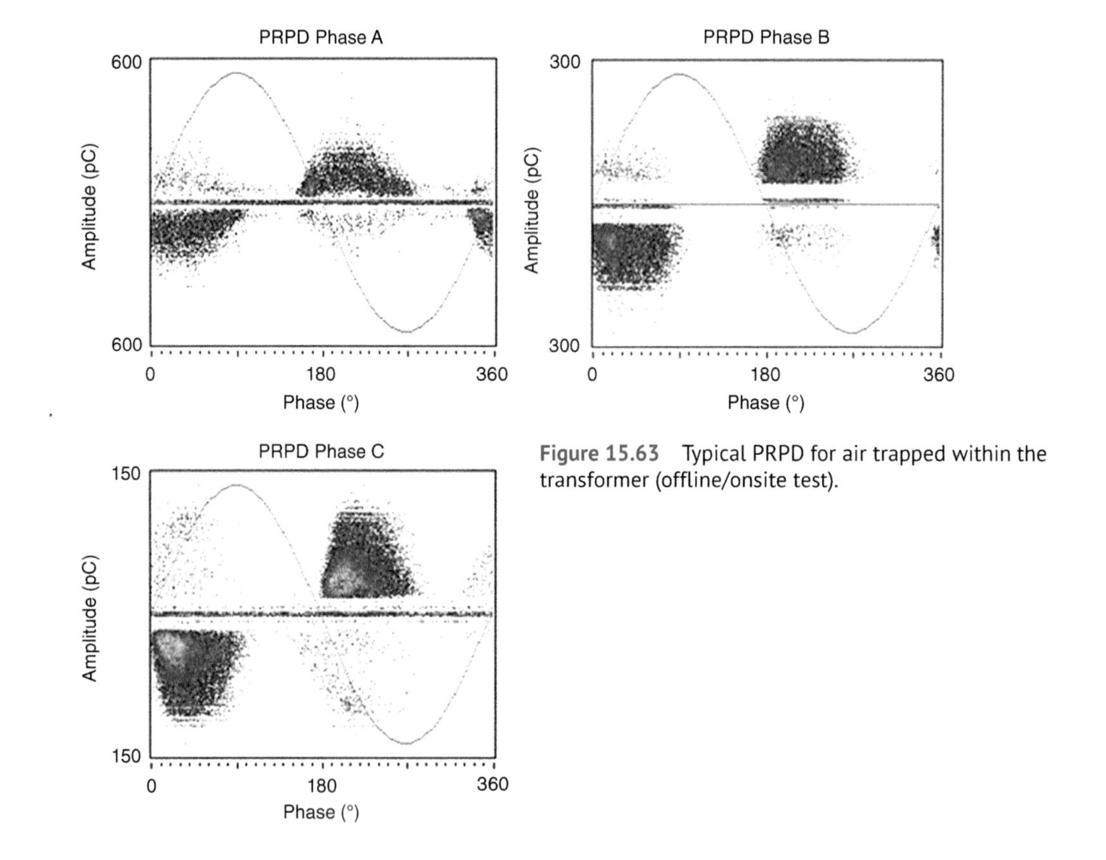

Figure 15.63 Typical PRPD for air trapped within the transformer (offline/onsite test).

15.13.5 PRPD Related to Probable Humidity in Paper Tapes

PD is more likely to occur at the top of the windings where the voltage is highest, and where the electric field may be intense due to the local geometry. As discussed in Section 15.4.6, high field regions are mitigated using metallic shields made of a core of robust organic materials covered by a sheet of thin conducting metal foil, which is then covered with a wrapped paper tape.

Figure 15.64 shows the effects of probable humidity trapped in the paper tape covering the HV shield of a HV winding. Figure 15.65 shows the PRPD pattern associated with this defect.

Figure 15.64 PD in a HV winding shield (red circle). *Source:* Courtesy of Camlin Power.

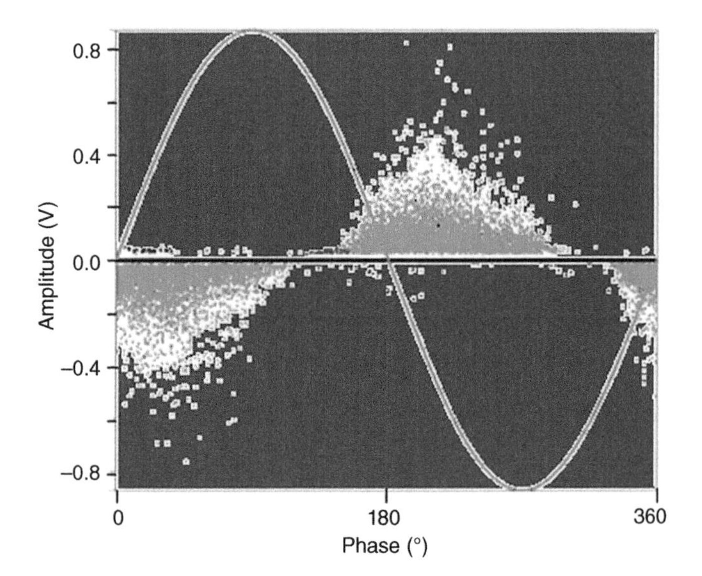

Figure 15.65 PRPD pattern from an online PD test where the paper tape of a HV winding shield that may have had moisture contamination. *Source:* Courtesy of Camlin Power.

15.13.6 PRPD Related to Turn-to-Turn PD

Turn to turn faults are quite rare. However, heavy electromechanical stresses (Section 15.4.5) as a consequence of a high through fault current occurring on the network, may cause tearing of the paper tape around the turn conductor (Figure 15.17). Likewise, thermal stress may cause brittleness of the paper tape (Figure 15.15). These types of damage may cause PD which are very dangerous because they quickly lead to winding failure. Normally the defect is first detected by a Buchholz relay trip. An offline PD measurement should then be performed before re-energizing the transformer. During the test, it is important to analyze the PRPD patterns in terms of PD intensity, repetition rate and, overall, the voltage threshold at which PD appears and disappears.

Normally, in the case of probable turn-to-turn PD, there will be a short circuit between turns. In any case, before the short circuit between turns occurs, the PRPD assumes a shape like that shown in Figure 15.66 (PRPD recognized during an offline test on 450 MVA, 230 kV phase shifter transformer). The red circle in Figure 15.67 shows the defect highlighted during the inspection of the winding.

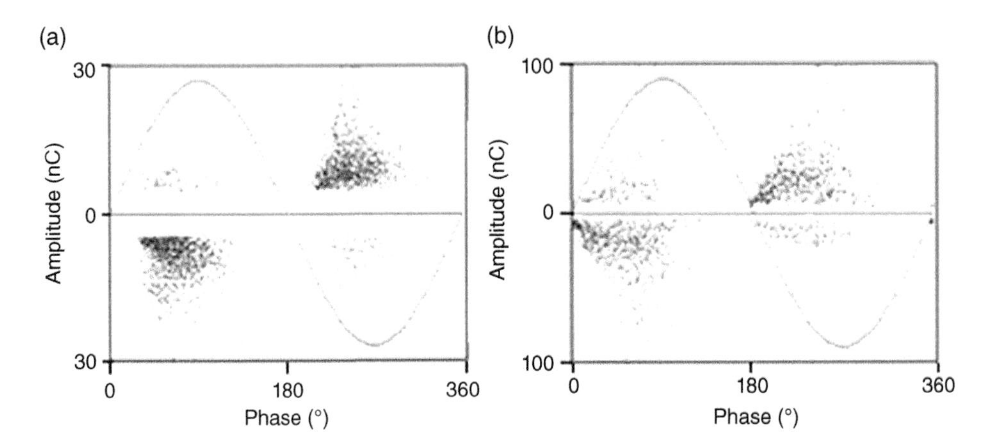

Figure 15.66 Increase of PD intensity in two PRPD recognized, during the observation time, in an off line test. Intensity in (a) is lower than in (b). PRPD in (a) is recognized earlier than that in (b).

Figure 15.67 Damage between two adjacent turns due to electrodynamic forces.

References

1 Ahuja, R. (2022). *Transformer Design Principles, Third Edition*. CRC Press.
2 Cimini, M., Bosetto, D., and Stevanato, F. (2020). *Il macchinario di trasformazione di potenza – Principi di funzionamento, Esercizio e Manutenzione*, 2nd edition, pp. 630 (in Italian).
3 Areva T&D (2008). *Power Transformers Vol. 1 Fundamentals*. Areva T&D.
4 Meador, J.R., Kaufman, R., and Brustle, H. (1966). Transformer corona testing. *IEEE Transactions on Power Apparatus and Systems* PAS-85: 893–900.

5 Narbut, P. (1965). Transformer corona measurement using condenser bushing tap and resonant measuring circuits. *IEEE Transactions on Power Apparatus and Systems* 84: 652–651.

6 Moser, H.P. and Dahinden, V. (1999). *Transformerboard II – Properties and Application of Transformerboard of Different Fibres*, 2e. Weidmann.

7 Trinh, N.G., Vincent, C., and Regis, J. (1982). Statistical Dielectric Degradation of Large-Volume Oil-Insulation. *IEEE Transactions on Power Apparatus and Systems* PAS-101 (10): 3712–3721. https://doi.org/10.1109/TPAS.1982.317056.

8 Scatiggio, F., Pompili, M., and Bartnikas, R. (2009). Oils with presence of corrosive sulphur: mitigation and collateral effects. *2009 IEEE Electrical Insulation Conference*, Montreal, QC (31 May to 3 June 2009).

9 IEC 60156 (2018). Insulating liquids – determination of the breakdown voltage at power frequency – test method.

10 Denat, A. (2011). Conduction and breakdown initiation in dielectric liquids. *2011 IEEE International Conference on Dielectric Liquids*, Trondheim, Norway, (26–30 June 2011).

11 Wang, Z.D., Matharage, S.Y., Liu, Q. et al. (2021). Research and development of ester filled power transformer. *Transformer Magazine* 8 (4): 24–33.

12 CIGRE (2013). *Oxidation Stability of Insulating Fluids*, Technical Brochure 526.

13 Wiklund, P. and Karlsson, C. (2011). Time-dynamic oxidation stability studies of mineral insulating oils. *2011 IEEE International Conference on Dielectric Liquids*, Trondheim, Norway (26–30 June 2001).

14 CIGRE (2010). *Experiences in Service with New Insulating Liquids*, Technical Brochure 436.

15 IEC 61294 (1993). Insulating liquids – determination of the partial discharge inception voltage (PDIV) – test procedure.

16 Clark, F.M. (1962). *Insulating Materials for Design and Engineering Practice*. Wiley.

17 CIGRE (2007). *Ageing of Cellulose in Mineral-Oil Insulated Transformers*, Technical Brochure 323.

18 McNutt, W.J. (1992). Insulation thermal life considerations for transformer loading guides. *IEEE Transactions on Power Delivery* 7 (1): 392–401.

19 Fallou, B. (1970). Synthèse des travaux effectués au LCIE sur le complexe papier-huile. RGE, Tome 79- No 8, September 1970.

20 Moser, H.P. (1981). *Future Transformer Insulations Systems*. WEIDMANN Publication 621.314.21.

21 CIGRE (2018). *Ageing of Liquid Impregnated Cellulose for Power Transformers*, Technical Brochure 738. CIGRE.

22 CIGRE (2016). *Partial Discharges in Transformers*, Technical Brochure 676.

23 IEC 60270:2015 (2015). High-voltage test techniques – partial discharge measurements.

24 IEC 62478:2016 (2016). *High Voltage Test Techniques – Measurement of Partial Discharges by Electromagnetic and Acoustic Methods*.

25 IEC 60076-3:2018 (2018). *Power Transformers: Insulation Levels, Dielectric Tests and External Clearances in Air*.

26 IEC 60076-11:2018 (2018). *Power Transformers – Part 11: Dry-Type Transformers*.

27 IEEE C57.12.90-2021 (2021). *IEEE Standard Test Code for Liquid-Immersed Distribution, Power, and Regulating Transformers*.

28 IEEE C57.113-2010 (2010). *IEEE Recommended Practice for Partial Discharge Measurement in Liquid-Filled Power Transformers and Shunt Reactors*.

29 IEEE C57.124-1991 (1991). *IEEE Recommended Practice for the Detection of Partial Discharge and the Measurement of Apparent Charge in Dry-Type Transformers*.

30 IEEE C57.127-2018 (2018). *IEEE Guide for the Detection, Location and Interpretation of Acoustic Emissions from Electrical Discharges in Power Transformers and Power Reactors*.

31 Hettiwatte, S.N., Wang, Z.D., Crossley, P.A. et al. (2002). Experimental investigation into the propagation of partial discharge pulses in transformers. *2002 IEEE Power Engineering Society Winter Meeting. Conference Proceedings (Cat. No.02CH37309)*, New York (27–31 January 2002), pp. 1372–1377, Vol. 2. https://doi.org/10.1109/PESW.2002.985240.

32 CIGRE (2010). *Guidelines for Unconventional Partial Discharge Measurements*, Technical Brochure 444.

33 Lundgaard, L.E. (1992). *Partial discharge. XIV. Acoustic partial discharge detection-practical application. IEEE Electrical Insulation Magazine* 8 (5): 34–43.

34 Hoek, S.M., Kraetge, A., Hummel, R. et al. (2012). Localizing partial discharge in power transformers by combining acoustic and different electrical methods. *2012 IEEE International Symposium on Electrical Insulation*, San Juan, PR (1–13 June 2012), pp. 237–241. https://doi.org/10.1109/ELINSL.2012.6251465.

35 CIGRE (2016). *Guidelines for Partial Discharge Detection Using Conventional (IEC 60270) and Unconventional Methods*, Technical Brochure 662. CIGRE.

36 Kuffel, E., Zaengl, W.S., and Kuffel, J. (2000). *High Voltage Engineering Fundamentals-Second Edition*, Chapter 4. Newnes.

37 CIGRE (2022). *Improvements to PD Measurements for Factory and Site Acceptance Tests of Power Transformers*. Technical Brochure 861.

38 Coenen, S., Tenbohlen, S., Markalous, S.M., and Strehl, T. (2008). *Sensitivity of UHF PD measurements in power transformers. IEEE Transactions on Dielectrics and Electrical Insulation* 15 (6): 1553–1558. https://doi.org/10.1109/TDEI.2008.4712657.

39 Meijer, S., Gulski, E., Smit, J.J., and Reijnders, H.F. (2004). Sensitivity check for UHF PD detection on power transformers. *Conference Record of the 2004 IEEE International Symposium on Electrical Insulation*, Indianapolis (19–22 September 2004), pp. 58–61. https://doi.org/10.1109/ELINSL.2004.1380448.

40 CIGRE (2019). *Advances in DGA Interpretation*, Technical Brochure 771.

41 IEC 60599 (2015). Mineral oil-impregnated electrical equipment in service – guide to the interpretation of dissolved and free gases analysis.

42 IEEE (2019). *IEEE C57.104-2019: Guide for the Interpretation of Gases Generated in Mineral Oil-Immersed Transformers*. IEEE https://doi.org/10.1109/IEEESTD.2019.8890040.

43 Duval, M. and Buchacz, J. (2022). *Identification of arcing faults in paper and oil in transformers—part i: using the Duval pentagons. IEEE Electrical Insulation Magazine* 38 (1): 19–23. https://doi.org/10.1109/MEI.2022.9648268.

44 Contin, A., Cavallini, A., Montanari, G.C. et al. (2002). *Digital detection and fuzzy classification of partial discharge signals. IEEE Transactions on Dielectrics and Electrical Insulation* 9 (3): 335–348. https://doi.org/10.1109/TDEI.2002.1007695.

45 Rethmeier, K., Kruger, M., Kraetge, A. et al. (2008). Experiences in on-site partial discharge measurements and prospects for PD monitoring. *2008 International Conference on Condition Monitoring and Diagnosis*, Beijing (21–24 April 2008), pp. 1279–1283. https://doi.org/10.1109/CMD.2008.4580210.

46 Harrold, R.T. (1985). Acoustical technology applications in electrical insulation and dielectrics. *IEEE Transactions on Electrical Insulation* EI-20 (1): 3–19.

47 CIGRE (2012). *High-Voltage On-site Testing with Partial Discharge Measurement*, Technical Brochure 502.

48 Unsworth, J., Kurusingal, J., and James, R.E. (1994). On-line partial discharge monitor for high voltage power transformers. *Proceedings of 1994 4th International Conference on Properties and Applications of Dielectric Materials (ICPADM)*, Brisbane, QLD (3–8 July 1994), pp. 729–732, Vol. 2. https://doi.org/10.1109/ICPADM.1994.414114.

49 Koch, M. and Kruger, M. (2008). A fast and reliable dielectric diagnostic method to determine moisture in power transformers. *International Conference on Condition Monitoring and Diagnosis*, 2008, pp. 467–470, https://doi.org/10.1109/CMD.2008.4580326.

50 Koch, M. and Krüger, M. (2012). A new method for on-line monitoring of bushings and partial discharges of power transformers. *2012 IEEE International Conference on Condition Monitoring and Diagnosis*, Bali (23–27 September 2012), pp. 1205–1208. https://doi.org/10.1109/CMD.2012.6416378.

16

Rotating Machine Stator Windings

16.1 Introduction

PD measurements are widely performed both on the individual coils used in stator windings and on complete stator windings in conventional electrical rotating machines, which includes motors, generators, and synchronous condensers. However, unlike for power cables, AIS, GIS, and transformers, PD tests are not required by either IEC or IEEE standards before the coils or windings are shipped from the factory. Furthermore, IEC and IEEE have no "acceptance criteria" for either PD magnitude, the PD extinction (*PDEV*), or the PD inception (*PDIV*) voltages [1–4]. Unless required by end users in a purchase contract, factory PD tests are optional for the machine manufacturer. However, even when they are not required by the end user, major manufacturers often use factory PD tests for stators rated 6 kV and above for process control purposes.

Instead of factory acceptance testing, the main application for PD tests has been for condition assessment, either using periodic onsite/offline tests or online monitoring. Here, the purpose is to help determine when winding maintenance is required, including rewinds.

This chapter is focused on the stator windings of machines rated 3 kV and above that use (in the case of motors) or generate (in the case of generators) 50 or 60 Hz AC. These tend to be large motors or generators in utility and industrial applications such as generating stations, petrochemical plants, pulp and paper mills, water and petrochemical pipelines, mining facilities, and some marine applications. The rotor windings in such machines are low voltage (typically <500 V DC) and thus do not produce PD. PD does sometimes occur in low voltage (<1 kV) windings in motors and generators that are fed from voltage source invertors. Such invertors may create voltage impulses up to a few thousand volts that can induce PD and PD-induced deterioration. This is a rapidly evolving field and is briefly discussed in Section 18.2 as well as in [5].

This chapter presents the main insulation system features of form-wound stator windings rated 3 kV and higher. There are many different insulation failure processes (aging mechanisms) that either produce PD as a symptom of insulation aging, or where PD is a direct cause of aging. These failure processes are identified and explained. The most common methods for factory testing, onsite/offline testing, and online PD monitoring are presented, as is the interpretation for each type of test. Establishing the root cause of any stator winding PD is critical, since if the root cause is properly identified, some repairs can be done for as little as a few percent of the cost of a stator rewind.

Practical Partial Discharge Measurement on Electrical Equipment, First Edition. Greg C. Stone, Andrea Cavallini, Glenn Behrmann, and Claudio Angelo Serafino.

16.2 Relevant Standards

IEEE 1434 was the first published international standard for performing PD tests on coils and windings [1]. It describes many electrical test methods for both offline and online testing. Although some guidance on interpretation is given, pass/fail criteria are not provided. As discussed in Section 6.6.3, this is mainly because PD measurements on complete stator windings cannot be "calibrated" into apparent charge (pC, nC) due to the inductive-capacitive nature of the windings (Section 6.6.3). This means that each combination of winding and test instrumentation will produce different results compared to other combinations. Using a different brand of PD measuring equipment, even on the same EUT, may yield different results that cannot be corrected for, and thus results from different brands of test equipment cannot be meaningfully compared. IEEE 1434 provides background information on the test, describes over a dozen test methods that have been used at least once, but provides only general advice on interpretation.

IEC 60034-27-1 (offline testing) and IEC 60034-27-2 (online testing) provide essentially the same information as IEEE 1434 [2, 3], with even less practical guidance on interpreting PD test results.

IEEE 1799 is a more recent standard that does present some specific interpretation advice [4]. This standard uses optical and acoustic methods in offline testing to locate PD occurring on the surface of coils and windings. These include the "black-out" or "lights-out" test, as well as the use of ultraviolet or acoustic imaging devices to identify and locate surface PD (Sections 5.3 and 5.4). IEEE 1799 provides a method of calibrating the sensitivity of surface PD measurements against the human eye, so that results among different devices can be compared. Since, in principle, no surface PD should be present in operation, IEEE 1799 sets the minimum voltages at which surface PD should not be detected. An important feature of the standard is testing the phase-to-ground insulation at a lower voltage than the phase-to-phase insulation.

16.3 Stator Winding Insulation Systems

The insulation system in stator windings rated 3 kV and above share the same basic features and materials, whether they are in motors or generators, or rated from a few hundred kW to 2000 MVA. Figure 16.1 shows a photograph of a typical stator from a large motor or small generator. The winding is composed of several coils that are connected in series between the high-voltage end of the winding and neutral, in a three-phase wye connection. The coils are installed in slots within the grounded, laminated steel core. The active area of the coil is the straight portion of the

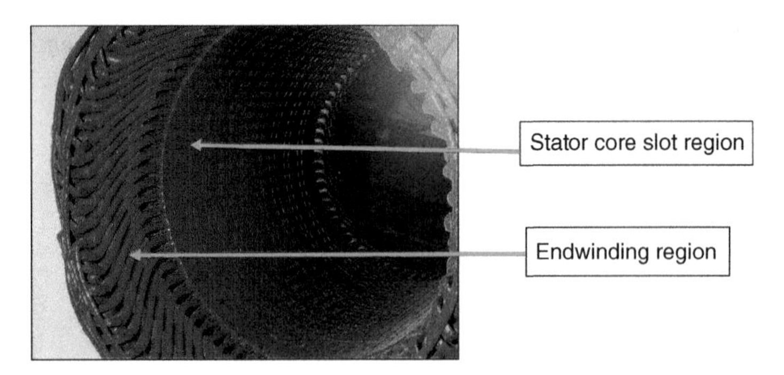

Figure 16.1 Photo of a 13.2 kV stator winding with multiturn coils. *Source:* Courtesy of Iris Power L.P.

coil in the stator core slots. The two regions of the coil outside of the stator core are called the endwindings (or sometimes end turn, end arm or overhang). PD can occur in either the slot or endwinding regions, and different PRPD patterns are found in the two regions (Section 16.12.1). Damage caused by PD in the endwinding region is usually fairly easy to repair, in comparison to damage caused by PD within the slot.

Much more information on stator winding insulation systems is given in [5].

Almost all motors and most generators rated about 50 MVA and below have a stator winding made from multiturn coils, i.e. each coil has two "legs" that are straight, and simultaneously placed in two different stator slots (Figure 16.2, upper). For stators rated more than about 50 MVA, the coils tend to be so large and heavy that there is a risk of insulation damage when inserting the two coil legs into the two stator slots. For such large stators, two separate half-turn "Roebel" bars are separately inserted into the stator slots and then connected together to form a (one-turn) coil. Since such bars (Figure 16.2, lower) have only one leg, they are much easier to install in the stator slots without cracking the insulation. Most bars have a half-turn design, so there is no turn insulation, and thus no turn insulation to fail. As discussed in Section 16.12.1, this has implications for interpreting PRPD patterns.

16.3.1 Insulation System Components

Figure 16.3 shows the cross sections of both a Roebel bar and a multiturn coil, together with the components of the insulation system. The same materials have been used in almost all motors and generators made since the 1960s. In multiturn coils, there are three insulation components:

- Strand insulation
- Turn insulation
- Groundwall (or mainwall) insulation

The strand insulation is usually a thin polymer film and/or a Dacron^{TM} plus fiberglass tape on each copper strand. It improves the efficiency of the winding by reducing eddy current and skin effect losses [5].

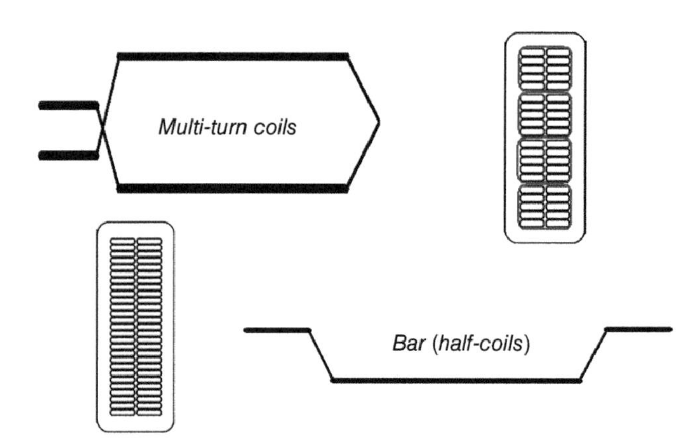

Figure 16.2 Differences between Roebel bars (lower) and a four-turn coil (upper). The Roebel bar is named after the engineer who developed a way to move copper strands to different locations within the copper stack to minimize copper losses. *Source:* Courtesy of Iris Power L.P.

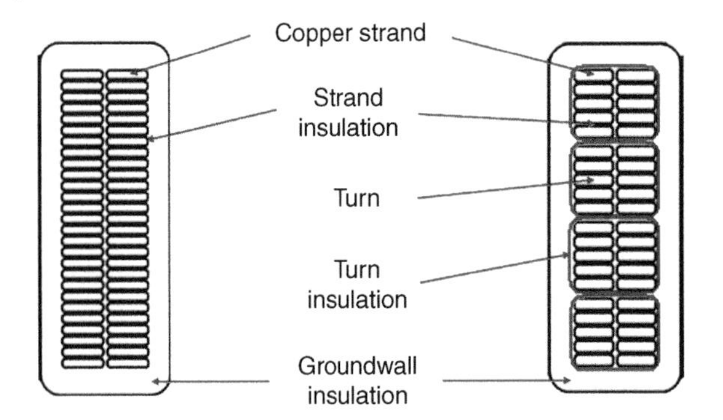

Figure 16.3 Cross-sections of a Roebel bar (left) and a multiturn coil (right). *Source:* Courtesy of Iris Power L.P.

The turn insulation is only present in multiturn coils and prevents short circuits between the turns. Although the power frequency voltage across the turn insulation is usually only a few tens of volts, if the turn insulation is shorted anywhere in the coil or winding, a very large 50/60 Hz current flows through the shorted turn. This current rapidly heats the copper, which then melts, usually burning through the groundwall insulation in seconds to minutes to result in a ground fault – i.e. machine failure [5]. Thus, the integrity of the stator winding turn insulation is essential for long winding life. In 50/60 Hz stator windings, PD cannot occur between the copper turns because the voltage (and the electric field stress) is not high enough to cause PD in any cavities that may be present between the copper turns (Figure 16.4). However, if cavities occur on the periphery of the turn stack (Figure 16.4), or between the groundwall and the turn insulation, then PD may occur in these cavities when the voltage across the insulation is sufficiently high. Although PD testing cannot detect defects between the turns since there is insufficient electrics stress, PD testing can detect problems at the copper stack interface with the groundwall insulation. PD at this location is relatively dangerous in multiturn coils since it may lead to turn insulation failure in just a few years. In contrast, the Roebel bars used in large generators do not have turn insulation, so PD at the copper stack and groundwall insulation interface are not as dangerous as the interface in multiturn coils. This has implications for the interpretation of PD on multiturn coil stators vs. Roebel bar windings, as discussed in Section 16.12.1.

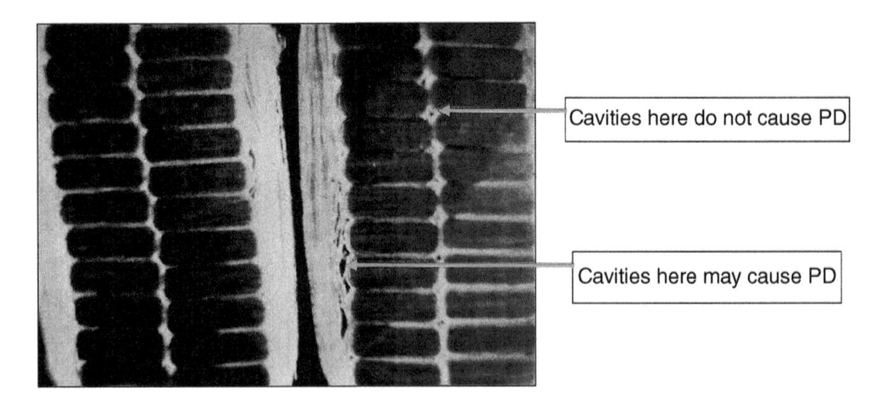

Figure 16.4 Cross-section of a 6.6 kV motor coil with cavities within the turn insulation stack (upper) and between the groundwall and turn insulation (lower). *Source:* [5]/with permission from IEEE.

The turn insulation in most stators rated less than 6 kV tends to be one or two layers of Dacron™ and fiberglass tape. In machines rated 6 kV or higher, most OEMs use one or two layers of "mica paper tape." Mica paper consists of very small mica platelets that are made into a layer using a process similar to making cellulose paper. The mica is then mechanically supported by either a polymer film or a Dacron and glass backing tape. Mica is an inorganic material that is almost universally used in stator winding insulation systems rated 3 kV and above because of its excellent resistance to PD and electrical treeing (Section 2.4.1). Mica also has a very high melting temperature, but poor mechanical properties (hence the need for a backing tape). Mica paper turn insulation makes the winding less likely to fail due to PD if any normal defects that may occur in manufacturing at the turn insulation/copper stack periphery.

Since the 1960s, virtually all stator windings use a groundwall composed of many layers of mica paper tapes that are bonded together with epoxy. The epoxy lowers the thermal resistance of the groundwall and prevents cavities, which can lead to PD. The epoxy is a thermosetting material that, in its simplest form, is made from mixing a liquid or paste resin (Part A) with a hardener (Part B). There are two main processes used to impregnate the mica tapes in coils or bars with epoxy:

- the resin rich, B-stage, or press-cure process
- the vacuum pressure impregnation (VPI) process

Both processes are still widely used for generator stators by OEMs and by service shops for rewinding.

The resin-rich process involves mixing the viscous forms of the resin and hardener together, and then impregnating the mica paper tapes themselves. These impregnated (B-stage) tapes are then cooled, leaving the epoxy uncured. When the groundwall is to be applied to the copper stack, the impregnated tapes are heated and helically wrapped around the conductors for the correct number of tape layers. Then, various methods are used to apply heat and pressure to the groundwall to squeeze out excess epoxy (ensuring a minimum of cavities) and curing the epoxy to a solid state.

The VPI process involves directly applying the dry (mainly unimpregnated) tapes to the copper stack. The taped coils or bars are then installed in a tank, air is removed with a vacuum pump to extract any air between the tape layers. Then, a low-viscosity liquid mixture of resin and hardener is pumped into the tank. After the coils and bars are covered with liquid, impregnation is aided by using a gas under pressure in the tank to drive the liquid into the groundwall. After impregnation, the coils/bars are removed from the VPI tank and put into an oven where the epoxy is cured at high temperature. Many more details of these processes are in [5, 6].

Since the 1970s a variation of the VPI process called the global VPI process (GVPI) has been used to impregnate the entire stator winding and core. Instead of impregnating each coil or bar individually, the GVPI process involves applying dry mica paper tapes to the coils or bars, then, while the coils/bars are still flexible, insert them into the stator slots and make all connections. Next, the entire stator with the windings is placed in a very large VPI tank and impregnated under pressure with low viscosity epoxy. Once impregnated, the stator is removed from the tank and placed in an oven to cure the epoxy. This is a very efficient process (once the manufacturer has climbed the learning curve) and is now used for virtually all motor stators, and many generators rated up to a few hundred MVA.

16.3.2 PD Suppression Coatings

The bars and coils in stators rated 6 kV and above almost always have special coatings on the surface of the groundwall to suppress surface PD. The coating in the slot is called the semiconductive or semicon coating (Figure 16.5). Confusingly, the same coating is also called the conductive

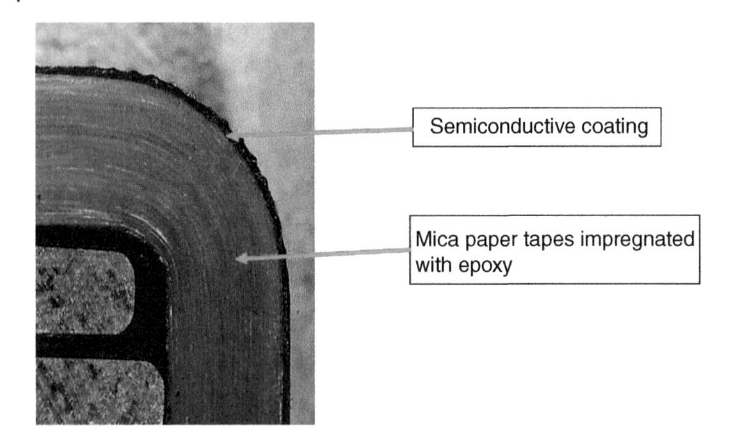

Figure 16.5 Cross-section of a stator coil showing the semiconductive (semicon) tape on the surface of the groundwall insulation to suppress PD between the coil and the stator core. *Source:* Courtesy of Iris Power L.P.

coating or OCP (for outer corona protection) coating. The coating is usually a tape or paint that is filled with graphite or carbon black to achieve a surface resistance on the order of a few thousand ohms per square.

In stator windings, there are always airgaps between the coil surface and the stator core somewhere along the length of the slot, since the width of the coils/bars must be slightly smaller than the width of the slot to insert them into the slots. This is even true in GVPI stators, since the low viscosity of the impregnating epoxy is not able to fill the relatively large gaps that will be present at some locations. As discussed in Section 6.4.1, the capacitance of the airgap in series with the groundwall capacitance will cause a large voltage to appear across the airgap. In coils operating at high voltage, the stress will be sufficient to breakdown the air in the gap, producing PD. If the surface of the coil has a semicon (short for semi-conducting) coating, and the coating is grounded to the stator core at many points along the slot, the semicon will be very close to ground potential. The result is that there will be no voltage across any airgap between the coil surface and the core, preventing PD at that location, which is sometimes referred to as slot discharge. This is the same reason that a semicon coating is extruded over the high-voltage insulation in power cables (Section 12.2.1).

The semicon coating is applied along the straight length portion of the coil/bar in stator slots and extends past the end of the stator core into the endwinding for a few centimeters. At each end of the semicon coating, unless special measures are taken, there will be a high electric field that may also lead to PD in this area. Therefore, as in power cable terminations (Section 12.2.2), special measures are taken to grade the electric field at the two ends of the semicon coating to prevent PD there. The grading coating (Figure 16.6) is usually a paint or tape filled with silicon carbide particles. More details on this coating and how it works are in [5]. The grading coating is sometimes called the semiconductive coating (even though that is applied by some to the slot coating) or the ECP (for endwinding corona protection).

16.3.3 Stator Winding Construction

Currents in the range of hundreds to many thousands of Amperes flow through the coils/bars in stator windings. These currents create magnetic fields from the coils that interact with one another and with the magnetic field from the rotor to produce twice AC frequency (i.e. 100 Hz or 120 Hz),

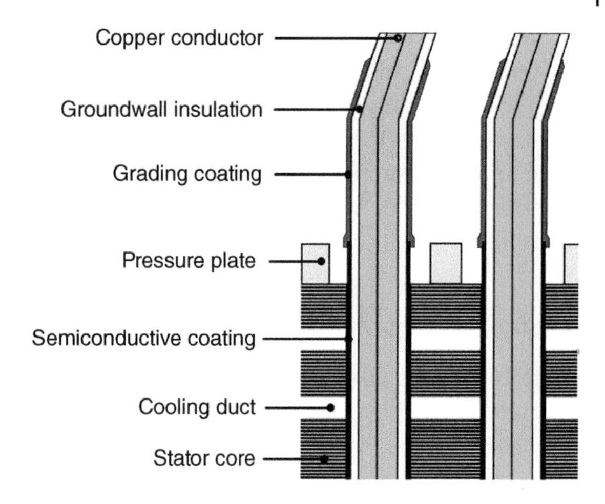

Figure 16.6 Location of the grading coating to suppress PD at each end of the semiconductive coating. *Source:* Courtesy of Iris Power L.P

mechanical forces for 50 Hz and 60 Hz machines, respectively. The vibrating mechanical forces are very large and require the coils to be securely held in place, both in the stator slots and in the endwindings. If they are not, then the groundwall insulation will abrade against the stator core in the slot and against bracing materials (which are usually insulating) in the endwinding. After years of relative movement, air spaces with voltage across them produce PD in the slot, or the endwinding abrasion can lead to exposed high-voltage copper conductors. Machine manufacturers have developed elaborate technologies to keep windings tight in the slot and in the endwinding. See [5] for more information.

16.4 Stator Winding Insulation Failure Processes

Reference [5] and IEEE 56:2016 [7] identify 20 or more insulation failure processes in stator windings. In stator windings rated 3 kV and above, most (but not all) of these failure processes are accompanied by PD. Some of the processes are a direct result of poor manufacturing; some are caused by aging stresses that occur during service; and some may be caused by either manufacturing or in-service operation. Factory PD tests help to identify issues caused during manufacturing that may lead to failure in service, and offline or online PD tests at site help to identify if any failure processes are occurring and track their progress over the years. Unlike most other types of HV equipment, PD may occur within and on the surface of the insulation for years or even decades before winding failure occurs. This is a consequence of the use of mica in the insulation system, which is extremely resistant to degradation by PD.

These different failure processes occur at different speeds, taking from a few years to decades to result in machine failure. Experience has shown that each of these processes produces different PRPD patterns and/or different PD responses as the machine operating condition changes (Section 16.12). Thus, if there is a single dominant failure process, PD testing can often identify the root cause. This will allow motor and generator owners to identify the best options to repair the machines and determine when the repairs should be done, to minimize repair cost and outage duration.

In general, air-cooled machines (virtually all motors and generators up to a few hundred MW) tend to have more PD pulses and higher magnitude PD than hydrogen-cooled machines with similar defects. This is because in hydrogen-cooled machines (large turbine generators and synchronous condensers), even though hydrogen has a breakdown strength about 50% lower than air at the same pressure (Section 2.6), the high-pressure hydrogen used in most hydrogen-cooled machines results in a much greater electric strength due to Paschen's Law (Section 2.3). Thus, high-pressure hydrogen cooling has a great impact on the interpretation of PD in hydrogen-cooled machines (Section 16.11.2).

The following summarizes the most common generic failure processes in windings that are associated with PD. More detailed descriptions of these processes, along with photos, are presented in [5].

16.4.1 Poor Epoxy Impregnation

Figure 16.4 shows a 6.6 kV, 8 MW motor coil cross-section, taken just outside of the stator slot of a GVPI stator. Clear cavities were formed, probably due to large wrinkles in the mica paper tapes, combined with the very low viscosity of the epoxy. Such large wrinkles cannot retain the epoxy when the stator is transferred to the oven during the GVPI process. Cavities can also occur due to processing problems such as poor resin viscosity control or poor temperature and pressure control. No matter the cause, the result will be PD in the cavities of coils/bars operating at high voltage.

In multiturn coils, if the cavities are adjacent to the turn insulation, the PD will gradually erode a path around the one or two layers of turn insulation (Section 2.4.1), eventually causing a turn insulation fault that rapidly morphs into a ground fault (Section 16.3.1). This can result in winding failure in as short as a few years, even with mica paper turn insulation, since erosion of only one or two tape layers is necessary. In contrast, such cavities adjacent to the strand insulation in a Roebel bar may lead to strand shorts, but these only cause minor local overheating if there are only a few shorts. To cause a winding failure, the PD will have to "tree" (Section 2.4.1) through all the groundwall mica-paper layers, which may take decades. Thus, the time to failure (and the time to plan for a rewind) depends on whether the winding is composed of multiturn coils or Roebel bars.

16.4.2 Inadequate PD Suppression Coatings

PD suppression coatings are usually present on most stator windings rated 6 kV and above (Section 16.3.2). Both the semicon and grading coatings may be incorrectly made or incorrectly applied during coil manufacture [5]. The carbon or silicon carbide particles may have too high or too low a concentration, or may be dispersed nonuniformly due to particle clumping. The paint or tape may also not be in intimate contact with the groundwall insulation in some areas. Furthermore, the resistance of the electrical connection between the semicon and grading coatings in the overlap region may be too high.

For the semicon coating, the result is isolated regions of PD occurring in the slot region between the groundwall and the stator core, even if the coils are not vibrating in the slot and in good contact with the core. Often the imperfection may have a diameter less than 1 cm, yet still cause PD between the imperfect semicon and the core. In air-cooled machines, the PD in air gives rise to ozone. Ozone combined with nitrogen in the air and some humidity creates nitric acid that then chemically attacks the remaining semicon, increasing gradually the area of the deterioration (Section 5.5.1). Ozone can be even more damaging to metal and rubbers within the machine than to the winding insulation. In the overlap region between the semicon and the grading coatings (Figure 16.6), if the resistance is too high, the capacitive current flowing through the overlap region creates excessive I^2R losses and local overheating, which further increases the resistance of the connection and, thus, the temperature. Ultimately, the overlap is so hot that it is fully oxidized and becomes insulating [5]. This leaves the grading coating disconnected from the semicon, and in the high-voltage coils, flashover occurs in the axial direction between the grading coating and the remaining semicon – which appears as very high intensity PD across the gap.

The PD from the semicon coating tends to produce a lot of pulses from all the sites that are deteriorating, however, the PD magnitude may be modest. In contrast, overlap discharging is of high intensity, but occurs in relatively few locations (producing fewer pulses). Both mechanisms may take many years or decades to cause failure, either due to its low intensity (semicon PD), or due to the discharges being parallel to the insulation surface (overlap PD) [5]. Since the PD occurs near to the grounded stator core, the PD is usually larger on the negative part of the AC cycle (i.e. positive PD > negative PD with the indirect measurement method in Figure 6.2).

This problem is much less likely to occur in hydrogen-cooled machines since oxygen is not present, thus, ozone will not be produced.

16.4.3 Loose Coils in the Stator Slots

In Section 16.3.2, we pointed out the need to keep coils well secured in the slot to prevent coil vibration due to the magnetic forces. In conventional stators (i.e. not GVPI), this involves the use of wedges, filler strips, ripple springs and/or silicone rubber, or other conformable materials. If the system to prevent movement was inadequately designed or poorly installed, then the coils will vibrate against the stator core in the slot, resulting in the steel laminations abrading through the semicon coating, leading to PD in coils operating at high voltage. If not corrected, the abrasion will also remove the groundwall insulation and eventually failure occurs. In addition to loose coils due to poor design/manufacture, this problem also occurs as a result of gradual shrinkage over time of the organic components (epoxy, filler strips, wedges) as well as relaxation of any ripple springs. Note that GVPI stators are very unlikely to experience this problem since the coils are bonded to the stator core.

This failure process can be relatively rapid, with ground faults occurring after just a few years in cases where the coils have been particularly loosely installed. How quickly failure occurs depends on the magnetic forces acting on the coils and how loose the coils are. PD is just a symptom of the process and only shortens the time to failure a little.

Although the physics is quite different, vibration sparking (also known as spark erosion) is also produced by loose coils in the slot, where the semicon coating is too conductive [5, 7].

16.4.4 Inadequate Separation of Coils and Bars in the Endwinding

Another design or manufacturing problem that can lead to PD, and ultimately winding failure, is caused by the coils in the endwinding, circuit ring busses, or connection leads being too close to one another. This is more likely to happen to air-cooled stators rated 11 kV and above. In most stators, usually in at least six locations at each end of the stator, a line-end (i.e. connected to the high voltage terminal) coil of one phase will be adjacent to a line end coil in a different phase. The copper conductors in these two coils will have the full phase-to-phase voltage across them. If the distance between the coil surfaces is too small (say <5 mm for an 11 kV winding), then PD may occur between the coil surfaces (Figure 10.1). If this phase-to-phase PD is not eliminated, the PD will eventually erode (tree) through the groundwall of both coils and lead to a phase-to-phase failure – which may be highly destructive. Since two groundwalls have to be penetrated by the PD, the failure process may take 15–20 years to cause a fault. Phase-to-ground PD caused by insufficient clearance may also occur, but it tends to be less likely since the voltage stress is 58% of the phase-to-phase voltage stress.

Since this problem is driven mainly by the phase-to-phase voltage, the PD is phase shifted from the phase-to-ground AC cycle phase position by either + or $-30°$. As discussed in Sections 10.5.4 and 16.12.1, this phase shift aids in identifying if this process is occurring.

16.4.5 Thermal Aging

Thermal aging is a common aging process in all types of equipment. For stator windings, the source of the heat is primarily the I^2R losses in the copper conductors within the coil and core losses in the stator. Depending on how aggressive the winding manufacturer has been, the copper cross-section and stator winding cooling system will result in a maximum coil insulation temperature (at the copper) of between 60 and 140 °C for industrial and utility machines using epoxy-mica groundwall insulation. The higher the temperature, the faster the insulation will age [5]. The presence of oxygen will also accelerate thermal aging. With epoxy mica insulation, the epoxy loses its bonding strength, and thus the mica tapes tend to separate under thermo-mechanical (heating/cooling) stress. Some call this process delamination of the groundwall. Delamination creates gas gaps between the tape layers, leading to PD in the high voltage coils. The PD, in turn, creates significant local heating and destroys the organic parts of the insulation system.

How rapidly this failure proceeds depends on how high the operating temperature of the winding is. Many windings that operate at moderate temperature (<120 °C at the copper) may take many decades to fail by this process. Since the insulation tapes near the copper are operating a few degrees warmer than the rest of the groundwall, the delamination process tends to occur closer to the copper conductors. This tends to result in a PRPD pattern with a negative PD predominance in the positive AC half cycle, assuming an indirect PD measurement method (Section 6.2.2).

16.4.6 Thermo-Mechanical Aging

Some machines tend to experience a lot of rapid load changes during operation. Examples include gas turbine generators, hydrogenerators and pump storage generator/motors. If such stators have cores longer than 2 m, then as the machine goes from low load to high load, the copper temperature will closely follow the load increase and due to the coefficient of thermal expansion, the copper may grow axially in the slot by a mm or so. Similarly, the copper shrinks as the load decreases. Since the coefficient of thermal expansion of the groundwall insulation is different from copper, a mechanical shear stress occurs between the copper stack and the groundwall. With many thermal cycles, the bond between the copper and the groundwall is fatigue-cracked, leading to an air space between the groundwall and the copper (even though there may be no thermal aging) [5, 7]. PD can then occur in the air (or gas) gap, which is a symptom of deterioration due to load cycling. The time to failure depends on the length of the stator slots, the difference in temperatures between high and low loads, and how rapidly the load changes. Failure will not occur until the PD tracks a path through the groundwall, or the copper stack vibrates enough to cause strand shorts and even higher local temperatures.

From a PD point of view, the PRPD patterns seem to be similar to those from thermal aging (Section 16.4.5).

16.4.7 Winding Contamination

Like thermal aging, coils being loose within the slots, and load cycling, another process that is primarily caused by operation and maintenance (or lack thereof) is winding contamination leading to electrical tracking over the endwindings. In air-cooled motors and generators, oil or moisture can sometimes enter the machine enclosure. When this liquid combines with almost any dust or dirt (e.g. cement dust, ground-up materials such as cloth, paper, brake dust, carbon dust from slip rings, and so on), a partly conductive coating (in the hundreds of kohm to the Mohm range) is deposited on the endwindings. Capacitive currents through the groundwall then flow through the contamination to ground or another phase. Wherever there is a dry (or high resistance) region, a high voltage may occur across the "dry" band, leading to a surface discharge (Section 3.7.2). The surface discharges burn and carbonize the organic materials on the groundwall surface. Eventually, carbonized tracks occur over the endwinding, sometimes burrowing into the groundwall, causing a ground fault or phase-to-phase fault. Depending on the conductivity of the contamination, the process may take from months to many decades to cause failure.

The PD that accompanies the electrical tracking process depends on the conductivity of the contamination, which in turn is dependent on the machine load (winding temperature) and the humidity. Tracking in the endwinding can be driven by both the phase-to-ground and the phase-to-phase voltages.

16.4.8 Metallic Debris

Like other types of equipment, sometimes isolated bits of metallic debris accumulate on the insulation surface in the endwindings, circuit ring busses, or connection leads. The debris could be shavings from drilling or grinding within the machine enclosure that was never cleaned up, metallic dust due to metal fretting, or tools or other items accidently left within the machine. If this debris is ferrous, magnetic forces may cause the debris to vibrate, work into the insulation, and eventually cause failure relatively quickly. If the debris has a sharp edge or point, then it may create a high electric stress (Section 2.1.3.4), which may be sufficient to create corona. Although the corona itself is usually relatively harmless, the existence of the metallic debris in a machine is always of concern and should be removed. Corona has a very distinctive phase position at the peaks of the AC cycle in the PRPD pattern (for example Figures 3.31 and 13.47).

16.5 PD Pulse Propagation in Stator Windings

A PD pulse within a stator winding must travel a complex path between the PD site and the PD sensor where the discharge pulse current is detected. Perhaps more than in any other type of electrical equipment, the three phases in a stator winding are physically close to each other, permitting significant mutual coupling of signals between phases. As a result the PD pulses can be attenuated, dispersed, and cross-coupled between winding elements as they propagate. Before the different types of detection methods are discussed, we first review the impact of pulse propagation on the detection sensitivity of the PD pulses in stator windings. This subject is also qualitatively discussed in the relevant standards [1–3].

16.5.1 Propagation Models

Work on models to predict how voltage pulses propagate in stator windings goes back to the 1930s. The initial concern was how voltage impulses from lightning strikes and switching created high turn-to-turn voltages in multiturn coils [5]. By the 1980s, analytical methods based on complex circuit models of windings were able to replicate experimental results on pulse propagation [8]. However, the models for switching impulse propagation assumed the shortest voltage risetime was

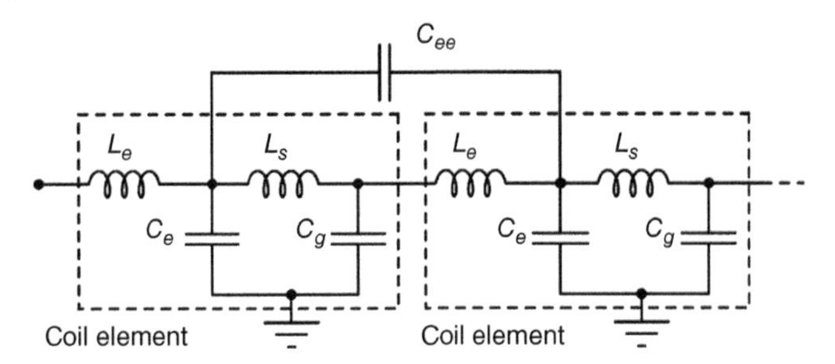

Figure 16.7 Simple LC model of each coil. The coil models are connected in series to simulate the propagation of a PD pulse from one end of a winding to the other. The model is complicated by the capacitance between coils, especially in the endwindings (C_{ee}).

about 100 ns, corresponding to a Fourier component up to about 3 MHz. PD has a much shorter risetime (typically a few ns), equivalent to over 100 MHz. Thus, the models and experimental results for surge voltage propagation in windings are not completely relevant for the propagation of PD pulses.

Figure 16.7 shows a very simple model for the first two coils in one parallel circuit between the line-end and neutral of one phase from a stator winding, which conveys the main elements of pulse propagation. Following on the work for surge voltage propagation in [8] and experiments by Zhu and Kemp [9], the coil consists of an inductance in the endwinding area (L_e) in series with an inductance for the slot area (L_s), together with a capacitance to ground in the endwinding area (C_e) and a capacitance to ground in the slot area C_g (where $C_g >> C_e$). In multiturn coils, there are also mutual inductances between the turns in the slot and endwinding areas, as well as turn-to-turn capacitances in the slot and endwinding area. These have been ignored for PD pulses since the capacitive impedance between turns is likely to be very low compared to the mutual inductance between turns at frequencies above a few tens of kHz. In fact, from the point of view of PD propagation, it is likely that the capacitive impedance between turns is so low, due to the thin turn insulation, that all the copper conductors in the slot and endwinding regions are at the same potential. With these simplifications, an equivalent circuit model for each coil is shown in Figure 16.7.

However, there is another propagation path for PD current pulses than just through the coils. In the endwinding area, two coils connected in series are often in two adjacent stator core slots. Thus, there is capacitive coupling in the endwinding area between the copper conductors in each coil (C_{ee} in Figure 16.7). For 3.3 and 4.1 kV stator windings, the coils in fact may be touching, creating a high coil-to-coil capacitance both in the endwinding and the slot regions (assuming no semiconductive coating is present in the slot area). If the windings are rated 6 kV and above, there is likely to be a grounded slot semiconductive coating, thus the coil-to-coil capacitance is limited to the endwinding region. For a 1 m long coil endwinding that has a copper stack about 50 mm high and a groundwall insulation thickness of 0.8 mm in each coil, this corresponds to a capacitance C_{ee} of about 1 nF. This results in a capacitive impedance that is 160 Ω at 1 MHz and only 0.16 Ω at 100 MHz. Especially at high frequencies, this capacitive impedance between coil/bar endwindings is likely to be much lower than the series impedance of the coil endwinding and slot regions. Thus, the high-frequency components of a PD pulse "skip" between the endwindings of adjacent coils/bars, rather than "travel" through the coils, encountering less attenuation. In contrast, the LF (<1 MHz) PD components will see higher coil-to-coil capacitive impedance, thus propagating through the coil inductance may be the preferred path with the least attenuation. This has led many researchers to conclude there are "fast" and "slow" PD propagation modes, each with its own attenuation [10, 11]. LF detectors are only sensitive to the "slow" mode, where VHF (30–300 MHz) detectors are primarily sensitive to the fast propagation mode.

This effect is more pronounced in lower voltage stators compared to higher voltage stators, since the spacing between coils in the endwinding tends to be greater at higher rated voltages, reducing C_{ee}. Similarly, the longer the endwindings (i.e. the higher the rotational speed), the greater is C_{ee}, and therefore, the lower the attenuation is between the PD site and the PD sensor.

Since the three phases are intermingled within the stator winding, there is also likely to be high capacitive coupling between the adjacent coils in different phases within the stator. This will lead to cross-coupling of PD in one phase to another phase, which often complicates the PRPD patterns from online tests on stator windings (Sections 10.5.5 and 16.12.1), especially with detection in the HF, VHF, and UHF ranges [12]. Generally, stator windings with short endwindings and higher voltage ratings experience less cross-coupling.

16.5.2 Experimental Findings

Experiments to measure how PD propagates in stator windings are very complex and require great skill. Two basic approaches have been taken:

- Create a PD source, such as a spark gap, in series with a small, high-voltage capacitance and insert it at different locations of the winding. Then, raise the winding voltage in an offline test until the spark gap is discharging, while detecting the PD at the machine terminal. As the PD source is moved to different positions in the winding, the signal attenuation and dispersion can be determined.
- Inject PD-like pulses from a pulse generator (typically 5 V, with a few nanosecond risetime) into various coils between the line-end and neutral while measuring the PD signal at the terminals. This test is performed with the machine shut down and does not involve high voltages.

Both methods have limitations. In the first method, the PD sources can only be installed at the series connections between coils or at the terminals, not in the slot region or just outside of the slot where most PD occurs. Not surprisingly, machine owners are reluctant to allow the needed winding modifications. The second method can normally only be done on stators that are to be rewound, since it is best to drill holes through the groundwall insulation (in the slot and endwinding regions) to allow galvanic injection of the pulse into the copper. Also, the pulse generator should have $<1\ \Omega$ source impedance and very short ground leads to the stator core, especially for detection in the HF and above ranges. Injecting pulses into the endwinding region may lead to problematic results due to the distance to the ground plane of the stator, which means that ground leads must be long.

Experimental results from reasonably careful experiments on turbine generators, hydrogenerators, and motors are in [9, 10, 12–14]. The signal from the PD sensor is measured either with an oscilloscope (which displays the PD signal in the time domain via a coupling capacitor or other PD sensor; or an oscilloscope probe) or with a commercial PD detector in different frequency bands. One of the earliest results is from three different 500 MVA turbine generators using VHF detection [13]. The PD sensor is connected to the high-voltage end of the winding. Figure 16.8 shows that there is a significant drop in output to an injected pulse between the line-end coil and the machine terminals, with the magnitude only 30–60% of the signal injected at the terminal. After the first coil, the signal attenuation somewhat stabilizes.

Lachance, in more recent measurements on a 234 MVA turbine generator, obtained similar results when injecting pulses at different locations in the winding and measuring the pulses with a 500 MHz analog bandwidth digital oscilloscope [14]. Figure 16.9 shows a large drop in signal along the circuit ring bus (from T2 to A), but relatively stable readings up to eight coils from the line end. This is similar to the pattern in Figure 16.8.

Significantly different results are measured when using a commercial PD detector operating in the LF and lower HF ranges on the same stator used in Figure 16.9. For this particular winding, the lower frequency ranges (120–280 and 100–400 kHz) show less attenuation in the first four coils (and no effect from the circuit ring bus) when compared to the three other measurement frequencies with upper cutoff frequencies of about 1, 2, and 10 MHz (Figure 16.10). This indicates that the attenuation can vary significantly depending on the measurement frequency in the LF and HF ranges and the distance of the PD site to the machine terminals. This result is similar to older work with conventional LF detectors. In contrast, the attenuation does not change significantly for

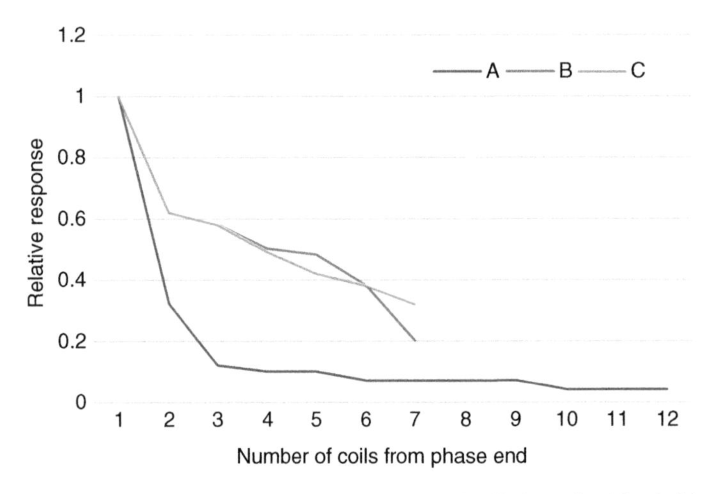

Figure 16.8 Attenuation versus the number of coils (or pair of Roebel bars) from the PD sensor at the three machine terminals measured with a 1 GHz analog oscilloscope for three different turbine generators. *Source:* Adapted from [13].

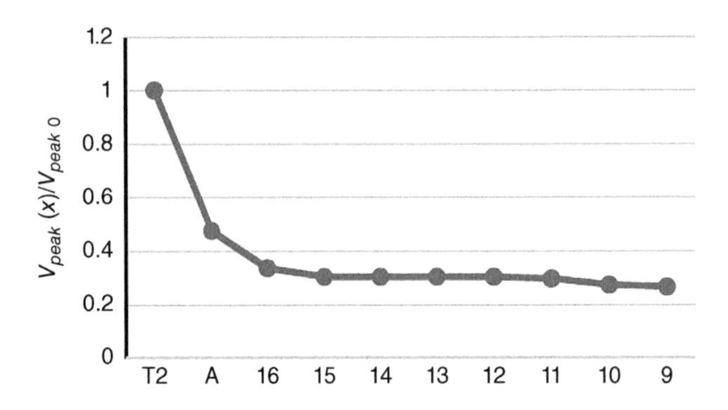

Figure 16.9 Attenuation of simulated PD pulses at different injection locations when measured at the winding terminal. T2 is at the winding terminal, A is where the circuit ring bus is connected to the first coil and 16 is one bar down, 15 is three bars down, etc. *Source:* [14].

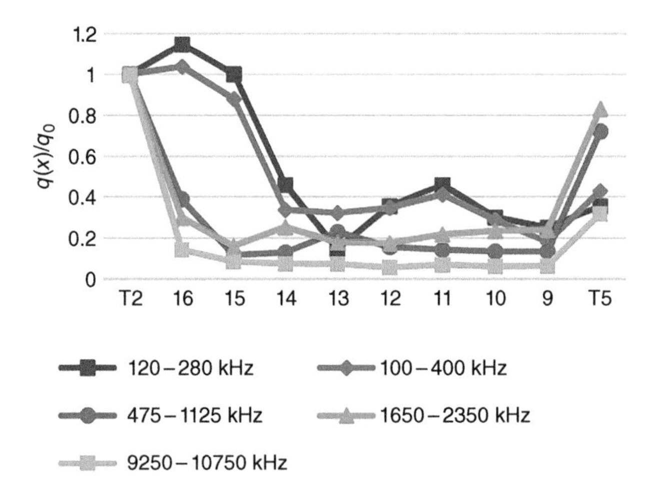

Figure 16.10 Relative response from a LF/HF commercial PD system in different frequency ranges where simulated PD is injected at different locations in one circuit parallel. T2 and T5 are the stator winding terminals at the high-voltage and neutral ends, respectively. *Source:* [14].

different PD injection sites in the VHF and UHF ranges (Figure 16.9). It also implies that one cannot compare PD magnitudes (in pC or mV) on the same winding when different instruments are used, different measurement frequency ranges are used, or both (see Section 6.6).

16.5.3 Impact of Coil Voltage on PD Signal

In Section 16.5.2, the detection frequency can result in different (usually lower) sensitivities to PD sites farther from the line-end terminal. This has an important impact in detecting PD from different winding locations in offline PD tests. In such tests, one phase of the winding is energized, and all the coils in that phase experience the same test voltage. That is, in an offline test, both the line-end and neutral-end coils are at the same voltage. Thus, PD is equally likely (assuming similar defects are present in all coils) to appear in any coil of the winding. If the PD sensor is located at the line end terminal, PD from the neutral end is likely to be detected as a much smaller signal due to pulse attenuation than PD in the line-end coil, even if the same defect size is present in both coils. Thus, to maximize sensitivity to PD occurring in as many coils as possible, it is advisable to measure the PD at both the line end and the neutral end when performing offline tests (Section 16.7.1).

The impact of pulse attenuation is less important in online tests. In an online test the voltage across the groundwall insulation depends on the position of the coil in the winding. A line-end coil has up to the full-rated phase-to-ground voltage across it (for example 8 kV rms for a 13.8 kV winding). In contrast, the neutral end coil will have close to 0 V across the insulation. PD cannot occur in a coil with close to 0 V across the groundwall insulation. Between the line-end coil and the neutral-end coil, the voltage across the groundwall insulation of coils in the parallel circuit drops linearly. As the voltage across a coil insulation decreases, so does the Q_{IEC} or Q_m (Figure 16.11). For the same size of defects, the PD magnitude measured at the machine terminal in an online test will primarily measure PD from defects in coils closest to the line end, even if there is no PD pulse attenuation between the defect site and the PD sensor, since the non-line end coils operate at lower voltage and, thus, have less PD activity. Anyone who has done many visual examinations of deteriorated stator windings can verify that PD deterioration is almost always on the line-end coils or within a few coils of the line end.

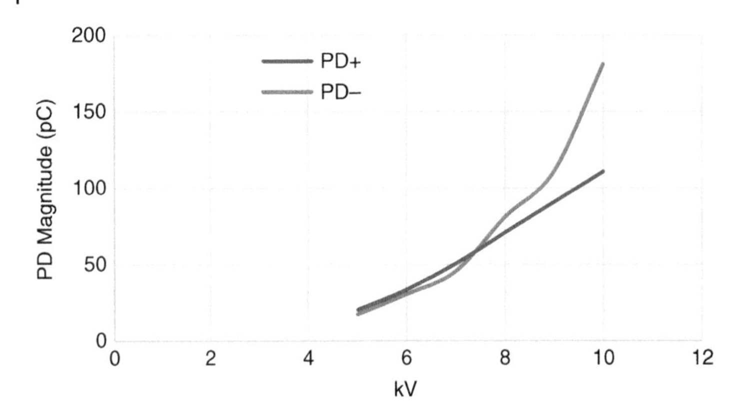

Figure 16.11 Tests on a 13.8 kV coil with known PD suppression coating problems and delamination, energized at different voltages above the *PDIV*. The PD was measured with a wideband IEC 60270 detector. *Source:* Adapted from [15].

16.6 PD Sensors

The types of sensors used for measuring PD on stator windings depend on whether the test is offline or online. For offline tests, the conventional IEC 60270 test in the LF range is the most common, which limits the range of PD sensors used. For online testing, an astonishing range of PD sensors have been used, although, in practice, either a large capacitance or a small capacitance sensor is most popular. For a general discussion on PD sensors, see Sections 6.2, 6.3, and 7.4.

16.6.1 Sensors for Offline Testing

For offline PD testing of coils and windings, the most popular way to detect the PD signal is an indirect measurement (Section 6.2.2) using a 1 nF or so high-voltage capacitor together with a detection impedance (or quadripole). Factory measurements tend to be made with IEC 60270-compliant measurement systems, and such systems are also common for onsite/offline tests. However, since many motor and generator factories and industrial plants tend to have many sources of electrical interference (sparking from overhead crane systems, switch-mode power supplies, welding systems, and power tool operation), LF measurements may have to be made when the manufacturing plant is not operating, i.e. during the night or on weekends. Alternatively, for factory PD testing, the PD measurements can be made in an electromagnetically shielded (Faraday) room (Section 9.3.1).

Especially in the Americas, some OEMs instead prefer to employ sensors for offline factory tests that are more commonly used for online PD detection, for example 80 pF sensors working in the VHF range. This usually permits PD measurements to be made in the normal factory environment during working hours, since VHF measurement is less susceptible to interference from the previously mentioned sources (Section 9.3). Similarly, generating stations and industrial plants may also use 80 pF sensors for onsite/offline tests, due to their higher interference immunity and additional means available to suppress interference, e.g. using time-of-flight methods (Section 9.3.6) and pulse shape analysis (Section 8.4.3).

16.6.2 Sensors for Online Testing

Over the past 70 years, virtually all types of PD sensors described in Sections 6.3 and 7.4 have been used for online PD detection in stator windings. Small capacitance sensors with a VHF measuring system and large capacitance sensors with a HF measuring system are the most common. Tens of thousands of machines have been equipped with permanently installed 80 pF or 1 nF (or so) sensors.

HFCTs installed on the generator neutral leads or the ground leads of surge capacitors on motor terminals have also been used, especially in combination with EMI monitoring (Sections 1.5.1 and 7.5). The Rogowski form of HFCT with an upper cutoff frequency of 100 kHz or so, installed around the unshielded portions of power supply cables, has been used for online PD detection in specialized situations in off-shore oil platforms in the UK.

UHF antennae are also used for PD detection, especially in large turbine generators that are typically hydrogen-cooled. In such machines, the PD levels are suppressed by the high-pressure hydrogen (Sections 16.4 and 16.11.2), at the same time the auxiliary systems tend to produce more interference, creating a low PD signal-to-noise ratio. As with GIS and power transformers, UHF PD detection tends to produce fewer false positive indications due to interference. The most common UHF sensor is the stator slot coupler or SSC (Section 7.4.5.4). Some generator OEMs also use UHF antenna installed within the generator enclosure and facing the endwinding region to detect endwinding PD in large generators.

When any type of capacitor is used for online PD testing of machines, special care is needed to ensure sensor reliability (Section 6.2). Such sensors are connected to the phase terminals of the machine and see the full-rated voltage of the stator winding. If the sensor breaks down or flashes over, the motor or generator protection system will take the machine out of service. The relevant IEEE and IEC standards have extensive discussions of the type and routine tests needed to assure machine operators that the PD sensors themselves will not lead to machine failure [1, 3].

Similarly, when capacitors are used in petrochemical plants, special testing is required to ensure that the sensor and connected termination panel does not spark, which could ignite an explosive gas that may be in the machine environment. This requires certification testing by various bodies according to ATEX or other hazardous area requirements [16, 17]. In addition, sensors used in nuclear applications such as reactor coolant pump motors require special tests for radiation, seismic vibration, and loss-of-coolant accident conditions [18].

16.7 Factory Acceptance Testing

Most motor and generator factories have a 50 or 60 Hz high-voltage supply that can energize the coil/bar or winding capacitance for a factory PD test. The reactive power volt-amps (VA) required by the test supply is:

$$VA = 2\Pi fCV^2$$

where f is the AC frequency, C is the capacitance of the test object, and the coupling capacitor and V is the applied voltage for the test.

For coil and bar tests, the capacitance is usually about 2 nF or less. At a 25 kV test voltage, the supply must be rated for 500 VA. A typical winding may have a capacitance per phase of 0.5 μF. For a maximum test voltage of 25 kV, the power transformer must be rated about 125 kVA, which is substantial, unless a resonant circuit is used to reduce the real power supplied. For motors with a "wye" connection at the neutral that is not accessible, the capacitance of all three phases must be energized at the same time, requiring a higher kVA rating for the AC supply. Since a large AC supply is needed for performing the AC withstand (hipot) tests required by IEC 60034-1 and IEEE C50.12 or C50.13, a suitable test supply is readily available in the factory.

Most major manufacturers of coils and windings rated 6 kV and above are likely already performing offline PD tests in the factory for quality control purposes. That is, manufacturers use the PD test to ensure that the *PDIV*/*PDEV* has not decreased from normal production or the Q_{IEC} or Q_m (Section 8.8) has not increased from the normal range, since this is an indicator that material properties have changed; or taping, impregnation, and coil insertion in the stator may not have been well performed. These manufacturers will have established a "normal" range for *PDEV*, Q_{IEC}, or Q_m based on experience.

PD tests in factories are almost always done in an open-air environment, even for coils and windings that will operate in high-pressure hydrogen. Since high-pressure hydrogen strongly suppresses PD (Sections 2.3, 3.8.2, and 16.11.2), the *PDIV* and *PDEV* will be much lower and the Q_{IEC} and Q_m will be much higher in air than in service. Thus, the results from a factory test cannot be directly compared to onsite/offline PD tests under pressure or online tests. In fact many hydrogen-cooled machines produce very little PD at all at operating hydrogen pressure and rated line-to-ground voltage.

Refer to Sections 10.2 and 10.3 for general test requirements to measure Q_m, Q_{IEC}, *PDIV*, and *PDEV* when performing offline tests.

As stated earlier, there is no formal requirement in relevant IEEE or IEC standards for coil and winding manufacturers to pass a PD test. IEEE 1434 and IEC 60034-27-1 only provide guidance on how to do offline PD tests and some very basic advice in interpretation. If end users want a factory PD test, they must stipulate this in the purchase contract. The purchase document must also identify the PD test method and the pass criteria. One utility did establish a 100 pC limit at rated line-to-ground voltage for coils when using a wideband IEC 60270 detector [19]. However, others have shown that even with coil/bar test objects, there may be up to a 300% difference in Q_{IEC} (in pC) using different brands of IEC 60270 detectors on the same coil or bar [20]. Levesque has found that some of this variation may be inherent in using Q_{IEC}, which tends to produce less repeatable results than measuring what she calls Q_p, which is the highest PD pulse over a several second interval that is clearly not associated with interference [21].

Levesque and her colleagues at Hydro Quebec have also explored using the "corona" or "TVA" probe (Section 16.13) for factory acceptance testing of stator bars and coils [22]. They seem to infer a limit of 5 mA for modern bars and coils at rated line-to-ground voltage.

16.7.1 LF Charge-Based Test Procedures

Commercial IEC 60270 charge-based measurement systems (Section 6.5) using an indirect circuit and a coupling capacitor (Section 6.2.2) seem to be the most widely employed method for both coils and complete windings (Sections 6.5.1 and 6.5.2) [23].

When PD is measured in the LF range (Section 6.5), a coil/bar behaves as a lumped capacitor, and thus, either the wide-band (few hundred kHz bandwidth) or the narrow band (a few tens of kHz bandwidth) methods in IEC 60270 can be used for coils and bars. Wide-band detection is most common unless there is significant interference at specific frequencies below 1 MHz. The results are calibrated in terms of apparent charge (pC, nC) using the process in Section 6.6.1. IEEE 2465 (currently in development) provides additional advice for performing PD tests on stator bars and coils [24].

For complete stator windings, a commercial IEC 60270 detector in the LF region should measure the PD in the wide-band mode (i.e. several hundred kHz bandwidth). Narrow band measurements should be avoided due to the high probability of natural frequencies in the stator winding in the LF range that can lead to resonance (Section 6.6.3). For example, Figure 6.29 shows the impedance versus frequency for a typical motor stator winding. Natural frequencies around 30, 50, 90, 150, and 600 kHz are noted. If a narrowband detector is centered on 30 kHz, a large PD response will be measured that will not likely be corrected for by the calibration process, whereas narrowband detection at 40 kHz will produce a smaller response. To reduce the variability caused by natural frequencies and resonance in the stator winding, IEC 60034-27-1 recommends LF measurements in the wide band mode to reduce the impact of any natural frequencies. Note, however, that even in the wideband mode, the effects of natural frequencies will affect the calibration procedure. Thus, tests must always be done in the same frequency range.

Calibration, in terms of apparent charge, using the procedure in IEC 60270 is not valid for the complex inductive-capacitive nature of a complete winding, since the test object is not a lumped capacitance (Section 6.6 and [23]), even in the LF range. In spite of this, it is common to use "apparent charge" for LF measurements, especially in Europe. Measurements on the same stator winding with different commercial IEC 60270 detectors at slightly different wideband frequencies may yield as much as a 30 to 1 variation in apparent charge (Q_{IEC}) values [15]. Users should be aware that the apparent charge readings from one detector are unlikely to be similar to the charge readings from another brand of IEC 60270 detector when testing the same stator winding. Comparisons of PD levels in Q_{IEC} between machines is only valid if the design of the windings is the same, and the same type of PD detector in the same frequency range and test setup is used.

A PD test at rated phase-to-ground voltage seems to be preferred in North America, whereas in Europe the preference is for a test at the rated phase-to-phase voltage, applied line-to-ground (i.e. at a 70% overvoltage for the groundwall insulation system). In addition to a three-phase test, if the phases can be isolated from one another at the neutral, it is preferable to test one phase at a time, with the other two phases and the stator frame grounded. This enables PD to be measured in the phase-to-ground insulation in the slot, as well as the phase-to-phase insulation in the endwinding. Note, however, that a test at rated phase-to-phase voltage will overstress the groundwall insulation and in particular the silicon carbide stress grading coating (Section 16.3.2), producing much higher PD than would occur in service. In contrast, a test at rated line-to-ground voltage will subject the insulation within the slot to the same stress level as would occur in service, but much less PD will occur in the endwinding due to the lower phase-to-phase voltage compared to normal operation.

If the phases can not be disconnected at the neutral, as is common in motors, then all phases will be at the same voltage, and thus, there will be no potential difference between adjacent coils in the endwinding. Hence, even severe insulation problems in the endwinding may not be detected by a three-phase test.

IEC 60034-27-1 gives an overview for all possible test connections for stator winding factory offline tests. Due to the attenuation of PD pulses as they propagate through a winding (Section 16.5), it is usual to test each phase with the PD sensor at the line-end terminals and then at the neutral-end terminal(s).

Electrical interference is sometimes present in coil and winding manufacturing plants due to power tool, welding, and crane operation. Sometimes this interference can be suppressed by judicious selection of the detection frequencies that can be easily varied in modern digital PD instruments (Section 9.3.4), combined with good grounding practice, power supply filtering, and/or the use of shielded rooms (Sections 9.3.1–9.3.3). If interference is still a problem, then some commercial detectors offer post-processing methods for IEC 60270 detectors such as time–frequency (TF) mapping (Section 8.9.2) or variations of it. However, the selection of clusters that are likely due to interference, just like observing the PRPD pattern directly, is somewhat subjective. Also, since IEC detectors operate at less than 1 MHz and thus the pulses are heavily modified by the integrator (Section 6.5), it is likely that the pulse frequency and time characteristics will be very similar for interference and test object PD.

To obtain stable, repeatable measurements, it is important to "condition" the coil or winding first (Sections 3.8.1 and 10.3.4). For coils and windings, the most common conditioning method involves raising the voltage to the test level and holding the voltage constant for at least five minutes before measuring the Q_m or Q_{IEC} and capturing the PRPD pattern. In fact, it may take up to 15 minutes for the PD level to stabilize. Thus, the five minute conditioning duration is a compromise between stability and testing speed, which can be lengthy if many coils/bars need to be tested. Another way to condition the coil or winding is to increase the voltage well above the rated line-to-ground voltage and hold it for a few tens of seconds, before determining the *PDIV*, *PDEV* and PD magnitude.

The outcome for factory LF PD tests is the *PDIV*, *PDEV*, Q_{IEC} (in nC), and the PRPD plot (Section 10.5) at rated line-to-ground or rated phase-to-phase voltage. If significant PD is detected in comparison to the windings of the same design tested with the same detection system, then sometimes the PD can be located within the coil or winding using the optical, acoustic, or RF methods described in Section 16.13, or the acoustic and optical methods in Sections 5.2, 5.3, and 5.4. IEEE 1434 suggests winding and coil manufacturers provide information for the "normal range" of *PDIV*/*PDEV* and Q_{IEC} for particular test object designs, but it seems these have not been published.

16.7.2 HF and VHF Methods

If significant levels of interference are present, then some manufacturers perform the factory test at higher frequencies than the upper limit of 1 MHz limit of IEC 60270 detectors. If LF detectors are used in the presence of interference, considerable experience is needed to subjectively separate test object PD from the interference.

The most common method to suppress interference in offline tests is to use VHF detection with two 80 pF capacitive couplers, and the time-of-flight method to suppress external interference from the power supply (Sections 8.4.2 and 9.3.6). As shown in Figure 8.5, one sensor is installed at the coil or winding under test, and the other sensor is connected close to the power supply. There must be a minimum distance between the sensors that depends on the PD instrument being used. For a VHF instrument that can resolve differences in arrival times of pulses from the two sensors of 6 ns or more, there must be 2 m of air-insulated bus or (better) about 1.5 m of shielded insulated cable between the sensors to determine if the signal is from the EUT or from the power supply. The advantage of this interference suppression method is that the suppression is automatic and does not require subjective judgment on what is test object PD and what is interference.

An HF method that is employed to suppress interference uses the T–F map approach to classify interference and test object PD (Section 8.9.2). Some skill is needed to correctly identify the interference and PD clusters on the T–F map.

Note that, as required by IEC 60270 and IEC 62478, the HF and VHF methods cannot be scientifically calibrated in terms of apparent charge, so the PD magnitudes should be measured in mV or some other units than pC. Otherwise, all other aspects of factory tests using HF and VHF detection is the same as in Section 16.7.1.

16.8 Onsite Offline Tests

Offline tests performed at the generating station or industrial plant are sometimes used during commissioning to verify that shipping and machine installation did not affect the insulation system. For large hydrogenerators that are often assembled at site, the test also serves to prove the coil installation and endwinding bracing have been adequate. The other purpose of onsite, offline tests is to determine when stator winding maintenance may be needed by trending the PD activity over time.

Onsite/offline PD can be measured either with a LF (IEC 60270 compliant) detector, or with HF/VHF detectors. The latter tend to be used if there are high levels of external interference. In any case, a formal calibration to apparent charge per IEC 60270 is not appropriate (Sections 6.6.3 and 16.7.1) since users may incorrectly feel the charge readings are absolute (not relative) and comparable without restriction. Instead, the PD magnitudes only have meaning in comparison to results from other very similar windings using the same detector and measurement frequencies and test set-up; or by trending the PD on a winding over time using the same detector and measurement frequency.

There are two key differences between factory tests and onsite, offline tests:

- a suitable portable AC test supply needs to be procured
- electrical interference is often more severe in generating stations and industrial plants.

A suitable 50/60 Hz power supply with the required kV and kVA ratings may be difficult to get to site and the required input power supply may not be available within the plant. A special-purpose mobile AC generator may be required to energize the test transformer. Note also that a means to gradually increase the voltage is needed, which normally requires a motor-driven variable autotransformer (Variac™ or similar). One practical solution is to use a resonant AC supply. Such supplies have a variable inductor in series or parallel with the capacitance of the winding. By adjusting the inductance to the capacitance of the winding, the circuit can be made resonant with a reasonable Q at 50 or 60 Hz. In this case, the AC supply only provides the losses to the circuit. For testing one phase of a stator winding, it is possible to supply 50 kVA from a 200 to 240 V output, 20 A circuit in the station or plant. Such resonant supplies are more expensive than conventional test transformers equipped with a variable autotransformer, but are smaller and lighter in weight. Unfortunately, some types of variable inductance produce considerable electrical interference, which makes noise suppression more critical. Another factor in using resonant supplies is that if the stator winding has significant PD activity, the losses from the PD will decrease the Q of the resonant circuit, and it may not be possible to energize the winding to the required test voltage. There are several manufacturers of resonant AC test sets, including Hipotronics (Hubbell), HighVolt, and Phenix Technologies (Doble).

An alternative to conventional or resonant AC supplies is to use a "very low frequency" or VLF AC supply. Such supplies usually provide a 0.1 Hz AC output. This results in a supply current to the EUT that is 1/600 of that needed for a 60 Hz AC supply. Even the largest windings can be supplied from a standard 100–220 V 50/60 Hz wall outlet. The basic concepts for this supply are in IEEE 433. To date, VLF supplies have not proved popular for stator winding PD testing. One reason for this is selecting the test voltage, and if it is equivalent to the 50 or 60 Hz voltage. Also, some types of VLF testers produce considerable interference which makes it difficult to extract PD from the interference pulses. In addition, the PD instrumentation must be able to synchronize to the 0.1 Hz frequency to produce a PRPD plot. Baur GmbH and High Voltage Inc. are two well-known suppliers of VLF test sets.

The second issue with offline, onsite PD testing is the higher interference environment. There is less ability to suppress the interference using power supply filtering, shielded rooms, and performing tests on the weekend, than may be available in factory tests. Thus, one or more of the following methods should be used to reduce the risk of false positive indications:

- Interpretation is done by a very experienced machine PD expert looking at the PRPD patterns
- Use of a time-of-flight interference separation method, which is only available when using detectors measuring in the VHF or higher frequency ranges (Sections 8.4.2 and 9.3.6)
- Use of T–F (time–frequency) mapping (or variations) to identify clusters of pulses that may be deduced as interference based on their PRPD pattern (Sections 8.9.2 and 9.3.8). This method is normally done in the HF range.

There are special considerations for hydrogen-cooled machines. Most PD test service providers perform the PD tests in air, as is done in factory testing (Section 16.7). Many owners of large hydrogen-cooled turbine generators have become unnecessarily alarmed when they see very high PD magnitudes and low *PDEVs* when tests are done in air. Yet in service, little or no PD may be present, and the stator winding insulation is likely to be in good condition. If possible, it is useful to also perform the offline, onsite test with the normal hydrogen pressure within the machine.

In most other respects, offline PD testing in a generating station or industrial plant is the same as tests done at the factory (Section 16.7).

16.9 Online Testing and Monitoring

The purpose of online PD testing or monitoring of stator windings is to help motor and generator owners determine when an action plan for stator winding insulation maintenance is needed. This technology can be traced back to the early 1950s (Section 1.5.1), and today there are tens of thousands of motors and generators either undergoing periodic online testing or equipped with continuous online monitoring. In both cases, the PD sensors must be installed (usually permanently, even for periodic testing) during a short outage or when the machine is being manufactured. In contrast with other types of equipment in this book (Chapters 11–15), most air-cooled machines rated 6 kV and above will have measurable PD from the first time they operate. Thus, recognizable PRPD patterns are usually present for the entire life of the machine. Also, as discussed later, the PD tends to be stable for years or decades until significant insulation aging starts to occur, which causes the PD magnitudes at operating voltage to increase significantly. The significant increase in PD magnitude is an indication that further tests or inspections are needed to confirm an insulation problem and/or plan appropriate maintenance. There are over a dozen vendors of online PD systems for stator windings, including systems from motor and generator manufacturers.

Since, by definition, motors and generators are connected to the power system for the online tests, there is usually considerable interference from the power system, some of which may be corona or PD in other connected apparatus. Generating stations and industrial plants usually also have many additional sources of electrical interference from other equipment in the plant. Thus, to have a low risk of false positive indications, it is essential to either have the results reviewed by a machine PD expert and/or use effective interference suppression methods. Until the 1990s, reliable online PD test results could be obtained by less than a dozen or so "experts" in the world.

The interpretation of PD test results is significantly different between machines and most other HV equipment (Chapters 11–15). This is because machine windings use PD-resistant mica, whereas most other HV apparatus uses only organic insulation, where any PD may lead to failure in days, weeks, or a few months.

16.9.1 Sources of Interference

Section 9.2.2 identifies the common sources for pulse-type electrical interference for all types of environments and types of electrical equipment. However, rotating machines have some interference sources that can be particularly troublesome for online PD testing/monitoring. Generators rated about 50 MVA and above are usually connected to the step-up transformer by isolated phase bus (IPB). In IPB (see also Section 14.8), for each phase there is usually a center hollow aluminum tube that is supported coaxially from an outer grounded shield by ceramic post insulators, as described in IEEE C37.23 [25]. To allow for axial expansion as the bus is loaded (which increases the bus temperature), there are usually flexible links between bus segments, and between the bus and the generator terminals. In large machines, there are usually dozens of bolted connections at the links. If even one of these bolts is not properly torqued, contact arcing can occur (Section 4.4). If the contact arcing is confined to just one or two bolts, there is little risk of IPB failure. However, this sparking produces PD-like pulses than can be orders of magnitude larger than stator winding PD, especially for hydrogen-cooled generators in which the winding PD is comparatively low [26]. This sparking can easily mislead machine owners to believe the winding has severe PD. Similarly, the ceramic insulators sometimes crack within the IPB due to thermomechanical forces, which can create very large PD, the unwary may think the discharging is due to stator winding insulation problems. Some types of IPB have a sliding contact between the top of the ceramic post insulator and the HV bus. This sliding contact, if not well maintained, can produce astonishingly large "floating potential" discharges that are harmless (Figure 4.1). Perhaps 30% of IPB connected to large generators experience this usually harmless discharging or sparking, and it has led to many false positive indications of serious stator winding insulation problems.

16.9.2 Periodic Testing vs Continuous Monitoring

Either periodic online testing or continuous monitoring can be effective to determine when stator winding insulation maintenance is needed. Unlike most other types of HV equipment that use purely organic insulation, PD progresses from a "low" level to a "high" level usually over three or more years. Hence, the relevant standards (IEC 60034-27-2 and IEEE 1434) suggest periodic testing every six months is usually sufficient to detect even relatively fast failure processes such as those described in Sections 16.4.1, 16.4.3, and 16.4.8. Six-month testing intervals allow enough time to verify high levels and identify an increasing trend, and for asset managers to plan corrective maintenance before an in-service failure is likely. From the 1950s to the 1990s, before commercial continuous PD monitors became available for stator windings, only periodic testing twice a year was done on machines – yet it was found to be effective.

The key features of periodic testing are:

- permanently or temporarily installed PD sensors (Sections 8.6.1 and 16.6.2)
- a method of synchronizing the PD to the machine's AC phase angle (Section 8.3.2)
- a portable instrument that can suppress interference as well as record the magnitude, number of pulses, phase position and polarity of the PD so that PRPD and other plots (Sections 8.7.3 and 10.5) useful for diagnostics can be created

- an interface to a computer for instrument control, data storage, and any post-processing after the raw data has been digitized
- a method for recording the operating voltage, winding temperature, load, hydrogen pressure (if applicable), ambient humidity; all of which can affect the PD activity.

Often this last step is done manually. A standard periodic test of a stator winding usually takes less than 30 minutes, unless the PD is to be recorded under different machine operating conditions.

The first commercial continuous PD monitors for machines were commissioned in 1994 [27]. Since then, probably more than 10,000 critical motors and generators have been equipped with continuous PD monitors by different vendors. Unlike the instruments needed for periodic monitoring, continuous monitoring technology changes rapidly due to advancements in computers, electronic communication, and ever-changing software platforms such as Windows™.

Continuous monitors for stator windings must have all the features of a periodic system above, plus:

- be permanently connected to the PD sensors, and incorporate protection against the voltage surges that typically occur in power systems
- automatically collect the motor or generator operating conditions either by a digital interface with the plant computer or directly by measurements from the machine's potential/current transformers and winding temperature indicators
- have a control system that initiates a measurement on a regular schedule such as once per day (measurement intervals less than this are not particularly useful since the time scale from the emergence of PD to indicating a high risk of failure is usually years, not days or months as is the case for power cables and transformers). Alternatively the control system can initiate measurements when some operating condition changes, for example, when the load has increased by 10%. Some control systems use a combination of the two approaches
- have algorithms to reduce the prodigious amounts of data collected, for example, by only permanently saving data when the PD activity has changed by some margin.

There is no purely technical reason to prefer periodic testing or continuous monitoring for stator windings. However, advantages of continuous PD monitoring are:

- As long as the system is operating properly, the data will always be collected and can be retroactively reviewed at any stage. Regrettably, many plants install PD sensors and procure a portable test instrument, but over time stop collecting data due to personnel changes or changing management priorities.
- Accurate machine operating data is collected and stored with each PD recording, which is essential for evaluating trends over time (Section 16.11.1) and identifying failure processes (Section 16.12.2).
- When the system is connected to the company intranet, data can be reviewed anywhere. Technicians no longer have to travel to the machine location, reducing ongoing testing costs. Also, instrument damage due to constant shipping is eliminated if the test instrument is shared between plants (as is common for periodic testing).
- The maximum possible warning time is given to plant maintenance engineers, since alerts of high PD or rapidly increasing PD activity are possible.
- For plants that connect to a company-wide central condition monitoring center, continuous PD data can usually be easily integrated to the center for trending and the application of artificial intelligence and statistical techniques to find abnormalities in the data that warrant further examination.

- If company cyber security rules permit this, the data can be reviewed over the internet or by cell/mobile phone connection by machine manufacturers, PD system suppliers or other third parties who may provide additional guidance on interpretation.

Another feature of continuous monitoring is that most continuous systems can provide “alerts” of “high” PD activity or a rising trend in PD activity to plant operators (Sections 16.11 and 16.12). Although such alerts for plant operators may be useful for equipment such as transformers, GIS, or power cable, they should usually be avoided in the case of motor and generator PD monitors. Plant operators generally get alarms for rapidly developing issues in the plant that often require immediate action. If operators get a PD alert for a stator winding, unless the motor or generator fails within their shift, which is highly unlikely, they tend to assume the alert is a false indication and ignore such alerts in future. Since stator winding PD is a slowly developing issue, it is far better to provide PD alerts only to maintenance personnel and not the plant control system. Such maintenance personnel should have training to recognize that they have time to do follow-up investigations.

The disadvantages of continuous PD monitoring on motors and generators compared to periodic monitoring are:

- High initial capital cost.
- Ongoing software update costs, as well as obsolescence as software platforms and communication protocols are no longer supported.
- Since motors and generators are often installed outdoors, the hardware may be damaged if the system is installed in direct sunlight, is supplied with power of poor quality, of where the ambient temperature exceeds 40 °C or so.
- Software platforms and hardware (especially the integrated circuits) tend to become obsolete relatively quickly, requiring the PD system vendor to frequently modify the product, even if the system specifications and features do not change. If continuous PD systems are installed in a plant or a company over many years, it is possible that later systems will not be interchangeable with systems installed earlier.
- Repairs often require the suppliers to visit the site to swap out boards, upgrade software and recalibrate the system, increasing repair costs. In contrast, damaged portable instruments can be shipped to the vendor for repair.

16.9.3 Common Testing/Monitoring Systems

Since the 1990s, in terms of machines monitored, the most common online method for both periodic testing and continuous monitoring uses 80 pF sensors together with VHF detection, allowing external interference suppression based on pulse shape analysis (Section 8.4.3) and/or time-of-flight discrimination (Section 8.4.2). The next most common online systems tend to use 1 or 2 nF PD sensors with instrumentation that works in the HF range. They use filters or post-processing methods such as time frequency maps (Section 8.9.2) or three-phase synchronous methods (Section 8.9.3) to identify known interference sources. For hydrogen-cooled machines, the UHF detection method using SSC-type sensors seems to be most common. This technology separates stator winding PD from both external (to the machine) and internal interference based on pulse shape analysis (Section 8.4.3) and the natural ability of UHF detection to suppress signals that are distant from the sensors (Section 7.4.5.4).

Conventional IEC 60270 detection systems (i.e. those operating in the 10 kHz to 1 MHz range) are rarely used for online PD detection. Most generating stations and industrial plants produce significant pulse-like interference in this frequency rage, from static excitation systems, variable speed drives, power line carrier communications, and/or battery charging systems. So this method will not be discussed further here.

16.9.3.1 VHF Methods

For machines connected to the power system by XLPE or EPR cables, the risetime and duration of the initial pulse lengthens as it propagates along the power cable (Section 12.6). Thus, PD-like interference from the power system and PD from equipment connected at the far end of the cable can be separated from stator winding PD by using a single VHF sensor per phase at the machine terminals (typically an 80 pF capacitor, Figure 16.12). Even PD in the switchgear that starts as a few nanosecond risetime pulse will distort as it propagates toward the stator winding along the power cable. After propagating 30 m or more, the risetime of the pulse from the power system is typically 6 ns or longer [28]. Pulses from the stator winding are largely unchanged in shape since the winding is only a few meters away. As discussed in Section 8.4.3, digital hardware can measure the risetime, pulse duration or other aspects of each pulse, and discriminate between stator PD and external pulses.

For machines where the power cable length is insufficient to significantly distort the pulse shape, or when the machine is connected to the power system by air-insulated isolated phase bus (IPB), which does not significantly distort the pulse shape, the time-of-flight method is used for distinguishing interference pulses (Section 8.4.2). For this method, two sensors per phase are used to separate stator PD from PD/sparking in the IPB along with other interference sources entering from the power system. The arrival times of a pulse at the two sensors can then be used to identify from which direction the pulse is coming. As discussed in Section 8.4.2, and shown in Figure 8.5, the minimum separation between the two sensors depends on the resolution of the electronics and the risetime of the pulses. For VHF instruments and 1–5 ns risetime pulses measured with 80 pF sensors, the two sensors should be 2 m or more apart to assure stator PD is separated from interference originating within the IPB or elsewhere in the power system. Figure 16.13 shows the 80 pF PD sensors installed at the machine terminals of a large turbine generator. There will be another set of sensors located 2 m or more along the bus, toward the step-up transformer. Figure 8.6 shows the

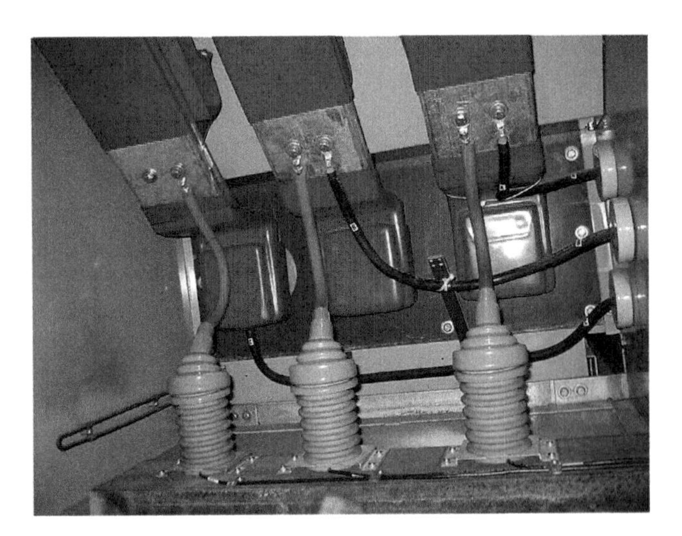

Figure 16.12 Photograph of 80 pF sensors (one per phase) installed on the terminals of a small generator. *Source:* Courtesy of Iris Power L.P.

Figure 16.13 Photograph of PD sensors installed at the end of the IPB near the generator output bushings. *Source:* Courtesy of Iris Power L.P.

stator PD activity separated from both the interference between the two sensors and from the power system, based on the time-of-flight method. Note that because the risetime of pulses is not 0 ns, and due to electronics processing time, there is about 1 m of uncertainty in the exact location of the signal source.

A variation of the time-of-flight interference suppression technique is used in large hydrogenerators that contain lap-wound stator windings (as opposed to wave wound stators). Lap wound stator windings almost always have relatively long circuit ring busses (2 m or more) connecting the individual parallel circuits to the machine terminals. Instead of mounting a pair of sensors on the output bus, the sensors are mounted at the end of the circuit ring bus where it connects to the parallel circuit (Figures 16.14 and 16.15). The length of the coaxial cables from the sensors are adjusted to ensure that electrical interference (noise) from the power system takes the same time to reach the two inputs of the PD instrument. Thus, if an interference pulse comes from the power system, the pulse will arrive at the two inputs of the instrument at the same time (within 6 ns of each other), and digital logic categorizes the pulse as interference. In contrast, PD located in a line-end coil/bar near (for example) the C1 sensor will arrive at the C1 instrument input more than 6 ns before it could arrive at the C2 input, as long as the circuit ring bus lengths to the terminals is greater than 1 m. More details about this popular method for hydrogenerator online PD detection are in [5, 29]. The advantage of this specific application of the time-of-flight method is that the PD activity in each parallel circuit can be identified separately, while also suppressing external interference.

All of these VHF interference suppression methods will also reduce sensitivity to PD pulses which have a long risetime (for example >6 ns).

16.9.3.2 HF Methods

Another popular method for online monitoring of PD in stator windings is to use high-frequency instrumentation, working in the 3–30 MHz range. The most common sensors used are 1 or 2 nF capacitors installed at the machine terminals (Figure 16.16), and sometimes at the neutral

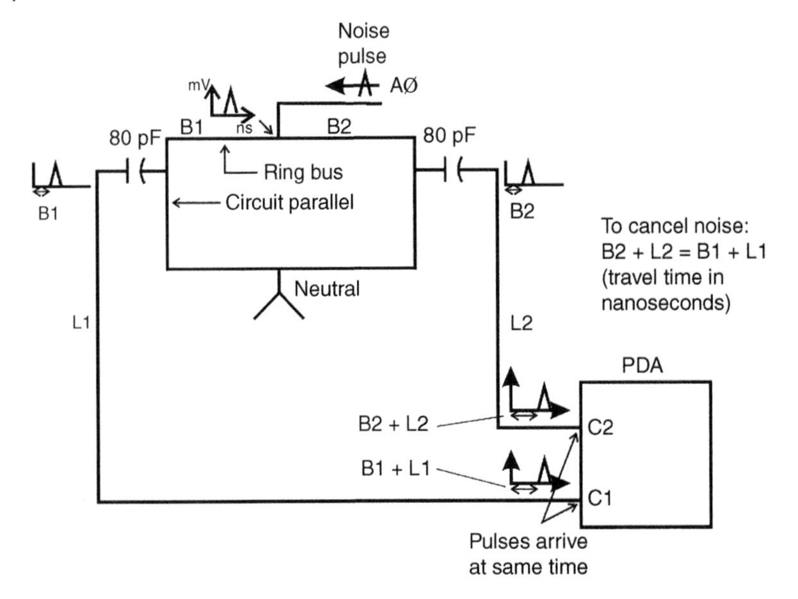

Figure 16.14 Schematic for the location of the pair of PD sensors at the ends of the circuit ring bus for one phase in a hydrogenerator. The length of the coaxial cables from the sensors are adjusted to ensure that electrical interference (noise) from the power system takes the same time to reach the two PD instrument inputs. *Source:* [5]/John Wiley & Sons.

Figure 16.15 Installation of 6 (of a total of 12) 80 pF PD sensors at the end of the circuit ring busses on the parallel circuits of a large hydrogenerator. *Source:* Courtesy of Iris Power L.P.

terminal as well. Some vendors have provided sensors as large as 9 nF in the past. Alternatively, an HFCT is installed on the ground side of surge capacitors (sometimes used in industrial motors) or the link between the power cable neutral and ground, if it is at the machine terminals. The signals from the sensors are directly fed to an analog-to-digital converter for processing, similar to the arrangements in Figures 8.2 or 8.3. Typically the A/D converter has a sampling frequency in the range of 20 MHz, corresponding to a 10 MHz analog bandwidth. Various PD system vendors use one or more of the following techniques to suppress electrical interference:

- Filtering with adjustable lower and upper cut-off frequencies (Section 9.3.4)
- Gating (Section 9.3.9)
- Time-frequency maps (Sections 8.9.2 and 9.3.8) [30]
- Three-phase synchronous analysis (Section 8.9.3) [31].

The latter method normally assumes that interference is common to all three phases. In addition, machine PD experts can review the PRPD patterns to determine (subjectively) which clusters of pulses are stator winding PD and what pulse clusters are likely to be due to interference.

The HF systems usually report the PD magnitudes in terms of mV, but since some of the systems can also be used as IEC 60270 charge-based detectors in the LF range, some will produce (erroneously – see Section 6.6.3) apparent charge magnitudes, even when used in the HF range.

Figure 16.16 A 1 nF PD sensor (outlined) connected to the output bus of a single phase of a large generator. *Source:* Courtesy of Iris Power L.P.

16.9.3.3 UHF Methods

Early experience with the time-of-flight interference suppression technique (Section 16.9.3.1) revealed a flaw in practical installations on large turbine generators. As shown in Figure 16.13, the machine-end PD sensor is often located on the IPB side of the flexible links that connect the IPB to the generator terminals. In large machines, there are dozens of bolts that connect the flexible links to the terminals. If a bolt is oxidized or improperly torqued, it can be a site for severe contact sparking (Sections 4.4, 4.5, and 16.9.1). Such contact sparking is common on older IPB, and if just one or two bolts are sparking, it poses little danger to the IPB. Yet due to its location, the time-of-flight method would digitally identify the signals as coming from the generator and erroneously be considered by the unwary as stator winding insulation problems.

The pervasiveness of this interference source led to the development of the SSC UHF sensor (Section 7.4.5.4) [32]. The SSC is a two-port directional electromagnetic coupler, which is installed in the slots containing bars connected to the machine high-voltage terminal (i.e. operating at the highest voltage), at the end of the stator core (Figure 16.17). It separates PD from all types of electrical interference, including contact sparking at the machine terminals, using pulse shape analysis (Section 8.4.3). The signal is detected from both ends of the SSC, which permits identifying if the PD is occurring in the endwinding or from the stator slot area, using time-of-flight analysis. The SSC is primarily used on hydrogen-cooled generators where the likelihood of interference from the terminals is higher, and where the PD magnitudes (due to the high pressure hydrogen) are lower. The relative effectiveness of the SSC to suppress interference is presented in [33]. The limitation of this UHF method is that it will only detect PD activity in the same slot in which it is installed, and perhaps only a portion of the slot, due to attenuation and pulse distortion effects of the PD pulse as it travels along the slot. PD in adjacent slots will be detected, but those signals will be classified as interference since they will have a slower risetime and considerably more ringing.

Figure 16.17 Photo of an SSC (black object) in the process of being installed under the stator wedges at the end of a stator slot in a large turbine generator. *Source:* Courtesy of Iris Power L.P.

16.10 Differences Between Online and Offline Tests

It seems reasonable to expect that online and offline tests would produce results (PD magnitude, PRPD plots) that are similar. Although this is often the case, in some situations significantly different results are obtained. Section 10.3 gives an overview of the differences between offline PD tests and those from online testing or monitoring for all types of equipment. But compared to transformers, power cables, and switchgear, there are more reasons for the offline and online results to differ in machines:

- In offline tests, all the coils in a phase will be at the test voltage (rated line-to-ground or rated phase-to-phase voltage applied line-to-ground). Thus since any coil in the phase can produce PD, offline tests sometimes have both more PD pulses, and higher magnitude PD, compared to online tests where only the phase-end coils will have enough voltage are to produce significant PD (Section 16.5.3).
- In offline tests there is either no voltage between coils in the endwinding (if all three phases are energized at the same time, as is typical in motors), or the voltage stress is significantly lower than in operation (if one phase is energized to rated line-to-ground voltage and the other two phases are grounded). This means phase-to-phase PD that occurs in service (Section 16.4.4) is unlikely to occur in an offline test. (If the rated phase-to-phase voltage is applied to one phase and the other phases are grounded in an offline test, the endwinding PD that is present in service will occur normally, but the phase-to-ground insulation in the stator slot is 70% higher than normal, which makes slot PD exceptionally high, and the endwinding PD is often not apparent.)
- As will be discussed in Section 16.12.2, windings in operation are at higher temperatures and high load (i.e. subject to higher magnetic forces) compared to offline tests. The higher winding temperatures during online tests may cause lower PD activity (from groundwall voids) or higher PD activity (due to defective PD suppression coatings or loose bars in the stator slots). PD caused by loose coils in the stator slots (Section 16.4.3) will be significantly lower in an offline test since the coils are not vibrating. See also Section 1.5.1 and the accompanying footnote.

In summary, compared to online tests, offline tests may result in higher PD activity (due to the voltage effect, or the temperature effect); whereas offline test PD may be lower (due to relative insensitivity to endwinding PD or the lower load or temperature).

These differences can make comparison between online and offline results a challenge. However, comparing offline and online results can also be an opportunity, since they may help to point to the root cause of the PD. For example:

- Endwinding PD is higher in online tests (Section 16.4.4)
- Loose coils in the stator slots causes higher PD in online tests (Section 16.4.3)
- Degraded PD suppression coatings causes higher PD in online tests (Section 16.4.2)
- PD due to thermal aging (Section 16.4.5) and thermomechanical aging (Section 16.4.6) is lower in online tests.

16.11 Interpretation

As mentioned in Section 16.7, IEEE 1434 and IEC 60034-27-1 standards do not have requirements for the minimum *PDIV* or *PDEV* for coils or windings and do not have requirements for the maximum PD magnitude allowed at rated voltage (or any other test voltage). IEEE 1434 does suggest the machine manufacturer compile a database of normal *PDIVs*/*PDEVs* and maximum PD magnitudes measured with the same PD detection system and the same test frequency range. When a new stator is tested, the results of the test can be compared with the machine manufacturer's database to determine if the *PDIV*, *PDEV* is lower than normal or the Q_m (or Q_{IEC}) are higher than normal. This can be accomplished using traditional statistical process control methods, which are familiar to most industrial manufacturers. It is likely that different databases need to be established for different voltage classes of machines (see Section 16.11.2). Depending on the PD measuring system used, different databases may be needed for different power ratings and types of machines (motor, hydro, or turbine generators). It would be ideal if the winding manufacturers share their database in the public domain, or at least with their customers. Unfortunately, there seems be a reluctance to do this.

IEEE 56:2016 does suggest that the *PDEV* be higher than 50% of the rated phase-to-ground AC voltage in new machines [7].

The rest of this interpretation section focuses on the second purpose of PD testing: using offline/onsite or online tests/monitoring to determine when winding maintenance is required. Note that PD testing cannot detect aging mechanisms that do not produce PD. Thus, two important failure processes: stator winding endwinding vibration and water leakage into the insulation in direct water-cooled machines cannot be detected by offline or online PD testing [5].

16.11.1 Trend Over Time

The trend of PD activity over time is the most powerful method to determine when winding maintenance may be needed. PD activity can be peak PD magnitude (Q_m or Q_{IEC}) or other integrated quantities such as PD current, PD power, NQs, NQN, etc. (Section 8.8.3). If the interference is largely suppressed, any PD system operating in any frequency range can be used for the trend. However, it is important that the same brand of test equipment using the same frequency range and test set-up be used for all tests on the same winding, for the reasons expressed in Section 16.5.

For a valid trend, the environmental conditions and test voltage (for offline tests) and the machine operating conditions (online tests) must be about the same from test to test. If the test voltage is higher in a test today than when it was tested a year ago, it will be unclear if the higher activity is due to aging over the past year or due to the increase in voltage, or both. Thus, the voltage must be the same for all tests over time for a valid trend to detect insulation aging. Offline tests should also use the same conditioning interval before the PD activity is recorded, since generally PD decreases after a winding is first excited (Section 10.3.4). For online trends, Sections 16.10 and 16.12.1 indicate that the PD activity for some failure processes depend on the winding temperature, load, ambient humidity, and/or hydrogen pressure. To establish a valid trend, IEEE 1434 suggests that the trend include tests that are collected within the following range of values:

- Voltage: ±1.5%
- Winding hotspot temperature: ±5 °C
- Real and reactive load: ±5%

In addition, the relative or absolute humidity should be recorded, since humidity can affect the surface PD activity in stator windings [34]. If the machine is hydrogen-cooled, increasing hydrogen pressure will suppress the PD activity; therefore, the hydrogen pressure should be between ±30 kPa for trending purposes. With the exception of voltage and gas pressure, all the other operating/test conditions will only be important if there is significant PD activity. If the activity is low, then wide variations in winding temperature and load will not significantly affect the activity.

An example of a trend plot of Q_m for a generator with online data periodically collected over a 17 year period is shown in Figure 8.12. There is a general rising trend where the Q_m increased over 10 times (i.e. 1000%) over 17 years. If PD were trended over just one day, it is likely that there would be ±25% variation in the hourly readings. Since the 1960s, therefore, most machine PD experts suggest a significant increase in PD activity is a doubling (100% increase) over a 6- to 12-month interval. This has been adopted in the relevant IEEE and IEC standards [1–3].

Although Figure 8.12 shows a continuously increasing Q_m, experience shows that once significant insulation aging starts, the PD will increase, but it does not increase until failure occurs [35]. Instead, over a winding's lifetime, the trend follows an evolution similar to Figure 16.18. In the first stage, both offline and online tests reveal that the PD will decrease (to as much as 33% of commissioning levels) in the first 10,000 hours or so of operation. It is speculated that this decrease is due to the epoxy completing its cure during machine operation, and/or the coil/slot semiconductive coating becoming better grounded. Once stage 1 is complete, if the winding was well made, the PD remains relatively stable (±25%) and low due to small voids and imperfections within the insulation system. Hopefully, this stage will last for many years or even decades. The third stage sees a significant increase in PD activity, usually due to thermal aging, coils becoming loose in the slot,

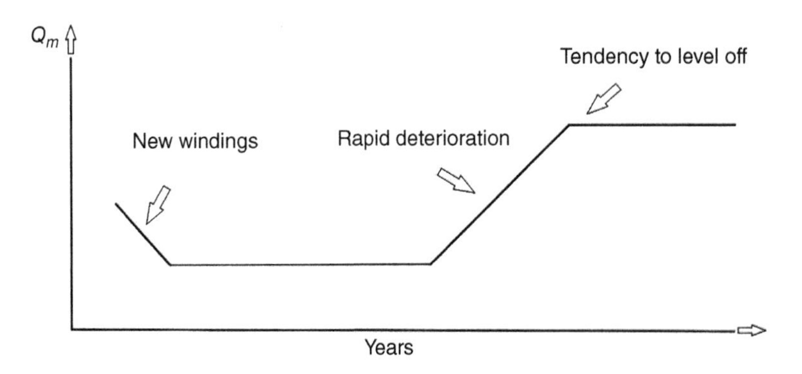

Figure 16.18 Generic trend of PD magnitude (Q_m) over a winding's lifetime. *Source:* [35].

contamination, etc. (Section 16.4). This is the stage when the PD may double in magnitude every six months to a year or so as the aging creates bigger voids and/or the number of locations having undergone deterioration increases.

Unfortunately, from a condition assessment point of view, stage 4 is where the PD activity (and especially the Q_m or Q_{IEC}) levels off, and even sometimes decreases. The reasons for this are not clear, even though it has been noted on many hundreds of machines. However, one can speculate that it is carbonization or a permanent space charge occurring on the interior walls of voids and delaminations, which limits the electric field within the voids. It is also probably aided by the excellent PD resistance of mica within the groundwall insulation, even if the organic epoxy has largely disappeared. Another possible cause can occur when coils are loose in the stator slots (Section 16.4.3); the semiconductive coating may essentially be re-established by debris from previous abrasion combining with any oil that is present within the machine.

The impact of the leveling-off in stage 4 is important to PD interpretation. If offline testing or online monitoring is only initiated once a winding has entered stage 4, the stable PD level that is evident after testing starts may lead the machine owner to conclude that the winding insulation is in good condition, even though aging is continuing without being apparent in the PD trend and failure will eventually still occur. Leveling-off also implies that one cannot predict when actual winding failure will occur.

16.11.2 Comparison to Similar Machines

It is not possible to set universally valid PD activity threshold levels (i.e. Q_m, Q_{IEC}, PD power, etc.) that indicate significant stator winding insulation problems. This is due to the complex interaction between each PD pulse and the inductive-capacitive nature of the stator winding, and that is further complicated by the interaction between the frequency dependence of the stator winding impedance and the PD detection system (Section 6.6.3). Also, PD pulses traveling from the PD sites to the PD sensor are attenuated and distorted. These effects cannot be compensated or corrected for by either calibration (IEC 60270 detectors) or sensitivity checks (detectors working in the HF, VHF and UHF range), as discussed in Section 6.6 and references of the chapter.

The only approach that may give a realistic indication about whether the PD activity is low or high is to compare the results of a test on a winding with tests on "similar" windings using the same brand of test equipment in the same frequency range tested at the same voltage and hydrogen pressure (if applicable). What constitutes a "similar" machine is the key factor. Of course if the machines are identical, i.e. are made by the same manufacturer, have the same model number and ratings, and were made within a few years of one another, then comparing results between machines is credible. The winding with the higher PD activity presumably has insulation that is more aged than the others.

But generalizing "similar" is tricky. Two organizations have published statistical analysis of their online PD test results (Q_m and NQN) collected with the same VHF instrumentation using 80 pF sensors [28, 36, 37]. In one, 750,000 test results from about 8500 machines were statistically analyzed [36]. Only the most recent test results from each machine were included, with the machines operating near or at full load and normal operating temperature. Each test result was tagged with the following information:

- Operating voltage
- Operating load
- Winding temperature
- Machine type (hydro, turbo, motor)

- Hydrogen pressure (if relevant)
- Year of machine (or rewind) manufacture
- Manufacturer

The database was then parsed by the above factors (and range of factors for voltage, load and pressure) and the mean and standard deviation of each subset was determined. Statistical tools were used to determine if there were significant differences between the mean and standard deviation of each subset. The following results were found:

- The mean Q_m (and other cumulative percentiles of the distribution) were significantly different between voltage ranges. Specifically, the 90th percentile of the Q_m distribution was significantly different between machines with a 6.6 kV versus a 13.8 kV operating voltage. In general (but not always), as the winding operating voltage increased, the mean and the 90th percentile increased. The conclusion is that, at least for this specific test method, one should not compare results from a 6.6 kV machine with those from a 13.8 kV machine.
- The machine power had no or a very weak correlation with the mean and 90th percentile of the Q_m and *NQN* distributions. That is, as long as the rated voltage is about the same, one can compare test results from machines rated, for example, from 1 to 100 MW.
- Hydrogen pressure had a strong influence on the statistical distributions. In general (although not always), as the pressure increases, the Q_m mean and 90th percentile show a statistically significant decrease.
- There are very large differences between the Q_m distributions depending on the age of the winding and the winding manufacturer.

Table 16.1 shows an example of the statistical distributions found in air-cooled machines as a function of voltage. The statistical distributions in Table 16.1 are similar to those obtained by other researchers using data from the same VHF method on a smaller collection of hydrogenerators [37]. These conclusions are only valid for the VHF system the results are based on. Similar tables for hydro generators and hydrogen-cooled machines are in [36]. The most recent tables can be found at [38].

Since visual examinations had been done on hundreds of the windings in the study, it seems that when the Q_m was higher than 90% of machines in the same operating voltage range, there was a

Table 16.1 Statistical distributions found in air-cooled machines as a function of voltage.

	Operating voltage (kV)					
Cumulative probability (%)	**2–5 kV**	**6–9 kV**	**10–12 kV**	**13–15 kV**	**16–18 kV**	**>19 kV**
25	7	21	32	45	42	45
50	24	55	78	111	85	90
75	71	141	175	239	186	191
90	208	308	368	488	346	507
95	393	476	587	730	506	798

Distribution of Q_m (mV) for air-cooled motors and generators with 80 pF sensors on the stator terminals, when the machine was operating near full load and the winding temperature is hot. The cells contain the Q_m levels in mV. For an 11 kV machine, 50% of stators have a $Q_m < 78$ mV, and 90% of machines have a $Q_m < 368$ mV. This data is based on results collected to the end of 2021 [38].

Source: https://irispower.com/online-partial-discharge-severity-tables/.

good correlation with significant insulation deterioration. An indicator of total PD activity, NQN (Section 8.8.3), seemed to have little correlation with the observed winding condition. Perhaps Q_m is a better indicator of damage because it tends to be correlated with the most serious damage, rather than the average condition of the insulation. Thus the Q_m 90th percentile has been established as the "PD Alert" level for continuous monitors using 80 pF sensors and the VHF method in [28, 36]. These findings illustrate that one should be careful of severity tables that do not distinguish between voltage class and hydrogen pressure.

The levels shown in Table 16.1 are drawn from stator windings that have one or more of the deterioration mechanisms in Section 16.4. However, it was pointed out in Section 16.4 that some mechanisms inherently produce higher PD magnitudes than others even though the risk of failure is the same or lower. One example is the deterioration of the overlap between the semicon and grading coatings, even though this mechanism is a very slow and relatively benign failure process (see Section 16.4.2), it produces very high PD magnitudes compared to PD adjacent to the turn insulation in a multiturn coil (Section 16.4.1 and 16.4.5). Thus PD activity alone is not a good predictor of how fast maintenance activity needs to be performed. To enable better prediction of the need for winding maintenance, tables such as Table 16.1 should be established for each failure process [35]. Unfortunately, this has not been possible to date.

16.11.3 Prediction of Remaining Winding Life

All relevant IEEE and IEC standards, as well as several publications, clearly indicate that PD testing cannot determine the remaining life of a winding using mica insulation, or predict when it will fail [35, 39]. Although some testing services and PD system vendors claim this is possible, it is extremely unlikely for the following reasons:

- For some important failure processes like loose windings in the slot and thermal aging, PD is only a symptom of the failure process, not the cause. While PD may indicate the problem is occurring, how quickly a winding fails depends on the winding temperature or how loose the coils are in the slots, not the PD activity itself.
- Even where PD is the main cause of aging (poor impregnation, poorly made PD suppression coatings, inadequate separation in the endwindings – see Section 16.4), the aging process is slow, typically taking many years to seriously degrade the insulation, and even then the time of failure is usually initiated by an external event such as a high voltage or high current transient, operating errors, etc.
- As discussed in Section 16.11.1, experience shows that PD activity does not continue to increase until failure occurs. Instead it levels off and sometimes decreases, even when aging is ongoing. Thus, increasing PD activity does not predict when the winding will fail.
- Some failure processes may not produce high PD, but, nevertheless, can cause failure quickly, and vice versa (see Section 16.11.2).

16.12 Root Cause Identification

There are two main reasons for identifying the root cause of any severe PD activity that is found by the trend over time and/or comparison to similar machines:

- Some aging processes (Section 16.4) lead to failure more-quickly, and some are much slower. By identifying the aging process and knowing if it is a fast or slow process, one can then plan maintenance in a timely manner to prevent in-service failure occurring.

- There is a different set of repair options for each possible aging process [5]. Thus, if the most likely aging process has been identified before the maintenance outage, competitive pricing for repair materials and labor can be arranged, with a good estimate of the duration of the repair work.

The root cause of failure processes can sometimes be identified by:

- PRPD pattern analysis (Section 10.5).
- Observing the effect of machine operation condition, and in particular winding temperature, load and humidity, on the PD activity level (Section 10.6).
- Visual inspection of the winding, sometimes aided by the use of corona probes (Section 16.13), acoustic and optical imaging devices to locate high PD sites.

The second can only be assessed with online PD testing/monitoring, and the third when the machine is shut down and at least partially dismantled.

Unlike some other types of apparatus, stator windings often have several different aging processes occurring at the same time. This is especially likely in older windings. Thus, the PRPD patterns often represent the combination of two or more processes, and the patterns can become very complicated. Furthermore, even though many papers have been published on the PRPD patterns associated with each failure process over the past 20 years, such patterns are sometimes found to result from entirely different mechanisms than expected when visual examination of the winding occurs [35, 40–42]. When expensive repair or rewind decisions must be made, online and offline PD tests should be confirmed with other types of tests, and most importantly, by visual examination of the winding [5, 7].

16.12.1 PRPD Pattern Analysis

As for other types of equipment, the PRPD pattern, i.e. the magnitude, number, and polarity (if available) of the PD pulses with respect to the 50/60 Hz AC cycle (Section 8.7.3), is a powerful tool both for noise suppression and for the identification of the cause of significant PD activity in a stator winding insulation system. For offline tests on windings, all the PD occurs with respect to the phase-to-ground voltage, because a single-phase power supply is usually employed. However, PRPD analysis for machines tested online should recognize that both the phase-to-ground AC voltage and the phase-to-phase voltage can create PD. Since there is a $\pm 30°$ phase-shift between the phase-to-ground and the phase-to-phase voltage (Section 10.5.4), this will result in different AC phase locations for the PD in the PRPD plot with respect to the AC phase-to-ground voltage (which is almost always used for synchronizing the PRPD plots).

Phase-to-ground PD in stator windings can be caused by (Section 16.4):

- Thermal aging (delamination)
- Thermomechanical aging (delamination)
- Slot discharge due to coil vibration in the slot
- Poor impregnation
- Poorly manufactured PD suppression coatings.

Phase-to-phase PD in online tests is caused by:

- Insufficient spacing between adjacent coils in the endwindings, circuit ring busses, and/or phase leads (Section 16.4.4)
- Partly conductive contamination leading to tracking (Section 16.4.7).

The first step in online PRPD analysis is to determine if phase-to-phase PD is occurring. This is discussed in Section 10.5.4. In addition, cross-coupling of PD from one phase to another tends to be significant in stator windings, especially in lower voltage machines where the spacing between the phases is usually small (Section 10.5.5). By using an appropriate phase-shifted three-phase PRPD plot, phase-to-ground PD, phase-to-phase PD, and irrelevant cross-coupling from another phase can be easily identified (Sections 10.5.2, 10.5.4, and 10.5.5, respectively). Figure 16.19 shows an example of the PRPD patterns for all three phases measured at the same time, with the start of the AC voltage for each phase shifted by 0°, 120°, and 240°, respectively. This means that a vertical line drawn between the three phases represents the same instant of time (Section 10.5.4). The software applications from several brands of PD systems are able to make such phase-shifted plots. The 3PARD plot (Section 8.9.3) can also be useful in separating phase-to-ground PD from interphasal PD in online tests.

The machine in Figure 16.19 has interphasal PD between phases B and C in the endwinding, with a small amount of cross-coupling into phase A. If B phase shows positive pulse predominance (i.e. the positive PD is more numerous and of higher magnitude), then C phase will show negative pulse predominance. The reason for this is described in Section 10.5.4. Thus, the "polarity predominance effect" is not useful for interphasal PD (Table 16.2).

Phase-to-ground PD occurs primarily in all three phases between 0° and 90° of the phase-to-ground AC cycle (negative pulses, when measured indirectly, Section 6.6.2) and 180° and 270° (positive pulses). Hudon and Belec have done careful laboratory experiments where they simulated some of the more important phase-to-ground failure processes on stator bars to determine their PRPD pattern [43]. Figure 16.20 shows the characteristics of the most common phase-to-ground PRPD patterns found in their laboratory testing. These were reproduced in IEEE 1434 and some parts of IEC 60034-27-1, and are valid in most cases for both offline and online testing. Table 16.2 provides a summary of the phase angles on which PD may most likely be centered, as well as the polarity effect associated with each winding deterioration process.

In reality, both offline and online PRPD plots are usually more complicated than those shown in Figures 16.19 and 16.20. This is because most stator windings have several different manufacturing defects and/or aging processes that each produce their own patterns (Section 10.5.6). Figure 16.21 shows a very complicated example. Pulses are occurring across the AC cycle (some of which must be cross-coupled from another phase or due to interference) and shows pulses of the wrong polarity for the AC phase position. Making sense of such a pattern requires a machine PD expert – and even then different experts rarely agree on the different processes occurring. Analyzing such plots can be aided using the T–F map and three-phase synchronous analysis tools (Sections 8.9.3 and 9.3.8, respectively). However, note that these tools (as do AI pattern recognition tools) assume the validity of the PRPD patterns in Figures 16.19 and 16.20 to deconvolute the complex patterns. References [30, 31] provide examples of such methods. In addition, examples are presented in IEEE 1434 and IEC 60034-27-2.

16.12.2 Effect of Operating Conditions on PD

After an initial PRPD pattern analysis, another way to determine the root cause of any stator winding PD, or to obtain additional confirmation about the root cause, is to perform special tests on the machine where the operating conditions are changed while observing the effect on PD activity during online PD testing/monitoring. That is, the winding load or temperature is manipulated, or the test is performed at times of different ambient humidity, to identify the aging processes. Apparently, the first publication on changing operating conditions to determine root causes of PD

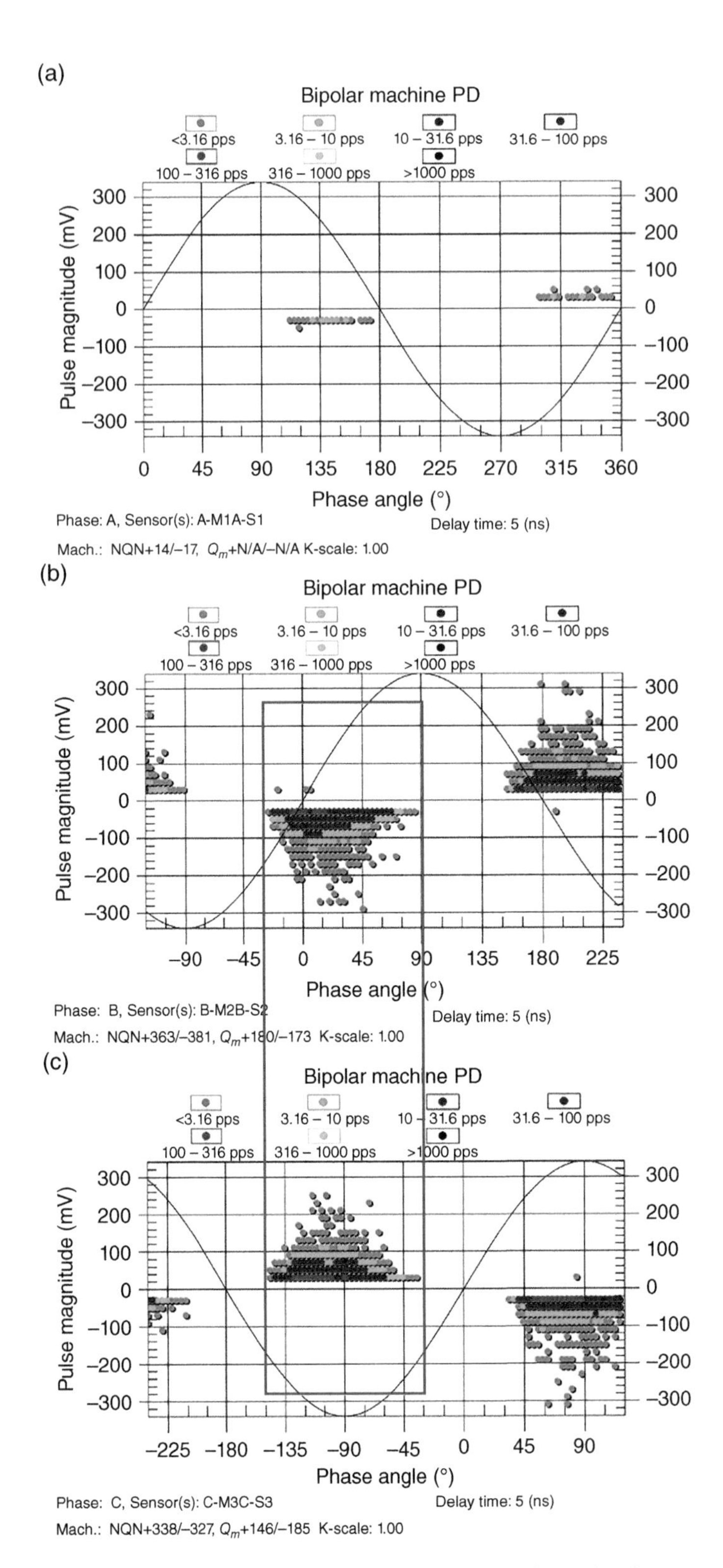

Figure 16.19 PRPD patterns in "phase shifted" plots (note the phase-to-ground AC voltage at the start of each plot is different). The interphasal PD is caused by insufficient spacing between B and C phases in the endwinding. Note that the PD is phase shifted either +30° or −30° from the normal phase-to-ground position, and the pulses at the same instant of time (red box) are of equal magnitude and reverse polarity. *Source:* Courtesy of Iris Power L.P.

Table 16.2 Characteristics of the PRPD features and the effect of operating conditions on the most common stator winding insulation deterioration processes.

		Operating condition effect			PRPD effect	
Aging process	**Section**	**Load**	**Temperature**	**Humidity**	**Phase angle modes**	**Polarity predominance**
Poor impregnation	16.4.1	None	Neg.	None	45° & 225°	None
Degraded PD coatings	16.4.2	Little	Pos.	Neg.	45° & 225°	Pos.
Loose coils in slots	16.4.3	Pos.	Little	Neg.	45° and 225°	Pos.
Endwinding separation	16.4.4	None	Little	Neg.	15°, 75°, 195° & 255°	Both or none
Thermal	16.4.5	None	Neg.	None	45° & 225°	None or neg.
Thermomechanical	16.4.6	None	Neg	None	45° and 225°	Neg.
Contamination	16.4.7	None	Little	Variable	All	None
Metallic debris	16.4.8	None	None	Neg.	90° & 270°	Either

"Pos." and "Neg." refers to the effect on PD activity by an increase in load, winding temperature, or humidity. "Phase angle mode" refers to the most commonly observed phase angle of either the positive or the negative PD, assuming an indirect measurement.

was published in 1984 [44]. It is now a standard tool for online data analysis and described in IEEE 1434 and IEC 60034-27-2. It is particularly easy to apply if data is available from continuous online PD monitors since arrangements for special testing is not needed.

The basic premise is that PD for some of the aging processes in Section 16.4 are sometimes affected by motor and generator operating conditions. The reasons for this dependency are described in [1, 3, 5], and were alluded to in Section 16.11, since such dependencies adversely affect collecting viable PD trend data. Table 16.2 shows how operating conditions can affect the PD activity. If only one condition is changed (and the others remain constant), the cells in Table 16.2 show:

- Positive if the PD increases with the condition (e.g. Q_m increases at higher winding temperatures)
- Negative if the PD decreases with the condition (e.g. Q_m decreases at higher temperatures)
- None if the temperature (or another condition) neither increases nor decreases with temperature
- Little if there may be a small increase or decrease with a changing condition

Since PD activity indicators such as Q_m vary ±25% over short periods of time when all operating conditions are stable (Section 16.11.1), the change in PD activity with load or temperature should be at least 50%, and similar for all three phases (since it is unlikely one phase will have one failure process, and another phase has a different process). The wider the temperature, load or humidity variations, the better to see clear effects. As a guide, the winding temperature difference should be at least 20 °C, and the load should be at least 40% different.

Separating load and winding temperature effects can be a challenge. If the load on the machine increases, then the current in the winding increases, and eventually so does the winding temperature. That is, over long periods of time, temperature and load are correlated. However, when a sudden load change occurs over a few minutes, it does take 10 minutes or more for the groundwall insulation temperature to change. Thus, for hydrogenerators and gas turbine generators, which can make significant (>50%) load changes in a few minutes, this method is especially useful. The PD can first be captured when the machine is either cold (low load) or hot (high load) for hours or days. Then the load is rapidly changed and the PD data is captured again

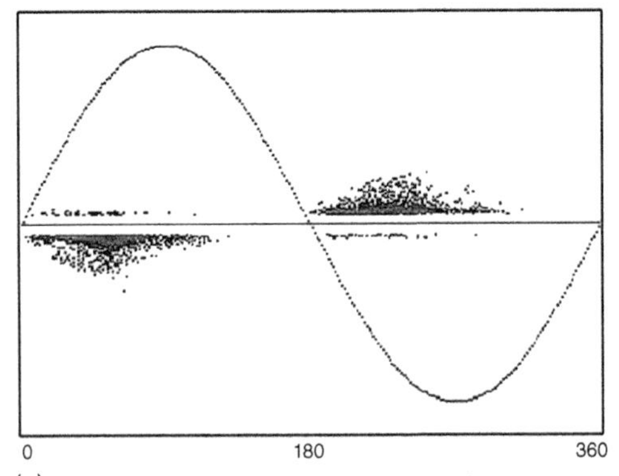

(a) Voids/delamination within the groundwall across the groundwall thickness:
- symmetrical pattern for negative and positive PD
- positive and negative pulse count rate and magnitude similar
- Gaussian pattern outline

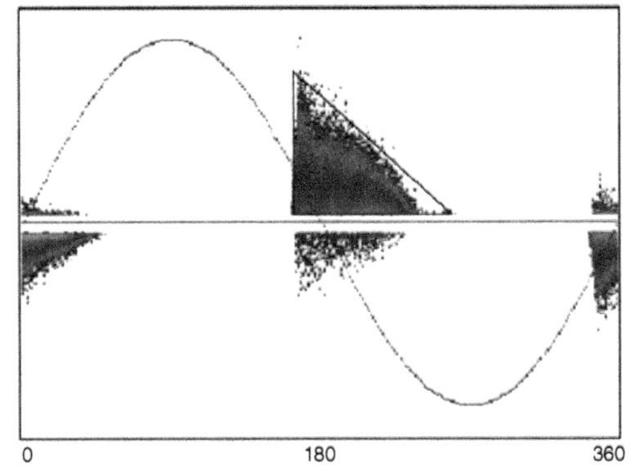

(b) Slot discharge/loose coils in the slot
- positive PD (when measured indirectly) is more numerous and higher than the negative PD
- PD tends to start near the AC cycle and sometimes the pattern has similar shape as a right-angled triangle

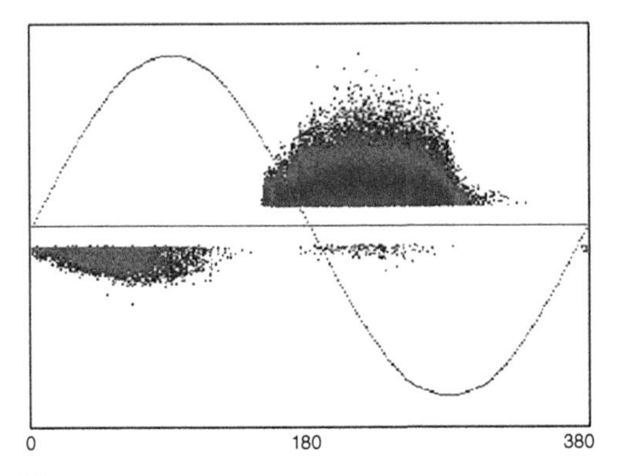

(c) Degraded semiconductive coating PD
- positive PD more numerous and higher than negative PD
- this pattern may also be due to loose coils in the slot or poorly made semiconductive coatings

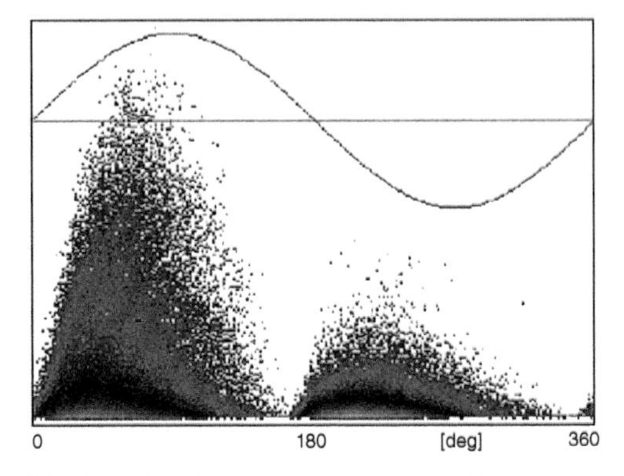

(d) Delamination adjacent to the copper conductors due to thermal aging or thermomechanical aging
- negative PD (occurring between 0° and 90° of the AC cycle) has a higher count rate and a higher magnitude than positive PD (between 180° and 270° of the AC cycle in this unipolar plot)
- usually a Gaussian pattern outline

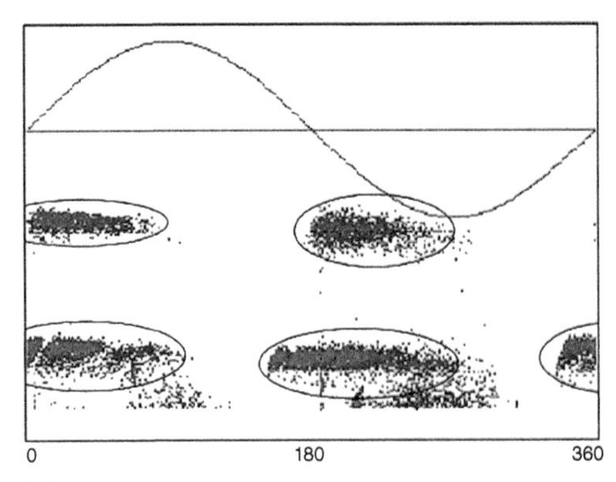

(e) Gap discharge due to bad semicon to grading coating deterioration, or insufficient stator core pressure finger to coil clearance, or other clearance issues to neutral end coils just outside of the slots.
- floating clouds of nearly equal magnitude PD across AC cycle

Figure 16.20 Examples of PRPD patterns from various types of phase-to-ground PD sources in laboratory tests on stator bars. Two different brands of IEC 60270 detectors were used. In (a) to (c) the PRPD is showing both polarities, with an indirect PD detection system (Section 6.2.2). Plots (d) and (e) only show one polarity (i.e. all the pulses have been converted by software to show positive pulses as negative, or negative pulses as positive pulses). *Source:* Adapted from [1, 43].

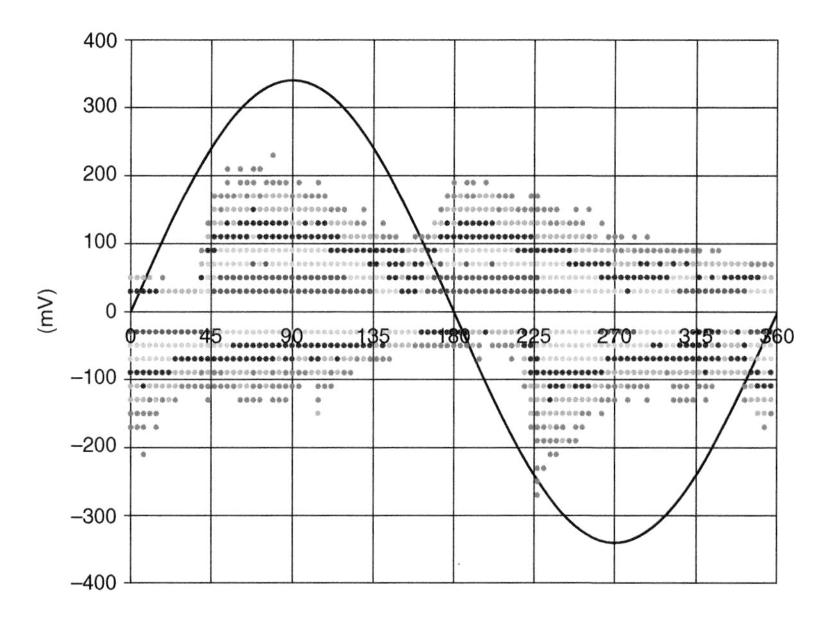

Figure 16.21 Complex PRPD plot from an online test on a stator winding.

before the winding temperature has changed much. In hydrogenerators and gas turbine generators, the data can be captured at:

1) full load, cold (after several hours of shutdown, and then a rapid load increase)
2) full load, hot winding (after a few hours at full load)
3) low load, hot (after many hours at full load and then a rapid load decrease)
4) low load, cold (after several hours of shutdown)

Only three of these load points are needed. IEEE 1434 and IEC 60034-27-2 provide more details.

For direct online start motors that only operate at full load, only tests 1 and 2 can be performed by capturing the PD immediately after starting and then again after a few hours when the winding temperature has stabilized. Thus, it is only possible to obtain the effect of winding temperature changes on the PD, and thus Table 16.2 is less useful.

The humidity cannot be controlled by an operator. Instead data is collected over several days under the same operating conditions while monitoring the humidity (preferably measured inside the machine enclosure).

Continuous PD monitors can automate this process by triggering the storage of test results when the four load points occur during the normal operation of the motor or generator over at least a year, assuming the ambient cooling water and/or air changes seasonally. Alternatively, a database search can pull out the continuously recorded results at the different load, temperature, and humidity points.

16.12.3 Interpretation Overview for Online PD Tests

There are two main stages for interpreting online PD data:

1) Determine the PD trend over time and compare levels to other "similar" windings to find which stator windings have increasing and/or high PD
2) Determine the causes of the high PD using PRPD analysis and effect of operating conditions on PD, but only if the machine has high and/or increasing PD in the first stage

For the first stage:

- From the most recent test results done near full load and with the stator winding hot, select the most sensitive scale (assuming there are multiple magnitude scales) with no over-magnitude warning flags.
- Find the phase with the highest PD activity (for example Q_m).
- Compare this activity against those from similar machines. If a statistical summary of the PD activity from a large number of machines is available (for example Table 16.1 for 80 pF sensors in air-cooled machines), determine the cumulative probability of the activity level.
- If the activity is less than the 50% level, and the trend is stable, stop the interpretation process, because the PD probably cannot be found in a winding visual inspection, and insulation problems (that PD can detect) are minor.
- If the trend is "increasing" – say doubling every year or so; and/or the phase exhibiting the highest activity is greater than 90% of similar machines – a significant issue exists and further careful examination of the data is warranted.

For the second stage, if there is a dominant deterioration process, the following can aid to identify it:

- If the activity is significant, review PRPD plots for all three phases using a phase-shifted presentation (Figure 16.19) to determine if the PD is caused by phase-to-ground or phase-to-phase activity while ignoring cross-coupled signals
- If the main source is interphasal PD, the cause is either insufficient spacing in the endwinding, or partly conductive contamination. In the case of interphasal PD, the polarity effect of the pulses conveys no useful information. On the other hand, if the PD is caused by phase-to-ground voltage, the polarity effect (Table 16.2) is useful to indicate whether the PD is near the copper (negative predominance) or the stator core (positive predominance).
- Only occasionally confirm the prognosis via special load point tests or a database search from a continuous monitor to determine the load, temperature, and humidity effects (Table 16.2).

Another tool to help confirm PD in air-cooled machines and to help identify the aging process is an ozone monitor (Section 5.5.1).

If there are multiple significant processes occurring at the same time, even most experts will disagree on the root causes. Offline PD and insulation resistance tests, together with the imaging methods in Sections 5.3 and 5.4, corona probe scans (Section 16.13.1), and perhaps the application of T–F maps (Section 16.8) may help. Of course, a careful visual inspection by an experienced person is the definitive way to assess the root causes of winding insulation deterioration.

16.13 Locating PD Sites

Some types of stator windings can be physically large, which can make visual inspection of the winding to find PD locations somewhat tedious to do. If the winding is exposed and can be energized with 50/60 Hz voltage to at least the rated line-to-ground voltage, the optical and acoustic imaging devices presented in Sections 5.3 and 5.4 can help identify the location of significant surface PD sites. These tools will not help to locate the PD sites if the PD is occurring within the groundwall insulation such as voids and delamination within the groundwall (Sections 16.4.1, 16.4.5, and 16.4.6). Instead radio-based probes call the corona probe or the insulation sniffer can be used.

16.13.1 Corona/Sniffer Probes

Each time a PD event occurs, some radio frequency (RF) signals in the range from a few tens of kHz to as high as a few GHz will be radiated, and these signals can be picked up with suitable antennas/receivers. This is discussed in extensive detail in Chapter 7. Over the years, various vendors have developed specialized, portable RF "probes" that can locate the sites of the PD activity on some types on stator windings and are sometimes used for PD location in IPB and AIS.

The first probe was developed by Dakin and was used widely by Westinghouse, especially in rotating machines that were partly disassembled [45]. The sensor was a ferrite core wrapped with several turns of wire to create a near-field resonant antenna tuned to about 5 MHz (using the capacitance of a coaxial cable that is connected to the measuring instrument). Originally, Dakin used a "noise meter" (see Section 1.5.1) to measure the RF signal from the antenna. However, Smith, while working for the American utility TVA, developed a small, portable peak pulse meter that could be connected to Dakin's antenna to measure the 5 MHz signals [46]. He decided to calibrate the device so that a PD pulse repetition rate of 10 pps[1] or more showed as the peak magnitude, measured in mA. Locating the PD source is possible since the RF radiation from PD activity attenuates the further the antenna is from the PD source; thus a higher signal is measured when the antenna is close to the PD site.

This probe has been widely used by North American and UK utilities to locate the PD in stator windings in offline testing, as well as for locating sources of discharging and arcing within a plant. This type of probe is often referred to as the TVA probe or the "corona probe." The latter term is misleading, since in most applications, the probe is detecting PD, not corona from an uninsulated HV conductor. A modern version of this device is still commercially produced by Qualitrol-Iris Power. Figure 16.22 shows the corona probe being used to locate coils with high PD activity in a hydrogenerator stator winding.

IEEE 1434, which is concerned with PD testing of stator windings, contains further detail on the TVA probe. This document also describes how to perform a test on a stator by sequentially

Figure 16.22 Photograph of the TVA probe test being performed on an energized hydrogenerator stator winding to detect coils with the highest PD activity. The probe tip is placed near the coil, bridging the stator core slot and the peak PD signal in mA is read on the instrument meter. *Source:* Courtesy of M. Sasic, Iris Power L.P.

---

1 The peak magnitude being determined at a repetition rate of 10 pps later found its way into conventional electrical PD instruments, since it is easily determined by digital electronics, as opposed to the normal IEC 60270 method of measuring the quasi-peak PD activity.

placing the probe across each slot in the stator and measuring the signal (Figure 16.22) when at least one phase of the stator winding is energized. For stator coils and bars that are insulated with mica tapes bonded together with epoxy, and with the winding excited to rated line-to-ground voltage, IEEE 1434 suggests that the readings are high if the indicator is greater than 20 mA. A problem with this test is that when energizing one phase at a time, sometimes a high response is measured in slots that do not contain energized coils/bars, which can lead to false positive indications of PD. These false readings are thought to occur because the PD from the energized phase can cross-couple into the other phases, usually via capacitive coupling in the end winding (Section 16.5.1).

At least one utility is now requiring the TVA probe to be used as a quality control test on stator coils and bars before they are inserted into a stator winding [47]. They suggest that acceptable bars/coils should have a reading of less than 5 mA at rated line-to-ground voltage. With such a reading, there will be no significant voids within the groundwall insulation or from imperfections in the surface PD suppression coatings.

A more flexible RF probe was developed by Lemke Diagnostics in the 1980s [48]. The LDP-5 is a compact, battery-operated, general purpose instrument that mainly works in the low-frequency (<1 MHz) range, although it could be connected to UHF sensors together with a suitable demodulator. This probe used either noncontact capacitive or inductive sensors, or it could be directly connected to conventional PD coupling capacitors. This probe was used on cable terminations and joints, as well as stator windings. It seems to be no longer commercially available.

Timperley, while with the utility AEP also popularized an electromagnetic interference (EMI) "sniffer" that is used to locate both PD and arcing in generating stations and substations [49]. Variations of the EMI sniffer are made by Radar Engineers (Model 246A) or as part of the Doble "Surveyor." The Model 246A measures both the electric field and the magnetic fields with separate sensors. It seems the electric field sensor operates in the 50 kHz to 6 MHz range and is most sensitive to PD; the H field sensor measures in the 35–450 kHz range which apparently is most sensitive to arcing (see Section 4.4). Figure 16.23 shows the sniffer being used to find PD and arcing sources in and around a turbine generator.

Figure 16.23 Photo showing the EMI sniffer being used to detect RF signals leaking from a turbine generator. *Source:* Courtesy of James E. Timperley, J.E. Timperley Consulting.

References

1 IEEE 1434. IEEE Guide for the Measurement of Partial Discharges in AC Electric Machinery.
2 IEC 60034-27-1. Rotating Electrical Machines – Part 27-1: Off-line Partial Discharge Measurements on the Winding Insulation.
3 IEC 60034-27-2. Rotating Electrical Machines – Part 27-2: On-line Partial Discharge Measurements on the Stator Winding Insulation of Rotating Electrical Machines.
4 IEEE 1799-2022. Recommended Practice for Quality Control Testing of External Discharges on Stator Coils, Bars, and Windings.
5 Stone, G.C., Culbert, I., Boulter, E.A., and Dhirani, H. (2014). *Electrical Insulation for Rotating Machines: Design, Evaluation, Aging, Testing, and Repair*, 2e. Wiley/IEEE Press.
6 Stone, G.C. and Miller, G.H. (2013). Progress in rotating-machine insulation systems and processing. *IEEE Electrical Insulation Magazine* 29 (4): 45–51. https://doi.org/10.1109/MEI.2013.6545259.
7 IEEE 56-2016. Guide for Insulation Maintenance of Electric Machines.
8 Narang, A., Gupta, B.K., Dick, E.P., and Sharma, D. (1989). Measurement and analysis of surge distribution in motor stator windings. *IEEE Transactions on Energy Conversion* 4 (1): 126–134. https://doi.org/10.1109/60.23163.
9 Zhu, H. and Kemp, I.J. (1992). Pulse propagation in rotating machines and its relationship to partial discharge measurements. *IEEE International Symposium on Electrical Insulation*, Baltimore, MD (7–10 June 1992), pp. 411–414.
10 Wilson, A., Jackson, R.J., and Wang, N. (1985). Discharge detection techniques for stator windings. *IEE Proceedings B* 132: 234–244.
11 Su, Q., Chang, C., and Tychsen, R.C. (1997). Travelling wave propagation of partial discharges along generator stator windings. *Proceedings of 5th International Conference on Properties and Applications of Dielectric Materials*, Seoul, Korea (25–30 May 1997), pp. 1132–1135, Vol. 2. https://doi.org/10.1109/ICPADM.1997.616648.
12 Stone, G.C., Campbell, S.R., and Sedding, H.G. (2011). Characteristics of noise and interphasal PD pulses in operating stator windings. *2011 Electrical Insulation Conference (EIC)*, Annapolis, MD (5–8 June 2011), pp. 15–19. https://doi.org/10.1109/EIC.2011.5996106.
13 Henriksen, M., Stone, G.C., and Kurtz, M. (1986). Propagation of partial discharge and noise pulses in turbine generators. *IEEE Transactions on Energy Conversion* EC-1 (3): 161–166. https://doi.org/10.1109/TEC.1986.4765750.
14 M. Lachance and F. Oettl (2020). A study of the pulse propagation behavior in a large turbo generator. *IEEE Electrical Insulation Conference (EIC)*, Knoxville, TN (22 June to 3 July 2020), pp. 434–439. https://doi.org/10.1109/EIC47619.2020.9158584.
15 Stone, G.C. (2000). Importance of bandwidth in PD measurement in operating motors and generators. *IEEE Transactions on Dielectrics and Electrical Insulation* 7 (1): 6–11. https://doi.org/10.1109/94.839335.
16 IEC 60079-0. Explosive Atmospheres – Part 0: Equipment – General Requirements.
17 IEC 60079-15. Explosive Atmospheres – Part 15: Equipment Protection by Type of Protection "n".
18 IEEE Standard 323. Standard for Qualifying Class 1E Equipment for Nuclear Power Generating Stations.
19 Stone, G.C., Gupta, B.K., Kurtz, M., and Sharma, D.K. (1984). Investigation of turn insulation failure mechanisms in large AC motors. *IEEE Transactions on Power Apparatus and Systems* PAS-103 (9): 2588–2595.
20 Petit, A. (2015). Comparison of PD amplitudes of stator bars taken with different instruments. *2015 IEEE Electrical Insulation Conference (EIC)*, Seattle, WA (7–10 June 2015), pp. 255–261. https://doi.org/10.1109/ICACACT.2014.7223548.

21 Levesque, M., Seol, Y.D., Hudon, C., and Provencher, H. (2022). Comparative study of phase resolved partial discharge patterns obtained on individual stator bars and coils. *Proceedings of the IEEE Electrical Insulation Conference*, Knoxville (June 2022).

22 Lévesque, M., Seol, Y.D., Hudon, C. et al. (2021). The correlation between PRPD patterns and dissection on individual stator coils. *2021 IEEE Electrical Insulation Conference (EIC)*, Denver, CO (7–28 June 2021), pp. 18–21. https://doi.org/10.1109/EIC49891.2021.9612355.

23 Stone, G.C. (1998). Partial discharge. XXV. Calibration of PD measurements for motor and generator windings-why it can't be done. *IEEE Electrical Insulation Magazine* 14 (1): 9–12.

24 IEEE 2465. Recommended practice for pulse-type partial discharge measurements on individual stator coils and bars. Document still in development at time of publication.

25 IEEE 37.23-2015, Standard for Metal-Enclosed Bus.

26 Timperley, J.E. (2022). Isolated-phase bus—design, operation, deterioration, and failure. *IEEE Electrical Insulation Magazine* 38 (3): 15–23. https://doi.org/10.1109/MEI.2022.9757854.

27 Lloyd, B.A., Campbell, S.R., and Stone, G.C. (1999). Continuous on-line partial discharge monitoring of generator stator windings. *IEEE Transactions on Energy Conversion* 14 (4): 1131–1138. https://doi.org/10.1109/60.815038.

28 Stone, G.C. and Warren, V. (2006). Objective methods to interpret partial-discharge data on rotating-machine stator windings. *IEEE Transactions on Industry Applications* 42 (1): 195–200. https://doi.org/10.1109/TIA.2005.861273.

29 Kurtz, M., Lyles, J.F., and Stone, G.C. (1983). Experience with the CEA generator insulation partial discharge test. *1983 EIC 6th Electrical/Electronical Insulation Conference,* Chicago, IL (3–6 October 1983), pp. 65–68. https://doi.org/10.1109/EEIC.1983.7465034.

30 Contin, A., Cavallini, A., Montanari, G.C., and Puletli, F. (2000). A novel technique for the identification of defects in stator bar insulation systems by partial discharge measurements. *Conference Record of the 2000 IEEE International Symposium on Electrical Insulation (Cat. No.00CH37075)*, Anaheim, CA (5 April 2000), pp. 501–505. https://doi.org/10.1109/ELINSL.2000.845558.

31 Koltunowicz, W., Gorgan, B., Broniecki, U. et al. (2020). Evaluation of stator winding insulation using a synchronous multi-channel PD technique. *IEEE Transactions on Dielectrics and Electrical Insulation* 27 (6): 1889–1897. https://doi.org/10.1109/TDEI.2020.009084.

32 Campbell, S.R., Stone, G.C., Sedding, H.G. et al. (1994). Practical on-line partial discharge tests for turbine generators and motors. *IEEE Transactions on Energy Conversion* 9 (2): 281–287. https://doi.org/10.1109/60.300147.

33 Stone, G.C., Chan, C., and Sedding, H.G. (2015). Relative ability of UHF antenna and VHF capacitor methods to detect partial discharge in turbine generator stator windings. *IEEE Transactions on Dielectrics and Electrical Insulation* 22 (6): 3069–3078. https://doi.org/10.1109/TDEI.2015.005180.

34 Fenger, M. and Stone, G.C. (2005). Investigations into the effect of humidity on stator winding partial discharges. *IEEE Transactions on Dielectrics and Electrical Insulation* 12 (2): 341–346. https://doi.org/10.1109/TDEI.2005.1430402.

35 Stone, G.C. (2012). A perspective on online partial discharge monitoring for assessment of the condition of rotating machine stator winding insulation. *IEEE Electrical Insulation Magazine* 28 (5): 8–13. https://doi.org/10.1109/MEI.2012.6268437.

36 Sedding, H.G., Stone, G.C., and Warren, V. (2016). Progress in Interpreting On-line PD Test Results from Motor and Generator Stator Windings. CIGRE Paper A1-202, August 2016.

37 Belec, M., Hudon, C., and Nguyen, D.N. (2006). Statistical analysis of partial discharge data. *Conference Record of the 2006 IEEE International Symposium on Electrical Insulation*, Toronto, ON (11–14 June 2006), pp. 122–125. https://doi.org/10.1109/ELINSL.2006.1665272.

38 IRIS Power (2022). Online partial discharge severity tables. https://irispower.com/online-partial-discharge-severity-tables/ (accessed June 2022).

39 Stone, G.C. and Culbert, I. (2010). Prediction of stator winding remaining life from diagnostic measurements. *2010 IEEE International Symposium on Electrical Insulation*, San Diego, CA (6–9 June 2010), pp. 1–4. https://doi.org/10.1109/ELINSL.2010.5549791.

40 Shahsavarian, T., Pan, Y., Zhang, Z. et al. (2021). A review of knowledge-based defect identification via PRPD patterns in high voltage apparatus. *IEEE Access* 9: 77705–77728. https://doi.org/10.1109/ACCESS.2021.3082858.

41 Stone, G.C. (2010). Relevance of phase resolved PD analysis to insulation diagnosis in industrial equipment. *2010 10th IEEE International Conference on Solid Dielectrics*, Potsdam, Germany (4–9 July 2010), pp. 1–5. https://doi.org/10.1109/ICSD.2010.5567971.

42 Stone, G., Sedding, H., and Veerkamp, W. (2021). What medium and high voltage stator winding partial discharge testing can – and can not – tell you. *2021 IEEE IAS Petroleum and Chemical Industry Technical Conference (PCIC)*, San Antonio, TX (13–16 September 2021), pp. 293–302. https://doi.org/10.1109/PCIC42579.2021.9728995.

43 Hudon, C. and Belec, M. (2005). Partial discharge signal interpretation for generator diagnostics. *IEEE Transactions on Dielectrics and Electrical Insulation* 12 (2): 297–319. https://doi.org/10.1109/TDEI.2005.1430399.

44 Kurtz, M., Lyles, J.F., and Stone, G.C. (1984). Application of partial discharge testing to hydro generator maintenance. *IEEE Transactions on Power Apparatus and Systems* PAS-103 (8): 2148–2157. https://doi.org/10.1109/TPAS.1984.318525.

45 Dakin, T.W., Works, C.N., and Johnson, J.S. (1969). An Electromagnetic Probe for Detecting and Locating Discharges in Large Rotating-Machine Stators. *IEEE Transactions on Power Apparatus and Systems* PAS-88 (3): 251–257.

46 Smith, L.E. (1970). A peak pulse ammeter-voltmeter for ionization (corona) measurement on electrical equipment. *Double Client Conference, Section 3-401*, Boston (April 1970).

47 Hudon, C., Lévesque, M., Bernier, S. et al. (2019). Scanning individual stator bars and coils with an antenna to detect localized partial discharges. *IEEE Electrical Insulation Conference (EIC)*, Calgary, AB (16–19 June 2019), pp. 457–460.

48 Lemke, E. (1989). PD probe measuring technique for on-site diagnosis tests on high voltage equipment. *International Symposium on High Voltage Engineering*, New Orleans, (1989), Paper 15.08.

49 Timperley, J.J. (2021). Field Guide: Detection of Generator, Exciter & Isolated Phase Bus EMI Using a Handheld Sniffer. EPRI Report 3002021509, Sept 2021.

17

PD Detection in DC Equipment

17.1 Why Is HVDC So Popular Now?

From 1880 to 1896, Tesla and Edison fought the war of currents. Eventually, Tesla won the war and AC became the choice for electric power systems for decades. The main reason AC won the war is that a transformer can change the voltage level very easily and with very low loss, enabling efficient long-distance transmission at high voltages and safe utilization at low voltages. Unfortunately for Edison, DC transformers do not exist.

Until the 1950s, high voltage alternating current (HVAC) transmission was undisputed. However, in 1951 the USSR built a DC link between Moscow and Kashira. In 1954, ASEA put into service a system linking Gotland to the mainland Sweden. The Swedish link was capable at 100 kV DC, transmitting 20 MW over 96 km. The change was marked by the progress in mercury-arc valves that enabled the conversion from AC to DC and *vice versa* at high voltages. The war between AC and DC was eventually over, and AC and DC gradually started to coexist. While the flexibility of AC is unrivalled, the advantages of DC for transmission over long distances are unquestioned. Among them:

1) The transmission line is simpler. Often it is possible to use terrestrial ground or the sea as the return path, so only one HV conductor needs to be installed.
2) With the same voltages and currents, a bipolar DC line transmits more power than an AC line.
3) The DC line does not require reactive power. Especially when cables are used, the large distributed capacitive reactance of the line will require extensive capital investments for reactive power compensation in AC applications, which are not needed for DC lines. The longest submarine HVAC cable lines are about 120 km [1], whereas the Sweden–Lithuania DC cable link, for example, is 453 km long [2].
4) The repetition rate of corona discharges is much lower in DC, reducing corona losses.
5) Skin effect does not exist in DC, so the cross-section of the conductors can be fully exploited.
6) Switching surges are less important in DC, so higher voltages can be reached.
7) DC lines do not suffer the power system stability problems inherent in long HVAC lines.
8) DC links enable interconnecting grids at different frequencies, creating a firewall for disturbances.

Compared to AC substations, the cost of a high-voltage direct current (HVDC) substations is large (although it is decreasing with time), so that HVDC transmission is suitable only above some critical transmission line lengths. See Section 13.13 that also discusses HVDC PD testing of gas-insulated switchgear and lines.

Practical Partial Discharge Measurement on Electrical Equipment, First Edition. Greg C. Stone, Andrea Cavallini, Glenn Behrmann, and Claudio Angelo Serafino.

17.2 Insulation System Design in DC

The design of the insulation system is more complex for DC applications. A key factor is that the electric field depends on resistivity of the insulation materials, not on permittivity as it was shown in Section 2.1.4 for AC systems. That is, the field in DC is resistively graded, whereas in AC it is capacitively graded. As an example, Figure 17.1, [3] shows the field distribution in the barriers of a transformer used for an HVDC substation. On the left, the figure shows the electric field under capacitive grading. Since the relative permittivity of the oil (≈2.2) and that of the pressboard (≈4) are close to one another, the electric field tends to be similar in the liquid and in the solid (although it is a little higher in the liquid). On the right, Figure 17.1 shows the resistive grading. As the pressboard can be 100 times more resistive than the oil, the field inside the pressboard tends to be 100 times larger than that in the oil. Due to the continuity of the displacement, elementary calculations show that a large negative charge (σ) can accumulate at the pressboard/oil interface:

$$\sigma = \varepsilon_{oil}E_{oil} - \varepsilon_{pressboard}E_{pressboard} \approx E_{pressboard}\left(\frac{2.2}{100} - 4.4\right) \approx -E_{pressboard}4.4 \tag{17.1}$$

where ε_{oil} and $\varepsilon_{pressboard}$ are the permittivities, E_{oil} and $E_{pressboard}$ are the electric fields in the oil and the pressboard, respectively.

These simple calculations illustrate two factors about insulation on DC: (i) the dependence of the field on the resistivity of the insulating components, a quantity that changes much more than permittivity, and (ii) the formation of space charge regions. Besides having a much larger range of variability compared with permittivity (going from graphene to polytetrafluoroethylene (PTFE), resistivity changes by 23 orders of magnitude!), resistivity is much more elusive than permittivity. The measurement of conductivity (γ) takes a very long time and often its final value is not

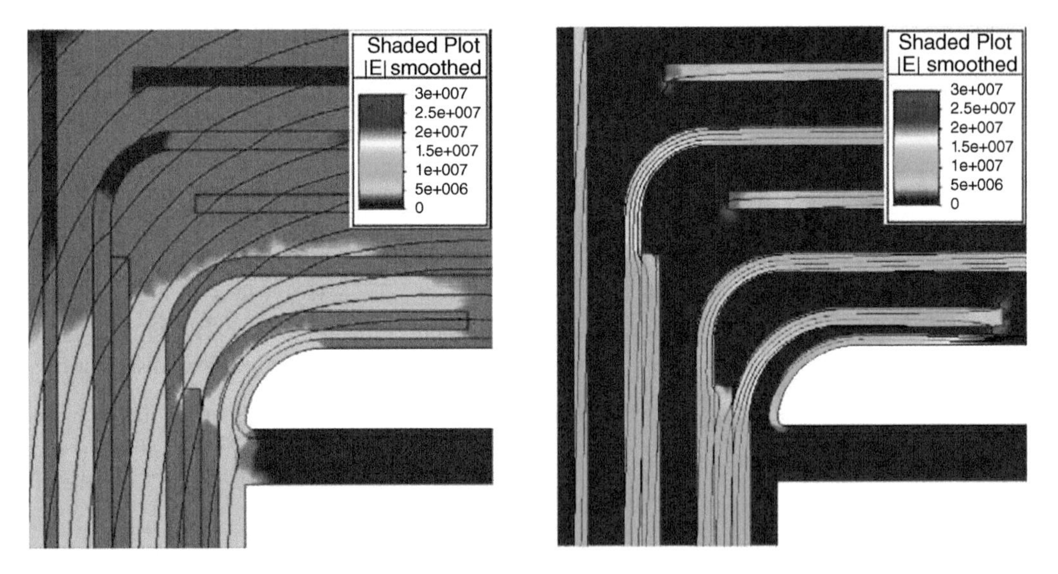

Figure 17.1 Electric field (V/m) in the barriers of a transformer used for an HVDC converter substation: (left) capacitive grading, (right) resistive grading.

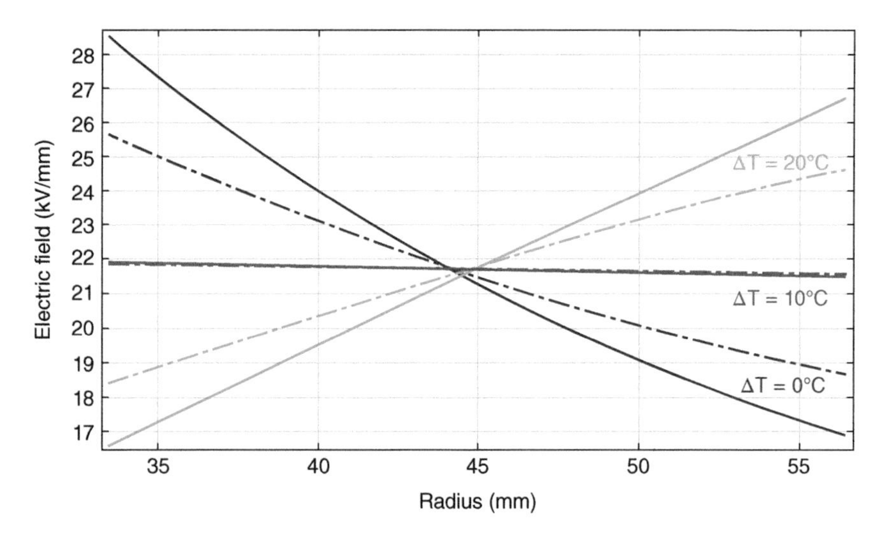

Figure 17.2 DC field distribution versus temperature gradient ΔT across the insulation in a power cable with (broken line) and without (solid lines) field dependency of the conductivity. *Source:* [4].

measured exactly. Conductivity also strongly depends on the insulation temperature (T) and the electric field (E) through equations such as the hopping equation:

$$\gamma = \gamma_0 \exp(-A/T)\frac{\exp(BE)}{E} \tag{17.2}$$

Accordingly, in operation, the profile of the field can change remarkably depending on the temperature profile of the system, as shown in Figure 17.2, [4]. The insulation system, however, should operate reliably also during transients (as an example, lightning impulses), when the field is capacitively graded. Thus, two different types of design must be carried out for DC equipment like power cables. Eventually, it should be noted that the presence of contaminants, for instance water, can change the conductivity and space charge (Section 3.6.2) in a remarkable way.

The impact of space charge (Section 3.6.2) will depend on its distribution with respect to the electrodes. Compared with the Laplacian case, if homocharge (i.e. space charge with the same polarity as the nearby conductor) is formed (the most common case), the field away from the electrodes increases above its Laplacian value (Section 2.1.2). If heterocharge is formed (space charge with the opposite polarity of the nearby electrode), the field at the involved electrode(s) can increase substantially above its Laplacian value. Since space charge is often trapped and stable over time, it can be an issue when transients take place, particularly with voltage polarity inversion transients. If the polarity is reversed too quickly in the presence of homocharge, the charge will eventually become heterocharge leading to a very high field in proximity of the electrodes. This can happen also in the transformer oil as transformers in HVDC substations can be exposed to DC voltages. If design does not account for this, or the oil is aged, a large discharge can occur in the oil leading to the degradation of solid barriers.

17.3 The Reasons for PD Testing Using DC

Partial discharges in DC systems are difficult to measure since there is no AC sinusoidal voltage, and thus creating a PRPD pattern (Section 10.5) is simply not possible. Thus, separation of PD from disturbances or PD source identification based on PRPD pattern analysis is not feasible.

Despite this, the interest in PD detection under DC voltages is high. The reasons are different from for AC. Under AC voltages, the results of PDIV can be very different from those observed under DC voltages. This is particularly relevant if:

- nonlinear stress control systems are used in the equipment, and/or
- space or trapped charge builds up in the region where PD takes place during AC tests.

In the first case, the stress control system might not operate as expected, thus leading to a lower PD inception. In the second case, the space charge might decrease the electric field thus increasing the PDIV.

As an example, the PDIV values achieved from measurements carried out on a MV cable supplied using 50 Hz and 0.1 Hz sinusoidal waveforms, as well as negative DC voltages are reported in Figure 17.3 [5]. The cable was terminated with silicone carbide tubes to control the field (Section 12.2.2). One of the tubes was not in contact with the cable screen to simulate a defect in the installation of the termination. As can be seen in Figure 17.3, some differences exist between the results achieved at 50 Hz and 0.1 Hz. These differences can be explained in part considering the different performance of the nonlinear silicon carbide stress relief, in part with the higher sensitivity achieved using a transformer at 50 Hz versus an electronic supply at 0.1 Hz. However, the large difference observed between AC and DC tests is striking. The much larger PDIV in DC was explained using a FEM model that confirmed that, using negative voltages, the electric field at the point where the silicon carbide material faces the cable screen was considerably lower in DC if compared with AC (for both 50 Hz and 0.1 Hz).

Besides being able to highlight the effect of nonlinear stress control systems, only tests in DC can reproduce the impact of space charge. As a matter of fact, transformers used in HVDC converter substations are also tested with DC sources since, in the case of monopole converter substations (see Figure 17.4) or bipole substations operated as monopolar due to the failure of one of the poles, the phase-to-ground insulation of the windings and the bushing is subjected to DC voltage components. The presence of space charge is an issue particularly for transformers used for

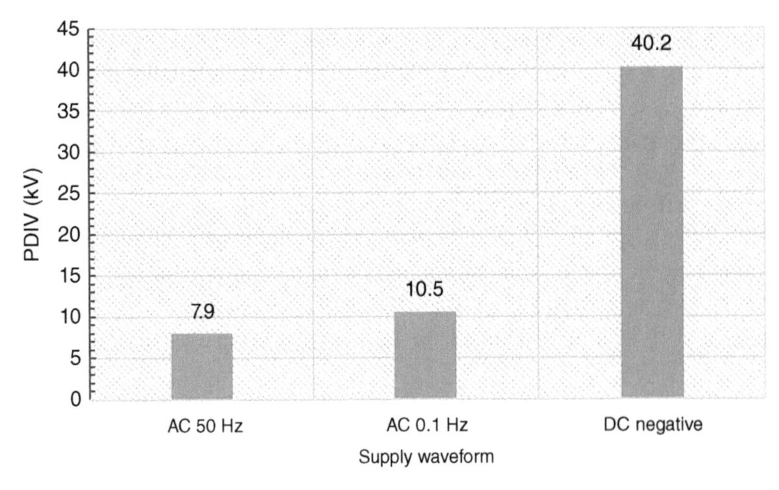

Figure 17.3 PDIV of a medium-voltage cable subjected to different types of voltage waveforms. *Source:* [5].

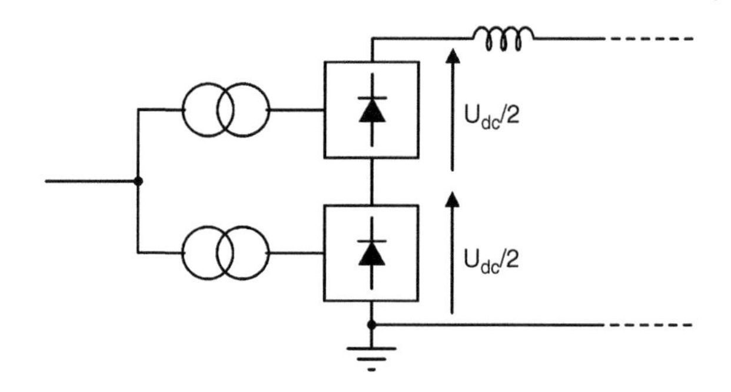

Figure 17.4 Transformers in a monopolar HVDC converter station.

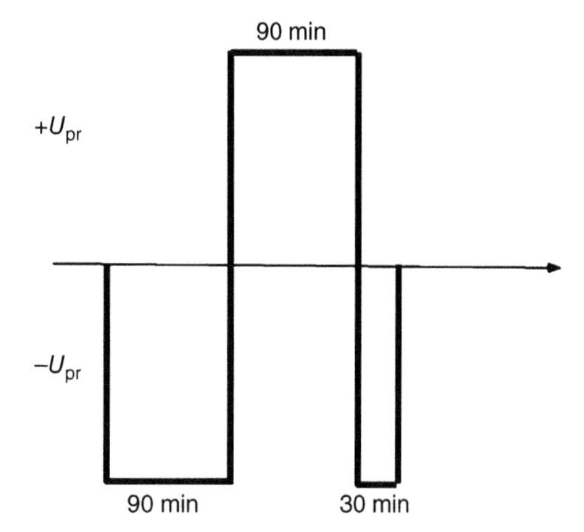

Figure 17.5 Double reversal test voltage profile for transformers used in HVDC substations. *Source:* [6].

line-commutated inverter (LCI) substations. With this type of converter, the power flow can be reversed only by reversing the polarity of the applied voltage. If the space charge injected during operation in one polarity is not removed quickly enough when the polarity is reversed, large disruptive discharges can take place.

For transformers used in HVDC substations, both withstand tests and polarity reversal tests require PD monitoring [6]. During withstand tests, the voltage is ramped up in one minute, kept constant for 120 minutes, and then reduced to zero in one minute. PDs are monitored during the test, and the pass/fail criterion is that no more than 30 pulses above 2 nC may be observed during the last 30 minutes of the test, with no more than 10 pulses above 2 nC observed in the last 10 minutes. For polarity reversal tests, the voltage waveforms are more complicated (see for example Figure 17.5), but the pass/fail criterion is basically the same: during the 30 minutes following the completion of each reversal, no more than 30 pulses above 2 nC may be detected, with no more than 10 pulses above 2 nC in the last 10 minutes [6].

As shown in Figure 17.2, the dependence of the electric field on temperature is remarkable under DC voltages. Heating large equipment is feasible by circulating current in the equipment with the leads shorted, but when high voltage needs to be applied for testing, the circulating current must be stopped. Thus, during the offline high-voltage test, the insulation cools down, and the HVDC PD test may not replicate the PD that occurs in service at operating temperature. Thus, online PD testing of HVDC systems would be advantageous. Besides, monitoring PD could be useful during transients as PD might be incepted more easily when the voltage changes rapidly.

17.4 Offline PD Testing with DC Excitation

IEC 60270 contains some information on how to measure PD under DC excitation and provides some guidance on how to display test results. The PD measurement uses the same couplers and instrumentation as used for AC PD measurements performed according to IEC 60270 (Chapter 6). Larger coupling capacitors could be used with DC excitation compared to AC excitation since they will not require reactive power. In practice, there are important differences between AC and DC excitation that need to be accounted for:

- The AC phase synchronization will not be available; thus, it will be impossible to generate a PRPD pattern (Section 8.7.3).
- The level of disturbance will be larger, due to transients from the power converter used in most high-voltage DC test supplies.

In general, PD detection sensitivity will be lower with DC, since distinguishing actual PD from background interference is difficult when there is no PRPD pattern to aid separation. Since the repetition rate of PD with DC is much lower than AC, the damage per unit time is lower with DC. It is customary to pay attention to very large PD magnitudes, in the nC range.

Indeed, IEC 60270 [7] suggests to use very simple indicators for PD activity, such as:

1) The largest PD recorded during the tests.
2) The accumulated apparent charge (see Figure 17.6)
3) The histogram (or count) of PD charges, as in Figure 17.7.
4) The cumulative PD pulse count exceeding specified limits for PD apparent charge, as in Figure 17.8

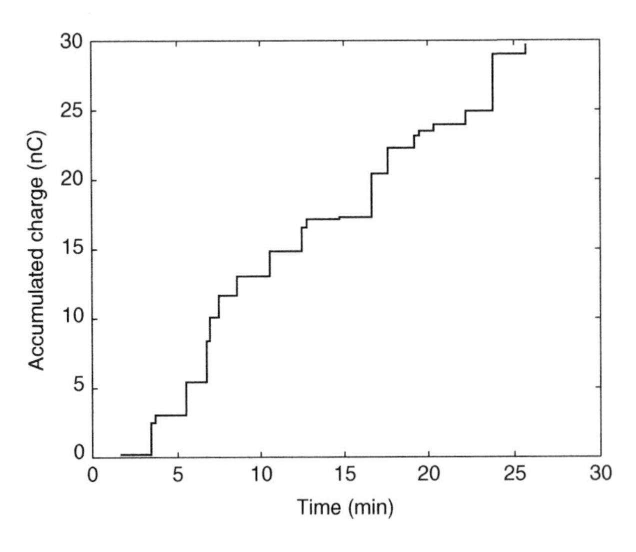

Figure 17.6 Accumulated apparent charge during a DC voltage test. Each "jump" correspond to a PD event having a magnitude equal to the jump itself.

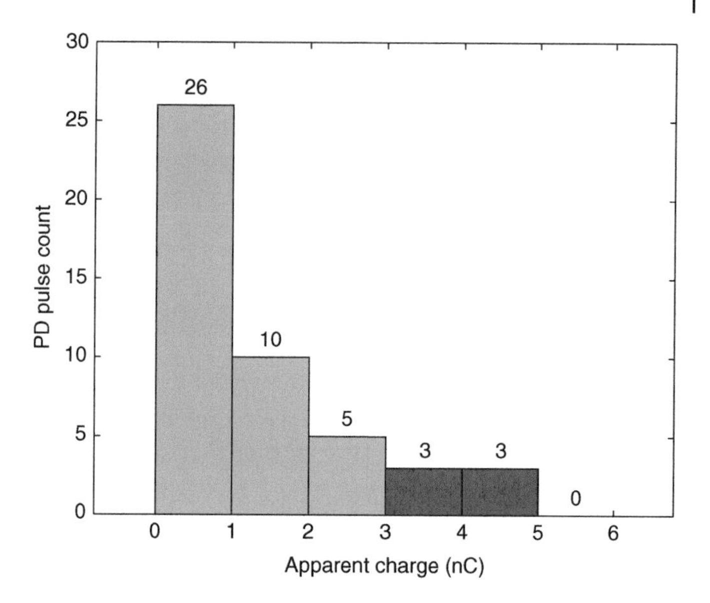

Figure 17.7 Histogram of the PD count vs apparent charge. In red: pulses exceeding a limit of 3 nC.

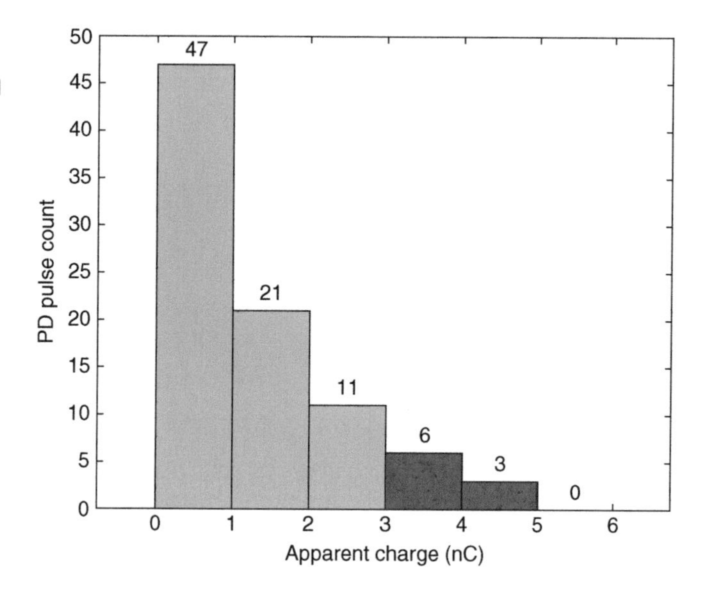

Figure 17.8 Histogram of cumulative PD pulse counts exceeding the indicated threshold (0 pC, 1 nC, etc.) for the data in Figure 17.7. In red are pulses exceeding 3 nC.

17.5 Interpretation of PD Measurements Under DC Excitation

As mentioned above, in contrast to AC PD testing, PD tests under DC cannot utilize a PRPD pattern. Attempts have been made to synchronize PD measurements with the voltage ripple. However, it is not clear whether PD are synchronized with the ripple in actual HVDC transmission systems, for instance, as it might be probably too low to influence PD activity.

17.5.1 Time Series Interpretation

Since under HVDC conditions, the periodic AC phase-reversal is not available, the only quantities available for recording are the PD magnitude and its arrival time. In 1995, Udo Fromm proposed other quantities derived from amplitude (q) and arrival time (t) [8]:

$$\begin{aligned}\Delta t_{pre}(k) &= t(k) - t(k-1)\\ \Delta t_{suc}(k) &= t(k+1) - t(k)\\ \Delta q_{pre}(k) &= q(k) - q(k-1)\\ \Delta q_{suc}(k) &= q(k+1) - q(k)\end{aligned} \tag{17.3}$$

These quantities are shown in Figure 17.9. Fromm's idea is based on the concept that the time between two discharges can be decomposed into:

1) A recovery time, t_R, which is the minimum time needed to remove part of the space charge from previous PD, so that the field at the defect site again becomes sufficient for PD inception;
2) A lag time, t_L, which is the time needed for a starting electron to arrive at the defect site once the field becomes larger than the minimum field for PD inception. During this time, the field at the defect site will tend to increase.

Figure 17.10 illustrates the origin and meaning of these quantities: (i) the recovery and lag phases for the kth PD, and (ii) the definition of two quantities: $t_L(k)$ and $t_R(k)$. The two latter quantities are both a function of the charge of the kth PD since:

- the longer $t_L(k)$ the larger the overvoltage at the time the kth PD is incepted, the higher the magnitude of the kth PD
- The larger the magnitude of the kth PD, the larger the space charge deployed by the PD, the longer the recovery time $t_R(k)$

Theoretically, these relationships could be accounted for by two functions that are uniquely associated with the nature of the PD source:

$$\begin{aligned}t_L &= f_1(q)\\ t_R &= f_2(q)\end{aligned} \tag{17.4}$$

Fromm suggested several ways to use the quantities in (17.3) to infer the functions in (17.4) and, thus, the type of PD source. Eventually, he advocated the use of the 3D histograms of $q(k)$ and

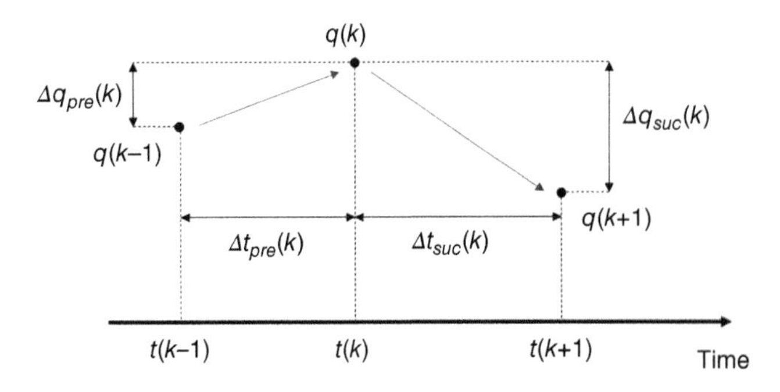

Figure 17.9 Definition of the quantities in Equation (17.3).

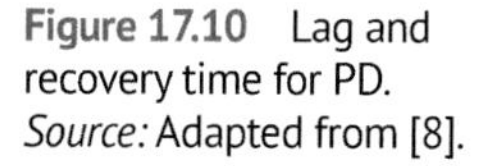

Figure 17.10 Lag and recovery time for PD. *Source:* Adapted from [8].

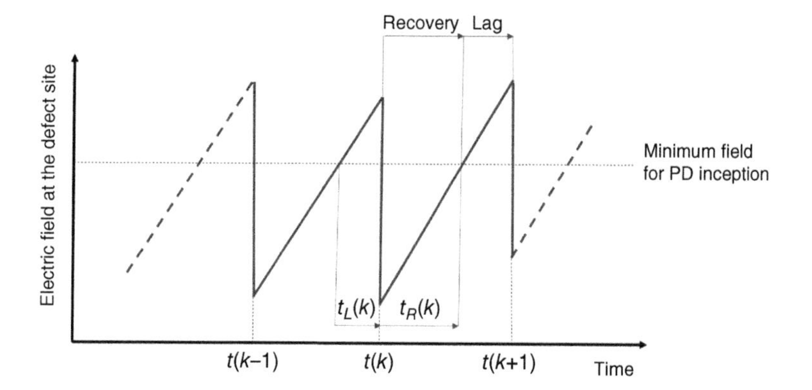

$\Delta t_{suc}(k)$ to identify different PD phenomena based on two classes: corona and internal discharges. Fromm found that the method provided effective results dealing with mass-impregnated cables and capacitors.

One of the limits in Fromm's proposal is that it requires that each event in the PD sequence be recorded. Actually, all the measurements reported in [8] were carried out in a shielded room using a PD-free high-voltage supply. For DC PD tests onsite, two different problems are likely to arise:

a) multiple PD phenomena, and
b) disturbance can be detected.

In both cases, the sequence of events necessary to calculate Equation (17.3) would be lost and the results would be inconsistent. Also, Fromm did not consider the category of surface discharges.

17.5.2 Magnitude Dispersion

In practice, many researchers tend to rely on the histograms of PD magnitudes, trying to separate the different type of defects based on their magnitude dispersion. In general, both positive and negative corona in DC tend to have a lower dispersion compared with internal and surface discharges. Figure 17.11 shows a statistical parameter, β_Q,[1] that indicates the dispersion of PD magnitudes: the lower β_Q the higher the dispersion. As can be seen, both positive and (to a larger extent) negative corona have statistically larger β_Q values, thus less dispersion of PD magnitudes.

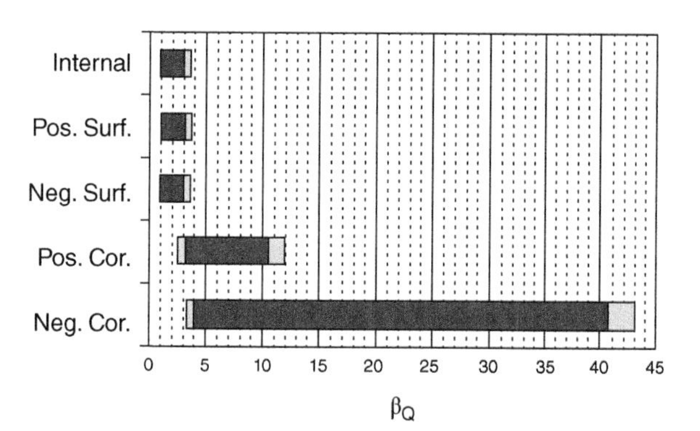

Figure 17.11 Range of β_Q for PD in a cavity (internal), positive and negative surface discharges (Pos. Surf., Neg. Surf.), positive and negative corona (Pos. Cor., Neg. Cor.). The figure reports, in dark gray, the intervals containing 90% of the experimental values as well as the minimum and maximum values (light gray). *Source:* [6].

1 β_Q is the shape parameter of the Weibull distribution fitting the empirical distribution of PD magnitudes.

Indeed, internal discharges and surface discharges (of both polarities) are perfectly overlapped, leaving no room for identification based on dispersion.

17.5.3 Effect of Operating Conditions on PD

Assuming that PD measurements can be carried out for a long enough time, or that the equipment is being monitored during operation, internal discharges (cavities, delamination, or other interface problems) can be distinguished from surface discharges on the outer boundaries of the insulation system, exploiting their dependence on environmental and operational quantities (Section 10.6).

For example, the temperature of the dielectric (thus, the load current) may have a large impact on internal discharges. Figure 17.12 shows the dependence of PD repetition rate on the temperature and on the applied voltage of the insulation in an artificial cavity [9]. It is clear that to achieve the same repetition rate (as an example, 10 PD pulses per minute), more than 25 kV is required at 25 °C, around 16 kV at 50 °C, and about 12 kV at 70 °C. Therefore, if the PD activity decreases with increasing temperature, this may be an indication that the PD activity is taking place within internal cavities. This is similar to the effect of temperature in AC PD tests on stator windings (Section 16.12.2).

The relative humidity of the air surrounding the equipment may also help to identify surface discharges. With surface discharges, the repetition rate can increase by an order of magnitude going from dry condition (30% relative humidity) to wet conditions (80% relative humidity), as shown in Figure 17.13. Another feature that seems associated with high humidity is the way PD are incepted. As shown in Figure 17.14, at low humidity, the repetition rate tends to be constant over time. In contrast, at high humidity, surface PD tend to come in bursts, probably associated with the drying and wetting cycles of the surface induced by the flow of high leakage currents.

In [9], the TF map (Section 8.9.2) was used to separate the PD caused by different phenomena, easing the task of identification. As an example, Figure 17.15 shows the decomposition of PD measurements in discharges due to positive corona, disturbances, and disturbance due to PD in a neighboring AC system (thus, synchronized with the 50 Hz voltage of the grid). The experience with using the separation of the pattern based on the TF map is insufficient, leaving the problem of identification uncertain.

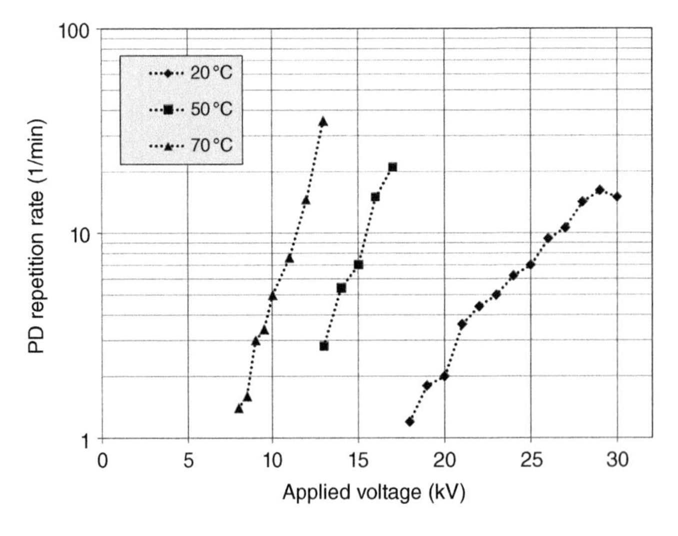

Figure 17.12 PD repetition rate vs temperature for the internal void-type PD under DC excitation. *Source:* [9].

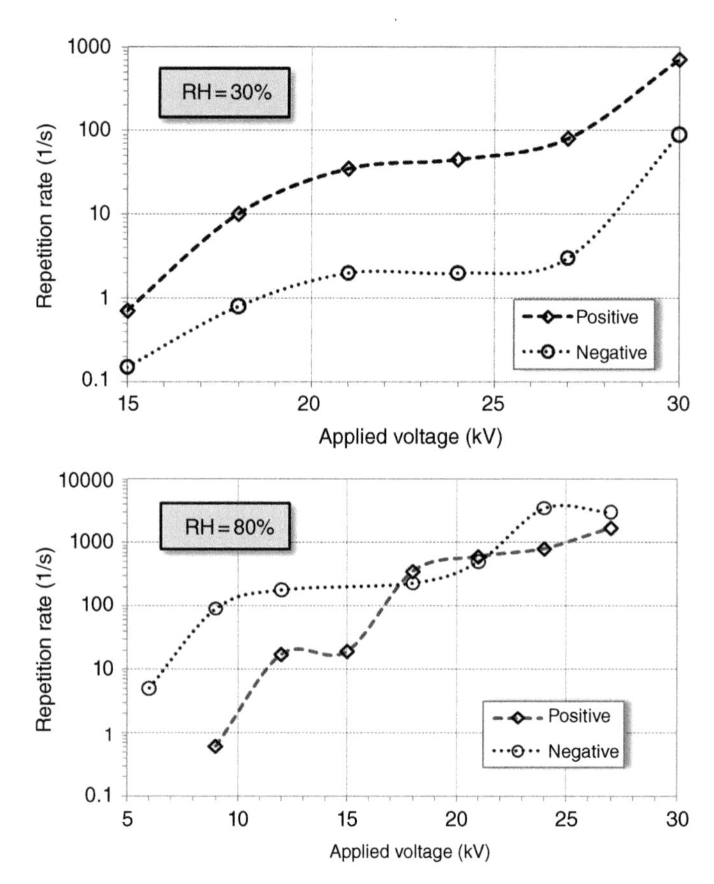

Figure 17.13 Repetition rate vs applied voltage under different humidity levels in DC PD tests on surface discharges. *Source:* [9].

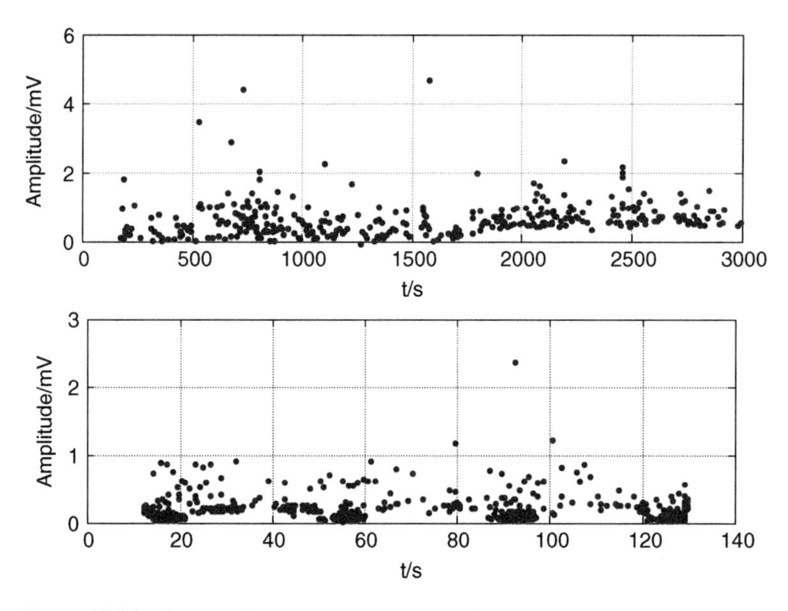

Figure 17.14 Surface discharge magnitude vs time, at different relative humidity (RH) levels. (Top: RH=30%. Bottom: RH=80%). *Source:* [9].

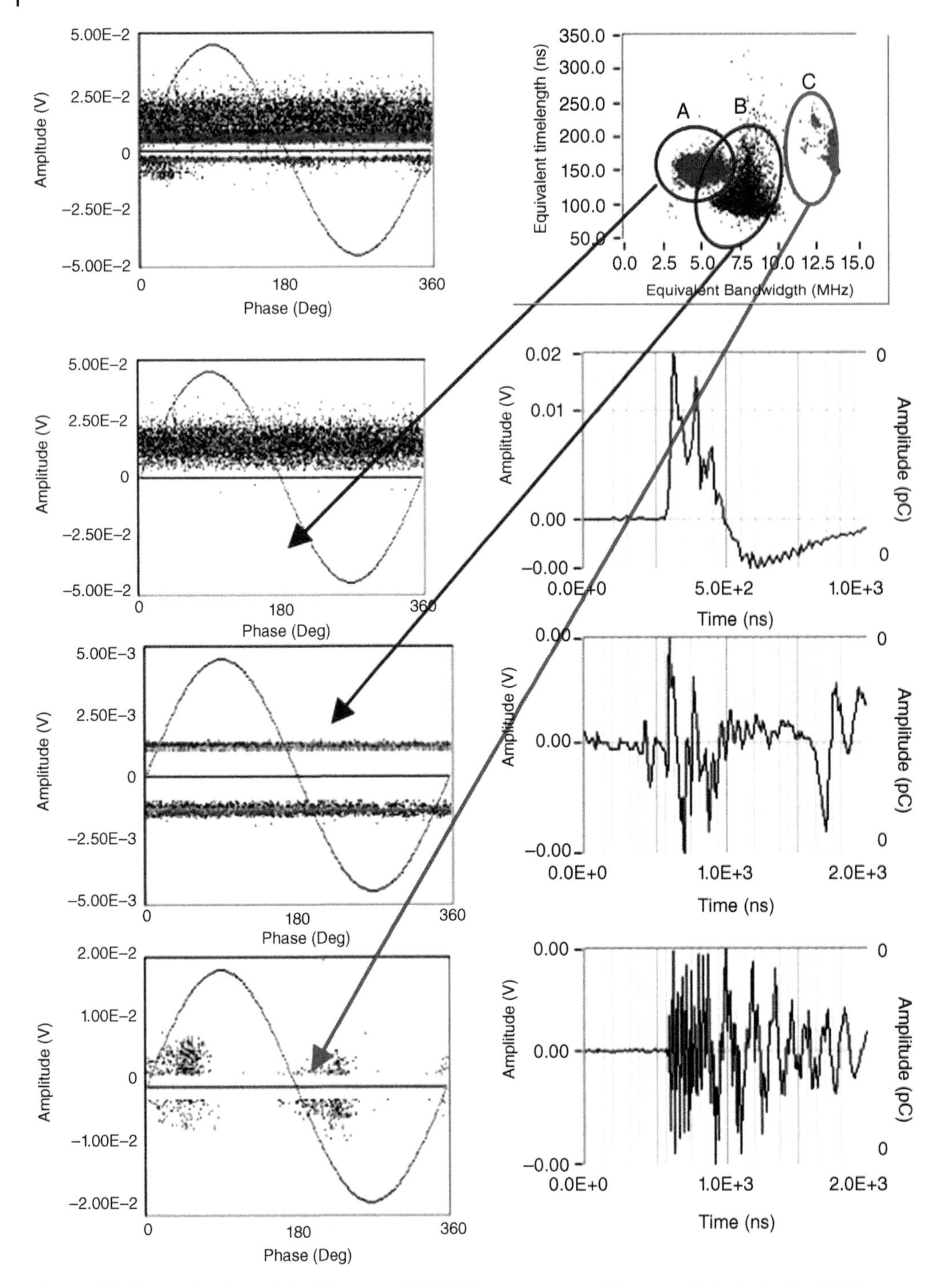

Figure 17.15 Application of the TF map to HVDC PD measurement. The upper left plot is the PRPD plot as recorded, with the TF map to its right showing three clusters. The arrows from the three clusters point to the PRPD plots for each cluster, and the associated pulse shapes (on the left) for each cluster. *Source:* [9].

17.6 Perspective

Measuring PD under DC excitation is already used for testing of transformers used in HVDC substation. Indeed, the PD magnitudes that are investigated are very large and polarity reversal tests are carried out in shielded rooms, which suppresses interference. Also, attempts were made recently to develop PD-based diagnosis of capacitors used in the HVDC lines for harmonic filtering [10]. Yet, testing PD in DC remains a niche activity due to its inherent difficulty (disturbance rejection, source identification). Besides the issues related to the absence of the PRPD pattern (interference rejection and PD source identification are very complex), PD pulse repetition rates under DC are very low, in the range of a few events per minute, as shown in Figures 17.12 and 17.13. Thus, very long test times are required to obtain a meaningful amount of PD data.

Online monitoring of PD in HVDC systems seems the most interesting proposition. Monitoring HVDC GIS substations was proposed in [11] and seems viable considering the large signal-to-noise ratio that can be achieved using sensors within the metallic enclosure. See also Section 13.13. However, the most critical asset that could benefit from PD monitoring in DC are transmission cables [12]. It is expected that, in the course of time, problems related to PD detection and analysis in DC will be solved, making polymeric HVDC cable PD monitoring common.

References

1 Lauria, S. and Palone, F. (2014). Optimal operation of long inhomogeneous AC cable lines: the Malta–Sicily interconnector. *IEEE Transactions on Power Delivery* 29 (3): 1036–1044. https://doi.org/10.1109/TPWRD.2013.2293054.

2 ABB (2022). Project NordBalt – 300 kV HVDC Light subm.pdf. https://library.e.abb.com/public/8b682feba2f4e10ec1257937003e299f/Project%20NordBalt%20-%20300%20kV%20HVDC%20Light%20subm.pdf (accessed 5 March 2022).

3 Fabian, J., Muhr, M., Jaufer, S., and Exner, W. (2012). Partial discharge behavior of mineral oil and oil-board insulation systems at HVDC. *2012 IEEE International Conference on Condition Monitoring and Diagnosis*, Bali, Indonesia (2012), pp. 285–288. https://doi.org/10.1109/CMD.2012.6416432.

4 Mauseth, F. and Haugdal, H. (2017). Electric field simulations of high voltage DC extruded cable systems. *IEEE Electrical Insulation Magazine* 33 (4): 16–21. https://doi.org/10.1109/MEI.2017.7956628.

5 Cavallini, A., Montanari, G.C., Boyer, L. et al. (2015). Partial discharge testing of XLPE cables for HVDC: challenges and opportunities. *Jicable International Conference on Insulated Power Cables*, Versailles, France (June 2015), p. A8.3.

6 ANSI (2017). IEC/IEEE International Standard – Power Transformers–Part 57-129: Transformers for HVDC Applications. *IECIEEE 60076-57-1292017*, pp. 1–58. https://doi.org/10.1109/IEEESTD.2017.8106674.

7 IEC Std. 60270 (2015). *High-Voltage Test Techniques – Partial Discharge Measurements*. IEEE.

8 Fromm, U. (1995). Interpretation of partial discharges at DC voltages. *IEEE Transactions on Dielectrics and Electrical Insulation* 2 (5): 761–770. https://doi.org/10.1109/94.469972.

9 Cavallini, A., Montanari, G.C., Tozzi, M., and Chen, X. (2011). Diagnostic of HVDC systems using partial discharges. *IEEE Transactions on Dielectrics and Electrical Insulation* 18 (1): 275–284. https://doi.org/10.1109/TDEI.2011.5704519.

10 Zhang, Z., Zeng, H., Su, M. et al. (2010). Research on DC partial discharge tentative of HVDC filter capacitors. *2010 International Conference on Electrical and Control Engineering*, Wuhan, China (26–28 June 2010), pp. 4081–4084. https://doi.org/10.1109/iCECE.2010.992.

11 Piccin, R., Mor, A.R., Morshuis, P. et al. (2015). Partial discharge analysis of gas insulated systems at high voltage AC and DC. *IEEE Transactions on Dielectrics and Electrical Insulation* 22 (1): 218–228. https://doi.org/10.1109/TDEI.2014.004711.

12 Winkelmann, E., Shevchenko, I., Steiner, C. et al. (2022). Monitoring of partial discharges in HVDC power cables. *IEEE Electrical Insulation Magazine* 38 (1): 7–18. https://doi.org/10.1109/MEI.2022.9648269.

18

PD Detection Under Impulse Voltage

18.1 Introduction

Most of this book has been focused on measurement of partial discharge when the test object is energized with 50 or 60 Hz sinusoidal AC voltage. However, measurement of PD when the test object is subject to short risetime voltage impulses has become an important commercial issue during the past 20 years. The widespread introduction of voltage source, pulse width modulated converters (VSCs) in wind turbines and for semiconductor-based motor speed control (also known as "drives") has resulted in many premature failures of motor stator windings, wind turbine windings, dry-type transformers, and the converters themselves, even on equipment rated as low as 400 V. Some of these failures were caused by PD initiated by the voltage impulses created by switching devices within the VSC. This has led to the development of PD measurement methods to assess the risk that electrical equipment supplied by VSCs will cause premature failure.

The conventional IEC 60270 test method, which detects PD in the 10 kHz to 1 MHz range, cannot be used when measuring PD caused by short risetime impulses. When the power semiconductor devices in the VSC switch the ± DC voltage on an off to create a simulated sine wave, powerful, short rise-time switching transients are created. If the DC bus voltage in the converter is 400 V, then impulses up to 800 V may be created with very short rise and fall times [1, 2]. Modern VSCs produce voltage impulses with risetimes as short as 20 ns if silicon carbide (SiC) or gallium nitride (GaN) switching devices are used. Even older VSCs that use silicon-based insulated gate bipolar transistors (IGBTs) create unfiltered risetimes of 50–100 ns. A 20 ns risetime corresponds to a frequency of about 15 MHz. If a conventional IEC 60270 detector is used, 800 V impulses at 15 MHz would be conducted through the (typical) 1 nF coupling capacitor (Chapter 6) and destroy the instrument electronics. Even if the instrument electronics could survive such impulses, the instrument will display switching transients that are hundreds of times larger than most reasonable PD, and thus the PD signal would be lost among the switching noise.

As a result, both offline and online electrical detection of PD under voltage impulses requires the use of VHF or UHF detection methods. IEC 62478 provides the basic requirements of VHF and UHF detection [3]. IEC 61934 describes PD detection methods for all types of test objects subject to VSC switching transients [4], and IEC TS 60034-27-5 provides offline PD test methods for rotating machine windings in particular [5]. There are no equivalent IEEE standards, since IEC has taken the lead in such applications.

It is important to note that current source inverters (CSI) and line-commutated inverters (LCI) produce relatively long risetime switching transients – 1 μs or more. Thus, the PD created by such

Practical Partial Discharge Measurement on Electrical Equipment, First Edition. Greg C. Stone, Andrea Cavallini, Glenn Behrmann, and Claudio Angelo Serafino.

converters can normally employ the PD detection methods used for sinusoidal 50/60 Hz equipment, if the PD instrument can synchronize to fundamental frequencies other than 50 or 60 Hz.

The detection of PD under short risetime voltage impulses is a rapidly developing field. Hundreds of technical papers are being written each year on this subject with many improvements in PD measurement technology.

18.2 Insulation Failure Due to Short Risetime Impulse Voltages

One of the first papers on voltage impulses leading to insulation failure of low-voltage motors was written by Persson [6]. He described turn insulation failures leading to stator winding failure in 600 V motors in pulp and paper mills. Subsequently, it was discovered that the failures were likely due to gradual aging of the insulation by PD. It was noted that low voltage motors use organic insulation materials such as polyamide-imide, polyesters and epoxies, which degrade relatively rapidly due to electron and ion bombardment when PD are occurring (Section 2.4.1). Persson's paper triggered a huge amount of research into the mechanisms where PD created by short risetime impulses can lead to equipment failure, with [7–9] being early examples.

The following describes how short risetime voltage impulses may lead to premature insulation failure. Of course the failure processes described in the other equipment chapters (Chapters 12–16) can also lead to PD.

18.2.1 High Peak Voltage

The peak output voltage impulses from voltage source converters are far higher than the rated peak line-to-ground voltage of equipment operated from the 50/60 Hz sinusoidal supplies. The reasons for the higher peak voltages are discussed in [1, 2]:

- The DC bus voltage within the VSC is higher than the peak line to ground voltage. For example, a two-level, three-phase convertor will have a DC bus voltage that is 1.35 times the rated RMS phase-to-phase voltage. The DC bus voltage is therefore 1.65 times the peak line to ground voltage under sinusoidal voltage. The DC bus voltage can be even larger when regenerative breaking occurs in electrical motors.
- If the VSC is remote from a load, which is common in industrial applications (although not for electric vehicle or aerospace applications), the VSC is connected to the load via often long power cables. Since the risetime of the impulse is some tens of nanoseconds, traveling wave phenomena occur that can result in voltage reflections at the load [1, 2, 6], depending on the characteristic impedances of the power cable, the impulse risetime, and the load impedance. This may result in a voltage doubling of the initial impulse (DC bus voltage) from the VSC. Thus the impulse may be 2×1.65 or 3.3 times the rated sinusoidal peak line-to-ground voltage at the load.

Reference [1] calculates the voltage impulse from a 2-level 500 V rated system may be as high as 1350 V line to ground, in the absence of regenerative braking. The peak line to ground voltage for a sinusoidal 500 V rated system is 408 V, only 30% of the peak voltage from the VSC. Since electrical breakdown and PD are driven by the peak voltage across the insulation system (Chapter 2), the higher peak voltage from the VSC may cause PD to incept in any voids that the peak line-to-ground voltage in a sinusoidal system will not. This can be made worse if the VSC manufacturer does not take measures to prevent double transitions – i.e. when two switching events (from say +Vdc to 0 V, and then 0 to –Vdc, or between phases) are not spaced apart in time.

Another aspect that is relevant is the impulse voltage repetition rate. Industrial medium voltage VSCs currently switch at 1–2 kHz, while in electric vehicle applications they may switch at 20–40 kHz. If each switching event produces a PD in a void, then many more PD events over the same time interval will occur with VSCs than with conventional 50/60 Hz sinusoidal voltage, leading to more rapid failure of the insulation, if it is organic.

18.2.2 Short Risetime Causing High Turn Voltages in Windings

Equipment that contains windings, such as motor stators, rotor windings in doubly-fed induction generators (wind turbines and pumped storage hydrogenerators) as well as transformers, have an increased risk of PD between the turns in such windings. It is well-known that short risetime impulses applied to windings cause a nonuniform distribution of the voltage across the turns in the winding [2]. As the impulse risetime decreases, a higher percentage of the impulse voltage appears across the turn-to-turn insulation in the coil that is connected to the phase terminal (and perhaps the neutral terminal [2]). Thus instead of a few tens of volts between turns anywhere in the winding for 50/60 Hz AC voltage, perhaps 75% of the impulse voltage will appear across the first turn in the line end coil if the impulse risetime is about 50 ns [1, 2]. In the 500 V motor example, 1000 V could appear across the turn insulation. If some free space is present between turns, then PD can result. Eventually a turn-to-turn failure may occur, which will rapidly morph into a ground fault.

18.2.3 Overheating of the Stress Relief Coatings

When VSCs started being applied to higher voltage motors, another failure process became apparent. The high-frequency content of the voltage impulses greatly increased the capacitive currents flowing through the PD suppression coatings of medium- and high-voltage motor coils (Section 16.3.2) as well as high voltage power cables. These high frequencies change the voltage profile along the coatings, and the higher currents overheat the coatings, destroying them if they were not designed for use with VSCs, creating very intense PD at the stator slot exits [2, 10, 11].

18.3 Electrical PD Detection

IEC 61934 and IEC TS 60034-27-5 present several methods for measuring PD during very high VSC switching transients. Some methods are mainly intended for offline testing where a voltage impulse generator energizes the test object. Some are more adapted for online PD detection when the electrical equipment is energized by a VSC. All work in the VHF or UHF ranges.

18.3.1 Directional Electromagnetic Couplers

Probably the first commercial method to measure PD from short risetime impulses used a modification of the SSC sensor for hydrogen-cooled turbine generators (Section 16.6.2) and works in the UHF range. The PDAlert was introduced by Iris Power L.P. in 1998 specifically to measure the PD in low voltage rotor and stator windings, as well as low voltage dry type transformers that are energized by commercial “surge testers.” Existing surge testers generate a few voltage impulses per second up to 40 kV, with risetimes as low as 100 ns (for example, see Baker Instruments or Schleich GmbH). The SSC PD sensor is similar in operation to a two-port directional coupler familiar in radio

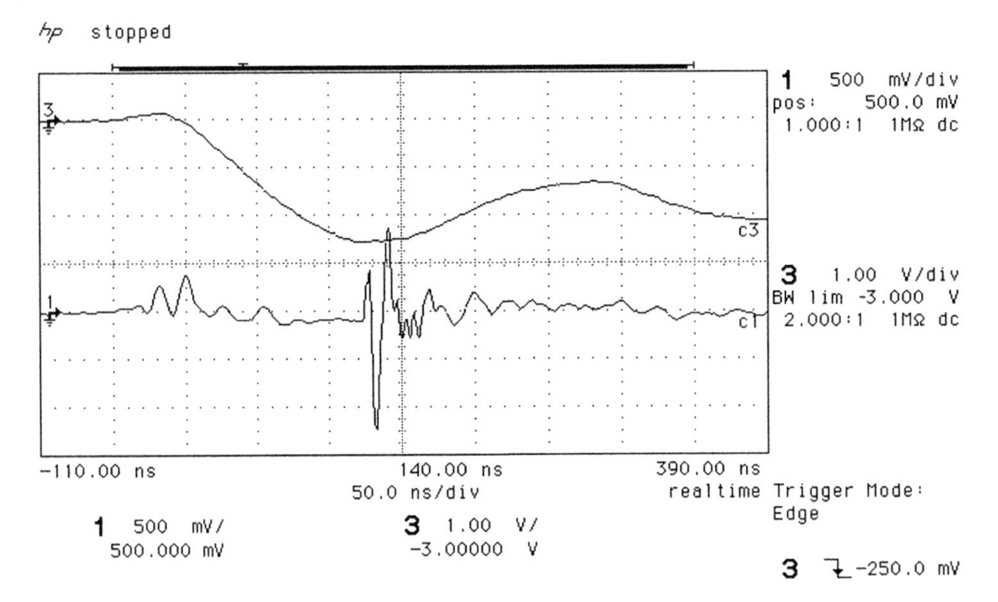

Figure 18.1 Example of a PD pulse measured on a 440 V stator winding supplied from a surge tester with a 100 ns risetime impulse. The surge voltage was −1200 V (top trace). The PD signal is the lower trace near the negative peak of the impulse. At the start of the impulse, there is still some residual signal from the surge tester switching device. *Source:* [13].

frequency and microwave applications. The electromagnetic coupler is connected between the surge tester output and the leads from the test object. Impulses from the surge tester are reduced in magnitude by 60–80 dB in the PD output, for impulse risetimes longer than 50 ns, while PD pulses from the test object are not attenuated [12]. The PD output signal is measured on a dual channel digital oscilloscope, with the second channel displaying the voltage impulse via a voltage divider incorporated into the PD sensor (Figure 18.1). The sensor preserves the polarity of the PD pulse.

The main advantage of this approach is that conducted PD pulses from the EUT are measured, not RF radiation as in antenna-based methods (Section 18.3.2). The PD can be measured from EUTs such as motors and transformers that are fully assembled and enclosed within their grounded metallic housings. In addition, it does not matter where the PD is occurring within the test object, since the UHF signals easily propagate throughout relatively small test objects. In general, this method is more sensitive for repetitive partial discharge inception voltage (RPDIV) and repetitive partial discharge extinction voltage (RPDEV) measurements than other approaches. This technique can be used for offline and online PD testing [13]. An important limitation of this technology is that it may not suppress the impulse voltage sufficiently if the impulse risetimes are less than about 50 ns, and thus may not be useful for surge testers using wide-band gap switching devices like silicon carbide. It is also limited to voltage impulses less than 5 kV, otherwise the sensor itself may produce PD.

18.3.2 VHF and UHF Antennas

The most common sensor for detecting PD in this application is an RF antenna. The antenna can be as simple as an RG 58 coaxial cable with a few cm of the metal outer shield removed to make a monopole antenna (Section 7.4.5.1). However, many types of antenna designs have been developed, with hundreds of papers published on their characteristics. See for example Section 7.4.5 or [14–22]. Sometimes antenna outputs are filtered to further suppress the residual switching transients with an external band-pass or high-pass filter (such as those made by RF Minicircuits). Using a filter is more common with VHF antennas than with narrowband microwave antennas. It is not yet clear what type of antenna is best suited for PD detection during short risetime voltage impulses; this area remains the subject of ongoing work.

The main advantage of an RF antenna is the low cost compared to other sensors. The main disadvantage of the antenna approach is that it only detects radiated RF from PD. If the test object is enclosed within a grounded metal housing with only small openings, for example a motor frame, then little or no signal will be detected due to Faraday shielding. This could be partly overcome if the antenna is installed within the motor frame or transformer. However, even then, PD occurring in the stator slots between turns deep within a large bundle of copper turns may not radiate much signal. Thus, the antenna approach is most sensitive when the PD is occurring near the surface. RF antennas also lose the polarity of the PD. In an online test, the PD from all three phases will be detected, making it difficult to determine what phase has the PD.

Some manufacturers of modern surge testers have incorporated what is believed to be VHF monopole antenna into a probe to measure the PD during voltage impulses (Figure 18.2). They display the PD signal on a second channel of the oscilloscope built into most surge testers to view the voltage impulse. The antenna sensor is suitable for offline tests, where the insulation of the test object is exposed (i.e. not enclosed within metal). To date, there are few reports of RF antenna being used for online testing.

18.3.3 Capacitive Couplers

Low-capacitance sensors connected to the test object terminals, in combination with high-pass filters have been used for online PD measurement in motors rated 3 kV and above fed by VSCs. The most popular sensors are 80 pF, low-inductance capacitors [22]. High-pass filters with a lower cutoff frequency of at least a few MHz are necessary to further suppress the residual switching transients from the VSC. The "special" coupler using in the surge tester in Figure 18.2 may be of this type [23].

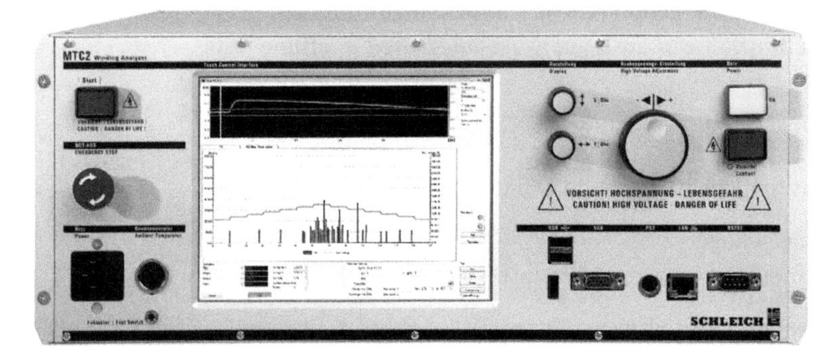

Figure 18.2 Photo of a commercial surge tester that has the ability to display PD on a second channel with either a UHF antenna or a "special" coupler that detects PD within an enclosed test object. The green trace shows the voltage level of the impulses (which first step up, and then step down, according to IEC 60034-27-5). The other traces show the PD detected by an RF antenna. *Source:* Courtesy of Schleich GmbH.

18.3.4 HFCT PD Sensors

As described in IEC 61934, HFCTs combined with a high-pass filter have been used as PD sensors in this application. Practical HFCTs have an upper cutoff frequency of about 100 MHz (with 30 MHz being more common). Due to the limited upper frequency response, HFCTs may produce significant residual switching transients if the impulse risetimes is below 200 ns. HFCTs may be installed on the lead between the impulse voltage supply and the test object. Alternatively the HFCT may be on the ground lead from the test object.

18.4 Nonelectrical Sensors

The extreme amounts of electrical interference caused by VSC switching transients makes the detection of insulation system PD challenging. Thus, many researchers have used optical or acoustic detection, which are not affected by switching transients.

For surface PD, photomultiplier tubes (PMTs) and other similar devices sensitive to visible or ultraviolet frequencies are the most sensitive method to measure individual PD pulses under impulse voltages (Section 5.3). PMTs produce an electrical pulse for each PD occurring on the test object. They must be used in the dark. As discussed in Section 5.3, UV imaging devices are more sensitive than devices working in the visible range (Section 5.3). Modern UV imaging devices can also record multiple PD locations on a test objects (Section 5.3).

The advantage of the optical methods is that they are completely immune to the EMI caused by the switching transients. However, they are mainly of use for research, since an optical line-of-sight is needed between the detector and the PD source, thus only surface PD is detected.

As discussed in Section 5.4, acoustic methods are also immune to switching transient EMI. Contact piezoelectric sensors can detect PD through a metal enclosure. Such sensors may be susceptable to the mechanical shock from the current pulses through the winding. If the PD is on the surface and direct line-of-sight to the PD site is possible, then acoustic imaging devices can also locate the PD.

18.5 PD Display and Quantities Measured

The outcome of a PD measurement usually includes either an oscilloscope image of the PD and the voltage impulse; or a PRPD plot (Section 8.7.3). The PD magnitudes at selected test voltages, and the inception and extinction voltages are usually also measured in offline tests.

18.5.1 PD Synchronized to the Voltage Impulse

There are two main methods to display captured PD pulses:

- With respect to the impulse voltage;
- With respect to the fundamental AC voltage waveform.

An example of the first display method is in Figure 18.1, where a two-channel oscilloscope shows the impulse voltage and the PD. The oscilloscope is triggered by the voltage impulse itself, which is very reliable in offline tests with a high-voltage pulse generator or a surge tester. The PD tends to occur just before the peak, at the peak (see Figure 18.1), or just after the peak of the impulse. Sometimes it may also occur on any high-magnitude oscillations produced by the impulse. On subsequent impulses, the PD location with respect to the impulse "dances" around the impulse waveform due to the need for initiatory electrons, and as a consequence of space charge or trapped charge from previous PD events (Section 3.6).

The use of this type of display is much more complicated when the impulse voltage is created by VSCs, since triggering of the oscilloscope is complicated by the high number of impulses at different DC bias levels (due to the fundamental frequency), and interactions between the three phases.

18.5.2 PD Synchronized to the Fundamental Frequency AC (PRPD)

The other type of display is similar to the phase-resolved PD (PRPD) plots measured with 50/60 Hz sinusoidal tests (Section 8.7.3). The display is suitable for test objects connected to VSCs with 2 or more levels of each polarity (when a simple surge tester is used for the impulse voltages, this display makes no technical sense since there is no AC waveform). For the PRPD plot, instead of 50 or 60 Hz AC, the phase angle of the fundamental frequency AC waveform from the VSC becomes the horizontal axis. There are two difficulties to address when creating PRPD plots on electrical equipment connected to VSCs [23, 24]:

- Obtaining a reliable fundamental frequency voltage waveform that can be used to synchronize the PD with the AC voltage;
- Changing the instrumentation to recognize AC frequencies other than 50 or 60 Hz, as the VSC may operate within a wide range of fundamental frequencies.

In most conventional PD measuring systems using capacitors as PD sensors, the AC voltage reference can be measured from the PD sensor itself because a small amount of 50/60 Hz current does come through the capacitor (Sections 6.3.2 and 8.3.2). However, in prototype installations involving VSCs, a reliable fundamental frequency voltage could not be extracted from the PD sensor due to the high-frequency switching transients, especially near the fundamental frequency AC voltage zero crossings.

Instead, a wide-band capacitive voltage divider can provide the fundamental frequency reference. A capacitive voltage divider produces a fixed ratio output independent of frequency, unlike a capacitor into a 50 Ω load. Thus, high-frequency transients do not dominate fundamental frequency detection and can be removed by simple low-pass filters. In one implementation, an 80 pF high-voltage capacitor is used for one half of the divider [23]. A low-voltage, high-capacitance capacitor is connected in series to the bottom end of the high-voltage capacitor to provide the required low-voltage output to the test instrument (Figure 18.3). The divider is normally installed near the EUT terminals (Figure 18.3).

Alternatively, a signal that represents the start of the fundamental frequency AC zero crossing may be available from the VSC.

Most PD instruments assume the synchronizing AC cycle is 50 or 60 Hz AC voltage. However, by definition, variable speed drives produce an output voltage and current that is variable over a wide range of frequencies. Conventional PD instrumentation use narrow band-pass filters at 50 and 60 Hz to eliminate the power frequency harmonics, combined with an AC voltage zero crossing detector to synchronize the PD to the AC cycle (Section 8.3.2). Thus conventional circuitry must be modified with a wider band-pass frequency range so that it could align the PD with respect to the AC waveform over a wide fundamental frequency range. Software changes also may be needed to allow frequencies other than 50 or 60 Hz.

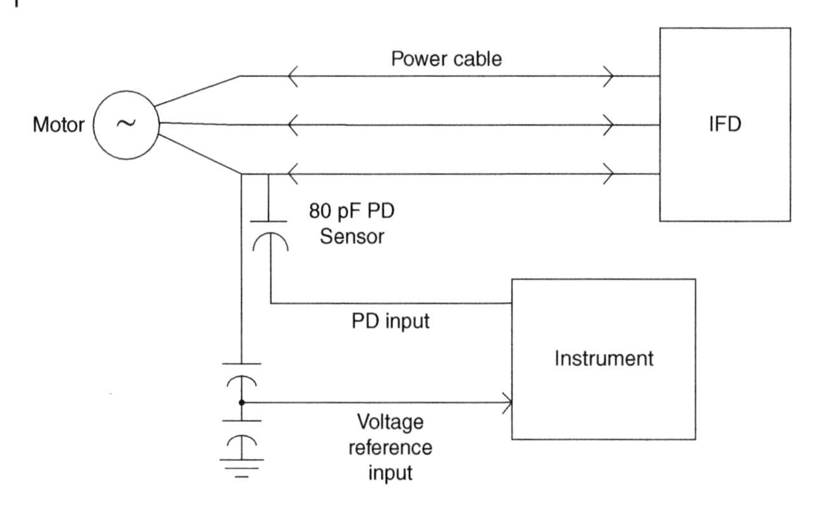

Figure 18.3 Schematic of the PD monitoring system intended for motors fed from a VSC (or IFD). There is one PD sensor per phase, and one capacitive voltage divider to provide a fundamental frequency AC synchronization signal. *Source:* [24].

18.5.3 PD Magnitude

The magnitude of the PD pulses from all types of sensors are measured in mV, consistent with IEC 62478, since they measure the signals in the VHF or UHF ranges. The number of PD pulses per second or per AC cycle may also be measured, as long as the residual switching transient is sufficiently low.

18.5.4 RPDIV and RPDEV

In offline tests with conventional power frequency PD detectors, the PDIV and PDEV are often measured (Sections 8.7.5 and 10.2). The PDIV is the voltage at which PD is detected above a specified Q_{IEC} (which is essentially 1 PD pulse per AC cycle) or Q_m (most often 10 PD pulses per second, whether the frequency is 50 or 60 Hz) – see Section 8.8. Since the fundamental AC frequency is variable with VSCs, these definitions are less relevant, and they are not relevant at all with surge testers. Thus in IEC 61934, the concept of repetitive PDIV and PDEV (i.e. RPDIV and RPDEV) was introduced. It is essentially the PDIV and PDEV where 50% of the voltage impulses create a PD pulse. The selection of 50% of the impulses having PD is arbitrary, but now standardized, although anecdotal experience shows that if a stream of impulses are applied, the PD may not occur at the beginning of the stream. The RPDIV and RPDEV have become very important for qualifying insulation systems for use on VSCs [1, 5].

18.6 Sensitivity and Interference Check

IEC 61934 describes the procedures to determine the sensitivity of the PD measuring system, and the amount of interference that occurs, with the latter almost always being determined by the residual interference from the surge tester or VSC. The procedure is applicable to both offline and online PD tests.

The sensitivity is measured by using an electronic pulse generator – typically outputting 5–10V pulses to the test object and the PD detection system. The pulse must have a risetime of just a few ns, much shorter than the pulse generators used for calibration of IEC 60270 detectors (Section 6.6). The output of the PD sensor is then observed, usually with an oscilloscope. The sensitivity is the lowest magnitude pulse (in mV) that can be clearly seen above the noise floor on the oscilloscope. The test is done with the VSC or surge tester connected to the circuit, but with the power turned off (otherwise the electronic pulse generator will be damaged).

For the background interference check, the electronic pulse generator is removed and the test object is replaced with a load of similar capacitance as the test object, but which is known to be PD free. The surge tester or VSC is turned on, and the level of interference at the desired test voltage is measured on the oscilloscope or the PD instrument. In most online tests, the interference is primarily from switching transients in the VSC.

There is no agreed standard for the maximum level of interference in mV allowed, or the minimum sensitivity in mV. These two numbers are merely presented in a test report.

18.7 Test Procedures

18.7.1 Offline Tests

In offline tests with a surge tester or square wave source, the RPDIV, RPDEV, and the PD magnitudes at selected impulse voltages are measured. Separating any residual switching transient from PD is critical. In a RPDIV test, if there is residual switching transient, this interference will grow linearly as the impulse voltage increases from zero. In contrast, PD is not present at low voltage, and then as the voltage is increased, it suddenly appears on the oscilloscope screen as the discharge inception voltage is attained. The PD will "dance" around the peak of the impulse voltage on subsequent impulses, as described in Section 18.5.1. Above the discharge inception voltage, the PD magnitude (in mV) may or may not increase as the impulse voltage further increases; however, more PD pulses are likely to appear on the oscilloscope screen. After a few offline tests, most observers can easily distinguish between insulation system PD and any residual transients from the impulse voltage supply.

To obtain repeatable RPDIV and RPDEV measurements, IEC 61934 and IEC 60034-27-5 recommend that the voltage be raised in steps, with at least five switching impulses occurring per step. The number of PD pulses per step is then assessed. The PRDIV is voltage at which 50% of the impulses are accompanied by PD. Similarly for the RPDEV. Some commercial instruments, such as the one shown in Figure 18.2, are available, which apply the required number of voltage impulses before the voltage is automatically raised (or lowered). Alternatively this can be approximated with normal commercial surge testers.

It should be noted that any surface PD is strongly affected by the humidity and test object temperature [25]. When twisted pairs of magnet wire are tested, human fingerprints or other contamination on the test object have been shown to lower the RPDIV and RPDEV. As would be expected from Paschen's law (Section 2.3), the ambient gas pressure also affects the measured quantities. This is especially important for insulation systems used in aircraft that are energized by VSCs [26].

With three-phase test objects like stator windings, there are many possible connections to the phase leads and ground [5]. By using different connections, it is sometimes possible to differentiate between turn-to-turn, phase-to-ground, and phase-to-phase discharges.

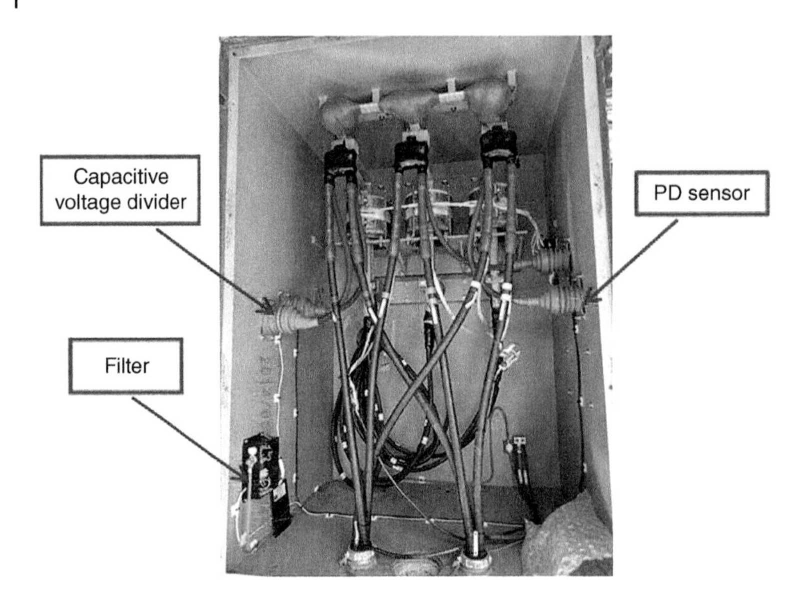

Figure 18.4 PD sensors and a capacitive voltage divider installed in the terminal box of a 4 kV motor connected to a VSC. The green objects are the 80 pF high-voltage capacitors used either as PD sensors or as part of a voltage divider. *Source:* Courtesy of Iris Power L.P.

18.7.2 Online Tests

To date there is very little published experience with online PD detection of test objects subject to VSCs during normal operation. However, it is known that RF antennas (Section 18.3.2) and electromagnetic couplers (Section 18.3.1) have been used on operating low-voltage motors.

One vendor has developed a system that has been deployed on many hundreds of operating motors rated 3–13.2 kV. The VSCs had multiple levels that limited the transient voltage magnitude and extended risetimes to a few hundred ns or longer [24]. Figure 18.3 shows the schematic of the system, and Figure 18.4 shows the installation of sensors on a 4 kV motor. The PD was measured with a portable VHF instrument that incorporated the pulse shape analysis inference suppression method (Section 8.4.3). The instrument used the AC signal from a capacitive voltage divider to synchronize the PD to the AC fundamental frequency to produce PRPD plots.

When performing online PD tests on both low-voltage and high-voltage test objects, it is important to measure the PD at different fundamental frequencies, since the severity of the voltage impulses may depend on load.

The usefulness and practicality of online PD testing on test objects connected to VSCs has still to be established and requires further technical development.

18.8 Interpretation

18.8.1 Type I (PD-Free) Insulation Systems

In test objects that are connected to voltage-source converters and where the insulation is purely organic in nature (e.g. epoxy, polyester, polyamide-imide, etc.), no PD should be present during operation. If PD is occurring, then the PD will lead to electrical treeing or electrical tracking

(Sections 2.4 and 3.7), which can rapidly cause failure. Therefore, the RPDIV and the RPDEV should be somewhat above expected voltage peak magnitude of the voltage impulses from the VSC. This is analogous to the specification for 50/60 Hz equipment where the PDIV and PDEV must be well above the expected operating voltage for power cables, GIS, AIS, and transformers (Chapters 12, 13, 14, and 15, respectively).

This is explicitly recognized in IEC 60034-18-41 for motor and generator windings that are fed from VSCs [1]. This standard defines a Type I machine insulation system as one that does **not** contain materials that are resistant to PD. This is usually windings rated less than 1000 V that do not contain mica. To avoid premature in-service failure, Type I windings are required to have an RPDIV that is essentially 25–60% higher than the expected maximum peak impulse voltage from the VSC during operation. The "enhancement factor" allows for likely variations in PD due to temperature and an assumed effect of in-service aging. The peak impulse voltage must be either measured in operation or calculated as described above from the DC bus voltage as described above, the number of converter levels, and the power cable length between the VSC and the winding [1].

For machine windings, the interpretation process is simply determining if the RPDIV and RPDEV exceed the requirements in IEC 60034-18-41. The requirements in [1], although intended for machines, are likely applicable to other types of electrical equipment that do not incorporate materials resistant to PD.

Periodic offline tests during outages of operating electrical equipment can be performed to establish if the RPDIV and RPDEV are decreasing over time. If these voltages are decreasing, then maintenance or replacement may be needed. Online PD testing is of limited use for Type I insulation systems, since there should be no PD occurring in operation. If PD is detected, the equipment is likely to fail soon.

18.8.2 Type II Insulation Systems

As yet there is no standard for interpreting PD test results for Type II insulation systems. Type II systems do have insulating materials that can withstand PD, such as mica, glass, or porcelain. IEC 60034-18-42 defines Type II insulation systems for machine windings where mica is often incorporated into the turn and groundwall insulation system [27]. To qualify Type II insulation systems for VSC applications, a set of voltage endurance tests must be performed [10, 27]. The offline partial discharge activity does not have to be measured, since it is expected that PD will be present in operation. If periodic offline tests are performed on equipment that has been operating, then any decrease over time in the RPDIV or RPDEV or any increase in the peak PD magnitude, Q_m, may indicate that aging is occurring.

Online PD testing may also provide insight on whether the insulation system is deteriorating, either directly from the PD or due to other aging processes such as thermal aging, load cycling, or contamination. Based on experience from all types of high-voltage equipment (Chapters 12–16), if the PD activity increases over time, then maintenance may be required. One expects this will also be the case for equipment energized by VSCs; however, there is little long-term experience with online PD in this application.

There is some preliminary online experience from motors fed from VSCs that have been equipped with VHF online PD systems. Examples of PRPD plots from online tests are shown in Figures 18.5 and 18.6. Figure 18.5 shows a classic PRPD pattern that could be expected for 50 or 60 Hz conventional motors, with positive PD predominance at the expected part of the fundamental frequency AC cycle. This pattern would be typical for surface PD just outside of the slot due to silicon carbide PD suppression coating deterioration (Sections 16.12.2 and 18.2.3). Recently, a sister motor in the same plant that had no PD monitoring did fail due to a stator winding insulation

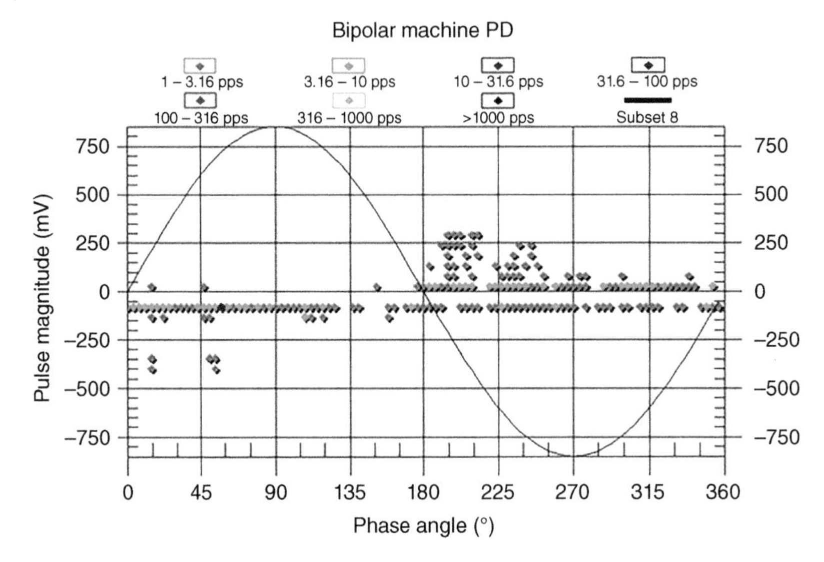

Figure 18.5 PRPD plot from one phase of a 45 MW, 7.2 kV motor fed by a VSC at 100 Hz. The vertical scale is the positive and negative PD magnitude. The horizontal scale is the AC phase angle of the fundamental frequency voltage. The color of the dots indicates the number of PD pulses per second. The PD shows a classic pattern indicating PD from degraded PD suppression coatings. *Source:* Image courtesy Iris Power L.P.

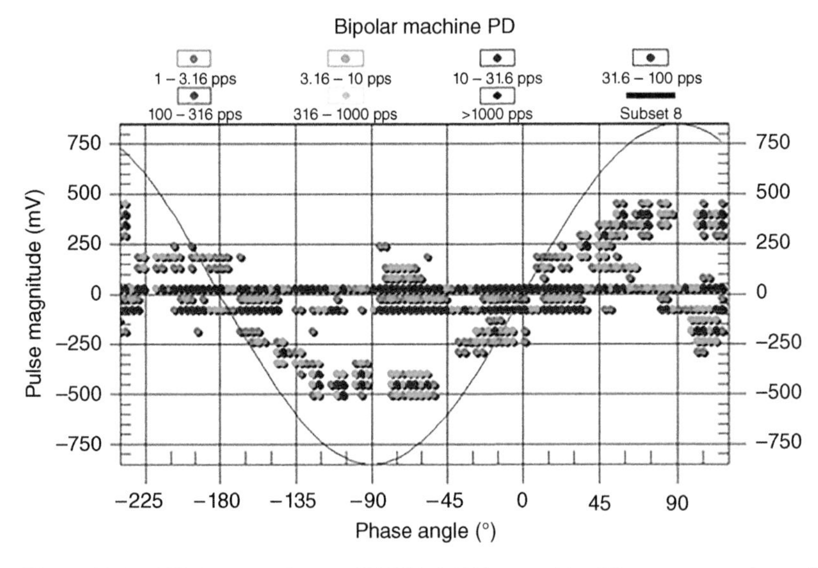

Figure 18.6 PRPD pattern from a 4000 HP, 4.1 kV motor (the PD sensors are shown in Figure 18.4). The pattern is complex and does not show a typical PD pattern. *Source:* Image courtesy Iris Power L.P.

deterioration. Figure 18.6 shows a much more complicated PRPD pattern. The root cause of this pattern has not been ascertained, but it may either be residual interference from VSC switching or perhaps PD occurring between the turns in the motor stator coils (Section 18.2.2). Such turn insulation PD would appear across the fundamental frequency AC cycle since it would be caused by the 2 kHz VSC switching frequency. The motor has not failed yet.

The effectiveness of online PD testing on equipment energized by VSCs will take more time to establish.

References

1 IEC (2014). IEC 60034-18-41. Rotating electrical machines – Part 18-41: partial discharge free electrical insulation systems (Type I) used in rotating electrical machines fed from voltage converters – qualification and quality control tests.

2 Stone, G.C., Culbert, I., Boulter, E.A., and Dhirani, H. (2014). *Electrical Insulation for Rotating Machines: Design, Evaluation, Aging, Testing, and Repair*, 2e. Wiley/IEEE Press.

3 IEC (2016). IEC 62478 TS, High voltage test techniques – measurement of partial discharges by electromagnetic and acoustic methods.

4 IEC (2011). IEC 61934 TS, Electrical insulating materials and systems – electrical measurement of partial discharges (PD) under short rise time and repetitive voltage impulses.

5 IEC (2021). IEC 60034-27-5 TS, Rotating electrical machines – Part 27-5: off-line measurement of partial discharge inception voltage on winding insulation under repetitive impulse voltage.

6 Persson, E. (1992). Transient effects in application of PWM inverters to induction motors. *IEEE Transactions on Industry Applications* 28 (5): 1095–1101. https://doi.org/10.1109/28.158834.

7 Stone, G.C., van Heeswijk, R.G., and Bartnikas, R. (1992). Investigation of the effect of repetitive voltage surges on epoxy insulation. *IEEE Transactions on Energy Conversion* 7 (4): 754–760. https://doi.org/10.1109/60.182659.

8 Kaufhold, M. (1995). Failure mechanism of the interturn insulation of low voltage electric machines fed by pulse controlled inverters. *Proceedings of 1995 Conference on Electrical Insulation and Dielectric Phenomena*, Virginia Beach (October 1995) pp. 254–257. https://doi.org/10.1109/CEIDP.1995.483711.

9 Stone, G.C., Campbell, S.R., and Sedding, H.G. (1998). Analysis of the effect of adjustable speed drives surges on motor stator windings. IEEE International Electric Machines and Drives Conference. IEMDC'99. Proceedings (Cat. No.99EX272), Seattle, WA (9–12 May 1999).

10 Stranges, M.K.W., Stone, G.C., and Bogh, D.L. (2009). Voltage endurance testing. *IEEE Industry Applications Magazine* 15 (6): 12–18. https://doi.org/10.1109/MIAS.2009.934439.

11 Sharifi, E., Jayaram, S.H., and Cherney, E.A. (2010). Analysis of thermal stresses in medium-voltage motor coils under repetitive fast pulse and high-frequency voltages. *IEEE Transactions on Dielectrics and Electrical Insulation* 17 (5): 1378–1384. https://doi.org/10.1109/TDEI.2010.5595539.

12 Stone, G.C., Campbell, S. R., and Susnik, M. (1999). New tools to determine the vulnerability of stator windings to voltage surges from IFDs. *Proceedings: Electrical Insulation Conference and Electrical Manufacturing and Coil Winding Conference (Cat. No.99CH37035)*, Cincinnati, OH (28 October 1999), pp. 149–153. https://doi.org/10.1109/EEIC.1999.826197.

13 Campbell, S.R. and Stone, G.C. (2000). Examples of stator winding partial discharge due to inverter drives. *Conference Record of the 2000 IEEE International Symposium on Electrical Insulation (Cat. No.00CH37075)*, Anaheim, CA (5 April 2000), pp. 231–234. https://doi.org/10.1109/ELINSL.2000.845495.

14 Hayakawa, N., Inano, H., and Okubo, H. (2007). Partial discharge inception characteristics by different measuring methods in magnet wire under surge voltage application. *Annual Report – Conference on Electrical Insulation and Dielectric Phenomena* 2007: 128–131. https://doi.org/10.1109/CEIDP.2007.4451515.

15 Kimura, K., Okada, S., and Hikita, M. (2008). Electromagnetic wave in GHz region of PD pulses under short rise time repetitive voltage impulses. *2008 International Symposium on Electrical Insulating Materials (ISEIM 2008)*, Yokkaichi (7–11 September 2008), pp. 633–636. https://doi.org/10.1109/ISEIM.2008.4664487.

16 Tozzi, M., Montanari, G.C., Fabiani, D. et al. (2009). Off-line and on-line PD measurements on induction motors fed by power electronic impulses. *IEEE Electrical Insulation Conference* 2009: 420–424. https://doi.org/10.1109/EIC.2009.5166383.

17 Kudo, Y., Yahara, N., Fujimoto, M. et al. (2011). Development and evaluation of small loop sensor for detecting electromagnetic field caused by partial discharges in simulated insulation system of inverter-fed rotating machine. *Proceedings of 2011 International Symposium on Electrical Insulating Materials*, Kyoto (6–10 September 2011), pp. 265-268. https://doi.org/10.1109/ISEIM.2011.6826283.

18 Billard, T., Fresnet, F., Makarov, M. et al. (2013). Using non-intrusive sensors to detect partial discharges in a PWM inverter environment: a twisted pair example. *2013 IEEE Electrical Insulation Conference (EIC)*, Ottawa, ON (2–5 June 2013), pp. 329–332. https://doi.org/10.1109/EIC.2013.6554260.

19 Chen, X., Cao, B., Wu, G. et al. (2014). Use of UHF method to measure partial discharge signal under square square-wave pulse. *Proceedings of 2014 International Symposium on Electrical Insulating Materials*, Niigata (1–5 June 2014), pp. 188–191. https://doi.org/10.1109/ISEIM.2014.6870750.

20 Nakamura, K., Uchimura, T., Kozako, M. et al. (2016). Comparison of sensor detection sensitivity in repetitive partial discharge inception voltage measurement for twisted pair placed in stator. *2016 IEEE Conference on Electrical Insulation and Dielectric Phenomena (CEIDP)*, Toronto, ON (16–19 October 2016), pp. 239-242. https://doi.org/10.1109/CEIDP.2016.7785620.

21 Wang, P., Zhou, W., Zhao, Z., and Cavallini, A. (2018). The limitation of partial discharge inception voltage tests at repetitive impulsive voltages using ultra-high frequency antenna and possible solutions. *2018 IEEE Electrical Insulation Conference (EIC)*, San Antonio, TX (17–20 June 2018), pp. 192–195. https://doi.org/10.1109/EIC.2018.8481044.

22 Zhou, W., Wang, P., Zhao, Z. et al. (2019). Design of an Archimedes spiral antenna for PD tests under repetitive impulsive voltages with fast rise times. *IEEE Transactions on Dielectrics and Electrical Insulation* 26 (2): 423–430. https://doi.org/10.1109/TDEI.2018.007738.

23 Stone, G.C., Culbert, I., and Campbell, S.R. (2014). Progress in on-line measurement of PD in motors fed by voltage source PWM drives. *2014 IEEE Electrical Insulation Conference (EIC)*, Philadelphia, PA (8–11 June 2014), pp. 172–175. https://doi.org/10.1109/EIC.2014.6869369.

24 Stone, G.C., Sedding, H.G., and Chan, C. (2018). Experience with online partial discharge measurement in high-voltage inverter-fed motors. *IEEE Transactions on Industry Applications* 54 (1): 866–872. https://doi.org/10.1109/TIA.2017.2740280.

25 Fenger, M. and Stone, G.C. (2005). Investigations into the effect of humidity on stator winding partial discharges. *IEEE Transactions on Dielectrics and Electrical Insulation* 12 (2): 341–346. https://doi.org/10.1109/TDEI.2005.1430402.

26 Gardner, R., Cotton, I., and Kohler, M. (2015). Comparison of power frequency and impulse based partial discharge measurements on a variety of aerospace components at 1000 and 116 mbar. *2015 IEEE Electrical Insulation Conference (EIC)*, Seattle, WA (7–10 June 2015), pp. 430–433. https://doi.org/10.1109/ICACACT.2014.7223588.

27 IEC (2017). IEC 60034-18-42: Partial discharge resistant electrical insulation systems (Type II) used in rotating electrical machines fed from voltage converters – qualification tests.

Index

Note: Page numbers in *italics* refer to figures and **bold** refers to tables respectively

Practical Partial Discharge Measurement on Electrical Equipment, First Edition. Greg C. Stone, Andrea Cavallini, Glenn Behrmann, and Claudio Angelo Serafino.

f

j

k

l

m

n

o

IEEE Press Series on Power and Energy Systems

Series Editor: Ganesh Kumar Venayagamoorthy, Clemson University, Clemson, South Carolina, USA.

The mission of the IEEE Press Series on Power and Energy Systems is to publish leading-edge books that cover a broad spectrum of current and forward-looking technologies in the fast-moving area of power and energy systems including smart grid, renewable energy systems, electric vehicles and related areas. Our target audience includes power and energy systems professionals from academia, industry and government who are interested in enhancing their knowledge and perspectives in their areas of interest.

1. *Electric Power Systems: Design and Analysis, Revised Printing*
   Mohamed E. El-Hawary
2. *Power System Stability*
   Edward W. Kimbark
3. *Analysis of Faulted Power Systems*
   Paul M. Anderson
4. *Inspection of Large Synchronous Machines: Checklists, Failure Identification, and Troubleshooting*
   Isidor Kerszenbaum
5. *Electric Power Applications of Fuzzy Systems*
   Mohamed E. El-Hawary
6. *Power System Protection*
   Paul M. Anderson
7. *Subsynchronous Resonance in Power Systems*
   Paul M. Anderson, B.L. Agrawal, J.E. Van Ness
8. *Understanding Power Quality Problems: Voltage Sags and Interruptions*
   Math H. Bollen
9. *Analysis of Electric Machinery*
   Paul C. Krause, Oleg Wasynczuk, and S.D. Sudhoff
10. *Power System Control and Stability, Revised Printing*
    Paul M. Anderson, A.A. Fouad
11. *Principles of Electric Machines with Power Electronic Applications, Second Edition*
    Mohamed E. El-Hawary
12. *Pulse Width Modulation for Power Converters: Principles and Practice*
    D. Grahame Holmes and Thomas Lipo
13. *Analysis of Electric Machinery and Drive Systems, Second Edition*
    Paul C. Krause, Oleg Wasynczuk, and S.D. Sudhoff

14. *Risk Assessment for Power Systems: Models, Methods, and Applications*
    Wenyuan Li
15. *Optimization Principles: Practical Applications to the Operations of Markets of the Electric Power Industry*
    Narayan S. Rau
16. *Electric Economics: Regulation and Deregulation*
    Geoffrey Rothwell and Tomas Gomez
17. *Electric Power Systems: Analysis and Control*
    Fabio Saccomanno
18. *Electrical Insulation for Rotating Machines: Design, Evaluation, Aging, Testing, and Repair*
    Greg C. Stone, Edward A. Boulter, Ian Culbert, and Hussein Dhirani
19. *Signal Processing of Power Quality Disturbances*
    Math H. J. Bollen and Irene Y. H. Gu
20. *Instantaneous Power Theory and Applications to Power Conditioning*
    Hirofumi Akagi, Edson H.Watanabe and Mauricio Aredes
21. *Maintaining Mission Critical Systems in a 24/7 Environment*
    Peter M. Curtis
22. *Elements of Tidal-Electric Engineering*
    Robert H. Clark
23. *Handbook of Large Turbo-Generator Operation and Maintenance, Second Edition*
    Geoff Klempner and Isidor Kerszenbaum
24. *Introduction to Electrical Power Systems*
    Mohamed E. El-Hawary
25. *Modeling and Control of Fuel Cells: Distributed Generation Applications*
    M. Hashem Nehrir and Caisheng Wang
26. *Power Distribution System Reliability: Practical Methods and Applications*
    Ali A. Chowdhury and Don O. Koval
27. *Introduction to FACTS Controllers: Theory, Modeling, and Applications*
    Kalyan K. Sen, Mey Ling Sen
28. *Economic Market Design and Planning for Electric Power Systems*
    James Momoh and Lamine Mili
29. *Operation and Control of Electric Energy Processing Systems*
    James Momoh and Lamine Mili
30. *Restructured Electric Power Systems: Analysis of Electricity Markets with Equilibrium Models*
    Xiao-Ping Zhang
31. *An Introduction to Wavelet Modulated Inverters*
    S.A. Saleh and M.A. Rahman
32. *Control of Electric Machine Drive Systems*
    Seung-Ki Sul

33. *Probabilistic Transmission System Planning*
Wenyuan Li
34. *Electricity Power Generation: The Changing Dimensions*
Digambar M. Tagare
35. *Electric Distribution Systems*
Abdelhay A. Sallam and Om P. Malik
36. *Practical Lighting Design with LEDs*
Ron Lenk, Carol Lenk
37. *High Voltage and Electrical Insulation Engineering*
Ravindra Arora and Wolfgang Mosch
38. *Maintaining Mission Critical Systems in a 24/7 Environment, Second Edition*
Peter Curtis
39. *Power Conversion and Control of Wind Energy Systems*
Bin Wu, Yongqiang Lang, Navid Zargari, Samir Kouro
40. *Integration of Distributed Generation in the Power System*
Math H. Bollen, Fainan Hassan
41. *Doubly Fed Induction Machine: Modeling and Control for Wind Energy Generation Applications*
Gonzalo Abad, Jesús López, Miguel Rodrigues, Luis Marroyo, and Grzegorz Iwanski
42. *High Voltage Protection for Telecommunications*
Steven W. Blume
43. *Smart Grid: Fundamentals of Design and Analysis*
James Momoh
44. *Electromechanical Motion Devices, Second Edition*
Paul Krause Oleg, Wasynczuk, Steven Pekarek
45. *Electrical Energy Conversion and Transport: An Interactive Computer-Based Approach, Second Edition*
George G. Karady, Keith E. Holbert
46. *ARC Flash Hazard and Analysis and Mitigation*
J.C. Das
47. *Handbook of Electrical Power System Dynamics: Modeling, Stability, and Control*
Mircea Eremia, Mohammad Shahidehpour
48. *Analysis of Electric Machinery and Drive Systems, Third Edition*
Paul C. Krause, Oleg Wasynczuk, S.D. Sudhoff, Steven D. Pekarek
49. *Extruded Cables for High-Voltage Direct-Current Transmission: Advances in Research and Development*
Giovanni Mazzanti, Massimo Marzinotto
50. *Power Magnetic Devices: A Multi-Objective Design Approach*
S.D. Sudhoff

51. *Risk Assessment of Power Systems: Models, Methods, and Applications, Second Edition*
    Wenyuan Li
52. *Practical Power System Operation*
    Ebrahim Vaahedi
53. *The Selection Process of Biomass Materials for the Production of Bio-Fuels and Co-Firing*
    Najib Altawell
54. *Electrical Insulation for Rotating Machines: Design, Evaluation, Aging, Testing, and Repair, Second Edition*
    Greg C. Stone, Ian Culbert, Edward A. Boulter, and Hussein Dhirani
55. *Principles of Electrical Safety*
    Peter E. Sutherland
56. *Advanced Power Electronics Converters: PWM Converters Processing AC Voltages*
    Euzeli Cipriano dos Santos Jr., Edison Roberto Cabral da Silva
57. *Optimization of Power System Operation, Second Edition*
    Jizhong Zhu
58. *Power System Harmonics and Passive Filter Designs*
    J.C. Das
59. *Digital Control of High-Frequency Switched-Mode Power Converters*
    Luca Corradini, Dragan Maksimoviæ, Paolo Mattavelli, Regan Zane
60. *Industrial Power Distribution, Second Edition*
    Ralph E. Fehr, III
61. *HVDC Grids: For Offshore and Supergrid of the Future*
    Dirk Van Hertem, Oriol Gomis-Bellmunt, Jun Liang
62. *Advanced Solutions in Power Systems: HVDC, FACTS, and Artificial Intelligence*
    Mircea Eremia, Chen-Ching Liu, Abdel-Aty Edris
63. *Operation and Maintenance of Large Turbo-Generators*
    Geoff Klempner, Isidor Kerszenbaum
64. *Electrical Energy Conversion and Transport: An Interactive Computer-Based Approach*
    George G. Karady, Keith E. Holbert
65. *Modeling and High-Performance Control of Electric Machines*
    John Chiasson
66. *Rating of Electric Power Cables in Unfavorable Thermal Environment*
    George J. Anders
67. *Electric Power System Basics for the Nonelectrical Professional*
    Steven W. Blume
68. *Modern Heuristic Optimization Techniques: Theory and Applications to Power Systems*
    Kwang Y. Lee, Mohamed A. El-Sharkawi
69. *Real-Time Stability Assessment in Modern Power System Control Centers*
    Savu C. Savulescu

70. *Optimization of Power System Operation*
Jizhong Zhu

71. *Insulators for Icing and Polluted Environments*
Masoud Farzaneh, William A. Chisholm

72. *PID and Predictive Control of Electric Devices and Power Converters Using MATLAB®/Simulink®*
Liuping Wang, Shan Chai, Dae Yoo, Lu Gan, Ki Ng

73. *Power Grid Operation in a Market Environment: Economic Efficiency and Risk Mitigation*
Hong Chen

74. *Electric Power System Basics for Nonelectrical Professional, Second Edition*
Steven W. Blume

75. *Energy Production Systems Engineering Thomas*
Howard Blair

76. *Model Predictive Control of Wind Energy Conversion Systems*
Venkata Yaramasu, Bin Wu

77. *Understanding Symmetrical Components for Power System Modeling*
J.C. Das

78. *High-Power Converters and AC Drives, Second Edition*
Bin Wu, Mehdi Narimani

79. *Current Signature Analysis for Condition Monitoring of Cage Induction Motors: Industrial Application and Case Histories*
William T. Thomson, Ian Culbert

80. *Introduction to Electric Power and Drive Systems*
Paul Krause, Oleg Wasynczuk, Timothy O'Connell, Maher Hasan

81. *Instantaneous Power Theory and Applications to Power Conditioning, Second Edition*
Hirofumi, Edson Hirokazu Watanabe, Mauricio Aredes

82. *Practical Lighting Design with LEDs, Second Edition*
Ron Lenk, Carol Lenk

83. *Introduction to AC Machine Design*
Thomas A. Lipo

84. *Advances in Electric Power and Energy Systems: Load and Price Forecasting*
Mohamed E. El-Hawary

85. *Electricity Markets: Theories and Applications*
Jeremy Lin, Jernando H. Magnago

86. *Multiphysics Simulation by Design for Electrical Machines, Power Electronics and Drives*
Marius Rosu, Ping Zhou, Dingsheng Lin, Dan M. Ionel, Mircea Popescu, Frede Blaabjerg, Vandana Rallabandi, David Staton

87. *Modular Multilevel Converters: Analysis, Control, and Applications*
Sixing Du, Apparao Dekka, Bin Wu, Navid Zargari

88. *Electrical Railway Transportation Systems*
Morris Brenna, Federica Foiadelli, Dario Zaninelli

89. *Energy Processing and Smart Grid*
James A. Momoh

90. *Handbook of Large Turbo-Generator Operation and Maintenance, 3rd Edition*
Geoff Klempner, Isidor Kerszenbaum

91. *Advanced Control of Doubly Fed Induction Generator for Wind Power Systems*
Dehong Xu, Dr. Frede Blaabjerg, Wenjie Chen, Nan Zhu

92. *Electric Distribution Systems, 2nd Edition*
Abdelhay A. Sallam, Om P. Malik

93. *Power Electronics in Renewable Energy Systems and Smart Grid: Technology and Applications*
Bimal K. Bose

94. *Distributed Fiber Optic Sensing and Dynamic Rating of Power Cables*
Sudhakar Cherukupalli, and George J Anders

95. *Power System and Control and Stability, Third Edition*
Vijay Vittal, James D. McCalley, Paul M. Anderson, and A. A. Fouad

96. *Electromechanical Motion Devices: Rotating Magnetic Field-Based Analysis and Online Animations, Third Edition*
Paul Krause, Oleg Wasynczuk, Steven D. Pekarek, and Timothy O'Connell

97. *Applications of Modern Heuristic Optimization Methods in Power and Energy Systems*
Kwang Y. Lee and Zita A. Vale

98. *Handbook of Large Hydro Generators: Operation and Maintenance*
Glenn Mottershead, Stefano Bomben, Isidor Kerszenbaum, and Geoff Klempner

99. *Advances in Electric Power and Energy: Static State Estimation*
Mohamed E. El-hawary

100. *Arc Flash Hazard Analysis and Mitigation, Second Edition*
J.C. Das

101. *Maintaining Mission Critical Systems in a 24/7 Environment, Third Edition*
Peter M. Curtis

102. *Real-Time Electromagnetic Transient Simulation of AC-DC Networks*
Venkata Dinavahi and Ning Lin

103. *Probabilistic Power System Expansion Planning with Renewable Energy Resources and Energy Storage Systems*
Jaeseok Choi and Kwang Y. Lee

104. *Power Magnetic Devices: A Multi-Objective Design Approach, Second Edition*
Scott D. Sudhoff

105. *Optimal Coordination of Power Protective Devices with Illustrative Examples*
Ali R. Al-Roomi

106. *Resilient Control Architectures and Power Systems*
Craig Rieger, Ronald Boring, Brian Johnson, and Timothy McJunkin

107. *Alternative Liquid Dielectrics for High Voltage Transformer Insulation Systems: Performance Analysis and Applications*
Edited by U. Mohan Rao, I. Fofana, and R. Sarathi

108. *Introduction to the Analysis of Electromechanical Systems*
Paul C. Krause, Oleg Wasynczuk, and Timothy O'Connell

109. *Power Flow Control Solutions for a Modern Grid using SMART Power Flow Controllers*
Kalyan K. Sen and Mey Ling Sen

110. *Power System Protection: Fundamentals and Applications*
John Ciufo and Aaron Cooperberg

111. *Soft-Switching Technology for Three-phase Power Electronics Converters*
Dehong Xu, Rui Li, Ning He, Jinyi Deng, and Yuying Wu

112. *Power System Protection, Second Edition*
Paul M. Anderson, Charles Henville, Rasheek Rifaat, Brian Johnson, and Sakis Meliopoulos

113. *High Voltage and Electrical Insulation Engineering, Second Edition*
Ravindra Arora and Wolfgang Mosch.

114. *Modeling and Control of Modern Electrical Energy Systems*
Masoud Karimi-Ghartemani

115. *Control of Power Electronic Converters with Microgrid Applications*
Armdam Ghosh and Firuz Zare

116. *Coordinated Operation and Planning of Modern Heat and Electricity Incorporated Networks*
Mohammadreza Daneshvar, Behnam Mohammadi-Ivatloo, and Kazem Zare

117. *Smart Energy for Transportation and Health in a Smart City*
Chun Sing Lai, Loi Lei Lai, and Qi Hong Lai

118. *Wireless Power Transfer: Principles and Applications*
Zhen Zhang and Hongliang Pang

119. *Intelligent Data Mining and Analysis in Power and Energy Systems: Models and Applications for Smarter Efficient Power Systems*
Zita Vale, Tiago Pinto, Michael Negnevitsky, and Ganesh Kumar Venayagamoorthy

120. *Introduction to Modern Analysis of Electric Machines and Drives*
Paul C. Krause and Thomas C. Krause

121. *Electromagnetic Analysis and Condition Monitoring of Synchronous Generators*
Hossein Ehya and Jawad Faiz

122. *Transportation Electrification: Breakthroughs in Electrified Vehicles, Aircraft, Rolling Stock, and Watercraft*
Ahmed A. Mohamed, Ahmad Arshan Khan, Ahmed T. Elsayed, Mohamed A. Elshaer

123. *Modular Multilevel Converters: Control, Fault Detection, and Protection*
Fuji Deng, Chengkai Liu, and Zhe Chen

124. *Stability-Constrained Optimization for Modern Power System Operation and Planning*
Yan Xu, Yuan Chi, and Heling Yuan

125. *Interval Methods for Uncertain Power System Analysis*
Alfredo Vaccaro

126. *Practical Partial Discharge Measurement on Electrical Equipment*
Greg C. Stone, Andrea Cavallini, Glenn Behrmann, and Claudio Angelo Serafino

Printed and bound by CPI Group (UK) Ltd, Croydon, CR0 4YY
16/04/2025
14658588-0004